UNITEXT for Physics

More information about this series at http://www.springer.com/series/13351

Riccardo D'Auria · Mario Trigiante

From Special Relativity to Feynman Diagrams

A Course in Theoretical Particle Physics for Beginners

Second Edition

 Springer

Riccardo D'Auria
Polytechnic University of Turin
Turin
Italy

Mario Trigiante
Polytechnic University of Turin
Turin
Italy

ISSN 2198-7882
UNITEXT for Physics
ISBN 978-3-319-34249-8
DOI 10.1007/978-3-319-22014-7

ISSN 2198-7890 (electronic)

ISBN 978-3-319-22014-7 (eBook)

Springer Cham Heidelberg New York Dordrecht London
© Springer International Publishing Switzerland 2012, 2016
Softcover re-print of the Hardcover 2nd edition 2016

Printed on acid-free paper

Springer International Publishing AG Switzerland is part of Springer Science+Business Media (www.springer.com)

To Anna Maria

Riccardo D'Auria

In loving memory
of my mother and my father

Mario Trigiante

Preface

This book has been developed from the lecture notes of a course in Advanced Quantum Mechanics given by the authors at the Politecnico of Torino for students of physical engineering, who, even though oriented towards applied physics and technology, were interested in acquiring a fair knowledge of modern fundamental physics. Although originally conceived for students of engineering, we have eventually extended the target of this book to also include students of physics who may be interested in a comprehensive and concise treatment of the main subjects of their theoretical physics courses. What underlies our choice of topics is the purpose of giving a coherent presentation of the theoretical ideas which have been developed since the very beginning of the last century, namely special relativity and quantum mechanics, up to the first consistent and experimentally validated quantum field theory, namely quantum electrodynamics. This theory provides a successful description of the interaction between photons and electrons and dates back to the middle of the last century.

Consistently with this purpose (and also for keeping the book within a reasonable size), we have refrained from dealing with the many important ideas that have been developed in the context of quantum field theory in the second part of the last century, although these are essential for a satisfactory understanding of the current status of elementary particle physics. A prominent example of such developments is the so-called standard model, in which for the first time all the (non-gravitational) interactions and the fundamental particles (quarks and leptons) were coherently described within a unified field theory framework. Looking at the past, however, one recognizes that this achievement has its very foundations in the two building blocks of any modern physical theory: special relativity and quantum mechanics, which have been left essentially unaffected by the later developments.

Quantum electrodynamics has provided a basic reference for the formulation of the standard model and in general for any field theory description of the fundamental interactions. In particular, a major role is played in quantum electrodynamics by the concept of gauge symmetry which is the guiding principle for the correct description of the interaction. Likewise, the standard model too, as a quantum field theory, is based on a suitable gauge symmetry, which is a non-abelian extension of that present in quantum electrodynamics.

On the basis of these considerations, we hope the concise account of quantum electrodynamics that we give at the end of our book can provide the interested reader with the necessary background to cope with more advanced topics on theoretical particle physics, in particular with the standard model.

The present book is intended to be accessible to students with only a basic knowledge of non-relativistic quantum mechanics.

We start with a concise, but (hopefully) comprehensive exposition of special relativity, for which we have added a chapter on the implications of the principle of equivalence. Here we have a principle whose importance can be hardly overestimated since it is at the very basis of the general theory of relativity, but whose discussion in a class, however, requires no more than a couple of hours. Nevertheless, this issue and its main implications are rarely dealt with even in graduate courses of physics. Can general relativity be totally absent from the scientific education of a student of physics or engineering? Of course it can be, as far as the full geometrical formulation of theory is concerned. However it is well known that many technological devices, mainly the GPS, require for their proper functioning to consider the corrections implied by the Einstein's theories of special and general relativity. Our account of the principle of equivalence and its main implications will allow us to derive in a rather non-rigorous but intuitive way the concepts of connection, curvature, geodesic lines, etc., emphasizing their intimate connection to gravitational physics.

Thereafter, in Chaps. 4 and 7, we give the basics of the theory of groups and Lie algebras, discussing the group of rotations, the Lorentz and the Poincaré group. We also give a concise account of representation theory and of tensor calculus, in view of its application to the formulation of relativistically covariant physical laws. These include the Maxwell's equations, which we discuss in their manifestly covariant form in Chap. 5.

In Chap. 6, anticipating part of the analysis which will be later developed, we discuss the quantization of the electromagnetic field in the radiation gauge. We thought it worth illustrating this important example early because it clarifies how the concepts of photon and of its spin emerge quite naturally from a straightforward application of special relativity and quantum theory in a field theoretical framework.

In Chap. 8 we review the essentials of the Lagrangian and Hamiltonian formalisms, first considering systems with a finite number of degrees of freedom, and then extending the discussion to fields. Particular importance is given to the relation between the symmetry properties of a physical system and conservation laws.

The last four chapters are devoted to the development of quantum field theory. In Chap. 9 we recall the basic construction of quantum mechanics in the Dirac notation. Eventually in Chap. 10, we study the quantum relativistic wave equations emphasizing their failure to represent the wave function evolution in a consistent way.

In Chap. 11 we perform the quantization of the free scalar, spin 1/2 and electromagnetic fields in the relativistically covariant approach. The final goal of this analysis is to provide the quantum relativistic description of fields in interaction, with particular reference to the interaction between spin 1/2 fields (like an electron) and the electromagnetic one (quantum electrodynamics). This is done in Chap. 12 in which the graphical description of interaction processes by means of Feynman diagrams is introduced. After the classical example of the tree-level processes, we start analysing the one-loop ones where infinities make their appearance. We then discuss how one can circumvent this difficulty through the process of renormalization, in order to obtain sensible results. We shall however limit ourselves to give only a brief preliminary account of the renormalization programme and its implementation at one-loop level.

As the reader can realize, there is scarcely any ambition on our side to develop various topics in an original way. Our goal, as pointed out earlier, is to give in a single one-year course the main concepts which are at the basis of contemporary theoretical physics.

A Note on the Bibliography

It is almost impossible to give even a short account of the many textbooks covering some of the topics which are dealt with in this book. Any textbook on relativity or elementary particle theory covers at least a part of the content in our book. We therefore limit ourselves to quote those excellent standard textbooks, which have been for us a precious guide for the preparation of the present work, referring them to the interested reader in order to deepen the understanding of the topics dealt with in this book.

Contents

Chapter 1
Special Relativity

1.1 The Principle of Relativity

The aim of physics is to describe the laws underlying physical phenomena.

This description would be devoid of a universal character if its formulation were different for different observers, that is for different frames of reference, and, as such, it could not deserve the status of an objective law of nature. Any physical theory should therefore fulfil the following requirement:

The laws of physics should not depend on the frame of reference used by the observer.

This statement is referred to as the *principle of relativity*, and is really at the heart of any physical theory aiming at the description of the physical world.

Actually, the physical laws are described in the language of mathematics, that is by means of one or more equations involving physical quantities, whose value in general will depend on the reference frame (RF) used for their measure. As a consequence of this, any change in the reference frame results in a change in the physical quantities appearing in the equations, so that in general these will satisfy new equations, called *transformed equations*. The requirement that the transformed equations be equivalent to the original ones, so that they describe the same physical law, allows us to give a more precise formulation of the principle of relativity:

The equations of a physical theory must preserve the same form under transformations induced by a change in the reference frame.

By *preserving the same form* we mean that if the physical law is given in terms of a single equation, the transformed equation will have exactly the same form, albeit in terms of the transformed variables. If we have a system of equations, we can allow the transformed system to be a linear combination of the old ones. Obviously, in both cases, the physical content of the original and transformed equations would be exactly the same.

Changes in the reference frame of an observer can be of different kinds: spatial translations, rotations, or any change in its state of motion. As we shall see in the

© Springer International Publishing Switzerland 2016
R. D'Auria and M. Trigiante, *From Special Relativity to Feynman Diagrams*,
UNITEXT for Physics, DOI 10.1007/978-3-319-22014-7_1

sequel, the latter transformations are the most relevant as far as the implications on the description of the physical world are concerned.

The simplest relative motion is of course the *uniform rectilinear motion* or *inertial motion*, and the requirement that the physical laws be independent of the particular inertial frame means that the theory satisfies the requirements of the principle of relativity only as far as inertial frames are concerned. We recall that *inertial frames* are those in which the Galilean principle of inertia holds, and that, given any inertial frame such as, for instance, the one attached to the center of mass of the solar system, with axes directed towards fixed stars (the *fixed star system*), all other inertial frames are in relative rectilinear uniform motion with respect to it.

In the following two chapters we shall refrain from considering accelerated (and thus non-inertial) frames of reference, restricting ourselves to the analysis of the implications of the principle of relativity only as far as inertial frames are concerned, which is the main subject of the *special theory of relativity*.

The extension of the principle of relativity to any kind of relative motion between observers, that is to accelerated reference frames, however, has a very deep impact on our ideas of space, time and matter and leads to a beautiful new interpretation of the gravitational force as a manifestation of the geometry of four-dimensional space-time. This analysis, which is the subject of Einstein's *general theory of relativity*, requires, for its understanding, a solid knowledge of differential geometry and goes beyond the scope of this book; in Chap. 3, however, we shall give a short introduction to general relativity by discussing the *principle of equivalence* and tidal forces. Furthermore an intuitive picture of the four-dimensional geometry of space-time and its relation to gravitation will be outlined.

1.1.1 Galilean Relativity in Classical Mechanics

In order to verify whether a theory satisfies the principle of relativity we need to know the transformation laws relating the measures of physical quantities obtained by different observers. When describing the motion of a system of bodies with respect to a reference frame all the quantities we need can be expressed in terms of length, time and mass.[1] It is therefore sufficient to find the transformation laws for these fundamental quantities.

The principle of relativity was first applied to classical mechanics; in this context, however, only the transformation law of the space intervals is relevant; indeed, as it is apparent from the formulation of Newton's second law, the *inertial mass of a point-like object is defined as the constant ratio between the strength of the force acting on it and the modulus of the resulting acceleration,* and *this constant value*

---

[1]The *reference frame* associated with an observer is defined by a coordinate system, which we shall choose to be a system of rectangular Cartesian coordinates (x, y, z) with origin O, with respect to which the observer is at rest. The frame also consists of all the instruments the observer needs for measuring the fundamental quantities: a ruler for lengths, a clock for time intervals and scales for masses.

is assumed to be independent of the actual value of the velocity of the body. Since a change in the state of motion of a reference frame results in a different velocity of the body as measured by the new observer, this implies that *the value of the mass is the same in all reference frames.*

As far as time intervals are concerned, they were also assumed to be independent of the particular inertial observer. In the words of Newton: *tempus est absolutum, spatium est absolutum.* The first statement about the *absolute* character of time means that time flows equally for all observers so that the same time-interval between two events is measured by any (inertial) observer; the second statement *space is absolute,* means that space-intervals, or lengths, do not depend on the reference frame in which they are measured, and, as we shall show presently, it is actually a consequence of the first.

To illustrate this we first need to derive the transformation law for the position vector $\mathbf{x}(t)$ of a material point due a change in the reference frame (for the sake of simplicity, here and in the following, unless differently stated, when speaking of reference frames or observers, we shall be always mean *inertial* frames and observers).

Let us denote by S and S' two inertial frames, as well as the observers associated with them, and let (x, y, z), O and (x', y', z'), O' be their coordinates and origins, respectively. Since S and S' are both assumed to be inertial, their relative motion, say of S' with respect to S, is of rigid translational type with constant velocity $\mathbf{V}$. Note that it is the same thing to say that S' moves with velocity $\mathbf{V}$ with respect to S or that S moves with velocity $-\mathbf{V}$ with respect to S'.

It is moreover convenient to make the following assumptions: The axes of S and S' are parallel and equally oriented, the x- and x'- axes coincide and have the same orientation as the velocity $\mathbf{V}$. If we denote by t and t' the times as measured by the two observers S and S' respectively, their common origin $t = t' = 0$ is chosen as the instant at which the two origins O and O' coincide: $O = O'$. With these assumptions the relative configuration of the two frames is referred to as the *standard configuration*, see Fig. 1.1. As we shall see in the following, all the main physical implications of the principle of relativity are already present in this simplified situation.

Fig. 1.1 Position vectors of P relative to two inertial frames, S, S', in standard configuration

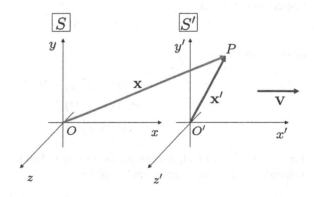

Let the two observers S and S' have identical instruments for measuring distances and time intervals. The assumption of absolute time implies that the times measured by S and S' are the same:

$$t = t'. \tag{1.1}$$

Next, suppose that the two inertial observers are describing a same event, say the position P at the time $t \equiv t'$, of a particle moving along a given trajectory. The position P with respect to O and O' is described by two different vectors $\mathbf{x}(t)$ and $\mathbf{x}'(t)$, whose components are:

$$\mathbf{x}(t) = (x(t), y(t), z(t)), \quad \mathbf{x}'(t) = (x'(t), y'(t), z'(t)). \tag{1.2}$$

By trivial geometrical considerations we derive the relation between $\mathbf{x}(t)$ and $\mathbf{x}'(t)$:

$$\mathbf{x}(t) = \mathbf{x}'(t) + \mathbf{OO}'(t) = \mathbf{x}'(t) + \mathbf{V}\,t. \tag{1.3}$$

In the standard configuration we have $\mathbf{V} = (V, 0, 0)$, and Eq. (1.3) can be written in components as follows:

$$\begin{aligned} x(t) &= x'(t) + V\,t, \\ y(t) &= y'(t), \\ z(t) &= z'(t). \end{aligned} \tag{1.4}$$

To obtain the relation between the velocities $\mathbf{v}$ and $\mathbf{v}'$ as measured with respect to S and S', respectively, one must differentiate $\mathbf{x}$ with respect to t in S and $\mathbf{x}'$ with respect to t' in S'. However, because of Eq. (1.1), we can simply differentiate both vectors with respect to the same variable $t = t'$, obtaining:

$$\mathbf{v} = \frac{d\mathbf{x}}{dt}, \quad \mathbf{v}' = \frac{d\mathbf{x}'}{dt'} = \frac{d\mathbf{x}'}{dt}. \tag{1.5}$$

Using the above definitions and differentiating both sides of Eqs. (1.3) or (1.4) with respect to t, we find:

$$\mathbf{v} = \mathbf{v}' + \mathbf{V}, \tag{1.6}$$

or in components:

$$\begin{aligned} v_x(t) &= v'_x(t) + V, \\ v_y(t) &= v'_y(t), \\ v_z(t) &= v'_z(t). \end{aligned} \tag{1.7}$$

Equation (1.6), or (1.7), defines the *composition law for velocities* and implies that velocities behave like vectors under addition.

A further differentiation of (1.6), or (1.7), with respect to t gives the relation between the accelerations as measured by the two observers. Taking into account that the relative velocity $\mathbf{V}$ is constant, we find:

$$\mathbf{a}(t) = \mathbf{a}'(t), \tag{1.8}$$

or, in components,

$$
\begin{aligned}
a_x &= a'_x, \\
a_y &= a'_y, \\
a_z &= a'_z.
\end{aligned}
\tag{1.9}
$$

Equations (1.3), (1.6), (1.8), or, in components, Eqs. (1.4), (1.7), (1.9), are called *Galilean transformations* and represent the relations between the measures of the kinematical quantities referred to two inertial reference frames in relative motion with constant velocity $\mathbf{V}$.

We are now ready to prove that the Newtonian statement about *absolute space* (*spatium est absolutum*) is a consequence of the analogous assumption about time. In other words, we verify that the *spatial distance between two points is the same for all inertial observers*. As an example, let us consider, as shown in Fig. 1.2, a rod placed along the x-axis whose endpoints A e B are at rest with respect to S'. The position vectors of A and B in S' are then $\mathbf{x}'_A = (x'_A, 0, 0)$ and $\mathbf{x}'_B = (x'_B, 0, 0)$, so that their distance L' in S', corresponding to the length of the rod, is:

$$L' = x'_B - x'_A. \tag{1.10}$$

We note that in S' the coordinates x'_B and x'_A are *time independent* and therefore can be measured at different times without affecting the value of their difference, that is the measure of the length of the rod. In S instead the coordinates of the endpoints depend on time, due to the relative motion of S' and S: $\mathbf{x}_A(t_A) = (x_A(t_A), y_A(t_A), z_A(t_A))$ and $\mathbf{x}_B(t_B) = (x_B(t_B), y_B(t_B), z_B(t_B))$. Their expression in terms of the coordinates of A and B in S' are given by (1.4):

Fig. 1.2 A rod at rest with respect to S'

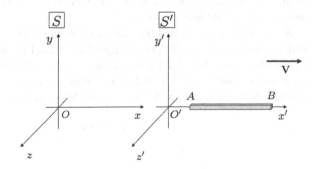

$$x_A(t_A) = x'_A + V\,t_A, \qquad x_B(t_B) = x'_B + V\,t_B,$$
$$y_A(t_A) = 0, \qquad\qquad\quad y_B(t_B) = 0, \qquad\qquad (1.11)$$
$$z_A(t_A) = 0, \qquad\qquad\quad z_B(t_B) = 0.$$

In order to compute the *length L* of the rod in S we must consider the coordinates of the endpoints A and B *at the same instant*, since evaluating them at different times would lead to a meaningless result. Setting $t = t_A = t_B$ we find:

$$x'_B - x'_A = (x_B(t_B) - V\,t_B) - (x_A(t_A) - V\,t_A) = x_B(t) - x_A(t). \qquad (1.12)$$

Equation (1.12) then implies:

$$L = L', \qquad\qquad (1.13)$$

that is, *the length of the rod is the same for both observers*. Note that, in defining the measure of the length L of the moving rod, we have used the notion of *simultaneity* of two events, $t_B = t_A$. This concept is, however, independent of the reference frame since, having assumed from the beginning the equality of time durations, that is $\Delta t = \Delta t'$, in different frames, simultaneity in S ($\Delta t = 0$) implies simultaneity in S' ($\Delta t' = 0$) for any two inertial frames S and S'. We have thus proven that invariance of the lengths (*absolute space*) is a consequence of invariance of the time intervals (*absolute time*).

In the previous discussion we have considered the rod lying along the x- axis, which is the direction of the relative motion. It is obvious that the distances along the y- or z-axes are also invariant since $y' = y$ e $z' = z$. This means that the vector describing the relative position of any two points in space is *invariant* under Galilean transformations. More specifically, if A and B are two points at rest in S', (not necessarily along the x-axis) with position vectors $\mathbf{x}'_A, \mathbf{x}'_B$ and relative position vector $\Delta\mathbf{x}' \equiv \mathbf{x}'_B - \mathbf{x}'_A$, and if $\mathbf{x}_A(t_A), \mathbf{x}_B(t_B)$ are the position vectors of the two points relative to S at different times, we define the relative position vector in S as the difference between the position vectors taken *at the same instant t*:

$$\Delta\mathbf{x}(t) \equiv \mathbf{x}_B(t) - \mathbf{x}_A(t) = (\mathbf{x}'_B + \mathbf{V}t) - (\mathbf{x}'_A + \mathbf{V}t) = \mathbf{x}'_B - \mathbf{x}'_A = \Delta\mathbf{x}'. \qquad (1.14)$$

We conclude that not only the spatial distance between A e B, but also the direction from A to B, i.e. the direction and orientation of the relative position vector, is invariant under Galilean transformations.

So far we have examined the change of inertial frames due to a relative motion with a constant velocity $\mathbf{V}$. The change of an inertial frame due to a rotation or to a rigid translation of the coordinate axes are in a sense trivial. They correspond to the *congruence* transformations of the Euclidean geometry leaving invariant the space relations between figures and objects. They have the form

$$\mathbf{x}' = \mathbf{R}\,\mathbf{x} + \mathbf{b},$$

where $\mathbf{R}$ denotes a 3×3 matrix which implements a generic rotation or reflection. Another trivial transformation is the change of the time origin, or *time translation* namely

$$t' = t + \beta.$$

In general one refers to the invariance of the laws of physics under rotations/reflections $\mathbf{R}$ and translations $\mathbf{b}$ as the properties of *isotropy and homogeneity of space*, respectively. Similarly the invariance under shifts in the time origin is referred to as *homogeneity in time*. Note that both transformations do not affect the Newton postulates of *absolute time* $\Delta t' - \Delta t$ and *absolute space*, $|\Delta \mathbf{x}'| = |\Delta \mathbf{x}|$. Including the congruence transformations and the time shift gives a more general form to the Galilean transformations, namely:

$$\mathbf{x}' = R\,\mathbf{x} + \mathbf{b} - \mathbf{V}t, \tag{1.15}$$
$$t' = t + \beta. \tag{1.16}$$

Unless explicitly mentioned, when referring to Galilean transformations we shall always refer to the simpler form given in Eq. (1.3) or (1.4), (1.6), (1.8).

1.1.2 Invariance of Classical Mechanics Under Galilean Transformations

We have seen that under the assumption of the invariance of time intervals, the Galilean transformations, expressed by Eqs. (1.3), (1.6), (1.8), or, in components, by Eqs. (1.4), (1.7), (1.9), provide the relations between the kinematical quantities as measured in any two inertial systems.

To verify that classical mechanics satisfies the principle of relativity, we need to transform the fundamental equations of the theory and see whether they keep the same form in the new reference frame.

Let us start from the *principle of inertia*: Suppose that in the frame S a free particle, that is not subject to interactions, moves at a constant velocity $\mathbf{v}$. From Eq. (1.6) we see that in S' its velocity $\mathbf{v}' = \mathbf{v} - \mathbf{V}$ is also constant, owing to the constancy of $\mathbf{V}$. Similarly if $\mathbf{v}'$ is constant, also $\mathbf{v}$ is and thus the law of inertia satisfies the principle of relativity.

Let us now examine the second law, namely the Newtonian equation of motion:

$$\mathbf{F} = m\,\mathbf{a}. \tag{1.17}$$

As already pointed out the mass appearing on the right-hand side of Eq. (1.17) is assumed to be the same in any reference frame; furthermore Eq. (1.8) implies that the acceleration has the same property. Thus the right-hand side is *invariant* under a change in the reference frame. In order for the principle of relativity to be satisfied, the force on the left-hand side must be invariant under Galilean transformations as well.

To ascertain this we recall that in classical mechanics a force[2] is defined as an *action at-a-distance* between two interacting particles with the following properties: Its direction coincides with the straight line connecting the particles, its strength only depends on their *distance* and it acts on each of them according to the *principle of action and reaction*. These properties define a conservative force. Explicitly, if Δx is the relative-position vector of the interacting particles and $|\Delta x|$ their distance, the force **F** acting on one of them has the following form:

$$\mathbf{F} = F(|\Delta \mathbf{x}|)\frac{\Delta \mathbf{x}}{|\Delta \mathbf{x}|}. \tag{1.18}$$

If we now recall that the vector Δx is left unchanged by Galilean transformations, we immediately conclude that the force itself is invariant.

Thus both sides of Eq. (1.17) are invariant[3] under a change in the reference frame and therefore the Newtonian equation of motion satisfies the principle of relativity. We refer to this propriety as the invariance of classical mechanics under Galileo transformations. We stress once again that this conclusion is valid only under the assumption that *the mass of a particle does not depend on its velocity*.

The study of electromagnetic phenomena has revealed the existence of fundamental forces of a different kind, not fitting the characterization given in classical (Newtonian) mechanics and described by Eq. (1.18). Think about the magnetic force exerted by an electric current in a segment of wire on the magnetized pointer of a compass. Its direction does not coincide with the straight-line connecting the wire to the compass, and moreover this force is non-conservative. Besides the action at-a-distance picture of classical mechanics turns out to be inadequate to describe electromagnetic interactions involving fast-moving charged particles. What these processes suggest is that the interaction between two particles should rather be described as mediated by a physical, propagating *field*, such as the electromagnetic field for interacting charges. In this new picture a force on one particle originates from an *action-by-contact* on it of the field electromagnetic field generated by another charged particle: instead the action of the gravitational field generated by one mass on another mass obeys the *action at-a-distance* principle in classical Newtonian mechanics.

Restricting ourselves, for the time being, to the purely mechanical case and to the Newtonian description of forces, it is interesting to examine the implications of the principle of relativity on the *conservation law of the total linear momentum of an isolated system of particles*.

This property is usually seen as a direct consequence of the second and third Newton laws. It is however well known that the law of conservation of linear

---

[2]Here we are referring to fundamental forces of the nature, like the gravitational force, not to phenomenological forces like elastic forces, friction etc.

[3]In general we call *covariant* an equation which takes the same form in different frames; if not just the form, but also the numerical values of the various terms are the same, we then say that the equation is *invariant*.

momentum can be taken, together with the principle of inertia, as a principle from which both Newton's law of motion and the action-reaction principle can be deduced. Indeed, from a modern point of view, the law of conservation of linear momentum is more fundamental than Newton's laws in that it retains its validity also in those situations where the concept of Newtonian forces is no longer applicable (provided its definition be appropriately extended).

Consider, with respect to some reference frame S, an interaction process in which two particles, not subject to external forces, with linear momenta $\mathbf{p}_1$ and $\mathbf{p}_2$, and masses m_1 and m_2, interact for a very short time (scattering), and give rise, in the final state, to two free particles with momenta $\mathbf{q}_1$ and $\mathbf{q}_2$, and masses μ_1 and μ_2 (which can be different from m_1, m_2, as it generally happens, for instance, in chemical reactions). Now suppose that the conservation of linear momentum is verified in S:

$$\mathbf{p}_1 + \mathbf{p}_2 = \mathbf{q}_1 + \mathbf{q}_2. \tag{1.19}$$

Denoting by $\mathbf{v}_1$, $\mathbf{v}_2$ and by $\mathbf{u}_1$, $\mathbf{u}_2$ the initial and final velocities of the two particles, respectively, Eq. (1.19) can be written as follows:

$$m_1\,\mathbf{v}_1 + m_2\,\mathbf{v}_2 = \mu_1\,\mathbf{u}_1 + \mu_2\,\mathbf{u}_2. \tag{1.20}$$

Let us now consider the same process in a new inertial frame S', related to S by a Galilean transformation. Substituting the old velocities in terms of the new ones and using the relation (1.6), Eq. (1.20) becomes:

$$m_1(\mathbf{v}_1' + \mathbf{V}) + m_2(\mathbf{v}_2' + \mathbf{V}) = \mu_1(\mathbf{u}_1' + \mathbf{V}) + \mu_2(\mathbf{u}_2' + \mathbf{V}). \tag{1.21}$$

Since $m_1\mathbf{v}_1'$, $m_2\mathbf{v}_2'$, $\mu_1\mathbf{u}_1'$, $\mu_2\mathbf{u}_2'$ are the initial and final linear momenta $\mathbf{p}_1'$, $\mathbf{p}_2'$, $\mathbf{q}_1'$, $\mathbf{q}_2'$ of the particles, as measured in S', (1.21) takes the following form:

$$\mathbf{p}_1' + \mathbf{p}_2' = \mathbf{q}_1' + \mathbf{q}_2' + (\mu_1 + \mu_2 - m_1 - m_2)\mathbf{V}. \tag{1.22}$$

This relation implies that the conservation of linear momentum satisfies the principle of relativity if, and only if, *the total mass is conserved*

$$\mu_1 + \mu_2 = m_1 + m_2. \tag{1.23}$$

Indeed, under this condition, Eq. (1.22) becomes:

$$\mathbf{p}_1' + \mathbf{p}_2' = \mathbf{q}_1' + \mathbf{q}_2', \tag{1.24}$$

which expresses the conservation of momentum also in the frame S', consistently with the principle of relativity.

We end this section with a few observations. From the above discussion, it follows that the conservation of the total mass is not an independent principle in classical mechanics, but rather a consequence of the law of conservation of linear momentum and the principle of relativity, a fact which is not always stressed in standard treatments of Newtonian mechanics.

Secondly, if we consider the more general Galileo transformations (1.15), the invariance of Newtonian mechanics with respect to spatial translations and time shifts is obvious. As far as the invariance under rotations of the coordinate frame is concerned it is sufficient to observe that the equations of classical mechanics can be written as *three-dimensional vector equations*; since vectors are geometrical objects (oriented segments) independent of the orientation of the coordinate frame, the same is true for the vector equations of the theory. As a third point it must be noted that the fact that a theory satisfies the principle of relativity *does imply that the same physical laws hold true in every inertial frame*, but it *does not imply that the actual description of the motion is the same in different frames*.

For example, if a ball is thrown vertically ($v_x = 0$) in S, in S' it will have an initial velocity $v'_x = -V \neq 0$. In S the trajectory is a vertical straight line, while in S' the trajectory is a parabola. Mathematically this follows from the fact that the laws of mechanics are 2-order differential equations whose solution depends on the initial conditions, which are different in different frames.

We also want to stress the different way the principle of relativity is implemented for Newton's second law, (1.17), and the conservation of linear momentum, (1.19): In the latter case the same law holds in the new frame, but the physical quantities, the momenta, have different values, while in the former case all the quantities, force, mass, acceleration, have exactly the same values in the two frames. Under Galilean transformations therefore the conservation of linear momentum is an example of a *covariant* law, while the Newtonian law of motion is *invariant*.

As a last point we observe that the independence of the laws of classical mechanics from the inertial frame, is easily verified in our everyday life. Everybody traveling by car, train, or ship and moving with uniform rectilinear motion with respect to the earth (considered as an inertial frame) can observe that the oscillation of a pendulum, the bouncing of a ball, the collisions of billiard-balls, etc., occur exactly in the same way as in the earth frame. These considerations (with reference to a ship in uniform motion) were actually clearly illustrated and discussed by Galilei in his celebrated book *Dialogue Concerning the Two Chief World Systems*, (1630). On the other hand if the moving frame is accelerated the laws of mechanics are violated, since the new frame is no longer inertial.

1.2 The Speed of Light and Electromagnetism

It is well known that many mechanical phenomena, such as vibrating strings, acoustic waves in a gas, ordinary waves on a liquid, can be described in terms of propagating waves. These *mechanical waves* describe the propagation through a given material medium of a perturbation originating from a *source* located in a point or a region in its interior (like for instance the impact of a stone on the surface of a pond), the propagation being due to the interactions among the molecules of the medium. If the medium is homogeneous and isotropic (which we shall always assume to be the case) the speed of propagation of a wave has the same constant value in every direction *with respect to the medium itself.*

For example, in the case of acoustic waves propagating through the atmosphere, the speed of sound is $v_{(s)} \simeq 330\,\text{m/s}$ with respect to the air (supposed still).

Let S be a frame at rest with respect to the air and S' another frame in relative uniform motion with velocity $\mathbf{V}$ with respect to the former (we assume the standard configuration between the two coordinate systems). By means of (1.6) we may compute the velocity $\mathbf{v}'_{(s)}$ of a sound signal with respect to S'.

$$\mathbf{v}'_{(s)} = \mathbf{v}_{(s)} - \mathbf{V}. \tag{1.25}$$

If the sound is emitted along the x-direction, that is in the same direction as the relative motion, $\mathbf{v}_{(s)} = (v_{(s)}, 0, 0)$ and one obtains:

$$
\begin{aligned}
v'_{(s)x} &= v_{(s)} - V, \\
v'_{(s)y} &= 0, \\
v'_{(s)z} &= 0,
\end{aligned}
\tag{1.26}
$$

so that the velocity of the sound measured in S' along the x-direction will be lower than in S, $v'_{(s)x} = v_{(s)} - V < v_{(s)}$. Viceversa, if the sound is emitted in the negative x-direction, that is $\mathbf{v}_s = (-v_{(s)}, 0, 0)$, then the modulus of the velocity measured in S' will be greater than in S; indeed

$$
\begin{aligned}
v'_{(s)x} &= -v_{(s)} - V, \\
v'_{(s)y} &= 0, \\
v'_{(s)z} &= 0,
\end{aligned}
\tag{1.27}
$$

implying $|v'_{(s)x}| = |v_{(s)} + V| > v_{(s)}$.

Let us now consider a sound wave propagating in S along a direction perpendicular to the that of the relative motion, say along the negative y-axis, $\mathbf{v}_{(s)} = (0, -v_{(s)}, 0)$ (see Fig. 1.3). In S' the velocity of the sound signal is:

$$\mathbf{v}'_{(s)} = \mathbf{v}_{(s)} - \mathbf{V} = (-V, -v_{(s)}, 0) \tag{1.28}$$

It follows that for the observer in S', the sound wave will propagate along a direction forming an angle α with respect to the y-axis given by (Fig. 1.4):

$$\tan \alpha = \frac{V}{v_{(s)}}, \tag{1.29}$$

Fig. 1.3 Sound wave propagating, with respect to S, along the y-axis

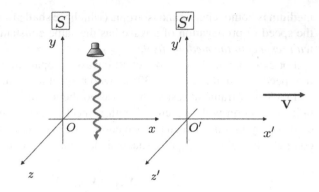

Fig. 1.4 Same sound wave as seen from S'

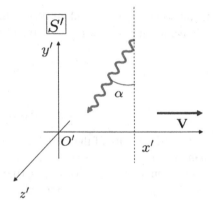

while the modulus of the velocity $v'_{(s)} \equiv |\mathbf{v}'_{(s)}|$ turns out to be:

$$v'_{(s)} = \sqrt{v^2_{(s)} + V^2} > v_{(s)}.\tag{1.30}$$

Note that, if $V \ll v_{(s)}$, the effect on the velocity of the motion of S' relative to the medium is an effect of order $\frac{V^2}{v^2_{(s)}}$ since:

$$\sqrt{v^2_{(s)} + V^2} \equiv v_{(s)}\sqrt{1 + \frac{V^2}{v^2_{(s)}}} \simeq v_{(s)}\left(1 + \frac{1}{2}\frac{V^2}{v^2_{(s)}}\right).\tag{1.31}$$

The example of a sound wave illustrates the general fact that the velocity of propagation of a mechanical wave is *isotropic*, that is the same in every direction, *only in a reference frame at rest with respect to the transmission medium*. In any other inertial frame, the wave velocity is not isotropic, but depends on the direction.

If we now consider the theory of electromagnetism, and in particular the propagation of electromagnetic waves, we immediately note some peculiarities with respect to ordinary material waves.

Electromagnetism, ignoring quantum processes, is described, with extremely good precision, by the *Maxwell equations*. Maxwell's theory predicts that electric and magnetic fields can propagate, in the form of electromagnetic waves, in the *vacuum*, that is apparently without a transmission medium, with a velocity, denoted by c, which is related to the parameters of the theory:

$$c = \frac{1}{\sqrt{\epsilon_0 \mu_0}} = 2.997925 \times 10^8 \, \text{ms}^{-1}.$$

We refer to this velocity as the *speed of light* since, as is well known, light is just an electromagnetic wave with wave-length in the approximate range between 380 and 780 nm. According to the principle of relativity, this velocity, being determined only by the parameters of the theory, should be the same for all the inertial observers. On the other hand, we have learned that the velocity of a wave should change by a change in the (inertial) reference frame. How can we resolve this apparent contradiction?

We note, first of all, that not only the velocity of electromagnetic waves changes under a Galilean transformation, but, as one can easily ascertain, also the Maxwell equations themselves are not left invariant by such transformations (in fact they are not even covariant). Since we can not give up the principle of relativity for electromagnetic phenomena, there are only two possibilities:

- Either the Maxwell equations and their consequences are valid only in a particular frame, and thus should change their form by a change in reference frame;
- or the Maxwell equations are valid in every inertial frame, but the principle of relativity should not be implemented by the Galilean transformations. Instead, the right transformation laws should be chosen in such a way as to keep the validity of this principle also for electromagnetism as expressed by the Maxwell equations.

Let us first discuss the former hypothesis. If there existed a privileged reference frame in which the Maxwell equations hold, and thus with respect to which light has velocity c, we should be able to experimentally detect it. In this respect, over the course of the nineteenth century, physicists made various hypotheses, among which we quote the following two.

A first hypothesis was that the frame with respect to which light has velocity c, is the frame of the light source; however this possibility was immediately ruled out because it would have led to consequences in sharp contrast with astronomical observations. Indeed, suppose we observe a binary system of stars, of comparable masses, revolving around their common center of mass; since the respective momenta are directed in opposite directions, there will be an instant when the motion of the stars will be in the direction of the terrestrial observer, one approaching and other withdrawing from it with velocities $\mathbf{v}$ and $-\mathbf{v}$ respectively. If the velocity of light were c with respect to the emitting source, with respect to the earth the velocities of the light signals emitted by the two stars would be $c + v$ and $c - v$, respectively, see

Fig. 1.5 Light emitted from
stars in a binary system

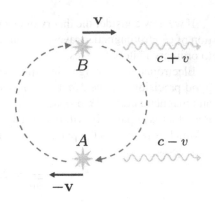

Fig. 1.5. Thus the two light waves would reach the terrestrial observer at different times and the motion observed from the earth would appear completely distorted with respect to that predicted by Newtonian mechanics. Needless to say that the motion we observe from the earth instead perfectly agrees with Newton's laws.[4]

A second hypothesis was based on the assumption that, in analogy with the mechanical waves, also electromagnetic waves propagate through a material medium, called *ether*; therefore, as it happens for the mechanical waves, the ether would be the privileged reference frame where the velocity of light is c and where the Maxwell equations take their usual form. If this were true, the ether, whose vibrations should propagate the electromagnetic waves, should fill the whole of space (thus allowing the light from the stars to reach the earth) and also penetrate the interior of material bodies. This hypothetical substance would actually have very unusual properties: It should be stiff enough to give light such an enormous velocity, but also light enough to allow for the motion of stars and planets through it.

If ether existed, it is reasonable to assume its rest frame to coincide, to a good approximation, with the frame of the fixed stars, which, as is well known, is the canonical reference frame where the principle of inertia and the whole of classical mechanics hold. Accordingly, one should be able to detect the change in the velocity of light (from the value c) as measured in a reference frame in motion with respect to the fixed stars, namely, with respect to the ether.

Actually, when we observe the light coming from a star, we are in a frame which, being attached to the earth, is moving with a velocity $V \approx 30$ km/s with respect to the fixed star system. The velocity of the light from such a star in the earth frame should be related to c by the Galilean transformations.

Let S be the fixed star frame, and S' the earth frame moving with velocity $\mathbf{V}$ along the x-direction. Suppose, for the sake of simplicity, that we are observing a light ray

---

[4]Another reason for ruling out the emitting source as the privileged frame where the Maxwell equations hold is the fact that, according to the laws of electromagnetism, an electric charge in an electric field $\mathbf{E}$ acquires an acceleration $\mathbf{a}$ which must vanish when $\mathbf{E} \to 0$; when the frame is accelerated, as it is generally the case for a moving source, $\mathbf{a}$ would preserve a non vanishing component equal to acceleration of the reference frame even in the limit of vanishing electric field.

coming from the negative direction of the y-axis in S; note that we are exactly in the same situation as previously described for the sound waves (see Figs. 1.3 and 1.4). Thus the same conclusions should hold provided we replace, in Eqs. (1.27)–(1.31), $\mathbf{v}_{(s)}$ by $\mathbf{c}$. In particular we find that the light ray will reach the telescope on earth at an angle α with respect to the y'-axis given by

$$\tan \alpha = \frac{V}{c},$$ (1.32)

and a speed

$$c' = \sqrt{c^2 + V^2} > c.$$ (1.33)

As far as the first effect is concerned, it implies that in order for the light ray to reach the observer, the telescope should be adjusted by an angle α with respect to vertical direction. Because of the revolution of the earth, in order to observe a same star over one year, the orientation of the telescope should be continuously adjusted, so that it describes a cone whose intersection with the sky defines a little ellipsis (Fig. 1.6). This phenomenon is in fact observed, and is called *aberration of starlight*. Every star on the sky is seen to describe an ellipsis over the course of one year. The angular half-width of the corresponding cone is given by (1.29), that is, taking into account of the numerical values of V and c, by (Fig. 1.6)

$$\tan \alpha \simeq 10^{-4} \quad \Rightarrow \quad \alpha = 20''.$$ (1.34)

The above value is consistent, in the limit of experimental errors, with the current astronomical observations, and this therefore seems to support the hypothesis that the velocity of the light is c only with respect to the ether.

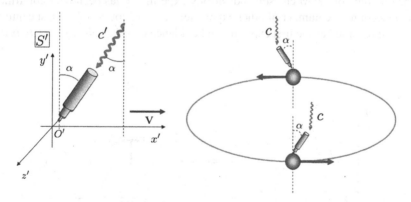

Fig. 1.6 Aberration of starlight: Figure to the left and to the right are referred to S' and S, respectively

This experimental evidence, however, relies on the measure of the change in the direction of the light ray, which is an effect of order $\frac{V}{c}$, but does not verify that its actual speed c' differs from c according to Eq. (1.33). To ascertain this we should be able to detect a non-vanishing difference $c' - c$, which, as observed in the analogous case of the sound waves, is an effect of order $\frac{V^2}{c^2}$, that is, in our case,

$$ c' = \sqrt{V^2 + c^2} = c\sqrt{1 + \frac{V^2}{c^2}} \quad \Rightarrow \quad \left(\frac{c' - c}{c}\right) \simeq \frac{V^2}{c^2} \approx 10^{-8}. $$

To verify so tiny an effect, it is therefore necessary to set up an experiment with a very high sensitivity. In 1886 Michelson and Morley realized such an experiment. The main idea behind it is that, if the ether existed, and if two light signals were emitted on the earth, one in the direction of its motion (with velocity **V** relative to the ether frame), and the other in the opposite direction, their velocities with respect to the earth would be $c - V$ and $-(c + V)$, respectively (Fig. 1.7). The experiment was so designed as to make two light rays, emitted by the same source, interfere after having traveled back and forth along two orthogonal paths. Under the ether hypothesis, a rotation of the interferometer by 90° would have changed the velocity of each light signal with respect to the earth, and this would have resulted in a shift of the interference fringes due to the change in the optical paths of the two beams.

The result of the experiment was negative, indicating that, with respect to the inertial frame tied to the earth, *the velocity of light is c in every direction*. Thus, either the earth reference frame is the ether frame where the velocity of light is the same in the every direction (this would give the earth back a central role in the Universe, four hundred years after Copernicus), or this result should be taken as the evidence that *the ether frame, and the ether itself as a physical substance, does not exist*.

The outcome of the Michelson and Morley experiment has been later confirmed by an unaccountable number of other experiments, in the course of the last century, and can be regarded as the first experimental evidence that *the velocity of light is the*

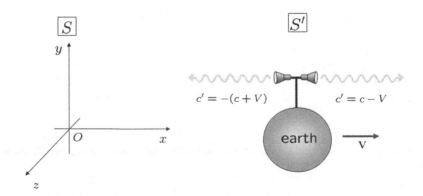

Fig. 1.7 Two light rays emitted from earth in opposite directions

same in every inertial frame. It is also consistent with the more general assumption that all the laws of physics, including electromagnetism, have the same form in every inertial frame, in agreement with the principle of relativity.

1.3 Lorentz Transformations

According to our discussion in the previous section, the apparent contradiction between the principle of relativity and the constancy of the speed of light, finds its natural solution in the possibility that the Galilean transformations, used for implementing the principle of relativity in classical mechanics, *are not the correct transformation laws relating the fundamental kinematic quantities x, y, z, t in different inertial frames.* Indeed, as we have seen, the use of the Galilean composition law for velocities leads to a speed of light which depends on the reference frame.

To find how the Galileo transformations must be amended, Einstein assumed as fundamental postulates the principle of relativity, which must apply to every law of physics (though restricted to inertial frames), and the following new proposition, based on the experimental evidence discussed in the previous section, and known as *the principle of the constancy of the speed of light*:

The speed of light in the vacuum is the same and is isotropic with respect any inertial reference frame, regardless of the motion of the source.

As we shall see in the following, this latter assumption is crucial in order to extend the validity of the principle of relativity from mechanics to electromagnetism and in general to all physical laws.

Starting from these two postulates Einstein developed his *theory of special relativity*[5] which led to a deep re-examination, from an operative point of view, of the very concepts of space and time and thus of the meaning of space and time intervals as well as of simultaneity. In particular the notions of absolute space and time, which the whole classical mechanics was founded on, were questioned.

Consistently with his two postulates (the principle of relativity and of constancy of the speed of light) Einstein proposed new transformation laws for space and time intervals which are known by the name of *Lorentz transformations* since they were originally formulated by the Dutch physicist Hendrik Lorentz, albeit with a different interpretation. These new transformation laws, together with all their consequences, represent the basis on which Einstein's theory is constructed. It is clear that the basic postulate $\Delta t = \Delta t'$ used in deriving the Galileo transformations must be given up, since, once taken for granted, the Galileo transformations are unavoidable.

To derive the new transformations let us start by writing down the most general relation between the coordinates *and times* of a given event as seen from two inertial frames S, S'. The validity of the principle of inertia in both frames $\mathbf{a} = 0 \Leftrightarrow \mathbf{a}' = 0$

---

[5]Here *special* refers to the fact that it is restricted to inertial frames.

requires that if the coordinates x, y, z depend linearly on time in S:

$$x = x_0 + v_x\,t,$$
$$y = y_0 + v_y\,t,$$
$$z = z_0 + v_z\,t,$$

also x', y', z' should depend linearly on t' in S', and this can only happen if the *transformation is linear*. We can therefore write:

$$
\begin{aligned}
x' &= a_{11}x + a_{12}y + a_{13}z + a_{14}t + a,\\
y' &= a_{21}x + a_{22}y + a_{23}z + a_{24}t + b,\\
z' &= a_{31}x + a_{32}y + a_{33}z + a_{34}t + c,\\
t' &= a_{41}x + a_{42}y + a_{43}z + a_{44}t + d.
\end{aligned}
\tag{1.35}
$$

Furthermore, for the sake of simplicity, we will also take the two frames, like in the Galilean case, in the *standard configuration*, see Fig. 1.1. We shall see in the following (see also Chap. 4) how the Lorentz transformations can be extended to more general configurations.

We shall show presently, working in the standard configuration, that the twenty undetermined coefficients appearing in (1.35) can be reduced by kinematical consideration, to just one.

First of all, having chosen the origin of times $t = t' = 0$ as the time at which the origins coincide $O = O'$, it immediately follows that, in (1.35), $a = b = c = d = 0$, so that $x = y = z = 0$ implies $x' = y' = z' = 0$.

Next we observe that the equation of the $z'y'$ plane, with respect to S has the form:

$$x = V t,$$

while for the observer S' the corresponding equation reads

$$x' = 0, \qquad \forall t'.$$

It then follows that

$$x' = \alpha(V)\,(x - Vt),\tag{1.36}$$

where the coefficient $\alpha(V)$ must be independent of the coordinates and time, and can then only depend on the "kinematic parameter" V.

Secondly, as the planes xz ed $x'z'$ coincide at all times, $y = 0$ should imply $y' = 0$ and this constrains the relation between y and y' to have the following form

$$y' = \beta(V)\,y.\tag{1.37}$$

Note that had we started with the opposite orientation of the x- and x'-axes we would have obtained:

$$y' = \beta(-V)\,y.\tag{1.38}$$

The simultaneous validity of Eqs. (1.37) and (1.38) requires β to be an even function of V:

$$\beta(V) = \beta(-V), \tag{1.39}$$

Furthermore the principle of relativity implies that nothing should change if we exchange the roles of the two observers, that is if we consider S in motion with respect to S' with velocity $-V$; in that case the primed coordinates become unprimed and viceversa, so that we may also write:

$$y = \beta(-V) y'. \tag{1.40}$$

Combining the Eqs. (1.37) with (1.40) we readily obtain:

$$y = \beta(-V) y' = \beta(-V) \beta(V) y = \beta^2(V) y \quad \Rightarrow \quad \beta^2(V) = 1.$$

which implies $\beta(V) = \pm 1$. On the other hand, since we have orientated y and y' in the same direction, we must have $\beta(V) \equiv 1$. By the same token we also find $z = z'$.

Thus the first three equations of the transformations (1.35) take the simple form:

$$x' = \alpha(V) (x - V t), \tag{1.41}$$
$$y' = y, \tag{1.42}$$
$$z' = z. \tag{1.43}$$

Let us now consider the fourth equation involving the time variable t'. Solving Eq. (1.41) with respect to t we find:

$$t = \frac{1}{V} \left(x - \frac{x'}{\alpha(V)} \right), \tag{1.44}$$

Using the same argument which led to Eq. (1.38), if we consider S in motion with velocity $-V$ with respect to S', the equation obtained from (1.44) by replacing t' with t, x with x' and V with $-V$ must also be true:

$$t' = -\frac{1}{V} \left(x' - \frac{x}{\alpha(-V)} \right)$$
$$= -\frac{1}{V} \left(\alpha(V)(x - V t) - \frac{x}{\alpha(-V)} \right)$$
$$= \alpha(V) t + \frac{1}{V} \left(\frac{1}{\alpha(-V)} - \alpha(V) \right) x. \tag{1.45}$$

We may then rewrite the transformation (1.45) as follows:

$$t' = \alpha(V) t + \delta(V) x. \tag{1.46}$$

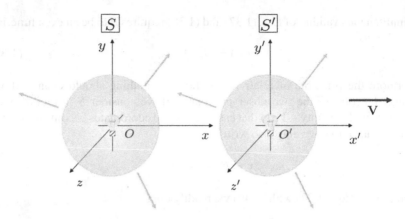

Fig. 1.8 Light signal as seen by S and S'

where we have set

$$\delta(V) = \frac{1}{V}\left(\frac{1}{\alpha(-V)} - \alpha(V)\right). \tag{1.47}$$

By simple considerations we have reduced the problem of determining all the coefficients in (1.35), to that of computing a single function $\alpha(V)$.

This coefficient will be now determined by implementing the principle of constancy and isotropy of the speed of light.

Let us suppose that at $t = t' = 0$, when $O \equiv O'$, a light (or electromagnetic wave) source emits a signal isotropically, see Fig. 1.8. According to this principle, the signal propagates isotropically with the same constant speed c for both observers S and S'. Thus with respect to the two frames the wave front of the electromagnetic signal will be described by spheres of radii $r = ct$ and $r' = ct'$ respectively. The equations for the wave front of the spherical wave are thus given by:

$$x^2 + y^2 + z^2 - c^2t^2 = 0, \tag{1.48}$$

for the observer S, and

$$x'^2 + y'^2 + z'^2 - c^2t'^2 = 0, \tag{1.49}$$

for the observer S'.

Since the four coordinates (x, y, z, t) and (x', y', z', t') refer to the same physical events, that is the locus of points reached by the signal at a fixed time, they must hold simultaneously. We must then have:

$$x^2 + y^2 + z^2 - c^2t^2 = \kappa\left(x'^2 + y'^2 + z'^2 - c^2t'^2\right). \tag{1.50}$$

where κ is a constant.

If we now substitute the expression of x', y', z', t' in terms of x, y, z, t as given by Eqs. (1.41), (1.42), (1.43) and (1.46), in Eq. (1.50), we obtain:

$$x^2 + y^2 + z^2 - c^2 t^2 = \kappa \left[\alpha^2 (x - Vt)^2 + y^2 + z^2 - c^2 (\alpha t + \delta x)^2 \right], \quad (1.51)$$

and this relation must be an identity in (x, y, z, t).

Comparing the coefficients of z and y on both sides, we immediately find $\kappa = 1$. Next, equating the coefficients of t^2, one finds:

$$-\alpha^2(V) \, V^2 = c^2 \left(1 - \alpha^2(V) \right) \quad \Rightarrow \quad \alpha(V) = \pm \frac{1}{\sqrt{1 - \frac{V^2}{c^2}}}.$$

Since at $t = t' = 0$, x and x' have the same orientation, we conclude that:

$$\alpha(V) = \alpha(-V) = \frac{1}{\sqrt{1 - \frac{V^2}{c^2}}}. \quad (1.52)$$

One can easily verify that, with the above value of $\alpha(V)$, also the coefficients of x^2 and $x\,t$ are equal. The transformation laws (1.41), (1.42), (1.43) and (1.46) now take the following final form

$$x' = \gamma(V) \, (x - V t), \quad (1.53)$$
$$y' = y, \quad (1.54)$$
$$z' = z, \quad (1.55)$$
$$t' = \gamma(V) \left(t - \frac{V}{c^2} x \right), \quad (1.56)$$

where

$$\gamma(V) \equiv \frac{1}{\sqrt{1 - \frac{V^2}{c^2}}} > 1. \quad (1.57)$$

Equations (1.53)–(1.56) are the *Lorentz transformations*. They represent the correct transformation laws connecting two inertial frames, which allow to extend the principle of relativity to electromagnetism, as we shall discuss in detail in Chap. 5. In the present chapter and in the following one we shall deal with the consequences of the Lorentz transformations in kinematics and dynamics.

One can verify, however, that the well established equations of classical mechanics, which are covariant under Galilean transformations, are not covariant under Lorentz transformations. It seems as if, by requiring the principle of relativity to hold for electromagnetism, we loose its validity in mechanics. In order to solve this apparent inconsistency, we should consider the fact that, until the beginning of the twentieth century, before the discovery of the subnuclear physics and of certain astrophysical phenomena, all the known physical processes involved bodies moving at speeds which are much lower than the speed of light. Now it is easy to show that the Lorentz transformations actually reduce to the Galilean transformations in the limit in which the velocity of the moving frame V is much smaller than c. Indeed, in this situation, applying the Taylor expansion to the factor $\gamma(V)$ in Eq. (1.53) and neglecting terms of order $\frac{V^2}{c^2}$,[6] we find

$$\gamma(V) = \frac{1}{\sqrt{1 - \frac{V^2}{c^2}}} \simeq 1 + \frac{1}{2}\frac{V^2}{c^2} + O\left(\frac{V^4}{c^4}\right) \simeq 1.$$

With the same approximation we may also set $t - \frac{V}{c^2}x \simeq t$. Thus in this limit the Lorentz transformations (1.53) reduce to the Galilean ones (1.4). The laws of classical mechanics should then be regarded as valid only in the limit in which velocities are much smaller than the speed of light (*non-relativistic limit*).

It is often useful to deduce from equations (1.53) the relation between the components of the relative position vector and the time lapse between two events occurring at points A and B and at different times. Let $(x_A, y_A, z_A), (x_B, y_B, z_B)$ be the coordinates of A and B, respectively, and t_A, t_B the times of the corresponding events, as measured in S, and let the primed symbols refer, as usual, to the same quantities relative to S'. Writing (1.53) for the two events in A e in B and subtracting the former from the latter we obtain:

$$\Delta x' = \gamma(V)(\Delta x - V\,\Delta t), \tag{1.58}$$

$$\Delta y' = \Delta y, \tag{1.59}$$

$$\Delta z' = \Delta z, \tag{1.60}$$

$$\Delta t' = \gamma(V)\left(\Delta t - \frac{V}{c^2}\,\Delta x\right), \tag{1.61}$$

where we have set:

$$\Delta x = x_B - x_A \ ; \ \Delta x' = x_B' - x_A',$$
$$\Delta t = t_B - t_A \ ; \ \Delta t' = t_B' - t_A'.$$

Equation (1.61) implies that, in contrast to Galilean transformations, the time lapse between two events is no longer invariant since $\Delta t \neq \Delta t'$. The postulate of absolute time (and thus of absolute space), as anticipated, are then inconsistent with the principles of relativity and of constancy of the speed of light and thus should be given up.

---

[6]To have an idea of this approximation, consider a very high velocity like, for instance, that of the earth around the sun, which is about $V \approx 3 \times 10^4$ m/s. In this case we have $V^2/c^2 \approx 10^{-8}$.

1.4 Kinematic Consequences of the Lorentz Transformations

Let us now discuss some properties and physical implications of the Lorentz transformations.

Reciprocity: We have already observed that, according to the principle of relativity, it is equivalent to say that S' is moving at velocity V with respect to S, or that S is moving with velocity $-V$ with respect to S'. This in particular implies that the inverse Lorentz transformations (1.53) expressing (x, y, z, t) in terms of (x', y', z', t') can be obtained by simply interchanging in (1.53) primed with unprimed coordinates, and V with $-V$. Indeed if we invert equations (1.53), we find:

$$\begin{cases} x = \gamma(V)\,(x' + V\,t'), \\ y = y', \\ z = z', \\ t = \gamma(V)\,\left(t' + \frac{V}{c^2}x'\right), \end{cases} \tag{1.62}$$

in accordance with the rule illustrated above.

Symmetry Between Space and Time Intervals: There is no doubt that the major difference between Lorentz and Galilean transformations is the fact that the former imply a non-trivial transformation of time intervals as opposed to the latter which are based on the assumption of absolute time. In the former, there is, moreover, a strong similarity between the transformation properties of space and time intervals which becomes apparent if we use as new time coordinate $x^0 \equiv c\,t$. In this notation the spatial and time coordinates all have the same physical dimensions of a length. Denoting x, y, z by x^1, x^2, x^3 and defining the dimensionless number $\beta = V/c$, the Lorentz transformations take the following form:

$$\begin{cases} x^{1'} = \gamma(V)(x^1 - \beta x^0), \\ x^{2'} = x^2, \\ x^{3'} = x^3, \\ x^{0'} = \gamma(V)\,(x^0 - \beta x^1), \end{cases} \tag{1.63}$$

where the symmetry between the transformation laws of spatial and time coordinates is evident as they all appear in the above equations on an equal footing.[7]

The Speed of Light as the Maximum Velocity: Let us first observe that if S' were moving at a velocity $V > c$ with respect to S, the factor $\gamma = \sqrt{1 - \frac{V^2}{c^2}}$ would be purely imaginary and thus the transformations (1.53) physically meaningless.

---

[7]The reader should not mistake the upper labels of the space-time coordinates x^0, x^1, x^2, x^3 as powers of a quantity x! The mathematical difference between quantities labeled by upper and lower indices will be extensively discussed in the following chapters.

Let us now show that if we require the *principle of causality* to be valid in any inertial reference frame, then *no physical signal can travel at a speed greater than c*.

Recall that the *principle of causality* states that if an event A causes a second event B to occur, then the event A should always *precede B* in time: $t_A < t_B$. If this principle were violated, no physical investigation would be possible, since no theory would be predictive.

Let us then consider two events A e B taking place in the reference frame S along the x-axis at the points x_A, x_B at the times t_A and t_B, respectively, and assume that in the frame S *the event in A precedes the event in B*, that is $t_B - t_A \equiv \Delta t > 0$. Now we ask whether is it possible to find a new reference frame S' where $\Delta t' < 0$, i.e. where *the event in B precedes the event in A*.

Suppose the answer is positive, so that if $\Delta t > 0$ in S, there exists a frame S' in which $\Delta t' < 0$. Using Eq. (1.61) we then find

$$\Delta t' < 0 \Rightarrow \frac{\Delta x}{\Delta t} > \frac{c^2}{V} > c, \tag{1.64}$$

and this can only happen if

$$\Delta t < \frac{V}{c^2} \Delta x < \frac{\Delta x}{c}, \tag{1.65}$$

where we have used $\frac{V}{c} < 1$.

On the other hand $\frac{\Delta x}{c}$ has the meaning of the time τ_{AB} that a light ray takes to cover the distance $\Delta x = x_B - x_A$; therefore the condition will be satisfied if:

$$\Delta t < \tau_{AB}. \tag{1.66}$$

When Eq. (1.66) holds one immediately finds that the velocity V of S' with respect to S must satisfy the inequality:

$$V > c \frac{\Delta t}{\tau_{AB}}, \tag{1.67}$$

what is certainly possible with $V < c$.

Having established under what condition it is possible to invert the chronological order of two events by a change of reference frame, let us now assume that the event A sends a *physical signal* at a velocity $c' > c$ and that this signal determines the occurrence of the event B. (For example, with reference to Fig. 1.9, the event A can be the pressing of a switch by the observer S at t_A with the emission of a hypothetical signal of velocity $c' > c$ whose effect in B is the lighting of a lamp at a later time t_B).

In this case the inequality (1.66) is certainly satisfied since $\Delta t = \frac{\Delta x}{c'}$ and $c' > c$.

We then reach the conclusion that, if a signal propagating at velocity $c' > c$ existed, that is if it were possible to transmit information at a velocity greater than c, then two causally related events in one reference frame would appear in a different frame in the inverse temporal sequence, thus violating the *principle of causality*,

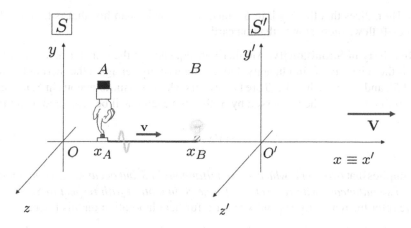

Fig. 1.9 Event A causing event B

since the effect would precede its cause. Since we can not give up the principle of causality, we conclude that:

No physical signal can propagate at a velocity greater than the speed of light.

Time Dilation: Equation (1.61) provides the explicit transformation law for the time intervals.

Consider a reference frame S' in motion at a constant speed V relative to another frame S (the standard configuration of the two frames is understood). Suppose an observer in S is measuring the time lapse $\Delta t = t_B - t_A$ between two events A and B which occur in S' at the same place but at different times, so that $t'_B > t'_A$ (for instance two successive positions of the second hand of a clock at rest in S'). If the events occur in S' along the x'-axis, we then suppose $\Delta x' = x'_B - x'_A = 0$. From Eq. (1.58) it then follows that $\Delta x = V \, \Delta t$. Substituting this relation in (1.61) we find

$$\Delta t' = \gamma(V) \left(\Delta t - \frac{V}{c^2} \Delta x \right) = \Delta t \, \gamma(V) \left(1 - \frac{V^2}{c^2} \right) = \frac{\Delta t}{\gamma(V)}. \qquad (1.68)$$

We conclude that the time lapse measured in S is

$$\Delta t = \gamma(V) \, \Delta t' > \Delta t'. \qquad (1.69)$$

This means that if an observer at rest in the frame S' measures a time interval $\Delta t'$, an observer in S will measure a lapse Δt greater than $\Delta t'$ by a factor $\gamma(V)$. As an example, let S' be a spacecraft traveling at a high velocity V relative a laboratory frame S on earth, and let time in S and S' be measured by two identical clocks. Suppose the observer in S measures the rate at which the clock inside the spacecraft ticks. He notices that the clock in S' runs more slowly than his, that is time on the spacecraft flows more slowly than on earth.[8]

Relativity of Simultaneity: Another consequence of the transformation law for time is the relativity of simultaneity. Indeed, let us consider again the inertial frames S and S', and suppose that A e B are two events which are simultaneous in S, namely $t_A = t_B$, $(\Delta t = 0)$. When observed by S' the two events will be separated by a time interval

$$\Delta t' = -\gamma(V) \frac{V}{c^2} \Delta x \neq 0. \tag{1.70}$$

This implies that *two events which are simultaneous in S, but occur at different points, are not simultaneous with respect to a frame S' in motion with respect to S.*

We refer the reader to Appendix H for a further elaboration on this issue.

Length Contraction: The fact that simultaneity between events is not an absolute concept implies that the distance between two points depends on the particular reference frame in which it is measured.

Let us consider the situation described in Sect. 1.1, in which a rod is placed at rest along the x'-axis of a frame S' moving with respect to S at velocity $\mathbf{V} = (V, 0, 0)$. Let the endpoints A and B of the rod be located in the points x'_B and x'_A. We can repeat one by one the arguments given in Sect. 1.1, from formula (1.10) to formula (1.12), using now the Lorentz transformations instead of the Galilean ones. In S' the length is defined as:

$$\Delta x' \equiv L' = x'_B - x'_A.$$

while the same length is measured in S as the difference between the coordinates of the endpoints taken at the same time, that is *simultaneously*:

$$\Delta x = L = x_B(t_B) - x_A(t_A) \equiv x_B(t) - x_A(t),$$

where we have set $t = t_B = t_A$. From Eq. (1.53) we then find:

$$L' = \Delta x' = \gamma(V)\,(\Delta x - V\,\Delta t) = \gamma(V)\,\Delta x = \gamma(V)\,L,$$

that is:

$$L = \gamma(V)^{-1}\,L'.$$

Since $\gamma(V)^{-1} = \left(1 - \frac{V^2}{c^2}\right)^{1/2} < 1$, the observer S in motion with respect to the rod will measure a length L contracted by the factor $\gamma(V)^{-1}$ with respect to L', which is the length of the object at rest. The conclusion is that:

---

[8]Note that the time dilation is a *relative effect*, that is if we have a clock at rest in S, from Eq. (1.61) it follows that $\Delta t' = \gamma(V)\,\Delta t$, that is time in S' is dilated with respect to S. The same observation applies to the length contraction to be discussed in the following.

The length L of an object in motion is contracted with respect its length L' at rest:

$$L = \sqrt{1 - \frac{V^2}{c^2}}\, L' < L'. \tag{1.71}$$

We note, instead, that lengths along the directions perpendicular to that of the relative motion are not affected by the motion itself $\Delta y = \Delta y'$, $\Delta z = \Delta z'$. This in particular implies that a volume $\Delta V = \Delta x \Delta y \Delta z$ transforms like the length of a rod parallel to the motion, namely:

$$\Delta V = \frac{1}{\gamma} \Delta V' < \Delta V'. \tag{1.72}$$

This in turn has the important consequence that the concept of *rigid body*, so useful un classical mechanics, looses its meaning in the framework of a relativistic theory, see Appendix H.

1.5 Proper Time and Space-Time Diagrams

We have learned, in the previous section, that both time and spatial intervals depend on the reference frame, that is, space and time are not *absolute* as they were in Newtonian mechanics, rather their transformation laws are combined in such a way that only the velocity of light is *absolute*. Note that, in the *standard configuration*, the transformation properties (1.58), (1.61) of Δx and Δt under Lorentz transformations are reminiscent of the way in which the components of a vector on the plane transform under a rotation of the corresponding coordinate axes. It is then natural to describe the effect of a change in the inertial frame as a kind of "rotation" of the space and time axes x, t. Considering also the other two coordinates y, z, which do not transform if the two inertial frames are in the *standard configuration*, one may regard a Lorentz transformation as a kind of "rotation" in a four-dimensional space, the fourth direction being spanned by the time variable.

A more precise definition of this kind of rotation will be given in Chap. 4; for the time being we call this four-dimensional space of points *space-time* or *Minkowski space*. Every point in space-time defines an *event* which occurs at a point in space of coordinates x, y, z, at a time t, and is labeled by the four coordinates t, x, y, z.

In three-dimensional Euclidean space $\mathbb{R}^3$ a rotation of the coordinate axes imply a transformation in the components $\Delta x, \Delta y, \Delta z$ of the relative position vector between two points, which however does not affect their squared distance $|\Delta \mathbf{x}|^2 = \Delta x^2 + \Delta y^2 + \Delta z^2$. In analogy to ordinary rotations in Euclidean space, a Lorentz transformation preserves a generalized "squared distance" between events in *Minkowski space* which generalizes the notion of distance between two points in space. To show this let us recall that, in determining the Lorentz transformations, we required the equality:

$$x^2 + y^2 + z^2 - c^2 t^2 = x'^2 + y'^2 + z'^2 - c^2 t'^2. \tag{1.73}$$

the left- and right-hand sides of this equation being separately zero, in accordance to the constancy of the speed of light in every inertial frame. The two events in A and B, in that case, were the emission of a spherical light wave in $O = O'$ at the time $t = t' = 0$ and the passage of the spherical wave-front through a generic point of coordinates x, y, z, t and x', y', z', t', respectively. Consider now a light wave which is emitted in a generic point, instead of the origin, at a generic time, and let the spherical wave propagate for a time Δt in S. Equation (1.73) can be written as:

$$
\begin{aligned}
|\Delta \mathbf{x}|^2 - c^2 \Delta t^2 &\equiv \Delta x^2 + \Delta y^2 + \Delta z^2 - c^2 \Delta t^2 \\
&= \Delta x'^2 + \Delta y'^2 + \Delta z'^2 - c^2 \Delta t'^2 = |\Delta \mathbf{x}'|^2 - c^2 \Delta t'^2. \tag{1.74}
\end{aligned}
$$

It is now a simple exercise to verify that equality (1.74) holds even if the two events do not refer to the propagation of a light ray. It is sufficient to express the primed quantities on the right-hand side in terms of the unprimed ones by using the Lorentz transformations (1.58)–(1.61). One then finds that Eq. (1.74) is identically satisfied. Defining the four-dimensional distance $\Delta \ell$, also called *proper distance* between two events, as

$$\Delta \ell^2 = |\Delta \mathbf{x}|^2 - c^2 \Delta t^2. \tag{1.75}$$

Equation (1.74) then implies that:

The proper distance between two events in space-time is invariant under Lorentz transformations. In particular, if there exists a frame where the two events are *simultaneous*, $\Delta t = 0$, the proper distance reduces in that frame to the ordinary distance $\Delta \ell = |\Delta \mathbf{x}|$.

While the proper distance has the dimension of a length, we may define an analogous *Lorentz-invariant* quantity, called *proper time interval* $\Delta \tau$, having dimension of a time, as follows:

$$\Delta \tau^2 = \Delta t^2 - \frac{1}{c^2} |\Delta \mathbf{x}|^2 = -\frac{1}{c^2} \Delta \ell^2. \tag{1.76}$$

Both $\Delta \ell$ and $\Delta \tau$, being proportional to each other, are referred to as *space-time intervals*. If we consider the reference frame, say S', where a body is at rest, then, since in this frame $\Delta \mathbf{x}' = 0$, we have:

$$\Delta \tau^2 = \Delta t'^2,$$

so that the physical meaning of the proper time interval $\Delta \tau$ is that of *the time interval between two events occurring at the same spatial point*. In any other frame S, being $\Delta \mathbf{x} \neq 0$, the time interval Δt will be different, their relation being given by Eq. (1.69).

Writing Eq. (1.76) in infinitesimal form (that is referring to infinitely close events), we find

$$d\tau^2 = dt^2 - \frac{1}{c^2}|d\mathbf{x}|^2 = dt^2 \left(1 - \frac{V^2}{c^2}\right),$$

since $\left|\frac{d\mathbf{x}}{dt}\right|$ is the velocity V of the frame S' attached to the particle. It follows

$$dt = \gamma(V)\,d\tau.$$

consistently with (1.69).

1.5.1 Space-Time and Causality

When studying the properties of the Lorentz transformations we have seen that if two events are *causally related*, namely if event A determines the occurrence of event B, then they must be connected by some physical signal, having a velocity $v \leq c$ and carrying to B the information of what happened in A (when A and B are the events describing the passage of a particle through two points along its trajectory, it is the very particle which carries the information).

Let $\Delta\mathbf{x}$ and Δt be the relative position vector and time lapse between the two events; then the velocity of the signal will be:

$$\mathbf{v} = \frac{\Delta\mathbf{x}}{\Delta t},$$

and we must have $|\mathbf{v}| \leq c$ for the two events to be causally related.

Looking at the definition of proper time, given in Eq. (1.76), we see that in this case we have:

$$\Delta\tau^2 \geq 0, \tag{1.77}$$

or, equivalently

$$\Delta\ell^2 \leq 0. \tag{1.78}$$

Indeed:

$$\Delta\tau^2 \geq 0 \iff c^2\,\Delta t^2 - |\Delta\mathbf{x}|^2 = \Delta t^2(c^2 - |\mathbf{v}|^2) \geq 0 \iff |\mathbf{v}| \leq c.$$

We conclude that *two events can be causally related, that is connected by a physical signal traveling at a velocity* $|\mathbf{v}| \leq c$, if and only if $\Delta\tau^2 \geq 0$ or, equivalently, $\Delta\ell^2 \leq 0$.

When (1.77) is *strictly* satisfied (i.e. when $\Delta\tau^2 > 0$) we say that the space-time interval between the two events is *time-like*, since we can always find a reference frame where the two events occur at the same point in space ($|\Delta\mathbf{x}| = 0$). Indeed,

referring to the Lorentz transformation between two frames in the standard configuration, $(\Delta y = \Delta y' = \Delta z = \Delta z' = 0)$ from Eq. (1.58) we see that

$$|\Delta \mathbf{x}'| \equiv \Delta x' = \gamma(V)\, \Delta t \left(\frac{\Delta x}{\Delta t} - V \right),$$

and since $\frac{\Delta x}{\Delta t} < c$ for a time-like interval, $\Delta x' = 0$ in a new Lorentz frame moving
at velocity $V = \frac{\Delta x}{\Delta t}$.

If $\Delta \tau^2 = 0$, the proper distance between the two events is zero and the corresponding space-time interval is called *light-like*, since the two points in space-time
can only be related by a light signal: $v = c$.

If instead the proper time interval between two events A and B is negative

$$\Delta \tau^2 < 0 \quad \Leftrightarrow \quad \Delta \ell^2 > 0 \tag{1.79}$$

the interval is called *space-like*. In this case the two events cannot be causally related,
since from Eq. (1.75) $|\Delta \mathbf{x}|/\Delta t \equiv v > c$, implying that no physical signal originating
from A, can ever reach the point B at a distance $|\Delta \mathbf{x}|$ during the time Δt.

In this case, however, it is possible to find a new reference frame where the two
events are simultaneous; indeed, if

$$\Delta \tau^2 < 0 \quad \Rightarrow \quad \frac{\Delta x}{\Delta t} > c, \tag{1.80}$$

from Eq. (1.61) it follows that we can choose a reference frame S', moving at a speed
$V < c$ relative to S given by

$$V = c^2 \frac{\Delta t}{\Delta x} < c, \tag{1.81}$$

with respect to which $\Delta t' = 0$. If, moreover, V satisfies the inequality $c > V > c^2 \frac{\Delta t}{\Delta x}$
we can find a frame in which the chronological order of the two events is inverted.

It is useful to give a geometric representation of space-time, (that is of Minkowski
space) supposing, for obvious graphical reasons, to have a two-dimensional Euclidean space spanned by the coordinates x, y instead of a three-dimensional one. The
time direction will be represented by an axis perpendicular to the xy-plane. It is
also convenient to measure time in units ct, so that all the coordinates of an event,
represented by a point in this space, share the same dimension. The origin O of this
coordinate system represents an event which has occurred in the point $x = y = 0$ at
$t = 0$, see Fig. 1.10.

All the points A whose proper distance ℓ^2 from O is time-like or light-like $c^2\tau^2 \equiv
-\ell^2 = c^2 t^2 - |\mathbf{x}|^2 \geq 0$ are enclosed within a cone in Minkowski space named *the
light-cone*. As discussed earlier, there can be a causality relation between the event in
O and any other point inside the light-cone. More precisely, referring to the figure,
O can determine the occurrence of A at $t > 0$ (A is said to belong to the *future
light-cone* of O), while it can have been determined by an event A' at $t < 0$ (A is
said to belong to the *past light-cone* of O); in any case the physical signal correlating
the two events travels at $v \leq c$ (Fig. 1.10).

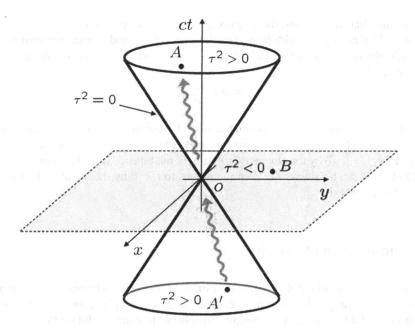

Fig. 1.10 Light-cone

Any event outside the light-cone instead, like the point B in the figure, is separated from O by a space-like interval $\tau^2 = -c^2\ell^2 = c^2t^2 - |\mathbf{x}|^2 < 0$, and thus can not be in any causal relation with it.

For the sake of simplicity we have just discussed the possible causal relations of events with a particular one located at origin of our space-time coordinate system. More generally we may associate with any event A its own light-cone dividing all events into two sets: Those in the interior of the cone, $\Delta\tau^2 \geq 0$, which can be causally related to A, and those outside the cone, $\Delta\tau^2 < 0$ which can not be correlated to A.

Let us focus on the plane described by the time- and x-axes and relabel the corresponding coordinates as follows: $x^0 \equiv ct$, $x^1 \equiv x$. We choose these axes to be orthogonal and their equations in this plane are $x^1 = 0$, $x^0 = 0$, respectively. Going to another reference frame S', by a Lorentz transformation, the old coordinates x^0, x^1 are related to the new ones x'^0, x'^1 as follows:

$$
\begin{aligned}
x'^1 &= \gamma(x^1 - \beta x^0), \\
x'^0 &= \gamma(x^0 - \beta x^1),
\end{aligned}
\tag{1.82}
$$

so that the new time and x'-axes ($x'^1 = 0$, $x'^0 = 0$, respectively) are described, in the old coordinates, by the equations:

$$
\begin{aligned}
x^1 &= \beta x^0, \\
x^0 &= \beta x^1.
\end{aligned}
\tag{1.83}
$$

This means that the ct'- and the x'-axes, describing space-time in the new frame, are rotated by an angle α with respect to the original ct- and x-axes, the former in the *clockwise direction* and the latter in the *counterclockwise direction*, the angle α being given by:

$$tg\alpha = \beta = \frac{v}{c} < 1,$$

so that $|\alpha| < \frac{\pi}{4}$. This explains geometrically why points inside the light-cone can always be brought, by means of a suitable Lorentz transformation, to the same point in space ($x = y = 0$) by the clockwise rotation of the time axis, while events outside the light-cone can be always made simultaneous to $t = 0$ by the counter-clockwise rotation of the x-axis.

1.6 Composition of Velocities

So far we have examined the implication of the Lorentz transformations as far as space and time intervals are concerned. Let us now consider how the velocities transform under Lorentz transformations. In contrast to what we did for the Galilean transformations, we can not simply differentiate both sides of Eqs. (1.58)–(1.60) with respect to time since $dt \neq dt'$. To find the correct relations we consider again two frames of reference S e S' as in Fig. 1.1, and a particle moving at velocity $\mathbf{v} = d\mathbf{x}/dt$ with respect to S and $\mathbf{v}' = d\mathbf{x}'/dt'$ with respect to S'. Restricting, as usual, to the standard configuration where the velocity of S' with respect to S is $(V, 0, 0)$, Eqs. (1.58)–(1.60), can be written in infinitesimal form:

$$dx' = \gamma(V)(dx - V\,dt),$$
$$dy' = dy,$$
$$dz' = dz,$$
$$dt' = \gamma(V)\left(dt - \frac{V}{c^2}\right).$$

From the above equations we easily find:

$$v'_x = \frac{dx'}{dt'} = \frac{dx - V\,dt}{dt - \frac{V}{c^2}dx} = \frac{v_x - V}{1 - \frac{V v_x}{c^2}},$$

$$v'_y = \frac{dy'}{dt'} = \frac{dy}{\gamma(V)\left(dt - \frac{V}{c^2}\,dx\right)} = \frac{v_y}{\gamma(V)\left(1 - \frac{V v_x}{c^2}\right)},$$

$$v'_z = \frac{v_z}{\gamma(V)\left(1 - \frac{V v_x}{c^2}\right)}.$$

We have thus derived the following *composition laws* for velocities:

$$v'_x = \frac{v_x - V}{1 - \frac{V\,v_x}{c^2}},$$

$$v'_y = \frac{v_y}{\gamma(V)\left(1 - \frac{V\,v_x}{c^2}\right)}, \tag{1.84}$$

$$v'_z = \frac{v_z}{\gamma(V)\left(1 - \frac{V\,v_x}{c^2}\right)}.$$

The different forms of the transformation of the x- component of the velocity $\mathbf{v}$, parallel to the relative velocity $\mathbf{V}$ between the two frames, with respect to the y- and z-components (orthogonal to $\mathbf{V}$) is obviously due to our choice of the *standard configuration*.

In the non-relativistic limit $V, |\mathbf{v}| \ll c$, we can neglect V^2/c^2 and $V\,v_x/c^2$ with respect to 1, so that:

$$\gamma(V) \simeq 1 + \frac{1}{2}\frac{V^2}{c^2} + \cdots \simeq 1,$$

$$\frac{1}{1 - \frac{V\,v_x}{c^2}} \simeq 1 + \frac{V\,v_x}{c^2} + \cdots \simeq 1,$$

and we retrieve the Galilean composition laws of velocities.

$$v'_x = v_x - V,$$

$$v'_y = v_y,$$

$$v'_z = v_z.$$

From Eqs. (1.84) we can easily verify that the composition of two velocities $|\mathbf{V}| \le c$ and $|\mathbf{v}| \le c$ can never result in a velocity $|\mathbf{v}'| \ge c$, in agreement with the fact that no signal or body can travel at a velocity greater than the speed of light. We can prove this property as follows:

$$|\mathbf{v}'|^2 = \frac{1}{\left(1 - \frac{V\,v_x}{c^2}\right)^2}\left[(v_x - V)^2 + \left(1 - \frac{V^2}{c^2}\right)(v_y^2 + v_z^2)\right]$$

$$= \frac{1}{\left(1 - \frac{V\,v_x}{c^2}\right)^2}\left[|\mathbf{v}|^2 - 2v_x\,V + V^2 - \frac{V^2}{c^2}(v_y^2 + v_z^2)\right]$$

$$= \frac{1}{\left(1 - \frac{V\,v_x}{c^2}\right)^2}\left[|\mathbf{v}|^2 - 2v_x\,V + V^2 - \frac{V^2}{c^2}(|\mathbf{v}|^2 - v_x^2)\right]$$

$$= \frac{1}{\left(1 - \frac{V v_x}{c^2}\right)^2} \left[c^2 - c^2 + |\mathbf{v}|^2 - 2v_x V + V^2 - \frac{V^2}{c^2}(|\mathbf{v}|^2 - v_x^2) \right]$$

$$= \frac{1}{\left(1 - \frac{V v_x}{c^2}\right)^2} \left[c^2 \left(1 - \frac{V v_x}{c^2}\right)^2 - c^2 \left(1 - \frac{|\mathbf{v}|^2}{c^2}\right) \left(1 - \frac{V^2}{c^2}\right) \right],$$

from which it follows that

$$\frac{|\mathbf{v}'|^2}{c^2} = 1 - \frac{1}{\left(1 - \frac{V v_x}{c^2}\right)^2} \left(1 - \frac{|\mathbf{v}|^2}{c^2}\right) \left(1 - \frac{V^2}{c^2}\right) \leq 1,$$

since $|\mathbf{v}| \leq c$ e $V \leq c$. In particular if $|\mathbf{v}| = c$, then also $|\mathbf{v}'| = c$, and we find that the velocity of light is the same in every inertial reference frame.

This concludes the examination of the kinematical effects of the Lorentz transformations. For the sake of simplicity we have used throughout *standard Lorentz transformations*, as defined in Sect. 1.3. More general transformations where the relative velocity $\mathbf{V}$ is not directed parallel to the x-axis do not affect the kinematical effects examined so far, as will be seen in the next chapter. Even the change of inertial frame due to rotations and translations of the reference frame as well as the change of the origin of the time coordinate do not affect the relativistic kinematics since they do not involve the relative velocity between the frames, on which all the kinematical effects depend. The explicit form of this more general change of frames will be discussed at the end of the next chapter, after the discussion of the relativistic dynamics. Indeed a physical theory obeying the principle of relativity must be covariant under any change of inertial frame and we shall see that the relativistic dynamics and the Maxwell theory satisfy this requirement.

1.6.1 Aberration Revisited

The fact that the speed of the light c is the same in every inertial system, does not imply that its direction of propagation is invariant under Lorentz transformations. We will illustrate this in the example of the aberration of starlight, which has already been discussed within the framework of the Galilean transformations where it seemed to find a natural explanation. Referring to the same configuration considered in that discussion, let us suppose that a light ray in the fixed-star frame S reaches a telescope on earth (frame S') with velocity: $\mathbf{c} = (0, -c, 0)$. Applying the composition laws of velocities (1.84) we find:

$$c'_x = \frac{0 - V}{1 - \frac{0 \cdot V}{c^2}} = -V,$$

$$c'_y = -\frac{1}{\gamma} \frac{c}{1 - \frac{0 \cdot V}{c^2}} = -\frac{c}{\gamma}, \qquad (1.85)$$

$$c'_z = 0, \qquad (1.86)$$

that is,

$$\mathbf{c}' = \left(-V, -\frac{c}{\gamma}, 0\right). \qquad (1.87)$$

It follows that in order for the light ray to be received by the observer in S', the telescope should be adjusted by an angle α with respect to the vertical x'-direction, given by:

$$\tan \alpha = \gamma \frac{V}{c} = \frac{1}{\sqrt{1 - \frac{V^2}{c^2}}} \frac{V}{c}. \qquad (1.88)$$

This formula, besides showing in a particular case how the direction of light changes by a change in the reference frame, also illustrates why the Galilean transformation laws of velocities give a fairly good account of the phenomenon of aberration. Indeed, comparing formula (1.88) with the Galilean expression (1.32), we see that the relativistic correction given by the factor $\gamma(V) \simeq 1 - \frac{1}{2}V^2/c^2 \simeq 1 - 10^{-8} \simeq 1$, is completely negligible, thus explaining why the observed phenomenon seemed to be in accordance with the ether hypothesis. On the other hand, note that the same formulae (1.88), give for the modulus of the velocity of the light signal

$$|\mathbf{c}'|^2 = \frac{c^2}{\gamma^2} + V^2 = c^2 \left(\frac{1}{\gamma^2} + \frac{V^2}{c^2}\right) = c^2.$$

in agreement with the principle of the constancy of the speed of light (Fig. 1.11).

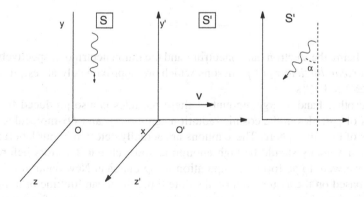

Fig. 1.11 Aberration using Lorentz transformations

1.7 Experimental Tests of Special Relativity

So far, apart from the aberration of starlight, we have never commented on the experimental tests of special relativity.

Einstein's special theory of relativity dates back to more than a century ago, over the course of which, considerable advances have been made in our understanding of the physical world, from the discovery of quantum mechanics to the developments of cosmology, and the formulation of the Standard Model for elementary particles. The latter theory, which receives almost daily confirmations from experiments carried out at the various particle accelerators all around the world, simply could not exist without a relativistic extension of the original quantum mechanics (called *relativistic quantum field theory*). In the absence of Einstein's theory, interaction processes involving high-energy elementary particles would simply appear incomprehensible while they are perfectly explained within the framework of relativistic kinematics, which has never been contradicted by experiments so far. The same can be said for our understanding of the universe as it results from the cosmological observations, which are perfectly described by Einstein's general theory of relativity, which extends the results of special relativity to non-inertial frames of reference, thereby including gravitation in the relativity principle (see late Chap. 3).

Nowadays the design of modern high-precision technological devices requires taking into account relativistic corrections for their correct functioning.

We postpone to the following chapters a more detailed analysis of the impact of special relativity on modern physics (restricting ourselves only to the quantum description of electromagnetic interactions), and on technology (e.g. GPS devices). For the time being it is interesting, from a historical point of view, to give a short account of one of the first experimental evidences of special relativity, which dates back to the thirties of the last century. In this experiment, which involved μ-mesons, the phenomenon of time dilation and length contractions were first observed.

The μ-meson particles (or muons), which are about 200 times as heavy as the electrons, can be produced in our laboratories, where they are observed to decay, in a very short time, into an electron and two neutrinos (very light neutral fermions):

$$\mu \to e + \bar{\nu}_e + \nu_\mu,$$

$\bar{\nu}_e$ and ν_μ being the electron (anti-)neutrino and the muon neutrino, respectively. The measured *mean lifetime* τ_μ of μ-mesons which are approximately at rest, turns out to be $\tau_\mu \simeq 2 \times 10^{-6}$ s.

On the other hand a large amount of these particles is also produced from the collisions of particles in the cosmic radiation against N_2- and O_2-molecules in the top layers of the atmosphere. These muons are actually detected in our laboratories, so that their velocity should be high enough as to reach our detectors before they decay. If we were to perform a computation using classical Newtonian mechanics, which is based on the assumption of absolute time, the mean lifetime of a muon is the same in every inertial frame. Therefore the minimum velocity v for a μ particle to reach the surface of the earth would be approximately given by the height h of the atmosphere divided its mean lifetime. In numbers:

$$v = \frac{h}{\tau_\mu} \approx 0.5 \times 10^{10} \, \text{m/s} > c.$$

Thus, according to Newtonian mechanics, they should have a velocity much greater than c, while the actual measure of their velocity turns out to be less than c.

This apparent contradiction, however, disappears when we reconsider the computation of v in the framework of special relativity. Indeed we know that a time interval, like the mean lifetime of a particle measured at rest, is not the same when measured in a different reference frame; in our case we must consider the mean lifetime $\tau_\mu^{(lab)}$ of the decaying muon as measured in the laboratory frame S tied to the earth, and the lifetime τ_μ measured in the frame S', moving at velocity v towards the earth, in which the particle is at rest. From Eq. (1.69) we deduce $\tau_\mu^{(lab)} = \gamma(v)\,\tau_\mu$. As a consequence the velocity is given by:

$$v = \frac{h}{\tau_\mu^{(lab)}} = \frac{1}{\gamma(v)} \frac{h}{\tau_\mu}. \tag{1.89}$$

Solving for v one finds:

$$v^2 = \left(1 - \frac{v^2}{c^2}\right)\left(\frac{h}{\tau_\mu}\right)^2 = \frac{c^2}{c^2 + \left(\frac{h}{\tau_\mu}\right)^2}\left(\frac{h}{\tau_\mu}\right)^2 = \frac{1}{1 + 3.6 \times 10^{-3}}\, c^2 < c^2.$$

Thus the velocity that the meson must have to reach the earth is $v \approx 0.998\, c < c$, in agreement with the experiments.

A possible objection to this result is the following: if we perform the computation from the point of view of an observer moving with the meson, then its lifetime would be measured at rest and so we should to use τ_μ instead of $\tau_\mu^{(lab)}$. Note, however, that the distance the meson should cover to reach the earth, as measured from its own frame S', would not be h, but rather

$$h' = \frac{1}{\gamma(v)}\, h,$$

since now the distance h is not at rest, but in motion with velocity v in the reference frame of the muon. Therefore, in S' the distance h' to cover is:

$$h' = \frac{1}{\gamma(v)}\, h.$$

And the corresponding velocity is

$$v = \frac{h'}{\tau_\mu} = \frac{1}{\gamma(v)} \frac{h}{\tau_\mu},$$

in agreement with the computation made in the earth frame S (1.89).

In conclusion, in both cases we obtain a result in agreement with experiment; in the former case by virtue of time dilation, in the latter case of length contraction.

As already stressed, uncountable phenomena where time dilation (or length contraction) is at work are observed in the elementary particle experiments. Indeed most of the particles created in high energy scattering processes have velocities close to the speed of light, so that the consequent time dilation can be easily observed.

In the next chapter, when discussing the implications of the Lorentz transformations on mechanics, other important consequences of the principle of relativity will be examined.

1.7.1 References

For further reading see Refs. [1, 11, 12].

Chapter 2
Relativistic Dynamics

2.1 Relativistic Energy and Momentum

In the previous chapter we have seen that a proper extension of the principle of relativity to electromagnetism necessarily implies that the correct transformation laws between two inertial frames are the Lorentz transformations. The price we have to pay, however, is that *the laws of classical mechanics are no longer invariant under changes in the inertial reference frame.*[1] We need therefore to re-examine the basic principles of the Newtonian mechanics and to investigate whether they can be made compatible with Einstein's formulation of the principle of relativity which, together with the principle of the constancy of the speed of light, requires invariance under the Lorentz, rather than the Galileo, transformations.

We have indeed learned, form the discussion in the last chapter, that the relativistic kinematics has an important bearing on the very concepts of space and time and, in particular, of *simultaneity*, which are no longer absolute. This fact is incompatible with some of the basic assumptions of classical mechanics. Let us recall that the fundamental force of this theory, the gravitational force, is described as acting at-a-distance. This gives rise to several inconsistencies from the point of view of special relativity:

1. The *instantaneous action* of a body on another, implies the transmission of the interaction at an *infinite velocity*. As we know, no physical signal can propagate with a velocity greater than c. To put it differently, in the action at-a-distance picture, the action of a body A and its effect, consisting in the consequent force applied to B, are simultaneous events localized at different points (corresponding to the positions of A and B respectively). Since simultaneity, in relativistic kinematics, is relative to the reference frame, there will in general exist an observer with respect to which the two events are no longer simultaneous, or in which the force is even seen to act on B before A exerts it, that is before A "knows" about B;

---

[1] Here by *classical mechanics* we refer to the Newtonian theory.

© Springer International Publishing Switzerland 2016
R. D'Auria and M. Trigiante, *From Special Relativity to Feynman Diagrams*,
UNITEXT for Physics, DOI 10.1007/978-3-319-22014-7_2

Fig. 2.1 Action by contact

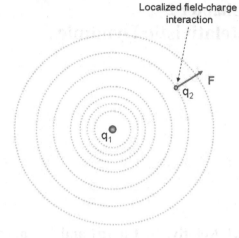

Localized field-charge interaction

2. By the same token also Newton's second law should be revisited. This equation indeed relates the acceleration of a point-mass to the total force exerted on it by all the other bodies, which is given by the sum of the individual forces taken at the same instant, that is *simultaneously*. These forces in turn will depend on the distances between the interacting objects. According to relativistic kinematics both simultaneity and spatial distances are relative to the inertial observer and thus, with respect to a different reference frame, the same forces will appear to be exerted at different times and distances.

The previous considerations imply that a proper formulation of mechanics (and in particular of dynamics) has to be given in terms of *localized interactions*, that is in terms of an interaction which takes place only when the two interacting parts are *in contact* and which is then localized in a certain point. This is in fact what happens when two point-charges interact through the electromagnetic field. The interaction is no longer represented as an action at-a-distance between the two charges but as mediated by the electromagnetic field, and can be divided into two moments (see the Fig. 2.1):

(a) a charge q_1 generates an electromagnetic field;
(b) The field, which is a physical quantity defined everywhere in space, propagates until it reaches the charge q_2 located at some point and acts on it by means of a force (the Lorentz force).

This mechanism is apparent when one of the two charges (say q_1) is moving at a very high speed. One then observes that the information about the position of the moving charge is transmitted to q_2 at the speed of light through the electromagnetic field, causing the force acting on it to be adjusted accordingly with a characteristic delay which depends on the distance between the two charges. In this *action-by-contact* picture the interacting parts are three instead of just two: the two charges and the field. The force acting on q_2 is the effect of the action of the field generated by q_1

on q_2. This implies that the action and the resulting force occur at the same time and place (the position of q_2) and this property is now Lorentz-invariant. Indeed if

$$\Delta t = |\Delta \mathbf{x}| = 0, \tag{2.1}$$

in a given frame, using the Lorentz transformations (1.58)–(1.61), we also have $\Delta t' = |\Delta \mathbf{x}'| = 0$ in any other frame. Thus *the action-by-contact representation is consistent with the principles of relativity and causality.*

As for the electromagnetic interaction, we would also expect the gravitational one to be mediated by a gravitational field. However, as we have mentioned earlier, a correct treatment of the gravitational interaction requires considering non-inertial frames of reference which goes beyond the framework of special relativity. In order to discuss how classical mechanics should be generalized in order to be compatible with Lorentz transformations (relativistic mechanics), we shall therefore refrain from considering gravitational interactions.

Even in classical mechanics we can consider processes in which the interaction is localized in space and time, so that the locality condition (2.1) is satisfied and we can avoid the inconsistencies discussed above, related to Newton's second law. These are typically *collisions* in which two or more particles interact for a very short time and in a very small region of space. Since the strength of the interaction is much higher than that of any other external force acting on the particles, the system can be regarded as isolated, so that the total linear momentum is conserved, and its initial and final states are described by free particles. Let us focus on this kind of processes in order to illustrate how one of the fundamental laws of classical mechanics, the conservation of linear momentum, can be made consistent with the principle of relativity, as implemented by the Lorentz transformations.

We shall first show that, *if we insist in defining the mass as independent of the velocity, then the conservation of momentum cannot hold in any reference frame,* thus violating the principle of relativity.[2]

Let us consider a simple process in which a mass m explodes into two fragments of masses $m_1 = m_2 = m/2$ (or equivalently a particle of mass m decays into two particles of equal masses, see Fig. 2.2). We shall assume the conservation of linear momentum to hold in the frame S in which the exploding mass is at rest:

$$\mathbf{v}_m = \mathbf{0} = \frac{m}{2}\mathbf{v}_1 + \frac{m}{2}\mathbf{v}_2 \quad \Rightarrow \quad \mathbf{v}_1 = -\mathbf{v}_2.$$

For the sake of simplicity we take the x-axis along the common direction of motion of the particles after the collision, so that $v_{1(y,z)} = v_{2(y,z)} = 0$. Let us now check whether the conservation of linear momentum also holds in a different frame S'. We choose S' to be the rest frame of fragment 1, which moves along the positive x-direction at a constant speed $V = v_{1(x)} \equiv v_1$ relative to S, and let the explosion

---

[2] Here and in the rest of this chapter, when referring to the conservation of the total linear momentum of an isolated system of particles, we shall often omit to specify that we consider the *total* momentum and that the system is *isolated*, regarding this as understood.

Fig. 2.2 Decay of a particle
into two particles of equal
masses

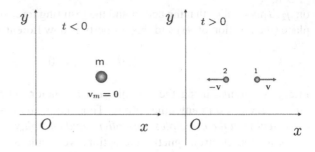

Fig. 2.3 Same decay in the
rest frame of particle 1

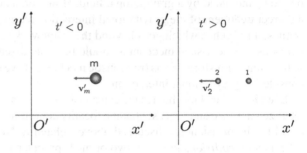

occur at the instant $t = t' = 0$, see Fig. 2.3. In the frame S', the velocity of the mass
m before the explosion is obtained by applying the relativistic composition law for
velocities (1.76):

$$v'_{(m)x} = \frac{0 - v_1}{1 - \frac{0 \cdot v_1}{c^2}} = -v_1 = -V, \quad v'_{(m)y,z} = 0.$$

Analogously, after the explosion, the velocities of the fragments in S' are given by:

$$v'_1 \equiv v'_{1x} = 0, \quad v'_2 \equiv v'_{2x} = \frac{-v_1 - v_1}{1 + \frac{v_1^2}{c^2}}, \quad v'_{2y} = v'_{2z} = 0.$$

Having computed the velocities in S' we may readily check whether the conservation
of linear momentum holds in this frame. It is sufficient to consider the components
of the linear momenta along the common axis $x = x'$; before and after the explosion
the total momenta in S' are given respectively by:

$$P'_{in} = m\,v'_{(m)} = -m\,v_1, \tag{2.2}$$

$$P'_{fin} = \frac{m}{2}\,v'_1 + \frac{m}{2}\,v'_{2x} = \frac{m}{2}\,v'_2 = -\frac{m\,v_1}{1 + \frac{v_1^2}{c^2}}. \tag{2.3}$$

Since

$$m\,v_1 \neq \frac{m\,v_1}{1+\frac{v_1^2}{c^2}}, \tag{2.4}$$

we conclude that, *in S' the total momentum is not conserved*. Or better, *the principle of conservation of linear momentum (as defined in classical mechanics) is not covariant under Lorentz transformations* thereby violating the principle of relativity. As such it cannot be taken as a founding principle of the new mechanics. It is clear that, just as the principle of relativity cannot be avoided in any physical theory, it would also be extremely unsatisfactory to give up the conservation of linear momentum; in the absence of it we would indeed be deprived of an important guiding principle for building up a theory of mechanics. To remediate this apparent shortcoming, it is important to trace back, in the above example, the origin of the non-conservation of the total momentum.

For this purpose we note the presence of the irksome factor $1+\frac{v^2}{c^2}$ on the right hand side of the inequality (2.4), which reduces to 1 in the non relativistic limit. This factor derives from the peculiar form of the composition law of velocities, which, in turn, originates from the non-invariance of time intervals under Lorentz transformations, namely:

$$dt' = \gamma(V)\left(dt - \frac{V}{c^2}dx\right) = \gamma(V)\,dt\left(1 - \frac{V\,v_x}{c^2}\right).$$

Thus we see that the non-trivial transformation property of dt is at the origin of the apparent failure of the conservation of momentum.

The same fact, however, gives us the clue to the solution of our problem: If we indeed replace, in the definition of the linear momentum **p** of a particle, the non-invariant time interval dt with the *proper time* $d\tau$, which is *invariant* under a change in the inertial frame, we may hope to have a conservation law of momentum compatible with the Lorentz transformations.

Let us then try to define the *relativistic linear momentum* of a particle as follows:

$$\mathbf{p} = m\frac{d\mathbf{x}}{d\tau}. \tag{2.5}$$

Recalling the relation between dt and $d\tau$, given by Eq. (1.76), and equations below, of the previous chapter,

$$d\tau = \frac{1}{\gamma(v)}dt = \sqrt{1 - \frac{v^2}{c^2}}\,dt, \tag{2.6}$$

we may write:

$$\mathbf{p} = m\frac{d\mathbf{x}}{d\tau} = m\,\gamma(v)\frac{d\mathbf{x}}{dt} = m(v)\,\mathbf{v}, \tag{2.7}$$

Fig. 2.4 Collision

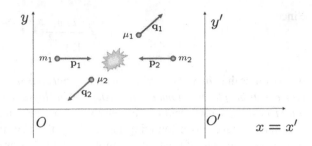

where

$$m(v) \equiv m\,\gamma(v) = \frac{m}{\sqrt{1 - \frac{v^2}{c^2}}}. \tag{2.8}$$

Note that the new definition of the relativistic momentum, (2.7), can be obtained from the classical one by *replacing the constant (classical) mass m*, with the velocity-dependent quantity $m(v)$, called the *relativistic mass*, so that the classical mass m coincides with the relativistic one only when the body is at rest: $m = m(v = 0)$. The mass m is then called *the rest mass* of the particle.

Let us now show that the conservation law of linear momentum is relativistic, provided we use (2.7) as the definition of the linear momentum of a particle. To prove the validity of this principle we would need to consider the most general process of interaction within an isolated system. For the sake of simplicity, we shall still restrict ourselves to collision processes, in order to deal with localized interactions, between two particles only. Consider then a process in which two particles of rest masses m_1, m_2 and linear momenta $\mathbf{p}_1$, $\mathbf{p}_2$ collide and two new particles are produced with rest masses and momenta μ_1, μ_2 and $\mathbf{q}_1$, $\mathbf{q}_2$, respectively, see Fig. 2.4.

We assume that, in a given frame S, the conservation of total linear momentum holds:

$$\mathbf{p}_1 + \mathbf{p}_2 = \mathbf{q}_1 + \mathbf{q}_2. \tag{2.9}$$

The above equation, using the definitions (2.7), can be rewritten in the following equivalent forms

$$m_1(v_1)\mathbf{v}_1 + m_2(v_2)\mathbf{v}_2 = \mu_1(u_1)\mathbf{u}_1 + \mu_2(u_2)\mathbf{u}_1,$$
$$m_1\frac{d\mathbf{x}_1}{d\tau_1} + m_1\frac{d\mathbf{x}_1}{d\tau_2} = \mu_1\frac{d\tilde{\mathbf{x}}_1}{d\tilde{\tau}_1} + \mu_2\frac{d\tilde{\mathbf{x}}_2}{d\tilde{\tau}_2}, \tag{2.10}$$

where we have marked with a tilde the quantities referring to the final state.

For the purpose of writing the conservation law in a new reference frame, we shall find it more useful to work with the second of Eqs. (2.10).

Let us consider now the process from a new frame S' moving with respect to S at constant speed, in the standard configuration. The two descriptions are related by a Lorentz transformation. In particular, if we apply the Lorentz transformation to

Eq. (2.10), we note first of all that the components of the same equation along the y- and z-axes do not change their form since the lengths along these directions are Lorentz-invariant ($dy' = dy, dz' = dz$), as well as the rest masses m_i, μ_i and the proper time intervals $d\tau_i$, $d\tilde{\tau}_i$. We can therefore restrict to the only component of Eq. (2.10) along the x-axis and prove that

$$m_1 \frac{dx_1}{d\tau_1} + m_1 \frac{dx_2}{d\tau_2} = \mu_1 \frac{d\tilde{x}_1}{d\tilde{\tau}_1} + \mu_2 \frac{d\tilde{x}_2}{d\tilde{\tau}_2}, \tag{2.11}$$

has the same form in the frame S', namely that it is *covariant* under a standard Lorentz transformation. This is readily done by transforming the differentials $dx_i, d\tilde{x}_i$ in Eq. (2.10), according to the inverse of transformation (1.58):

$$\sum_{i=1}^{2} m_i \left(\frac{dx_i'}{d\tau_i} + V \frac{dt'}{d\tau_i} \right) \gamma(V) = \sum_{i=1}^{2} \mu_i \left(\frac{d\tilde{x}_i'}{d\tilde{\tau}_i} + V \frac{dt'}{d\tilde{\tau}_i} \right) \gamma(V).$$

Let us now perform, using Eq. (2.7), the following replacement

$$\frac{d}{d\tau} = \frac{dt}{d\tau} \frac{d}{dt} = \gamma(v) \frac{d}{dt},$$

v being the velocity of the particle, so that, recalling the definition of the relativistic momentum, (2.11) takes the following form[3]:

$$\left(\sum_i p_{xi}' - \sum_i q_{xi}' \right) = \gamma(V) V \sum_i \left(m_i \gamma(v_i') - \mu_i \gamma(u_i') \right),$$

where v_i' and u_i', as usual, denote the velocities of the particles before and after the collision in the frame S'. The above relation can also be written in vector form as follows:

$$\left(\sum_i \mathbf{p}_i' - \sum_i \mathbf{q}_i' \right) \propto \sum_i \left(m_i(v_i') - \mu_i(u_i'), 0, 0 \right). \tag{2.12}$$

---

[3]Note that $\gamma(V) = \frac{1}{\sqrt{1-\frac{V^2}{c^2}}}$ is the relativistic factor associated with the motion of S' relative to S, while $\gamma(v_i') = \frac{1}{\sqrt{1-\frac{v_i'^2}{c^2}}}$ and $\gamma(u_i') = \frac{1}{\sqrt{1-\frac{u_i'^2}{c^2}}}$ are the relativistic factors depending on the velocities of each particle and relate the time dt' in S' to the proper times $d\tau_i, d\tilde{\tau}_i$ referred to the rest-frames of the various particles, according to

$$\begin{cases} d\tau_i = \sqrt{1 - \frac{v_i'^2}{c^2}} dt' \\ d\tilde{\tau}_i = \sqrt{1 - \frac{u_i'^2}{c^2}} dt'. \end{cases}$$

Since the right hand side contains only the difference between the sum of the *relativistic masses* before and after the collision, it follows that, in order for the conservation of linear momentum to hold in the new reference frame

$$\sum_i \mathbf{p}_i' = \sum_i \mathbf{q}_i', \tag{2.13}$$

we must have:

$$\sum_i m_i(v_i') = \sum_i \mu_i(u_i'), \tag{2.14}$$

that is the total relativistic mass must be conserved. From this analysis we can conclude that:

Given the new definition of linear momentum, (2.7), the conservation of momentum is consistent with the principle of relativity, i.e. covariant under Lorentz transformations, if and only if the total relativistic mass is also conserved.

Let us emphasize the deep analogy between our present conclusion and the analogous result obtained when studying the covariance of the conservation law of momentum under Galilean transformations in Newtonian mechanics (see Sect. 1.1.2).

2.1.1 Energy and Mass

We have seen that the concept of force as an action at a distance on a given particle looses its meaning in a relativistic theory. However nothing prevents us from *defining* the force acting on a particle as the time derivative of its relativistic momentum:

$$\mathbf{F} = \frac{d\mathbf{p}}{dt}. \tag{2.15}$$

Recalling the definition of $\mathbf{p}$, namely, $\mathbf{p} = m(v)\,\mathbf{v}$, we find:

$$\mathbf{F} = \frac{d}{dt}(m(v)\mathbf{v}) = \frac{dm(v)}{dt}\mathbf{v} + m(v)\frac{d\mathbf{v}}{dt},$$

Note that $\mathbf{F}$ is in general no longer proportional to the acceleration $\mathbf{a} = \frac{d\mathbf{v}}{dt}$. Writing $\mathbf{v} = v\,\mathbf{u}$, where $\mathbf{u}$ is the unit vector in the direction of motion, we obtain

$$\mathbf{a} = \frac{d\mathbf{v}}{dt} = \left(\frac{dv}{dt}\right)\mathbf{u} + \frac{v^2}{\rho}\mathbf{n},$$

where, as is well known, the unit vector $\mathbf{n}$ is normal to $\mathbf{u}$ and oriented towards the concavity of the trajectory, ρ being the radius of curvature.

Computing the time derivative of the relativistic mass we find

$$\frac{dm(v)}{dt} = m \frac{d}{dt} \left(\frac{1}{\sqrt{1 - \frac{v^2}{c^2}}} \right) = \frac{m}{\left(1 - \frac{v^2}{c^2}\right)^{3/2}} \frac{v}{c^2} \frac{dv}{dt} = \frac{m(v)}{c^2} \frac{v}{1 - \frac{v^2}{c^2}} \frac{dv}{dt},$$

so that:

$$\mathbf{F} = m(v) \left(\frac{1}{1 - \frac{v^2}{c^2}} \frac{v^2}{c^2} \frac{dv}{dt} + \frac{dv}{dt} \right) \mathbf{u} + m(v) \frac{v^2}{\rho} \mathbf{n} = \frac{m(v)}{1 - \frac{v^2}{c^2}} \frac{dv}{dt} \mathbf{u} + m(v) \frac{v^2}{\rho} \mathbf{n}.$$

We are now ready to determine the relativistic expression for the kinetic energy of a particle by computing the work done by the total force $\mathbf{F}$ acting on it. For an infinitesimal displacement $d\mathbf{x} = \mathbf{v}\, dt$ along the trajectory, the work reads:

$$dW = \mathbf{F} \cdot d\mathbf{x} = \mathbf{F} \cdot \mathbf{v}\, dt = \frac{m(v)}{1 - \frac{v^2}{c^2}} v \frac{dv}{dt} dt = \frac{1}{2} \frac{m(v)}{1 - \frac{v^2}{c^2}} d(v^2).$$

Integrating along the trajectory Γ (and changing the integration variable into $x = 1 - \frac{v^2}{c^2}$), we easily find:

$$W = \int_\Gamma \mathbf{F} \cdot d\mathbf{x} = \int_\Gamma \frac{1}{2} \frac{m}{\left(1 - \frac{v^2}{c^2}\right)^{3/2}} dv^2 = -\frac{c^2}{2} m \int \frac{dx}{x^{3/2}} = mc^2 \left(\frac{1}{x^{1/2}} \right)$$

$$= \frac{m}{\sqrt{1 - \frac{v^2}{c^2}}} c^2 + \text{const.} = m(v)c^2 + \text{const.}$$

If we define the kinetic energy, as in the classical case, to be zero when the particle is at rest, then the constant is determined to be $-m(0)c^2$, so that, the kinetic energy E_k acquired by the particle will be given by:

$$E_k(v) = m(v)\, c^2 - m\, c^2, \qquad (2.16)$$

where, from now on, $m = m(0)$ will always denote the rest mass. Note that in the non-relativistic limit $v^2/c^2 \ll 1$, we retrieve the Newtonian result:

$$E_k(v) = \frac{mc^2}{\sqrt{1 - \frac{v^2}{c^2}}} - mc^2 \simeq mc^2 \left(1 + \frac{1}{2} \frac{v^2}{c^2} \right) - mc^2 = \frac{1}{2} mv^2. \qquad (2.17)$$

where we have neglected terms of order $O(v^4/c^4)$.

Let us define the *total energy* of a body as:

$$E = m(v) c^2, \tag{2.18}$$

The kinetic energy is then expressed, in (2.16), as the difference between the total energy and the *rest energy*, which is the amount of energy a mass possesses when it is at rest:

$$E_{rest} = E(v = 0) = m c^2. \tag{2.19}$$

To motivate the definition of the total energy of a particle given in (2.18), we prove that the total energy in a collision process, defined as the sum of the total energies of each colliding particle, is always conserved. This immediately follows from the conservation of the total (relativistic) mass, which we have shown to be a necessary requirement for the conservation law of momentum to be covariant. Indeed by multiplying both sides of

$$m_1(v_1) + m_2(v_2) + \cdots + m_k(v_k) = \text{const.}$$

by c^2 and using the definition (2.18), we find

$$E_1(v_1) + E_2(v_2) + \cdots + E_k(v_k) = \text{const.}$$

This fact has no correspondence in classical mechanics where we know that, as opposed to the total linear momentum, which is always conserved in collision processes, the conservation of mechanical energy only holds in elastic collisions. This apparent clash between the classical and the relativistic laws of energy conservation is obviously a consequence of the fact that the rest energy can be transformed into other forms of energy, like kinetic energy, etc. We can give a clear illustration of this by considering again the collision of two particles with rest masses m_1, m_2 and velocities $\mathbf{v}_1$ e $\mathbf{v}_2$. Suppose that the collision is perfectly inelastic, so that the two particles stick together into a single one of rest mass M. It is convenient to describe the process in the center of mass frame, in which the final particle is at rest. Let us first describe the collision in the context of Newtonian mechanics. The conservation of momentum reads:

$$\mathbf{p}_1 + \mathbf{p}_2 = \mathbf{P} = 0,$$

or, equivalently

$$m_1 \mathbf{v}_1 + m_2 \mathbf{v}_2 = 0.$$

Moreover conservation of the *classical mass* is also assumed.

$$M = m_1 + m_2. \tag{2.20}$$

The initial and final mechanical (kinetic) energies are however different since

$$E_k^i = m_1 \frac{v_1^2}{2} + m_2 \frac{v_2^2}{2}, \quad E_k^f = 0.$$

and thus

$$\Delta E_k = E_k^f - E_k^i = -\left(m_1 \frac{v_1^2}{2} + m_2 \frac{v_2^2}{2} \right) \neq 0.$$

In Newtonian physics, the interpretation of this result is that the kinetic energy of the particles in the initial state is not conserved, while, from the thermodynamical point of view it has been converted into heat, increasing the thermal energy of the final body, that is the disordered kinetic energy of the constituent molecules.

Let us now describe the same process from the relativistic point of view.

Conservation of momentum and energy give the following two equations:

$$m_1 \gamma(v_1)\mathbf{v}_1 + m_2 \gamma(v_2)\mathbf{v}_2 = 0$$
$$m_1(v_1)c^2 + m_2(v_2)c^2 = M(0)c^2$$

where we have set $M(v = 0) = M(0)$. Using Eq. (2.16) to separate the rest masses from the (relativistic) kinetic energies, we obtain

$$E_k(v_1) + m_1 c^2 + E_k(v_2) + m_2 c^2 = E_k^f(0) + M(0)c^2. \qquad (2.21)$$

where on the right hand side $E_k^f(0) \equiv E_k^f(v = 0) = 0$. It follows[4]

$$c^2 \Delta M(0) \equiv c^2(M(0) - m_1 - m_2) = -(0 - E_k(v_1) - E_k(v_2)) = -\Delta E_k. \quad (2.22)$$

From the above relation we recognize that the loss of kinetic energy has been transformed in an increase of the final rest mass $M = M(0)$; thus M is not the sum of the rest masses of the initial particles (as it was instead assumed in the classical case, see Eq. (2.20)).

If we consider the inverse process in which a particle of rest mass M decays, in its rest frame, into two particles of rest masses m_1 and m_2, we see that part of the initial rest mass is now converted into the kinetic energy of the decay products. The importance of this effect obviously depends on the size of the ratio (v^2/c^2).

These examples illustrate an important implication of relativistic dynamics: The rest mass m of an object can be regarded as a form of energy, the rest energy $m\,c^2$, which can be converted into other forms of energy (kinetic, potential, thermal etc.). Let us illustrate this property in an other example. Consider a body of mass M at some given temperature: M will be given by the sum of the *relativistic masses* of its constituent molecules, and its temperature is related to their thermal motion. If we

---

[4]Note that at order $O(v^2/c^2)$ Eq. (2.21) can be written $c^2 \Delta M(0) = (\frac{1}{2}m_1 v_1^2 + \frac{1}{2}m_2 v_2^2)$.

now transfer an amount of energy E, in the form of heat, to the body, the total kinetic energy of its molecules will increase by E, thereby implying an increase in the mass by E/c^2, $M \to M + E/c^2$. Since all forms of energy can be transformed into one another and in particular into heat, we see that we can associate an equivalent amount of energy E with any mass, and, in particular, with the rest mass $m = m(0)$. Vice versa to each form of energy there corresponds an equivalent amount of mass given by $m(v) = E/c^2$.

The *equivalence between mass ed energy*, which is expressed by Eq. (2.18) or, equivalently, by

$$\Delta E = \Delta m(v) \, c^2, \tag{2.23}$$

is one of the major results of Einstein's theory of relativity. As a consequence, for a system made of interacting parts we define the total energy as:

$$E_{tot} = E_0 + E_k + U + \ldots, \tag{2.24}$$

where the sum is made over all the forms of energy which are present in the system: total rest energy, kinetic energy, potential energy and so on.

As a further example, let us consider a *bound system*. By definition a bound system is a system of interacting bodies such that the sum of the kinetic and potential energies is negative:

$$E_k + U < 0, \tag{2.25}$$

provided we fix the potential energy U to be zero when all the components are at infinite distance from each other and thus non-interacting: $U_\infty = 0$. If we think of the bound system as a single particle of rest mass M, in its rest frame we can write its energy $E_0 = M c^2$ as the total energy of the system, namely as the sum of the rest energies of its constituents and their total kinetic and potential energies, according to Eq. (2.24):

$$E_0 = M c^2 = \sum_i E_{0i} + E_k + U = \sum_i m_i c^2 + E_k + U, \tag{2.26}$$

m_i being the rest masses of the constituent particles. Equations (2.25) and (2.26) imply that, in order to disassemble the system bringing its elementary parts to infinite distances from one another (non-interacting configuration), we should supply it with an amount of energy (called the *binding energy* of the system) given by

$$\Delta E = -(E_k + U) > 0.$$

Note that, being $E_k + U$ a negative quantity, from Eq. (2.26) it immediately follows that the rest mass of the bound state is smaller than the sum of the rest masses of its constituents

$$M = \sum_i m_i - \frac{\Delta E}{c^2} < \sum_i m_i, \tag{2.27}$$

the "missing" rest mass being the equivalent in mass of the binding energy, as it follows from Eq. (2.26)

$$\Delta M \equiv \sum_i m_i - M = \frac{\Delta E}{c^2}. \tag{2.28}$$

Therefore when a bound state of two or more particles is formed starting from a non-interacting configuration, the system looses part of its total rest mass which, being the total energy conserved, is converted into an equivalent amount of energy $\Delta E = \Delta M \, c^2$ and released as, for instance, radiation.

An example of bound state is the hydrogen atom. It consists of a positively charged proton and a negatively charged electron, the two being bound together by the electric force. The rest masses of the two particles are respectively:

$$m_p \cong 938.3 \text{ MeV}/c^2; \qquad m_e \cong 0.5 \text{ MeV}/c^2,$$

where, taking into account the equivalence between mass and energy, we have used for the masses the unit MeV/c^2.[5] The corresponding binding energy

$$\Delta E = 1\mathbb{R}_y \cong 13.5 \text{ eV},$$

is called a *Rydberg*. Since the rest energy of the hydrogen atom is

$$M c^2 = m_e c^2 + m_p c^2 - \Delta E = \left(938.3 \times 10^6 + 0.5 \times 10^6 - 13.5\right) \text{ eV},$$

it follows that

$$\frac{\Sigma_i m^i c^2 - M c^2}{\Sigma_i m^i c^2} = \frac{\Delta M}{m_e + m_p} = \frac{13.5}{938.8 \times 10^6} \cong 10^{-8}. \tag{2.29}$$

Thus we see that, in this case, where the force in play is the electric one, the rate of change in rest mass, $\Delta M/M$, is quite negligible.

---

[5]We recall that $1 \text{ MeV} = 10^6 \text{ eV}$, where 1 eV is the energy acquired by an electron (whose charge is $e \cong 1.6 \times 10^{-19}$ C) crossing an electric potential difference of 1 V:

$$1 \text{ eV} = 1.6 \times 10^{-19} \text{ J}.$$

Another commonly used unit, when considering energy exchanges in atomic processes, is the atomic mass unit u, that is defined as $1/12$ the rest mass M_C of the isotope ^{12}C of the carbon atom at rest; this unit is more or less the proton mass. Precisely we have: $1\,u = 1.660\,538\,782(83) \times 10^{-24}$ g $= \frac{1}{N_A}$ g, where N_A is the Avogadro number. Taking into account the equivalence mass-energy we also have

$$1\,u = \frac{1}{12} M_C \simeq 931.494 \text{ MeV}/c^2.$$

Let us now consider a two-body system bound by the *nuclear force*, The ratio of the strength of the nuclear force to that of the electric one is of order 10^5. An example is the deuteron system which is a bound state of a proton and a neutron. In this case we may expect a much larger binding energy and, consequently, a greater rest mass variation. Using the values of the proton and neutron masses,

$$c^2 m_p \cong 938.272 \, \text{MeV} \simeq 1.00728 \, u,$$
$$c^2 m_n \cong 939.566 \, \text{MeV} \simeq 1.00867 \, u,$$
$$\Delta E \simeq 2.225 \, \text{MeV},$$

it turns out that the corresponding loss of rest mass is:

$$\frac{-\Delta M}{m_p + m_n} = \frac{2.225}{1877.838} = 1.18 \times 10^{-3},$$

that is, five orders of magnitudes greater than in the case of the hydrogen atom. This missing rest mass (times the square of the speed of light) results in an amount of energy which is released when the bound state is created. A similar mass defect is present in all atomic nuclei. In fact, as the reader can easily verify, the atomic mass of an atom, which can be read off the Mendeleev table, is always smaller than the sum of the masses of the protons and neutrons entering the corresponding nuclei, since they form a bound state.

2.1.2 Nuclear Fusion and the Energy of a Star

Taking into account that life on earth depends almost exclusively on the energy released by the sun, it is of outmost importance to realize that the source of such energy is the continuous conversion of the solar rest mass into radiation energy and heat that we receive on earth, through the so-called *nuclear fusion*; just as for the reaction discussed above, leading to the creation of the deuteron, nuclear fusion essentially amounts to the formation of a bound state of nucleons (protons and neutrons) with a consequent reduction of rest mass which is released in the form of energy (radiation). The fact that the solar energy, or more generally, the energy of a star, could not originate from chemical reactions, can be inferred from the astronomical observation that the mean life of a typical star, like the sun, is of the order of 10^9–10^{10} years. If the energy released by the sun were of chemical origin, one can calculate that the mean life of the sun would not exceed 10^5–10^6 years. It is only through the conversion of mass into energy, explained by the theory of special relativity, that the lifetime of stars can be fully explained in relation to their energy emission.

Without entering into a detailed description of the sequences of nuclear processes taking place in the core of a burning star (which also depend on the mass of the star), we limit ourselves to give a qualitative description of the essential phenomenon.

We recall that after the formation of a star, an enormous gravitational pressure is generated in its interior, so that the internal temperature increases to typical values of 10^6–10^7 K. At such temperatures nuclear fusion reactions begin to take place, since the average kinetic energy of nucleons is large enough to overcome the repulsive (electrostatic) potential barrier separating them. At sufficiently short distances, the interaction between nucleons is dominated by the attractive nuclear force and nucleon bound states can form. The fundamental reaction essentially involves four protons which give rise, after intermediate processes, to a nucleus of Helium, ^4_2He, together with two *positrons and neutrinos*:

$$4 \times {}^1_1H \rightarrow {}^4_2He + 2e^+ + 2v_e. \tag{2.30}$$

where e^+ denotes the positron (the anti-particle of an electron) and v_e the (electronic) neutrino, their masses being respectively: $m_{e^+} = m_{e^-} \simeq 0.5\,\text{MeV}$, $m_{v_e} \simeq 0$.
(Note that ionized hydrogen, that is protons, comprise most of the actual content of a star.)

The reaction (2.30) is the aforementioned nuclear fusion taking place in the interior of a typical star. To evaluate the mass reduction involved in this reaction we use the value of the mass of a ^4He nucleus, and obtain:

$$\Delta M \cong 0.0283\,u = 0.0283 \times 931.494\,\text{MeV}/c^2 \simeq 26.36\,\text{MeV}/c^2.$$

This implies that every time a nucleus of ^4He is formed out four protons, an amount of energy of about 26.36 MeV is released.

Consider now the fusion of 1 kg of ionized hydrogen. Since 1 mole of ^1_1H, weighting about 1 g, contains $N_A \simeq 6.023 \times 10^{23}$ (Avogadro's number) particles, there will be a total of $\sim 1.5 \times 10^{26}$ reactions described by (2.30), resulting in an energy release of[6]:

$$\Delta E(1\,\text{kg}) = 26.36 \times 1.5 \times 10^{26}\,\text{MeV} \simeq 3.97 \times 10^{27}\,\text{MeV} \approx 6.35 \times 10^{14}\,\text{J}.$$

On the other hand, we know that a star like the sun fuses H^1_1 at a rate of about $5.64 \times 10^{11}\,\text{kg}\,\text{s}^{-1}$, the total energy released every second by our star amounts approximately to:

$$\frac{\Delta E}{\Delta t} = 6.35 \times 10^{14} \times 5.64 \times 10^{11} \approx 3.58 \times 10^{26}\,\text{J}\,\text{s}^{-1},$$

This implies a reduction of the solar mass at a rate of:

---

[6]Note that if we had a chemical reaction instead of a nuclear one, involving just the electrons of two hydrogen atoms ($H + H \rightarrow H_2$) we would obtain an energy release of $E \simeq 2 \times 10^6$ J, which is eight orders of magnitude smaller.

$$\frac{\Delta m}{\Delta t} = \frac{1}{c^2} \frac{\Delta E}{\Delta t} \cong 3.98 \times 10^9 \, \text{kg s}^{-1}.$$

Since the energy emitted over one year (=3.2×10^7 s) is $\Delta E_{year} \simeq 1.1 \times 10^{34}$ J, the corresponding mass lost each year by our sun is $\Delta M_{year} \simeq 1.3 \times 10^{17}$ kg. If this loss of mass would continue indefinitely,[7] using the present value of the solar mass, $M_\odot \simeq 1.9 \times 10^{30}$ kg, its mean life can be roughly estimated to be of the order of

$$T = \frac{M_\odot}{\Delta M} \text{(years)} \simeq 1.5 \times 10^{13} \text{ years}.$$

2.2 Space-Time and Four-Vectors

It is useful at this point to introduce a mathematical set up where all the kinematic quantities introduced until now and their transformation properties have a natural and transparent interpretation.

To summarize our results so far, the energy and momentum of a particle of rest mass m moving at velocity $\mathbf{v}$ in a given frame S, are defined as:

- *energy* : $E = m(v)c^2 = m\gamma(v)c^2 = m\frac{dt}{d\tau}c^2$,
- *momentum*: $\mathbf{p} = m\frac{d\mathbf{x}}{d\tau} = m(v)\mathbf{v}$,

where $\mathbf{v} = \frac{d\mathbf{x}}{dt}$, and $m(v) = \frac{m}{\sqrt{1-\frac{v^2}{c^2}}} = m\frac{dt}{d\tau}$.

From the above definitions we immediately realize that the four quantities:

$$\left(\frac{E}{c}, \mathbf{p}\right) \equiv \left(m\frac{dt}{d\tau}, m\frac{d\mathbf{x}}{d\tau}\right), \tag{2.31}$$

transform exactly as $(c\,dt, d\mathbf{x})$ under a Lorentz transformation, since both m and $d\tau$ are invariant. Thus, using a standard configuration for the two frames in relative motion with velocity V, we may readily compute the transformation law of E, $\mathbf{p}$:

$$p'_x = m\frac{dx'}{d\tau} = m\gamma(V)\frac{dx - V\,dt}{d\tau} = \gamma(V)\left(m\frac{dx}{d\tau} - V\,m\frac{dt}{d\tau}\right),$$

$$= \gamma(V)\left(p_x - V\frac{E}{c^2}\right). \tag{2.32}$$

---

[7]This does not happen however, because the nuclear fusion of hydrogen ceases when there is no more hydrogen, and after that new reactions and astrophysical phenomena begin to take place.

where we have used that $m\,dt/d\tau = m\,\gamma(v) = E/c^2$. Furthermore we also have

$$p'_y = p_y, \tag{2.33}$$

$$p'_z = p_z, \tag{2.34}$$

$$\frac{E'}{c} = mc\frac{dt'}{d\tau} = m\,\gamma(V)\frac{c\,dt - \frac{V}{c}dx}{d\tau} = \gamma(V)\left(\frac{E}{c} - \frac{V}{c}p_x\right), \tag{2.35}$$

where we have used the property that the proper time interval, as defined in Eq. (1.76), is Lorentz-invariant: $d\tau' = d\tau$. Comparing the transformation laws for the time and spatial coordinates with those for energy and momentum, given by the Eqs. (2.32)–(2.35), we realize that, given the correspondences $(p_x, p_y, p_z) \rightarrow (dx, dy, dz)$ and $E/c \rightarrow c\,dt$, they are identical:

$$\begin{cases} p'_x = \gamma(V)\left(p_x - V\frac{E}{c^2}\right) \\ p'_y = p_y \\ p'_z = p_z \\ \frac{E'}{c} = \gamma(V)\left(\frac{E}{c} - \frac{V}{c}p_x\right) \end{cases} \leftrightarrow \begin{cases} dx' = \gamma(V)(dx - V\,dt) \\ dy' = dy \\ dz' = dz \\ c\,dt' = \gamma(V)\left(c\,dt - \frac{V}{c}dx\right) \end{cases} \tag{2.36}$$

We now recall the expression of the Lorentz-invariant proper time interval, as defined in Eq. (1.76):

$$d\tau^2 = dt^2 - \frac{1}{c^2}|d\mathbf{x}|^2. \tag{2.37}$$

From the above correspondence it follows that the analogous quantity

$$\frac{E^2}{c^2} - |\mathbf{p}|^2 = m^2\gamma(v)^2c^2 - m^2\gamma(v)^2v^2 = \frac{m^2c^2}{1 - \frac{v^2}{c^2}}\left(1 - \frac{v^2}{c^2}\right) = m^2c^2,$$

is Lorentz-invariant as well, being simply proportional to the rest mass of the particle.
Note that the relativistic relation between energy and momentum given by

$$\frac{E^2}{c^2} - |\mathbf{p}|^2 = m^2c^2 \quad \Rightarrow \quad E = \sqrt{\mathbf{p}^2c^2 + m^2c^4}, \tag{2.38}$$

separating the relativistic kinetic energy from the rest mass, can be rewritten as follows:

$$mc^2 + E_k^{rel.} = mc^2\left(\sqrt{1 + \frac{\mathbf{p}_{rel}^2}{m^2c^2}}\right). \tag{2.39}$$

In the non-relativistic limit, neglecting higher order terms in v^2/c^2, Eq. (2.39) becomes:

$$m c^2 + \frac{1}{2} m v^2 \simeq m c^2 \left(1 + \frac{\mathbf{p}_{rel}^2}{2m^2 c^2}\right) \quad \Rightarrow \quad \frac{1}{2} m v^2 = \frac{\mathbf{p}_{class.}^2}{2m}, \tag{2.40}$$

in agreement with the standard relation between kinetic energy and momentum in classical mechanics.

2.2.1 Four-Vectors

In the previous chapter we have seen that the time and space coordinates of an *event* may be regarded as coordinates (ct, x, y, z) of a four-dimensional space-time called *Minkowski space*, for which we shall use the following short-hand notation

$$(x^\mu) = (x^0, x^1, x^2, x^3) = (ct, x, y, z); \quad (\mu = 0, 1, 2, 3),$$

The time coordinate $x^0 = ct$ has been defined in such a way that all the four coordinates x^μ share the same dimension. These coordinates can be viewed as the orthogonal components of the position vector of an event relative to the origin-event $O(x^\mu \equiv 0)$.

Given two events A, B labeled by

$$x_A^\mu = (ct_A, x_A, y_A, z_A), \quad x_B^\mu = (ct_B, x_B, y_B, z_B),$$

we may then define a relative position vector of B with respect to A:

$$\Delta x^\mu = x_B^\mu - x_A^\mu = (\Delta x^0, \Delta x^1, \Delta x^2, \Delta x^3) = (c\Delta t, \Delta x, \Delta y, \Delta z).$$

Using this notation, the Lorentz transformation of the four coordinate differences Δx^μ (or their infinitesimal form dx^μ) is given, in the standard configuration, by (see also Eqs. (1.58)–(1.61) and (1.63)):

$$\begin{cases} \Delta x'^0 = \gamma(V) \left(\Delta x^0 - \frac{V}{c} \Delta x^1\right), \\ \Delta x'^1 = \gamma(V) \left(\Delta x^1 - \frac{V}{c} \Delta x^0\right), \\ \Delta x'^2 = \Delta x^2, \\ \Delta x'^3 = \Delta x^3, \end{cases} \tag{2.41}$$

which, in matrix form, can be rewritten as:

$$\begin{pmatrix} \Delta x'^0 \\ \Delta x'^1 \\ \Delta x'^2 \\ \Delta x'^3 \end{pmatrix} = \begin{pmatrix} \gamma & -\gamma\beta & 0 & 0 \\ -\gamma\beta & \gamma & 0 & 0 \\ 0 & 0 & 1 & 0 \\ 0 & 0 & 0 & 1 \end{pmatrix} \begin{pmatrix} \Delta x^0 \\ \Delta x^1 \\ \Delta x^2 \\ \Delta x^3 \end{pmatrix}, \qquad (2.42)$$

where, as usual, $\beta = V/c$. Restricting ourselves to a *standard configuration* we shall provisionally call *four-vector* any set of four quantities that, under a *standard Lorentz transformation*, undergoes the transformation (2.41) in Minkowski space. In particular, recalling Eq. (2.36), we see that the four quantities[8]

$$p^\mu = (p^0, p^1, p^2, p^3) \equiv (E/c, p_x, p_y, p_z),$$

are the components of a four-vector, the *energy-momentum* vector, which transforms by the same matrix (2.42) as (Δx^μ). Since $p^0 = E/c = m\,\gamma(v)\,c$, recalling Eq. (2.7), the *energy-momentum* vector can also be written as

$$p^\mu = m \frac{dx^\mu}{d\tau} = m\,\gamma(v)(c, v_x, v_y, v_z) = m\,U^\mu, \qquad (2.43)$$

where U^μ, called *four-velocity*, is also a four-vector, since the rest mass m is an invariant.

Recall that, in Eqs. (1.75) and (1.76), we defined as *proper distance* in Minkowski space the Lorentz-invariant quantity

$$\Delta\ell^2 = (\Delta x^1)^2 + (\Delta x^2)^2 + (\Delta x^3)^2 - (\Delta x^0)^2 \equiv -c^2\,\Delta\tau^2. \qquad (2.44)$$

which is the natural extension to Minkowski space of the Euclidean three-dimensional distance in Cartesian coordinates. However in the following we shall mostly use as *space-time* or *four-dimensional distance*[9] in Minkowski space the quantity $\Delta s^2 = c^2\,\Delta\tau^2 = -\Delta\ell^2$, that is the *negative* of the proper distance. This choice is dictated by the conventions we shall introduce in the following chapters when discussing the geometry of Minkowski space. Thus, for example, the square of the *four-dimensional distance or norm* of the four-vector Δx^μ is defined as

$$\|\Delta x^\mu\|^2 = (\Delta x^0)^2 - (\Delta x^1)^2 - (\Delta x^2)^2 - (\Delta x^3)^2 = c^2\,\Delta\tau^2 = \Delta s^2. \qquad (2.45)$$

Note, however, that the square of the Lorentzian norm is not *positive definite*, that is, it is not the sum of the squared components of the vector (see Eq. (2.44)) as

---

[8]As for Δx^μ we define $p^0 = E/c$ so that all the four components of p^μ share the same physical dimension.

[9]Alternatively also the denominations Lorentzian or Minkowskian distance are used.

the Euclidean norm $|\Delta \mathbf{x}|^2$ is. Consequently a non-vanishing four-vector can have a vanishing norm.

In analogy with the relative position four-vector, we define the norm of the energy-momentum vector as

$$\| p^\mu \|^2 = (p^0)^2 - (p^1)^2 - (p^2)^2 - (p^3)^2 = (p^0)^2 - |\mathbf{p}|^2.$$

From Eq. (2.38) it follows that this norm is precisely the (Lorentz-invariant) squared rest mass of the particle times c^2: $\| p^\mu \|^2 = m^2 c^2$. Using the notation of four-vectors, we may rewrite the results obtained so far in a more compact way.

Consider once again a collision between two particles with initial energies and momenta E_1, E_2 and $\mathbf{p}_1, \mathbf{p}_2$, respectively, from which two new particles are produced, with energies and momenta E_3, E_4, $\mathbf{p}_3$, $\mathbf{p}_4$. The conservation laws of energy and momentum read:

$$E_1 + E_2 = E_3 + E_4,$$
$$\mathbf{p}_1 + \mathbf{p}_2 = \mathbf{p}_3 + \mathbf{p}_4. \tag{2.46}$$

If we now introduce the four-vectors p_n^μ, $n = 1, 2, 3, 4$ associated with the initial and final particles

$$p_n^\mu = \begin{pmatrix} E_n/c \\ p_{nx} \\ p_{ny} \\ p_{nz} \end{pmatrix},$$

and define the *total energy-momentum* as the sum of the corresponding four-vectors associated with the two particles before and after the process, we realize that the conservation laws of energy and momentum are equivalent to the statement that the total energy momentum four-vector is conserved. To show this we note that Eqs. (2.46) can be rewritten in a simpler and more compact form as the *conservation law of the total energy-momentum four-vector*:

$$p_{tot}^\mu = p_1^\mu + p_2^\mu = p_3^\mu + p_4^\mu. \tag{2.47}$$

Indeed the 0th component of this equation expresses the conservation of energy, while the components $\mu = 1, 2, 3$ (spatial components) express the conservation of linear momentum. Note that for each particle the norm of the energy-momentum four-vector gives the corresponding rest mass:

$$\| p_n^\mu \|^2 = \left(\frac{E_n}{c} \right)^2 - |\mathbf{p}_n|^2 = m_n^2 c^2.$$

Until now we have restricted ourselves to Lorentz transformations between frames in *standard configuration*. For the next developments it is worth generalizing our setting to Lorentz transformations with generic relative velocity vector $\mathbf{V}$, however keeping, for the time being, the three coordinate axes parallel and the origins coincident at the time $t = t' = 0$. Consider two events with relative position four-vector $\Delta x \equiv (\Delta x^\mu) = (c\,\Delta t, \,\Delta \mathbf{x})$ with respect to a frame S. We start decomposing the three-dimensional vector $\Delta \mathbf{x}$ as follows

$$\Delta \mathbf{x} = \Delta \mathbf{x}_\perp + \Delta \mathbf{x}_\|,$$

where $\Delta \mathbf{x}_\perp$ and $\Delta \mathbf{x}_\|$ denote the components of $\Delta \mathbf{x}$ orthogonal and parallel to $\mathbf{V}$, respectively. Consider now the same events described in a RF S' moving with respect to S at a velocity $\mathbf{V}$. It is easy to realize that the corresponding Lorentz transformation can be written as follows

$$\Delta \mathbf{x}' = \Delta \mathbf{x}_\perp + \gamma(V) \left(\Delta \mathbf{x}_\| - \mathbf{V}\Delta t \right), \tag{2.48}$$

$$\Delta t' = \gamma(V) \left(\Delta t - \frac{\Delta \mathbf{x} \cdot \mathbf{V}}{c^2} \right). \tag{2.49}$$

Indeed they leave invariant the fundamental Eq. (1.50) or, equivalently, the proper time (and thus the proper distance):

$$c^2\,\Delta t'^2 - |\Delta \mathbf{x}'|^2 = c^2\,\Delta t^2 - |\Delta \mathbf{x}|^2. \tag{2.50}$$

Writing $\Delta \mathbf{x}_\perp = \Delta \mathbf{x} - \Delta \mathbf{x}_\|$, $\gamma(V) \equiv \gamma$ and using the variables $\Delta x^0 = c\,\Delta t$ and $\boldsymbol{\beta} = \frac{\mathbf{V}}{c}$, Eqs. (2.48) and (2.49) become:

$$\Delta \mathbf{x}' = \Delta \mathbf{x} + (\gamma - 1)\,\Delta \mathbf{x}_\| - \gamma\,\boldsymbol{\beta}\,\Delta x^0, \tag{2.51}$$

$$\Delta x'^0 = \gamma \left(\Delta x^0 - \Delta \mathbf{x} \cdot \boldsymbol{\beta} \right). \tag{2.52}$$

Recalling that the four-vector $p \equiv (p^\mu) = (\frac{E}{c}, \mathbf{p})$ transforms as $x \equiv (x^\mu) = (ct, \mathbf{x})$, we also obtain

$$\mathbf{p}' = \mathbf{p} + (\gamma - 1)\,\mathbf{p}_\| - \gamma\,\mathbf{V}\,\frac{E}{c^2}, \tag{2.53}$$

$$E' = \gamma\,(E - \mathbf{p} \cdot \mathbf{V}), \tag{2.54}$$

and since $\mathbf{p} = m(v)\mathbf{v} = \frac{E}{c^2}\mathbf{v}$, the energy transformation (2.54) can be written as follows:

$$E' = \gamma \left(E - \frac{\mathbf{v} \cdot \mathbf{V}}{c^2} E \right). \tag{2.55}$$

Observing that the vector $\Delta \mathbf{x}_\parallel$ can also be written as $\frac{\boldsymbol{\beta} \cdot \Delta \mathbf{x}}{|\boldsymbol{\beta}|^2} \boldsymbol{\beta}$, the matrix form corresponding to Eqs. (2.51) and (2.52) is

$$
\begin{pmatrix} \Delta x_0' \\ \Delta x_1' \\ \Delta x_2' \\ \Delta x_3' \end{pmatrix} = \begin{pmatrix} \gamma & -\gamma \beta^1 & -\gamma \beta^2 & -\gamma \beta^3 \\ -\gamma \beta^1 & 1 + \frac{(\gamma-1)}{|\beta|^2} \beta^1 \beta^1 & \frac{(\gamma-1)}{|\beta|^2} \beta^1 \beta^2 & \frac{(\gamma-1)}{|\beta|^2} \beta^1 \beta^3 \\ -\gamma \beta^2 & \frac{(\gamma-1)}{|\beta|^2} \beta^2 \beta^1 & 1 + \frac{(\gamma-1)}{|\beta|^2} \beta^2 \beta^2 & \frac{(\gamma-1)}{|\beta|^2} \beta^2 \beta^3 \\ -\gamma \beta^3 & \frac{(\gamma-1)}{|\beta|^2} \beta^3 \beta^1 & \frac{(\gamma-1)}{|\beta|^2} \beta^3 \beta^2 & 1 + \frac{(\gamma-1)}{|\beta|^2} \beta^3 \beta^3 \end{pmatrix} \begin{pmatrix} \Delta x_0 \\ \Delta x_1 \\ \Delta x_2 \\ \Delta x_3 \end{pmatrix}.
$$

$$(2.56)$$

In the sequel we shall use the following abbreviated notation for the matrix (2.56):

$$
\Lambda'^{\mu}{}_{\nu} = \begin{pmatrix} \gamma & -\beta^j \gamma \\ -\beta^i \gamma & \delta^{ij} + (\gamma - 1) \frac{\beta^i \beta^j}{|\beta|^2} \end{pmatrix} \tag{2.57}
$$

where $i, j = 1, 2, 3$ label the rows and columns of the 3×3 matrix acting on the spatial components x^1, x^2, x^3, and

$$
\beta^i = \frac{v^i}{c} \Rightarrow \gamma = \frac{1}{\sqrt{1 - \beta^2}},
$$

where we have defined $\beta \equiv |\boldsymbol{\beta}|$. The symbol δ^{ij} is the Kronecker delta defined by the property:

$$
\delta^{ij} = 1 \text{ if } i = j, \qquad \delta^{ij} = 0 \text{ if } i \neq j.
$$

In Chap. 4 it will be shown that the most general Lorentz transformation $\Lambda^{\mu}{}_{\nu}$, $\mu\nu = 0, 1, 2, 3$ is obtained by multiplying the matrix $\Lambda'^{\mu}{}_{\nu}$ by a matrix $\mathbf{R} \equiv (R^{\mu}{}_{\nu})$

$$
\mathbf{R} = \begin{pmatrix} 1 & 0 \\ 0 & R^i_j \end{pmatrix} \tag{2.58}
$$

where the 3×3 matrix R^i_j describes a generic rotation of the three axes (x, y, z), so that $\boldsymbol{\Lambda} = \boldsymbol{\Lambda}' \mathbf{R}$. It is in terms of this general matrix that the notion of *four-vector* is defined: *A four-vector is a set of four quantities that under a general Lorentz transformation transform with the matrix $\Lambda^{\mu}{}_{\nu}$.* For example Δx^{μ} and p^{μ} are both four-vectors; indeed they have the same transformation properties under a general Lorentz transformation

$$
p'^{\mu} = \sum_{\nu=0}^{3} \Lambda^{\mu}{}_{\nu} p^{\nu},
$$

$$
\Delta x'^{\mu} = \sum_{\nu=0}^{3} \Lambda^{\mu}{}_{\nu} \Delta x^{\nu}.
$$

$$(2.59)$$

To simplify our notation, let us introduce the *Einstein summation convention*: Whenever in a formula a same index appears in upper and lower positions,[10] summation over that index is understood and the two indices are said to be *contracted (or dummy) indices*.[11] Using this convention, when we write for instance $\Lambda^\mu{}_\nu \, p^\nu$, summation over the repeated index ν will be understood, so that:

$$\Lambda^\mu{}_\nu \, p^\nu \equiv \sum_{\nu=0}^{3} \Lambda^\mu{}_\nu \, p^\nu.$$

Using (2.59) it is now very simple to show in a concise way that the conservation of total momentum **P** implies the conservation of the total energy E and viceversa.

Let us consider the collision of an isolated system of N particles each having a linear momentum p_n^i, $i, j = 1, 2, 3$ and let us denote by $P^i \equiv \sum_{n=1}^{N} p_n^i$ their total momentum. For each $i = 1, 2, 3$ the change ΔP^i of the total momentum $P^i \equiv \sum_{n=1}^{N} p_n^i$ occurring during the collision will be:

$$\Delta P^i \equiv \left(\sum_n p_n^i \right)_{fin} - \left(\sum_n p_n^i \right)_{in}. \tag{2.60}$$

We assume that the total momentum is conserved in a certain frame, say S, that is:

$$\Delta P^i = 0, \qquad \forall i = 1, 2, 3, \tag{2.61}$$

and carry out a Lorentz transformation to a new frame S'. Taking into account that the momentum of each particle transforms as

$$p_n'^i = \Lambda^i{}_j \, p_n^j + \Lambda^i{}_0 \, p_n^0, \tag{2.62}$$

in the new frame S' the change of the total momentum is:

$$\Delta P'^i = \left(\sum_n p_n'^i \right)_{fin} - \left(\sum_n p_n'^i \right)_{in} = \Lambda^i{}_j \left[\left(\sum_n p_n^j \right)_{fin} - \left(\sum_n p_n^j \right)_{in} \right]$$
$$+ \Lambda^i{}_0 \left[\left(\sum_n p_n^0 \right)_{fin} - \left(\sum_n p_n^0 \right)_{in} \right], \tag{2.63}$$

where Einstein's summation convention is used and summation over the repeated index $j = 1, 2, 3$ is understood. Since the first term in square brackets on the right

---

[10] So far the position of indices in vector components and matrices has been conventionally fixed. We shall give it a meaning in the next chapters.

[11] We observe that contracted indices, *being summed over*, can be denoted by arbitrary symbols, for example $\Lambda^\mu{}_\nu \, p^\nu \equiv \Lambda^\mu{}_\rho \, p^\rho$.

hand side of Eq. (2.63) is zero by hypothesis, $\Delta P^i = 0$, requiring conservation of the total momentum in S', $\Delta P'^i = 0$, implies:

$$\left(\sum_n p_n^0\right)_{fin} = \left(\sum_n p_n^0\right)_{in}, \qquad (2.64)$$

that is, since

$$\sum_n p_n^0 = c \sum_n m_n(v) = \frac{E_{tot}}{c}, \qquad (2.65)$$

the total mass, or equivalently the total energy, must be also conserved.

Viceversa, if we start assuming the conservation of energy $\sum_n p_n^0 = \sum_n E_n/c$, in S and write, for each n,

$$p_n^{0'} = \Lambda^0{}_i \, p_n^i + \Lambda^0{}_0 \, p_n^0, \qquad (2.66)$$

the same Lorentz transformation gives:

$$\frac{1}{c}\Delta E'_{tot.} = \left(\sum_n p_n'^0\right)_{fin} - \left(\sum_n p_n'^0\right)_{in} = \Lambda^0{}_j\left[\left(\sum_n p_n^j\right)_{fin} - \left(\sum_n p_n^j\right)_{in}\right]$$
$$+ \Lambda^0{}_0\left[\left(\sum_n p_n^0\right)_{fin} - \left(\sum_n p_n^0\right)_{in}\right]. \qquad (2.67)$$

The second term in square brackets on right hand side is $\Delta E_{tot}/c$ and is zero, by assumption; Being $\Lambda^0{}_j$ the three components of an arbitrary vector for generic relative motions between the two frames, each of their coefficients must vanish separately. We then conclude that the energy is conserved in S' if and only if also the total linear momentum is.

The notion of four-vector can be also used to give generalize the relativistic vector Eq. (2.15) in a four-vector notation. Recalling the invariance of the proper time interval, we may also define the *force four-vector* or *four-force* as:

$$f^\mu \equiv \frac{dP^\mu}{d\tau}. \qquad (2.68)$$

To understand the content of this equation, let us consider, at a certain instant, an inertial reference frame S' moving at the same velocity as the particle. In this frame the particle will thus appear *instantaneously* at rest (the reason for not considering the rest frame of the particle, namely the frame in which the particle is *constantly at rest* is that such frame is, in general, accelerated, and thus *not inertial*). We know from our discussion of proper time that $d\tau$ coincides with the time dt' in the particle frame S' so that Eq. (2.68) becomes:

$$f'^0 = m \frac{d^2 x'^0}{dt'^2} = mc \frac{d^2 t'}{dt'^2} \equiv 0, \qquad (2.69)$$

$$f'^i = m \frac{d^2 x'^i}{dt'^2} \equiv F^i, \qquad (2.70)$$

where we have used $dt' = d\tau$. It follows that in S' $f'^0 \equiv 0$ so that Eq. (2.70) becomes the ordinary Newtonian equation of classical mechanics. To see what happens in a generic inertial frame S, we perform a Lorentz transformation from S' to S and find:

$$f^\mu = \sum_{\nu=0}^{3} \Lambda^\mu{}_\nu f'^\nu = \sum_{i=1}^{3} \Lambda^\mu{}_i f'^i \equiv \sum_{i=1}^{3} \Lambda^\mu{}_i F^i. \qquad (2.71)$$

The content of this equation is better understood by writing its time ($\mu = 0$) and spatial ($\mu = i$) components separately (here, for the sake of simplicity, we neglect in the Lorentz transformation the rotation part):

$$f^0 = \Lambda^0{}_i F^i = \frac{\gamma}{c} \mathbf{v} \cdot \mathbf{F}, \quad f^i = \sum_{j=1}^{3} \Lambda^i{}_j f'^j = \sum_{j=1}^{3} \Lambda^i{}_j F^j$$

$$= \sum_{j=1}^{3} \left(\delta^i_j + (\gamma - 1) \frac{v^i v^j}{v^2} \right) F^j \equiv F^i + (\gamma - 1) \frac{v^i}{v^2} \mathbf{v} \cdot \mathbf{F}, \qquad (2.72)$$

$\mathbf{v}$ being the velocity of the particle.

We observe that the expression $\mathbf{v} \cdot \mathbf{F}$ on the right hand side of the first of Eq. (2.72), is the *power* of the force $\mathbf{F}$ acting on the moving particle. The time component of Eq. (2.68) then reads

$$f^0 = \frac{dP^0}{d\tau} = \frac{\gamma}{c} \frac{dE}{dt} = = \frac{\gamma}{c} \mathbf{v} \cdot \mathbf{F}, \qquad (2.73)$$

and is the familiar statement that the rate of change of the energy in time equals the power of the force.

If no force is acting on the particle the equation of the motion reduces to:

$$\frac{dP^\mu}{d\tau} = 0, \qquad (2.74)$$

or, using Eq. (2.43),

$$\frac{dU^\mu}{d\tau} = \frac{d^2 x^\mu}{d\tau^2} = 0, \qquad (2.75)$$

Being $d\tau = \sqrt{1 - \frac{v^2}{c^2}} \, dt$, it is easy to see that this equation implies $\mathbf{v} = \text{const.}$ Thus Eq. (2.75) is the Lorentz covariant way of expressing the principle of inertia.

2.2.2 Relativistic Theories and Poincarè Transformations

We have seen that the conservation of the four-momentum and Eq. (2.68) defining the four-force are relations between four-vectors and therefore they automatically satisfy the principle of relativity being covariant under the Lorentz transformations implemented by the matrix $\Lambda \equiv (\Lambda^\mu{}_\nu)$. It follows that the laws of mechanics discussed in this chapter, *excluding the treatment of the gravitational forces*, satisfy the principle of relativity, implemented in terms of general Lorentz transformations.

We may further extend the covariance of relativistic dynamics by adding transformations corresponding to constant shifts or translations

$$x'^\mu = x^\mu + b^\mu, \tag{2.76}$$

b^μ being a constant four-vector. This transformation is actually the four-dimensional transcription of time shifts and space translations already discussed for the extended Galilean transformations (1.15). However, differently from the Galilean case, there is no need in the relativistic context to add three-dimensional rotations, since, as mentioned before, they are actually part of the general Lorentz transformations implemented by the matrix Λ.

It is easy to realize that the four-dimensional translations do not affect the proper time or proper distance definitions, nor the fundamental equations of the relativistic mechanics, Eqs. (2.47) and (2.68).

We conclude that relativistic dynamics is covariant under the following set of transformations

$$x'^\mu = \Lambda^\mu{}_\nu x^\nu + b^\mu, \tag{2.77}$$

which are referred to as *Poincaré transformations*. Both the Lorentz and Poincaré transformations will be treated in detail in Chap. 4. Furthermore in Chap. 5 it will be shown that also the Maxwell theory is covariant under the transformation (2.77), thus proving that the whole of the relativistic physics, namely relativistic dynamics and electromagnetism, is invariant under Poincaré transformations.

One could think that the invariance under translations and time shifts should not play an important role on the interpretation of a physical theory. On the contrary we shall see that such invariance *implies* the conservation of the energy and momentum in the Galilean case and of the four-momentum in the relativistic case (see Chap. 8).

2.2.3 References

For further reading see Refs. [1, 11, 12].

Chapter 3
The Equivalence Principle

3.1 Inertial and Gravitational Masses

In this section we discuss the *principle of equivalence*. We shall see that, besides allowing the extension of the principle of relativity to *a generic*, not necessarily inertial, frame of reference, it allows to define gravity, in a relativistic framework, as a property of the four-dimensional space-time geometry.

The *principle of equivalence*, in the so-called *weak* form, asserts the *exact equivalence between the inertial mass m_I and the gravitational one m_G*.

We recall that the inertial mass m_I is defined through Newton's second law of dynamics:

$$\mathbf{F} = m_I \, \mathbf{a}. \tag{3.1}$$

Its physical meaning is, as is well known, that of *inertia* of a body, that is its reluctance to be set in motion or, more generally, to change its velocity.

The gravitational mass m_G, on the other hand, enters the definition of Newton's universal law of gravitation according to:

$$\mathbf{F} = -G \, \frac{m_G M_G}{r^2} \, \mathbf{u_r} \tag{3.2}$$

where $G = 6.6732 \times 10^{-11} \, \mathrm{N \, m^2/kg^2}$ is the gravitational constant, m_G and M_G are the gravitational masses of the two attracting bodies.[1] For definiteness we refer to the situation where the mass M_G attracts the mass m_G. We see that the gravitational mass sets the strength of the gravitational force that, *ceteris paribus,* a given body exerts on another one. In this sense it would better deserve the name of *gravitational charge*,

---

[1] It goes without saying that we are referring to two spherical bodies or to bodies whose dimensions are negligible with respect to their distance r.

© Springer International Publishing Switzerland 2016
R. D'Auria and M. Trigiante, *From Special Relativity to Feynman Diagrams*,
UNITEXT for Physics, DOI 10.1007/978-3-319-22014-7_3

in analogy with Coulomb's law of the electrostatic interaction, where the electric charge sets the strength of the electric interaction, the mathematical structures of the two laws being exactly the same.

However, besides the enormous *quantitative* difference between the *strengths* of *gravitational* and *electrostatic* forces[2] there is a major qualitative difference between them: While any two bodies have mass, they do not necessarily have charge. As a consequence the gravitational force is *universal*, while the electric force is not.

It is important to note that the equality between inertial and gravitational masses, expressed by the principle of equivalence, is in some sense surprising, given the substantial difference between these two physical concepts, which reflects into different operational definitions of their respective measures.

It is, however, one of the best established results from the experimental point of view.

Indeed a large number of experiments were devised, since Newton's times, to ascertain the validity of this unexpected coincidence; among them, of particular importance from a historical point of view is the *Eötvos experiment*, of which we give a short description in Appendix A. The precision reached in this experiment is such that $\frac{\Delta m}{m_I} \equiv \frac{|m_G - m_I|}{m_I}$, is less than 2×10^{-3}. Thanks to the advanced technological features of modern experimental physics, the above ratio has been pushed to less than 10^{-15}.

This justifies the theoretical assumption of the exact equality between inertial and gravitational mass:

$$m_I = m_G, \tag{3.3}$$

that is the *principle of equivalence in its weak form*, can be safely assumed as one of the experimentally best established principles of theoretical physics.

The great intuition Einstein had at the beginning of last century, was to realize the considerable importance of this seemingly curious "coincidence", since it implies that *locally, it is impossible to distinguish between the effects of a gravitational field and those of an accelerated frame of reference.*

To justify such a conclusion we illustrate a so-called "Gedankenexperiment", that is a conceptual experiment, originally formulated by Einstein to be performed in a frame of reference attached to an elevator; in our era of space journeys, it seems however more appropriate to update this experiment by replacing Einstein's elevator with a spaceship. Note that the time duration of the experiment inside the spaceship, together with its spatial extension, define a four-dimensional region of space-time. In the following we shall moreover restrict ourselves to the framework of the Newtonian theory of gravitation.

Let us now describe this conceptual experiment, in the following four steps:

(i) Suppose the spaceship, to be simply referred to by $\mathcal{A}$, be initially placed, with the engines turned off, in a region of space which is far enough from any celestial

---

[2] We recall that the ratio between the electric and gravitational forces between two protons is of the order of 10^{38}.

body for the net gravitational force acting on it to be negligible. By definition this is an *inertial frame of reference*. If a physicist performs experiments within the spaceship, he will find that all bodies (if not subject to other kind of forces) move in a rectilinear uniform motion, according to Galileo's principle of inertia.

(ii) Let us assume next that, at some stage, the spaceship reaches the proximity of a planet, and thus becomes subject to a gravitational acceleration of the form:

$$\mathbf{g}(\mathbf{r}) = -G\frac{M}{r^2}\mathbf{u_r}, \tag{3.4}$$

M being the (gravitational) mass of the planet.

In the presence of such an attraction $\mathcal{A}$ starts orbiting around the planet, thus becoming an *accelerated frame of reference*. The accelerated frame of reference defined by *any* massive body which is not subject to forces other than gravity is usually referred to as a *free falling frame*. Our spaceship $\mathcal{A}$ orbiting around the planet, is an example of a free falling frame. In this situation an observer inside $\mathcal{A}$ still finds that all bodies describe *inertial motions*: Indeed, the same acceleration $\mathbf{g}(\mathbf{r})$, acting on the frame $\mathcal{A}$, also acts on each body inside of it. It follows that the motion of bodies in $\mathcal{A}$ is inertial with respect to the spaceship itself, since their relative acceleration with respect to $\mathcal{A}$ is zero.

Since in both cases (i) and (ii) the motion inside $\mathcal{A}$ is of the same kind, that is inertial, there is no way an observer in the spaceship can distinguish between the two situations. We conclude that:

It is not possible through experiments performed in the interior of $\mathcal{A}$ to tell whether it is a free falling frame or an inertial one.[3]

(iii) Consider again the spaceship $\mathcal{A}$ in a region far from any celestial body, and suppose that now, in contrast to case (i), its engines are turned on, thus producing an acceleration $\mathbf{a_R}$ which is uniform all over the spaceship and constant in time. Since now $\mathcal{A}$ is accelerated, all bodies inside of it fall with the same acceleration $\mathbf{a'} = -\mathbf{a_R}$, where $\mathbf{a'}$ and $\mathbf{a_R}$ are the accelerations with respect to $\mathcal{A}$ and the relative acceleration of $\mathcal{A}$ with respect to an inertial system, respectively.

(iv) Finally consider the situation in which the spaceship is at rest on a massive body (for example the earth) which creates a gravitational field of acceleration $\mathbf{g}$; because of the equality between inertial and gravitational masses, *all bodies inside $\mathcal{A}$ fall with the same acceleration $\mathbf{g}$.*

Comparing these two last examples, we see the an observer inside the spaceship cannot tell whether he is in case (iv), where the observed acceleration is due to a gravitational field and his frame of reference is inertial, or in case (iii) where his frame of reference is accelerated by the engines of the spacecraft with acceleration $\mathbf{a_R} = -\mathbf{g}$ with respect to an inertial frame. We conclude that:

It is not possible, through experiments performed in the interior of $\mathcal{A}$, to

---

[3]The inertial motion inside a free falling system is a well known fact nowadays, think about the absence of weight of astronauts inside orbiting spacecrafts.

distinguish an inertial frame in the presence of a gravitational field from an accelerated system of reference.

The previous discussion seems to lead to the conclusion that a perfect equivalence exists between the effects of a gravitational field as observed in an inertial frame of reference and those observed in an accelerated frame. This may seem very strange: A gravitational field after all has its sources in the massive bodies where the field lines converge, and its intensity is a decreasing function of the distance from them, vanishing at infinity. The lines of force of the acceleration field $\mathbf{a}'$, on the contrary, are not converging and do not fall down to zero at infinity.

In fact, as we shall now show, a more accurate and quantitative discussion of the "Gedankenexperiment" examined above, shows that the assumed *equivalence between gravitational field and non-inertial frames, holds only if we restrict ourselves to events taking place in a small space region and during a small time interval, that is, mathematically, in an infinitesimal region of space-time.*

To clarify this point, let us examine the situations described in points (i) and (ii) in some more detail. It is not difficult to realize that the supposed equivalence between the free falling frame and the inertial system holds only in the limit where the gravitational acceleration $\mathbf{g}(\mathbf{r})$ can be regarded as uniform in the interior of $\mathcal{A}$, that is only if:

$$\mathbf{g}(\mathbf{r}) \simeq \mathbf{g}(\mathbf{r}_0), \tag{3.5}$$

where $\mathbf{r}_0$ is the barycenter of the spaceship.

Suppose indeed we have a system of particles of masses $m_1, m_2, \ldots, m_N$ in a gravitational field, $\mathbf{g}$, each obeying its own equation of motion; let us now rewrite the equation of motion of, say, the kth-particle (3.1), in an accelerated frame S' having a relative acceleration $\mathbf{a}_R$ with respect to an inertial frame S.[4] One has:

$$m_{G(k)}\, \mathbf{g}(\mathbf{r}_k) = m_{I(k)} \left(\mathbf{a}'_k + \mathbf{a}_R \right) \tag{3.6}$$

where $\mathbf{a}'_k$ is the acceleration of the particle in the frame S'. In particular, if S' is the free falling system, we have:

$$\mathbf{a}_R = \mathbf{g}(\mathbf{r}_0), \tag{3.7}$$

so that Eq. (3.6) becomes:

$$m_{G(k)}\, \mathbf{g}(\mathbf{r}_k) - m_{I(k)}\, \mathbf{g}(\mathbf{r}_0) = m_{I(k)}\, \mathbf{a}'_k. \tag{3.8}$$

The discussion made in points (i) and (ii) holds provided we use the approximation (3.5), $\mathbf{g}(\mathbf{r}_k) \simeq \mathbf{g}(\mathbf{r}_0)$, that is *we consider a sufficiently small neighborhood* of the center of mass of $\mathcal{A}$, such that the gravitational acceleration can be approximated by a constant vector and higher order effects in the distance $|\mathbf{r}_k - \mathbf{r}_0|$ can be neglected.

---

[4]We recall that the relative acceleration $\mathbf{a}_R$ is given in general by the sum of four terms, the translation acceleration $\mathbf{a}'$, the centripetal acceleration $\mathbf{a}_{(centr)}$, the Coriolis acceleration $\mathbf{a}_{(Cor)}$ and a further term proportional to the angular acceleration of S' with respect to S.

In this case, taking into account the principle of equivalence, $m_{G(k)} = m_{I(k)}$, from (3.8) it follows:

$$\mathbf{a}'_k \simeq 0. \tag{3.9}$$

This more accurate analysis implies that the equivalence between the free falling frame and the inertial frame described in the former two experiments only holds *locally*, that is in an infinitesimal neighborhood where, up to higher order terms, Eq. (3.5) is valid. The previous discussion can be made more general by assuming that, besides the interaction with the gravitational field, the particles are also subjected to reciprocal *non gravitational forces* $\mathbf{F}_{kl}$. In this case let us write the equation of motion of the kth particle in an inertial frame where a gravitational field $\mathbf{g}$ is present:

$$m_{Gk}\, \mathbf{g}(\mathbf{r}_k) + \sum_{l \neq k} \mathbf{F}_{kl}(|\mathbf{r}_k - \mathbf{r}_l|) = m_{Ik}\, \frac{d^2\, \mathbf{r}_k}{d\, t^2}. \tag{3.10}$$

We may write the corresponding equation in the free falling system through the coordinate transformation

$$\mathbf{r}' = \mathbf{r} - \frac{1}{2}\mathbf{g}t^2. \tag{3.11}$$

In the same hypotheses made before, namely assuming Eq. (3.5), which implies $\mathbf{g} = \text{const.}$ and (3.3), Eq. (3.10) takes the following form:

$$\sum_{l} \mathbf{F}_{kl}(|\mathbf{r}'_k - \mathbf{r}'_l|) = m_{Ik}\, \frac{d^2\, \mathbf{r}'_k}{d\, t^2}, \tag{3.12}$$

which is the equation of motion of classical mechanics in the absence of a gravitational field.

As far as the other two situations (iii) and (iv) are concerned, it is clear that also in this case the equivalence only holds if the approximation (3.5) is used, that is only *locally*, since the gravitational field would in general be a function of the point inside the spaceship, while the field of accelerations is, at each instant, exactly uniform.

Let us recall that our analysis so far has been made within the framework of classical Newtonian mechanics, without any reference to the implications of Einstein's special relativity. If we now assume that the statement of the *local equivalence* between an inertial frame of reference in a gravitational field and an accelerated one also holds in the framework of relativistic mechanics, described in Chap. 2, then we may reformulate the conclusions of points (i) and (ii), as follows:

In the presence of a gravitational field, locally, the physical laws observed in a free falling frame are those of special relativity in the absence of gravity.

As explained above, *locally* means, mathematically, in an infinitesimal neighborhood or, more physically, in a sufficiently small neighborhood of a point such that, up to higher order terms, the approximation (3.5) holds.

The previous statement is referred to as the *equivalence principle in its strong form* or simply *strong equivalence principle.*

In particular we see that when gravitation is included in the description of physical systems, if the *strong equivalence principle* is assumed to hold, the special role played by *inertial frames* is lost, since frames of reference which are rigorously inertial only exist in infinitesimal regions of space-time. Moreover, since acceleration and gravitation are locally indistinguishable it is reasonable to assume that *the physical laws should take the same form also in non-inertial frames*, or, in other words, to require that *the principle of relativity be valid not only in inertial frames, but more generally, in every frame of reference, described by generic four-dimensional coordinate systems.* Recall indeed that we restricted ourselves, in the first chapter, to coordinate transformations which were linear, since we were only interested in inertial frames of references (endowed with three-dimensional Cartesian coordinate systems).[5] Anticipating concepts to be introduced in next chapter, this characterizes the four-dimensional coordinate system associated with inertial frames of reference to be Cartesian or *rectilinear.* Extending our analysis to non-inertial frames implies considering general (four-dimensional) coordinate systems, related to one another by non-linear transformations: Transformations to *arbitrary accelerated frames of reference* are described by *arbitrary transformations of the four coordinates labeling space-time.* It follows that in order to implement the principle of relativity on the laws of physics, in the presence of gravitation, we must require these to have the same form in *any frame of reference*, so that they be covariant under *general coordinate transformations*, i.e. arbitrary changes of coordinates with a non-vanishing, coordinate-dependent, Jacobian:

$$x^{\alpha\prime} = f^\alpha(x^0 \equiv ct, \, x^1, \, x^2, \, x^3). \tag{3.13}$$

Summarizing, while in the special theory of relativity, where the gravitational interaction is not taken into account, the physical laws were required to be covariant only under the linear Lorentz transformations, relating inertial frames, in the presence of gravity the implementation of the principle of relativity requires that:

The laws of the Physics be covariant under general coordinate transformations (3.13).

Note that *covariance under general coordinate transformations* means, as we discussed in the case of the Lorentz transformations, that the equations describing the physical laws have exactly the same form, albeit in the transformed variables, in every coordinate system.

In other words: A theory including a treatment of the gravitational field must be *generally covariant.*

---

[5]Linearity was then a consequence of the requirement that the principle of inertia holds in both the old and the transformed frames: A motion which is uniform with respect to one of them cannot be seen as accelerated with respect to the other.

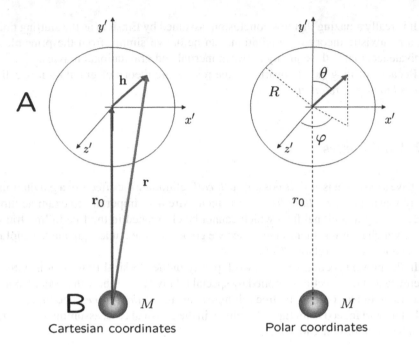

Fig. 3.2 Coordinates

Let us compute, to the first order in h^k, the ith component of the gravitational acceleration in the free falling frame. From Eq. (3.14) it follows:

$$a^{i\prime} = \sum_{k=1}^{3} \left.\frac{\partial g^i}{\partial x^k}\right|_{h=0} h^k + O(|\mathbf{h}|^2) \equiv -\sum_{k=1}^{3} \left.\frac{\partial^2 \phi}{\partial x^k \partial x^i}\right|_{h=0} h^k + O(|\mathbf{h}|^2). \quad (3.15)$$

where $i = 1, 2, 3$ and ϕ denotes the gravitational potential.

Taking into account that

$$g^i = -\frac{\partial \phi}{\partial x^i} = -G\frac{M\,x^i}{r^3}, \quad (3.16)$$

Equation (3.15) becomes to first order in h^i:

$$a^{i\prime} = -G\frac{M}{r_0^3} \sum_{k=1}^{3} \left(\delta^{ik} - 3\frac{x_0^i\, x_0^k}{r_0^2}\right) h^k. \quad (3.17)$$

The acceleration field $\mathbf{a}' \equiv (a^{i\prime})$, defined in Eq. (3.17), is *the remnant, to order* $O(|\mathbf{h}|)$, of the gravitational field in the free falling frame. We see that it is essentially given by the *gradient of the gravitational field* and is called the *tidal field*.

Correspondingly, the force $\mathbf{f} = m\mathbf{a}'$ of the tidal field, acting on a given mass m, is the *tidal force*.

We stress that all these considerations have been obtained using the classical Newtonian formula for the gravitational field which is both *static and non-relativistic*.

In order to show what the effect of tidal forces on an extended body is, let us consider two bodies, say A and B, subject to their mutual gravitational interaction, and let us assume that A is free falling in the gravitational field of B, like for instance an orbiting satellite. We also assume, for the sake of simplicity, that the free falling body A is spherical. We call S' the frame of reference attached to A and S the one attached to B. Let the origin of S' coincide with the barycenter $\mathbf{r}_0$ of A, the z'-axis coincide with direction joining $\mathbf{r}_0$ to the center of mass of the attracting body B and the x' and y'-axes lie, as usual, in the plane orthogonal to the z'-axis (see Fig. 3.2).

With reference to this configuration we observe that $x_0^3 \equiv z_0 \equiv r_0$, so that Eq. (3.17) gives:

$$f^{z'} = 2G \frac{Mm}{r_0^3} h^{z'}, \tag{3.18}$$

$$f^{x'} = -G \frac{Mm}{r_0^3} h^{x'}, \tag{3.19}$$

$$f^{y'} = -G \frac{Mm}{r_0^3} h^{y'}, \tag{3.20}$$

that is, in matrix notation:

$$\begin{pmatrix} f^{x'} \\ f^{y'} \\ f^{z'} \end{pmatrix} = m \begin{pmatrix} -G\frac{M}{r_0^3} & 0 & 0 \\ 0 & -G\frac{M}{r_0^3} & 0 \\ 0 & 0 & 2G\frac{M}{r_0^3} \end{pmatrix} \begin{pmatrix} h^{x'} \\ h^{y'} \\ h^{z'} \end{pmatrix}, \tag{3.21}$$

where the entries of the matrix on the right hand side are $\frac{\partial g^i}{\partial x^k}$ computed in $\mathbf{r}_0$. From (3.18) it follows that $f^{z'}$ is attractive or repulsive according to the sign of $h^{z'}$; that is, referring to Fig. 3.1, it points downwards on the part of the body facing B ($z' > 0$), and upwards on the opposite side ($z' < 0$). The two horizontal components $f^{x'}$ e $f^{y'}$, instead, are always attractive, that is they are directed towards the origin of S'. The net result is an outward stress acting along the line joining A and B and an inward stress on the horizontal planes $z' = $ const.

Tidal forces can be very strong in the astrophysical phenomena; for example the tidal forces exerted by the greatest planets of the solar system on their satellites induce a tidal heating due to the consequent internal friction; in the case of the Jupiter satellite Io this results in dramatic volcanic eruptions.[6]

---

[6]In the proximity of the event horizon of a black hole tidal forces are so strong as to completely disintegrate any body falling inside.

In the case of a deformable body, tidal forces deform a spherical body to the shape of an ellipsoid, the major axis lying along the $A-B$ direction. This is in fact what happens in the case of the earth where the corresponding phenomenon, induced by the moon,[7] gives rise to the ordinary oceanic tides, whence the denomination *tidal forces* has its origin.

We may indeed think of the earth as the body A in free fall on the gravitational field of the moon, the body B. The "thin" layer represented by the oceans covering the earth's surface is indeed deformable. It follows that on the side facing the moon and on the opposite side tidal forces give rise to high tides, while in the directions perpendicular to the line earth-moon tidal forces produce low tides (Fig. 3.1). Because of the bipolar character of these bulges and compressions, the periodicity of tides is of 12 h.

We may make a crude estimation of the tidal size on the earth.

First we observe that tidal forces are conservative. Indeed from Eqs. (3.18)–(3.20) it follows that the associated tidal potential energy is:

$$\mathcal{E}_P = -\frac{G\,M_L m_T}{r_0^3}\left(z'^2 - \frac{1}{2}x'^2 - \frac{1}{2}y'^2\right) = -\frac{G\,M_L m_T}{r_0^3}\left(-\frac{1}{2}R^2 + \frac{3}{2}z'^2\right)$$

$$= -\frac{G\,M_L m_T}{r_0^3}\,R^2\left(\frac{3\cos^2\theta - 1}{2}\right),\tag{3.22}$$

where we have introduced spherical coordinates with origin on the center of the earth (see Fig. 3.2), R being its radius, θ the latitude and we have denoted by M_L, m_T, r_0, the moon's mass, the earth's mass and the earth-moon distance, respectively.

The total potential energy of a water particle is obtained as the sum $\mathcal{E}_P + m\,g\,h$, where $m\,g\,h$ is the potential energy of the terrestrial attraction. At the equilibrium the form of the oceans' equipotential surface is determined by the condition:

$$E_P^{tot} = m\,g\,h - \frac{G\,M_L m_T}{r_0^3}\,R^2\left(\frac{3\cos^2\theta - 1}{2}\right) = \text{const.}\tag{3.23}$$

Equation (3.23) implies $h = h(\theta)$; the difference in height between high ($\theta = 0$) and low ($\theta = \frac{\pi}{2}$) tides turns out to be:

$$\Delta h \equiv h(0) - h(\pi/2) = \frac{3GM_l}{2g\,r_0^3}\,R^2 \simeq 53\,\text{cm}.\tag{3.24}$$

If the gravitational pull of the sun is also taken into account, one finds an additional contribution about half as large as the previous one.[8]

---

[7]More correctly one should also consider the contribution of the sun which is not much less than that of the moon. For the sake of simplicity we just illustrate the main contribution due to the moon.

[8]This crude estimate must be considered, together with the correction due to the sun, just a mean value. It does not take into account resonance phenomena due to the earth rotation and the shape

3.3 The Geometric Analogy

In this section we try to explain how the equivalence principle discussed in the previous section naturally leads to a modification of the space-time geometry, described in special relativity by the Minkowski space. To this end we need to anticipate part of the discussion of next chapter.

We begin by recalling that in Euclidean geometry the distance between two points P_1 e P_2, which is *invariant* under rotations and translations of the Cartesian coordinate system, can be written as[9]

$$\Delta \ell^2 = (\Delta \xi^1)^2 + (\Delta \xi^2)^2 + (\Delta \xi^3)^2 = \sum_{i,j} \delta_{ij} \, \Delta \xi^i \, \Delta \xi^j, \qquad (3.25)$$

if and only if we are using Cartesian (rectangular) coordinates ξ^i. By Cartesian (rectangular) coordinates we mean rectangular (or rectilinear) coordinates, which, in studying Euclidean geometry, are defined throughout the space. This is possible only if the geometry is Euclidean, that is the properties of figures, the notion of parallelism and so on are those derived by Euclid's axioms.

In Eq. (3.25) we have adopted the following notation (see Sect. 2.2.1):

$$\delta_{ij} = 1 \qquad i = j,$$
$$\delta_{ij} = 0 \qquad i \neq j, \qquad (3.26)$$

that is, in matrix notation:

$$(\delta_{ij}) = \begin{pmatrix} 1 & 0 & 0 \\ 0 & 1 & 0 \\ 0 & 0 & 1 \end{pmatrix}. \qquad (3.27)$$

The Kronecker symbol δ_{ij} defines the so called *metric tensor*.

If we adopt the Einstein convention that repeated indices are summed over, then Eq. (3.25) can be written as follows:

$$\Delta \ell^2 = \sum_{i,j} \delta_{ij} \, \Delta \xi^i \, \Delta \xi^j \equiv \delta_{ij} \, \Delta \xi^i \, \Delta \xi^j, \qquad (3.28)$$

or, if the two points are infinitesimally apart,

$$d\ell^2 = \delta_{ij} \, d\xi^i d\xi^j. \qquad (3.29)$$

(Footnote 8 continued)
of the oceanic depths which can locally alter in a sensible way the rough computation leading to (3.24).

[9]In this section only we denote by ξ^i the Cartesian coordinates, while the notation x^i is used for generic curvilinear coordinates.

It is clear that the same considerations and formalism hold if the Euclidean space has a generic number D of dimensions; in this case the indices $i, j, \ldots$ would run over D values instead of only three.

As we have already remarked in Sect. 1.5 of the first chapter, there is a close analogy between the *four-dimensional distance* in Minkowski space-time and the Euclidean distance (3.28) in $D = 4$ dimensions. Indeed the *four-dimensional distance* between two events[10] labeled by the four coordinates $\xi^0, \xi^1, \xi^2, \xi^3$ was written as[11]:

$$\Delta s^2 = c^2(\Delta\tau)^2 = (\Delta\xi^0)^2 - (\Delta\xi^1)^2 - (\Delta\xi^2)^2 - (\Delta\xi^3)^2 = \sum_{\alpha,\beta} \eta_{\alpha\beta}\, \Delta\xi^\alpha \Delta\xi^\beta$$

$$\equiv \eta_{\alpha\beta}\, \Delta\xi^\alpha \Delta\xi^\beta \quad (\alpha, \beta = 0, 1, 2, 3) \tag{3.30}$$

that is in a way which is strictly analogous to the four-dimensional Euclidean distance, the only difference being the replacement of the metric δ_{ij} with minus the Minkowski metric $\eta_{\alpha\beta}$:

$$\eta_{\alpha\beta} = \begin{pmatrix} 1 & 0 & 0 & 0 \\ 0 & -1 & 0 & 0 \\ 0 & 0 & -1 & 0 \\ 0 & 0 & 0 & -1 \end{pmatrix}, \tag{3.31}$$

where the extra fourth coordinate is related to time, $\xi^0 = ct$. We stress that these simple expressions of distance in Euclidean space and of proper distance in Minkowski case, are only valid if we use the three-dimensional Cartesian rectangular or the analogous four-dimensional Cartesian rectangular (also referred to as Minkowskian[12]) coordinates ξ^α, the latter being defined as the coordinates used to describe inertial frames in terms of the spatial Cartesian rectangular coordinates and the usual time coordinate t. In any other coordinate system, not related by a three-dimensional rotation or a Lorentz transformation, respectively, the Euclidean distance (3.28) or the Minkowski proper distance (3.30) would take a more complicated form.

Suppose indeed that in the Euclidean case we want to use an arbitrary system of curvilinear coordinates x^i, $i = 1, 2, 3$ (an example would be the spherical polar coordinates); we would then have:

$$\xi^i = \xi^i(x^j). \tag{3.32}$$

---

[10]Recall that the space-time (four-dimensional) distance was conventionally defined as the *negative* of the *proper distance*, see Eq. (2.45).

[11]From now on, we use Greek indices to label four dimensional space-time coordinates and Latin ones for the coordinates in Euclidean space.

[12]We shall call Minkowskian the Cartesian rectangular coordinates in the four-dimensional Minkowski space-time, with metric $\eta_{\alpha\beta}$.

In these new coordinates the infinitesimal distance (3.30) becomes:

$$d\ell^2 = \delta_{ij} \frac{\partial \xi^i}{\partial x^k} \frac{\partial \xi^j}{\partial x^\ell} dx^k dx^\ell \doteq g_{k\ell} dx^k dx^\ell, \tag{3.33}$$

where

$$g_{k\ell} = \delta_{ij} V_k^i V_\ell^j, \tag{3.34}$$

being:

$$V_k^i \doteq \frac{\partial \xi^i}{\partial x^k}. \tag{3.35}$$

The dimensionless quantity $g_{k\ell}(x)$, replacing δ_{ij} in the formula for the squared distance, is called *metric tensor* or, more simply, *metric*[13] in curvilinear coordinates.

It is obvious that all the geometric quantities of Euclidean geometry (lengths, angles, areas,etc) do not depend of the particular coordinates used for their description. However it is well known that in general it is much simpler to compute geometric quantities using Cartesian coordinates, rather than the curvilinear ones.[14]

The same considerations would of course apply to Minkowski space-time in special relativity, if we were to use arbitrary "curvilinear" four-dimensional coordinate frames. The physical interpretation in this case would be the following: Since an arbitrary change of coordinates would correspond to arbitrary functions of the original Minkowski coordinates $\xi_0 \equiv ct$, ξ_1, ξ_2, ξ_3, the new frame of reference cannot be inertial since the transformation is, in general, not linear as it is the case for Lorentz transformations. It then follows that the new coordinate system (x^μ) must correspond to an *accelerated frame of reference*. Moreover, as it happens in the Euclidean case, instead of a constant Minkowski metric $\eta_{\alpha\beta}$ we would end up with a metric tensor depending on the four coordinates $x^\mu = x^0, x^1, x^2, x^3$. This, of course, would not change the physical laws, but only describe them in a more general, albeit cumbersome, way.

Given these preliminaries we now define a space, or a space-time as *flat* if there exists a *special class of coordinates* such that the metric assumes a constant value *in a finite or infinite domain of the space or space-time*. In Euclidean and Minkowski spaces this special class is given by the Cartesian coordinates (in particular the Minkowski coordinates of special relativity), any two elements of this class being related by a combination of rotations and translations in Euclidean space or of Lorentz

---

[13] Here and in the following of this chapter we use the word *tensor*, whose precise meaning will be given in Chap. 4, in a loose sense, that is as a quantity carrying indices and whose transformation properties are fixed in terms of the change of coordinates (or of reference frame). In this chapter the transformation of coordinates considered are either cartesian orthogonal, or Lorentzian or even arbitrary, as explained below.

[14] When dealing with problems which exhibit some degree of symmetry it may, however, be more useful to use curvilinear coordinates, like spherical, cylindrical, etc.

transformations and translations in Minkowski space. Note that the computation in generic curvilinear coordinates of geometric quantities or of physical laws in Euclidean or Minkowskian flat space, respectively, would involve the use of the metric $g_{\mu\nu}(x^\rho)$, that is of a matrix whose elements are function of the coordinates. However, in the special class of frames, the corresponding metric is simply given by the constant matrices (3.27) or (3.31).

If, however, such a *special class of coordinates cannot be found in a large domain,*[15] we then say that the space or space-time is *curved,* that is it exhibits *curvature,* a concept which we dwell on in the next section.

3.4 Curvature

Suppose now we have a space (finite or infinite) which possesses a curvature, that is in which it is *not* possible to introduce in a large domain Cartesian (Minkowskian) coordinates. For the sake of simplicity, and to have a help from our intuition, we suppose for the moment that the space in consideration has two space dimensions, though it may not necessarily be the plane $\mathbb{R}^2$, but rather an arbitrary surface Σ. It is well known that on a generic surface it is not possible to introduce *in a finite domain* Cartesian coordinates, but only *curvilinear ones* x^i, $i = 1, 2$.

Restricting our attention to an *infinitesimal neighborhood* of a point P, however, we can approximate the surface by the tangent plane to Σ at P. We say that the *local geometry* of the surface at P is identified with that of the corresponding tangent plane, the local coordinates on Σ, in the close vicinity of P, coinciding with the Cartesian ones ξ^i, $(i = 1, 2)$ on the plane[16] (see Fig. 3.3).

In an infinitely small neighborhood of P we can then use Eq. (3.30) to compute the infinitesimal distance between two points, and thus we conclude that *locally the metric can be always reduced to the form* δ_{ij}.[17] However to compute geometric quantities in a *finite domain* it is necessary to use general curvilinear coordinates by transforming the local coordinates through Eq. (3.32).[18] This means that the geometry in a large, possibly infinite, domain is determined by the metric tensor $g_{ij}(x^1, x^2)$, function of the curvilinear coordinates. For example, the length of a curve $\gamma = \gamma(\tau)$ can be computed as:

---

[15]By *large* domain we mean a domain whose extension is finite or infinite.

[16]Being the plane flat, we can describe it by Cartesian coordinates.

[17]Here and in the following by *locally* we mean that our statement is valid in an infinitesimal neighborhood of a point where higher order terms can be neglected.

[18]Alternatively one can use Cartesian coordinates at each point using the local tangent plane; in this case, however, one needs a quantity, called *connection*, which relates the local geometries associated with different tangent planes.

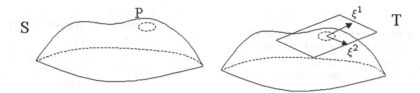

Fig. 3.3 Local geometry and tangent space

$$\ell = \int_\gamma d\ell = \int_\gamma \left(g_{ij}(x)dx^i dx^j\right)^{\frac{1}{2}} = \int_\gamma \left(g_{ij}(x)\frac{dx^i}{d\tau}\frac{dx^j}{d\tau}\right)^{\frac{1}{2}} d\tau, \quad (3.36)$$

τ being any parameter on the curve.

We conclude that on a surface with curvature *the metric cannot be reduced to the constant matrix δ_{ij} in a finite domain;* its geometry is therefore described by the metric tensor $g_{k\ell}(x)$ or, equivalently, by the matrix V_k^i defined by Eq. (3.35).

Even if these considerations have been made in the special case of a two-dimensional surface, they can be straightforwardly extended to three- or N-dimensional curved spaces[19] that is to spaces where Cartesian coordinates cannot be introduced in large domains and, also, to spaces *where the local metric is not δ_{ij}, but $\eta_{\alpha\beta}$,* as it happens for the Minkowskian space-time of special relativity.

We can now give a precise geometric interpretation of the *strong equivalence principle* of the previous section: Saying that in a free falling frame the laws of special relativity hold, means that the space-time geometry in such frame is *locally* the same as that of the four-dimensional (hyper)-plane tangent to space-time at the point in which the frame is located.

This (four-dimensional) tangent plane is of course the Minkowski space of special relativity. Indeed we have learned that in a free falling frame, *locally*, that is in an infinitesimal neighborhood of a point $P = (x^0, \mathbf{x})$ in its interior, up to higher order terms, gravitation is absent, so that the metric tensor reduces to the constant metric tensor $\eta_{\alpha\beta}$ of special relativity.[20] If instead we use a general frame of reference, which is not free falling, the metric tensor $g_{\mu\nu}(x)$ must describe the presence of the gravitational field.[21]

We may thus establish the following correspondence between the presence of curvature in a four-dimensional space which is locally Euclidean (metric δ_{ij}) and a space which is locally Minkowskian (metric $\eta_{\alpha\beta}$).

---

[19] A more precise definition of curvature will be given in the next section.

[20] Note that this implies that the free falling frames are the inertial frames described by special relativity. However, in the presence of a gravitational field, they can be only defined locally, since only locally the effects of gravity can be canceled.

[21] As we shall see in the sequel of this chapter the space-time metric $g_{\mu\nu}(x)$ is related to the gravitational *potential* rather than to the gravitational *field*.

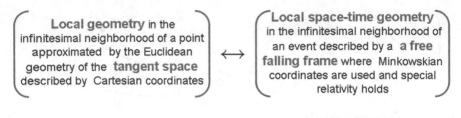

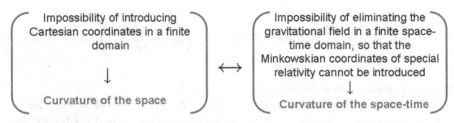

Summarizing, the geometry of space-time is the geometry of a four-dimensional space with local Minkowski metric; this geometry cannot be flat in the presence of a gravitational field. Space-time geometry must then be described by a metric $g_{\mu\nu}(x)$ which depends on the presence of a gravitational field, each entry being a function of the space-time coordinates x^0, x^1, x^2, x^3. The metric can be reduced to the special relativity form $\eta_{\alpha\beta}$ only locally, that is in an infinitely small neighborhood of an event P, but not in a large region of space-time.

Generalizing Eqs. (3.33) and (3.34) to a four-dimensional space-time, where ξ^α ($\alpha = 0, 1, 2, 3$) are the local Minkowskian coordinates and x^μ ($\mu = 0, 1, 2, 3$) are the general coordinates parametrizing a large domain of the space, we have:

$$ds^2 = \eta_{\alpha\beta} \, V_\mu^\alpha V_\nu^\beta dx^\mu dx^\nu = g_{\mu\nu} \, dx^\mu dx^\nu, \tag{3.37}$$

where

$$V_\mu^\alpha \Big|_P = \frac{\partial \xi^\alpha}{\partial x^\mu}\Big|_P, \tag{3.38}$$

and ξ^α are the *local Minkowskian* (i.e. inertial) coordinates in an infinitesimal neighborhood of P.

3.4.1 An Elementary Approach to the Curvature

We recall that the Euclidean geometry is based on Euclid's eleventh postulate which states the uniqueness of the straight line passing through a given point P and parallel to a given straight line. This postulate, in particular, implies that the three interior angles of a triangle sum up to 180°, see Fig. 3.4:

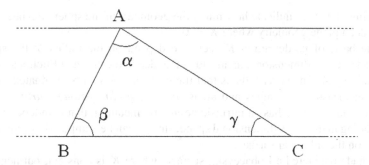

Fig. 3.4 Euclidean geometry

Fig. 3.5 Spherical geometry

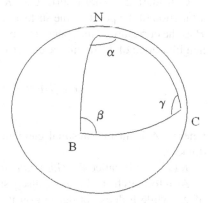

$$\alpha + \beta + \gamma = \pi. \tag{3.39}$$

Let us take again, for the sake of simplicity and intuition, a two-dimensional space, more specifically a 2-sphere (see Fig. 3.5). Let us then consider a spherical triangle defined by joining along maximal circles three arbitrary points of the sphere.[22]

In spherical geometry one can show that the following relation holds:

$$\alpha + \beta + \gamma = \pi + \frac{A}{R^2}, \tag{3.40}$$

where A is the area of the spherical surface enclosed by the triangle and R is the radius of the sphere.

Let us define the *curvature K* of the sphere as:

$$K = \frac{1}{R^2}. \tag{3.41}$$

---

[22]A maximal circle is the shortest path joining two points on the sphere. It can be obtained by intersecting the spherical surface with a plane determined by the two points and the center of the sphere.

the meaning of K is to indicate how much the geometry of the sphere deviates from the Euclidean plane geometry where $K = 0$.

On the basis of its definition K seems to depend on the radius of the sphere considered as a two dimensional manifold embedded in the ambient Euclidean space $\mathbb{R}^3$. Equation (3.40) however tells us that the curvature K can be evaluated by just performing measures of angles and areas *on the two-dimensional surface of the sphere*. It follows that K has an intrinsic geometric meaning, that is independent of its embedding in the flat 3-dimensional space, since it can be computed by only using measures on the spherical surface.

If we had considered a Lobacevskij surface, where K is constant, but negative, Eq. (3.40) is also valid.

Considering a *generic* surface Σ, K is of course no longer constant, but becomes a function of the point on the surface. Indeed we may characterize the value of K at P as the curvature of the sphere that best approximates the generic surface in a small neighborhood of P. In this case Eq. (3.40) can be generalized by writing:

$$\alpha + \beta + \gamma - \pi = \int\limits_{\Delta} K(x^1, x^2)\, dA, \tag{3.42}$$

where dA is an infinitesimal element of the surface Σ and Δ is the integration domain.

$K(x^1, x^2)$ is called the *Gaussian curvature* of S at $P \equiv (x^1, x^2)$.

As it follows from the above discussion, K only depends on the intrinsic geometry of Σ, while it does not depend on its particular representation in $\mathbb{R}^3$, in terms of parametric equations of the form:

$$x = x(x^1, x^2), \quad y = y(x^1, x^2), \quad z = z(x^1, x^2), \tag{3.43}$$

where x^1, x^2, are curvilinear coordinates on Σ and x, y, z are Cartesian coordinates on $\mathbb{R}^3$. In fact one can show that K is invariant if S is flexed without stretching or tearing; for example, since K is zero on a plane, it will also be zero on a cone or a cylinder or in general on any surface which can be unfolded on a plane. If instead we are to map a portion of the terrestrial globe on a plane, stretching is necessary, since we have to change the value of K from a positive constant to zero. In this case the map can be considered a good approximation only if the area considered is much smaller than $\frac{1}{K} = R^2$.

3.4.2 Parallel Transport

There is an equivalent way of describing the curvature of a sphere.

Consider a cannon at the north pole (which may metaphorically represent a tangent vector) and let us carry it along a meridian till it reaches the equator at a point A,

Fig. 3.6 Parallel transport

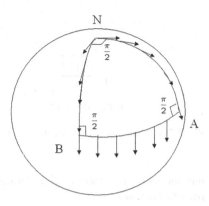

displacing it in such a way as to always keep it *parallel to itself,* so that the angle it forms with the meridian remains constant (for example, in Fig. 3.6, the vector forms an angle zero with the arcs *NA* and *BN* and $\pi/2$ with the arc *AB*). This is what it is meant by *parallel transport.* Eventually, the same kind of *parallel transport* is performed along an arc *AB* of the equator whose length is $\frac{1}{4}$ the circumference, and finally we carry the cannon back to the north pole along the meridian *BN*, see Fig. 3.6.

We easily realize that after this tour the cannon arrives back *rotated* by an angle of $\frac{\pi}{2}$, which is exactly the angular excess described by the curvature:

$$\Delta\theta = \alpha + \beta + \gamma - \pi = \frac{3}{2}\pi - \pi = \frac{\pi}{2} = K\,A. \tag{3.44}$$

Indeed the ratio between the angular excess $\Delta\theta$ and the area of the spherical octant, $\frac{\pi R^2}{2}$, gives exactly the value of the curvature, namely $\frac{1}{R^2} \equiv K$.

Although this result was obtained in a very particular case, it can be shown to hold for a *parallel transport* along any closed path γ, not necessarily along geodesic triangles, enclosing an area Δ and on any surface. Indeed one can expect and actually show that in the general case the rotation angle is found by first applying Eq. (3.44) to an infinitesimal area dA and then integrating over the whole area Δ:

$$\Delta\theta = \int_{\Delta} K(x)\,dA. \tag{3.45}$$

where Δ is the area enclosed by the curve γ.

This alternative way of defining the curvature can be easily extended to manifolds with any number of dimensions. Since our goal is to define the curvature for the four-dimensional space-time with local Minkowski metric $\eta_{\alpha\beta}$, we consider the *parallel transport* of a vector v^μ, ($\mu = 0, 1, 2, 3$) along a closed path γ in a four-dimensional manifold. After the trip, the vector will get back to the initial point "rotated" with respect to the original direction. However, as we have previously learned, such a "rotation" is a "four-dimensional rotation" in a space endowed with a metric which

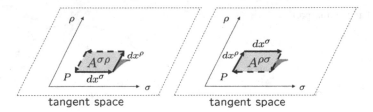

Fig. 3.7 Parallel transports along infinitesimal contours with opposite orientations

has the same signature as the Minkowski one $\eta_{\alpha\beta}$,[23] and therefore is a *Lorentz transformation*.

In particular, if the closed path γ is infinitesimal, we can obtain the analogous of the *Gaussian curvature* as in the case of a two-dimensional surface.

However, while in two dimensions we just have one orientation for the infinitesimal area dA enclosed by the path γ, (that of the plane tangent to the surface), in four dimensions dA can have several different orientations. More specifically, if we represent γ as an infinitesimal parallelogram, we have $\binom{4}{2} = 6$ orientations, which we may label by the ordered couple of indices of the two infinitesimal displacements dx^ρ, dx^σ defining the parallelogram dA: $(\rho, \sigma) = (01), (02), (03), (12), (13), (23)$.

We can thus write:

$$dA \to dA^{\rho\sigma} = -dA^{\sigma\rho} \equiv dx^\rho \wedge dx^\sigma, \tag{3.46}$$

where the antisymmetry in the couple $\rho\sigma$, denoted by the symbol $\wedge$, is due to the orientation of $dA^{\rho\sigma}$, which is related to the orientation of the curve γ (in other words the orientation of dA depends on whether the vector is transported along γ in one direction or in the other, namely if the displacement dx^ρ precedes or follows dx^σ, see Fig. 3.7).

In conclusion, performing a parallel transport of a Lorentz vector v^μ[24] along an infinitesimal path in space-time the vector undergo an (infinitesimal) Lorentz transformation given by:

$$\delta v^\mu = R^\mu{}_{\nu\rho\sigma} v^\nu dx^\rho \wedge dx^\sigma. \tag{3.47}$$

where Einstein convention of repeated indices is applied and we *define* $R^\mu{}_{\nu\rho\sigma}$ to be the curvature of the space-time manifold at a point P. We see that in four dimensions, the curvature is actually described by an object carrying four four-dimensional indices,

---

[23] It can be shown that the signature of the metric, that is the number of positive and negative eigenvalues of the matrix $g_{\mu\nu}(x)$, is the same at each point of a manifold. Since at a given point the metric can be taken to coincide with $\eta_{\alpha\beta}$ this explains the meaning of the statement in the text.

[24] Note that since the closed path is infinitesimal we are allowed to use the in the tangent hyper-plane the usual flat geometry of special relativity and therefore the reference to v^μ as a *Lorentz* vector is appropriate.

called *Riemann curvature tensor*, or simply *Riemann tensor*, providing the natural generalization of the two-dimensional *Gaussian curvature*.

One can show that the number of *independent* components of $R^{\alpha}{}_{\beta\gamma\delta}$ is 20 (instead of 4^4).

3.4.3 Tidal Forces and Space-Time Curvature

From the discussion in the previous section on the geometric analogy and from the above geometric definition of the curvature, it follows that the *Riemann tensor* $R^{\mu}{}_{\nu\rho\sigma}$ describes the deviation of the actual space-time geometry from the *flat* Minkowski geometry, in an intrinsic, coordinate independent way. On the other hand, we have seen that, in the Newtonian approximation, tidal forces acting in the free falling frame, also describe the deviation from the Minkowskian geometry of special relativity. It is therefore obvious that there must be a close relation between curvature and tidal forces.

To establish this relation, we recall that Eq. (3.21) gives the tidal force in the *classical Newtonian approximation*, that is in the limit where the description of the gravitational field is static, non-relativistic and to the first order in the displacement vector **h**.

We also note that the matrix in Eq. (3.21), which is the explicit form of $\frac{\partial g^i}{\partial x^k}|_{\mathbf{r}=\mathbf{r}_0}$, has the physical dimensions of $[T^{-2}]$, while curvature has dimension $[L^{-2}]$. Thus if a relation between tidal forces and curvature exists, the two quantities must be related by a factor c^2. At first sight this could seem impossible, since $\frac{\partial g^i}{\partial x^k}$ is a 3×3 matrix with only spatial indices, $i, k = 1, 2, 3$, while the Riemann tensor has four four-dimensional indices, $\alpha, \beta, \ldots = 0, 1, 2, 3$.

This different structure of indices, however, simply means that *only some components of the Riemann tensor survive when we take the non-relativistic, static limit of the full relativistic expression of the (gradient of the) gravitational field as given by the Riemann tensor in the general theory*. Indeed, from the exact formula of the Riemann tensor of the general theory of relativity, one can see that, in the non relativistic limit $c \to \infty$, denoting by Latin letters the space indices, to lowest order in $1/c$ one obtains:

$$R^k{}_{0\ell 0} = -\frac{1}{2\,c^2}\frac{\partial g^k}{\partial x^\ell}, \tag{3.48}$$

all the other components of the Riemann tensor being of higher order in $1/c$. We may therefore write:

$$f^k = -2\,mc^2\,R^k{}_{0\ell 0}\,h^\ell \quad (k, \ell = 1, 2, 3). \tag{3.49}$$

Comparing Eqs. (3.48) and (3.49) with Eq. (3.21) one obtains:

$$R^k{}_{0\ell 0} = -\frac{1}{2}\begin{pmatrix} 2\frac{GM}{r_0^3 c^2} & 0 & 0 \\ 0 & -\frac{GM}{r_0^3 c^2} & 0 \\ 0 & 0 & -\frac{GM}{r_0^3 c^2} \end{pmatrix}, \tag{3.50}$$

We conclude that this matrix describes the curvature of space-time in the classical non-relativistic limit.

3.5 Motion of a Particle in Curved Space-Time

To determine the trajectory of a particle in space-time we take advantage of the principle of equivalence, by first writing the equation of motion in the free falling frame and then transforming to a general frame.

Locally, in the free falling frame attached to the particle, there is no gravitational field and the motion is purely *inertial* in the local special relativistic coordinates ξ^α ($\alpha = 0, 1, 2, 3$). On the other hand, in Chap. 2, the special relativistic inertial motion was described by Eq. (2.75), namely:[25]

$$U^\alpha = \frac{d\xi^\alpha}{d\tau} = \text{const.} \longrightarrow \frac{d^2\xi^\alpha}{d\tau^2} = \frac{dU^\alpha}{d\tau} = 0, \tag{3.51}$$

We stress that this equation is valid in an infinitesimal four-dimensional space-time neighborhood of the particle. Switching to the "laboratory" system where the general coordinates x^μ are used, and using the general relation between the two coordinates ξ^α and x^μ:

$$\xi^\alpha = \xi^\alpha(x^\mu), \tag{3.52}$$

we find

$$\frac{d\xi^\alpha}{d\tau} = \frac{\partial\xi^\alpha}{\partial x^\mu}\frac{dx^\mu}{d\tau} = V^\alpha_\mu\frac{dx^\mu}{d\tau}, \tag{3.53}$$

$$\frac{d^2\xi^\alpha}{d\tau^2} = \frac{d}{d\tau}\left(V^\alpha_\mu\frac{dx^\mu}{d\tau}\right) = V^\alpha_\mu\frac{d^2x^\mu}{d\tau^2} + \frac{\partial V^\alpha_\mu}{\partial x^\nu}\frac{dx^\mu}{d\tau}\frac{dx^\nu}{d\tau} = 0. \tag{3.54}$$

We solve this equation with respect to the second derivatives, by multiplying both sides by the inverse matrix $(V^{-1})^\mu_\alpha$, that we denote by

---

[25] With respect to the notations used in Chap. 2, we have changed notation for the locally inertial coordinates from x^μ to ξ^α ($\alpha = 0, 1, 2, 3$) since in the present setting the latter describe the locally inertial coordinates of special relativity, while the former describe a general frame, for example the coordinates used in the "laboratory" frame, where the gravitational field is present.

$$V_\alpha^\mu \equiv (V^{-1})_\alpha^\mu; \quad V_\alpha^\mu V_\nu^\alpha = \delta_\nu^\mu. \tag{3.55}$$

Equation (3.54) takes then the following form:

$$\frac{d^2 x^\mu}{d\tau^2} + \Gamma_{\nu\rho}^\mu \frac{dx^\nu}{d\tau} \frac{dx^\rho}{d\tau} = 0, \tag{3.56}$$

where we have set:

$$\Gamma_{\nu\rho}^\mu \equiv \frac{1}{2} V_\alpha^\mu \left(\frac{\partial V_\nu^\alpha}{\partial x^\rho} + \frac{\partial V_\rho^\alpha}{\partial x^\nu} \right). \tag{3.57}$$

The number of independent components of $\Gamma_{\nu\rho}^\mu$, taking into account the symmetry in the two lower indices, is 40.

Note that the second term on the left hand side of Eq. (3.56) can be given the meaning of the *gravitational acceleration* impressed to the particle of coordinates x^μ; thus $\Gamma_{\nu\rho}^\mu$, called *affine connection*, represents the relativistic generalization of the gravitational field.

The solution to Eq. (3.56) for the spatial coordinates x^i, ($i = 1, 2, 3$), gives the trajectory of the particle in the gravitational field while the solution for $x^0 = ct$ gives the relation between the local time t and the proper time τ.

One can show that $\Gamma_{\nu\rho}^\mu$ can be expressed in terms of the metric and its derivatives. The quickest way do so is to observe that Eq. (3.56), from the four-dimensional point of view, describes a *free inertial motion* in the curved space-time since the gravitational field has been expressed in terms of the affine connection which is a geometric property of space-time. From this point of view there is no force driving the particle; instead, the very presence of a non-trivial geometry, characterized by a non-vanishing Riemann tensor, implies that the free motion must be described by Eq. (3.56).

This interpretation is corroborated by the observation that a free motion in a *curved space-time* is the analogue of the inertial motion in flat space-time, given by a straight line. We must therefore expect that the solution to Eq. (3.56) must represent the analogue, in a curved space, of a straight line in flat space. Such curve is called a *geodesic* and is defined as the shortest line joining two points.

Let us consider two points A and B (events) in the four-dimensional space-time and let γ be a generic curve joining them. Its four-dimensional length $s(\gamma)$ is given by:

$$s(\gamma) = \int_{\gamma[A \to B]} ds = \int_{\gamma[A \to B]} (g_{\mu\nu} dx^\mu dx^\nu)^{\frac{1}{2}}. \tag{3.58}$$

The analogue, in curved space, of the straight line in flat space can be obtained by requiring the curve γ to be such that its length $s(\gamma)$, as a functional of γ, has a minimum, as it is the case for the straight line in flat space.

Solving the variational problem one precisely finds Eq. (3.56) with:

$$\Gamma^{\mu}_{\nu\rho} = \frac{1}{2} g^{\mu\delta} \left(-\frac{\partial}{\partial x^{\delta}} g_{\nu\rho} + \frac{\partial}{\partial x^{\rho}} g_{\nu\delta} + \frac{\partial}{\partial x^{\nu}} g_{\rho\delta} \right). \tag{3.59}$$

Note that since the relativistic gravitational field is given in terms of $\Gamma^{\mu}_{\nu\rho}$, which contains the first derivatives of the metric, the ten components of the metric $g_{\mu\nu}(x)$ are the relativistic generalization of the Newtonian gravitational potential.

To simplify formulas, in the rest of the book we shall often use the following short-hand notation for partial derivatives:

$$\partial_{\mu} \equiv \frac{\partial}{\partial x^{\mu}}; \quad \partial_i \equiv \frac{\partial}{\partial x^i}, \tag{3.60}$$

where $\mu = 0, 1, 2, 3$ and $i = 1, 2, 3$.

3.5.1 The Newtonian Limit

Since the geodesic equation (3.56) describes the trajectory of a particle in a gravitational field, it must reduce to the usual Newtonian formula in the non-relativistic limit. Recalling that the metric is related to the gravitational potential, we define the classical Newtonian limit as that in which, besides the non-relativistic condition $v \ll c$, we also require the gravitational field to be weak and static.

On the metric this implies:

$$g_{\mu\nu} = \eta_{\mu\nu} + h_{\mu\nu} + O(h^2), \tag{3.61}$$

$$\frac{\partial g_{\mu\nu}}{\partial t} = c \frac{\partial g_{\mu\nu}}{\partial x^0} = 0, \tag{3.62}$$

where $h_{\mu\nu}(x)$ is the first order deviation from the flat Minkowski space corresponding to the absence of gravitational field. In Appendix B we show that, in this case, the only non-vanishing components of the affine connection are:

$$\Gamma^i_{00} = \frac{1}{2} g^{i\mu}(-\partial_{\mu} g_{00}) \simeq -\frac{1}{2} \eta^{ij} \partial_j h_{00} = \frac{1}{2} \partial_i h_{00}, \tag{3.63}$$

where i is a three-dimensional space index and we have used $\eta^{ij} = -\delta^{ij}$.

With these approximations the geodesic equation for the index $\mu = 0$ gives $\frac{dt}{d\tau} \simeq 1$, while for the spatial index $\mu = i$ we have:

$$\frac{1}{c^2} \frac{d^2 x^i}{dt^2} = -\frac{1}{2} \partial_i h_{00}. \tag{3.64}$$

Equation (3.64) then coincides with the Newton equation if we set:

$$\frac{\phi}{c^2} = \frac{1}{2} h_{00}. \tag{3.65}$$

where ϕ is the classical gravitational potential. With this identification, Eq. (3.64) becomes:

$$\frac{d^2 x^i}{dt^2} = -\partial_i \phi. \tag{3.66}$$

In particular from Eq. (3.61) we find:

$$g_{00} = 1 + h_{00} = 1 + 2\frac{\phi}{c^2}. \tag{3.67}$$

3.5.2 Time Intervals in a Gravitational Field

Equation (3.67) allows us to evaluate how time intervals are affected by the presence of a gravitational field. Let us suppose that we are in a free falling frame, where the Minkowskian coordinates ξ^α can be used. According to the principle of equivalence, a clock at rest in such a system measures a time interval which coincides with the proper time in the absence of gravity:

$$d\tau^2 = = \frac{1}{c^2} \eta_{\alpha\beta} d\xi^\alpha d\xi^\beta = \eta_{00} \, dt^2 = dt^2 \quad (\alpha, \beta = 0, 1, 2, 3), \tag{3.68}$$

since, for a clock at rest, $\frac{d\xi^i}{dt} = 0$.

In any other frame of reference with coordinates x^μ, like our laboratory, the gravitational field is present and the proper time interval will take the following form:

$$d\tau^2 = \frac{1}{c^2} g_{\mu\nu} dx^\mu dx^\nu. \tag{3.69}$$

If in this frame the clock has four-velocity $dx^\mu/dt = v^\mu$, then the time interval dt between two consecutive (infinitely close) ticks satisfies the relation:

$$\left(\frac{d\tau}{dt}\right)^2 = \frac{1}{c^2} g_{\mu\nu} \frac{dx^\mu}{dt} \frac{dx^\nu}{dt}. \tag{3.70}$$

In particular, if the clock is at rest in the laboratory frame, that is if $v^i = 0$, we obtain:

$$\frac{dt}{d\tau} = (g_{00})^{-\frac{1}{2}}.\tag{3.71}$$

The dilation factor on the right hand side of (3.71), however, cannot be observed, since the gravitational field affects in the same way the ticks of the standard clock and those of the clock being studied. However, the difference between dt_1 e dt_2 in two different points $\mathbf{x}_1$, $\mathbf{x}_2$ can be observed; indeed Eq. (3.71) implies:

$$dt_1 = d\tau \, (g_{00}(\mathbf{x}_1))^{-\frac{1}{2}},\tag{3.72}$$

$$dt_2 = d\tau \, (g_{00}(\mathbf{x}_2))^{-\frac{1}{2}},\tag{3.73}$$

so that:

$$\frac{dt_2}{dt_1} = \left[\frac{g_{00}(\mathbf{x}_1)}{g_{00}(\mathbf{x}_2)}\right]^{\frac{1}{2}}.\tag{3.74}$$

In particular, in the classical Newtonian limit, we may use (3.67) and, recalling Eq. (3.67), we obtain

$$\frac{dt_2}{dt_1} = \left[\frac{1 + 2\frac{\phi_1}{c^2}}{1 + 2\frac{\phi_2}{c^2}}\right]^{\frac{1}{2}} \approx \left(1 + 2\frac{\phi_1}{c^2}\right)^{\frac{1}{2}} \left(1 - 2\frac{\phi_2}{c^2}\right)^{\frac{1}{2}} \approx 1 - \frac{\phi_2 - \phi_1}{c^2} = 1 - \frac{\Delta\phi}{c^2},\tag{3.75}$$

where we have defined $\phi_1 = \phi(\mathbf{x}_1)$ and $\phi_2 = \phi(\mathbf{x}_2)$ and used that $\frac{\phi}{c^2}$ is very small in most situations.[26]

For example, if a clock is placed at the point $\mathbf{x}_1$ far away from other bodies, so that no gravitational field is present, we have $\phi(\mathbf{x}_1) = 0$, and therefore:

$$dt_1 = d\tau,\tag{3.76}$$

since $g_{00}(\mathbf{x}_1) = \eta_{00} = 1$. The same clock placed at a point $\mathbf{x}_2$, e.g. on the earth's surface, will tick time intervals dt_2 such that:

$$\frac{dt_2}{dt_1} = \frac{dt_2}{d\tau} \approx 1 - \frac{\Delta\phi}{c^2} = 1 - \frac{\phi_2}{c^2}.\tag{3.77}$$

Since $\phi_2 < 0$ we find $dt_2 > dt_1$, that is *time intervals are dilated in a gravitational field* by a factor $(1 - \frac{\phi_2}{c^2})$. In the case of the terrestrial gravitational field we have:

---

[26] As usual the value at infinity of ϕ is set equal to zero.

$$dt_2 = \left(1 + \frac{GM}{c^2 r}\right) d\tau. \tag{3.78}$$

where M is the earth's mass.

In particular a same clock will tick at a different rate, depending on whether it is placed at sea level or at height h. Indeed, since

$$\phi_1 = -\frac{GM}{R}; \quad \phi_2 = -\frac{GM}{R+h}, \tag{3.79}$$

we obtain:

$$dt_2 = \left(1 - \frac{\phi_2 - \phi_1}{c^2}\right) dt_1 = \left[1 - \frac{GM}{c^2}\left(\frac{1}{R} - \frac{1}{R+h}\right)\right] dt_1. \tag{3.80}$$

In general we may say that the more negative the gravitational potential is, (or the greater its absolute value is), the more dilated time intervals are.

Because of the factor $1/c^2$, these effects are generally small, however gravitational time dilation has been experimentally measured in various different situations.

The first verification by a experiment on earth was performed by Pound and Rebka combining the gravitational dilation with Doppler effect in the emission and absorption of photons in Fe^{57}. Further experimental evidence can be inferred from astrophysical observations (especially from the light spectra of white dwarfs). In this case we can safely set to zero the earth's gravitational potential being much smaller in absolute value than the potential on the surface of a star. Thus, in this case, we have:

$$\frac{dt_2}{dt_1} = \frac{\nu_1}{\nu_2} > 1, \tag{3.81}$$

where ν denotes the light frequency and the suffix 1 and 2 are referred to the earth and to the star, respectively. The frequency of the emitted light is then higher than that observed on earth and we have a measurable shift towards the red of the wavelength.

In the eighties further confirmations were gained by experiments with time signals sent to and from Viking 1 Mars lander and from time measurements using atomic clocks on airplanes; the clocks that traveled aboard the airplanes, upon return, were slightly faster than those on the ground.

However, the most spectacular evidence of the gravitational red-shift is nowadays given by its technological application to GPS devices.

A GPS gives the absolute position on the surface of the earth to within 5–10 m of precision; this requires the clock ticks on the GPS satellite to be known with an accuracy of 20–30 ns. Such an accuracy cannot be reached if we neglected the special and general relativity effects on time intervals . To compute these effects we first observe that the transmitting clock is subject to the special relativistic time dilation due to the satellite orbital speed, compared to an identical clock on the earth.

From our discussion of time dilation in special relativity the clock on the satellite would run slower, compared to a clock on the earth, by the factor:

$$\sqrt{1 - \frac{v^2}{c^2}} \simeq 1 - \frac{v^2}{2c^2},$$ (3.82)

to first order in $\frac{v^2}{c^2}$, since v, the satellite speed, is much smaller than c.

Suppose the satellite is orbiting at a distance from the center of the earth of about four times the earth's radius R. Using the classical result

$$\frac{v^2}{r} = \frac{GM}{r^2},$$ (3.83)

with $r = 4R$, we obtain a time loss of about $-7\,\mu s/day$.

We now compute the general relativistic effect. First we recall that a clock in a greater gravitational potential runs *faster* than one on earth. Then, calling Δt_E and Δt_s the time intervals on earth and on the satellite, respectively, we find:

$$1 - \frac{\Delta t_E}{\Delta t_s} = \frac{\Delta \phi}{c^2} = \frac{1}{c^2}\left[\frac{GM}{R+h} - \frac{GM}{R}\right]$$ (3.84)

and if $R + h \simeq 4R$ this gives a gain in time of $\simeq 45\,\mu s/day$, six times larger than the special relativistic effect.

Summing up the two effects we find that the clock on the satellite runs faster $\simeq 38\,\mu s/day \equiv 38 \times 10^3\,ns/day$. We see that neglecting the relativistic effects, would imply errors three order of magnitude higher than the necessary accuracy of 20–$30\,ns!$[27]

3.5.3 The Einstein Equation

Until now we have been discussing some consequences of the principle of equivalence in relation to the motion of a particle in a given gravitational field. We have seen that the motion is essentially a *free motion* in a curved space-time, the generalized gravitational potential being described by the metric $g_{\mu\nu}(x)$ which was supposed to be a known function of the space-time coordinates.

The knowledge of the metric field is, however, not known *a priori*, and the description of the gravitational field requires the knowledge of the equation of motion of the metric field.

To arrive to a rigorous determination of this equation the principle of equivalence is no more sufficient and we must address the full geometric formalism of general relativity.

Nevertheless, in this section, with no ambition of being rigorous or complete, we shall try to develop some heuristic considerations to justify the actual form of the gravitational equations.

---

[27] We also note that the given error increases day by day, since it is a *cumulative* effect.

We recall that in the Newtonian non-relativistic theory the gravitational potential is static, that is, non-propagating, and, given a distribution of mass in space, it can be determined through the Poisson equation:

$$\nabla^2 \phi = -4\pi G \, \rho(\mathbf{x}), \tag{3.85}$$

where ρ is the density of matter and, as usual, $\phi(\mathbf{x})$ the gravitational potential. On the basis of the discussions given in the previous sections, it is clear that the relativistic extension we are looking for must be *covariant under general coordinate transformations*. Furthermore the ten components of the metric field $g_{\mu\nu}(x)$ must describe a *propagating* field generalizing the single static component of the Newtonian gravitational potential ϕ. Indeed, we know that already at the level of special relativity, dependence on the spatial coordinates implies dependence on the time.

It then follows that the source term ρ must also be generalized, in a covariant setting,[28] in terms of ten quantities depending on the four space-time coordinates. As we will show in Chaps. 5 and 8, such relativistic extension is given in terms of the so-called *energy momentum tensor* $T_{\mu\nu}$, symmetric in $\mu\nu$, which describes the density of matter four-momentum and of its current and can be shown to reduce, to lowest order in v/c, to the single non-vanishing component $T_{00} = \rho c = \epsilon/c$, ϵ being the energy density.

On the other hand, on the left hand side, the Laplace operator in the classical Poisson equation must be replaced by an expression satisfying the following requirements:

1. It must have the same index structure as $T_{\mu\nu}$;
2. It must be a second order differential expression on the metric field $g_{\mu\nu}(x)$;
3. It must be covariant under general coordinate transformations;
4. It must reduce to the left hand side of Poisson equation in the non-relativistic limit.

Let us denote such unknown expression by $G_{\mu\nu}$. Then the equation we are looking for should have the form:

$$G_{\mu\nu} = \alpha T_{\mu\nu}, \tag{3.86}$$

where α is a constant. On the other hand our previous discussion on the geometric properties of space-time in the presence of gravitation tells us that the Riemann tensor $R^{\mu}{}_{\nu\rho\sigma}$, the relativistic extension of the tidal forces, determines the geometric properties of space–time associated with the presence of gravitation. Thus we expect that $G_{\mu\nu}$ must be related to the Riemann tensor. Actually one can show that, in Riemannian geometry, all the requirements enumerated before are satisfied if we set:

$$G_{\mu\nu} = R^{\alpha}{}_{\mu\alpha\nu} - \frac{1}{2} g_{\mu\nu} R^{\tau}{}_{\rho\tau\sigma} g^{\rho\sigma}, \tag{3.87}$$

---

[28] Here covariant means with respect to general coordinate transformations.

and set the unknown constant α of Eq. (3.86) $\alpha = -8\pi\, G$ in order to reproduce in the non relativistic static weak-field limit the Poisson equation (3.85). Thus we see that the equation for the metric field, also known as Einstein's equation, is a second order differential equation whose source is $T_{\mu\nu}$, which, as we shall see later on, describes the distribution of energy and momentum in space-time. If such distribution is known and the initial Cauchy data are given, we can determine the solution for the metric field $g_{\mu\nu}$. In other words: *The energy-momentum distribution, that is the content of matter-energy and of its current, determines the geometry of space-time.*

3.5.4 References

For further reading see Refs. [1, 11, 12].

Chapter 4
The Poincaré Group

In this chapter, after briefly reviewing the notions of linear vector spaces, inner product of vectors and metric in (three-dimensional) Euclidean space, we shall focus on coordinate transformations, namely maps between different descriptions of the same points in space. This will allow us to introduce *covariant* and *contravariant* vectors, as well as tensors, characterized by specific transformation properties under coordinate transformations. Though we shall be mainly concerned with Cartesian coordinate transformations, which are implemented by linear relations between the old and new coordinates, the formalism is readily extended to more general transformations relating curvilinear coordinate systems, and thus also to curved spaces where Cartesian coordinates cannot be defined. We shall then study rotations in Euclidean space and show that they close an object called a *Lie group*, whose properties are locally captured by a *Lie algebra*. This will lead us to the important concept of covariance of an equation of motion with respect to rotations. The generalization of all these notions from Euclidean to Minkowski space will be straightforward. As anticipated in an earlier chapter, points in Minkowski space are described by a Cartesian system of four coordinates x^0, x^1, x^2, x^3 and the distance between two points is defined by a metric with Minkowskian (or Lorentzian) signature. Poincaré transformations will then be introduced as Cartesian coordinate transformations which leave the coordinate dependence of the distance between two points invariant. These linear transformations include, as the homogeneous part, the Lorentz transformations, which generalize the notion of rotation to Minkowski space. Poincaré transformations also comprise, as their inhomogeneous part, the space-time translations, and close a Lie group called the *Poincaré group*. The principle of special relativity is now restated as the condition that the equations of motion be covariant with respect to Poincaré transformations.

© Springer International Publishing Switzerland 2016
R. D'Auria and M. Trigiante, *From Special Relativity to Feynman Diagrams*,
UNITEXT for Physics, DOI 10.1007/978-3-319-22014-7_4

4.1 Linear Vector Spaces

Let us briefly recall some basic facts about vector spaces. Consider the three-dimensional Euclidean space E_3. We can associate with each couple of points A, B in E_3 a vector $\overrightarrow{AB}$ originating in A and ending in B. If we arbitrarily fix an *origin O* in E_3, any other point A in E_3 will be uniquely identified by its *position vector* $\mathbf{r} \equiv \overrightarrow{OA}$.

We can then consider the collection of vectors, each associated with couples of points in E_3, which constitutes a *vector space* associated with E_3 and denoted by V_3. Indeed the space V_3 is endowed with a linear structure, which means that an operation of sum and multiplication by real numbers is defined on its elements: We can consider a generic linear combination of two or more vectors $\mathbf{V}_1, \ldots, \mathbf{V}_k$ in V_3 with real coefficients and the result $\mathbf{V}$:

$$\mathbf{V} = a_1 \mathbf{V}_1 + \cdots a_k \mathbf{V}_k, \tag{4.1}$$

is still a vector, namely an element of V_3, that is there is a couple of points A, B in E_3 such that $\mathbf{V} = \overrightarrow{AB}$. If A and B coincide the corresponding vector is the *null vector* $\mathbf{0} \equiv \overrightarrow{AA}$.

A set of three linearly independent vectors $\{\mathbf{u}_1, \mathbf{u}_2, \mathbf{u}_3\} = \{\mathbf{u}_i\}$[1] defines a *basis* for V_3 and any vector $\mathbf{V}$ can be expressed as a *unique* linear combination of $\{\mathbf{u}_i\}$:

$$\mathbf{V} = V^1 \mathbf{u}_1 + V^2 \mathbf{u}_2 + V^3 \mathbf{u}_3 = \sum_i V^i \mathbf{u}_i, \tag{4.2}$$

where V^1, V^2, V^3 (with the upper index), are the components of $\mathbf{V}$ in the basis $\{\mathbf{u}_i\}$.

It is useful to describe the vectors $\{\mathbf{u}_i\}$ as column vectors in the following way:

$$\mathbf{u}_1 \equiv \begin{pmatrix} 1 \\ 0 \\ 0 \end{pmatrix}; \quad \mathbf{u}_2 \equiv \begin{pmatrix} 0 \\ 1 \\ 0 \end{pmatrix}; \quad \mathbf{u}_3 \equiv \begin{pmatrix} 0 \\ 0 \\ 1 \end{pmatrix}. \tag{4.3}$$

This allows to describe a generic vector $\mathbf{V}$ as a column vector having as entries the components of $\mathbf{V}$ with respect to the basis $\{\mathbf{u}_i\}$.

$$\mathbf{V} \equiv V^1 \begin{pmatrix} 1 \\ 0 \\ 0 \end{pmatrix} + V^2 \begin{pmatrix} 0 \\ 1 \\ 0 \end{pmatrix} + V^3 \begin{pmatrix} 0 \\ 0 \\ 1 \end{pmatrix} = \begin{pmatrix} V^1 \\ V^2 \\ V^3 \end{pmatrix}. \tag{4.4}$$

This formalism will allow us to reduce all operations among vectors to matrix operations. We shall also use the boldface to denote the matrix representation of a given quantity.

---

[1] Recall that this property means that $a_1 \mathbf{u}_1 + a_2 \mathbf{u}_2 + a_3 \mathbf{u}_3 = 0$ if and only if $a_1 = a_2 = a_3 = 0$.

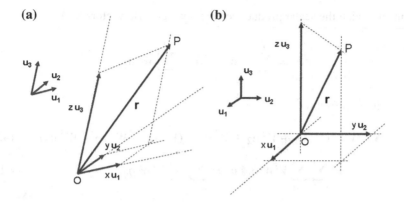

Fig. 4.1 **a** Generic Cartesian coordinate system; **b** Cartesian rectangular coordinate system

A *Cartesian coordinate system* in E_3 is defined by an origin O and a basis $\{\mathbf{u}_i\}$ of V_3 and it allows to uniquely describe each point P in E_3 by means of three coordinates x, y, z, which are the components of the corresponding position vector $\mathbf{r} = \overrightarrow{OP}$, that is the parallel projections along $\mathbf{u}_i$, see Fig. 4.1a:

$$\mathbf{r} = \overrightarrow{OP} = x\mathbf{u}_1 + y\mathbf{u}_2 + z\mathbf{u}_3 \equiv \begin{pmatrix} x \\ y \\ z \end{pmatrix}. \tag{4.5}$$

It will be convenient to rename the coordinates as follows: $x^1 = x, x^2 = y, x^3 = z$. This will allow us to use the following short-hand description of the position vector:

$$\mathbf{r} \equiv \sum_{i=1}^{3} x^i \mathbf{u}_i \equiv \{x^i\}. \tag{4.6}$$

A frame of reference (RF) in Euclidean space will be defined by a Cartesian coordinate system. In V_3 a *scalar product* is defined which associates with each couple of vectors $\mathbf{V}$, $\mathbf{W}$ a real number $\mathbf{V} \cdot \mathbf{W} \in \mathbb{R}$ and which satisfies the following properties:

(a) $\mathbf{V} \cdot \mathbf{W} = \mathbf{W} \cdot \mathbf{V}$ (*symmetry*),
(b) $(a\mathbf{V}_1 + b\mathbf{V}_2) \cdot \mathbf{W} = a(\mathbf{V}_1 \cdot \mathbf{W}) + b(\mathbf{V}_2 \cdot \mathbf{W})$ (*distributivity*), (4.7)
(c) $\mathbf{V} \cdot \mathbf{V} \geq 0; \quad \mathbf{V} \cdot \mathbf{V} = 0 \Rightarrow \mathbf{V} = \mathbf{0}$ (*positive definiteness*)

With respect to a basis $\mathbf{u}_i$, $i = 1, 2, 3$, a scalar product can be described by means of a symmetric non singular matrix called *metric*:

$$\mathbf{g} = (g_{ij}) \equiv (\mathbf{u}_i \cdot \mathbf{u}_j) \quad i, j = 1, 2, 3, \tag{4.8}$$

in terms of which the scalar product between two generic vectors **V**, **W**

$$\mathbf{V} = \sum_{i=1}^{3} V^i \mathbf{u}_i, \quad \mathbf{W} = \sum_{i=1}^{3} W^i \mathbf{u}_i,$$

can be written as follows:

$$\mathbf{V} \cdot \mathbf{W} = (V^1 \mathbf{u}_1 + V^2 \mathbf{u}_2 + V^3 \mathbf{u}_3) \cdot (W^1 \mathbf{u}_1 + W^2 \mathbf{u}_2 + W^3 \mathbf{u}_3) \tag{4.9}$$

$$= \sum_{i=1}^{3} \sum_{j=1}^{3} V^i W^j \mathbf{u}_i \cdot \mathbf{u}_j = \sum_{i=1}^{3} \sum_{j=1}^{3} V^i W^j g_{ij} \tag{4.10}$$

$$= (V^1, V^2, V^3) \begin{pmatrix} g_{11} & g_{12} & g_{13} \\ g_{21} & g_{22} & g_{23} \\ g_{31} & g_{32} & g_{33} \end{pmatrix} \begin{pmatrix} W^1 \\ W^2 \\ W^3 \end{pmatrix} = \mathbf{V}^T \mathbf{g} \mathbf{W}, \tag{4.11}$$

where we have applied properties (*a*) and (*b*). Property (*c*) is specific to Euclidean space and expresses the positive definiteness of its metric, namely that for any vector $\mathbf{V} \in V_3$, different from the null vector $\mathbf{0} = (0, 0, 0)$, the quantity $\|\mathbf{V}\|^2 \equiv \mathbf{V} \cdot \mathbf{V} = V^i g_{ij} V^j$, called the *norm* squared of **V**, is positive. This in turn implies that the symmetric matrix g_{ij} has only positive eigenvalues. This property will not hold for the metric in Minkowski space, which has three negative and one positive eigenvalues.

As an example let us consider a basis in which the scalar product is described by the following metric:

$$\mathbf{g} = (g_{ij}) = \begin{pmatrix} 1 & 0 & 0 \\ 0 & 3 & 2 \\ 0 & 2 & 3 \end{pmatrix}.$$

Given two vectors:

$$\mathbf{V} = 3\mathbf{u}_1 + 4\mathbf{u}_2 \equiv (3, 4, 0) = \mathbf{V},$$
$$\mathbf{W} = 5\mathbf{u}_1 + 2\mathbf{u}_3 \equiv (5, 0, 2) = \mathbf{W},$$

their scalar product can be expressed in terms of the following matrix operation:

$$\mathbf{V} \cdot \mathbf{W} = \mathbf{V}^T \mathbf{g} \mathbf{W} = (3, 4, 0) \begin{pmatrix} 1 & 0 & 0 \\ 0 & 3 & 2 \\ 0 & 2 & 3 \end{pmatrix} \begin{pmatrix} 5 \\ 0 \\ 2 \end{pmatrix} = 31.$$

It is very useful, in writing this kind of formulae, to use the Einstein summation convention introduced in Chap. 2: Whenever in a formula a same index appears in upper and lower positions, summation over that index is understood. We say that a *contraction* is performed over that index, which is also called *dummy index*. For instance the formula (4.11) for the scalar product can be written as follows:

$$\mathbf{V} \cdot \mathbf{W} = \sum_{i=1}^{3} \sum_{j=1}^{3} V^i W^j g_{ij} \equiv V^i W^j g_{ij}, \tag{4.12}$$

contraction being over the indices i and j. In what follows we shall always use this convention in order to make formulae simpler and more transparent. Starting from the notion of scalar product on V_3 we can define a *distance* in E_3: The distance $d(A, B)$ between two points A and B, described by the position vectors $\mathbf{r}_A = x_A^i \mathbf{u}_i$, $\mathbf{r}_B = x_B^i \mathbf{u}_i$ respectively, is defined as the norm of the *relative position vector* $\mathbf{r}_A - \mathbf{r}_B$:

$$d(A, B) \equiv \|\mathbf{r}_A - \mathbf{r}_B\| = \sqrt{(\mathbf{r}_A - \mathbf{r}_B) \cdot (\mathbf{r}_A - \mathbf{r}_B)} = \sqrt{(x_A^i - x_B^i) g_{ij} (x_A^j - x_B^j)}, \tag{4.13}$$

where we have used the fact that

$$\mathbf{r}_A - \mathbf{r}_B = \left(x_A^1 - x_B^1\right) \mathbf{u}_1 + \left(x_A^2 - x_B^2\right) \mathbf{u}_2 + \left(x_A^3 - x_B^3\right) \mathbf{u}_3. \tag{4.14}$$

From property (c) of Euclidean metric it follows that, if two points have vanishing distance, they coincide. Indeed $d(A, B) = 0$ means that $\|\mathbf{r}_A - \mathbf{r}_B\| = 0$, which is the case only if the relative position vector $\mathbf{r}_A - \mathbf{r}_B$ equals the 0-vector $(0, 0, 0)$, i.e. if $\mathbf{r}_A = \mathbf{r}_B$, that is $A = B$. This will not hold in Minkowski space where two distinct points (i.e. two different events) can have vanishing four-dimensional distance.

In V_3 we can always choose a basis of vectors $\{\mathbf{u}_i\}$ which are *orthonormal* (defining a Cartesian *rectangular* coordinate system, see Fig. 4.1b), namely satisfy the condition:

$$g_{ij} = \mathbf{u}_i \cdot \mathbf{u}_j = \delta_{ij} = \begin{cases} 1 & i = j \\ 0 & i \neq j \end{cases} \tag{4.15}$$

The unit vectors $\{\mathbf{u}_i\}$ define three mutually orthogonal axes: X, Y, Z.[2] The metric matrix, in this case, reads:

$$\mathbf{g} = \begin{pmatrix} 1 & 0 & 0 \\ 0 & 1 & 0 \\ 0 & 0 & 1 \end{pmatrix}. \tag{4.16}$$

The scalar product between two vectors in this basis acquires the following simple form:

---

[2]When referring to the collection of Cartesian rectangular coordinates in our Euclidean three-dimensional space we shall often use, as we did in Chaps. 1 and 2, the symbol $\mathbf{x}$ instead of $\mathbf{r}$: $\mathbf{x} \equiv (x, y, z)$.

$$\mathbf{V} \cdot \mathbf{W} = V^i \, \delta_{ij} \, W^j = V^1 W^1 + V^2 W^2 + V^3 W^3. \tag{4.17}$$

The squared norm of a non-null vector reads: $\|\mathbf{V}\|^2 = (V^1)^2 + (V^2)^2 + (V^3)^2 > 0$. Also the expression (4.14) of the distance between two points A and B simplifies considerably:

$$d(A, B) = \sqrt{(x_A - x_B)^2 + (y_A - y_B)^2 + (z_A - z_B)^2}. \tag{4.18}$$

The above formula could have been deduced directly using Pythagoras' theorem.

4.1.1 Covariant and Contravariant Components

Consider a transformation of the Cartesian coordinate system which leaves the origin fixed but brings a basis $\{\mathbf{u}_i\}$ into a new one $\{\mathbf{u}'_j\}$ and let us see how the components V'^i of a vector $\mathbf{V}$ in the new basis are related to those (V^i) in the old basis. Each vector $\mathbf{u}'_i$ can be expressed in terms of its components relative to the old basis $\{\mathbf{u}_i\}$:

$$\mathbf{u}'_j = M^i{}_j \, \mathbf{u}_i, \tag{4.19}$$

where $\mathbf{M} = (M^i{}_j)$, i labelling the rows, j the columns, has to be an invertible matrix in order for $\mathbf{u}'_i$ to be linearly independent. For the sake of convenience let us denote by $\mathbf{D}$ the inverse of $\mathbf{M}$, so that $\mathbf{M} = \mathbf{D}^{-1}$. Notice that in the expression on the right hand side of (4.19) the summation is taken over the row-index of $\mathbf{M}$. If we arrange the basis elements $\mathbf{u}_i$ in a row vector, Eq. (4.19) can be written in a matrix form:

$$(\mathbf{u}'_1, \, \mathbf{u}'_2, \, \mathbf{u}'_3) = (\mathbf{u}_1, \mathbf{u}_2, \mathbf{u}_3) \, \mathbf{D}^{-1}, \tag{4.20}$$

namely the row vector $(\mathbf{u}_i)$ transforms by acting on it with the matrix $\mathbf{D}^{-1}$ from the right. Alternatively, thinking of $(\mathbf{u}_i)$ as a column vector, it transforms by the action of $(\mathbf{D}^{-1})^T$ to the left. The components V^i of the vector

$$\mathbf{V} = V^i \mathbf{u}_i, \tag{4.21}$$

will then transform with the matrix $\mathbf{D}$. Indeed, if V'^i and V^i are different descriptions of a *same* vector, we have

$$\mathbf{V} = V^i \, \mathbf{u}_i = V'^i \, \mathbf{u}'_i, \tag{4.22}$$

which implies:

$$V^j \mathbf{u}_j = V'^i D^{-1j}{}_i \mathbf{u}_j. \tag{4.23}$$

Being $\mathbf{u}_i$ independent we find

$$V'^i D^{-1j}{}_i = V^j. \tag{4.24}$$

Using the matrix formalism, we can describe the *same abstract vector* in Eq. (4.21) in terms of two column vectors $\mathbf{V}' = (V'^i)$ and $\mathbf{V} = (V^i)$ consisting of the corresponding components in the new and old bases respectively. Equation (4.24) can then be recast in a matrix form:

$$\mathbf{D}^{-1} \mathbf{V}' = \mathbf{V}. \tag{4.25}$$

We can solve Eq. (4.25) in $\mathbf{V}'$ multiplying both sides by the matrix $\mathbf{D}^{-1}$:

$$\mathbf{V}' = \mathbf{D}\mathbf{V} \quad \Leftrightarrow \quad V'^i = D^i{}_j V^j. \tag{4.26}$$

Compare now the two Eqs. (4.20) and (4.26). If the elements of a basis (which are labeled by a lower index), as elements of a row vector, transform with a matrix $\mathbf{D}^{-1}$ from the right (or $\mathbf{D}^{-1T}$ to the left if seen as a column vector), the corresponding components of a vector (labeled by an upper index), as elements of a column vector, transform with the matrix $\mathbf{D}$ from the left. We say that the elements of a basis transform as a *covariant vector* (having a lower index), while the components of a vector transform as a *contravariant vector* (having an upper index). In our conventions we will often represent covariant and contravariant quantities, in matrix notation, as components of row and column vectors respectively. Let us now consider the scalar product (4.12) and define the quantities $V_i \equiv g_{ij} V^j$. The presence of a lower index suggests that it should transform as a *covariant* vector, as we presently show. Indeed, from the definition (4.8) of metric and from (4.20) we can deduce its transformation property:

$$\mathbf{g}' = (g'_{ij}) = (\mathbf{u}'_i \cdot \mathbf{u}'_j) = (D^{-1k}{}_i D^{-1\ell}{}_j g_{k\ell}) = \mathbf{D}^{-T} \mathbf{g} \mathbf{D}^{-1}, \tag{4.27}$$

where we have used the distributive property of the scalar product and defined $\mathbf{D}^{-T}$ as $(\mathbf{D}^{-1})^T$, namely the transpose of the inverse. The scalar product between two vectors can be expressed in the following simple form:

$$\mathbf{V} \cdot \mathbf{W} = V_i W^i = V^i W_i, \tag{4.28}$$

or, in matrix notation

$$\mathbf{V} \cdot \mathbf{W} = (V_1, V_2, V_3) \begin{pmatrix} W^1 \\ W^2 \\ W^3 \end{pmatrix} = (W_1, W_2, W_3) \begin{pmatrix} V^1 \\ V^2 \\ V^3 \end{pmatrix}, \tag{4.29}$$

It is useful to define the *inverse metric* $\mathbf{g}^{-1}$, whose components are denoted by $g^{ij} \equiv g^{-1\,ij}$, so that $g^{ik}\,g_{kj} = \delta^i_j$. From Eqs. (4.27) and (4.15) it follows that the inverse metric transforms as follows:

$$\mathbf{g}^{-1\prime} = (g'^{ij}) = (D^i{}_k\,D^j{}_\ell\,g^{k\ell}) = \mathbf{D}\,\mathbf{g}^{-1}\,\mathbf{D}^T. \tag{4.30}$$

From (4.27) we deduce the transformation property of V_i:

$$V'_i = g'_{ij}\,V^{j\prime} = D^{-1\,k}{}_i\,D^{-1\,\ell}{}_j\,g_{k\ell}\,D^j{}_s\,V^s = D^{-1\,k}{}_i\,g_{k\ell}\,\delta^\ell_s\,V^s = D^{-1\,k}{}_i\,V_k, \tag{4.31}$$

where we have used the definition of inverse matrix: $D^{-1\,\ell}{}_j\,D^j{}_s = \delta^\ell_s$. Comparing (4.31) with (4.20) we conclude that V_i transform as the basis elements $\mathbf{u}_i$, namely as components of a covariant vector. We say that V_i are the covariant components of the vector $\mathbf{V}$, since they transform covariantly with the basis $\{\mathbf{u}_i\}$. To define them we needed the notion of metric g_{ij}. Equivalently we can write the contravariant components in terms of the covariant ones by contracting the latter with the inverse metric $V^i = g^{ij}\,V_j$. We conclude that a vector $\mathbf{V}$ can be characterized either in terms of its covariant or of its contravariant components, and that we can lower a contravariant index or raise a covariant one by contracting it with the metric or the inverse metric, respectively.

From (4.31) we also conclude that the scalar product between two vectors is invariant under a change of basis, as we would expect since the result of this product is a number (scalar):

$$\mathbf{V}' \cdot \mathbf{W}' = V'_i\,W'^i = V_k\,D^{-1\,k}{}_i\,D^i{}_\ell\,W^\ell = V_k\,\delta^k_\ell\,W^\ell = V_i\,W^i = \mathbf{V} \cdot \mathbf{W}.$$

Geometrically the covariant components of a vector are its orthogonal projections along the coordinate axes. Indeed we can write:

$$V_i = g_{ij}\,V^j \equiv \mathbf{u}_i \cdot \mathbf{u}_j\,V^j \equiv \mathbf{u}_i \cdot (\mathbf{u}_j\,V^j) = \mathbf{u}_i \cdot \mathbf{V}, \tag{4.32}$$

recalling the geometric meaning of the scalar product between two vectors, V_i is the orthogonal projection of $\mathbf{V}$ along $\mathbf{u}_i$ (provided $\mathbf{u}_i$ has unit length), while the contravariant component is obviously the parallel projection, as it follows form (4.22).

Clearly if $\{\mathbf{u}_i\}$ is an orthonormal basis, namely if $\mathbf{u}_i \cdot \mathbf{u}_j = \delta_{ij}$ the covariant and contravariant components of a vector coincide: $V_i = \delta_{ij}\,V^j = V^i$.

Let $\mathbf{r} = (x^i)$ and $\mathbf{r}' = (x'^i)$ denote the coordinate vectors of a point P with respect to the two coordinate systems. By Eq. (4.26) we find the following relation between the two:

$$\mathbf{r}' = \mathbf{D}\,\mathbf{r} \quad \Leftrightarrow \quad x'^i = D^i{}_j\,x^j. \tag{4.33}$$

Let us now consider the most general transformation relating two Cartesian coordinate systems. It is an *affine* transformation which acts not only on the basis of vectors but also on the origin, by means of a translation. Let O, $(\mathbf{u}_i)$ and O', $(\mathbf{u}'_i)$ denote the origins and the bases of the two systems. The two bases are related as in Eq. (4.20). A point P is described by the vector $\overrightarrow{OP} = x^i \mathbf{u}_i$ with respect to the former coordinate system, and by $\overrightarrow{O'P} = x'^i \mathbf{u}'_i$ with respect to the latter. Let $\overrightarrow{OO'} = x^i_0 \mathbf{u}'_i$ be the position vector of O' relative to O in the new basis. From the relation:

$$\overrightarrow{O'P} = \overrightarrow{OP} - \overrightarrow{OO'}, \tag{4.34}$$

we derive the following relation between the new and old coordinates of P

$$x'^i \mathbf{u}'_i = x^i \mathbf{u}_i - x^i_0 \mathbf{u}'_i = x^j D^i_{\ j} \mathbf{u}'_i - x^i_0 \mathbf{u}'_i. \tag{4.35}$$

Equating the components of the vectors on the right and left hand side we find

$$x'^i = D^i_{\ j} x^j - x^i_0, \tag{4.36}$$

or, as a relation between coordinate vectors,

$$\mathbf{r}' = (\mathbf{D}, \mathbf{r}_0) \cdot \mathbf{r} \equiv \mathbf{D}\,\mathbf{r} - \mathbf{r}_0, \tag{4.37}$$

where $\mathbf{r}_0 \equiv (x^i_0)$. The most general transformation of a Cartesian coordinate system is then implemented by a linear relation (4.36) between the old and the new coordinates. In (4.37) this relation has been described as the action on $\mathbf{r}$ of a couple $(\mathbf{D}, \mathbf{r}_0)$ consisting of an invertible matrix $\mathbf{D}$ and a vector $\mathbf{r}_0$ defining the homogeneous and inhomogeneous part of the transformation, respectively. *Homogeneous transformations* are those considered at the beginning of the present section, which do not affect position of the origin, $O \equiv O'$, and thus are just characterized by the matrix $\mathbf{D}$, being $\mathbf{r}_0 \equiv \mathbf{0}$. If, on the other hand, the homogeneous component of the transformation is trivial, $\mathbf{D} = \mathbf{1}$, the affine transformation $(\mathbf{1}, \mathbf{r}_0)$ only describes a rigid *translation* of the frame of reference: $x'^i = x^i - x^i_0$. Let us stress here that the matrix elements $D^i_{\ j}$ and the parameters x^i_0 are constant, namely coordinate-independent. We can consider the relative position vector between two infinitely close points. Its components $d\mathbf{r} = (dx^i)$, $d\mathbf{r}' = (dx'^i)$, with respect to the two coordinate systems, are the infinitesimal differences between the coordinates of the two points, i.e. the coordinate differentials. Their relation is obtained from (4.36) by differentiating both sides:

$$dx'^j = D^j_{\ i}\, dx^i = \frac{\partial x'^j}{\partial x^i}\, dx^i. \tag{4.38}$$

The matrix $\mathbf{D}$ thus represents the coordinate-independent Jacobian matrix of the coordinate transformation.

General coordinate transformations involve non-Cartesian, i.e. curvilinear coordinate systems, and are typically described by non-linear coordinate relations[3] $x'^i = x'^i(x) \equiv x'^i(x^1, x^2, x^3)$, as anticipated in Chap. 3. In this case the Jacobian matrix $\mathbf{D} = \left(\frac{\partial x'^j}{\partial x^i}\right)$ in Eq. (4.38) will no longer be coordinate-independent (think about the relation between Cartesian orthogonal coordinates x, y, z and spherical polar coordinates r, θ, φ).[4] We shall come back to this point at the end of this section.

All that have been said about three-dimensional Euclidean space E_3 can be easily extended its n-dimensional version E_n. It is sufficient to take the indices $i, j, \ldots$ to run from 1 to n instead of taking only three values.

So far we have been considering the transformation properties of the components of a (covariant or contravariant) vector as the basis of the reference frame is changed. In physics (and geometry) one in general has to deal with vectors which are functions of the point in space through its coordinates $\mathbf{V}(\mathbf{r}) = \mathbf{V}(x^1, x^2, \ldots, x^n)$, namely with *vector fields*. Nothing changes in the transformation rule of the (covariant or contravariant) components of the vector field, since, if we perform a Cartesian coordinate transformation (4.36) $(\mathbf{D}, \mathbf{r}_0)$, at a given point P we will have:

$$V'^i(P) = D^i{}_j V^j(P), \quad \text{contravariant vector,}$$
$$V'_i(P) = (D^{-1})^k{}_i V_k(P), \quad \text{covariant vector.} \qquad (4.39)$$

However the same point P, in the two reference frames, will be described by two different sets of coordinates: $\mathbf{r} \equiv (x^i)$ and $\mathbf{r}' \equiv (x'^i)$ respectively, $i = 1, \ldots, n$. Therefore the dependence of the components of the vector field on the coordinates will in general change as a consequence of the transformation:

$$V'^i(\mathbf{r}') = V'^i(\mathbf{D}\mathbf{r} - \mathbf{r}_0) = D^i{}_j V^j(\mathbf{r})$$
$$V'_i(\mathbf{r}') = V'_i(\mathbf{D}\mathbf{r} - \mathbf{r}_0) = (D^{-1})^k{}_i V_k(\mathbf{r}), \qquad (4.40)$$

where we have used (4.36). In what follows we shall, for the sake of simplicity, talk about vectors even when dealing with vector fields, omitting their explicit coordinate dependence, whenever this is not required by the context.[5]

The vector space V_n, with a positive definite scalar product, will capture all the geometric properties of E_n. In particular we can describe all the points in E_n in terms of a Cartesian coordinate system defined by an origin and a basis $\{\mathbf{u}_i\}_{i=1,\ldots,n}$ of V_n. This is a feature of *flat spaces* in general (the Euclidean space being an example of flat space) and in the following of this book we shall restrict to this kind of spaces only. Let us just mention that *non-flat* spaces have been considered in Chap. 3 and

---

[3] Such relations are, by definition, invertible, namely the Jacobian matrix $\left(\frac{\partial x'^j}{\partial x^i}\right)$ is non-singular.

[4] We shall use $\mathbf{r} = (x^i)$ to denote the collection of Cartesian coordinates. Generic coordinates will also be collectively denoted by $x = (x^i)$.

[5] This remark will also apply to tensors and tensor–fields, to be introduced in next section.

their features have been described in a non-rigorous way by the introduction of the concept of curvature. In particular we have seen that if the space is not flat (consider a sphere in the three-dimensional Euclidean space E_3), its geometric properties are no longer captured by a vector space (take two vectors in E_3 connecting two couples of points on the sphere, their sum in general does not connect two points on the sphere). One can show, however, as was described in an intuitive way in the previous chapter, that infinitesimal displacements in the neighborhood of any point P of the space, do close a vector space, called *tangent space at P*. The latter therefore captures only the *local* properties of the space, just as the tangent plane to a sphere at a point P approximates the sphere in the immediate vicinity of P. As anticipated in Chap. 3, curved spaces can not be described, in a finite or infinite region, in terms of Cartesian but only by means of curvilinear coordinates. If x^i are coordinates in this space, an infinitesimal displacement is a vector having, as components, the differentials dx^i of the coordinates. All that has been defined for the vector space V_n associated with a flat space, such as the metric, covariant and contravariant vectors etc. can now be defined on the tangent space to a curved space at a generic point, the matrix $\mathbf{D}$ representing the coordinate-dependent Jacobian matrix of the general coordinate transformation, see the end of this section. Since in this more general situation, the coordinates x^i are no longer components of vectors, it is correct to associate with the differentials dx^i, rather than with the coordinates x^i themselves, contravariant transformation properties. If the space is flat the tangent spaces at all points coincide and the geometry is captured by a single vector space.

We end this section by giving a more general definition of contravariant and covariant vectors, which holds also for non-linear coordinate transformations, and thus extends the definition given earlier to generic coordinate transformations and, in the light of our previous remark, to transformations on curved spaces. If we effect a coordinate transformation:

$$x^i \longrightarrow x'^i = x'^i(x^1, x^2, \ldots, x^n) \quad i = 1 \ldots, n, \tag{4.41}$$

the coordinate differentials dx^i transform through the (coordinate-dependent) Jacobian matrix:

$$dx'^i = \frac{\partial x'^i}{\partial x^j} dx^j. \tag{4.42}$$

We shall call contravariant a vector V^i whose components transform as the coordinate differentials dx^i:

$$V^i \longrightarrow V'^i = \frac{\partial x'^i}{\partial x^j} V^j. \tag{4.43}$$

In case the transformation (4.41) connects two Cartesian coordinate systems, it is linear, of the form (4.36), and the Jacobian matrix coincides with the constant matrix $\mathbf{D}$ relating the two bases, so that we retrieve the previous definition (4.26).

Consider now the following differential operators:

$$\frac{\partial}{\partial x^i} : \quad f \longrightarrow \frac{\partial f}{\partial x^i}, \tag{4.44}$$

where $\frac{\partial f}{\partial x^i}$ are the components of the *gradient vector* ∇f of a function $f(\mathbf{r}) = f(x^1, x^2, \ldots, x^n)$. These quantities transform under (4.41) according to the rule of derivatives of composite functions:

$$\frac{\partial}{\partial x'^i} = \frac{\partial x^j}{\partial x'^i} \frac{\partial}{\partial x^j}. \tag{4.45}$$

We shall call *covariant* any vector whose components transform as a gradient vector, namely as (4.45). For an affine transformation (4.36) we then find

$$x^j = D^{-1j}{}_i x'^i \quad \Longrightarrow \quad \frac{\partial x^j}{\partial x'^i} \equiv D^{-1j}{}_i, \tag{4.46}$$

so that the components of the gradient vector transform as the basis elements $\{\mathbf{u}_i\}$ of the Cartesian coordinate system:

$$\frac{\partial}{\partial x'^i} = D^{-1j}{}_i \frac{\partial}{\partial x^j}, \tag{4.47}$$

consistently with the earlier characterization of covariant vector.

In the following we shall restrict to Cartesian coordinate systems and thus will only consider affine transformations, unless explicitly stated.

4.2 Tensors

Consider now the set of all quantities of the form $V^i W^j$, namely expressible as the product of the contravariant components of two vectors. Under a change of basis (4.20), resulting from a Cartesian coordinate transformation (4.36), we have:

$$V'^i W'^j = D^i{}_k D^j{}_\ell V^k W^\ell. \tag{4.48}$$

A collection of n^2 numbers F^{ij} $(i, j = 1, 2, \ldots, n)$ is a contravariant tensor of order 2 and type $(2, 0)$, if, under a change of basis, it transforms as the product of two contravariant vectors, namely as in (4.48):

$$F^{ij'} = D^i{}_k D^j{}_\ell F^{k\ell}. \tag{4.49}$$

The set of all such objects form a vector space (i.e. a linear combination of two type $(2, 0)$-tensors is again a type $(2, 0)$-tensor) which is denoted by $V_n \otimes V_n$. Let us recall at this point, that given two vector spaces V_n, V_m, the tensor product $V_n \otimes V_m$ is a vector space containing the tensor products $\mathbf{V} \otimes \mathbf{W}$ of vectors $\mathbf{V} \in V_n$ and $\mathbf{W} \in V_m$. The tensor product operation $\otimes$ is bilinear in its two arguments: $(\alpha \, \mathbf{V}_1 + \beta \, \mathbf{V}_2) \otimes \mathbf{W} = \alpha \, \mathbf{V}_1 \otimes \mathbf{W} + \beta \, \mathbf{V}_2 \otimes \mathbf{W}$ and $\mathbf{V} \otimes (\alpha \, \mathbf{W}_1 + \beta \, \mathbf{W}_2) = \alpha \, \mathbf{V} \otimes \mathbf{W}_1 + \beta \, \mathbf{V} \otimes \mathbf{W}_2$. Therefore, if $\{\mathbf{u}_i\}_{i=1,\ldots,n}$ and $\{\mathbf{w}_\alpha\}_{\alpha=1,\ldots,m}$ are bases of V_n and V_m, respectively, all products $\mathbf{V} \otimes \mathbf{W}$ can be expanded in the basis $\{\mathbf{u}_i \otimes \mathbf{w}_\alpha\}$ consisting of mn (linearly independent) elements, their components being the product of the components of the two vectors:

$$\mathbf{V} \otimes \mathbf{W} = (\mathbf{V} \otimes \mathbf{W})^{i\alpha} \, \mathbf{u}_i \otimes \mathbf{w}_\alpha = V^i \, W^\alpha \, \mathbf{u}_i \otimes \mathbf{w}_\alpha,$$

where the summation over i and α is understood. The tensor product space $V_n \otimes V_m$ is defined as the vector space spanned by $\{\mathbf{u}_i \otimes \mathbf{w}_\alpha\}$ and has therefore dimension nm. A generic element of it has the following form:

$$\mathbf{F} \in V_n \otimes V_m, \quad \mathbf{F} = F^{i\alpha} \, \mathbf{u}_i \otimes \mathbf{w}_\alpha.$$

Notice that $\mathbf{F}$ is in general *not* the tensor product of two vectors: $F^{i\alpha} \neq V^i \, W^\alpha$. We can generalize the above construction to define the tensor product of three or more spaces: $V_n \otimes V_m \otimes V_k \equiv (V_n \otimes V_m) \otimes V_k$ (so that $\mathbf{V}_1 \otimes \mathbf{V}_2 \otimes \mathbf{V}_3 \equiv (\mathbf{V}_1 \otimes \mathbf{V}_2) \otimes \mathbf{V}_3$, for any $\mathbf{V}_1 \in V_n$, $\mathbf{V}_2 \in V_m$, $\mathbf{V}_3 \in V_k$) and so on. Given ℓ vector spaces V_{n_k}, $k = 1, \ldots, \ell$, of dimension n_k each, the tensor product $V_{n_1} \otimes V_{n_2} \otimes \cdots \otimes V_{n_\ell}$ is the vector space spanned by the ℓ-fold tensor product $\mathbf{u}_{i_1}^{(1)} \otimes \mathbf{u}_{i_2}^{(2)} \otimes \cdots \otimes \mathbf{u}_{i_\ell}^{(\ell)}$, where $\{\mathbf{u}_{i_k}^{(k)}\}_{i_k=1,\ldots,n_k}$ is a basis of V_{n_k}. The notion of tensor product of spaces is important not just for the definition of tensors but also when describing, in quantum mechanics, the quantum states for a system of non-interacting particles (see Chap. 9).

The n^2 entries $F^{k\ell}$ can either be arranged in a $n \times n$ matrix or can be viewed as the components of a n^2-dimensional "vector" which transform linearly under a change of basis. Indeed the quantities $D^i{}_k \, D^j{}_\ell$ on the right hand side of (4.49) can be thought of as entries of a single matrix $\mathbf{M} = (M^{ij}{}_{k\ell})$ in which the row and column indices are represented by the couples (i, j), (k, ℓ), respectively, running over the n^2 different combinations. This matrix would act on the column vector $\mathbf{F} \equiv (F^{k\ell}) = (F^{12}, F^{13}, \ldots, F^{n\,n-1}, F^{nn})$, whose components are labeled by the couple (k, ℓ). We shall denote the matrix $\mathbf{M}$ by $\mathbf{D} \otimes \mathbf{D}$, also called Kronecker product the two $\mathbf{D}$ matrices, so that we can rewrite Eq. (4.49) in the following form:

$$F^{ij\prime} = D^i{}_k \, D^j{}_\ell \, F^{k\ell} = (\mathbf{D} \otimes \mathbf{D})^{ij}{}_{k\ell} \, F^{k\ell}. \tag{4.50}$$

As anticipated, F^{kl}, having two contravariant indices, is called *contravariant tensor of order (or rank) 2*, or simply a $(2, 0)$-tensor, the latter notation indicating that it has two contravariant (upper) indices and no covariant ones.

Similarly we can define a *covariant tensor of order 2, namely of type* $(0, 2)$ as a quantity having two lower indices and transforming as the product of the covariant components of two vectors $V_i W_j$:

$$F'_{ij} = D^{-1k}{}_i D^{-1\ell}{}_j F_{k\ell} \equiv \left(\mathbf{D}^{-1} \otimes \mathbf{D}^{-1}\right)^{k\ell}{}_{ij} F_{k\ell}. \tag{4.51}$$

Clearly $(0, 2)$-tensors form a vector space as well. Finally we can consider objects whose components have the form $F^i{}_j$ and transform as the product of a covariant and a contravariant vector $V^i W_j$:

$$V'^i W'_j = D^i{}_k D^{-1\ell}{}_j V^k W_\ell. \tag{4.52}$$

Such objects are order 2 tensors called type $(1, 1)$-tensors and are therefore collections of entries $F^i{}_j$ transforming as:

$$F'^i{}_j = D^i{}_k D^{-1\ell}{}_j F^k{}_\ell \equiv \left(\mathbf{D} \otimes \mathbf{D}^{-T}\right)^{i;\,\ell}{}_{j;\,k} F^k{}_\ell. \tag{4.53}$$

Of this kind are the non-singular matrices $\mathbf{A} = (A^i{}_j)$ defining linear transformations on V_n, i.e. linear mappings of (contravariant) vectors $\mathbf{V} = (V^i)$ into (contravariant) vectors $\mathbf{W} = (W^i)$:

$$\mathbf{V} \longrightarrow \mathbf{W} = \mathbf{A}\,\mathbf{V} \;\Leftrightarrow\; W^i = A^i{}_j V^j, \tag{4.54}$$

To show that $A^i{}_j$ is a $(1, 1)$-tensor, let us consider the effect on it of a change of basis. The column vectors $\mathbf{V}$ and $\mathbf{W}$ are mapped into $\mathbf{V}'$ and $\mathbf{W}'$, which still satisfy a relation of the form:

$$\mathbf{W}' = \mathbf{A}'\,\mathbf{V}'. \tag{4.55}$$

Expressing the transformed (primed) quantities in terms of the old ones we can write:

$$\mathbf{A}'\,\mathbf{D}\,\mathbf{V} = \mathbf{D}\,\mathbf{W} \;\Rightarrow\; \left(\mathbf{D}^{-1}\mathbf{A}'\,\mathbf{D}\right)\mathbf{V} = \mathbf{W}. \tag{4.56}$$

Being $\mathbf{V}$ and $\mathbf{W}$ generic, the above relation implies:

$$\mathbf{A}' = \mathbf{D}^{-1}\mathbf{A}\,\mathbf{D}, \tag{4.57}$$

or, in components,

$$A'^i{}_j = D^i{}_k D^{-1\ell}{}_j A^k{}_\ell, \tag{4.58}$$

which shows that the matrix $\mathbf{A} = (A^i{}_j)$ is a type-$(1, 1)$ tensor.

The transformation property (4.27) implies that the metric is a covariant tensor of order 2, namely a type-(0, 2) tensor called *metric tensor*. Similarly, Eq. (4.27) implies that the inverse metric g^{ij} is a type (2, 0) tensor.

The Kronecker symbol δ^i_j is a (1, 1)-tensor which has the property of being *invariant*, namely to have the same form in whatever coordinate system:

$$\delta'^i_j = D^i{}_k D^{-1}{}^\ell{}_j \, \delta^j_\ell = D^i{}_k D^{-1}{}^k{}_j = \delta^i_j. \tag{4.59}$$

It is natural now to define tensors with more than two indices. We define a (p, q)-tensor, an object having p upper and q lower indices transforming as the product of q covariant and p contravariant vector components:

$$T'^{a_1...a_p}{}_{b_1...b_q} = D^{a_1}{}_{c_1} \cdots D^{a_p}{}_{c_p} D^{-1}{}^{l_1}{}_{b_1} \cdots D^{-1}{}^{l_q}{}_{b_q} T^{c_1...c_p}{}_{l_1...l_q}. \tag{4.60}$$

or, using the obvious extension of the notation used for rank two tensors

$$T'^{a_1...a_p}{}_{b_1...b_q} = \left(\mathbf{D} \otimes \cdots \otimes \mathbf{D} \otimes \mathbf{D}^{-T} \otimes \cdots \otimes \mathbf{D}^{-T} \right)^{a_1...a_p;l_1...l_q}_{b_1...b_q;c_1...c_p} T^{c_1...c_p}{}_{l_1...l_q}, \tag{4.61}$$

with p factors $\mathbf{D}$ and q factors $\mathbf{D}^{-T}$.

We can convert covariant indices into contravariant ones using the metric tensor. A typical example is the definition, given earlier, of the covariant components of a vector, obtained from the contravariant ones V^i by contraction with the metric tensor: $V_i = g_{ij} V^j$. To start with, let us consider, as an example, a type (2, 1) tensor $T^{ij}{}_k$. Multiplying this quantity by the metric tensor and contracting over one index, we obtain a new 3-index tensor:

$$T_i{}^j{}_k \equiv g_{i\ell} T^{\ell j}{}_k, \tag{4.62}$$

which is of type (1, 2): The first index, which used to be contravariant, has become covariant due to the contraction by g_{ij}. In general this procedure allows us to map a type-(p, q) tensor into a type-$(p - 1, q + 1)$ one.

Similarly we can use the tensor g^{jk} to convert a covariant index into a contravariant one and thus to map a type-(p, q) tensor into a type-$(p + 1, q - 1)$ one. For example:

$$T^i{}_k = g^{ij} T_{jk}. \tag{4.63}$$

4.3 Tensor Algebra

We have previously pointed out that rank 2 tensors of the same type ((2, 0), (1, 1) or (0, 2)) can be considered as elements of a linear vector space: It is straightforward to show that the linear combination of two rank 2 tensors of the same type is again a tensor of the same type. The same property holds for generic type (p, q) tensors,

It is really amazing that this conclusion, assumed by Einstein as the starting point for a relativistic theory of gravitation, can be drawn simply from the principle of equivalence, that is the equality between inertial and gravitational masses.

Because of its general covariance the relativistic theory of gravitation is called *general theory of relativity.*

3.2 Tidal Forces

We have seen that it is always possible to *locally* eliminate the effects of a gravitational field by using a *free falling frame*. It is then extremely important to examine those effects of a gravitational field which cannot be eliminated in the free falling frame, that is which manifest themselves when we go beyond the crude equivalence implied by the approximate relation (3.5).

In the present section we shall work purely in the classical limit, that is with no reference to the corrections implied by special relativity, and show that what remains of a gravitational force in the free falling frame are the *tidal forces*, see Fig. 3.1.

Let us start indeed from Eq. (3.8) which, in the classical case, assuming $m_I = m_G$, contains no approximations:

$$\mathbf{g}(\mathbf{r}) - \mathbf{g}(\mathbf{r_0}) = \mathbf{a}', \qquad (3.14)$$

the position vector $\mathbf{r}$ defining a generic point in a neighborhood of $\mathbf{r_0}$. We shall also denote by $\mathbf{h} = \mathbf{r} - \mathbf{r_0} = (h^i)$ the relative position vector between the two points, with components $h^k \equiv x^k - x_0^k$, where $k = 1, 2, 3$ and $(x^k) \equiv (x, y, z)$, $(x_0^k) \equiv (x_0, y_0, z_0)$ (see Fig. 3.2).

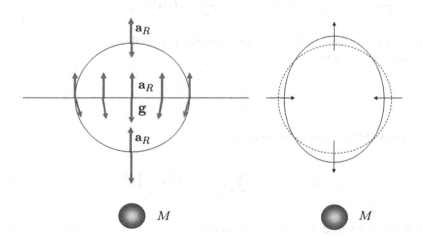

Fig. 3.1 Tidal forces

which form, for given values of p and q a linear vector space: The linear combination
of two type (p, q) tensors F and G:

$$S^{a_1 \ldots a_p}{}_{b_1 \ldots b_q} = \alpha\, F^{a_1 \ldots a_p}{}_{b_1 \ldots b_q} + \beta\, G^{a_1 \ldots a_p}{}_{b_1 \ldots b_q}, \tag{4.64}$$

is again a type (p, q) tensor.

Moreover we can multiply tensors of different type to obtain a new tensor. Consider two tensors F and G of type (p, q) and (r, s) respectively. We define the tensor
product $F \otimes G$ of the two tensors, the following type $(p + r, q + s)$ tensor:

$$(F \otimes G)^{a_1 \ldots a_p\, a_{p+1} \ldots a_{p+r}}{}_{b_1 \ldots b_q\, b_{q+1} \ldots b_{q+s}} = F^{a_1 \ldots a_p}{}_{b_1 \ldots b_q}\, G^{a_{p+1} \ldots a_{p+r}}{}_{b_{q+1} \ldots b_{q+s}}. \tag{4.65}$$

Take for example a type $(2, 0)$ tensor F^{ij} and a type $(0, 1)$ tensor G_k. We can construct
a type $(2, 1)$ tensor from the tensor product of the two: $T^{ij}{}_k = (F \otimes G)^{ij}{}_k \equiv F^{ij}\, G_k$.
Indeed, from the transformation properties of F and G:

$$F'^{ij} = D^i{}_k\, D^j{}_\ell\, F^{k\ell}; \quad G'_i = D^{-1j}{}_i\, G_j, \tag{4.66}$$

it follows that:

$$T'^{ij}{}_k = F'^{ij}\, G'_k = D^i{}_\ell D^j{}_m (D^{-1})^n{}_k\, F^{\ell m}\, G_n = D^i{}_\ell D^j{}_m (D^{-1})^n{}_k\, T^{\ell m}{}_n.$$

The generalization of the above proof to tensors of generic rank is straightforward.
The set of all tensors, endowed with the tensor product operation, is called *tensor
algebra*.

Another operation defined within a tensor algebra is the *contraction* or *trace*,
which maps a type (p, q) tensor into a type $(p - 1, q - 1)$ one, and which consists
in taking the entries of a tensor with the same values of an upper (contravariant) and
a lower (covariant) index and summing them over these common values. We say
that the upper and lower indices are contracted with one another. This is what we do
when we compute the trace of a matrix with entries $a^i{}_j$: We consider the entries with
equal values of i and j (i.e. the diagonal entries) and we sum them up, namely we
compute $\mathrm{tr}(a^i{}_j) = a^i{}_i \equiv \sum_{i=1}^n a^i{}_i$. In computing the trace of $(a^i{}_j)$, in other words,
we are *contracting* the index i with the index j. Let us consider as an example the
tensor $T^{ij}{}_k$, which transforms as follows:

$$T'^{ij}{}_k = D^i{}_\ell\, D^j{}_m\, D^{-1s}{}_k\, T^{\ell m}{}_s. \tag{4.67}$$

If we contract j with k, namely we set $j = k$ and sum over j from 1 to n, we obtain:

$$T'^{ij}{}_j = D^i{}_\ell\, D^j{}_m\, D^{-1s}{}_j\, T^{\ell m}{}_s = D^i{}_\ell\, \delta^s_m\, T^{\ell m}{}_s = D^i{}_\ell\, T^{\ell m}{}_m. \tag{4.68}$$

We observe that $T^{\ell m}{}_m$ transforms as a contravariant vector, namely as a $(1, 0)$ tensor.
In particular, if we have a tensor, or a product of tensors, with all indices contracted,

the result is a $(0, 0)$ tensor, which is a *scalar*, namely a quantity which does not depend on the chosen coordinate system (an example is the trace $a^i{}_i$ of the matrix $(a^i{}_j)$). Consider, for instance, the transformation property of the product $T^{ij} U_{ij}$ of a $(2, 0)$ and a $(0, 2)$-tensors:

$$T'^{ij} U'_{ij} = = D^i{}_k D^j{}_\ell D^{-1}{}^m{}_i D^{-1}{}^n{}_j T^{kl} U_{mn} = \delta^m_k \delta^n_\ell T^{kl} U_{mn} = T^{k\ell} U_{k\ell}.$$

An other example is the scalar product itself $V_i W^i$.

Just as we did for vectors, we may define a *scalar field*, that is a $(0, 0)$-tensor as a scalar quantity defined in each point is space, i.e. a function over space. As such, its value at any point does not depend on the coordinates used to describe it:

$$f'(P) = f(P). \tag{4.69}$$

This implies that the scalar function will in general have a different dependence on the chosen coordinates, namely that, under a change of coordinates $x^i \to x'^i \equiv x'^i(x)$ it will be described by a new function $f'(x')$ related to $f(x)$ as follows:

$$f'(x') = f(x). \tag{4.70}$$

If the *functional* dependence of f on the new and old coordinates does not change, that is if:

$$f'(x') = f(x'), \tag{4.71}$$

the scalar function f is said to be *invariant*.[6]

A *tensor field* is a tensor quantity which depends on the coordinates of a point P in space. A change in coordinates, besides transforming the tensor components, will also transform the coordinate dependence of the tensor, as we have shown for the vector and scalar fields. Take for instance a $(2, 1)$ tensor field described by a set of functions $T^{ij}{}_k(x)$ in a given coordinate system. Under a coordinate transformation we have:

$$T'^{ij}{}_k(x') = D^i{}_\ell D^j{}_m D^{-1}{}^s{}_k T^{\ell m}{}_s(x). \tag{4.72}$$

Using the explicit form (4.36) of a Cartesian coordinate transformation, we find:

$$T'^{ij}{}_k(\mathbf{r}') = D^i{}_\ell D^j{}_m D^{-1}{}^s{}_k T^{\ell m}{}_s(\mathbf{D}^{-1} \mathbf{r}' + \mathbf{D}^{-1} \mathbf{r}_0), \tag{4.73}$$

where, in the argument on the right hand side, we have expressed the old coordinate vector $\mathbf{r}$ in terms of the new one $\mathbf{r}'$ by inverting Eq. (4.37).

---

[6]Of course Eq. (4.70) can be also written $f'(x) = f(x)$, since x is a variable.

The notion of invariance, which was given for scalar fields, can be extended to more general tensor fields. Let us still take, for the sake of simplicity, the type (2, 1) tensor field $T^{ij}{}_k(\mathbf{r})$. We will say that $T^{ij}{}_k(x)$ is invariant, if it transforms, under a coordinate transformation, as follows:

$$T'^{ij}{}_k(x') = D^i{}_\ell D^j{}_m D^{-1}{}^s{}_k T^{\ell m}{}_s(x) \equiv T^{ij}{}_k(x'). \tag{4.74}$$

The above invariance condition has an obvious generalization to tensors of type (p, q). An example of invariant tensor is the Kronecker symbol, as it was shown in the previous section.

Let us now define a (2, 0)-tensor F^{ij} symmetric if $F^{ij} = F^{ji}$ and antisymmetric if $F^{ij} = -F^{ji}$. Considering a generic type (2, 0) tensor F^{ij}, it can be decomposed into a symmetric and an anti-symmetric part, with respect to the exchange of the two indices, by writing the following trivial identity:

$$F^{ik} = \frac{1}{2}(F^{ik} + F^{ki}) + \frac{1}{2}(F^{ik} - F^{ki}) \doteq F_S{}^{ik} + F_A{}^{ik}, \tag{4.75}$$

where $F_S{}^{ik} = F_S{}^{ki}$ and $F_A{}^{ik} = -F_A{}^{ki}$ define the symmetric and anti-symmetric parts of F^{ij}.

This decomposition does not depend on the coordinate basis we use, since under a coordinate transformation a symmetric tensor $F_S{}^{ik}$ is mapped into a symmetric tensor and similarly for the anti-symmetric ones:

$$F_S'^{ij} = D^i{}_\ell D^j{}_m F_S^{\ell m} = D^i{}_\ell D^j{}_m F_S^{m\ell} = D^j{}_m D^i{}_\ell F_S^{m\ell} = F_S'^{ji},$$
$$F_A'^{ij} = D^i{}_\ell D^j{}_m F_A^{\ell m} = -D^i{}_\ell D^j{}_m F_A^{m\ell} = -D^j{}_m D^i{}_\ell F_A^{m\ell} = -F_A'^{ji}. \tag{4.76}$$

We conclude that the vector space of type (2, 0)-tensors can be decomposed into the direct sum of two disjoint subspaces spanned by symmetric and antisymmetric tensors. The same decomposition can be performed on the space of (0, 2)-tensors, by writing a generic covariant rank 2 tensor F_{ij} into the sum of its symmetric and anti-symmetric components: $F_{ij} = F_{S\,ij} + F_{A\,ij}$. It is straightforward to prove that *the contraction over all indices of a type (2, 0) and a type (0, 2) tensors with opposite symmetry (i.e. one symmetric and the other anti-symmetric) is zero*. Consider, for instance, the contraction of a symmetric (2, 0)-tensor with an anti-symmetric (0, 2) one:

$$F_S{}^{ik} F_{A\,ik} = F_S{}^{ki} F_{A\,ki} = -F_S{}^{ik} F_{A\,ik} = 0. \tag{4.77}$$

By the same token we would have $F_A{}^{ik} F_{S\,ik} = 0$. As a consequence of this property, any rank 2 tensor contracted with a symmetric or an anti-symmetric tensor gets projected into its symmetric or anti-symmetric component. Consider, for instance, a tensor T^{ij} with a definite symmetry property (i.e. it is either symmetric or anti-symmetric) and let U_{ij} be a generic type (0, 2) tensor, which has symmetric ($U_{S\,ij}$)

and anti-symmetric ($U_{A\,ij}$) components. Contracting the two tensors over all indices we find

$$T^{ij}\,U_{ij} = T^{ij}\left[\frac{1}{2}\left(U_{ij} + U_{ji}\right) + \frac{1}{2}\left(U_{ij} - U_{ji}\right)\right] = T^{ij}\left(U_{S\,ij} + U_{A\,ij}\right). \qquad (4.78)$$

Recall now that, according to (4.77), if T^{ij} is symmetric $T^{ij}\,U_{A\,ij} = 0$ and thus $T^{ij}\,U_{ij} = T^{ij}\,U_{S\,ij}$, whereas if T^{ij} is anti-symmetric, $T^{ij}\,U_{S\,ij} = 0$ and so $T^{ij}\,U_{ij} = T^{ij}\,U_{A\,ij}$.

We note that the previous decompositions into symmetric and antisymmetric part cannot be performed for $(1, 1)$ tensors, since the two indices transform differently and therefore the symmetry or antisymmetry properties are not preserved by coordinate transformations.

Let us finally introduce the operation of *differentiation* over tensor fields. By definition, tensor fields depend on coordinates, and thus can be differentiated with respect to them. The partial derivative with respect to the coordinate x^k of a type-(p, q) tensor field is a type-$(p, q+1)$ tensor, whose structure differs from the original one by one additional lower (covariant) index k. Consider, for instance, a type-$(2, 1)$ tensor field $T^{ij}{}_k(\mathbf{r})$. Differentiating with respect to x^ℓ we find a new quantity $U^{ij}{}_{\ell k}$:

$$\frac{\partial}{\partial x^\ell} : \quad T^{ij}{}_k(\mathbf{r}) \longrightarrow U^{ij}{}_{\ell k} \equiv \frac{\partial}{\partial x^\ell} T^{ij}{}_k(\mathbf{r}), \qquad (4.79)$$

which transforms, under a coordinate transformation, as follows:

$$U'^{ij}{}_{\ell k}(\mathbf{r}') = \frac{\partial}{\partial x'^\ell} T'^{ij}{}_k(\mathbf{r}') = \frac{\partial}{\partial x'^\ell}\left[D^i{}_m D^j{}_n D^{-1}{}^p{}_k T^{mn}{}_p(\mathbf{r})\right]$$

$$= D^i{}_m D^j{}_n D^{-1}{}^p{}_k \frac{\partial x^s}{\partial x'^\ell} \frac{\partial}{\partial x^s} T^{mn}{}_p(\mathbf{r}). \qquad (4.80)$$

On the other hand $\left(\frac{\partial x^s}{\partial x'^\ell}\right)$ is the (constant) inverse Jacobian matrix of (4.36), that is $(D^{-1\,s}{}_\ell)$. Substituting this in Eq. (4.80) we find:

$$U'^{ij}{}_{\ell k}(\mathbf{r}') = D^i{}_m D^j{}_n D^{-1}{}^p{}_k D^{-1\,s}{}_\ell U^{mn}{}_{ps}(\mathbf{r}),$$

that is the quantity $U^{ij}{}_{\ell k}(\mathbf{r}) \equiv \frac{\partial}{\partial x^\ell} T^{ij}{}_k(\mathbf{r})$ is a tensor field, and, more specifically, a type $(2, 2)$ tensor. The operator $\frac{\partial}{\partial x^k}$, to be also denoted by the symbol ∂_k, behaves, by definition, as a type-$(0, 1)$ tensor, i.e. as a covariant vector:

$$\partial'_\ell \equiv \frac{\partial}{\partial x'^\ell} = D^{-1\,s}{}_\ell \frac{\partial}{\partial x^s} = D^{-1\,s}{}_\ell \partial_s. \qquad (4.81)$$

4.4 Rotations in 3-Dimensions

As we have previously pointed out, the scalar product associates with a couple of vectors a number which does not depend on the basis we use to describe the vectors. However, its explicit expression in terms of the vector components is basis-dependent, since the metric tensor changes: $g'_{ij} \neq g_{ij}$.

Suppose now the change of basis is such that the metric is invariant, that is $g'_{ij} = g_{ij}$. We will then have:

$$\mathbf{V} \cdot \mathbf{W} = V^i \, g_{ij} \, W^j = V'^i \, g_{ij} \, W'^j, \qquad (4.82)$$

that is the *functional dependence* of $\mathbf{V} \cdot \mathbf{W}$ over the old and new components of the two vectors is the *same*. Let us denote by

$$\mathbf{R} \equiv (R^i{}_j) = \begin{pmatrix} R^1{}_1 & R^1{}_2 & R^1{}_3 \\ R^2{}_1 & R^2{}_2 & R^2{}_3 \\ R^3{}_1 & R^3{}_2 & R^3{}_3 \end{pmatrix}, \qquad (4.83)$$

the matrix implementing such transformation: $V'^i = R^i{}_j \, V^j$, $W'^i = R^i{}_j \, W^j$ (or, in matrix notation $\mathbf{V}' = \mathbf{R}\,\mathbf{V}$, $\mathbf{W}' = \mathbf{R}\,\mathbf{W}$). Expressing in (4.82) the new components in terms of the old ones we find:

$$V^i \, g_{ij} \, W^j = V^k \, R^i{}_k \, g_{ij} \, R^j{}_\ell \, W^\ell. \qquad (4.84)$$

Requiring the above invariance to hold for any couple of vectors (V^i) and (W^i), we conclude that:

$$R^i{}_k \, g_{ij} \, R^j{}_\ell = g_{k\ell}. \qquad (4.85)$$

In matrix notation Eq. (4.85) reads

$$\mathbf{R}^T \mathbf{g} \mathbf{R} = \mathbf{g}, \qquad (4.86)$$

where $\mathbf{g} \equiv (g_{ij})$ is the matrix whose components are the entries of the metric tensor g_{ij}. Recalling that $g_{ij} = \mathbf{u}_i \cdot \mathbf{u}_j$, the above relation is telling us that scalar products among the basis elements are invariant under $\mathbf{R}$. It is now convenient to use an ortho-normal basis $(\mathbf{u}_i)$ to start with:

$$\mathbf{u}_i \cdot \mathbf{u}_j = g_{ij} \equiv \delta_{ij}, \qquad (4.87)$$

since the ortho-normality property of a basis is clearly preserved by all the transformations $\mathbf{R}$ which leave the metric invariant. In the ortho-normal basis the relations (4.85) and (4.86) become:

$$R^i{}_k \, \delta_{ij} \, R^j{}_\ell = \sum_{i=1}^{n} R^i{}_k \, R^i{}_\ell = \delta_{k\ell}, \qquad (4.88)$$

and, in matrix form:

$$\mathbf{R}^T \, \mathbf{1} \, \mathbf{R} = \mathbf{R}^T \, \mathbf{R} = \mathbf{1}, \qquad (4.89)$$

where

$$\mathbf{1} \equiv (\delta_{ij}) = \begin{pmatrix} 1 & 0 & 0 \\ 0 & 1 & 0 \\ 0 & 0 & 1 \end{pmatrix}. \qquad (4.90)$$

Transformation matrices satisfying Eqs. (4.88), or equivalently (4.89), are called *orthogonal*. Orthogonal transformations can be alternatively characterized as the most general Cartesian coordinate transformations in Euclidean space mapping two orthonormal bases into one another, leaving the origin fixed, i.e. the most general homogeneous transformations between Cartesian rectangular coordinate systems.[7] Recalling from Eq. (4.13) that the distance squared between two points is defined as the squared norm of the relative position vector, an orthogonal transformation leaves its coordinate dependence invariant. Vice versa, if an affine transformation $x^i \to x'^i = R^i{}_j \, x^j - x_0^i$ of the Cartesian coordinates x^i leaves the distance between any two points, as a function of their coordinates, invariant, its homogeneous part, described by the matrix $\mathbf{R}$ and defining the transformation of the relative position vector, is an invariance of the metric tensor. This means that, starting from an ortho-normal basis in which $g_{ij} = \delta_{ij}$, $\mathbf{R}$ is an orthogonal matrix. To illustrate the above implication, note that the invariance of the coordinate dependence of the distance $d(A, B)$ between any two points translates into the invariance of the norm of any vector as a function of its components. This latter property amounts to stating that, if $\mathbf{V} = (V^i)$ and $\mathbf{V}' = (V'^i)$ are the components of a same vector in the old and new bases, related by the transformation $\mathbf{R}$, then $\|\mathbf{V}\|^2 = \mathbf{V}^T \, \mathbf{V} = \|\mathbf{V}'\|^2 = \mathbf{V}'^T \, \mathbf{V}'$. Applying this property to the squared norm $\|\mathbf{V} + \mathbf{W}\|^2$ of the sum of two generic vectors $\mathbf{V}$, $\mathbf{W}$, one easily finds that the scalar product $(\mathbf{V}, \mathbf{W}) = \mathbf{V}^T \, \mathbf{W}$ is functionally invariant under $\mathbf{R}$, namely that $\mathbf{V}^T \, \mathbf{W} = \mathbf{V}'^T \, \mathbf{W}' = \mathbf{V}^T \, (\mathbf{R}^T \mathbf{R}) \, \mathbf{W}$. From the arbitrariness of $\mathbf{V}$ and $\mathbf{W}$, property (4.89) follows. Rotations about an axis and reflections in a plane are examples of orthogonal transformations in E_3.

Since Eq. (4.89) implies $(\mathbf{R}^T)^{-1} = \mathbf{R}$, there is no distinction between the transformation properties of the covariant and contravariant components of a vector under orthogonal transformations, as it is apparent from the fact that, being the metric δ_{ij}

---

[7] In what follows, when referring to Cartesian coordinate systems, the specification *rectangular* will be understood, unless explicitly stated, since we shall mainly restrict ourselves to coordinate systems of this kind.

Fig. 4.2 Rotation about the
X axis by an angle θ

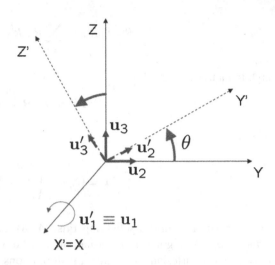

invariant, the two kinds of components coincide $V_i = \delta_{ij} V^j = V^i$ in any Cartesian
coordinate system.

A simple example of orthogonal transformation is a rotation by an angle θ about
the X axis , see Fig. 4.2.[8] The relation between the new and the old basis reads

$$
\begin{aligned}
\mathbf{u}_1' &= \mathbf{u}_1, \\
\mathbf{u}_2' &= \cos\theta\,\mathbf{u}_2 + \sin\theta\,\mathbf{u}_3, \\
\mathbf{u}_3' &= -\sin\theta\,\mathbf{u}_2 + \cos\theta\,\mathbf{u}_3.
\end{aligned}
\tag{4.91}
$$

Being $\mathbf{u}_i' = R_x^{-1j}{}_i\,\mathbf{u}_j$, from Eq. (4.91) we can read the form of the inverse of the
rotation matrix $\mathbf{R}_x$:

$$
\mathbf{R}_x^{-1} = (R_x^{-1j}{}_i) = \begin{pmatrix} 1 & 0 & 0 \\ 0 & \cos\theta & -\sin\theta \\ 0 & \sin\theta & \cos\theta \end{pmatrix},
\tag{4.92}
$$

from which we derive:

$$
\mathbf{R}_x = (R_x^j{}_i) = \begin{pmatrix} 1 & 0 & 0 \\ 0 & \cos\theta & \sin\theta \\ 0 & -\sin\theta & \cos\theta \end{pmatrix},
\tag{4.93}
$$

---

[8]In our conventions, the rotation angle θ, on any of the three mutually orthogonal planes
XY, XZ, YZ, is positive if its orientation is related to that of the axis orthogonal to it (i.e. Z, Y, X)
by the *right-hand rule*.

The new components V'^i of a vector are related to the old ones V^i according to: $V'^i = R_x{}^i{}_j V^j$, that is

$$V'^1 = V^1,$$
$$V'^2 = \cos\theta\, V^2 + \sin\theta\, V^3,$$
$$V'^3 = -\sin\theta\, V^2 + \cos\theta\, V^3. \qquad (4.94)$$

The matrix $\mathbf{R}_x$ in (4.93), which describes this rotation, depends on the continuous parameter θ: $\mathbf{R}_x = \mathbf{R}_x(\theta)$. The reader can easily verify that Eq. (4.89) is satisfied by $\mathbf{R}_x$. Let us observe that $\det(\mathbf{R}_x) = 1$. This is a common feature of all the rotation matrices and can be deduced by computing the determinant of both sides of Eq. (4.89) and using the known properties of the determinant: $\det(A^T) = \det(A)$, $\det(AB) = \det(A)\det(B)$:

$$\det(\mathbf{R})\det(\mathbf{R}^T) = \det(\mathbf{R})^2 = 1 \quad\Rightarrow\quad \det(\mathbf{R}) = \pm 1. \qquad (4.95)$$

Orthogonal transformations with $\det(\mathbf{R}) = +1$ are called *proper rotations*, or simply *rotations*, while those with $\det(\mathbf{R}) = -1$ also involve reflections and are called *improper*. A matrix $\mathbf{R}$ having this property is called *improper rotations*. A typical example of improper rotation is given by a pure reflection, that is a transformation changing the orientation of one or all the coordinate axes, e.g.

$$\begin{pmatrix} -1 & 0 & 0 \\ 0 & -1 & 0 \\ 0 & 0 & -1 \end{pmatrix}. \qquad (4.96)$$

Let us now perform two consecutive rotations, represented by the matrices $\mathbf{R}_1$, $\mathbf{R}_2$. Starting from a basis $(\mathbf{u}_i)$, the components V^i of a generic vector will transform as follows:

$$V^i \xrightarrow{R_1} V'^i = R_1{}^i{}_j V^j \xrightarrow{R_2} V''^i = R_2{}^i{}_j V'^i = R_2{}^i{}_j R_1{}^j{}_k V^k = R_3{}^i{}_j V^j,$$

or, in matrix form: $\mathbf{V} \to \mathbf{V}'' = \mathbf{R}_3\,\mathbf{V}$, where $\mathbf{R}_3 \equiv \mathbf{R}_2\,\mathbf{R}_1$. Let us show now that the resulting transformation, implemented by $\mathbf{R}_3$ is still a rotation, namely that it is orthogonal (i.e. $\mathbf{R}_3^T\,\mathbf{R}_3 = 1$) and has unit determinant:

$$\mathbf{R}_3^T\,\mathbf{R}_3 = (\mathbf{R}_2\,\mathbf{R}_1)^T\,(\mathbf{R}_2\,\mathbf{R}_1) = \mathbf{R}_1^T\,(\mathbf{R}_2^T\,\mathbf{R}_2)\,\mathbf{R}_1 = \mathbf{R}_1^T\,\mathbf{R}_1 = 1,$$
$$\det(\mathbf{R}_3) = \det(\mathbf{R}_2\,\mathbf{R}_1) = \det(\mathbf{R}_2)\,\det(\mathbf{R}_1) = 1. \qquad (4.97)$$

This proves that *the product of two rotations is still a rotation*.

In general the product of two rotations is not commutative:

$$\mathbf{R}_2\,\mathbf{R}_1 \neq \mathbf{R}_1\,\mathbf{R}_2. \qquad (4.98)$$

We can easily understand this by a simple example: If we rotate a system of Cartesian axes first about the X axis by $90°$ and then about the Z axis by the same angle or we perform the two rotations in opposite order, we end up with two different configurations of axes.

Any orthogonal matrix is *invertible*, having a non vanishing determinant, and its inverse is still orthogonal. Indeed let $\mathbf{R}$ be an orthogonal matrix and $\mathbf{R}^{-1}$ its inverse. We can multiply both sides of $\mathbf{R}^T \mathbf{R} = \mathbf{1}$ by $(\mathbf{R}^{-1})^T$ to the left and by $\mathbf{R}^{-1}$ to the right, obtaining:

$$\mathbf{R}^{-1T} \mathbf{R}^{-1} = I, \tag{4.99}$$

which proves that $\mathbf{R}^{-1}$ is still orthogonal. Clearly, if $\mathbf{R}$ is a rotation, namely $\det(\mathbf{R}) = 1$, also its inverse is, since: $\det(\mathbf{R}^{-1}) = 1/\det(\mathbf{R}) = 1$.

Also the identity matrix $\mathbf{1}$ defines a rotation since it is orthogonal and has unit determinant. It represents the trivial rotation leaving the system of axes invariant.

We have thus deduced, from their very definition (4.89), the following properties of orthogonal matrices:

 (i) The product of two orthogonal matrices is still an orthogonal matrix;
 (ii) The identity matrix $\mathbf{1}$ represents the orthogonal transformation such that, given any other orthogonal transformation $\mathbf{R}$: $\mathbf{R}\,\mathbf{1} = \mathbf{1}\,\mathbf{R} = \mathbf{R}$;
 (iii) For any orthogonal transformation $\mathbf{R}$ one can define its inverse $\mathbf{R}^{-1}$: $\mathbf{R}\,\mathbf{R}^{-1} = \mathbf{R}^{-1}\,\mathbf{R} = \mathbf{1}$. $\mathbf{R}^{-1}$ is still is orthogonal;
 (iv) Let us add the *associative* property of the product of orthogonal transformations, which actually holds for any transformation which is realized by matrices: Given any 3 matrices $\mathbf{R}_1\,(\mathbf{R}_2\,\mathbf{R}_3) = (\mathbf{R}_1\,\mathbf{R}_2)\,\mathbf{R}_3$.

The above properties define a *group* called O(3), where O stands for *orthogonal*, namely for the defining property (4.89) of the transformations, and 3 refers to the dimensionality of the space on which they act. The group O(3) contains the set of all matrices describing rotations. This set is itself a group, since it satisfies the above properties and thus is a *subgroup* of O(3), denoted by SO(3), where the additional S stands for *special*, namely having unit determinant. Therefore SO(3) is the *rotation group* in three dimensional Euclidean space.

4.5 Groups of Transformations

The orthogonal group is just an instance of the more general notion of *group of transformations*. In general any set of elements G among which a product operation $\cdot$ is defined and which satisfies the same properties (i), (ii), (iii), (iv) as the orthogonal transformations, is called a *group*.

Consider general coordinate transformations and define the product $A \cdot B$ of two such transformations A, B, as the transformation resulting from the consecutive action of B and A on the initial coordinate system S: If B transforms S, of coordinates x^i, into a system S', of coordinates $x'^i = x'^i(x)$, A maps S' into S'', of coordinates $x''^i = x''^i(x')$, $A \cdot B$ will transform S into S'' and will be defined by the coordinate relations $x''^i = x''^i(x'(x))$. Given a transformation A which maps S into S', defined by the relations $x'^i(x)$, its inverse A^{-1} is the unique transformation mapping S' into S, and is defined by the inverse relations $x^i(x')$. The identity transformation I is the trivial transformation mapping a coordinate system S into itself. Finally we can convince ourselves that the product of transformations is associative, namely that, if A, B, C are three transformations, $A \cdot (B \cdot C) = (A \cdot B) \cdot C$. This proves that the set of all coordinate transformations satisfy the same properties (i), (ii), (iii), (iv) as the rotations, and thus close a group called the group of coordinate transformations.

We can generalize the concept of orthogonal transformations and of rotations to the n-*dimensional Euclidean space* E_n, namely to a n-dimensional space endowed with a positive definite metric g_{ij}, $i, j = 1, \ldots, n$. Orthogonal transformations in n are those which leave this metric tensor invariant, and are represented, in an orthonormal basis in which $g_{ij} = \delta_{ij}$, by $n \times n$ matrices $\mathbf{R}$ satisfying the orthogonality property: $\mathbf{R}^T \mathbf{R} = \mathbf{1}$, $\mathbf{1}$ being the $n \times n$ identity matrix. These transformations close themselves a group (i.e. satisfy axioms (i), (ii), (iii), (iv)), denoted by O(n), which contains, as a subgroup, the group of rotations SO(n) over the n-dimensional space, described of orthogonal matrices with unit determinant.

Let us consider the set of all Cartesian (not necessarily rectangular) n-dimensional linear coordinate transformations, i.e. the *affine transformations* (4.36) and show that they close a group. To this end, let us consider the effect, on a coordinate vector $\mathbf{r}$ of two consecutive affine transformations $(\mathbf{D}_1, \mathbf{r}_1)$, $(\mathbf{D}_2, \mathbf{r}_2)$:

$$\mathbf{r} \xrightarrow{1} \mathbf{r}' = \mathbf{D}_1 \mathbf{r} - \mathbf{r}_1 \xrightarrow{2} \mathbf{r}'' = \mathbf{D}_2 \mathbf{r}' - \mathbf{r}_2 = \mathbf{D}_2 \mathbf{D}_1 \mathbf{r} - \mathbf{D}_2 \mathbf{r}_1 - \mathbf{r}_2$$
$$= (\mathbf{D}_2 \mathbf{D}_1, \mathbf{D}_2 \mathbf{r}_1 + \mathbf{r}_2) \cdot \mathbf{r}.$$

The result of the two transformations defines their product, which is still an affine transformation:

$$(\mathbf{D}_3, \mathbf{r}_3) \equiv (\mathbf{D}_2, \mathbf{r}_2) \cdot (\mathbf{D}_1, \mathbf{r}_1) = (\mathbf{D}_2 \mathbf{D}_2, \mathbf{D}_2 \mathbf{r}_1 + \mathbf{r}_2). \qquad (4.100)$$

The identity element and the inverse of an affine transformation have the following form:

$$I = (\mathbf{1}, \mathbf{0}), \quad (\mathbf{D}, \mathbf{a})^{-1} = (\mathbf{D}^{-1}, -\mathbf{D}^{-1}\mathbf{a}). \qquad (4.101)$$

This proves that the affine transformations close a group, called the *affine group*. A subset of affine transformations are the homogeneous transformations $(\mathbf{D}, \mathbf{0})$ which do not shift the origin of the Cartesian system, but describe the most general transformation on the basis elements, and are defined by an invertible matrix $\mathbf{D}$. They close themselves a group, as the reader can easily verify, which is the group of non-singular $n \times n$ matrices, called *general linear group*, and denoted by GL(n). We say that GL(n) is a *subgroup* of the affine group. In general if a subset G' of a group G is itself a group with respect to the product defined on G, then G' is a subgroup of G. For instance the rotation group SO(3) is a subgroup of GL(3), since all its elements are invertible 3×3 matrices. Similarly the rotation group in a n-dimensional Euclidean space SO(n) is a subgroup of the general linear group on the same space GL(n). The most general transformation relating two Cartesian rectangular coordinate systems is an affine transformation of the form (**R**, $\mathbf{r}_0$):

$$\mathbf{r}' = (\mathbf{R}, \ \mathbf{r}_0) \cdot \mathbf{r} = \mathbf{R}\,\mathbf{r} - \mathbf{r}_0, \tag{4.102}$$

where **R** is an orthogonal matrix, since the two bases $\{\mathbf{u}_i\}$, $\{\mathbf{u}'_i\}$ are both ortho-normal, and we allowed for a translation of the origin $O \rightarrow O'$. Since this translation does not affect the actual value of the relative position vector between two points, Eq. (4.102) defines the most general Cartesian coordinate transformation leaving the distance between two points, as a function of their coordinates, invariant. The reader can verify that these transformations close a group, called the *Euclidean* or *congruence* group E(n), which is therefore a subgroup of the affine one.[9]

We can now refine the notion of *tensor*, relating it to a certain group of transformations. We have introduced tensors as quantities with definite transformation properties relative to the most general homogeneous linear transformations, i.e. relative to the group GL(n). We can consider the transformation property of tensors with respect to the subgroup SO(n) of GL(n). A tensor which is invariant with respect to the latter, such as δ^i_j, is *a fortiori*, invariant under any of its subgroups, including SO(n). However, a tensor which is invariant with respect to SO(n) is not in general invariant under GL(n). As an example consider the Ricci tensor ϵ_{ijk}, $i, j, k = 1, 2, 3$, which is SO(3)-invariant but not GL(3)-invariant. Such tensor is defined as follows: It is completely anti-symmetric in its three indices[10] and therefore vanishes if any couple of indices have equal value; Its value is $+1$ or -1 depending on whether (i, j, k) is an even or odd permutation of $(1, 2, 3)$ (for instance $\epsilon_{123} = +1$). Under a SO(3) transformation:

$$\epsilon'_{ijk} = R^{-1\,m}{}_i \, R^{-1\,n}{}_j \, R^{-1\,\ell}{}_k \, \epsilon_{mn\ell} = \det(\mathbf{R}^{-1}) \, \epsilon_{ijk} = \epsilon_{ijk}. \tag{4.103}$$

This proves that ϵ_{ijk} is SO(3)-invariant. It clearly is not GL(3)-invariant since transformations in GL(3) may in general have a determinant which is not 1. One can

---

[9]Let us recall that Euclidean geometry can be fully characterized by the invariance under the corresponding congruence group.

[10] *Complete antisymmetrization* in the three indices μ, ν, ρ on a generic tensor $U_{\mu\nu\rho}$, is defined as follows:

$$U_{[\mu\nu\rho]} = \frac{1}{3!}(U_{\mu\nu\rho} + U_{\nu\rho\mu} + U_{\rho\mu\nu} - U_{\mu\rho\nu} - U_{\nu\mu\rho} - U_{\rho\nu\mu}).$$

It amounts to summing over the even permutations of μ, ν, ρ with a plus sign and over the odd ones with a minus sign, the result being normalized by dividing it by the total number 6 of permutations. (see Chap. 5).

verify that for ϵ_{ijk} the following properties hold:

$$\epsilon_{ijk}\epsilon_{ijk} = 3!,$$
$$\epsilon_{ijk}\epsilon_{ljk} = 2!\delta_{il},$$
$$\epsilon_{ijk}\epsilon_{lnk} = (\delta_{il}\delta_{jn} - \delta_{in}\delta_{jl}).$$

Another tensor which is invariant under SO(3) (more generally with respect to O(3)) but not with respect to GL(3) is the metric $g_{ij} = \delta_{ij}$ (note that this differs from the tensor δ^i_j in that both indices are covariant). This follows from the very definition of orthogonal matrices (4.88). For the same reason δ_{ij}, $i, j = 1, \ldots, n$, is in general O(n)-invariant but not GL(n)-invariant. Note that δ^{ij}, inverse of δ_{ij}, clearly coincides with δ_{ij}, and thus is still O(n)-invariant.

Let us now consider the decomposition (4.75) for tensors transforming under the subgroup O(n) $\subset$ Gl(n). We have shown that the two vector spaces spanned by the symmetric F^{ij}_S and anti-symmetric F^{ij}_A components of rank 2 tensors F^{ij} are *invariant*, in the sense that a symmetric (anti-symmetric) tensor is mapped by any element of GL(n) into a tensor with the same symmetry property. It is easy to show that, if we consider transformations of tensors with respect to O(n), we can use the O(n) invariant tensor δ_{ij} to decompose the symmetric component F^{ij}_S in Eq. (4.75) into a *trace* part $\delta_{ij} F^k_k$, where

$$F^k_k \equiv \delta_{ij} F^{ij} = \delta_{ij} F^{ij}_S, \tag{4.104}$$

and a *traceless* part $\tilde{F}^{ij}_S$ defined as:

$$\tilde{F}^{ij}_S = \frac{1}{2}(F^{ij} + F^{ji}) - \frac{1}{n}\delta^{ij} F^k_k. \tag{4.105}$$

As the reader can easily verify from the definition of trace, $\tilde{F}^{ij}_S$ is indeed a symmetric traceless tensor, namely: $\tilde{F}^{ij}_S \delta_{ij} = 0$. We can now decompose F^{ij} as follows

$$F^{ij} = \left(\tilde{F}^{ij}_S + D^{ij}\right) + F^{ij}_A, \tag{4.106}$$

where

$$D^{ij} = \frac{1}{n}\delta^{ij} F^k_k, \tag{4.107}$$

is the trace part, while, as usual

$$F^{ij}_A = \frac{1}{2}(F^{ij} - F^{ij}), \tag{4.108}$$

is the anti-symmetric component. Let us show that the components $\tilde{F}_S^{ij}$ and D^{ij} of all the type $(2, 0)$ tensors span two invariant vector spaces with respect to $O(n)$. We need first to show that the $O(n)$-transformed of $\tilde{F}_S^{ij}$ is still symmetric traceless:

$$\tilde{F}_S^{\prime\, ij} = R^i{}_k R^j{}_\ell \tilde{F}_S^{k\ell} \implies \delta_{ij} \tilde{F}_S^{\prime\, ij} = \delta_{ij} R^i{}_k R^j{}_\ell \tilde{F}_S^{k\ell} = \delta_{k\ell} \tilde{F}_S^{k\ell} = 0. \qquad (4.109)$$

Finally the trace part $D^{ij} = \frac{1}{n} \delta^{ij} \delta_{k\ell} F^{k\ell}$ is also invariant being δ^{ij} $O(n)$-invariant.

4.5.1 Lie Algebra of the SO(3) Group

Let us consider some properties of the rotation group $SO(3)$. This group has *dimension* 3, which means that the most general rotation in the three dimensional Euclidean space is parametrized by three angles, such as for instance the Euler angles defining the relative position of two Cartesian systems of orthogonal axes:

$$\mathbf{R} = \mathbf{R}(\boldsymbol{\theta}) \equiv \mathbf{R}(\theta^1, \theta^2, \theta^3) \qquad \boldsymbol{\theta} \equiv (\theta^i). \qquad (4.110)$$

The Euler angles are often denoted by (θ, ϕ, ψ) and correspond to describing a generic rotation as a sequence of three elementary ones: A first rotation about the Z axis by an angle θ, followed by a rotation about the new Y axis by an angle ϕ, and a final rotation about the new Z axis by an angle ψ. The entries of the rotation matrix $\mathbf{R}(\boldsymbol{\theta})$ are continuous functions of the three angles.

In general the dependence of the group elements on their parameters θ^i is *continuous* and the parameters are chosen so that

$$\mathbf{R}(\theta^i \equiv 0) = \mathbf{1}. \qquad (4.111)$$

We also know that the product of two rotations is still a rotation and one can verify that the parameters defining the resulting rotation are analytic functions of those defining the first two:

$$\mathbf{R}(\boldsymbol{\theta}_1) \cdot \mathbf{R}(\boldsymbol{\theta}_2) = \mathbf{R}(\boldsymbol{\theta}_3), \qquad (4.112)$$

where $\theta_3^i = \theta_3^i(\boldsymbol{\theta}_1, \boldsymbol{\theta}_2)$ are analytic functions. In general a group of continuous transformations satisfying the above properties is called a *Lie group*.

Since rotation matrices are continuous functions of angles, we can consider rotations which are infinitely close to the identity element. These transformations, called *infinitesimal* rotations, are defined by very small (i.e. infinitesimal) angles θ^i. We can expand the entries of an infinitesimal rotation matrix $\mathbf{R}(\theta_1, \theta_2, \theta_3)$ in Taylor series with respect to its parameters and write, to first order in the angles:

$$\mathbf{R}(\theta_1, \theta_2, \theta_3) = \mathbf{1} + \left. \frac{\partial \mathbf{R}}{\partial \theta^i} \right|_{\theta^i = 0} \theta^i + O(|\boldsymbol{\theta}|^2). \qquad (4.113)$$

Introducing the matrices $\mathbf{L}_i \equiv \frac{\partial R}{\partial \theta^i} |_{\theta^i=0}$, called *infinitesimal generators* of rotations, the above expansion, to first order, reads:

$$\mathbf{R}(\theta) = 1 + \theta^i \mathbf{L}_i + O(|\boldsymbol{\theta}|^2) \simeq 1 + \theta^i \mathbf{L}_i. \tag{4.114}$$

Let us consider, as an example, a rotation about the X axis, described by the matrix $\mathbf{R}_x$ in (4.93), by an angle θ and let us expand it, for small θ, up to fist order in the angle:

$$\mathbf{R}_x = \begin{pmatrix} 1 & 0 & 0 \\ 0 & \cos\theta & \sin\theta \\ 0 & -\sin\theta & \cos\theta \end{pmatrix} = \begin{pmatrix} 1 & 0 & 0 \\ 0 & 1 & 0 \\ 0 & 0 & 1 \end{pmatrix} + \theta \begin{pmatrix} 0 & 0 & 0 \\ 0 & 0 & 1 \\ 0 & -1 & 0 \end{pmatrix} + O(\theta^2)$$

$$\simeq 1 + \theta \mathbf{L}_1. \tag{4.115}$$

From this equation we can read the expression of the first infinitesimal generator $\mathbf{L}_1$, associated with rotations about the X axis:

$$\mathbf{L}_1 = \begin{pmatrix} 0 & 0 & 0 \\ 0 & 0 & 1 \\ 0 & -1 & 0 \end{pmatrix}. \tag{4.116}$$

Similarly, expanding infinitesimal rotation matrices about the Y and Z axes we find:

$$\mathbf{R}_y(\theta) \equiv \begin{pmatrix} \cos\theta & 0 & -\sin\theta \\ 0 & 1 & 0 \\ \sin\theta & 0 & \cos\theta \end{pmatrix} \simeq 1 + \theta \mathbf{L}_2, \tag{4.117}$$

$$\mathbf{R}_z(\theta) = \begin{pmatrix} \cos\theta & \sin\theta & 0 \\ -\sin\theta & \cos\theta & 0 \\ 0 & 0 & 1 \end{pmatrix} \simeq 1 + \theta \mathbf{L}_3, \tag{4.118}$$

from which we can derive the corresponding infinitesimal generators:

$$\mathbf{L}_2 = \begin{pmatrix} 0 & 0 & -1 \\ 0 & 0 & 0 \\ 1 & 0 & 0 \end{pmatrix}, \quad \mathbf{L}_3 = \begin{pmatrix} 0 & 1 & 0 \\ -1 & 0 & 0 \\ 0 & 0 & 0 \end{pmatrix}. \tag{4.119}$$

In a more compact notation we may write the three matrices $\mathbf{L}_i$ as follows[11]:

$$(\mathbf{L}_i)^j{}_k = \epsilon_{ijk}. \tag{4.120}$$

---

[11] Recall that the orthogonal group makes no difference between upper and lower indices.

Since a generic rotation $\mathbf{R}(\theta^i)$ can be written as a sequence of consecutive rotations about the three axes:

$$\mathbf{R}(\theta^1, \theta^2, \theta^3) \equiv \mathbf{R}_z(\theta^3)\, \mathbf{R}_y(\theta^2)\, \mathbf{R}_x(\theta^1), \qquad (4.121)$$

expanding the right hand side for small θ^i, up to the first order, we find

$$\mathbf{R}(\theta^1, \theta^2, \theta^3) \equiv \mathbf{1} + \theta^1\, \mathbf{L}_1 + \theta^2\, \mathbf{L}_2 + \theta^3\, \mathbf{L}_3, \qquad (4.122)$$

that is the infinitesimal generator of a generic rotation is expressed as a linear combination (whose parameters are the rotation angles) of the three matrices $\mathbf{L}_i$ given in Eqs. (4.116) and (4.119). In other words, any linear combination of infinitesimal generators is itself an infinitesimal generator, that is *infinitesimal generators span a linear vector space*, of which the matrices $(\mathbf{L}_i)$ define a basis. From Eq. (4.120) it follows that the effect of an infinitesimal rotation $\mathbf{R}(\delta\theta)$, by infinitesimal angles $\delta\theta^i \approx 0$, can be described in terms of the following displacement of the coordinates:

$$x'^{\,i} = x^i - \epsilon_{ijk}\, \delta\theta^j\, x^k \quad \Leftrightarrow \quad \mathbf{r}' = \mathbf{R}(\delta\theta)\,\mathbf{r} \simeq \mathbf{r} - \delta\theta \times \mathbf{r}, \qquad (4.123)$$

where $\times$ denotes the external product between two vectors: $\delta\theta \times \mathbf{r} \equiv (\epsilon_{ijk}\, \delta\theta^j\, x^k)$. The reader can easily verify the following *commutation* relation between the infinitesimal generators:

$$\left[\mathbf{L}_i, \mathbf{L}_j\right] \equiv \mathbf{L}_i\, \mathbf{L}_j - \mathbf{L}_j\, \mathbf{L}_i = C_{ij}{}^k\, \mathbf{L}_k, \qquad (4.124)$$

where

$$C_{ij}{}^k = -\epsilon_{ijk}. \qquad (4.125)$$

In other words the commutator [,] of two infinitesimal generators is still in the same vector space. As a consequence of this, in virtue of the linearity property of the commutator with respect to its two arguments, the commutator of any two matrices in the vector space belongs to the same vector space. The commutator then provides a composition law on the vector space of infinitesimal generators which promotes it to an *algebra*. Equations (4.124) and (4.125) define the *structure* of this algebra and the constant entries of the SO(3)-tensor $C_{ij}{}^k$ are called *structure constants*. From (4.125) it follows that $C_{ij}{}^k$ is a SO(3)-invariant tensor.

Let us now show how, from the explicit form of the infinitesimal generators $\mathbf{L}_i$, we can derive the matrix defining a generic finite rotation. Consider a rotation $\mathbf{R}(\theta)$, parametrized by some finite angles θ^i. We can think of performing it through a sequence of a very large number $N \gg 1$ of infinitesimal rotations $\mathbf{R}(\delta\theta)$ by angles $\delta\theta^i \equiv \frac{\theta^i}{N} \ll 1$. For large N, each infinitesimal rotation reads: $\mathbf{R}(\delta\theta) \approx \mathbf{1} + \delta\theta^i\, \mathbf{L}_i = \mathbf{1} + \frac{1}{N}\, \theta^i\, \mathbf{L}_i$. The finite rotation will therefore be approximated as follows:

$$\mathbf{R}(\theta^i) \approx [\mathbf{R}(\delta\boldsymbol{\theta})]^N = \left(1 + \frac{1}{N}\theta^i\,\mathbf{L}_i\right)^N, \quad N \gg 1. \tag{4.126}$$

Intuitively, the larger N the better the above approximation is. Therefore we expect, in the limit $N \to \infty$, to obtain an exact representation of the finite rotation:

$$\mathbf{R}(\boldsymbol{\theta}) = \lim_{N\to\infty} \left(1 + \frac{1}{N}\theta^i\,\mathbf{L}_i\right)^N. \tag{4.127}$$

Recalling that, if x is a number, we can express its exponential e^x as the limit $e^x = \lim_{N\to\infty}\left(1 + \frac{x}{N}\right)^N$, in a similar way it can be shown that the limit on the right hand side of (4.127) is the *exponential of the matrix* $\theta^i\,\mathbf{L}_i$:

$$\mathbf{R}(\boldsymbol{\theta}) = \exp(\theta^i\,\mathbf{L}_i), \tag{4.128}$$

where the exponential of a matrix $\mathbf{A}$ is defined by the same infinite series defining the exponential of a number:

$$\exp(\mathbf{A}) \equiv \sum_{n=0}^{\infty} \frac{1}{n!}\,(\mathbf{A})^n. \tag{4.129}$$

Therefore, knowing the infinitesimal generators of the rotation group (and, as we shall see in Chap. 7, the same is true for any Lie group), we can express any rotation as the exponential of an element of the infinitesimal generator algebra:

$$\mathbf{R}(\boldsymbol{\theta}) = e^{\theta^i\,\mathbf{L}_i}. \tag{4.130}$$

Obviously the determinant of the rotation matrix $\mathbf{R}(\theta^i)$, being a continuous function of its entries, will be a continuous function of the three angles as well. Since orthogonal matrices can only have determinant ± 1, and the matrix in (4.130) at $\theta^i \equiv 0$, has determinant $+1$, in virtue of its continuity, the value of $\det(\mathbf{R}(\boldsymbol{\theta}))$ cannot jump to -1 for some values of the angles. We conclude that the exponential in (4.130) has determinant $+1$ and thus that only rotations can be expressed as exponentials. Therefore transformations in O(3) involving also reflections, which have determinant -1, cannot be written in that form. As opposed to rotations, we will say that these transformations of O(3) are not in the *neighborhood* of the origin in which the exponential representation holds. We can however write a generic orthogonal matrix with determinant -1 as the product of a rotation times a given reflection $\mathbf{O}$, e.g. $\mathbf{O} = \mathrm{diag}(-1, 1, 1)$.

From the physical point of view the infinitesimal generators of rotations have an important meaning in quantum mechanics. Let us define the following matrices:

$$\mathbf{M}_i = -i\,\hbar\,\mathbf{L}_i, \tag{4.131}$$

which, in virtue of Eqs. (4.124) and (4.125), satisfy the following commutation relations:

$$[\mathbf{M}_i, \mathbf{M}_j] = i\hbar\,\epsilon_{ijk}\,\mathbf{M}_k. \tag{4.132}$$

These are the commutation relations between the components of the *angular momentum* operator in quantum mechanics. Aside from the new normalization, Eq. (4.131) expresses the fact that the angular momentum components can be identified with the infinitesimal generators of the rotation group SO(3). Similarly, when dealing with symmetries in Hamiltonian (classical) mechanics, we will learn that we can associate with any continuous symmetry transformation of the Hamiltonian, a conserved quantity on the phase space. This quantity will be identified with the infinitesimal generator of such transformations. In particular invariance under rotations will imply the conservation of the corresponding infinitesimal generators, which we shall show to be the components of angular momentum (see Chap. 8).

Let us observe that the infinitesimal generators $\mathbf{L}_i$ are represented by antisymmetric matrices, as it is apparent from Eqs. (4.116) and (4.119). This is not accidental, but follows from the defining property of the rotation group. Consider an infinitesimal rotation $\mathbf{R}(\delta\theta^i) = \mathbf{1} + \delta\theta^i\,\mathbf{L}_i \in$ SO(3), $\delta\theta^i \approx 0$. Let us write for $\mathbf{R}(\delta\theta)$ the orthogonality condition:

$$\mathbf{1} = \mathbf{R}^T(\delta\theta)^T\mathbf{R}(\delta\theta) = (\mathbf{1} + \delta\theta^i\,\mathbf{L}_i^T)\,(\mathbf{1} + \delta\theta^j\,\mathbf{L}_j) = \mathbf{1} + \delta\theta^i\,(\mathbf{L}_i^T + \mathbf{L}_i),$$

where we have neglected orders in $\delta\theta^i$ higher than the first. Form the above condition it then follows that:

$$\mathbf{L}_i = -\mathbf{L}_i^T. \tag{4.133}$$

that is *the infinitesimal generators of rotations, with respect to an ortho-normal basis, are represented by anti-symmetric matrices.*

Let us end this section by giving the explicit form of a generic rotation in terms of the Euler angles:

$$\mathbf{R}(\theta, \phi, \psi) = e^{\theta\,\mathbf{L}_3}e^{\phi\,\mathbf{L}_2}e^{\psi\,\mathbf{L}_3}. \tag{4.134}$$

To construct the infinitesimal generators we have used the parametrization of a rotation in terms of $\theta_1, \theta_2, \theta_3$ and not the Euler angles. This is due to the fact that the latter define a parametrization which is singular at the origin where infinitesimal generators are defined, while this is not the case for the parametrization we used. If we indeed expand the matrix (4.134) for infinitesimal Euler angles, we do not find the complete basis of generators.

4.6 Principle of Relativity and Covariance of Physical Laws

The tensorial formalism is particularly convenient since it allows to easily tell whether physical laws are written in a form which does not depend on the frame of reference we use, namely if the principle of relativity holds for the group of transformations with respect to which the tensor quantities are defined. Indeed consider a group G of Cartesian coordinate transformations (like the rotation group), subgroup of the affine one.

If an equation is expressed as an equality between two tensors of the same type with respect to G, if it holds in a RF, it will hold in any other RF related to it by a transformation of the group G.

To prove this property, let us consider an equation which holds in a basis defining a certain RF and which is written as an equality between tensors of the same type, with respect to G:

$$T^{i_1 \cdots i_k}{}_{j_i \cdots j_p} = U^{i_1 \cdots i_k}{}_{j_1 \cdots j_p}, \tag{4.135}$$

and define $A^{i_1 \cdots i_k}{}_{j_i \cdots j_p} = T^{i_1 \cdots i_k}{}_{j_i \cdots j_p} - U^{i_1 \cdots i_k}{}_{j_i \cdots j_p}$. In the original RF Eq. (4.135) can also be written as follows:

$$A^{i_1 \cdots i_k}{}_{j_i \cdots j_p} = 0. \tag{4.136}$$

In a new RF obtained from the original one by means of G-transformation, using Eq. (4.60), we will have a new tensor A', related to A as follows:

$$A'^{i_1 \cdots i_k}{}_{j_i \cdots j_p} = D^{i_1}{}_{n_1} \cdots D^{i_k}{}_{n_k} D^{-1 m_1}{}_{j_1} \cdots D^{-1 m_p}{}_{j_p} A^{n_1 \cdots n_k}{}_{m_1 \cdots m_p}. \tag{4.137}$$

$A'^{i_1 \cdots i_k}{}_{j_i \cdots j_p}$, the same equation as seen in a G-related reference frame, is still vanishing due to Eq. (4.136); Indeed the action of the tensor (or Kronecker) product of D-matrices on the right hand side is an invertible transformation, being the D-matrices themselves invertible by assumption. This can be proven using the following general property of the Kronecker product of matrices: $(\mathbf{A} \otimes \mathbf{B}) \cdot (\mathbf{C} \otimes \mathbf{D}) = \mathbf{AC} \otimes \mathbf{BD}$, which can be generalized to an n-fold tensor product. As a consequence of this, the identity transformation on a tensor is the tensor product of identity matrices acting on each index and, moreover, the inverse of a tensor product of invertible matrices exists and is the tensor product of the inverses of each factor: If A and B are invertible, $(\mathbf{A} \otimes \mathbf{B})^{-1} = \mathbf{A}^{-1} \otimes \mathbf{B}^{-1}$.

The physical law expressed by Eq. (4.135), will then hold also in the new RF, obtained from the original one through a G-transformation. We say that this equation is *manifestly covariant* with respect to the transformation group G.

As an example let us show that the fundamental law of dynamics does not depend on the orientation of the Cartesian orthogonal axes of the chosen RF, namely that it is covariant with respect to O(3). Let us consider the simple case of Newton's second law in the presence of conservative forces $\mathbf{F} = m\,\mathbf{a}$. Being the force conservative, its contravariant components F^i are expressed in terms of the gradient of a potential energy U, which is a function of the position of the point particle. Being the components of a gradient covariant, this relation should involve the metric tensor: $F^i = -g^{ij}\,\partial_j U$, where $g^{ij} = \delta^{ij}$. Also the acceleration is described by contravariant components (a^i), since it is expressed as the second derivative with respect to time of the position vector $\mathbf{r}$, which is described by contravariant components x^i: $a^i \equiv \frac{d^2 x^i}{dt^2}$. Newton's second law is then written as an equality between two type $(1, 0)$ $O(3)$-tensors (contravariant vectors):

$$m\,a^i = F^i = -g^{ij}\,\partial_j U. \tag{4.138}$$

If we act on the original RF by a transformation in $O(3)$ (rotations and reflections), we find:

$$F'^i - m\frac{d^2 x'^i}{dt^2} = R^i{}_j \left(F^j - m\frac{d^2 x^j}{dt^2} \right) = 0, \tag{4.139}$$

which shows that the fundamental law of dynamics is covariant with respect to $O(3)$, namely with respect to rotations and reflections of the RF.

When we shall consider four-dimensional space-time instead of the three-dimensional Euclidean space, among all the possible transformations on a RF, of particular interest are the Lorentz transformations, on which Einstein's principle of relativity is based. We shall show, at the end of this chapter, that Lorentz transformations close a group, *the Lorentz group*. If we also include space-time translations, this group enlarges to the *Poincaré group. If physical laws are expressed as an equality between tensors of the same type with respect to the Lorentz group, we will be guaranteed that the principle of relativity holds.*

4.7 Minkowski Space-Time and Lorentz Transformations

In discussing special relativity, we have seen that space-time can be regarded as a *four dimensional space M^4* whose points are described by a set of four Cartesian coordinates

$$(x^\mu) = (x^0, x^1, x^2, x^3), \quad \mu = 0, 1, 2, 3, \tag{4.140}$$

three of which $(x^i) = (x^1, x^2, x^3) = (x, y, z)$ are spatial coordinates of our Euclidean space E_3, and one $x^0 = c\,t$ is related to time. A point on M^4 describes an event taking place at the point (x, y, z), at the time t. Just as for the Euclidean space, we can define vectors connecting couples of points in M^4, like the infinitesimal displacement vector connecting two infinitely close events:

$$dx \equiv (dx^\mu) = (dx^0, dx^1, dx^2, dx^3). \tag{4.141}$$

These vectors span a four-dimensional linear vector space on which a symmetric scalar product is defined by means of the metric $g_{\mu\nu} = \eta_{\mu\nu}$, where[12]:

$$
\eta_{\mu\nu} = \begin{pmatrix} 1 & 0 & 0 & 0 \\ 0 & -1 & 0 & 0 \\ 0 & 0 & -1 & 0 \\ 0 & 0 & 0 & -1 \end{pmatrix}. \tag{4.142}
$$

Given two 4-vectors $P \equiv (P^\mu)$ and $Q \equiv (Q^\mu)$, their scalar product reads:

$$
P \cdot Q = P^\mu \, \eta_{\mu\nu} \, Q^\nu = P^0 Q^0 - \sum_{i=1}^{3} P^i Q^i. \tag{4.143}
$$

This scalar product, in contrast to the one defined on the Euclidean space, is not positive definite, namely does not satisfy property c) of (4.7), since the corresponding metric has one positive and three negative diagonal entries (indefinite or Minkowskian signature). As a consequence of this the squared *norm* of a 4-vector $P \equiv (P^\mu)$, defined using this scalar product:

$$
\|P\|^2 \equiv P \cdot P \equiv P^\mu \, \eta_{\mu\nu} \, P^\nu = (P^0)^2 - \sum_{i-1}^{3}(P^i)^2, \tag{4.144}
$$

can vanish even if P is not zero. In particular a non-vanishing 4-vector can have positive, zero or negative squared norm, in which cases we talk about a *time-like*, *null* or *space-like* 4-vector, respectively. We can take, as 4-vector, the displacement vector dx, whose squared norm measures the squared *space-time* distance ds^2 between two infinitely close events:

$$
ds^2 = \|dx\|^2 = dx^\mu \, \eta_{\mu\nu} \, dx^\nu = (dx^0)^2 - (dx^1)^2 - (dx^2)^2 - (dx^3)^3. \tag{4.145}
$$

As pointed out when discussing about relativity, the distance ds in (4.145) should be interpreted as the infinitesimal proper-time interval times (square of) the velocity of light: $ds = c\, d\tau$. A four-dimensional space on which the metric (4.142) is defined, is called *Minkowski space* (or better space-time).

Let us now consider linear coordinate transformations $x^\mu \to x'^\mu = x'^\mu(x)$ which do not affect the position of the origin of the coordinate system (i.e. the origins of the two Euclidean coordinate systems $O(x = 0, y = 0, z = 0)$ and $O'(x' = 0, y' = 0, z' = 0)$ coincide at some common initial instant $t = t' = 0$). Such transformations

---

[12]Some authors alternatively define the Lorentzian metric η as diag$(-1, +1, +1, +1)$. This notation is common in the general relativity literature and has the advantage of yielding the Euclidean metric when restricted to the spatial directions 1, 2, 3.

are defined by homogeneous relations between old and new coordinates

$$x'^{\mu} = \Lambda^{\mu}{}_{\nu} x^{\nu} \quad \Rightarrow \quad dx'^{\mu} = \Lambda^{\mu}{}_{\nu} dx^{\nu}. \tag{4.146}$$

Just as we defined orthogonal transformations in Euclidean space, we can consider homogeneous transformations (4.146) which leave the distance ds, in (4.145), between two events, as a function of their coordinates, invariant (*invariance of ds*). This condition defines the *Lorentz transformations*, which are thus implemented by a 4×4 invertible matrix $\Lambda = (\Lambda^{\mu}{}_{\nu})$. We can alternatively characterize a Lorentz transformation by requiring that its action on two generic 4-vectors $P \equiv (P^{\mu})$ and $Q \equiv (Q^{\mu})$, which transform as dx^{μ} in (4.146), namely

$$P^{\mu} \rightarrow P'^{\mu} = \Lambda^{\mu}{}_{\nu} P^{\nu},$$
$$Q^{\mu} \rightarrow Q'^{\mu} = \Lambda^{\mu}{}_{\nu} Q^{\nu}, \tag{4.147}$$

leaves their scalar product $P \cdot Q$ invariant:

$$P'^{\mu} \eta_{\mu\nu} Q'^{\nu} = P^{\mu} \eta_{\mu\nu} Q^{\nu}. \tag{4.148}$$

Substituting the expressions in (4.147) into the above equation, and requiring the equality to hold for any choice of the two 4-vectors, we derive the following general condition defining the matrix Λ

$$\Lambda^{\rho}{}_{\mu} \Lambda^{\sigma}{}_{\nu} \eta_{\rho\sigma} = \eta_{\mu\nu}, \tag{4.149}$$

or, in matrix notation, setting $\eta \equiv (\eta_{\mu\nu})$:

$$\Lambda^T \eta \Lambda = \eta. \tag{4.150}$$

Lorentz transformations are thus the linear homogeneous coordinate transformations which leave the metric $\eta_{\mu\nu}$ invariant. Physically they represent the most general coordinate transformation relating two inertial frames of reference, whose four-dimensional origins $x^{\mu} = 0$ and $x'^{\mu} = 0$ coincide. Comparing Eq. (4.149) with Eq. (4.88) we see that Lorentz transformations play in Minkowski the role that orthogonal transformation have in Euclidean space. The reader can easily verify that the set of all matrices Λ, solution to Eq. (4.149), i.e. the Lorentz transformations, satisfy axioms (*i*), (*ii*), (*iii*), (*iv*) of Sect. 4.4 which define a group structure. Lorentz transformations therefore form a group called *the Lorentz group*. The elements of this group depend on a set of continuous parameters, which are the entries $\Lambda^{\mu}{}_{\nu}$ of the matrix Λ, subject to the condition (4.149). The Lorentz group is therefore another example of continuous groups, together with the rotation group, which we have characterized in the previous sections as Lie groups. The identity transformation $\mathbf{1} \equiv (\delta^{\mu}_{\nu})$ is in particular a Lorentz transformation corresponding to a particular choice of the continuous parameters: $\Lambda^{\mu}{}_{\nu} = \delta^{\mu}_{\nu}$.

Consider now the component $\mu = \nu = 0$ of Eq. (4.149):

$$(\Lambda^0{}_0)^2 - (\Lambda^i{}_0)^2 = 1 \quad \Rightarrow \quad (\Lambda^0{}_0)^2 \geq 1, \tag{4.151}$$

the above property implies that we can either have $\Lambda^0{}_0 \geq 1$ or $\Lambda^0{}_0 \leq -1$. Moreover, from Eq. (4.149), it also follows that

$$\det(\Lambda)^2 = 1 \quad \Rightarrow \quad \det(\Lambda) = \pm 1. \tag{4.152}$$

Lorentz transformations are then divided in the following four classes:

(i) $\Lambda^0{}_0 \geq 1, \quad \det(\Lambda) = 1$ (proper transformations);
(ii) $\Lambda^0{}_0 \geq 1, \quad \det(\Lambda) = -1$;
(iii) $\Lambda^0{}_0 \leq -1, \quad \det(\Lambda) = -1$;
(iv) $\Lambda^0{}_0 \leq -1, \quad \det(\Lambda) = +1$.

Lorentz transformations in the first class are called *proper* and, as the reader can easily verify, close a group. An example of a Lorentz transformation of the second kind is the *parity* P, which is implemented by the matrix $\Lambda_P = \eta = \mathrm{diag}(+1, -1, -1, -1)$. Its effect is to reverse the orientation of the three Cartesian axes X, Y, Z:

$$x^\mu \xrightarrow{\;P\;} x'^\mu = \Lambda_P{}^\mu{}_\nu x^\nu \quad \Leftrightarrow \quad \begin{cases} t \longrightarrow t' = t \\ \mathbf{x} \longrightarrow \mathbf{x}' = -\mathbf{x} \end{cases}. \tag{4.153}$$

A transformation of the kind (*iii*) is the *time reversal* T, which consists in reversing the orientation of time while leaving the space-coordinates inert. It is implemented by the matrix $\Lambda_T = -\eta = \mathrm{diag}(-1, +1, +1, +1)$

$$x^\mu \xrightarrow{\;T\;} x'^\mu = \Lambda_T{}^\mu{}_\nu x^\nu \quad \Leftrightarrow \quad \begin{cases} t \longrightarrow t' = -t \\ \mathbf{x} \longrightarrow \mathbf{x}' = \mathbf{x} \end{cases}. \tag{4.154}$$

Finally a representative of last class is the product of the parity and time reversal transformations, implemented by the matrix $\Lambda_P \Lambda_T = -\mathbf{1}$. Its effect is to reverse the orientation of the space and time Cartesian axes in Minkowski space-time.

Transformations with $\Lambda^0{}_0 \geq 1$ are called *orthochronous* since they do not involve time reversal. Let us now prove an important property of Lorentz transformations:

Orthochronous transformations leave the sign of the time-component of time-like (or in general non-space-like) four-vectors invariant, while non-orthochronous ones reverse it.

To prove it let us consider a non-space-like four vector $P \equiv (P^\mu) = (P^0, \mathbf{P})$:

$$\|P\|^2 \geq 0 \quad \Leftrightarrow \quad \frac{|\mathbf{P}|}{|P^0|} \leq 1. \tag{4.155}$$

Let $\boldsymbol{\Lambda}$ be a Lorentz transformation which maps P^μ into $P'^\mu = \Lambda^\mu{}_\nu P^\nu$. The time-component of the transformed vector reads:

$$P'^0 = \Lambda^0{}_0 P^0 + \Lambda^0{}_i P^i. \tag{4.156}$$

The second term on the right hand side has the form of a scalar product $\boldsymbol{\Lambda}^0 \cdot \mathbf{P}$ between the vectors $\boldsymbol{\Lambda}^0 \equiv (\Lambda^0{}_i)$ and $\mathbf{P}$, which can be written as the product of their norms times the cosine of the angle between them: $\Lambda^0{}_i P^i = |\boldsymbol{\Lambda}^0||\mathbf{P}|\cos(\theta)$. Dividing both sides of (4.156) by P^0, we find:

$$\frac{P'^0}{P^0} = \Lambda^0{}_0 \left(1 + \frac{|\boldsymbol{\Lambda}^0|}{\Lambda^0{}_0} \frac{|\mathbf{P}|}{P^0} \cos(\theta) \right) = \Lambda^0{}_0 \left(1 + A \right). \tag{4.157}$$

From (4.151) we find that $|\boldsymbol{\Lambda}^0| = \sqrt{(\Lambda^0{}_0)^2 - 1} < |\Lambda^0{}_0|$. This property and Eq. (4.155) imply that the constant A in (4.157) is, in modulus, smaller than one: $|A| < 1$, so that $1 + A$ is positive and P'^0/P^0 has the same sign as $\Lambda^0{}_0$. This proves the property stated above, namely that P'^0 and P^0 have the same sign if, and only if, the transformation is orthochronous ($\Lambda^0{}_0 \geq 1$).

Let us now consider the product of two Lorentz transformations $\boldsymbol{\Lambda}_3 = \boldsymbol{\Lambda}_1 \boldsymbol{\Lambda}_2$ and, in particular, the four-vector defined by the components $\Lambda_3{}^\mu{}_0 = \Lambda_1{}^\mu{}_\nu \Lambda_2{}^\nu{}_0$. It is expressed as the transformed through $\boldsymbol{\Lambda}_1$ of the four vector $\Lambda_2{}^\nu{}_0$. Both $\Lambda_3{}^\mu{}_0$ and $\Lambda_2{}^\mu{}_0$ are time-like since, by virtue of Eq. (4.151), $\|\Lambda_2{}^\nu{}_0\|^2 = \|\Lambda_3{}^\nu{}_0\|^2 = 1$. As a consequence of the previously proven property:

$$\text{sign}(\Lambda_3{}^0{}_0) = \text{sign}(\Lambda_1{}^0{}_0)\,\text{sign}(\Lambda_2{}^0{}_0), \tag{4.158}$$

namely the product of two orthochronous or two non-orthochronous transformations is orthochronous, while the product of an orthochronous transformation and a non-orthochronous one is non-orthochronous. Since the product of two Lorentz transformations in each class always has unit determinant, we conclude that *the product of any two transformations in each of the above four classes is a proper Lorentz transformation.* Consider now the inverse $\boldsymbol{\Lambda}^{-1}$ of a Lorentz transformation $\boldsymbol{\Lambda}$. Since $\boldsymbol{\Lambda}^{-1}\boldsymbol{\Lambda} = \mathbf{1}$, and $\mathbf{1}$ is orthochronous, $\boldsymbol{\Lambda}^{-1}$ is orthochronous if and only if $\boldsymbol{\Lambda}$ is. This implies that any two representatives of each of the above classes are connected, through the right (or left) multiplication, by a proper Lorentz transformation. Consider indeed two transformations $\boldsymbol{\Lambda}_1$, $\boldsymbol{\Lambda}_2$ within a same class. From our previous discussion it follows that $\boldsymbol{\Lambda} = \boldsymbol{\Lambda}_1 \boldsymbol{\Lambda}_2^{-1}$ is a proper Lorentz transformation such that $\boldsymbol{\Lambda}_1 = \boldsymbol{\Lambda} \boldsymbol{\Lambda}_2$. We conclude that any representative of the classes (*ii*), (*iii*), (*iv*) can be written as the product of a proper Lorentz transformation times $\boldsymbol{\Lambda}_P$, $\boldsymbol{\Lambda}_T$, $\boldsymbol{\Lambda}_P \boldsymbol{\Lambda}_T$, respectively.

We shall be mainly interested in those transformations $\Lambda^\mu{}_\nu$ which are continuously connected to the identity transformation $\mathbf{1}$. Since $\delta_0^0 = 1$ and $\det(\mathbf{1}) = 1$, these transformations, by continuity, should be the proper Lorentz transformations. They close a group denoted by SO(1, 3), which differs from the group SO(4) of rotations in a 4-dimensional Euclidean space in that the corresponding invariant metric, instead of being $\delta_{\mu\nu}$ with diagonal entries $(+1, +1, +1, +1)$, is the matrix $\eta_{\mu\nu}$ defined in (4.142), with diagonal entries $(+1, -1, -1, -1)$. The argument $(1, 3)$ in the symbol of the group refers then to the signature of the corresponding invariant metric.[13]

Just as we did for the rotation group and the general linear group, we define vectors (V^μ) which are *contravariant* and vectors (V_μ) which are *covariant* with respect to the Lorentz group as quantities transforming as (dx^μ) and as the gradient $(\frac{\partial}{\partial x^\mu} f(x))$ of a function $f(x)$, respectively:

$$V^\mu \to V'^\mu = \Lambda^\mu{}_\nu V^\nu, \tag{4.159}$$

$$V_\mu \to V'_\mu = \Lambda^{-1\nu}{}_\mu V_\nu. \tag{4.160}$$

Using the metric tensor $g_{\mu\nu} = \eta_{\mu\nu}$, we can *raise* or *lower* indices, that is we can map a covariant into a contravariant vector and vice-versa, as we have seen in the more general case

$$V^\mu \to V_\mu = \eta_{\mu\nu} V^\nu,$$

$$V_\mu \to V^\mu = \eta^{\mu\nu} V_\nu,$$

where we have used $\eta^{\mu\nu} \eta_{\nu\sigma} = \delta^\mu_\sigma$. This is proven in the same way as in the general case, this time using Eq. (4.149). Notice that, since $\eta_{00} = 1$, raising or lowering a time component will not alter its sign, while, being $\eta_{ij} = -\delta_{ij}$, the same operation will invert the sign of the spatial components.

All the general properties that we have learned for tensors with respect to GL(4), clearly apply to SO(1, 3)-tensors as well. For instance we can define a Lorentz tensor of type (p, q), that is with p contravariant and q covariant indices

$$T^{\mu_1 \ldots \mu_p}{}_{\nu_1 \ldots \nu_q}, \tag{4.161}$$

as a quantity transforming under Λ as follows:

$$T'^{\mu_1 \ldots \mu_p}{}_{\nu_1 \ldots \nu_q} = \Lambda^{\mu_1}{}_{\rho_1} \ldots \Lambda^{\mu_q}{}_{\rho_p} \Lambda^{-1\sigma_1}{}_{\nu_1} \ldots \Lambda^{-1\sigma_q}{}_{\nu_q} T^{\rho_1 \ldots \rho_p}{}_{\sigma_1 \ldots \sigma_q}. \tag{4.162}$$

Tensors of the same type (p, q) form a linear vector space and the collection of all possible tensors form an algebra with respect to the tensor product operation and contraction.

Anticipating some concepts which will be introduced and discussed in Chap. 7, tensors of a given type (p, q) *form a basis of a representation of the Lorentz group*, on which the group action is defined by (4.162). Such property means that the effect on a type (p, q) tensor of two consecutive Lorentz transformations Λ_1, Λ_2 is the transformation induced by the product of the two $\Lambda_2 \Lambda_1$. This follows from the

---

[13]If the invariant metric were diagonal with entries $(+1, +1, -1, -1)$, the corresponding group would have been SO(2, 2).

definition of the Kronecker product of matrices and in particular from the property $(\mathbf{A} \otimes \mathbf{B}) \cdot (\mathbf{C} \otimes \mathbf{D}) = (\mathbf{A} \, \mathbf{C}) \otimes (\mathbf{B} \, \mathbf{D})$, see Sect. 4.6. This representation is in general *reducible*, that is the vector space spanned by type (p, q) tensors may decompose into the direct sum of orthogonal subspaces each of which are stable under the action of the Lorentz group, and therefore define themselves bases of representations of the group. As an example let us consider a Lorentz tensor with two contravariant indices $F^{\mu\nu}$, transforming according to (4.162). Similarly to what happened in the case of the rotation group, see Eqs. (4.104) and (4.109), we can decompose this tensor into three components which transform into themselves under the action of SO(1, 3). Let us define the trace operation:

$$F^{\rho}_{\rho} \equiv \eta_{\mu\nu} \, F^{\mu\nu}, \qquad (4.163)$$

and decompose $F^{\mu\nu}$ as follows

$$F^{\mu\nu} = \left(\tilde{F}^{\mu\nu}_{S} + D^{\mu\nu} \right) + F^{\mu\nu}_{A}. \qquad (4.164)$$

The first term within brackets denotes the symmetric traceless component of $F^{\mu\nu}$:

$$\tilde{F}^{\mu\nu}_{S} = \frac{1}{2}(F^{\mu\nu} + F^{\mu\nu}) - \frac{1}{4} \eta^{\mu\nu} \, F^{\rho}_{\rho}, \quad \tilde{F}^{\mu\nu}_{S} \, \eta_{\mu\nu} = 0.$$

The second term within brackets in (4.164) represents the trace part:

$$D^{\mu\nu} = \frac{1}{4}\eta^{\mu\nu} \, F^{\rho}_{\rho},$$

and, finally,

$$F^{\mu\nu}_{A} = \frac{1}{2}(F^{\mu\nu} - F^{\mu\nu}).$$

is the anti-symmetric component. With the above definitions the proof that each of these components, under a Lorentz transformation, is mapped into the corresponding component of the transformed tensor, is the same as the one given for the rotation group. We conclude that antisymmetric, symmetric traceless and the trace each span three orthogonal subspaces of the total space of type (2, 0) tensors, which are stable under the action of the Lorentz group. Since they cannot be further reduced, we say that they define the bases of three *irreducible representations* of SO(1, 3). The same result applies to (0, 2)-tensors as well.

We can now apply the discussion of Sect. 4.6 to the case in which the group G is the Lorentz group and conclude that:

If a physical law is written as an equality between Lorentz tensors of a same type, in a given RF, it will hold in any other RF connected to the original one by a Lorentz transformation.

Since Lorentz transformations are the most general homogeneous transformations relating two inertial RFs. in relative motion, we conclude that *a physical law which can be expressed as an equality between Lorentz tensors of the same type, that is in a manifestly Lorentz covariant way, is consistent with the principle of special relativity.*

The principle of relativity in other words requires all physical laws to be written in the following general form:

$$F^{\mu_1\cdots\mu_p}{}_{\nu_1\ldots\nu_q} = G^{\mu_1\cdots\mu_p}{}_{\nu_1\ldots\nu_q}, \qquad (4.165)$$

where F and G are Lorentz-tensors. As we shall see, this is indeed the case for Maxwell's equations. A tensorial equation of the form (4.165) is said to be *manifestly covariant* under Lorentz transformations.

4.7.1 General Form of (Proper) Lorentz Transformations

In our discussion about special relativity in Chap. 1, we limited ourselves to reference frames with parallel axes and whose relative constant velocity vector was oriented along the common X axis (standard configuration). In Sect. 2.2.1 of Chap. 2, however, we have also given the form of the Lorentz transformation when, keeping the three axes parallel, the velocity has an arbitrary direction. In this subsection we shall construct the most general proper Lorentz transformation through the construction of its infinitesimal generators, just as we did in the case of the rotation group, showing that for parallel axes it coincides with Eq. (2.57) and then generalizing to the case where the axes of the two frames S and S' are rotated with respect to each other.

Let us start considering an infinitesimal Lorentz transformation, i.e. a Lorentz transformation which is infinitely close to the identity $\mathbf{1}$:

$$\Lambda^{\mu}{}_{\nu} \simeq \delta^{\mu}_{\nu} + \omega^{\mu}{}_{\nu}, \qquad (4.166)$$

where $\omega = (\omega^{\mu}{}_{\nu})$ is the infinitesimal generator of the transformation. It has infinitesimal entries, for which we use the first order approximation. Substituting Eq. (4.166) into (4.149) we find

$$\eta_{\mu\sigma}\, \omega^{\sigma}{}_{\nu} = -\eta_{\nu\sigma}\, \omega^{\sigma}{}_{\mu} \quad \Leftrightarrow \quad \eta\,\omega = -\omega^{T}\,\eta, \qquad (4.167)$$

Defining the matrix $\omega_{\mu\nu} = \eta_{\mu\sigma}\, \omega^{\sigma}{}_{\nu}$, Eq. 4.167 implies

$$\omega_{\mu\nu} = -\omega_{\nu\mu}, \qquad (4.168)$$

namely the infinitesimal generator of the most general proper Lorentz transformation, upon lowering one index by means of the metric, is represented by a 4×4 anti-symmetric matrix. An anti-symmetric matrix has $4 \times (4-1)/2 = 6$ independent entries, i.e. all entries above the main diagonal:

$$(\omega_{\mu\nu}) = \begin{pmatrix} 0 & \omega_{01} & \omega_{02} & \omega_{03} \\ -\omega_{01} & 0 & \omega_{12} & \omega_{13} \\ -\omega_{02} & -\omega_{12} & 0 & \omega_{23} \\ -\omega_{03} & -\omega_{13} & -\omega_{23} & 0 \end{pmatrix} = \frac{1}{2} \omega_{\rho\sigma} (L^{\rho\sigma})_{\mu\nu},$$

where $(L^{\rho\sigma})_{\mu\nu} = \delta^{\rho}_{\mu} \delta^{\sigma}_{\nu} - \delta^{\sigma}_{\mu} \delta^{\rho}_{\nu} = -(L^{\sigma\rho})_{\mu\nu}$ is an ortho-normal basis of anti-symmetric matrices labeled by the (anti-symmetric) couple of indices $(\rho\sigma)$, ρ, $\sigma = 0, \ldots, 3$. For notational convenience, we shall denote the entries $\omega_{\rho\sigma}$ in the above equation by $\delta\theta_{\rho\sigma}$ and write $\omega_{\mu\nu} = \frac{1}{2} \delta\theta_{\rho\sigma} (L^{\rho\sigma})_{\mu\nu}$. The generic infinitesimal generator is obtained by raising one index of $\omega_{\mu\nu}$ and has therefore the following form:

$$\boldsymbol{\omega} \equiv (\omega^{\mu}{}_{\nu}) = (\eta^{\mu\sigma} \omega_{\sigma\nu}) = \frac{1}{2} \delta\theta_{\rho\sigma} ((L^{\rho\sigma})^{\mu}{}_{\nu}) = \frac{1}{2} \delta\theta_{\rho\sigma} \mathbf{L}^{\rho\sigma}, \quad (4.169)$$

where the matrices $\mathbf{L}^{\rho\sigma}$ read:

$$\mathbf{L}^{\rho\sigma} = ((L^{\rho\sigma})^{\mu}{}_{\nu}) = \left(\eta^{\mu\delta} (L^{\rho\sigma})_{\delta\nu}\right) = (\eta^{\rho\mu}\delta^{\sigma}_{\nu} - \eta^{\sigma\mu}\delta^{\rho}_{\nu}). \quad (4.170)$$

The matrices $\mathbf{L}^{\rho\sigma}$ play for the Lorentz group the same role that the matrices $\mathbf{L}_i$ had for the rotation group: They form a basis for the six-dimensional vector space spanned by the infinitesimal generators of Lorentz transformations (recall that the infinitesimal generators of the rotation group spanned a three-dimensional vector space of which the matrices $\mathbf{L}_i$ represented a basis). The parameters $\delta\theta_{\rho\sigma} = -\delta\theta_{\sigma\rho}$ (only six of which are independent!) play then the same role of the 3 angles $\delta\theta^i$ in the SO(3) case. A generic Lorentz transformation depends then on six independent continuous parameters, and therefore we say that *the Lorentz group has dimension* 6.

Using Eqs. (4.166) and (4.169), we can write an infinitesimal proper Lorentz transformation as follows:

$$\Lambda^{\mu}{}_{\nu} = \delta^{\mu}_{\nu} + \frac{1}{2} \delta\theta_{\rho\sigma}(L^{\rho\sigma})^{\mu}{}_{\nu} \quad \Leftrightarrow \quad \Lambda = 1 + \frac{1}{2} \delta\theta_{\rho\sigma} \mathbf{L}^{\rho\sigma}, \quad (4.171)$$

for infinitesimal $\delta\theta_{\rho\sigma}$. After some algebra the reader can show that the following commutation relations among the infinitesimal generators hold:

$$\left[\mathbf{L}^{\mu\nu}, \mathbf{L}^{\rho\sigma}\right] = \eta^{\nu\rho} \mathbf{L}^{\mu\sigma} + \eta^{\mu\sigma} \mathbf{L}^{\nu\rho} - \eta^{\mu\rho} \mathbf{L}^{\nu\sigma} - \eta^{\nu\sigma} \mathbf{L}^{\mu\rho}. \quad (4.172)$$

In the sequel, we shall define a more general action of the Lorentz group on objects which are not 4-vectors. In other words we shall consider different *matrix representations* of the same Lorentz group. However the commutation relations (4.172) between its infinitesimal generators, namely the structure constants, will not depend on the particular matrix representation considered. For this reason they will characterize the properties of the abstract Lorentz group in a neighborhood of the origin (proper transformations). Let us observe that, aside from δ^{μ}_{ν}, there exist two invariant tensors with respect to (proper) Lorentz transformations: The metric $\eta_{\mu\nu}$ (which is invariant under generic Lorentz transformations) and the Levi-Civita tensor $\epsilon_{\mu\nu\rho\sigma}$, defined as follows:

$$\epsilon_{\mu\nu\rho\sigma} = 1 \quad (\mu\nu\rho\sigma) \text{ even permutation of } (0, 1, 2, 3), \tag{4.173}$$

$$\epsilon_{\mu\nu\rho\sigma} = -1 \quad (\mu\nu\rho\sigma) \text{ odd permutation of } (0, 1, 2, 3), \tag{4.174}$$

so that $\epsilon_{0123} = +1$. Indeed, in virtue of Eq. (4.149)

$$\eta_{\mu\nu} \to \Lambda^\rho{}_\mu \Lambda^\sigma{}_\nu \eta_{\rho\sigma} = \eta_{\mu\nu},$$

$$\epsilon_{\mu\nu\rho\sigma} \to \Lambda^{\mu'}{}_\mu \Lambda^{\nu'}{}_\nu \Lambda^{\rho'}{}_\rho \Lambda^{\sigma'}{}_\sigma \epsilon_{\mu'\nu'\rho'\sigma'} = \det(\Lambda)\, \epsilon_{\mu\nu\rho\sigma} = \epsilon_{\mu\nu\rho\sigma}.$$

Let us return to the basis of infinitesimal generators $\mathbf{L}^{\rho\sigma}$. As previously pointed out, only six of them are independent. It is therefore convenient to reorganize them as follows:

$$\mathbf{L}_1 \equiv -\mathbf{L}^{23}, \quad \mathbf{L}_2 \equiv -\mathbf{L}^{31}, \quad \mathbf{L}_3 \equiv -\mathbf{L}^{12}, \tag{4.175}$$

or, equivalently:

$$\mathbf{L}_i = -\frac{1}{2} \epsilon_{ijk} \mathbf{L}^{jk}, \quad i, j, k = 1, 2, 3, \tag{4.176}$$

and

$$\mathbf{K}_i = \mathbf{L}^{0i}, \quad i = 1, 2, 3. \tag{4.177}$$

The generators $\mathbf{K}_i$ are given the name of *boost generators*. From (4.172) we may deduce the commutation relations between the six generators $\mathbf{L}_i$, $\mathbf{K}_i$. For instance, let us write the following commutator:

$$[\mathbf{L}_1, \mathbf{L}_2] = \left[\mathbf{L}^{23}, \mathbf{L}^{31}\right] = \eta^{33} \mathbf{L}^{21} = -\mathbf{L}^{21} = -\mathbf{L}_3. \tag{4.178}$$

Similarly we can show that:

$$\left[\mathbf{L}_i, \mathbf{L}_j\right] = -\epsilon_{ijk} \mathbf{L}_k, \tag{4.179}$$

$$\left[\mathbf{L}_i, \mathbf{K}_j\right] = -\epsilon_{ijk} \mathbf{K}_k, \tag{4.180}$$

$$\left[\mathbf{K}_i, \mathbf{K}_j\right] = \epsilon_{ijk} \mathbf{L}_k. \tag{4.181}$$

By comparing Eq. (4.179) with Eqs. (4.124) and (4.125), we conclude that $\mathbf{L}_i$ are the generators of rotations since they satisfy the corresponding commutation relations.

They generate Lorentz transformations of the form

$$(\Lambda_R{}^\mu{}_\nu) \equiv \begin{pmatrix} 1 & 0 \\ 0 & R^i{}_j \end{pmatrix} \in SO(3), \tag{4.182}$$

which clearly leave $\eta_{\mu\nu}$ invariant. The Lorentz group therefore contains the rotation group $SO(3)$ as a subgroup.

Consider now the generators $\mathbf{K}_i$. As opposed to $\mathbf{L}_i$, they do not close an algebra. From Eqs. (4.170) and (4.177) we may deduce their matrix form:

$$\mathbf{K}_1 = \begin{pmatrix} 0 & 1 & 0 & 0 \\ 1 & 0 & 0 & 0 \\ 0 & 0 & 0 & 0 \\ 0 & 0 & 0 & 0 \end{pmatrix} ; \quad \mathbf{K}_2 = \begin{pmatrix} 0 & 0 & 1 & 0 \\ 0 & 0 & 0 & 0 \\ 1 & 0 & 0 & 0 \\ 0 & 0 & 0 & 0 \end{pmatrix} ; \quad \mathbf{K}_3 = \begin{pmatrix} 0 & 0 & 0 & 1 \\ 0 & 0 & 0 & 0 \\ 0 & 0 & 0 & 0 \\ 1 & 0 & 0 & 0 \end{pmatrix} ,$$

note that $\mathbf{K}_i$ are symmetric matrices, as opposed to $\mathbf{L}_i$. Let us see what a finite transformation $\Lambda^\mu{}_\nu$ generated by the set of $\mathbf{K}_i$ looks like. Since the rotation generators $\mathbf{L}_i$ are not involved, this transformation will not affect the orientation of the Cartesian axes and thus the two RFs will keep their axes parallel.

Let us denote by σ^i the parameters of the transformation. According to our previous analysis, to obtain a finite Lorentz transformation we need to exponentiate the infinitesimal generators (in this case $\sigma^i \mathbf{K}_i$). We find:

$$\Lambda = e^{\sigma^i \mathbf{K}_i} = \sum_{n=0}^{\infty} \frac{1}{n!} (\sigma^i \mathbf{K}_i)^n. \tag{4.183}$$

where the parameters σ^i are identified with ω_{0i}. Let us now define the norm σ and the unit vector $\mathbf{u} = (u^i)$ associated with the vector (σ^i):

$$\sigma = \|(\sigma^i)\| = \sqrt{\sum_i (\sigma^i)^2}, \quad u^i = \frac{\sigma^i}{\sigma}, \quad \sum_{i=1}^{3} u^i u^i = 1, \quad \sigma^i = \sigma\, u^i.$$

Let us compute the following matrices:

$$u^i \mathbf{K}_i = \begin{pmatrix} 0 & u^j \\ u^i & \varnothing^{ij} \end{pmatrix} ; \quad (\varnothing^{ij}) = \begin{pmatrix} 0 & 0 & 0 \\ 0 & 0 & 0 \\ 0 & 0 & 0 \end{pmatrix} ,$$

$$(u^i \mathbf{K}_i)^2 = \begin{pmatrix} 1 & 0 \\ 0 & u^i u^j \end{pmatrix} ; \quad (u^i \mathbf{K}_i)^3 = (u^i \mathbf{K}_i); \quad (u^i \mathbf{K}_i)^4 = (u^i \mathbf{K}_i)^2 \ldots$$

$$\ldots (u^i \mathbf{K}_i)^{2k} = (u^i \mathbf{K}_i)^2; \quad (u^i \mathbf{K}_i)^{2k+1} = (u^i \mathbf{K}_i).$$

We can now compute the exponential in Eq. (4.183):

$$\Lambda = 1 + \sigma \left(u^i \, \mathbf{K}_i\right) + \frac{\sigma^2}{2}(u^i \, \mathbf{K}_i)^2 + \frac{\sigma^3}{3!}\left(u^i \, \mathbf{K}_i\right) + \frac{\sigma^4}{4!}(u^i \, \mathbf{K}_i)^2 + \cdots$$

$$= 1 + \left(\sigma + \frac{\sigma^3}{3!} + \frac{\sigma^5}{5!} + \cdots\right)(u^i \, \mathbf{K}_i) + \left(\frac{\sigma^2}{2} + \frac{\sigma^4}{4!} + \cdots\right)(u^i \, \mathbf{K}_i)^2$$

$$= 1 - (u^i \, \mathbf{K}_i)^2 + \sinh \sigma \, (u^i \, \mathbf{K}_i) + \cosh \sigma (u^i \, \mathbf{K}_i)^2$$

$$= \begin{pmatrix} \cosh \sigma & \sinh \sigma \, u^j \\ \sinh \sigma \, u^i & \delta^{ij} + (\cosh \sigma - 1) \, u^i u^j \end{pmatrix}.$$

Let us take, for instance:

$$(\sigma^i) = (\sigma, 0, 0) \quad \rightarrow \quad \mathbf{u} = (u^i) = (1, 0, 0), \tag{4.184}$$

The corresponding transformation reads:

$$(\Lambda^\mu{}_\nu) = \begin{pmatrix} \cosh \sigma & \sinh \sigma & 0 & 0 \\ \sinh \sigma & \cosh \sigma & 0 & 0 \\ 0 & 0 & 1 & 0 \\ 0 & 0 & 0 & 1 \end{pmatrix}. \tag{4.185}$$

If we set in (4.185) $\sinh \sigma = -\frac{v}{c}\gamma$ and $\cosh \sigma = \gamma$, which is consistent with the property $\cosh^2 \sigma - \sinh^2 \sigma = 1$ provided $\gamma = \frac{1}{\sqrt{1 - \frac{v^2}{c^2}}}$, the transformation Λ becomes:

$$\Lambda^\mu{}_\nu = \begin{pmatrix} \gamma & -\frac{v}{c}\gamma & 0 & 0 \\ -\frac{v}{c}\gamma & \gamma & 0 & 0 \\ 0 & 0 & 1 & 0 \\ 0 & 0 & 0 & 1 \end{pmatrix}. \tag{4.186}$$

The transformation $\Delta x'^\mu = \Lambda^\mu{}_\nu \Delta x^\nu$ is precisely the one which maps a RF S onto a RF S' in uniform motion with respect to the former with a velocity $\mathbf{v} = (v, 0, 0)$, which we have derived in Chap. 1.

If, more generally, we define the following vector:

$$(\beta^i) = \beta \, (u^1, u^2, u^3) = \left(\frac{v^i}{c}\right) = \frac{\mathbf{v}}{c}, \tag{4.187}$$

where $\beta = v/c$, so that we can write $\gamma = \frac{1}{\sqrt{1-\beta^2}}$, and set $\sinh \sigma = -\beta\gamma$, $\cosh \sigma = \gamma$, the most general proper Lorentz transformation generated by $\mathbf{K}_i$ reads:

$$\Lambda^{\mu}{}_{\nu} = \begin{pmatrix} \gamma & -\beta^{j}\gamma \\ -\beta^{i}\gamma & \delta^{ij} + (\gamma - 1)\frac{\beta^{i}\beta^{j}}{\beta^{2}} \end{pmatrix}. \tag{4.188}$$

This is the Lorentz transformation which connects two frames of reference S and S', the latter moving with respect to the former with a translational uniform motion with constant velocity vector $\mathbf{v} = (v^{i})$. It was also derived in Chap. 2, see Eq. (2.57). We say that this transformation *boosts* the RF S onto S' and is therefore called a *boost transformation*, to be denoted by Λ_{B}. Consequently the $\mathbf{K}_{i}$ are called infinitesimal generators of Lorentz boosts and their parameters are related to the relative velocity vector.

The most general Lorentz transformation can be written as the product of a *boost* $\Lambda_{B} = \exp(\sigma^{i}\,\mathbf{K}_{i})$ times a rotation $\Lambda_{R} = \exp(\theta^{i}\mathbf{L}_{i})$:

$$\Lambda = \Lambda_{B}\,\Lambda_{R} = \exp(\sigma^{i}\,\mathbf{K}_{i})\,\exp(\theta^{i}\mathbf{L}_{i}), \tag{4.189}$$

or, alternatively, as the exponential of a finite combination of the infinitesimal generators $\mathbf{L}^{\rho\sigma}$: $\Lambda = \exp(\frac{1}{2}\,\theta_{\rho\sigma}\,\mathbf{L}^{\rho\sigma})$.

It is useful at this point to give the explicit matrix form of the Lorentz boost Λ_{p} which connects the rest frame S_{0} of a massive particle, in which $\mathbf{p} = \mathbf{0}$ and thus the corresponding four-momentum is $\bar{p} = (mc, \mathbf{0})$, to a generic RF S in which the particle has momentum $p \equiv (p^{\mu}) = (E/c, \mathbf{p})$:

$$p^{\mu} = \Lambda_{p}{}^{\mu}{}_{\nu}\,\bar{p}^{\nu}.$$

The energy and the linear momentum of the particle in S are related by Eq. (2.38) of Chap. 2: $E^{2} - |\mathbf{p}|^{2}c^{2} = m^{2}c^{4}$. Moreover the velocity of the particle in S is $\mathbf{p}c/E$. Since S moves relative to S_{0} (in the standard configuration) with a velocity $\mathbf{v}$ which is the opposite of that of the particle, we have to set in (4.188) $\mathbf{v}/c^{2} = -\mathbf{p}/E$. Using the relation $\gamma(v) = E/(mc^{2})$ we find for Λ_{p} the following matrix expression:

$$\Lambda_{p} = \begin{pmatrix} \frac{E}{mc^{2}} & \frac{p^{j}}{mc} \\ \frac{p^{i}}{mc} & \delta^{ij} + \frac{p^{i}p^{j}}{m(E+mc^{2})} \end{pmatrix}. \tag{4.190}$$

4.7.2 The Poincaré Group

We want now to write the most general coordinate transformation which leaves the four-dimensional distance ds, as a coordinate function, invariant. It will generalize the Lorentz transformation in (4.146) by allowing the four-dimensional origins of the two systems of coordinates not to coincide. It will therefore be described by an affine transformation (Λ, x_{0}):

$$x'^{\mu} = \Lambda^{\mu}{}_{\nu}x^{\nu} - x_{0}^{\mu} \quad \Rightarrow \quad dx'^{\mu} = \Lambda^{\mu}{}_{\nu}\,dx^{\nu}, \tag{4.191}$$

whose homogeneous part Λ is a Lorentz transformation acting on the directions of the space-time axes, while the inhomogeneous part $x_0 = (x_0^\mu)$ describes a global space-time translation. The reader can easily show, along the lines of Sect. 4.5, that these transformations, called *Poincaré transformations*, close a group, named the *Poincaré group*. A generic Poincaré transformation depends analytically on the six parameters of the Lorentz part and the four parameters x_0^μ associated with the space-time translations. The Poincaré group is therefore a *ten*-parameter Lie group.

In this chapter we have been dealing with matrix representations of transformation groups acting on component vectors. In order to characterize the algebra associated with the Poincaré group, we would need to work out a basis of infinitesimal generators. Such basis would comprise the six generators $\mathbf{L}^{\rho\sigma}$ of the Lorentz subgroup and the four generators $\mathbf{P}_\gamma$ of the space-time translations. It is useful work with a matrix realization of a generic group element. This is done by associating with a transformation (Λ, x_0) the following 5×5 matrix

$$(\Lambda, x_0) \rightarrow \begin{pmatrix} \Lambda^\mu{}_\nu & -x_0^\mu \\ \varnothing_\nu & 1 \end{pmatrix}, \quad [(\varnothing_\nu) \equiv (0, 0, 0, 0)], \tag{4.192}$$

acting on the coordinate vector, extended by an additional entry 1: $(x^\mu, 1) \equiv (x^0, x^1, x^2, x^3, 1)$. The first four components of the resulting 5-vector are the transformed coordinates:

$$\begin{pmatrix} x^\mu \\ 1 \end{pmatrix} \rightarrow \begin{pmatrix} \Lambda^\mu{}_\nu & -x_0^\mu \\ \varnothing_\nu & 1 \end{pmatrix} \begin{pmatrix} x^\nu \\ 1 \end{pmatrix} = \begin{pmatrix} \Lambda^\mu{}_\nu\, x^\nu - x_0^\mu \\ 1 \end{pmatrix} = \begin{pmatrix} x'^\mu \\ 1 \end{pmatrix}. \tag{4.193}$$

This matrix construction applies to a generic affine transformation. We wish now to write the matrix representation of the infinitesimal Poincaré generators. To this end let us write an infinitesimal Poincaré transformation to first order in its small parameters $\delta\theta_{\rho\sigma}$, δx_0^μ:

$$\begin{pmatrix} \Lambda^\mu{}_\nu & -\delta x_0^\mu \\ \varnothing_\nu & 1 \end{pmatrix} \approx \begin{pmatrix} \delta^\mu{}_\nu + \frac{1}{2}\,\delta\theta_{\rho\sigma}\,(L^{\rho\sigma})^\mu{}_\nu & -\delta x_0^\mu \\ \varnothing_\nu & 1 \end{pmatrix} = \mathbf{1} + \frac{1}{2}\,\delta\theta_{\rho\sigma}\,\mathbf{L}^{\rho\sigma} + \delta x_0^\gamma\,\mathbf{P}_\gamma,$$

where $\mathbf{1}$ is the 5×5 identity matrix and $\mathbf{L}^{\rho\sigma}$ are now represented by 5×5 matrices:

$$\mathbf{L}^{\rho\sigma} \equiv \begin{pmatrix} (L^{\rho\sigma})^\mu{}_\nu & \varnothing^\mu \\ \varnothing_\nu & 0 \end{pmatrix}. \tag{4.194}$$

The reader can verify that the commutation relations (4.172) still hold. The four matrices $\mathbf{P}_\gamma$ generate the space-time translations and read:

$$\mathbf{P}_\gamma \equiv \begin{pmatrix} \varnothing^\mu{}_\nu & -\delta^\mu_\gamma \\ \varnothing_\nu & 0 \end{pmatrix}. \tag{4.195}$$

The effect of the infinitesimal transformation on x^μ is the following:

$$x^\mu \to x'^\mu = x^\mu + \delta x^\mu, \quad \delta x^\mu = \delta\theta^\mu{}_\nu x^\nu - \delta x_0^\mu, \tag{4.196}$$

where we have used the property $\frac{1}{2}\delta\theta_{\rho\sigma}(L^{\rho\sigma})^\mu{}_\nu = \delta\theta^\mu{}_\nu$ which follows from Eq. (4.170). A finite Poincaré transformation, with a proper Lorentz component Λ, can be expressed in terms of exponentials of finite combinations of the infinitesimal generators:

$$\begin{pmatrix} \Lambda^\mu{}_\nu & -x_0^\mu \\ \varnothing_\nu & 1 \end{pmatrix} = e^{x_0^\gamma \mathbf{P}_\gamma} \cdot e^{\frac{1}{2}\theta_{\rho\sigma}\mathbf{L}^{\rho\sigma}}. \tag{4.197}$$

As an example consider the subset consisting of pure translations $(\mathbf{1}, x_0)$. The reader can verify that it is a subgroup. Moreover the result of two consecutive translations does not depend on the order in which they are effected: $(\mathbf{1}, x_0) \cdot (\mathbf{1}, x_1) = (\mathbf{1}, x_1) \cdot (\mathbf{1}, x_0)$. This property makes the group of translations *commutative* or *abelian*. Let us verify that a finite translation $(\mathbf{1}, x_0)$ is indeed represented by the 5×5 matrix $e^{x_0^\gamma \mathbf{P}_\gamma}$:

$$e^{x_0^\gamma \mathbf{P}_\gamma} \cdot \begin{pmatrix} x^\mu \\ 1 \end{pmatrix} = \begin{pmatrix} \delta^\mu{}_\nu & -x_0^\mu \\ \varnothing_\nu & 1 \end{pmatrix} \begin{pmatrix} x^\mu \\ 1 \end{pmatrix} = \begin{pmatrix} x^\mu - x_0^\mu \\ 1 \end{pmatrix}, \tag{4.198}$$

where we have used the definition (4.129) of the exponential of a matrix, and the property that powers of $x_0^\gamma \mathbf{P}_\gamma$ higher than one vanish: $(x_0^\gamma \mathbf{P}_\gamma)^n = \mathbf{0}, n \geq 2$.

Let us compute the commutation relations between the Poincaré generators. We clearly have $[\mathbf{P}_\gamma, \mathbf{P}_\sigma] = \mathbf{0}$. This represents the fact that the group of translations is commutative. Let us now compute $[\mathbf{L}^{\rho\sigma}, \mathbf{P}_\gamma]$. Clearly $\mathbf{P}_\gamma \mathbf{L}^{\rho\sigma} = \mathbf{0}$, while:

$$[\mathbf{L}^{\rho\sigma}, \mathbf{P}_\gamma] = \mathbf{L}^{\rho\sigma} \mathbf{P}_\gamma = \begin{pmatrix} \varnothing^\mu{}_\nu & -\eta^{\rho\mu}\delta^\sigma_\gamma + \eta^{\sigma\mu}\delta^\rho_\gamma \\ \varnothing_\nu & 0 \end{pmatrix} = (\eta^{\rho\nu}\delta^\sigma_\gamma - \eta^{\sigma\nu}\delta^\rho_\gamma)\mathbf{P}_\nu.$$

Let us now summarize the commutation relations among the Poincaré generators:

$$\left[\mathbf{L}^{\mu\nu}, \mathbf{L}^{\rho\sigma}\right] = \eta^{\nu\rho}\mathbf{L}^{\mu\sigma} + \eta^{\mu\sigma}\mathbf{L}^{\nu\rho} - \eta^{\mu\rho}\mathbf{L}^{\nu\sigma} - \eta^{\nu\sigma}\mathbf{L}^{\mu\rho}, \tag{4.199}$$

$$\left[\mathbf{L}^{\mu\nu}, \mathbf{P}_\rho\right] = \mathbf{P}^\mu \delta^\nu_\rho - \mathbf{P}^\nu \delta^\mu_\rho, \tag{4.200}$$

$$\left[\mathbf{P}_\mu, \mathbf{P}_\nu\right] = \mathbf{0}. \tag{4.201}$$

4.7.3 References

For further reading see Refs. [2, 5, 14].

Chapter 5
Maxwell Equations and Special Relativity

5.1 Electromagnetism in Tensor Form

As we have already noted in Chap. 1, Maxwell's electromagnetic theory is by definition a relativistic theory, since it implies in particular the constancy of the speed of
light in every RF. As such it must be covariant under the group of Lorentz transformations, or, using the terminology of the previous chapter, covariant under the group
SO(1, 3).

In this chapter, using the tensor formalism developed for the Lorentz group, we
shall establish the covariance of the electromagnetic theory under the Lorentz group
by formulating the Maxwell equations as tensor equations, namely as equalities
between Lorentz-tensors of the same kind.

The use of the Lorentz-tensors notation, besides making the relativistic nature of
Maxwell's theory manifest, will also be useful for deriving some consequences of
the electromagnetic theory in a simpler and more transparent way.

To begin with, let us write the Maxwell equations in the usual vector notation,
which, by definition, is manifestly covariant under the three-dimensional rotation
group SO(3):

$$\nabla \cdot \mathbf{E} = \rho, \tag{5.1}$$

$$\nabla \times \mathbf{B} = \frac{1}{c} \frac{\partial \mathbf{E}}{\partial t} + \frac{\mathbf{j}}{c}, \tag{5.2}$$

$$\nabla \cdot \mathbf{B} = 0, \tag{5.3}$$

$$\nabla \times \mathbf{E} = -\frac{1}{c} \frac{\partial \mathbf{B}}{\partial t}, \tag{5.4}$$

where $\mathbf{E}(x)$ and $\mathbf{B}(x)$ denote as usual the electric and the magnetic field, respectively, and we define $x \equiv (x^\mu) = (ct, x^1, x^2, x^3)$.[1] In the following we shall use

---

[1] We are using the so called Heaviside-Lorentz (HL) system of units, the most useful for theoretical considerations. It amounts to considering the electric charge as a quantity whose physical

© Springer International Publishing Switzerland 2016
R. D'Auria and M. Trigiante, *From Special Relativity to Feynman Diagrams*,
UNITEXT for Physics, DOI 10.1007/978-3-319-22014-7_5

the compact notation (3.60) for partial differentiation with respect to Minkowski and spatial coordinates. We first translate the vector notation into a 3-dimensional tensor notation. For example, using the SO(3)-tensor notation, the three-dimensional divergence and curl operators can be written as follows:

$$\partial_i E^i \equiv \nabla \cdot \mathbf{E}, \quad \epsilon_{ijk}\partial_j B_k \equiv (\nabla \times \mathbf{B})_i.$$

In the same notation, Eqs. (5.1)–(5.4) are recast in the following equivalent form:

$$\partial_i E_i = \rho, \tag{5.7}$$

$$\epsilon_{ijk}\partial_j B_k = \frac{1}{c}\frac{\partial E_i}{\partial t} + \frac{j_i}{c}, \tag{5.8}$$

$$\partial_i B_i = 0, \tag{5.9}$$

$$\epsilon_{ijk}\partial_j E_k = -\frac{1}{c}\frac{\partial}{\partial t}B_i. \tag{5.10}$$

Of course, in this formalism, only covariance with respect to three-dimensional rotations is manifest. Recall that, with respect to the orthogonal group SO(3) (or in general SO(n) for a n-dimensional Euclidean space), there is no difference between covariant (lower) and contravariant (upper) indices, since they transform in the same way. Indices are raised and lowered by contraction with the identity matrix, which

---

(Footnote 1 continued)
dimensions are derived from the basic dimensional quantities $[M, L, T]$ (the corresponding units being [kilogram, meter, second]), by writing Coulomb's law without additional physical constants, namely in the following form:

$$F = \frac{1}{4\pi}\frac{q_1 q_2}{r^2}. \tag{5.5}$$

In this way the electric charge has the physical dimensions $[M^{\frac{1}{2}}L^{\frac{3}{2}}T^{-1}]$, and the electric field $[M^{\frac{1}{2}}L^{-\frac{1}{2}}T^{-1}]$. (Note that the presence of the factor $\frac{1}{4\pi}$ in Coulomb's law means that the HL system is *rationalized*, that is there are no factors 4π explicitly appearing in Maxwell's equations). Moreover the electric and magnetic fields are defined so as to have the same dimension, so that the Lorentz force reads:

$$\mathbf{F} = q\left(\mathbf{E} + \frac{\mathbf{v}}{c} \times \mathbf{B}\right). \tag{5.6}$$

The quickest way to translate formulae written in the international system of units (S.I.) into the HL one, is to redefine the electric charge as follows: Let $\tilde{e}$ and e be the measures of the electric charge in the S.I. and the HL systems respectively. We then have:

$$e = \frac{\tilde{e}}{\sqrt{\varepsilon_0}} \Rightarrow \rho = \frac{\tilde{\rho}}{\sqrt{\varepsilon_0}}, \quad \mathbf{j} = \frac{\tilde{j}}{\sqrt{\varepsilon_0}}.$$

Moreover the electric and magnetic fields $\tilde{\mathbf{E}}, \tilde{\mathbf{B}}$ in the SI system are related to the analogous quantities $\mathbf{E}$ e $\mathbf{B}$ in the HL system as follows: $\sqrt{\varepsilon_0}\tilde{\mathbf{E}} = \mathbf{E};\quad \mathbf{B} = \frac{1}{\sqrt{\mu_0}}\tilde{\mathbf{B}}$.

For example, the energy density takes the form: $\rho_E = \frac{1}{2}\left(\varepsilon_0|\tilde{\mathbf{E}}|^2 + \frac{|\tilde{\mathbf{B}}|^2}{\mu_0}\right) = \frac{1}{2}\left(|\mathbf{E}|^2 + |\mathbf{B}|^2\right)$.

does not affect the value of the corresponding components. For this reason, when writing SO(3)-tensors like the Euclidean vectors E_i, B_i or the Levi-Civita symbol ϵ_{ijk}, we shall not care about the position of their indices: $E_i = E^i$, $B_i = B^i$ and so on.

In order to make covariance with respect to SO(1, 3), that is Lorentz transformations, manifest, we introduce a 4×4 *antisymmetric matrix* $F^{\mu\nu}$ whose entries are defined as follows:

$$F^{0i} = -F^{i0} = E^i = E_i, \tag{5.11}$$

$$F^{ij} = \epsilon_{ijk}B_k \Leftrightarrow B_i = \frac{1}{2}\epsilon_{ijk}F^{jk}, \tag{5.12}$$

that is:

$$F^{\mu\nu} = \begin{pmatrix} 0 & E^1 & E^2 & E^3 \\ -E^1 & 0 & B^3 & -B^2 \\ -E^2 & -B^3 & 0 & B^1 \\ -E^3 & B^2 & -B^1 & 0 \end{pmatrix}. \tag{5.13}$$

The above quantity will be characterized as a Lorentz (i.e. a SO(1, 3))-tensor. The position of its indices can be changed only with the Lorentz metric (4.142) $\eta_{\mu\nu}$ ($\eta_{00} = 1$, $\eta_{ij} = -\delta_{ij}$, $\eta_{0i} = 0$). As remarked above, when three-dimensional indices $i, j, \ldots$ belong to SO(3)-tensor quantities, like the electric and magnetic fields, their position is irrelevant since they are raised or lowered with the metric δ_{ij}. Instead when indices $i, j, \ldots$ are a subset of the four-dimensional ones $\mu, \nu, \ldots$, namely label components of SO(1, 3)-tensors, we must use the Lorentz metric, so that the raising, or lowering, of three-dimensional indices implies a change of sign of the corresponding components, while the same operation on time components $\mu = 0$ leaves their sign unchanged.

Therefore if we lower the two upper indices of $F^{\mu\nu}$ with the Minkowski metric $\eta_{\mu\nu}$, we obtain:

$$F_{\mu\nu} = \eta_{\mu\sigma}\eta_{\nu\rho}F^{\sigma\rho} = \begin{pmatrix} 0 & -E^1 & -E^2 & -E^3 \\ E^1 & 0 & B^3 & -B^2 \\ E^2 & -B^3 & 0 & B^1 \\ E^3 & B^2 & -B^1 & 0 \end{pmatrix}. \tag{5.14}$$

We shall prove in the sequel that $F^{\mu\nu}$ (and $F_{\mu\nu}$) are actually *contravariant (and covariant) antisymmetric tensors* of the Lorentz group.

We further define the *electromagnetic four-current* or, in short, the *four-current* as:

$$J^\mu = \left(\rho, \frac{1}{c}j^k\right) \rightarrow J^0 = \rho, \ J^k = \frac{1}{c}j^k. \tag{5.15}$$

Note that we have denoted J^μ as a Lorentz four-vector. The proof that the four components of J^μ actually transform as a contravariant four-vector will be given in Sect. 5.4.

We now show that Eqs. (5.7) and (5.8) can be written in the following compact form:

$$\partial_\nu F^{\nu\mu} = -J^\mu. \tag{5.16}$$

Considering first the $\mu = 0$ component of Eq. (5.16) and, taking into account the antisymmetry of $F^{\nu\mu}$ ($F^{00} = 0$), we have:

$$\partial_i F^{i0} \equiv -\partial_i F^{0i} = -\rho \Rightarrow \partial_i E^i = \rho,$$

which coincides with Eq. (5.7).

Setting instead $\mu = i$ in Eq. (5.16) one obtains:

$$\partial_0 F^{0i} + \partial_j F^{ji} = -\frac{j^i}{c},$$

and since $\epsilon_{ijk} = -\epsilon_{jik}$,

$$\partial_0 E^i - \epsilon_{ijk}\partial_j B_k = -\frac{j^i}{c} \to \epsilon_{ijk}\partial_j B_k = \frac{1}{c}\frac{\partial E_i}{\partial t} + \frac{j^i}{c},$$

which coincides with Eq. (5.8).

Thus Eq. (5.16), written in terms of four-dimensional indices, is equivalent to the two non-homogeneous Maxwell equations, (5.7) and (5.8).

Coming next to the homogeneous Maxwell equations (5.9) and (5.10), we show that, using four-dimensional Minkowski indices, they can also be written in terms of the following single covariant equation:

$$\partial_{[\mu} F_{\nu\rho]} \equiv \frac{1}{3}(\partial_\mu F_{\nu\rho} + \partial_\nu F_{\rho\mu} + \partial_\rho F_{\mu\nu}) = 0. \tag{5.17}$$

The symbol $[\mu\nu\rho]$ denotes the *complete antisymmetrization* in the three indices μ, ν, ρ. On a generic tensor $U_{\mu\nu\rho}$, this operation is defined as follows:

$$U_{[\mu\nu\rho]} = \frac{1}{3!}(U_{\mu\nu\rho} - U_{\mu\rho\nu} + U_{\nu\rho\mu} - U_{\nu\mu\rho} + U_{\rho\mu\nu} - U_{\rho\nu\mu}).$$

In words, it consists in summing over the even permutations of μ, ν, ρ with a plus sign and over the odd ones with a minus sign, the result being normalized by dividing it by the total number 6 of permutations. Since $F_{\mu\nu} = -F_{\nu\mu}$, this definition applied to $\partial_\mu F_{\nu\rho}$, gives Eq. (5.17).

Let us write Eq. (5.17) choosing one time-index and two spatial indices, that is $\mu = 0; \nu = i; \rho = j$:

$$\partial_0 F_{ij} + \partial_i F_{j0} + \partial_j F_{0i} = 0 \quad \Leftrightarrow \quad \epsilon_{ijk}\partial_0 B_k + \partial_i E_j - \partial_j E_i = 0,$$

where we have used $E^i = E_i = -F_{0i}$, since $F_{0i} = \eta_{0\mu}\eta_{i\nu}F^{\mu\nu} = \eta_{00}\eta_{ij}F^{0j} = -E_i$. This equation can be easily identified with one of the homogeneous Maxwell equations; it is sufficient to multiply it by $\epsilon_{ij\ell}$, summing over i, j and using the formula: $\epsilon_{ijk}\epsilon_{ij\ell} = 2\delta_{k\ell}$. We find:

$$2\partial_0 B_\ell + 2\epsilon_{\ell ij}\partial_i E_j = 0 \quad \Leftrightarrow \quad \epsilon_{\ell ij}\partial_i E_j = -\frac{1}{c}\frac{\partial B_\ell}{\partial t},$$

which coincides with Eq. (5.10).

If, instead, in Eq. (5.17) we consider three spatial indices, namely, $\mu = i$, $\nu = j, \rho = k$, we find:

$$\partial_i F_{jk} + \partial_j F_{ki} + \partial_k F_{ij} = 0.$$

In this case we multiply the above equation by ϵ_{ijk} and sum over i, j, k, obtaining:

$$3\,\epsilon_{ijk}\partial_i F_{jk} = 3\,\epsilon_{ijk}\epsilon_{jk\ell}\partial_i B_\ell = 6\,\delta_\ell^i \partial_i B_\ell = 6\,\partial_i B_i = 0.$$

so that Eq. (5.9) is retrieved.

Summarizing: We have defined two quantities $F^{\mu\nu}$ and J^μ such that the Maxwell equations are written as:

$$\partial_\mu F^{\mu\nu} = -J^\nu, \tag{5.18}$$

$$\partial_{[\mu}F_{\nu\rho]} = 0. \tag{5.19}$$

In particular, if we compute the *four-dimensional divergence* ∂_ν of Eq. (5.18) and take into account that $\partial_\nu\partial_\mu F^{\mu\nu} = 0$, which follows from the fact that $\partial_\mu\partial_\nu = \partial_\nu\partial_\mu$ is symmetric while $F^{\mu\nu}$ is antisymmetric, we obtain the equation:

$$\partial_\nu J^\nu = 0, \tag{5.20}$$

which, in three-dimensional notation, reads

$$\partial_0\rho + \frac{1}{c}\partial_i j^i = 0 \quad \Leftrightarrow \quad \frac{\partial\rho}{\partial t} + \nabla\cdot\mathbf{j} = 0. \tag{5.21}$$

We recognize the above equation as the well known *continuity equation* of the electric current expressing, in local form, the conservation of the electric charge. Since this property has been verified so far with no exception in different inertial systems, it is natural to expect Eq. (5.20) to be a Lorentz *covariant* equation,[2] namely independent of the particular inertial system. This implies that J^μ *must transform as a Lorentz four-vector.* In any case we will explicitly verify the four-vector nature of J^μ from its very definition in Sect. 5.4.

Assuming, for the time being, J^μ to be a four-vector, we may readily show that $F^{\mu\nu}$, introduced as a matrix in Eq. (5.13), is actually a *(contravariant antisymmetric)* tensor with respect to the group of Lorentz transformations, and that consequently the inhomogeneous Maxwell equations are SO(1, 3)-covariant.

To show this, let us assume Eq. (5.16) to hold in a certain inertial RF S':

$$\partial'_\mu F'^{\mu\nu} = -J'^\nu. \tag{5.22}$$

Since ∂_μ and J^μ are covariant and contravariant vectors, respectively, in a new RF S, related to S' by a Lorentz transformation (i.e. an SO(1, 3) rotation), we have:

$$\Lambda^{-1\rho}{}_\mu \partial_\rho F'^{\mu\nu} = -\Lambda^\nu{}_\sigma J^\sigma,$$

Multiplying by the matrix $\Lambda^{-1\tau}{}_\nu$ and summing over ν we obtain:

$$\Lambda^{-1\rho}{}_\mu \Lambda^{-1\tau}{}_\nu \partial_\rho F'^{\mu\nu} = -J^\tau$$

Therefore in the RF S we may write:

$$F^{\rho\tau} = \Lambda^{-1\rho}{}_\mu \Lambda^{-1\tau}{}_\nu F'^{\mu\nu}.$$

Finally, solving with respect to $F'^{\mu\nu}$, we conclude:

$$F'^{\mu\nu} = \Lambda^\mu{}_\rho \Lambda^\nu{}_\sigma F^{\rho\sigma}, \tag{5.23}$$

expressing the fact that *the matrix $F^{\mu\nu}$ is indeed a (contravariant) tensor of order two.*

It follows that the Maxwell equation (5.19) is also Lorentz covariant owing to the four-vector nature of the differentiation operator ∂_μ.

In conclusion, *the theory of electromagnetism, described by Mawxwell's equations, is covariant under Lorentz transformations,* a fact which is consistent with our discussion about the principle of invariance of the velocity of light given in Chap. 1. Moreover, recalling the definition of the *Poincaré* group given in Chap. 4 and the fact that Maxwell's equations are obviously *invariant* under four-dimensional translations, we may assert that the electromagnetic theory is invariant under the full *Poincaré group* as it is the case for relativistic mechanics, see discussion in Chap. 2.

---

[2] Actually, since Eq. (5.20) contains no free indices, it is a scalar equation, namely $\partial_\mu J^\mu(x) = \partial'_\mu J'^\mu(x')$.

5.2 The Lorentz Force

We recall that in the Maxwell theory the Lorentz force acting on a given charge e is given by

$$\mathbf{F} = e\left(\mathbf{E} + \frac{\mathbf{v}}{c} \times \mathbf{B}\right), \tag{5.24}$$

so that its equation of motion reads

$$\mathbf{F} = \frac{d\mathbf{p}}{dt}. \tag{5.25}$$

Equation (5.24), as it stands, is not written in an explicit tensor form. However we shall prove that it has the same form in all inertial frames (we wish to remark here that the tensor form of a physical law is a sufficient, though not necessary condition for its validity in every RF). To show that Eq. (5.25) holds in every RF, we use the covariant form of the equation of motion in Eq. (2.68)

$$f^\mu = \frac{d}{d\tau} p^\mu, \tag{5.26}$$

and define the *four-force* f^μ acting on the charge as follows:

$$f^\mu = -\frac{e}{c} F^{\mu\nu} \frac{dx_\nu}{d\tau} = -\frac{e}{c} F^{\mu\nu} \frac{dx_\nu}{dt} \frac{dt}{d\tau}. \tag{5.27}$$

Note that Eq. (5.27) is a covariant equation.

Let us examine both sides of Eq. (5.26) in components. Considering the time-component, we have:

$$f^0 = \frac{e}{c} F^{0i} v^i \frac{dt}{d\tau} = \frac{e}{c} \mathbf{E} \cdot \mathbf{v} \frac{dt}{d\tau}, \qquad \frac{dp^0}{d\tau} = \frac{1}{c} \frac{d\mathcal{E}}{dt} \frac{dt}{d\tau}, \tag{5.28}$$

respectively, where we have denoted by $\mathcal{E}$ the energy of the charged particle. Equating the two expressions we find that the $\mu = 0$ component of Eq. (5.26) becomes:

$$\frac{d\mathcal{E}}{dt} = e\,\mathbf{E} \cdot \mathbf{v} = \frac{d\mathcal{W}}{dt},$$

where $\mathcal{W}$ is the work of the force. Thus we retrieve the general result given by Eq. (2.73): The rate of change of the energy of a particle in time equals the power of the force (in our case of the electric force only).

Let us now consider the $\mu = i$ component of Eq. (5.26); on the left hand side we find:

$$f^i = e\,E^i \frac{dt}{d\tau} + \frac{e}{c} \epsilon^{ijk} v^j B^k \frac{dt}{d\tau} = e\left(E^i + \epsilon^{ijk} \frac{v^j}{c} B^k\right) \frac{dt}{d\tau},$$

while, on the right hand side, we may write:

$$\frac{dp^i}{d\tau} = \frac{dp^i}{dt}\frac{dt}{d\tau},$$

Therefore the spatial components of Eq. (5.26) become:

$$e\left(\mathbf{E} + \frac{\mathbf{v}}{c} \times \mathbf{B}\right)\frac{dt}{d\tau} = \frac{d\mathbf{p}}{dt}\frac{dt}{d\tau}.$$

Erasing the common factor $\frac{dt}{d\tau} = \gamma$, we obtain Eq. (5.24), corresponding to the spatial part of Eq. (5.26).

Thus we conclude that Eq. (5.24), even if not written in a manifestly covariant form, is covariant under Lorentz transformation and therefore valid in every RF.

5.3 Behavior of E and B Under Lorentz Transformations

Once we know the transformation properties of the electromagnetic tensor under Lorentz transformations, we may easily find the corresponding laws for its three-dimensional components $\mathbf{E}$ and $\mathbf{B}$.

As $F^{\mu\nu}$ is a Lorentz tensor its transformation under a change of RF is given by:

$$F^{\mu\nu} \to F'^{\mu\nu} = \Lambda^\mu{}_\rho \Lambda^\nu{}_\sigma F^{\rho\sigma}, \tag{5.29}$$

where the Lorentz transformation matrix has been computed in the previous chapter. For a generic boost it is given by Eq. (4.188), or Eq. (2.57), that is:

$$\Lambda^\mu{}_\nu = \begin{pmatrix} \gamma & -\beta^j\gamma \\ -\beta^i\gamma & \delta_{ij} + (\gamma - 1)\frac{\beta^i\beta^j}{\beta^2} \end{pmatrix}; \quad i, j = 1, 2, 3; \quad \beta^i \equiv \frac{V^i}{c},$$

$\mathbf{V} \equiv (V^i)$ being the velocity of S' relative to S and V its norm. Recalling the relations (5.11)–(5.12) and specializing Eq. (5.29) to the components $(\mu, \nu = 0, i)$ and $(\mu, \nu = i, j)$, a simple computation yields the following transformation laws for $\mathbf{E}$ and $\mathbf{B}$:

$$\mathbf{E}' = \gamma(\mathbf{E} + \boldsymbol{\beta} \times \mathbf{B}) + \frac{(1 - \gamma)}{\beta^2}(\boldsymbol{\beta} \cdot \mathbf{E})\,\boldsymbol{\beta}, \tag{5.30}$$

$$\mathbf{B}' = \gamma(\mathbf{B} - \boldsymbol{\beta} \times \mathbf{E}) + \frac{(1 - \gamma)}{\beta^2}(\boldsymbol{\beta} \cdot \mathbf{B})\,\boldsymbol{\beta}, \tag{5.31}$$

where, as usual, we have set $\boldsymbol{\beta} = \frac{\mathbf{V}}{c}$ and denoted by β its length. An equivalent, and somewhat simpler, way to write the previous transformations is to decompose the electric and magnetic fields into components $\mathbf{E}_\parallel$, $\mathbf{B}_\parallel$ which are *parallel* and

$\mathbf{E}_\perp = \mathbf{E} - \mathbf{E}_\|$, $\mathbf{B}_\perp = \mathbf{B} - \mathbf{B}_\|$ which are *transverse* to **V**. It is not difficult to see that in this case Eq. (5.30) take the following form:

$$\mathbf{E}'_\| = \mathbf{E}_\|; \qquad \mathbf{B}'_\| = \mathbf{B}_\|, \tag{5.32}$$

$$\mathbf{E}'_\perp = \gamma(V)\left(\mathbf{E}_\perp + \frac{1}{c}\mathbf{V} \times \mathbf{B}_\perp\right), \tag{5.33}$$

$$\mathbf{B}'_\perp = \gamma(V)\left(\mathbf{B}_\perp - \frac{1}{c}\mathbf{V} \times \mathbf{E}_\perp\right). \tag{5.34}$$

As an example, we compute the electromagnetic field of a charge e in uniform motion with velocity **v** in a frame S.

Let the charge e be at rest at the origin of a RF S'. An observer in S' will observe a Coulombian field:

$$\mathbf{E}' = \frac{e}{4\pi}\frac{\mathbf{x}'}{r'^3}, \tag{5.35}$$

and no magnetic field, $\mathbf{B}' = 0$.

Let the RF S be in standard configuration with respect to S' so that $\mathbf{V} \equiv \mathbf{v} = (v, 0, 0)$. To find the fields in S in terms of those in S', it is sufficient to exchange the role of the two observers in Eq. (5.30), what amounts to exchange the primed quantities with the unprimed ones and to change the sign of the velocity. One obtains:

$$\mathbf{E} = \gamma(\mathbf{E}' - \beta \times \mathbf{B}') + \frac{(1-\gamma)}{v^2}(\mathbf{v} \cdot \mathbf{E}')\,\mathbf{v}, \tag{5.36}$$

$$\mathbf{B} = \gamma(\mathbf{B}' + \beta \times \mathbf{E}') + \frac{(1-\gamma)}{v^2}(\mathbf{v} \cdot \mathbf{B}')\,\mathbf{v}. \tag{5.37}$$

Taking into account that in S' we have $\mathbf{B}' = 0$, Eq. (5.36) becomes:

$$\mathbf{E} = \gamma(v)\,\mathbf{E}' + (1-\gamma)\,E'_x\frac{\mathbf{v}}{v}, \tag{5.38}$$

$$\mathbf{B} = \gamma(v)\,\beta \times \mathbf{E}'. \tag{5.39}$$

Writing **E** in components we find:

$$E_x = E'_x, \tag{5.40}$$
$$E_y = \gamma\,E'_y, \tag{5.41}$$
$$E_z = \gamma\,E'_z. \tag{5.42}$$

Moreover, substituting these values of **E** into the expression for **B** given by Eq. (5.38), one finds that in the frame S the following relation holds:

$$\mathbf{B} = \frac{\mathbf{v} \times \mathbf{E}}{c}, \tag{5.43}$$

which gives the value of the magnetic field generated by a charge moving at a constant velocity $\mathbf{v}$ in S.

Since $\mathbf{E}'$ depends on $\mathbf{x}'$, to obtain the value of $\mathbf{E}$ in S it is still necessary to express the position vector $\mathbf{x}'$, as measured in S', in terms of the one $(\mathbf{x})$ measured in S.

We may suppose, with no loss of generality, the field $\mathbf{E}'$ to lie in the xy-plane of the frame S'; then, taking into account the contraction of lengths along the direction of motion, namely the x-axis, we have:

$$x' = \gamma x; \quad y' = y. \tag{5.44}$$

Hence Eq. (5.40) can be rewritten as follows:

$$E_x = \frac{\gamma e}{4\pi} \frac{x}{r'^3}, \tag{5.45}$$

$$E_y = \frac{\gamma e}{4\pi} \frac{y}{r'^3}, \tag{5.46}$$

$$E_z = 0,$$

that is:

$$\mathbf{E} = \frac{\gamma e}{4\pi} \frac{\mathbf{x}}{r'^3}. \tag{5.47}$$

We can then express r' in terms of r by using the following relation:

$$r'^2 = x'^2 + y'^2 = \frac{x^2}{1 - \frac{v^2}{c^2}} + y^2 = \frac{r^2 - \frac{v^2}{c^2}y^2}{1 - \frac{v^2}{c^2}} = \gamma^2 r^2 \left(1 - \frac{v^2}{c^2} \sin^2 \theta \right), \tag{5.48}$$

where $\sin \theta \equiv y/r$, θ being the angle between the direction of $\mathbf{x}$ and the y-axis. From Eq. (5.48) it follows:

$$r'^3 = r^3 \gamma^3 \left(1 - \frac{v^2}{c^2} \sin^2 \theta \right)^{\frac{3}{2}}, \tag{5.49}$$

and substituting in Eq. (5.47) we obtain the final result:

$$\mathbf{E} = \frac{1}{4\pi} \frac{e}{r^3} \mathbf{x} \left[\frac{1 - \frac{v^2}{c^2}}{\left(1 - \frac{v^2}{c^2} \sin^2 \theta \right)^{\frac{3}{2}}} \right]. \tag{5.50}$$

Moreover, substitution of Eq. (5.50) into Eq. (5.43) gives the value of the magnetic field.

The formula (5.50) tells us that when a charge is moving with constant velocity **v**, the electric field differs from the electrostatic value by the relativistic factor:

$$f(v, \theta) \equiv \frac{1 - \frac{v^2}{c^2}}{\left(1 - \frac{v^2}{c^2} \sin^2 \theta\right)^{\frac{3}{2}}}. \tag{5.51}$$

When the velocity v of the charge is much smaller than the speed of light, $v \ll c$, we can set $f \approx 1$. However, when the velocity of the charge is close to c the modulus of **E** changes according to its direction, thus breaking spherical symmetry. Indeed, since $f(v, \theta)$ is θ-dependent, the strength of the field will be larger when $\sin \theta \approx \pm 1$, that is when $\theta \approx \pm \frac{\pi}{2}$, while it will be smaller when $\theta \approx 0$.

Note, however, that the electric field is always *radial*, as in the static case.

5.4 The Four-Current and the Conservation of the Electric Charge

In Sect. 5.1 the quantity defined in Eq. (5.15) was assumed to be a contravariant four-vector under Lorentz transformations. An argument in favor of this was based on the requirement that the conservation of the electric charge hold in any inertial frame.

In this section we shall construct the explicit expression of J^μ from which its nature of Lorentz four-vector will be manifest.

Let us consider, in a given RF, a system of moving point-like charges e_k ($k = 1, \ldots n$), and denote by $\mathbf{x}_k(t)$ their positions at a given instant t. We want to derive the explicit expressions for the charge density $\rho(\mathbf{x}, t)$ and the current density $\mathbf{j}(\mathbf{x}, t)$.[3] In the three-dimensional notation they can be respectively written as follows:

$$\rho(\mathbf{x}, t) = \sum_k e_k \, \delta^3(\mathbf{x} - \mathbf{x}_k(t)), \tag{5.52}$$

$$j^i(\mathbf{x}, t) = \sum_k e_k \frac{dx_k^i}{dt} \delta^3(\mathbf{x} - \mathbf{x}_k(t)),$$

where $\delta^3(\mathbf{x} - \mathbf{x}')$ is *the three-dimensional Dirac delta function*, defined by the property[4]:

$$\int d^3\mathbf{x}' \, f(\mathbf{x}') \delta^3(\mathbf{x} - \mathbf{x}') = f(\mathbf{x}).$$

---

[3] Note that $\mathbf{x}_k(t)$ is a kinematical variable referred to the kth particle, while $(x^\mu) = (ct, \mathbf{x})$ are space-time labels.

[4] In Cartesian coordinates, if $\mathbf{x} = (x, y, z)$ and $\mathbf{x}_k(t) = (x_k(t), y_k(t), z_k(t))$, the three-dimensional Dirac delta function reads: $\delta^3(\mathbf{x} - \mathbf{x}_k) = \delta(x - x_k(t)) \, \delta(y - y_k(t)) \, \delta(z - z_k(t))$.

To show that $J^\mu = (\rho, \frac{j}{c})$ is a four-vector, we associate with each charge the coordinate four-vector

$$x_k^\mu(t) \equiv (ct, \mathbf{x}_k(t)),$$

so that J^μ takes the following form:

$$J^\mu = \frac{1}{c} \sum_k e_k \frac{dx_k^\mu}{dt} \delta^3(\mathbf{x} - \mathbf{x}_k(t)),$$

or, equivalently:

$$J^\mu(\mathbf{x}, t) = \frac{1}{c} \sum_k e_k \int dt' \frac{dx_k^\mu(t')}{dt'} \delta^3(\mathbf{x} - \mathbf{x}_k(t')) \, \delta(t - t').$$

Since t' is an integration variable, it can be replaced by any other variable. In particular, in the kth-term of the sum, we may replace t' with the *proper time* τ_k of the kth-particle, thus obtaining:

$$J^\mu(\mathbf{x}, t) = \sum_k e_k \int d\tau_k \frac{dx_k^\mu}{d\tau_k} \delta^4(x^\mu - x_k^\mu(\tau_k)), \qquad (5.53)$$

where

$$x_k^\mu(\tau_k) = (c\tau_k, \mathbf{x}_k(\tau_k)), \qquad (5.54)$$

and[5]

$$\delta^4(x^\mu - x_k^\mu(\tau_k)) \equiv \delta^3(\mathbf{x} - \mathbf{x}_k(\tau_k)) \, \delta\left(c(t - \tau_k)\right)$$

$$= \frac{1}{c}\delta^3(\mathbf{x} - \mathbf{x}_k(\tau_k))\delta(t - \tau_k). \qquad (5.55)$$

We now observe that given a four-vector W_μ, $\delta^4(W^\mu)$ is a Lorentz scalar, independently of W^μ. Indeed, by well known properties of the Dirac δ-function we have[6]:

$$\delta^4(W'^\mu) = \delta^4(\Lambda^\mu{}_\nu W^\nu) = \frac{1}{|\det(\Lambda)|}\delta^4(W^\mu) = \delta^4(W^\mu), \qquad (5.56)$$

---

[5]We used $\delta(\alpha x) = \delta(x)/\alpha$, a particular case of the incoming formula (5.56).

[6]This property is easily proven on test functions $f(x) = f(x^\mu)$. Indeed we can write $\int d^4x \, \delta^4$ $(\Lambda \cdot x) f(x) = \int d^4x \, \delta^4(x') f(\Lambda^{-1} \cdot x') = \int \frac{d^4x'}{|\det(\Lambda)|} \delta^4(x') f(\Lambda^{-1} \cdot x') = \frac{f(0)}{|\det(\Lambda)|}$, where we have changed the integration variable from $x \equiv (x^\mu)$ to $x' \equiv \Lambda \cdot x$. Recalling that $f(0) = \int d^4x \, \delta^4(x) f(x)$, and being $f(x)$ generic, we conclude that $\delta^4(\Lambda \cdot x) = \frac{1}{|\det(\Lambda)|} \delta^4(x)$.

since the determinant of a Lorentz transformation is ± 1. It is then apparent that, since in Eq. (5.53) both τ_k and $\delta^4(x^\mu - x_k^\mu(\tau_k))$ are Lorentz scalars, J^μ will transform as dx_k^μ, that is as a four-vector.

We can also derive from the previous expression the continuity equation (5.21) leading to the conservation of the electric charge. To this end, let us compute the divergence of the current density $\mathbf{j} = (j^i)$:

$$
\begin{aligned}
\nabla \cdot \mathbf{j} = \partial_i j^i &= \sum_k e_k \frac{dx_k^i}{dt} \frac{\partial}{\partial x^i} \delta^3(\mathbf{x} - \mathbf{x}_k(t)) \\
&= -\sum_k e_k \frac{dx_k^i}{dt} \frac{\partial}{\partial x_k^i} \delta^3(\mathbf{x} - \mathbf{x}_k(t)) \\
&= -\sum_k e_k \frac{\partial}{\partial t} \delta^3(\mathbf{x} - \mathbf{x}_k(t)) = -\frac{\partial}{\partial t} \rho(\mathbf{x}, t).
\end{aligned}
\tag{5.57}
$$

Thus we retrieve the continuity equation of the electric current:

$$
\partial_i j^i + \frac{\partial}{\partial t} \rho = 0 \Leftrightarrow \partial_\mu J^\mu = 0.
$$

Let us recall how the conservation of the electric charge is obtained from this equation. Let

$$
Q = \int_V d^3x \rho(\mathbf{x}, t),
$$

be the total electric charge contained in the volume V. Then

$$
\frac{dQ}{dt} = \int_V d^3x \frac{\partial}{\partial t} \rho(\mathbf{x}, t) = -\int_V d^3x \, \partial_i j^i(\mathbf{x}, t) = -\int_S dS \, \mathbf{n} \cdot \mathbf{j}(\mathbf{x}, t),
\tag{5.58}
$$

where S is the surface enclosing the volume V, and $\mathbf{n}$ is the unit vector normal to dS. If V represents a *finite domain* of the space, then Eq. (5.58) expresses the fact that the variation of the charge inside V is compensated by the flux of current through its boundary S, which is one way of characterizing the conservation of electric charge.

If, instead, V extends over the whole three-dimensional space, $V \equiv \mathbb{R}^3$, then $S = S_\infty$ is a sphere located at infinity, and, since there is no current at the spatial infinity, the last term on the left hand side of Eq. (5.58) is zero. It then follows that:

$$
\frac{dQ}{dt} = 0,
$$

that is, *the total electric charge in the whole space is conserved.*

5.5 The Energy-Momentum Tensor

The procedure of assembling together the charge density and its current density into the four-vector J^μ can be also used to construct a tensor quantity describing, together with the energy and momentum densities associated with a system of electric charges, the corresponding currents.

Let us denote by $T^{\mu 0}_{\text{part.}}$ the density of the total energy-momentum $P^\mu_{\text{part.}}$ of the system of charges, defined as:

$$T^{\mu 0}_{\text{part.}} = \sum_k p^\mu_k(t)\delta^3(\mathbf{x} - \mathbf{x}_k(t)), \tag{5.59}$$

where p^μ_k is the four-momentum of the single charge e_k. Upon integration over the whole space V, we find:

$$\int_V d^3\mathbf{x}\, T^{\mu 0}_{\text{part.}} = \sum_k p^\mu_k = P^\mu_{\text{part.}},$$

which is the *total four-momentum* of the system of charges. In particular its (00) component reads: $T^{00}_{\text{part}} = \frac{1}{c}\rho_E$ where ρ_E is the energy density of the system of charges.

We now define the *current density* of P^μ as the following three-vector:

$$T^{\mu i}_{\text{part.}} = \frac{1}{c}\sum_k p^\mu_k \frac{dx^i_k}{dt}\delta^3(\mathbf{x} - \mathbf{x}_k(t)).$$

In particular $T^{ji}_{\text{part.}}$ is the ith component of the current density associated with the jth component of the total momentum of the system.

We may now set together $T^{\mu 0}_{\text{part.}}$ and $T^{\mu i}_{\text{part.}}$ to build a 4×4 matrix $T^{\mu\nu}_{\text{part.}}$:

$$T^{\mu\nu}_{\text{part.}} = \frac{1}{c}\sum_k p^\mu_k \frac{dx^\nu_k}{dt}\delta^3(\mathbf{x} - \mathbf{x}_k(t)). \tag{5.60}$$

If we use the property that for each massive particle $p^\mu_k = m(v_k)\frac{dx^\mu_k}{dt}$, so that we can write:

$$\frac{p^\mu_k}{m(v_k)} = \frac{p^\mu_k}{E_k}c^2,$$

Equation (5.60) can be recast as follows:

$$T^{\mu\nu}_{\text{part.}} = c\sum_k \frac{p^\mu_k p^\nu_k}{E_k}\delta^3(\mathbf{x} - \mathbf{x}_k(t)) = T^{\nu\mu}_{\text{part.}} \tag{5.61}$$

showing that the matrix $T^{\mu\nu}_{\text{part.}}$ is manifestly *symmetric*.

We now prove that $T_{\text{part.}}^{\mu\nu}$ is a *tensor* under Lorentz transformations. Indeed, following the same steps used to prove that J^μ is a four-vector, we can rewrite Eq. (5.60) as follows:

$$T_{\text{part.}}^{\mu\nu} = \frac{1}{c} \int dt' \sum_k p_k^\mu \frac{dx_k^\nu}{dt'}(t') \, \delta^3(\mathbf{x} - \mathbf{x}_k(t')) \, \delta(t - t') \tag{5.62}$$

$$= \sum_k \int d\tau_k \, p_k^\mu \frac{dx_k^\nu}{d\tau_k}(\tau_k) \, \delta^4(x^\rho - x_k^\rho(\tau_k)). \tag{5.63}$$

where we have used (5.54) and (5.55).

Since $T_{\text{part.}}^{\mu\nu}$ transforms as the product of the two four-vectors p_k^μ and dx_k^ν it is, by definition, a rank two tensor, symmetric in the two upper indices. It is called the *energy-momentum* tensor of the charge system.[7]

Let us now compute its four-dimensional divergence $\partial_\mu T_{\text{part.}}^{\mu\nu}$. We first compute its three-dimensional counterpart from the definition of $T_{\text{part.}}^{\mu i}$ given above:

$$\partial_i T_{\text{part.}}^{\mu i} = \frac{1}{c} \sum_k p_k^\mu \frac{dx_k^i}{dt} \frac{\partial}{\partial x^i} \delta^3(\mathbf{x} - \mathbf{x}_k(t))$$

$$= -\frac{1}{c} \sum_k p_k^\mu \frac{dx_k^i}{dt} \frac{\partial}{\partial x_k^i} \delta^3(\mathbf{x} - \mathbf{x}_k(t)) = -\frac{1}{c} \sum_k p_k^\mu \frac{\partial}{\partial t} \delta^3(\mathbf{x} - \mathbf{x}_k(t))$$

$$= -\frac{\partial}{\partial t} \left(\frac{1}{c} \sum_k p_k^\mu \delta^3(\mathbf{x} - \mathbf{x}_k(t)) \right) + \frac{1}{c} \sum_k \left(\frac{d}{dt} p_k^\mu \right) \delta^3(\mathbf{x} - \mathbf{x}_k(t)). \tag{5.64}$$

On the other hand, from the definition (5.59), we have:

$$\frac{\partial}{\partial t} \left(\frac{1}{c} \sum_k p_k^\mu \, \delta^3(\mathbf{x} - \mathbf{x}_k(t)) \right) = \frac{\partial}{\partial x^0} T_{\text{part.}}^{0\mu}, \tag{5.65}$$

so that Eq. (5.64) becomes:

$$\partial_i T_{\text{part.}}^{i\mu} + \frac{\partial}{\partial x^0} T_{\text{part.}}^{0\mu} = \frac{1}{c} \sum_k \left(\frac{d}{dt} p_k^\mu \right) \delta^3(\mathbf{x} - \mathbf{x}_k(t)), \tag{5.66}$$

that is:

$$\partial_\mu T_{\text{part.}}^{\mu\nu} = \frac{1}{c} G^\nu, \tag{5.67}$$

---

[7] Note that since $T_{\text{part.}}^{\mu\nu}$ is a tensor, and being the product $p^\mu p^\nu$ a rank-two tensor as well, from Eq. (5.61) it follows that $\delta^3(\mathbf{x} - \mathbf{x}_k(t))/E_k$ transforms as a rank 0-tensor, that is a scalar quantity.

where the four-vector G^ν defines the *density of the total force* acting on the system:

$$G^\mu = \sum_k \frac{dp_k^\mu}{dt} \delta^3(\mathbf{x} - \mathbf{x}_k(t)). \tag{5.68}$$

It is important to note that, differently from the case of the four-current, where $\partial_\mu J^\mu = 0$, here we find:

$$\partial_\mu T_{\text{part.}}^{\mu\nu} \neq 0.$$

Actually, this was to be expected since $\partial_\mu T_{\text{part.}}^{\mu\nu} = 0$ would imply the conservation of the total four-momentum $P_{\text{part.}}^\mu$ of the system of charged particles, and this cannot be true since the system is *not isolated* being in interaction with the electromagnetic field.

We may however expect that, since the *total* system consisting not only of the particles, but also of the electromagnetic field, is isolated, *its total four momentum is conserved*.

Let us show this in detail. We first observe that in the expression of the Lorentz force-density, given by Eq. (5.68), we may replace on the right hand side $\frac{dp_k^\mu}{dt}$ with $f^\mu \frac{d\tau}{dt}$. For our system of charges in interaction with the electromagnetic field we may therefore write:

$$\begin{aligned} G^\mu &= \sum_k f^\mu \frac{d\tau}{dt} \delta^3(\mathbf{x} - \mathbf{x}_k(t)) \\ &= -\frac{1}{c} \sum_k e_k \frac{d\tau}{dt} F^\mu{}_\nu \frac{dx_k^\nu}{d\tau} \delta^3(\mathbf{x} - \mathbf{x}_k(t)) \\ &= -\frac{1}{c} F^\mu{}_\nu \sum_k e_k \frac{dx_k^\nu}{dt} \delta^3(\mathbf{x} - \mathbf{x}_k(t)) = -F^\mu{}_\nu J^\nu. \end{aligned} \tag{5.69}$$

where we have used the dynamic definition (5.27) of the four-force exerted by the electromagnetic field on a charge. Using the Maxwell equation $\partial_\rho F^{\rho\nu} = -J^\nu$, G^μ can be rewritten as follows:

$$\begin{aligned} G^\mu &= F^\mu{}_\nu \partial_\rho F^{\rho\nu} = \partial_\rho(F^\mu{}_\nu F^{\rho\nu}) - F^{\rho\nu}(\partial_\rho F^\mu{}_\nu) \\ &= \partial_\rho(F^\mu{}_\nu F^{\rho\nu}) - \frac{1}{2} F^{\rho\nu}(\partial_\rho F^\mu{}_\nu - \partial_\nu F^\mu{}_\rho) \\ &= \partial_\rho(F^\mu{}_\nu F^{\rho\nu}) - \frac{1}{2} F_{\rho\nu}(\partial^\rho F^{\mu\nu} + \partial^\nu F^{\rho\mu}), \end{aligned} \tag{5.70}$$

where we have replaced $\partial_\rho F^\mu{}_\nu$ with its antisymmetric part in (ρ, ν), since it is contracted with $F^{\rho\nu}$.

Next, by using the homogeneous Maxwell equation $\partial^{[\rho} F^{\mu\nu]} = 0$, we can make the following replacement in the last term of the previous equation:

$$\partial^\rho F^{\mu\nu} + \partial^\nu F^{\rho\mu} = -\partial^\mu F^{\nu\rho} = \partial^\mu F^{\rho\nu},$$

and we find:

$$G^\mu = \partial_\rho (F^\mu{}_\nu F^{\rho\nu}) - \frac{1}{2} F_{\rho\nu} \partial^\mu F^{\rho\nu} = \partial_\rho (F^\mu{}_\nu F^{\rho\nu}) - \frac{1}{4} \partial^\mu (F_{\rho\nu} F^{\rho\nu})$$

$$= \partial_\rho \left(F^\mu{}_\nu F^{\rho\nu} - \frac{1}{4} \eta^{\mu\rho} F_{\sigma\nu} F^{\sigma\nu} \right) = -c \, \partial_\rho T_{em}^{\rho\mu},$$

where we have defined the *energy-momentum tensor of the electromagnetic field*:

$$T_{em}^{\mu\nu} = -\frac{1}{c} \left(F^\mu{}_\rho F^{\nu\rho} - \frac{1}{4} \eta^{\mu\nu} F_{\rho\sigma} F^{\rho\sigma} \right). \tag{5.71}$$

We note that, as for the particle energy- momentum tensor, $T_{em}^{\rho\mu}$ defines *the distribution in space of the energy and momentum, and of their currents, associated with the electromagnetic field.*[8]

We now substitute the expression of f^μ into (5.67), to obtain:

$$\partial_\mu T_{\text{part.}}^{\mu\nu} = \frac{1}{c} G^\nu \Leftrightarrow \partial_\mu \left(T_{\text{part.}}^{\mu\nu} + T_{em}^{\nu\mu} \right) = 0.$$

Defining the sum of the energy-momentum tensors of particles and electromagnetic field as the *total energy-momentum tensor* $T^{\mu\nu}$ of the system, we obtain the result:

$$\partial_\mu T^{\mu\nu} = 0. \tag{5.72}$$

As in the case of the four dimensional current, the vanishing of the four-dimensional divergence (5.72) implies the *conservation of the total four-momentum* P_{tot}^μ, that is of the total energy and linear momentum of the *isolated* system consisting of the charges *and* the electromagnetic field.

Indeed, following the same steps as in the derivation of the charge conservation from the continuity equation, and setting $V = \mathbb{R}^3$, we have:

$$\int_V d^3\mathbf{x} \, \partial_0 T^{\mu 0} = -\int_V d^3\mathbf{x} \, \partial_i T^{\mu i} = -\int_{S_\infty} dS \, T^{\mu i} \, n_i. \tag{5.73}$$

---

[8]Note that in our conventions all the components of $T_{em}^{\mu\nu}$ have the physical dimensions of a momentum density.

Since there is no energy or momentum density at spatial infinity, the last integral vanishes and we obtain:

$$\frac{\partial}{\partial t} \int_V d^3\mathbf{x} \, T^{\mu 0} = 0 \quad \Rightarrow \quad \frac{dP^\mu_{tot}}{dt} = 0. \tag{5.74}$$

Coming back to the expression (5.71) of the electromagnetic energy-momentum tensor, we now show how to retrieve the familiar definitions of energy, linear momentum, and Poynting vector from our four-dimensional formalism.

If we compute the density $T^{0\mu}_{em}$ in terms of the fields $\mathbf{E}$ and $\mathbf{B}$, we find:

$$T^{00}_{em} = -\frac{1}{c} \left(F^0{}_i F^{0i} - \frac{1}{2} F_{0i} F^{0i} - \frac{1}{4} F_{ij} F^{ij} \right) \tag{5.75}$$

$$= -\frac{1}{c} \left(-E_i E_i + \frac{1}{2} E_i E_i - \frac{1}{2} B_i B_i \right) \tag{5.76}$$

$$= \frac{1}{2c} \left(|\mathbf{E}|^2 + |\mathbf{B}|^2 \right) = \frac{1}{c} \rho_E, \tag{5.77}$$

where:

$$\rho_E = \frac{1}{2} \left(|\mathbf{E}|^2 + |\mathbf{B}|^2 \right), \tag{5.78}$$

is the *energy density* of the electromagnetic field.

Moreover:

$$T^{i0}_{em} = -\frac{1}{c} F^{ij} F^0{}_j = \frac{1}{c} F^{ij} F^{0j} = \frac{1}{c} \epsilon_{ijk} B^k E^j$$

$$= \frac{1}{c} (\mathbf{E} \times \mathbf{B})^i = \pi^i_{em} = \frac{1}{c^2} S^i, \tag{5.79}$$

where:

$$\pi^i_{em} = \frac{1}{c} (\mathbf{E} \times \mathbf{B})^i, \tag{5.80}$$

is the *momentum density* of the electromagnetic field and $\mathbf{S} = c\,\mathbf{E} \times \mathbf{B}$ is the *Poynting vector* measuring the *energy current density* carried by the electromagnetic field. Note that $|\mathbf{S}| = \rho_E\, c$.

We can rephrase the previous results, by stating that the energy $\mathcal{E}$ and the linear momentum $\mathcal{P}$ associated with an electromagnetic field, in a given volume V, are given by:

$$\mathcal{E} = c \int_V d^3\mathbf{x} \, T^{00}_{em} = \int_V d^3\mathbf{x} \, \rho_E, \tag{5.81}$$

$$\mathcal{P}^i = \int_V d^3\mathbf{x}\, T_{em}^{i0} = \frac{1}{c^2} \int_V d^3\mathbf{x}\, S^i. \tag{5.82}$$

As an example, suppose we consider a region where no charges are present. In this case, $T_{part}^{\mu\nu} = 0$, and Eq. (5.72) reduces to $\partial_\mu T_{em}^{\mu\nu} = 0$. Separating the $\nu = 0$ and the $\nu = i$ components, we have:

$$\frac{\partial}{\partial t} \rho_E + \partial_i S^i = 0,$$

which expresses the *conservation of energy* in its local form. Upon integration over a volume V we have:

$$\frac{d}{dt} \int_V d^3\mathbf{x}\, \rho_E \equiv \frac{d}{dt} \mathcal{E} = -\int_V d^3\mathbf{x}\, \partial_i S^i = -\int_S d\sigma\, \mathbf{n} \cdot \mathbf{S}. \tag{5.83}$$

As is well known, the physical interpretation of this equation is the following: The positive (negative) rate of change of electromagnetic energy inside the volume is compensated by the incoming (outgoing) flux of energy across the boundary S. In particular, when the integration volume is infinite, owing to the vanishing of the surface integral, the equation implies the conservation of the electromagnetic energy, in its global form:

$$\frac{d\mathcal{E}}{dt} = 0.$$

Similarly, when $\nu = i$, we obtain:

$$\frac{1}{c} \frac{\partial}{\partial t} \pi^i + \partial_j T_{em}^{ji} = 0,$$

and, upon integration over V, we find:

$$\frac{1}{c} \frac{d}{dt} \mathcal{P}^i = -\int_V d^3\mathbf{x}\, \partial_j T_{em}^{ji} = -\int_S dS\, T^{ji} n^j, \tag{5.84}$$

where, as usual, we have applied the divergence theorem (over index j) to the volume integral on the right hand side. As in the previous case, Eq. (5.84) means that the positive (negative) rate of change of electromagnetic linear momentum inside the volume V is compensated by the incoming (outgoing) flux of momentum across S. In particular, when the integration volume is infinite, the equation implies the conservation of the electromagnetic linear momentum:

$$\frac{d}{dt} \mathcal{P}^i = 0.$$

5.6 The Four-Potential

We now observe that *the homogeneous Maxwell equation* (5.19) *can be immediately solved introducing a four-vector* $A_\mu = (A_0, A_i)$, $i = 1, 2, 3$, *and setting*:

$$F_{\mu\nu} = \partial_\mu A_\nu - \partial_\nu A_\mu. \tag{5.85}$$

Indeed, since on any function $\partial_\mu \partial_\nu$ is a symmetric tensor, the total antisymmetrization of (5.19), using the definition (5.85), gives identically zero:

$$\partial_{[\mu} F_{\nu\rho]} = \partial_{[\mu} \partial_\nu A_{\rho]} - \partial_{[\mu} \partial_\rho A_{\nu]} = 2\partial_{[\mu} \partial_\nu A_{\rho]} = 0.$$

The four-vector A_μ is given the name of *four-potential*.

We may readily express **E** e **B** in terms of A_μ, using Eqs. (5.14) and (5.85):

$$E^i = -F_{0i} = -\partial_0 A_i + \partial_i A_0 = -\frac{1}{c}\frac{\partial}{\partial t} A_i + \partial_i A_0, \tag{5.86}$$

$$B_i = \frac{1}{2}\epsilon_{ijk} F_{jk} = \epsilon_{ijk}\partial_j A_k. \tag{5.87}$$

These formulae allow us to identify $-A_0(\mathbf{x}, \mathbf{t})$ and $A_i(\mathbf{x}, \mathbf{t})$ with the electric potential V and the vector potential $\mathbf{A}$, respectively; indeed, as is well known in the electromagnetic theory, one has:

$$\mathbf{E} = -\frac{1}{c}\frac{\partial}{\partial t}\mathbf{A} - \nabla V. \tag{5.88}$$

and

$$\mathbf{B} = \nabla \times \mathbf{A}, \tag{5.89}$$

which are the vector form of the previous Eqs. (5.86) and (5.87). The four-potential, however, *is not uniquely defined*; if we redefine A_μ by adding the four-dimensional gradient of a scalar field,

$$A_\mu \rightarrow A'_\mu = A_\mu + \partial_\mu\varphi, \tag{5.90}$$

where $\varphi(x) \equiv \varphi(x^\mu)$ is an arbitrary function of the space-time coordinates, then $F_{\mu\nu}$, and therefore **E** and **B**, remains *unchanged*.[9] Indeed:

---

[9]While in the classical theory the only *measurable physical quantities* are **E** and **B**, so that the four-potential seems not necessary for a complete description of the electromagnetic field, in quantum-mechanics the *Aharonov-Bohm effect* shows that the **E** and **B** fields are not sufficient for describing the electromagnetic field in interaction with matter, and that for its full description the four-potential $A_\mu(x)$ is necessary.

$$F_{\mu\nu} \longrightarrow F'_{\mu\nu} = \partial_\mu A'_\nu - \partial_\nu A'_\mu = \partial_\mu A_\nu - \partial_\nu A_\mu$$
$$+ \partial_\mu \partial_\nu \varphi - \partial_\nu \partial_\mu \varphi = \partial_\mu A_\nu - \partial_\nu A_\mu = F_{\mu\nu}. \tag{5.91}$$

From the invariance of $F_{\mu\nu}$ under the change (5.90), it follows that Maxwell's equations (5.18) and (5.19) are invariant as well. The transformation (5.90) is called *gauge transformation* and the corresponding invariance of the Maxwell equations is referred to as *the gauge invariance of electromagnetism.*

One can exploit the gauge invariance of electromagnetism to simplify Eq. (5.18). Indeed one can choose the arbitrary scalar field $\varphi(x)$ in such a way that the transformed four-potential satisfy an auxiliary condition. We may for example require:

$$\partial_\mu A'^\mu \equiv \frac{1}{c}\frac{\partial}{\partial t} A'_0 - \partial_i A'_i = 0. \tag{5.92}$$

Using Eq. (5.90), we see that we can always construct a four-potential A'_μ satisfying the above equation starting from one (A_μ) which does not, by choosing $\varphi(x)$ in such a way as to solve the equation:

$$\partial_\mu A'^\mu = \partial_\mu A^\mu + \partial_\mu \partial^\mu \varphi = 0 \rightarrow \partial_\mu A^\mu + \Box \varphi = 0 \tag{5.93}$$

where we have introduced the (*Lorentz-invariant*) d'Alembertian operator:

$$\Box \equiv \partial_\mu \partial^\mu = (\partial_0)^2 - \sum_{i=1}^{3} (\partial_i)^2 = \frac{1}{c^2}\frac{\partial^2}{\partial t^2} - \nabla^2.$$

Indeed, as is well known, given $\partial_\mu A^\mu$ and suitable Cauchy data, Eq. (5.93) always admits a solution in the unknown function $\varphi(x)$.

Thus *gauge invariance* implies that we can always choose a four-potential $A_\mu(x)$ satisfying

$$\partial_\mu A^\mu = 0. \tag{5.94}$$

As the above condition fixes (though not completely) the gauge function $\varphi(x)$, see Eq. (5.111) and below, it is called a gauge-fixing condition. The corresponding choice of φ, such that (5.94) holds, is referred to as the *Lorentz gauge*.[10] Note that the Lorentz gauge fixing condition (or simply Lorentz gauge condition) is a scalar under Lorentz transformations.

When the Lorentz gauge is used, Eq. (5.18) simplifies considerably. Indeed, writing on the left hand side of the inhomogeneous Maxwell equation (5.18) $F_{\mu\nu}$ in terms

---

[10]This is not the only possible gauge condition. Several other choices are possible. In particular, when discussing the quantization of the electromagnetic field in Chap. 6, we shall use the more convenient Coulomb gauge $\nabla \cdot \mathbf{A}(x) = 0$. The Lorentz gauge has the advantage of being Lorentz covariant. In Chap. 11, the same quantization will be performed using the covariant Lorentz gauge.

of A_μ and using the Lorentz gauge condition (5.94) we obtain:

$$\partial_\mu F^{\mu\nu} = \partial_\mu \partial^\mu A^\nu - \partial_\mu \partial^\nu A^\mu = \Box A^\nu - \partial^\nu(\partial_\mu A^\mu) = \Box A^\nu,$$

that is, each component of the four-potential A_μ satisfies the well known *wave equation* in the presence of a source:

$$\Box A^\mu = -J^\mu. \tag{5.95}$$

Let us consider the case where, in some space domain, the source is absent, $J^\mu(x) \equiv 0$; then Eq. (5.95) becomes:

$$\Box A_\mu = \frac{1}{c^2} \frac{\partial^2 A_\mu}{\partial t^2} - \nabla^2 A_\mu = 0. \tag{5.96}$$

It is well known that the general solution to the homogeneous wave equation can be written as a superposition of plane waves whose polarization is orthogonal to the direction of propagation. Let us retrieve this result in our Lorentz- covariant formalism.

We start solving Eq. (5.96) within a bounded region shaped as a parallelepiped, with sides L_A, L_B, L_C along the three Cartesian axes and volume $V = L_A L_B L_C$, requiring periodic boundary conditions on the solution in each coordinate. This allows us to expand the field $A_\mu(\mathbf{x}, t)$ in a triple *Fourier series* with respect to the coordinates $\mathbf{x} = (x, y, z)$:

$$A_\mu(\mathbf{x}, t) = \sum_{k_1 k_2 k_3} \tilde{A}_{\mathbf{k}, \mu}(t) \, e^{i\,\mathbf{k}\cdot\mathbf{x}}. \tag{5.97}$$

where the components of the wave number vector $\mathbf{k} = (k_1, k_2, k_3)$ have the following discrete values:

$$k_1 = \frac{2\pi n_1}{L_A}; \quad k_2 = \frac{2\pi n_2}{L_B}; \quad k_3 = \frac{2\pi n_3}{L_C}, \tag{5.98}$$

n_1, n_2, n_3 being integers. Reality of $A_\mu(x)$ further imposes that: $\tilde{A}_{-\mathbf{k}, \mu}(t) = \tilde{A}_{\mathbf{k}, \mu}(t)^*$. Inserting the expansion (5.97) in Eq. (5.96), in virtue of linearity of Maxwell's equations in the vacuum, $A_\mu(x)$ is a solution if and only if each of its Fourier components are. Equation (5.96) on the generic $\mathbf{k}$-component reads:

$$\frac{d^2}{dt^2} \tilde{A}_{\mathbf{k}, \mu}(t) + c^2 \,|\mathbf{k}|^2 \, \tilde{A}_{\mathbf{k}, \mu}(t) = 0, \tag{5.99}$$

where we have used the property that

$$\partial_i A_\mu(x) = \sum_{\mathbf{k}} i \, k_i \, \tilde{A}_{\mathbf{k}, \mu}(t) \, e^{i\,\mathbf{k}\cdot\mathbf{x}} \;\Rightarrow\; \nabla^2 \tilde{A}_\mu(x) = -\sum_{\mathbf{k}} |\mathbf{k}|^2 \, \tilde{A}_{\mathbf{k}, \mu}(t) \, e^{i\,\mathbf{k}\cdot\mathbf{x}}.$$

$$\tag{5.100}$$

From Eq. (5.99) we see that each component of $\tilde{A}_{\mathbf{k},\,\mu}(t)$ has to satisfy the equation of a harmonic oscillator of angular frequency

$$\omega_k = c\,|\mathbf{k}| \tag{5.101}$$

and can thus be written in the form:

$$\tilde{A}_{\mathbf{k},\,\mu}(t) = \epsilon_{\mathbf{k},\,\mu}(\omega_k)\,e^{-i\omega_k t} + \epsilon_{\mathbf{k},\,\mu}(-\omega_k)\,e^{i\omega_k t}, \tag{5.102}$$

where the reality condition on $A_\mu(x)$ further requires that:

$$\epsilon_{-\mathbf{k},\,\mu}(-\omega_k) = \epsilon_{\mathbf{k},\,\mu}(\omega_k)^*. \tag{5.103}$$

The solution (5.97) to Maxwell's equation in the vacuum then reads:

$$
\begin{aligned}
A_\mu(\mathbf{x}, t) &= \sum_{\mathbf{k}} \epsilon_{\mathbf{k},\,\mu}(\omega_k)\,e^{-i(\omega_k t - \mathbf{k}\cdot\mathbf{x})} + \sum_{\mathbf{k}} \epsilon_{\mathbf{k},\,\mu}(-\omega_k)\,e^{i(\omega_k t + \mathbf{k}\cdot\mathbf{x})} \\
&= \sum_{\mathbf{k}} \epsilon_{\mathbf{k},\,\mu}(\omega_k)\,e^{-i(\omega_k t - \mathbf{k}\cdot\mathbf{x})} + \sum_{\mathbf{k}} \epsilon_{-\mathbf{k},\,\mu}(-\omega_k)\,e^{i(\omega_k t - \mathbf{k}\cdot\mathbf{x})} \\
&= \sum_{\mathbf{k}} \left(\epsilon_{\mathbf{k},\,\mu}\,e^{-i(\omega_k t - \mathbf{k}\cdot\mathbf{x})} + \epsilon_{\mathbf{k},\,\mu}^*\,e^{i(\omega_k t - \mathbf{k}\cdot\mathbf{x})} \right) \\
&= \sum_{\mathbf{k}} \left(\epsilon_{\mathbf{k},\,\mu}\,e^{-i k\cdot x} + \epsilon_{\mathbf{k},\,\mu}^*\,e^{i k\cdot x} \right), \tag{5.104}
\end{aligned}
$$

where we have defined $\epsilon_{\mathbf{k},\,\mu} \equiv \epsilon_{\mathbf{k},\,\mu}(\omega_k)$ referred to as the *polarization four-vector* and used Eq. (5.103). Moreover we have also defined the *wave-number* four-vector as:

$$k \equiv (k^\mu) = (k^0, k^i); \quad k^0 \equiv \frac{\omega_k}{c}, \tag{5.105}$$

so that

$$k \cdot x \equiv k^\mu x_\mu = k^\mu \eta_{\mu\nu} x^\nu = k^0 x^0 - \mathbf{k}\cdot\mathbf{x},$$

Equation (5.104) represents an expansion of the electromagnetic potential $A_\mu(x)$ in plane waves, progressing in the direction of the wave number vector $\mathbf{k}$ with angular frequency $\omega_k \equiv \frac{2\pi}{T} = c\,|\mathbf{k}|$. We shall also write the solution (5.104) in the more implicit form:

$$A_\mu(\mathbf{x}, t) = \sum_{\mathbf{k}} \left(A_{\mathbf{k},\,\mu}(t)\,e^{i\,\mathbf{k}\cdot\mathbf{x}} + A_{\mathbf{k},\,\mu}(t)^*\,e^{-i\,\mathbf{k}\cdot\mathbf{x}} \right), \tag{5.106}$$

where $A_{\mathbf{k},\,\mu}(t) \equiv \epsilon_{\mathbf{k},\,\mu}\,e^{-i\omega_k t}$ and $A_{\mathbf{k},\,\mu}(t)^*$ are the two independent solutions to the harmonic oscillator equation (5.99).

From its definition and taking into account (5.101), we see that $k \equiv (k^\mu)$ is light-like

$$k^2 \equiv k^\mu k_\mu = (k^0)^2 - |\mathbf{k}|^2 = \frac{\omega_k^2}{c^2} - |\mathbf{k}|^2 = 0.$$

Anticipating part of the discussion of next chapter, we may give a quantum interpretation to our result; indeed, from the quantum theory point of view, a monochromatic plane wave can be interpreted as the *relativistic wave function of a particle of energy* $E = \hbar \omega_k$ *and momentum* $\mathbf{p} = \hbar \mathbf{k}$, that is having a four-momentum[11]:

$$p^\mu = \hbar k^\mu = \hbar \left(\frac{\omega_k}{c}, \mathbf{k} \right). \tag{5.107}$$

If we now recall the general relation between energy and momentum given in Eq. (2.38) of Chap. 2, we see that, from the quantum point of view, the Maxwell equations imply:

$$m^2 c^2 = p^2 = \hbar^2 k^2 = 0, \tag{5.108}$$

that is *the rest mass of a particle associated with the plane wave solution to Eq. (5.96), called photon, is exactly zero.*

5.6.1 The Spin of a Plane Wave

We must still require $A_\mu(x)$ to satisfy, besides the wave equation, the Lorentz gauge condition (5.94).

As we are going to show, this requirement implies that, at each point $\mathbf{x}$, the physical degrees of freedom of a freely propagating electromagnetic field are just two.

Indeed, if we apply condition (5.94) to the generic plane wave superposition (5.104) and separately equate each Fourier component $\mathbf{k}$ to zero, we easily find:

$$\partial_\mu A^\mu = 0 \quad \Leftrightarrow \quad k^\mu \epsilon_{\mathbf{k},\,\mu} = 0, \qquad \forall \mathbf{k}. \tag{5.109}$$

Now we observe that, since k^μ is a light-like vector, i.e. $k^2 = 0$, it is always possible to find a RF where it takes the form $k^\mu \equiv (\kappa, \kappa, 0, 0)$ and thus $k_\mu = (\kappa, -\kappa, 0, 0)$. In this RF, using Eq. (5.109), the polarization four-vector has the following form:

$$\epsilon_{\mathbf{k},\,\mu} \equiv (\epsilon_{\mathbf{k}}, -\epsilon_{\mathbf{k}}, \epsilon_2, \epsilon_3). \tag{5.110}$$

One may conclude that, for each term of the expansion, the degrees of freedoms, that is the independent components of the polarization four-vector are three: $\epsilon_{\mathbf{k}}, \epsilon_2, \epsilon_3$.

---

[11] This interpretation was already proposed by Einstein in 1905 for the description of the photoelectric effect.

However this is not the end of the story: As previously anticipated, *the choice of the Lorentz gauge does not completely fix the gauge freedom.* We may indeed perform on $A_\mu(x)$ a further gauge transformation

$$A_\mu \to A_\mu + \partial_\mu \varphi, \tag{5.111}$$

still preserving $\partial_\mu A^\mu = 0$, provided:

$$\Box \varphi = 0. \tag{5.112}$$

This is easily shown by implementing such transformation on the four-potential in the Lorentz gauge condition. Since it implies:

$$\partial_\mu A^\mu \longrightarrow \partial_\mu (A^\mu + \partial^\mu \varphi), \tag{5.113}$$

if condition (5.112) holds, the transformed field will still be in the Lorentz gauge. A solution φ to the wave equation (5.112), describing the residual gauge symmetry, can be expressed by the same Fourier expansion (5.104) as $A_\mu(x)$:

$$\varphi(x) = \sum_{\mathbf{k}} \left(\xi_{\mathbf{k}} \, e^{-i\,k\cdot x} + \xi_{\mathbf{k}}^* e^{i\,k\cdot x} \right); \quad k^2 = 0,$$

so that:

$$\partial_\mu \varphi(x) = \sum_{\mathbf{k}} \left(-i\,k_\mu \xi_{\mathbf{k}} \, e^{-ik\cdot x} + i\,k_\mu \xi_{\mathbf{k}}^* \, e^{ik\cdot x} \right).$$

Therefore under the transformation (5.111), supplemented by condition (5.112), the solution (5.104) takes the form:

$$A'_\mu(x) = \sum_{\mathbf{k}} \left(\epsilon'_{\mathbf{k}, \mu} e^{-i\,k\cdot x} + \epsilon'^{*}_{\mathbf{k}, \mu} e^{i\,k\cdot x} \right), \tag{5.114}$$

where

$$\epsilon'_{\mathbf{k}, \mu} = \epsilon_{\mathbf{k}, \mu} - ik_\mu \, \xi_{\mathbf{k}} \equiv (\epsilon_{\mathbf{k}} - i\kappa \xi_{\mathbf{k}}, -\epsilon_{\mathbf{k}} + i\kappa \xi_{\mathbf{k}}, \epsilon_2, \epsilon_3). \tag{5.115}$$

Being φ, and therefore $\xi_{\mathbf{k}}$, arbitrary, we may fix $\xi_{\mathbf{k}}$ in such a way that the first two components of the polarization four-vector both vanish. In particular, setting $\xi_{\mathbf{k}} = -i\frac{\epsilon_{\mathbf{k}}}{\kappa}$, we obtain:

$$\epsilon'_{\mathbf{k}, \mu} = (0, 0, \epsilon_2, \epsilon_3). \tag{5.116}$$

We see that, once the gauge freedom has been completely fixed by Eqs. (5.93) and (5.112), *the independent components of the polarization four-vector are only two,* and precisely those *transverse* to the propagation vector $\mathbf{k}$, namely ϵ_2, ϵ_3.

Since there are only two physical components of the four-dimensional polarization vector for each wave vector $\mathbf{k}$, we conclude that a generic electromagnetic wave $A_\mu(x)$ has, at each point in space $\mathbf{x}$, only two *physical* degrees of freedom, both transverse to the direction of propagation.

This fact leads us to the concept of *spin of a plane wave*, or better, from the quantum point of view, of *spin of a photon*.

Let us define the *spin group* of a particle as the residual subgroup of the Lorentz group which remains once we fix our RF to be attached to the particle itself. Quite generally when our particle has a non-vanishing rest mass this RF is defined as its rest frame, in which $p^\mu = (mc, \mathbf{p} = \mathbf{0}) \rightarrow m^2 = \frac{p^2}{c^2} > 0$. In this case the residual group, that is the spin group, is the rotation group $\tilde{SO}(3)$, subgroup of the Lorentz group, which leaves the p^0 component of the four-momentum invariant.

If, on the other hand, the particle has *vanishing rest mass*, there exists no RF where the particle is at rest, its velocity being c. In this case only *rotations around the propagation direction of the particle, corresponding to two-dimensional rotations on the transverse plane, can be properly defined as the relevant spin group*.

This latter is clearly the case of the electromagnetic field, the relevant particle being the *photon*. We want to see how its physical degrees of freedom ϵ_2, ϵ_3 transform under the SO(2) rotations in the transverse plane, in our case the $y - z$ plane. This can be easily found by recalling that $A_\mu(x)$ is a covariant vector field, thus transforming, under Lorentz transformations, as follows:

$$A_\mu(x) \rightarrow A'_\mu(x') = \Lambda^{-1\,\nu}{}_\mu\, A_\nu(x),$$

where $x' = \Lambda \cdot x \Rightarrow x = \Lambda^{-1} \cdot x'$. In our case, the subgroup of the Lorentz group leaving the components $p^0 \equiv \hbar k^0$ and $p^1 \equiv \hbar k^1$ (along the x-direction) of the photon momentum invariant, apparently coincides with the SO(2) rotation subgroup whose generic element $\Lambda^{(0)}$ has the following matrix form:

$$\Lambda^{(0)} = \begin{pmatrix} 1 & 0 & 0 & 0 \\ 0 & 1 & 0 & 0 \\ 0 & 0 & \cos\theta & \sin\theta \\ 0 & 0 & -\sin\theta & \cos\theta \end{pmatrix}. \tag{5.117}$$

Since in this RF the polarization four-vector has the form: $\epsilon_\mu = (0, 0, \epsilon_2, \epsilon_3)$, the spin group acts on the physical degrees of freedom as follows:

$$\begin{pmatrix} \epsilon'_2 \\ \epsilon'_3 \end{pmatrix} = \begin{pmatrix} \cos\theta & \sin\theta \\ -\sin\theta & \cos\theta \end{pmatrix} \begin{pmatrix} \epsilon_2 \\ \epsilon_3 \end{pmatrix}. \tag{5.118}$$

It is convenient to use the complex basis $\epsilon_2 \pm i\epsilon_3$, so that the rotation matrix takes a diagonal form and the transformation (5.118) becomes:

$$\epsilon_2 \pm i\epsilon_3 \rightarrow \epsilon'_2 \pm i\epsilon'_3 = e^{\mp i\theta}(\epsilon_2 \pm i\epsilon_3).$$

In general, when the two polarization states of a massless particle transform, under a rotation about the propagation direction, with a factor $e^{\mp in\theta}$, one defines $n\hbar$ to be the *spin* of the particle (the particle is simply said to have spin n). In our case we have found $n = 1$, so that we conclude that *the photon*, the massless particle associated with an electromagnetic plane wave, has spin one.

5.6.2 Large Volume Limit

In this section we have solved Maxwell's equations in the vacuum within a finite "box" of sides L_A, L_B, L_C. This had the advantage of allowing us to work with Fourier series instead of integrals, when expanding the solution $A_\mu(x)$ as a super-position of plane waves. The components of the wave number vector $\mathbf{k}$ are indeed discrete quantities, the step Δk_i between two successive values of one of them k_i ($i = 1, 2, 3$) being $\Delta k_i = 2\pi/L_i$, $L_i = L_A, L_B, L_C$. An *elementary cell* in the space parametrized by k_1, k_2, k_3 has then volume:

$$\Delta^3 \mathbf{k} \equiv \Delta k_1 \Delta k_2 \Delta k_3 = \frac{(2\pi)^3}{V}. \tag{5.119}$$

We can write the discrete sum over the Fourier modes $\mathbf{k}$ of a function $f(\mathbf{x})$ as:

$$f(\mathbf{x}) = \sum_{\mathbf{k}} f_{\mathbf{k}} \, e^{i\,\mathbf{k}\cdot\mathbf{x}} = \sum_{\mathbf{k}} \frac{\Delta^3 \mathbf{k}}{(2\pi)^3} \, V \, f_{\mathbf{k}} \, e^{i\,\mathbf{k}\cdot\mathbf{x}}. \tag{5.120}$$

In the *large volume limit*, in which the size of the box becomes infinite, $L_A, L_B, L_C \to \infty$, $\Delta k_i \to dk_i$, the k_i's become continuous variables and the Fourier sum in (5.120) is replaced by an integral in $d^3\mathbf{k} \equiv dk_1 dk_2 dk_3$:

$$f(\mathbf{x}) = \sum_{\mathbf{k}} \frac{\Delta^3 \mathbf{k}}{(2\pi)^3} \, V \, f_{\mathbf{k}} \, e^{i\,\mathbf{k}\cdot\mathbf{x}} \;\to\; \int \frac{d^3\mathbf{k}}{(2\pi)^3} \, V \, f(\mathbf{k}) \, e^{i\,\mathbf{k}\cdot\mathbf{x}}, \tag{5.121}$$

where the notation $f(\mathbf{k}) \equiv f_{\mathbf{k}}$ emphasizes the fact that we are now treating $\mathbf{k}$ as a continuous variable rather than a discrete label.

Thus the passage to the large volume limit is effected by replacing, in the triple Fourier series:

$$\sum_{\mathbf{k}} \to \int \frac{d^3\mathbf{k}}{(2\pi)^3} \, V, \tag{5.122}$$

which amounts to passing from a *Fourier series* expansion to a *Fourier integral*. The general expansion (5.104) of the solution $A_\mu(x)$, in this limit, reads

$$A_\mu(\mathbf{x}) = \int \frac{d^3\mathbf{k}}{(2\pi)^3} V \left(\epsilon_\mu(\mathbf{k}) e^{-i\,k\cdot x} + \epsilon_\mu(\mathbf{k})^* e^{i\,k\cdot x}\right). \qquad (5.123)$$

Let us finally note that the delta-function $\delta_{\mathbf{k},\mathbf{k}'}$, defined on discrete momenta as follows

$$\delta_{\mathbf{k},\mathbf{k}'} = \begin{cases} 1 & \mathbf{k} = \mathbf{k}' \\ 0 & \mathbf{k} \neq \mathbf{k}' \end{cases}, \qquad (5.124)$$

so that

$$\sum_{\mathbf{k}'} \delta_{\mathbf{k},\mathbf{k}'} f_{\mathbf{k}'} = f_{\mathbf{k}}, \qquad (5.125)$$

in the continuum limit becomes

$$\delta_{\mathbf{k},\mathbf{k}'} \longrightarrow \frac{(2\pi)^3}{V} \delta^3(\mathbf{k} - \mathbf{k}'). \qquad (5.126)$$

Indeed we find:

$$\sum_{\mathbf{k}'} \delta_{\mathbf{k},\mathbf{k}'} f_{\mathbf{k}'} \to \int \frac{d^3\mathbf{k}}{(2\pi)^3} V \frac{(2\pi)^3}{V} \delta^3(\mathbf{k} - \mathbf{k}') f(\mathbf{k}') = f(\mathbf{k}). \qquad (5.127)$$

5.6.3 References

For further reading see Ref. [8] (Vol. 2).

Chapter 6
Quantization of the Electromagnetic Field

In this section we shall analyze the general solution $A_\mu(x)$ to Maxwell's equations
in the vacuum by choosing as gauge fixing condition the so called *Coulomb gauge*
differently from the *Lorentz gauge* condition used in the general discussion of the
electromagnetic field given in Chap. 5. In this new framework we shall be able to
describe the electromagnetic field as a collection of infinitely many decoupled harmonic oscillators. This will pave the way for the quantization of the electromagnetic
field and the consequent introduction of the notion of photon.

6.1 The Electromagnetic Field as an Infinite System of Harmonic Oscillators

Let us now still consider an electromagnetic field, described by the vector potential
$A_\mu(x) \equiv A_\mu(\mathbf{x}, t)$ in a region which is "far away" from any charge and current. It is
a solution to the following Maxwell's equations:

$$\partial_\mu F^{\mu\nu} = 0; \quad F_{\mu\nu} = \partial_\mu A_\nu - \partial_\nu A_\mu. \tag{6.1}$$

As shown in the previous chapter, Equations (6.1) are invariant under gauge transformations:

$$A_\mu \to A_\mu + \partial_\mu \phi. \tag{6.2}$$

We have also shown that, upon using the Lorentz gauge:

$$\partial_\mu A^\mu = 0, \tag{6.3}$$

and by suitably fixing ϕ, we can set to zero, for each term in the Fourier expansion, the time-component and the longitudinal component, proportional to $\mathbf{k}$, of the
polarization four-vector $\epsilon_{\mathbf{k}, \mu}$, that is two components of the four-potential $A_\mu(x)$.

© Springer International Publishing Switzerland 2016 17
R. D'Auria and M. Trigiante, *From Special Relativity to Feynman Diagrams*,
UNITEXT for Physics, DOI 10.1007/978-3-319-22014-7_6

For example, given an arbitrary wave propagating, say, in the x^1 direction we can set to zero the components A_0, A_1. It is however possible to choose other gauge-fixing conditions, as we are going to do in the present chapter. In particular we are going to use the *Coulomb gauge*, which corresponds to imposing the condition:

$$\mathbf{\nabla} \cdot \mathbf{A}(x) = 0 \quad \Leftrightarrow \quad \sum_{i=1}^{3} \partial_i A_i(x) = 0. \tag{6.4}$$

Let us notice that, in contrast to the Lorentz gauge (6.3), this gauge choice is not Lorentz covariant, since if it is satisfied in a frame S in a new frame S' we find $\mathbf{\nabla}' \cdot \mathbf{A}'(x) \neq 0$.

In a region in which $J^\mu \equiv 0$, the wave equation (6.1), which can be written in the form:

$$\partial_\rho \partial^\rho A_\mu - \partial_\nu \partial_\mu A^\nu = 0, \tag{6.5}$$

when decomposed in the components $\mu = 0$ and $\mu = i$, yields the following equations[1]:

$$(\partial_0 \partial_0 - \partial_i \partial_i)A_0 - \partial_0(\partial_0 A_0 - \partial_i A_i) = 0, \tag{6.6}$$
$$(\partial_0 \partial_0 - \partial_j \partial_j)A_i - \partial_i \partial_0 A_0 + \partial_i(\partial_j A_j) = 0. \tag{6.7}$$

Using the gauge choice (6.4) the first equation becomes:

$$\nabla^2 A^0 = 0, \tag{6.8}$$

whose solution is the electrostatic potential $A^0 = -V$ in the absence of charges, which can be set to zero, $A^0 = 0$. The second Eq. (6.7), in the Coulomb gauge, becomes:

$$\Box A_i = 0, \tag{6.9}$$

or, equivalently:

$$\left(\frac{1}{c^2}\partial_t^2 - \sum_{i=1}^{3} \partial_i \partial_i\right)\mathbf{A}(x) = 0. \tag{6.10}$$

Equations (6.9) are wave equations for each component $A_i(x)$ of $\mathbf{A}(x)$ whose solutions describe electromagnetic waves. Let us solve these equations, as we did in Sect. 5.6,

[1]Recall that summation over repeated Euclidean indices (i, j or k), independently of their relative position, is understood.

within a box with sides L_A, L_B, L_C and volume $V = L_A L_B L_C$. This allows us to expand the field $\mathbf{A}(\mathbf{x}, t)$ in the form of a triple Fourier series with respect to the coordinates $\mathbf{x}$ and write the solution to Maxwell's equations in the vacuum in the form (5.106):

$$\mathbf{A}(\mathbf{x}, t) = \sum_{k_1 k_2 k_3} \left(\mathbf{A_k}(t) e^{i\mathbf{k}\cdot\mathbf{x}} + \mathbf{A_k}(t)^* e^{-i\mathbf{k}\cdot\mathbf{x}} \right), \tag{6.11}$$

The choice of the Coulomb gauge $\nabla \cdot \mathbf{A} = 0$ then implies, on each single Fourier component, the *transversality condition*: $\mathbf{k} \cdot \mathbf{A_k} = 0$. Since we are working within a finite size domain, the periodic boundary conditions on the surface of the cube require the components of the wave number vector $\mathbf{k} = (k_1, k_2, k_3)$ to have the discrete values in Eq. (5.98). Substituting the expansion (6.11) into Eq. (6.9), and using the linearity property of Maxwell's equations in the vacuum, we find that each Fourier component $\mathbf{A_k}(t)$ satisfies the harmonic oscillator equation (5.99):

$$\frac{d^2 \mathbf{A_k}}{dt^2} + c^2 |\mathbf{k}|^2 \mathbf{A_k} = 0, \tag{6.12}$$

having used the property (5.100). The vectors $\mathbf{A_k}(t)$ and $\mathbf{A_k}(t)^*$ were defined in Sect. 5.6 to correspond to the two independent solutions to Eq. (6.12):

$$\mathbf{A_k}(t) \equiv \boldsymbol{\epsilon_k} \, e^{-i\omega_k t}, \quad \mathbf{A_k}(t)^* \equiv \boldsymbol{\epsilon_k^*} \, e^{i\omega_k t}, \tag{6.13}$$

so that Eq. (6.11) represents an expansion in plane waves progressing along the direction of $\mathbf{k}$ with angular frequency

$$\omega_k \equiv \frac{2\pi}{T} = c \, |\mathbf{k}|. \tag{6.14}$$

The following relations then hold:

$$\dot{\mathbf{A}}_\mathbf{k}(t) = -i\omega_k \, \mathbf{A_k}(t) \; ; \quad \dot{\mathbf{A}}_\mathbf{k}(t)^* = i\omega_k \, \mathbf{A_k^*}(t). \tag{6.15}$$

The Fourier expansion (6.11) can then be also written in the form (see Eq. (5.104)):

$$\mathbf{A}(\mathbf{x}, t) = \sum_{k_1 k_2 k_3} \left(\boldsymbol{\epsilon_k} \, e^{-i\mathbf{k}\cdot\mathbf{x}} + \boldsymbol{\epsilon_k^*} \, e^{i\mathbf{k}\cdot\mathbf{x}} \right). \tag{6.16}$$

Let us now consider the electric field vector:

$$\mathbf{E} = -\frac{1}{c}\frac{\partial}{\partial t}\mathbf{A} + \nabla A_0 = -\frac{1}{c}\frac{\partial}{\partial t}\mathbf{A}, \tag{6.17}$$

and expand its components in Fourier series as we did for $\mathbf{A}(x)$:

$$\mathbf{E} = \sum_k \left(\mathbf{E_k}(t)\, e^{i\mathbf{k}\cdot\mathbf{x}} + \mathbf{E_k}(t)^*\, e^{-i\mathbf{k}\cdot\mathbf{x}} \right), \tag{6.18}$$

where, in virtue of Eqs. (6.17) and (6.13), $\mathbf{E_k}(t)$ are related to $\mathbf{A_k}(t)$ as follows:

$$\mathbf{E_k}(t) = \frac{i}{c}\, \omega_k\, \mathbf{A_k}(t) = i\,|\mathbf{k}|\, \mathbf{A_k}(t). \tag{6.19}$$

Similarly we can Fourier-expand the magnetic field $\mathbf{B} = \nabla \times \mathbf{A}$:

$$\mathbf{B} = \sum_k \left(\mathbf{B_k}(t)\, e^{i\mathbf{k}\cdot\mathbf{x}} + \mathbf{B_k}(t)^*\, e^{-i\mathbf{k}\cdot\mathbf{x}} \right), \tag{6.20}$$

and find

$$\mathbf{B_k} = i\,\mathbf{k} \times \mathbf{A_k}(t) = \mathbf{n_k} \times \mathbf{E_k}, \tag{6.21}$$

where $\mathbf{n_k} \equiv \frac{\mathbf{k}}{|\mathbf{k}|}$ is the unit vector along the direction of propagation. Our aim is now to compute the wave number expansion of the energy (5.81) of the electromagnetic field enclosed in the box:

$$\mathcal{E} = \int_V d^3\mathbf{x} \frac{1}{2} \left(|\mathbf{E}|^2 + |\mathbf{B}|^2 \right). \tag{6.22}$$

Let us first compute $|\mathbf{E}|^2$. Using the expansion (6.18) we can write:

$$|\mathbf{E}|^2 = \frac{1}{c^2} \sum_{\mathbf{k},\mathbf{k}'} \left[\mathbf{E_k} \cdot \mathbf{E_{k'}}\, e^{i(\mathbf{k}+\mathbf{k}')\cdot\mathbf{x}} + \mathbf{E_k^*} \cdot \mathbf{E_{k'}^*}\, e^{-i(\mathbf{k}+\mathbf{k}')\cdot\mathbf{x}} \right.$$
$$\left. + \mathbf{E_k} \cdot \mathbf{E_{k'}^*}\, e^{i(\mathbf{k}-\mathbf{k}')\cdot\mathbf{x}} + \mathbf{E_k^*} \cdot \mathbf{E_{k'}}\, e^{-i(\mathbf{k}-\mathbf{k}')\cdot\mathbf{x}} \right],$$

where, for the sake of simplicity, we have suppressed the dependence of the Fourier components on time. A similar expansion can be written for $|\mathbf{B}|^2$. Note that, in the above expression for $|\mathbf{E}|^2$, the spatial coordinates $\mathbf{x}$ only appear in the complex exponentials. It then follows that for computing the integral of $|\mathbf{E}|^2$ over the volume V we just need to integrate these exponentials. Consider the following integral:

$$\int_V d^3\mathbf{x}\, e^{i\,(\mathbf{k}+\mathbf{k}')\cdot\mathbf{x}} = \int_0^{L_A} dx \int_0^{L_B} dy \int_0^{L_C} dz\; e^{i(k_1+k_1')x}\, e^{i(k_2+k_2')\,y}\, e^{i(k_3+k_3')\,z}.$$

It is the product of three integrals of the same kind. Let us evaluate for instance the one in dx:

$$\int_0^{L_A} dx\, e^{i(k_1+k_1')x} = \frac{1}{i(k_1+k_1')} e^{i(k_1+k_1')x}\Big|_0^{L_A} = \frac{1}{i(k_1+k_1')} e^{2i\pi \frac{(n_1+n_1')}{L_A}x}\Big|_0^{L_A}.$$

(6.23)

Since $n_1 + n_1'$ is an integer, the above integral is always zero unless $k_1 + k_1' = 0$, in which case $\int_0^{L_A} dx\, e^{i(k_1+k_1')x} = \int_0^{L_A} dx = L_A$. We then find:

$$\int_V d^3\mathbf{x}\, e^{i\,(\mathbf{k}+\mathbf{k}')\cdot\mathbf{x}} = L_A\,L_B\,L_C\,\delta_{\mathbf{k},-\mathbf{k}'} = V\,\delta_{\mathbf{k},-\mathbf{k}'},$$

(6.24)

where

$$\delta_{\mathbf{k},-\mathbf{k}'} = \begin{cases} 1 & \text{if } \mathbf{k} = -\mathbf{k}' \\ 0 & \text{if } \mathbf{k} \neq -\mathbf{k}' \end{cases}.$$

(6.25)

Similarly:

$$\int_V d^3\mathbf{x}\, e^{i(\mathbf{k}-\mathbf{k}')\cdot\mathbf{x}} = V\,\delta_{\mathbf{k},\mathbf{k}'},$$

(6.26)

We can now perform the volume integrals in the expression for $|\mathbf{E}|^2$ and the analogous ones for $|\mathbf{B}|^2$ and then find the expansion in $\mathbf{k}$ of the energy $\mathcal{E}$ of the electromagnetic field within V:

$$\mathcal{E} = \frac{1}{2}\int_V d^3\mathbf{x}\, \left(|\mathbf{E}|^2 + |\mathbf{B}|^2\right) = \frac{V}{2}\sum_k \big[(\mathbf{E}_\mathbf{k}\cdot\mathbf{E}_{-\mathbf{k}} + \mathbf{E}_\mathbf{k}\cdot\mathbf{E}_\mathbf{k}^* + c.c.)$$
$$+ (\mathbf{B}_\mathbf{k}\cdot\mathbf{B}_{-\mathbf{k}} + \mathbf{B}_\mathbf{k}\cdot\mathbf{B}_\mathbf{k}^* + c.c.)\big],$$

(6.27)

where $c.c.$ denotes the complex conjugate of the previous terms. The terms $\mathbf{E}_\mathbf{k}\cdot\mathbf{E}_{-\mathbf{k}}$ and $\mathbf{B}_\mathbf{k}\cdot\mathbf{B}_{-\mathbf{k}}$ cancel since:

$$\mathbf{B}_\mathbf{k}\cdot\mathbf{B}_{-\mathbf{k}} = (\mathbf{n}_\mathbf{k}\times\mathbf{E}_\mathbf{k})\cdot(\mathbf{n}_{-\mathbf{k}}\times\mathbf{E}_{-\mathbf{k}}) = -\epsilon_{ij\ell}\,n_\mathbf{k}^j\,E_\mathbf{k}^\ell\,\epsilon_{ipq}\,n_\mathbf{k}^p\,E_{-\mathbf{k}}^q$$
$$= -|\mathbf{n}_\mathbf{k}|^2\mathbf{E}_\mathbf{k}\cdot\mathbf{E}_{-\mathbf{k}} + (\mathbf{n}_\mathbf{k}\cdot\mathbf{E}_\mathbf{k})(\mathbf{n}_\mathbf{k}\cdot\mathbf{E}_{-\mathbf{k}}) = -\mathbf{E}_\mathbf{k}\cdot\mathbf{E}_{-\mathbf{k}},$$

where we have set $\mathbf{n}_\mathbf{k} = -\mathbf{n}_{-\mathbf{k}}$ and have used the transversality condition $\mathbf{n}_\mathbf{k}\cdot\mathbf{E}_{\pm\mathbf{k}} = 0$ and the contraction properties of two ϵ_{ijk} symbols (see Sect. 4.5).

We also find:

$$|\mathbf{B_k}|^2 = |\mathbf{E_k}|^2 = \frac{\omega_k^2}{c^2}(\mathbf{A_k} \cdot \mathbf{A_k^*}),\tag{6.28}$$

which allows us to rewrite Eq. (6.27) in the form:

$$\mathcal{E} = 2V\sum_k |\mathbf{E_k}|^2 = \frac{V}{c^2}\sum_k \omega_k^2\left((\mathbf{A_k}\cdot\mathbf{A_k^*}) + (\mathbf{A_k^*}\cdot\mathbf{A_k})\right)$$

$$\equiv 2\frac{V}{c^2}\sum_k \omega_k^2(\mathbf{A_k}\cdot\mathbf{A_k^*}).\tag{6.29}$$

(The symmetric form of the sum on the right hand side in the first line will be seen to be convenient for the purpose of the quantization (see next section).)
Let us now introduce the following variables[2]:

$$\mathbf{Q_k} = \frac{\sqrt{V}}{c}(\mathbf{A_k} + \mathbf{A_k^*}); \quad \mathbf{P_k} = -i\omega_k\frac{\sqrt{V}}{c}(\mathbf{A_k} - \mathbf{A_k^*}).\tag{6.30}$$

Taking into account the time-dependence of $\mathbf{A_k}(t)$, see Eq. (6.15), it is straightforward to verify that $\mathbf{P_k} = \dot{\mathbf{Q}}_{\mathbf{k}}$. Equations (6.30) can be easily inverted to express $\mathbf{A_k}$ and $\mathbf{A_k^*}$ in terms of $\mathbf{Q_k}$, $\mathbf{P_k}$:

$$\mathbf{A_k} = \frac{c}{2\omega_k\sqrt{V}}(i\,\mathbf{P_k} + \omega_k\mathbf{Q_k}); \quad \mathbf{A_k^*} = \frac{c}{2\omega_k\sqrt{V}}(-i\,\mathbf{P_k} + \omega_k\mathbf{Q_k}),\tag{6.31}$$

Using the above relations we can rewrite the energy in the new variables:

$$\mathcal{E} = \mathcal{H} = \frac{1}{2}\sum_k(|\mathbf{P_k}|^2 + \omega_k^2|\mathbf{Q_k}|^2) = \sum_k \mathcal{E}_\mathbf{k},\tag{6.32}$$

where we have identified the energy $\mathcal{E}$ with the Hamiltonian $\mathcal{H}$ of the system of infinitely many degrees of freedom, described by $\mathbf{Q_k} = (Q_\mathbf{k}^i)$, each labeled by a wave number vector $\mathbf{k}$ and a polarization index i (as a consequence of the transversality condition, not all these polarizations are independent, as we shall discuss below). We can easily verify that $\mathbf{P_k}$, $\mathbf{Q_k}$ are indeed the canonical variables corresponding to the Hamiltonian $\mathcal{H}$ by showing that they satisfy Hamilton's equations[3]:

$$\dot{Q}_\mathbf{k}^i = \frac{\partial\mathcal{H}}{\partial P_k^i} = P_\mathbf{k}^i, \quad \dot{P}_\mathbf{k}^i = -\frac{\partial\mathcal{H}}{\partial Q_k^i} = -\omega_k^2 Q_\mathbf{k}^i,$$

---

[2]Notice that the $\mathbf{P_k}$ here have dimension $(Energy)^{\frac{1}{2}}$.

[3]We are anticipating the Hamiltonian formulation of the equations of motion of a mechanical system, which will be fully discussed in Chap. 8. However we assume the reader to have a basic knowledge of the Hamilton formalism which is propaedeutical to elementary quantum mechanics.

where, as usual, the dot represents the time-derivative. These equations can also be written in the second order form:

$$\ddot{Q}_{\mathbf{k}}^i + \omega_k^2 Q_{\mathbf{k}}^i = 0, \tag{6.33}$$

which, given the relation (6.30), are equivalent to the Maxwell equations (6.12) for each component $\mathbf{A_k}$. We realize that the above equation in the variable $Q_{\mathbf{k}}^i$, for each polarization component i and wave-number vector $\mathbf{k}$, is the equation of motion of a harmonic oscillator with angular frequency ω_k. Note now that the vectors $\mathbf{P_k}$ and $\mathbf{Q_k}$ are orthogonal to $\mathbf{k}$ in virtue of the transversality property $\mathbf{A_k}$ and $\mathbf{A_k^*}$:

$$\mathbf{k} \cdot \mathbf{P_k} = \mathbf{k} \cdot \mathbf{Q_k} = 0. \tag{6.34}$$

This allows us, for a given direction of propagation $\mathbf{n_k}$, to decompose $\mathbf{P_k}$ and $\mathbf{Q_k}$ along an ortho-normal basis $\mathbf{u}_{\mathbf{k},\alpha}$, $\alpha = 1, 2$, on the plane transverse to $\mathbf{n}$:

$$\mathbf{P_k} = \sum_1^2 P_{\mathbf{k}\alpha} \mathbf{u}_{\mathbf{k},\alpha}, \quad \mathbf{Q_k} = \sum_1^2 Q_{\mathbf{k}\alpha} \mathbf{u}_{\mathbf{k},\alpha}.$$

The index α labels the two polarizations of the plane wave. Taking into account the ortho-normality of the $(\mathbf{u}_{\mathbf{k},\alpha})$ we can write the Hamiltonian as follows:

$$\mathcal{H} = \frac{1}{2} \sum_{\mathbf{k}} \sum_{\alpha=1,2} (P_{\alpha\mathbf{k}}^2 + \omega_k^2 Q_{\alpha\mathbf{k}}^2) = \sum_{\mathbf{k}} \mathcal{H}_{\mathbf{k}} = \sum_{\mathbf{k}} \mathcal{E}_{\mathbf{k}} = \sum_{\mathbf{k}} \sum_{\alpha=1,2} \mathcal{H}_{\mathbf{k},\alpha}, \tag{6.35}$$

which describes a system of infinitely many, decoupled, harmonic oscillators, each described by the conjugate variables $P_{\alpha\mathbf{k}}$, $Q_{\alpha\mathbf{k}}$ and Hamiltonian function $\mathcal{H}_{\mathbf{k},\alpha}$.

Summarizing we have shown that *the electromagnetic field, far from charges and currents, can be represented by a system of decoupled harmonic oscillators, each associated with a wave-number vector $\mathbf{k}$ and polarization α and characterized by an angular frequency $\omega_k = c\,|\mathbf{k}|$.*

We can also compute the momentum $\mathcal{P} \equiv (\mathcal{P}^i)$ associated with the electromagnetic field, given by the formula:

$$\mathcal{P} = \frac{1}{c} \int_V d^3\mathbf{x} \, \mathbf{E} \times \mathbf{B}, \tag{6.36}$$

Using the mode expansion of the electric and magnetic fields we find:

$$\mathcal{P} = \frac{V}{c} \sum_{\mathbf{k}} \left[\mathbf{E_k} \times \mathbf{B_{-k}} + (\mathbf{E_k} \times \mathbf{B_{-k}})^* + \mathbf{E_k} \times \mathbf{B_k^*} + \mathbf{E_k^*} \times \mathbf{B_k} \right]. \tag{6.37}$$

The first term in the above sum can be recast in the following form:

$$\left[\mathbf{E_k} \times \mathbf{B_{-k}}\right]_i = -\left[\mathbf{E_k} \times (\mathbf{n_k} \times \mathbf{E_{-k}})\right]_i = -\epsilon_{ijl}\,\epsilon_{lpq}\,E_k^j\,n_k^p\,E_{-k}^q = -n_k^i\,(\mathbf{E_k} \cdot \mathbf{E_{-k}}),$$

which changes sign as $\mathbf{k} \to -\mathbf{k}$, implying that $\sum_{\mathbf{k}} \mathbf{E_k} \times \mathbf{B_{-k}} = 0$.

On the other hand we have:

$$\mathbf{E_k} \times \mathbf{B_k^*} = \mathbf{n_k}|\mathbf{E_k}|^2. \tag{6.38}$$

Using Eq. (6.29) we then find:

$$\mathcal{P} = \frac{2V}{c}\sum_{\mathbf{k}} \mathbf{n_k}\,|\mathbf{E_k}|^2 = \sum_{\mathbf{k}} \frac{\mathbf{n_k}}{c}\,\mathcal{E}_{\mathbf{k}}. \tag{6.39}$$

6.2 Quantization of the Electromagnetic Field

We have described in the previous section an electromagnetic field in a box, far from charges and currents, as a collection of infinitely many, decoupled, harmonic oscillators, each described by a couple of conjugate canonical variables. The field itself is then a system having infinite degrees of freedom, its physical state being described by infinitely many canonical variables:

$$\{P_{\mathbf{k}\,\alpha},\ Q_{\mathbf{k}\,\alpha}\}, \quad \mathbf{k} = \left(2\pi\frac{n_1}{L_A}, 2\pi\frac{n_2}{L_B}, 2\pi\frac{n_3}{L_C}\right), \tag{6.40}$$

where $\alpha = 1, 2$ labels the physical components of $\mathbf{P_k}$ and $\mathbf{Q_k}$, which are transverse to the direction of propagation of the corresponding plane-wave. The dynamics of the field is encoded in the Hamiltonian $\mathcal{H}$ given in (6.32). Having described the degrees of freedom of our system in the canonical formalism, we can now proceed to its quantization[4]: The canonical variables $P_{\mathbf{k}\,\alpha}$, $Q_{\mathbf{k}\,\alpha}$ now become *linear operators* $\hat{P}_{\mathbf{k}\,\alpha}$, $\hat{Q}_{\mathbf{k}\,\alpha}$ in the *space of states* of the system, and the Poisson bracket between classical variables is replaced by the commutator between the corresponding operators according to the rule $\{\cdot, \cdot\}_{P.B.} \to \frac{1}{i\hbar}[\cdot, \cdot]$. Since the Poisson bracket between conjugate variables p, q corresponding to the same degree of freedom is $\{\hat{q}, \hat{p}\}_{P.B.} = 1$, while that computed between variables associated with different degrees of freedom vanishes, the *operators* $\hat{Q}_{\mathbf{k}\,\alpha}$, $\hat{P}_{\mathbf{k}'\,\alpha'}$ satisfy the following commutation relations

$$[\hat{Q}_{\mathbf{k},\alpha},\ \hat{P}_{\mathbf{k}',\alpha'}] = i\,\hbar\,\delta_{\mathbf{k},\mathbf{k}'}\,\delta_{\alpha,\alpha'}. \tag{6.41}$$

---

[4]Although we assume the reader to have a basic knowledge of non-relativistic quantum mechanics, the relevant notions will be reviewed in Chap. 9. We refer the reader to that chapter for the notations used here.

To compute the Hamiltonian operator let us first define the *operators* $\hat{\mathbf{A}}_{\mathbf{k}}$ in terms of the canonical operators $\hat{\mathbf{P}}_{\mathbf{k}}, \hat{\mathbf{Q}}_{\mathbf{k}}$ using the same relations (6.31):

$$\hat{\mathbf{A}}_{\mathbf{k}} \equiv \frac{c}{2\omega_k \sqrt{V}} \left(i\hat{\mathbf{P}}_{\mathbf{k}} + \omega_k \hat{\mathbf{Q}}_{\mathbf{k}} \right). \qquad (6.42)$$

Next we expand $\hat{\mathbf{A}}_{\mathbf{k}}$ along the two transverse directions and define the dimensionless operators $a_{\mathbf{k},\alpha}$ as follows:

$$\hat{\mathbf{A}}_{\mathbf{k}} = c \sqrt{\frac{\hbar}{2\omega_k V}} \sum_{\alpha=1}^{2} a_{\mathbf{k},\alpha} \, \mathbf{u}(\mathbf{k}, \alpha). \qquad (6.43)$$

where

$$a_{\mathbf{k}\,\alpha} = \frac{1}{\sqrt{2\hbar\omega_k}} \left(i\hat{P}_{\mathbf{k},\alpha} + \omega_k \hat{Q}_{\mathbf{k},\alpha} \right). \qquad (6.44)$$

As it is well known, when passing from classical quantities to quantum operators, the complex conjugation operation is replaced by hermitian conjugation. The hermitian conjugate of $a_{\mathbf{k}\,\alpha}$ is:

$$a_{\mathbf{k},\alpha}^{\dagger} = \frac{1}{\sqrt{2\hbar\omega_k}} \left(-i\hat{P}_{\mathbf{k},\alpha} + \omega_k \hat{Q}_{\mathbf{k},\alpha} \right). \qquad (6.45)$$

Using Eq. (6.41), we find that the operators $a_{\mathbf{k}\,\alpha}, a_{\mathbf{k}\,\alpha}^{\dagger}$ satisfy the following commutation relations:

$$[a_{\mathbf{k},\alpha}, a_{\mathbf{k}',\alpha'}] = [a_{\mathbf{k},\alpha}^{\dagger}, a_{\mathbf{k}',\alpha'}^{\dagger}] = 0,$$
$$[a_{\mathbf{k},\alpha}, a_{\mathbf{k}',\alpha'}^{\dagger}] = \delta_{\mathbf{k},\mathbf{k}'} \, \delta_{\alpha,\alpha'}. \qquad (6.46)$$

The operator $\hat{\mathbf{A}}(\mathbf{x}, t)$ associated with the vector potential of the electromagnetic field is then expressed by the Fourier series:

$$\hat{\mathbf{A}}(\mathbf{x}, t) = \sum_{\mathbf{k}} (\hat{\mathbf{A}}_{\mathbf{k}}(t) \, e^{i\mathbf{k}\cdot\mathbf{x}} + \hat{\mathbf{A}}_{\mathbf{k}}(t)^{\dagger} \, e^{-i\mathbf{k}\cdot\mathbf{x}})$$

$$= \sum_{\alpha=1}^{2} \sum_{\mathbf{k}} \frac{c}{\omega_k \sqrt{V}} \left[\omega_k \hat{Q}_{\mathbf{k}\,\alpha} \cos(\mathbf{k}\cdot\mathbf{x}) - \hat{P}_{\mathbf{k}\,\alpha} \sin(\mathbf{k}\cdot\mathbf{x}) \right] \mathbf{u}_{\mathbf{k},\alpha}.$$

We can now use the expansions (6.18) and (6.20) as well as Eqs. (6.19) and (6.21) to define the electric and magnetic field operators $\hat{\mathbf{E}}$ and $\hat{\mathbf{B}}$ in terms of the operators $\hat{\mathbf{A}}_{\mathbf{k}}, \hat{\mathbf{A}}_{\mathbf{k}}^{\dagger}$ and thus of $a_{\mathbf{k}\,\alpha}, a_{\mathbf{k}\,\alpha}^{\dagger}$:

$$\hat{\mathbf{E}} = \sum_k \left(\hat{\mathbf{E}}_k e^{i\mathbf{k} \cdot \mathbf{x}} + \hat{\mathbf{E}}_k^{\dagger} e^{-i\mathbf{k} \cdot \mathbf{x}} \right) ; \quad \hat{\mathbf{B}} = \sum_k \left(\hat{\mathbf{B}}_k e^{i\mathbf{k} \cdot \mathbf{x}} + \hat{\mathbf{B}}_k^{\dagger} e^{-i\mathbf{k} \cdot \mathbf{x}} \right) . \quad (6.47)$$

with:

$$\hat{\mathbf{E}}_k = \sum_{\alpha=1}^{2} \hat{E}_{k\,\alpha} \, \mathbf{u}_{\mathbf{k},\alpha} = \sum_{\alpha=1}^{2} i \sqrt{\frac{\hbar \omega_k}{2V}} \, a_{k\,\alpha} \, \mathbf{u}_{\mathbf{k},\alpha}; \quad \hat{\mathbf{B}}_k = \sum_{\alpha=1}^{2} \hat{B}_{k\,\alpha} \mathbf{u}_{\mathbf{k},\alpha} = \mathbf{n}_k \times \hat{\mathbf{E}}_k .$$

$$(6.48)$$

We are now able to compute the Hamiltonian operator $\hat{\mathcal{H}}$ following the same derivation as in the classical case. Care, however, has to be used in deriving the operator
versions of Eqs. (6.28) and (6.29) from (6.27) since, as opposed to the corresponding
classical quantities which were just numbers, the operators $\hat{E}_k$ and $\hat{E}_k^{\dagger}$, as well as
their magnetic counterparts, no longer commute. As a consequence of this, in writ-
ing the expression for the Hamiltonian, we should keep the order of factors in each
product and thus, instead of translating in operatorial form the term in the second
line of Eq. (6.29) we use the symmetric expression $(\hat{\mathbf{A}}_k \hat{\mathbf{A}}_k^{\dagger} + \hat{\mathbf{A}}_k^{\dagger} \hat{\mathbf{A}}_k)$ in the first line
of the same equation since $\hat{\mathbf{A}}_k$ and $\hat{\mathbf{A}}_k^{\dagger}$ do not commute. The Hamiltonian operator
then reads:

$$\hat{\mathcal{H}} = \sum_k \sum_\alpha \frac{\hbar \omega_k}{2} \left(a_{\mathbf{k},\alpha} a_{\mathbf{k}\,\alpha}^{\dagger} + a_{\mathbf{k}\,\alpha}^{\dagger} a_{\mathbf{k}\,\alpha} \right) = \sum_k \sum_\alpha \left(\hat{N}_{\mathbf{k},\alpha} + \frac{1}{2} \right) \hbar \omega_k, \quad (6.49)$$

where

$$\hat{N}_{\mathbf{k},\alpha} \equiv a_{\mathbf{k},\alpha}^{\dagger} a_{\mathbf{k},\alpha}, \quad (6.50)$$

The operator $\hat{\mathcal{H}}$, in terms of the canonical operators, has the same form (6.35):

$$\hat{\mathcal{H}} = \frac{1}{2} \sum_k \sum_\alpha \left[(\hat{P}_{\mathbf{k},\alpha})^2 + \omega_k (\hat{Q}_{\mathbf{k},\alpha})^2 \right] . \quad (6.51)$$

Equations (6.49) and (6.51) describe the Hamiltonian operator associated with the
system of infinitely many quantum harmonic oscillators $(\mathbf{k}, \alpha)$ defined in the previ-
ous section. The quantities $a_{\mathbf{k},\alpha}$ and $a_{\mathbf{k},\alpha}^{\dagger}$ are indeed nothing but the annihilation and
creation operators associated with the quantum oscillator $(\mathbf{k}, \alpha)$, which are useful in
constructing the corresponding quantum states. It is now straightforward to determine
the expression for the momentum operator, by using Eqs. (6.39) and (6.49):

$$\hat{\mathcal{P}} = \frac{1}{c} \int_V d^3 \mathbf{x} \, \mathbf{E} \times \mathbf{B} = \sum_k \sum_\alpha \hbar \mathbf{k} \left(\hat{N}_{\mathbf{k},\alpha} + \frac{1}{2} \right) . \quad (6.52)$$

Both $\mathcal{H}$ and $\hat{\mathcal{P}}$ are expressed in terms of the *occupation number* operators $\hat{N}_{\mathbf{k},\alpha}$ (6.50) associated with the quantum oscillators $(\mathbf{k}, \alpha)$. We know from elementary quantum theory that the states of each oscillator can be described as eigenstates of the occupation number operator. These eigenstates of the $(\mathbf{k}, \alpha)$-oscillator at the time t can then be written in the form $|N_{\mathbf{k},\alpha}, t\rangle$ and satisfy:

$$\hat{N}_{\mathbf{k},\alpha}|N_{\mathbf{k},\alpha}, t\rangle = N_{\mathbf{k},\alpha}|N_{\mathbf{k},\alpha}, t\rangle, \tag{6.53}$$

where $N_{\mathbf{k},\alpha}$, eigenvalue of $\hat{N}_{\mathbf{k},\alpha}$, is a positive integer. The energy and momentum of this state is

$$\mathcal{E}_{\mathbf{k},\alpha} = \hbar\omega_k \left(N_{\mathbf{k},\alpha} + \frac{1}{2}\right), \quad \mathcal{P}_{\mathbf{k},\alpha} = \hbar\mathbf{k}\left(N_{\mathbf{k},\alpha} + \frac{1}{2}\right). \tag{6.54}$$

Note that the operators $\hat{\mathbf{A}}_{\mathbf{k}\alpha}(t)$, and thus also $a_{\mathbf{k}\alpha}(t)$, depend on time through a factor $e^{-i\omega_k t}$. It is apparent however that neither the Hamiltonian and the momentum operators, nor the commutation relations, depend on time. We choose to use the Schrodinger representation in which states depend on time while operators are time-independent: $a_{\mathbf{k},\alpha} \equiv a_{\mathbf{k},\alpha}(0)$. The state $|N_{\mathbf{k},\alpha}, t\rangle$ is constructed by applying $N_{\mathbf{k},\alpha}$-times the creation operator $a_{\mathbf{k},\alpha}^{\dagger}$ to the ground state $|0\rangle$:

$$|N_{\mathbf{k},\alpha}, t\rangle = \frac{1}{\sqrt{N_{\mathbf{k},\alpha}!}}(a_{\mathbf{k},\alpha}^{\dagger})^{N_{\mathbf{k},\alpha}}|0, t\rangle. \tag{6.55}$$

where $\dfrac{1}{\sqrt{N_{\mathbf{k},\alpha}!}}$ is a normalization factor, the states being normalized to one, and the ground state satisfies the relation:

$$a_{\mathbf{k},\alpha}|0, t\rangle = 0. \tag{6.56}$$

Equations (6.54) are telling us that the energy and momentum of a state are quantized in units $\hbar\omega_k$ and $\hbar\mathbf{k}$ respectively.

The electromagnetic field, being a collection of decoupled harmonic oscillators, is described by a state which is the tensor product of the states associated with each oscillator. It will be then characterized by all the occupation numbers $\{N_{\mathbf{k},\alpha}\} \equiv \{N_{\mathbf{k}_1,\alpha_1}, N_{\mathbf{k}_2,\alpha_2}, \dots\}$ of the constituent states:

$$|\{N_{\mathbf{k},\alpha}\}, t\rangle \equiv |N_{\mathbf{k}_1,\alpha_1}, t\rangle|N_{\mathbf{k}_2,\alpha_2}, t\rangle\cdots . \tag{6.57}$$

In particular the ground state of the system is the direct product of the oscillator ground states. The energy and momentum of the field are the sum of the energies and momenta associated with each oscillator state, as we easily find by applying the operators in (6.49) and (6.52) to the state (6.57):

$$\hat{\mathcal{H}} |\{N_{\mathbf{k},\alpha}\}, t\rangle = \left[\sum_k \sum_\alpha \left(N_{\mathbf{k},\alpha} + \frac{1}{2} \right) \hbar \omega_k \right] |\{N_{\mathbf{k},\alpha}\}, t\rangle,$$

$$\hat{\mathcal{P}} |\{N_{\mathbf{k},\alpha}\}, t\rangle = \left[\sum_k \sum_\alpha \left(N_{\mathbf{k},\alpha} + \frac{1}{2} \right) \hbar \mathbf{k} \right] |\{N_{\mathbf{k},\alpha}\}, t\rangle, \qquad (6.58)$$

We observe that the above eigenvalues exhibit an infinite term, which is the ground state energy $\mathcal{E}_0$ and momentum $\mathcal{P}_0$, sum over all the oscillators $(\mathbf{k}, \alpha)$, of the corresponding ground state energies $\frac{\hbar \omega_k}{2}$ and momenta $\frac{\hbar \mathbf{k}}{2}$:

$$\mathcal{E}_0 = \sum_k \sum_\alpha \frac{1}{2} \hbar \omega_k = \infty,$$

$$\mathcal{P}_0 = \sum_k \sum_\alpha \frac{1}{2} \hbar \mathbf{k} = \infty. \qquad (6.59)$$

These terms have no physical meaning and make the energy and momentum operators, as given in (6.49) and (6.52), ill defined. In order to correctly define these operators in terms of a and $a^\dagger$, let us introduce the notion of "normal ordering" : : for a generic bosonic field (such as the electromagnetic one, as we shall see), as the operation by which all the operators a and $a^\dagger$, in a product, are reordered so that the $a^\dagger$ are moved to the left and the a to the right:

$$: a^\dagger a := a^\dagger a, \qquad : a a^\dagger := a^\dagger a. \qquad (6.60)$$

For instance:

$$: a_1 a_1^\dagger a_2 a_3 a_2^\dagger := a_1^\dagger a_2^\dagger a_1 a_2 a_3. \qquad (6.61)$$

We then give the prescription that all the operators associated with physical observables, should be defined as normal ordered products of the field operators. The Hamiltonian operator $\hat{\mathcal{H}}$, for instance, should be defined as follows:

$$\hat{\mathcal{H}} = V \sum_k : |\hat{\mathbf{E}}_\mathbf{k}|^2 := \sum_k \sum_\alpha \frac{\hbar \omega_k}{2} : (a_{\mathbf{k},\alpha} a_{\mathbf{k},\alpha}^\dagger + a_{\mathbf{k},\alpha}^\dagger a_{\mathbf{k},\alpha}) :$$

$$= \sum_k \sum_\alpha \hbar \omega_k (a_{\mathbf{k},\alpha}^\dagger a_{\mathbf{k},\alpha}) = \sum_k \sum_\alpha \hbar \omega_k \hat{N}_{\mathbf{k},\alpha}. \qquad (6.62)$$

Similarly, the correct definition of the momentum operator is:

$$\hat{\mathcal{P}} = \frac{2V}{kc} \sum_k \mathbf{k} : |\hat{\mathbf{E}}_\mathbf{k}|^2 := \sum_k \sum_\alpha \hbar \mathbf{k} \, \hat{N}_{\mathbf{k},\alpha}. \qquad (6.63)$$

Let us note that using the normal ordering in the definition of $\hat{\mathcal{H}}$ and $\hat{\mathcal{P}}$ amounts to subtracting to their eigenvalues the infinite un-physical contribution associated with their ground state in the previous definitions (6.49), (6.52).

Having set the energy and momentum of the ground state $|\{0\}, t\rangle$ to zero, the energy and momentum of a generic state $|\{N_{k,\alpha}\}, t\rangle$ of the electromagnetic field is now simply given by the sum of *quanta* $\hbar\omega_k$ and $\hbar\mathbf{k}$:

$$\mathcal{E} = \sum_k \sum_\alpha N_{k,\alpha} \hbar\omega_k, \quad \mathcal{P} = \left(\sum_k \sum_\alpha N_{k,\alpha} \hbar\mathbf{k}\right), \tag{6.64}$$

We associate with each oscillator $(\mathbf{k}, \alpha)$, i.e. with each plane wave, a *state* of a particle called *photon* and denoted by the symbol γ, carrying the quantum of momentum, $\hbar\mathbf{k}$, and of energy, $\hbar\omega_k$, and having polarization α. The state $|N_{k,\alpha}, t\rangle$ of the $(\mathbf{k}, \alpha)$-oscillator is then then interpreted as describing $N_{k,\alpha}$ photons in the state $(\mathbf{k}, \alpha)$. Its energy and momentum are $N_{k,\alpha} \hbar\omega_k$ and $N_{k,\alpha} \hbar\mathbf{k}$ respectively, namely the sum of the $N_{k,\alpha}$ quanta of the two quantities associated with each photon. The state $|\{N_{k,\alpha}\}, t\rangle$ of the whole electromagnetic field then describes $N_{k,\alpha}$ photons in each state $(\mathbf{k}, \alpha)$ and its energy and momentum, as given in (6.64), is the sum of the energy and momenta of the photons in the various states. A photon with energy $E = \hbar\omega_k$ and momentum $\mathbf{p} = \hbar\mathbf{k}$ has a rest mass, given by:

$$m_\gamma^2 = \frac{1}{c^4} E^2 - \frac{1}{c^2} |\mathbf{p}|^2 = \frac{1}{c^4} \hbar^2 (\omega_k^2 - c^2|\mathbf{k}|^2) = 0, \tag{6.65}$$

where we have used the definition of ω_k, (6.14). As was anticipated in Chap. 5, the photon is therefore a massless particle. Its momentum four-vector p^μ is thus $\hbar$ times the wave number four-vector k^μ associated with the corresponding plane wave and defined in Eq. (5.105).

The action of $a_{k,\alpha}^\dagger$ or of $a_{k,\alpha}$ on a state amounts to "creating" or "destroying" a $(\mathbf{k}, \alpha)$-photon since they increase or decrease the energy and momentum of the corresponding oscillator state by one quantum respectively. This can be seen by recalling, from elementary quantum mechanics, the following relations which hold for the $(\mathbf{k}, \alpha)$-oscillator:

$$a_{k,\alpha}^\dagger |N_{k,\alpha}, t\rangle = \sqrt{N_{k,\alpha} + 1} |N_{k,\alpha} + 1, t\rangle,$$
$$a_{k,\alpha} |N_{k,\alpha}, t\rangle = \sqrt{N_{k,\alpha}} |N_{k,\alpha} - 1, t\rangle. \tag{6.66}$$

Expressing the canonical operators in terms of $a_{k,\alpha}$, $a_{k,\alpha}^\dagger$,

$$\hat{P}_{k,\alpha} = -i\sqrt{\frac{\hbar\omega_k}{2}} \left(a_{k,\alpha} - a_{k,\alpha}^+\right),$$
$$\hat{Q}_{k,\alpha} = \sqrt{\frac{\hbar}{2\omega_k}} \left(a_{k,\alpha} + a_{k,\alpha}^+\right), \tag{6.67}$$

we find the following relations:

$$\langle N_{\mathbf{k},\alpha}|Q_{\mathbf{k},\alpha}|N_{\mathbf{k},\alpha}-1\rangle = \langle N_{\mathbf{k},\alpha}-1|Q_{\mathbf{k},\alpha}|N_{\mathbf{k},\alpha}\rangle = \sqrt{\frac{\hbar N_{\mathbf{k},\alpha}}{2\omega_k}},$$

$$\langle N_{\mathbf{k},\alpha}|P_{\mathbf{k},\alpha}|N_{\mathbf{k},\alpha}-1\rangle = -\langle N_{\mathbf{k},\alpha}-1|P_{\mathbf{k},\alpha}|N_{\mathbf{k},\alpha}\rangle = i\sqrt{\frac{\hbar \omega_k N_{\mathbf{k},\alpha}}{2}}. \quad (6.68)$$

The representation of the states of the electromagnetic field in terms of occupation number eigenstates associated with the constituent harmonic oscillators, is called *occupation number representation* or *second quantization*. In this construction each state is obtained by applying the $a_{\mathbf{k},\alpha}^{\dagger}$ operators to the ground state.

We have been using, so far, the Schroedinger representation in which quantum states evolve in time while operators are constant. The time evolution of a quantum state $|a, t\rangle$ is described in this picture by the Schroedinger equation (see Chap. 9 for a general review of the subject):

$$\hat{\mathcal{H}}|a, t\rangle = i\hbar \frac{\partial}{\partial t}|a, t\rangle, \quad (6.69)$$

where $|a, t\rangle$ describes the electromagnetic field at a time t and is a generic linear combination of the basis elements $|\{N_{\mathbf{k},\alpha}\}, t\rangle$. Let us recall that, if the Hamiltonian operator, as in our case, does not explicitly depend on time, a solution to (6.69) at a time t can be expressed in terms of a time evolution operator of the form $e^{-\frac{i}{\hbar}\hat{\mathcal{H}}t}$ acting on the state at a given initial time $t = 0$

$$|a, t\rangle = e^{-\frac{i}{\hbar}\hat{\mathcal{H}}t}|a, t = 0\rangle.$$

On the basis elements $|\{N_{\mathbf{k},\alpha}\}\rangle$ we have:

$$|\{N_{\mathbf{k},\alpha}\}, t\rangle = e^{-\frac{i}{\hbar}\hat{\mathcal{H}}t}|\{N_{\mathbf{k},\alpha}\}, t = 0\rangle = e^{-i(\sum_{\mathbf{k},\alpha} N_{\mathbf{k},\alpha}\omega_k)t}|\{N_{\mathbf{k},\alpha}\}, 0\rangle. \quad (6.70)$$

On the other hand operators are all computed at $t = 0$. In this representation Lorentz covariance is not manifest.

If we adopt the Heisenberg picture (or representation) instead, see Chap. 9, the dependence on time is associated with operators, quantum states being time-independent. We can easily obtain such representation by writing the matrix element of the operator $\hat{A}_{\mathbf{k}}(\mathbf{x})$, in the Schroedinger representation, between two states at a time t and equating it to the matrix element of a time-dependent operator $\hat{A}(\mathbf{x}, t)$ between the same states computed at $t = 0$:

$$\langle\{N_{\mathbf{k}',\alpha'}\}, t|\hat{A}(\mathbf{x})|\{N_{\mathbf{k},\alpha}\}, t\rangle = \langle\{N_{\mathbf{k}',\alpha'}\}, 0|e^{\frac{i}{\hbar}\hat{\mathcal{H}}t}\hat{A}(\mathbf{x})e^{-\frac{i}{\hbar}\hat{\mathcal{H}}t}|\{N_{\mathbf{k},\alpha}\}, 0\rangle$$

$$= \langle\{N_{\mathbf{k}',\alpha'}\}, 0|\hat{A}(\mathbf{x}, t)|\{N_{\mathbf{k},\alpha}\}, 0\rangle. \quad (6.71)$$

In Heisenberg's representation we act on constant states by means of the time-dependent operator $\hat{\mathbf{A}}(\mathbf{x}, t)$:

$$\hat{\mathbf{A}}_{\mathbf{k},\alpha}(\mathbf{x}, t) = e^{\frac{i}{\hbar}\hat{\mathcal{H}}t}\,\hat{\mathbf{A}}_{\mathbf{k}}(\mathbf{x})\,e^{-\frac{i}{\hbar}\hat{\mathcal{H}}t}. \tag{6.72}$$

The resulting expression is manifestly Lorentz-covariant, as we can verify by computing the matrix elements of the Fourier components $\hat{\mathbf{A}}_{\mathbf{k}}(t)$ at a time t between two states:

$$\langle N_{\mathbf{k},\alpha}|\hat{\mathbf{A}}_{\mathbf{k}}(t)|N_{\mathbf{k},\alpha}+1\rangle = \langle N_{\mathbf{k},\alpha}|e^{\frac{i}{\hbar}\hat{\mathcal{H}}t}\,\hat{\mathbf{A}}_{\mathbf{k}}\,e^{-\frac{i}{\hbar}\hat{\mathcal{H}}t}|N_{\mathbf{k},\alpha}+1\rangle$$
$$= e^{-i\omega_k t}\langle N_{\mathbf{k},\alpha}|\hat{\mathbf{A}}_{\mathbf{k}}|N_{\mathbf{k},\alpha}+1\rangle \Rightarrow \hat{\mathbf{A}}_{\mathbf{k}}(t) = \hat{\mathbf{A}}_{\mathbf{k}}e^{-i\omega_k t}. \tag{6.73}$$

In the Heisenberg representation we can then write the electromagnetic field operator on the space of states in the form:

$$\hat{\mathbf{A}}(\mathbf{x}, t) = \sum_k \left(\hat{\mathbf{A}}_{\mathbf{k},\alpha}e^{-i(\omega_k t - \mathbf{k}\cdot\mathbf{x})} + \hat{\mathbf{A}}_{\mathbf{k},\alpha}e^{i(\omega_k t - \mathbf{k}\cdot\mathbf{x})}\right)$$

$$= c\sqrt{\frac{\hbar}{2\omega_k V}}\sum_k\sum_\alpha \left[a_{\mathbf{k},\alpha}\mathbf{u}_{\mathbf{k},\alpha}e^{-ik\cdot x} + a^\dagger_{\mathbf{k},\alpha}\mathbf{u}^*_{\mathbf{k},\alpha}e^{ik\cdot x}\right]. \tag{6.74}$$

From the above expansion it is apparent that the operator $\hat{\mathbf{A}}(\mathbf{x}, t)$ depends on the space-time coordinates, just as in the classical case, through the Lorentz-invariant product: $k \cdot x \equiv k_\mu x^\mu = k^0 x^0 - \mathbf{k}\cdot\mathbf{x}$, where, as usual, $(k^\mu) \equiv (\frac{\omega_k}{c}, \mathbf{k})$.

6.3 Spin of the Photon

We have learned, from our previous discussion, that each plane wave component

$$\mathbf{A}(\mathbf{x}, t) = \boldsymbol{\epsilon}_{\mathbf{k}}\,e^{-ik\cdot x} + \text{c.c.} \equiv \boldsymbol{\epsilon}_{\mathbf{k}}\,e^{-\frac{i}{\hbar}p\cdot x} + \text{c.c.}, \tag{6.75}$$

in the expansion (6.16) of a generic solution to Maxwell's equation in the vacuum, is associated with the quantum state of a photon of energy $E = \hbar\omega_k$, momentum $\mathbf{p} = \hbar\mathbf{k}$ and polarization $\boldsymbol{\epsilon}_{\mathbf{k}}$. It can thus be interpreted as the *wave function* of the corresponding photon.

We know, however, that the photon is a massless particle and, as such, there exists no RF in which its linear momentum vanishes: $\mathbf{p} = \mathbf{0}$. This implies that there is no RF in which the total angular momentum $\mathbf{J} \equiv \mathbf{M} + \mathbf{S} = \mathbf{x} \times \mathbf{p} + \mathbf{S}$, where $\mathbf{M}$ is the orbital part and $\mathbf{S}$ is the spin (see Chap. 9), coincides with $\mathbf{S}$ and thus acts on the internal degrees of freedom only. The only component of $\mathbf{J}$ which acts only on the internal degrees of freedom of the photon and which thus can be taken as a definition of its spin, is its component along $\mathbf{p}$, called the "helicity" and denoted by Γ:

$$\Gamma \equiv \mathbf{J} \cdot \frac{\mathbf{p}}{|\mathbf{p}|} = (\mathbf{x} \times \mathbf{p}) \cdot \frac{\mathbf{p}}{|\mathbf{p}|} + \mathbf{S} \cdot \frac{\mathbf{p}}{|\mathbf{p}|} = \mathbf{S} \cdot \mathbf{n_k}. \tag{6.76}$$

The helicity Γ generates rotations about the direction $\mathbf{n_k}$ of $\mathbf{p}$:

$$\Lambda_R(\theta) = e^{\frac{i}{\hbar} \Gamma \theta}. \tag{6.77}$$

On the internal components (polarization) of the photon, which are components of a four-vector $(\epsilon_\mu(\mathbf{k})) = (0, \ \epsilon_\mathbf{k})$ (transverse components of A_μ), this transformation acts as a particular Lorentz transformation. Let us choose a RF in which $\mathbf{p}$ is aligned to the x-direction, $\mathbf{p} = (p, 0, 0) = \hbar\, \mathbf{k}$. The infinitesimal generator of rotations about the x axis is represented, on the four-vector k^μ, by the matrix J_1:

$$\Gamma = J_1 = -i\hbar \begin{pmatrix} 0 & 0 & 0 & 0 \\ 0 & 0 & 0 & 0 \\ 0 & 0 & 0 & 1 \\ 0 & 0 & -1 & 0 \end{pmatrix}. \tag{6.78}$$

Since $\epsilon_\mathbf{k}$ is transverse to the direction x of motion, we have: $(\epsilon_\mu(\mathbf{k})) = (0, 0, \epsilon_2, \epsilon_3)$, we easily find that Γ has two eigenvalues $i\,(\mp i\,\hbar) = \pm\hbar$ with eigenvectors:

$$\epsilon_\mu^{(+)}(\mathbf{k}) = \begin{pmatrix} 0 \\ 0 \\ 1 \\ i \end{pmatrix} \quad \text{and} \quad \epsilon_\mu^{(-)}(\mathbf{k}) = \begin{pmatrix} 0 \\ 0 \\ 1 \\ -i \end{pmatrix}. \tag{6.79}$$

We define the spin of a massless particle as the number s such that its states are eigenstates of Γ to the eigenvalues $\pm\hbar\, s$. It then follows that the photon has spin $s = 1$. Note that the transformation $\Lambda_R(\theta)$ precisely coincides with the transformation $\Lambda^{(0)}$ given in Eq. (5.117), so that the definition of spin of a photon given here corresponds to the definition of spin of a plane wave given in Sect. 5.6.1.

6.3.1 References

For further reading see Refs. [8] (Vol. 4), [9].

Chapter 7
Group Representations and Lie Algebras

7.1 Lie Groups

As already mentioned in Chap. 4 several properties of the rotation group SO(3) and of the Lorentz group SO(1, 3) are actually valid for any *Lie group G* and do not depend of the particular representation of their elements in terms of matrices. Such representation independent features are encoded in the notion of an *abstract group*.

In this chapter we give the definition of an abstract group, restricting to *Lie groups* only. Without any pretension to rigour or completeness, we define the general concept of *representation* and that of a *Lie algebra*. This will be essential for showing the deep relation, existing in classical and quantum field theories, between symmetry and/or invariance properties of a system, to be described in group theoretical language, and conservation laws of physical quantities. These interrelations will be discussed in the next chapters.

Let us first give the general axioms defining an *abstract group*.

Def: An abstract group G is a set of elements within which a law of composition $\cdot$ (to be characterized as a "product") is defined, such that, given any two elements in it $g_1, g_2 \in G$, their product is an element of G as well: $g_1 \cdot g_2 \in G$.

The following conditions are to be satisfied:

(1) Associative law: $g_1 \cdot (g_2 \cdot g_3) = (g_1 \cdot g_2) \cdot g_3$;
(2) There exists an element g_0, called the *identity*[1] which leaves any g unaltered by the group composition: $g_0 \cdot g = g \cdot g_0 = g$;
(3) For each $g \in G$ there exists an element called the *inverse* and denoted g^{-1} such that: $g \cdot g^{-1} = g^{-1} \cdot g = g_0$.

---

[1] The identity element is also called the *unit* element and is sometimes denoted by e.

© Springer International Publishing Switzerland 2016
R. D'Auria and M. Trigiante, *From Special Relativity to Feynman Diagrams*,
UNITEXT for Physics, DOI 10.1007/978-3-319-22014-7_7

In general, given two group elements g_1, g_2, $g_1 \cdot g_2 \neq g_2 \cdot g_1$. If for any g_1, $g_2 \in G$, $g_1 \cdot g_2 = g_2 \cdot g_1$, the group is called *commutative* or *abelian*.[2] As shown in Chap. 4, the set of all non-singular $n \times n$ matrices close a group with respect to the matrix multiplication, which is denoted by $GL(n, \mathbb{C})$, for complex matrices, and $GL(n, \mathbb{R})$, or simply $GL(n)$, for real ones.

In order to define a *Lie group*, we first define a *continuous group*. In general a *q-parameter continuous group* G has its elements labeled by q continuously varying parameters $(\theta^r) \equiv (\theta^1, \ldots, \theta^q)$:

$$g \in G : g = g(\theta^r) \equiv g(\theta^1, \ldots, \theta^q), \tag{7.1}$$

where continuity is expressed in terms of a (squared) "distance" d^2 in parameter space, $d^2 = \sum_r (\theta^r - \theta'^r)^2$.

A *Lie group* is a continuous group such that the dependence of its elements on the parameters θ^r satisfies the following requirement: If $g(\theta_1^r)$, $g(\theta_2^r)$ are two generic elements of it, the parameters $(\theta_3^r) = (\theta_3^1, \ldots, \theta_3^q)$ defining their product

$$g(\theta_1^r) \cdot g(\theta_2^r) = g(\theta_3^r), \tag{7.2}$$

are *q analytic* functions $\theta_3^r = \theta_3^r(\theta_1^s, \theta_2^s)$ of (θ_1^s) and (θ_2^s) (here the lower index on the parameters refers to the corresponding group element). Moreover the dependence of the group elements on the parameters is conventionally fixed so that

$$g(\theta^r \equiv 0) = g_0.$$

For example, in Chap. 4, we defined the three-dimensional rotation group $SO(3)$ as the group of 3×3 matrices $R^i{}_j$ acting on the three-dimensional Euclidean space and leaving the metric $g_{ij} = \delta_{ij}$ invariant. However the same group could have been defined abstractly, that is independently of its matrix realization, as the group of continuous transformations, depending on three parameters, and obeying a given composition law $\boldsymbol{\theta}_3 = \boldsymbol{\theta}_3(\boldsymbol{\theta}_1, \boldsymbol{\theta}_2)$ or, equivalently, as a Lie group described in the neighborhood of the identity by an algebra of generators whose structure is defined by Eqs. (4.124) and (4.125).

---

[2]We could have used a different notation and characterize the composition law as a "sum" $+$: g_1, $g_2 \in G$, $g_3 = g_1 + g_2 \in G$. In this case the identity element is called the *zero-element* and denoted by $\mathbf{0}$: $\forall g \in G$, $g + \mathbf{0} = \mathbf{0} + g = g$. The inverse of $g \in G$ is denoted by $-g$. This is clearly just a notation since in general the $+$ composition law has nothing to do with the ordinary sum of numbers. As an example the real numbers form an abelian group with respect to the ordinary sum, the zero-element clearly being number 0.

7.2 Representations

The notion of an n-dimensional vector space V_n over the real numbers (real vector space), introduced in Chap. 4, readily generalizes to that of an n-dimensional vector space V_n over the complex numbers (complex vector space). The elements of a complex vector space are uniquely defined by a collection of n complex numbers representing their components relative to a given basis $(\mathbf{u}_i)$: $\mathbf{V} = V^i \mathbf{u}_i \equiv (V^i)$, $V^i \in \mathbb{C}$. Three dimensional rotations, Lorentz transformations and homogeneous transformations of Cartesian coordinate systems, discussed in Chap. 4, are examples of *linear, homogeneous* transformations on vector spaces (three-dimensional rotations act on vectors in E_3, Lorentz transformations on four-vectors in M_4 an so on). A *linear function*, or *operator*, A on a vector space V_n is in general defined as mapping of V_n into itself, which associates with any vector $\mathbf{V} \in V_n$ a vector $A(\mathbf{V})$ in the same space, and which satisfies the linearity condition: Given any two vectors $\mathbf{V}, \mathbf{W} \in V_n$ and two numbers a, b (real or complex depending on whether V_n is defined over the real or complex numbers):

$$A(a\mathbf{V} + b\mathbf{W}) = a A(\mathbf{V}) + b A(\mathbf{W}). \tag{7.3}$$

Suppose now A is invertible, so that one can define the inverse linear transformation A^{-1} on V_n, then A is called a *linear transformation*. Being A invertible, if $\mathbf{V}$, $\mathbf{W}$ are linearly independent, also $A(\mathbf{V})$, $A(\mathbf{W})$ are. A therefore maps a basis $(\mathbf{u}_i)$ of V_n into a new basis $(\mathbf{u}'_i) \equiv (A(\mathbf{u}_i))$. We have dealt in Chap. 4 with linear transformations on vectors when describing the correspondence between Cartesian coordinate systems with a common origin (homogeneous linear coordinate transformations). In that case we have adopted a *passive point of view* and made transformations act on the base elements $(\mathbf{u}_i)$ of the coordinate system only and not on vectors in space. We have then considered the relation between the components of a same vector $\mathbf{V}$ in the two bases. In this perspective the action of A is uniquely defined by the $n \times n$ invertible matrix $\mathbf{A} \equiv (A^i{}_j)$ defining the components of the old basis relative to the new one.

$$\mathbf{u}_i = A^j{}_i \mathbf{u}'_j. \tag{7.4}$$

The components $\mathbf{V}' = (V'^i)$ and $\mathbf{V} = (V^i)$ of the same geometrical vector relative to the new and old bases, respectively, are related by the action of $\mathbf{A}$:

$$V'^i = A^i{}_j V^j \quad \Leftrightarrow \quad \mathbf{V}' = \mathbf{A}\mathbf{V}. \tag{7.5}$$

With an abuse of notation we shall denote the array vector $\mathbf{V}'$ of the new components by $A(\mathbf{V})$.

The same relation is obtained if we use the *active* description of transformations and view them as correspondences between different vectors (and in general points) in space. Then if $\mathbf{V} = V^i \mathbf{u}_i$ is a vector in V_n, the active action of a linear transformation

A will map it into a different vector $\mathbf{V}' = A(\mathbf{V})$. If we now define the matrix elements $A^i{}_j$ as the components of the new basis element $\mathbf{u}'_j = A(\mathbf{u}_j)$ along $\mathbf{u}_i$:

$$\mathbf{u}'_j = A^i{}_j \, \mathbf{u}_i, \tag{7.6}$$

using the linearity property of A we can write:

$$\mathbf{V}' = A(\mathbf{V}) = V^i A(\mathbf{u}_i) = V^i A^j{}_i \, \mathbf{u}_j = V'^j \, \mathbf{u}_j. \tag{7.7}$$

Although we find the same relation (7.5), the quantities involved have a different interpretation: V'^i and V^i in the passive description are the components of the *same* vector in the new and old bases, while in the active representation they represent the components of the new and old vectors with respect to the *same* basis. We shall use the active description when describing the effect of a coordinate transformation on the quantum states (which are vectors in a complex vector space). From now on we shall represent each vector by the array of its components $\mathbf{V} \equiv (V^i)$ with respect to a given basis, so that the effect of a transformation A, in both the complementary descriptions, is then described by the same matrix relation (7.5): $\mathbf{V} \to \mathbf{V}' = A(\mathbf{V}) \equiv \mathbf{A}\,\mathbf{V}$.

If we have two linear transformations A, B on V_n, their product $A \cdot B$ is the linear transformation resulting from their consecutive action on each vector: If B maps $\mathbf{V}$ into $\mathbf{V}' = B(\mathbf{V}) = \mathbf{B}\,\mathbf{V}$ and A maps $\mathbf{V}'$ into $\mathbf{V}'' = A(\mathbf{V}') = \mathbf{A}\,\mathbf{V}'$, then $A \cdot B$ is the transformation which maps $\mathbf{V}$ into $\mathbf{V}'' = A(B(\mathbf{V})) = A(\mathbf{B}\,\mathbf{V}) = (\mathbf{A}\,\mathbf{B})\,\mathbf{V}$. The product of two transformations is thus represented by the product of the matrices associated with each of them, in the same order.[3] The identity transformation I is the linear transformation which maps any vector into itself and it is represented by the identity $n \times n$ matrix $\mathbf{1}$. For any linear transformation A we trivially have $A \cdot I = I \cdot A = A$. Finally, being a linear transformation invertible, we can define its inverse A^{-1} such that, if A maps $\mathbf{V}$ into $\mathbf{V}' = A(\mathbf{V})$, A^{-1} is the linear transformation mapping $\mathbf{V}'$ into the unique vector $\mathbf{V} = A^{-1}(\mathbf{V}')$ which corresponds to $\mathbf{V}$ through A. It follows that A^{-1} is represented by the inverse $\mathbf{A}^{-1}$ of the matrix $\mathbf{A}$ associated with A. Finally the product of linear transformations is associative, the argument being substantially the same as the one used for coordinate transformations in Sect. 4.5. Linear transformations on vector spaces close therefore a group. Given the identification of linear transformations on V_n with $n \times n$ non singular matrices, the group of all such transformations can be identified with the group $GL(n, \mathbb{C})$, if V_n is complex, or $GL(n)$ if V_n is real (the symbol GL stands indeed for *General Linear transformations*).

---

[3] The action of a non-invertible operator A is also represented by a matrix $\mathbf{A}$, its definition being analogous to the one given for transformations. Such matrix, however, is singular. The product of two generic operators A and B is defined as for transformations and is represented by the product of the corresponding matrices in the same order. Examples of non-invertible operators appear among the hermitian operators representing observables in quantum mechanics, V_n being in this case the infinite dimensional vector space of quantum states. Another example of not necessarily invertible operators are the infinitesimal generators of continuous transformations, to be introduced below, which are indeed related, as we shall discover in the next chapters, to observables in quantum mechanics.

An *n-dimensional representation* D (or representation of degree n) consists in associating with each element $g \in G$ a *linear* transformation $D(g)$ on a linear vector space V_n in such a way that:

$$D(g) \cdot D(g') = D(g \cdot g'). \qquad (7.8)$$

Since linear transformations on V_n are uniquely defined by $n \times n$ invertible matrices, with respect to a given basis, Eq. (7.8) characterizes a representation as *a homomorphic map of G into the set (group) of $n \times n$ invertible matrices.*[4] Introducing a *basis* $(\mathbf{u}_i)$, $i = 1, \ldots n$, on V_n, $D(g)$ acts as an $n \times n$ matrix $\mathbf{D}(g) \equiv (D(g)^i{}_j)$ on the components of a vector $\mathbf{V} = (V^i, \ldots, V^n)$ according to the law

$$V'^i = D(g)^i{}_j\, V^j \quad \Leftrightarrow \quad \mathbf{V}' = \mathbf{D}(g)\,\mathbf{V},$$

We shall denote by the bold symbol $\mathbf{D}$ the representation of a group in terms of matrices. The vector space V_n is called the *carrier of the representation, or representation space*. In the case of the rotation group, for example, the three dimensional Euclidean space V_3 is the carrier of the representation studied in Chap. 4:

$$g(\theta_1, \theta_1, \theta_2) \in SO(3) \xrightarrow{\ \mathbf{D}\ } D(g)^i{}_j = R(\theta_1, \theta_1, \theta_2)^i{}_j, \quad i, j = 1, 2, 3.$$

For a general representation the matrix $\mathbf{D}(g)$ is an element of $GL(n, \mathbb{C})$ or of $GL(n, \mathbb{R})$, depending on whether the base space is a complex or real vector space.

In the active picture, for any $g \in G$, $D(g)$ maps vectors into vectors, all represented with respect to a same basis $(\mathbf{u}_i)$. On replacing the original basis $(\mathbf{u}_i)$ by a new one $(\mathbf{u}'_i)$, related to it through a non singular matrix $\mathbf{A}$, as in Eq. (7.4), the matrix $\mathbf{D}(g)$ gets replaced by the matrix $\mathbf{D}'(g) = \mathbf{A}\,\mathbf{D}(g)\,\mathbf{A}^{-1}$ which represents the action of $D(g)$ in the new basis. This is easily shown starting from the matrix relation between the components of a vector $\mathbf{V}_1$ and its transformed $\mathbf{V}_2$ in the old basis: $\mathbf{V}_2 = \mathbf{D}(g)\,\mathbf{V}_1$. Being the components $\mathbf{V}'_1$ and $\mathbf{V}'_2$ of the two vectors in the new basis given by $\mathbf{V}'_1 = \mathbf{A}\,\mathbf{V}_1$, $\mathbf{V}'_2 = \mathbf{A}\,\mathbf{V}_2$, we find:

$$\mathbf{V}'_2 = \mathbf{A}\,\mathbf{V}_2 = \mathbf{A}\,\mathbf{D}(g)\,\mathbf{V}_1 = \mathbf{A}\,\mathbf{D}(g)\,\mathbf{A}^{-1}\,\mathbf{V}'_1 = \mathbf{D}'(g)\,\mathbf{V}'_1. \qquad (7.9)$$

It is easily verified that the mapping $\mathbf{D}'$ of a generic group element g into $\mathbf{D}'(g)$ is still a representation, also denoted by $\mathbf{D}' = \mathbf{A}\,\mathbf{D}\,\mathbf{A}^{-1}$.

The representations $\mathbf{A}\,\mathbf{D}\,\mathbf{A}^{-1}$ and $\mathbf{D}$ are then said *equivalent*, and we write:

$$\mathbf{D} \sim \mathbf{A}\,\mathbf{D}\,\mathbf{A}^{-1}. \qquad (7.10)$$

---

[4]It is obvious that the identity $g_0 = e$ element of the group is represented by the unit n-dimensional matrix that we will denote by $\mathbf{1}$ or else by I.

If the homomorphic mapping:

$$g \longrightarrow D(g), \tag{7.11}$$

is *isomorphic*, namely it is one-to-one and onto, then the representation is *faithful*, otherwise it is *unfaithful*. A trivial, but important, unfaithful representation is obtained by the mapping:

$$g \longrightarrow 1, \quad \forall g \in G, \tag{7.12}$$

and is called the *identity, or trivial representation*, simply denoted by **1**.

Coming back to the general case, let us assume that there exists a *subspace* $V_m \subset V_n$ of dimensions $m < n$ such that every element of the subspace V_m is transformed into an element of the same subspace under all the transformations of the group G:

$$\forall g \in G \quad D(g) : V_m \to V_m.$$

If such a subspace exists it is called *invariant* under the action of G and the representation D acting on V_n is said to be *reducible* in V_n. A representation is *irreducible* if it is not reducible, that is if there is no proper invariant subspace of the carrier space. If a representation is *reducible*, we may find a *basis* in V_n in which all matrices $\mathbf{D}(g)$, with $g \in G$, can be *simultaneously* brought to the form

$$\mathbf{D}(g) = (D(g)^i{}_j) = \begin{pmatrix} \mathbf{A} & \mathbf{0} \\ \mathbf{B} & \mathbf{C} \end{pmatrix}, \tag{7.13}$$

where $\mathbf{A}$, $\mathbf{B}$, $\mathbf{C}$ are matrices of dimensions $(n-m) \times (n-m)$, $m \times (n-m)$ and $m \times m$ and $\mathbf{0}$ is the $(n-m) \times m$ matrix whose elements are all zero. The corresponding basis $(\mathbf{u}_i)$ is chosen so that its last m elements $(\mathbf{u}_\ell)$, $\ell = 1, \ldots, m$, form a basis of V_m, while the first $m - n$ elements $(\mathbf{u}_a)$, $a = 1, \ldots, n - m$, generate the complement V_{n-m} of V_m in V_n. The components of a generic column vector then split accordingly: $\mathbf{V} = (V^a, V^\ell)$, and transform as follows:

$$V'^a = A^a{}_b V^b , \quad V'^\ell = C^\ell{}_{\ell'} V^{\ell'} + B^\ell{}_a V^a. \tag{7.14}$$

Therefore, if $\mathbf{V} \in V_m$, $V^a \equiv 0$ and thus $V'^a \equiv 0$, that is $\mathbf{V}' = D(g)(\mathbf{V}) \in V_m$.

If it is possible to find a basis in which all the matrices of the representation assume the form (7.13), *but with* $\mathbf{B} = \mathbf{0}$, we say that the representation is *fully reducible* or *decomposable*. In this case both V_{n-m} and V_m are *invariant subspaces*.[5] The space V_n, as a vector space, is the *direct sum* of V_{n-m} and V_m, $V_n = V_{n-m} \oplus V_m$, and the representation $\mathbf{D}$ is said to be the *direct sum* of $\mathbf{D}_{n-m}$ and $\mathbf{D}_m$

$$\mathbf{D}_n = \mathbf{D}_{n-m} \oplus \mathbf{D}_m,$$

---

[5]It can be proven that, for groups which admit finite-dimensional representations in terms of unitary or orthogonal matrices (like the group of rotations in the Euclidean space), all reducible representations are completely reducible.

where $\mathbf{D}_{n-m}$ and $\mathbf{D}_m$ are the two representations of G defined, for any $g \in G$, by the upper and lower diagonal blocks of $\mathbf{D}(g)$

$$\mathbf{D}(g) = \begin{pmatrix} \mathbf{D}_{n-m}(g) & \mathbf{0} \\ \mathbf{0} & \mathbf{D}_m(g) \end{pmatrix}. \tag{7.15}$$

The representations $\mathbf{D}_m$, $\mathbf{D}_{n-m}$ may still be completely reducible, and thus may be further decomposed into lower dimensional representations. We can iterate the above procedure until we end up with *irreducible representations*: $\mathbf{D}_{k_1}, \ldots, \mathbf{D}_{k_\ell}$, where $\sum_{i=1}^{\ell} k_i = n$. This corresponds to finding a basis in which the matrix representation under $\mathbf{D}$ of a generic element $g \in G$ has the following block structure:

$$\mathbf{D}(g) = \begin{pmatrix} \mathbf{D}_{k_1}(g) & \mathbf{0} & \cdots & \mathbf{0} \\ \mathbf{0} & \mathbf{D}_{k_2}(g) & \cdots & \mathbf{0} \\ \vdots & & \ddots & \vdots \\ \mathbf{0} & \mathbf{0} & \cdots & \mathbf{D}_{k_\ell}(g) \end{pmatrix}. \tag{7.16}$$

We say that the original representation $\mathbf{D}$ is completely reducible into the irreducible representations $\mathbf{D}_{k_i}$ and write:

$$\mathbf{D} = \bigoplus_{i=1}^{\ell} \mathbf{D}_{k_i} \equiv \mathbf{D}_{k_1} \oplus \mathbf{D}_{k_2} \oplus \ldots \mathbf{D}_{k_\ell}. \tag{7.17}$$

Correspondingly the representation space V_n of $\mathbf{D}$ has been decomposed into the direct sum of spaces V_{k_i} on which $\mathbf{D}_{k_1}[g]$ act:

$$V_n = V_{k_1} \oplus V_{k_2} \oplus \ldots V_{k_\ell}. \tag{7.18}$$

As a simple example we may consider the group SO(2) of rotations in the (x, y) plane. The three-dimensional representation acting on a generic vector of components (x, y, z) has the following form:

$$\begin{pmatrix} \cos \theta & \sin \theta & 0 \\ -\sin \theta & \cos \theta & 0 \\ 0 & 0 & 1 \end{pmatrix}. \tag{7.19}$$

We see that the representation is fully reducible into a two-dimensional representation acting on the components x, y and a one-dimensional representation acting on the component z (which leaves it invariant).

The simplest (faithful) representation of GL(n) is given in terms of the set of matrices acting on the components of a vector $\mathbf{V} \in V_n$ and is called the *defining representation*.

However, while studying the tensor algebra, we have emphasized that the (p, q)-tensors can be thought of as *vectors* in a representation space of $GL(n)$. More precisely, the set of $n^{k+\ell}$ components $T^{i_1 \cdots i_k}{}_{j_1 \cdots j_\ell}$ of a tensor of type (k, ℓ), can be understood as the components of a vector in the representation space $V_{n^{k+\ell}}$ on which the linear action of an element $g \in GL(n)$ is defined by Eqs. (4.60) and (4.61), that is by the *tensor product* $\mathbf{D} \otimes \cdots \otimes \mathbf{D} \otimes \mathbf{D}^{-T} \cdots \otimes \mathbf{D}^{-T}$ of k matrices $\mathbf{D}$ and ℓ matrices $\mathbf{D}^{-T}$. As anticipated in Chap. 4, using the properties of the Kronecker product of matrices, one can easily verify that the action of the group on tensors satisfies Eq. (7.8) and thus defines a representation.

As an example, let us recall from Sect. 4.3, that a generic tensor F^{ij} can be split into a symmetric and antisymmetric component (F_S^{ij}, F_A^{ij}, respectively), see Eq. (4.75). F^{ij} belongs to the vector space V_{n^2} of dimension n^2, its n^2 components can be thought of as the independent entries of the $n \times n$ matrix (F^{ij}). This vector space is the base space of a representation of $GL(n)$, each tensor F^{ij} transforming according to Eq. (4.49). The symmetric and antisymmetric components F_S^{ij}, F_A^{ij} span orthogonal subspaces $V_{(S)}$, $V_{(A)}$ of V_{n^2}, such that:

$V^{(S)}$ contains as elements the symmetric tensors, $F_S^{ij} = F_S^{ji}$;
$V^{(A)}$ contains as elements the antisymmetric tensors, $F_A^{ij} = -F_A^{ji}$.

The dimensions of $V_{(S)}$ and $V_{(A)}$ are $\frac{n(n+1)}{2}$ and $\frac{n(n-1)}{2}$, so that their sum matches the dimension of V_{n^2}:

$$\frac{n(n+1)}{2} + \frac{n(n-1)}{2} = n^2. \qquad (7.20)$$

In other words V_{n^2} is the direct sum of $V_{(S)}$ and $V_{(A)}$:

$$V^{n^2} = V^{(S)} \oplus V^{(A)}.$$

Since symmetric (antisymmetric) tensors are transformed into symmetric (antisymmetric) tensors, see Eqs. (4.76), both $V_{(S)}$ and $V_{(A)}$ are invariant subspaces of V_{n^2}, and thus that the representation $\mathbf{D} \otimes \mathbf{D}$ is fully reducible into the direct sum of a representation $\mathbf{D}_{(S)}$ acting on symmetric tensors and a representation $\mathbf{D}_{(A)}$ acting on antisymmetric ones:

$$\mathbf{D} \otimes \mathbf{D} = \mathbf{D}_{(S)} \oplus \mathbf{D}_{(A)}. \qquad (7.21)$$

It must be observed that if we restrict the transformations of a group G to those of a subgroup $G' \subset G$, a representation which was irreducible with respect to the G may become reducible with respect the smaller group G'. This is what happens, for example, when we restrict the transformations of $GL(n)$ to those of the subgroup $O(n)$ as it was observed at the end of Sect. 4.5. In fact, with reference to Eqs. (4.106) and (4.105), we see that if we restrict to $O(n)$ transformations only, the space of symmetric tensors, which was irreducible with respect to $GL(n)$, becomes now a *direct sum* of the subspace of the *symmetric and traceless* tensors and of the one-dimensional subspace of tensors proportional to δ^{ij}. It then follows that the n^2-dimensional space of rank-two $O(n)$-tensors F^{ij} can now be reduced into the direct sum of three subspaces according to the decompositions of tensors described in Eqs. (4.106) and (4.105) that here we rewrite, for the sake of completeness:

$$F^{ij} = \tilde{F}^{ij}_S + F^{ij}_A + D^{ij},$$

where

$$\tilde{F}^{ij}_S = \frac{1}{2}(F^{ij} + F^{ji}) - \frac{1}{n}(\delta_{pq} F^{pq})\,\delta^{ij},$$

$$F^{ij}_A = \frac{1}{2}(F^{ij} - F^{ji}), \quad D^{ij} = \frac{1}{n}(\delta_{pq} F^{pq})\delta^{ij}.$$

As it was shown in Chap. 4 each of the three subspaces is *invariant* under $O(n)$ transformations, elements of each subspace being transformed into elements of the same subspace. It follows that the n^2-dimensional representation of $O(n)$ is fully reducible into three irreducible representations $\mathbf{D}_{(S)}$, $\mathbf{D}_{(A)}$, $\mathbf{D}_{\text{Tr}} = 1$ of dimensions $\frac{n(n+1)}{2} - 1$, $\frac{n(n-1)}{2}$ and 1, respectively:

$$\mathbf{D} \otimes \mathbf{D} = \mathbf{D}_{(S)} \oplus \mathbf{D}_{(A)} \oplus \mathbf{D}_{\text{Tr}}, \tag{7.22}$$

where $\mathbf{D}_{(S)}$ act on symmetric traceless matrices and $\mathbf{D}_{\text{Tr}}$ on the tensors proportional to δ^{ij} (traces).

The same decompositions hold if instead of the group $O(n)$ we have a non-compact form like the Lorentz group $SO(1, 3)$ when $n = 4$. The only difference is that the one-dimensional subspace is now proportional to the Minkowski metric $\eta_{\mu\nu}$.

Let us now discuss a property in group theory which has important applications in physics.

Schur's Lemma: *Let* $\mathbf{D}$ *be an irreducible n-dimensional representation of a group* G. *A matrix* $\mathbf{T}$ *which commutes with all matrices* $\mathbf{D}(g)$, *for any* $g \in G$, *is proportional to the identity matrix* $\mathbf{1}_n$.

In formulas, if

$$\forall g \in G: \quad \mathbf{T}\mathbf{D}(g) = \mathbf{D}(g)\,\mathbf{T}, \tag{7.23}$$

there exists a number λ such that:

$$\mathbf{T} = \lambda\,\mathbf{1}_n \quad \Leftrightarrow \quad T^i{}_j = \lambda\,\delta^i_j, \quad i, j = 1, \ldots, n. \tag{7.24}$$

To show this, let λ be an eigenvalue of $\mathbf{T}$ in V_n (which always exists) and $\mathbf{V}$ the corresponding eigenvector:

$$\mathbf{T}\mathbf{V} = \lambda\,\mathbf{V}. \tag{7.25}$$

Let $V_\lambda = \{ \mathbf{V}' \in V_n \mid \mathbf{T}\mathbf{V}' = \lambda\,\mathbf{V}' \}$ be the eigenspace of the matrix $\mathbf{T}$ corresponding to the eigenvalue λ. This space is non-empty since $\mathbf{V} \in V_\lambda$. It can be easily verified that V_λ is invariant under the action of G. Indeed for any $\mathbf{V}' \in V_\lambda$ and $g \in G$ the vector $\mathbf{D}(g)\mathbf{V}'$ is still in V_λ since:

$$\mathbf{T}\mathbf{D}(g)\mathbf{V}' = \mathbf{D}(g)\mathbf{T}\mathbf{V}' = \lambda\,\mathbf{D}(g)\mathbf{V}', \tag{7.26}$$

where we have used the hypothesis (7.23) of Schur's lemma that $\mathbf{T}$ commutes with the action of G on V_n defined by the representation $\mathbf{D}$. Since V_λ is a non-empty invariant subspace of V_n and being $\mathbf{D}$ an irreducible representation by assumption, V_λ can only coincide with V_n. We conclude that $\mathbf{T}$ acts on V_n as λ times the identity matrix.

An important consequence of Schur's lemma is that, *if $\mathbf{D}$ is a n-dimensional representation of a group G and if there exists a matrix $\mathbf{T}$ which commutes with all matrices $\mathbf{D}(g)$, for any $g \in G$, and which is not proportional to the identity matrix $\mathbf{1}_n$, then $\mathbf{D}$ is reducible.*

This property provides us with a powerful criterion for telling if a representation is reducible and, in some cases, to determine its irreducible components: Suppose we find an operator T on V_n which commutes with all the transformations $D(g)$ representing the action of a group G on the same space. The matrix representation $\mathbf{T}$ of T will then have the form:

$$\mathbf{T} = \begin{pmatrix} c_1\,\mathbf{1}_{k_1} & & \mathbf{0} \\ & \ddots & \\ \mathbf{0} & & c_s\,\mathbf{1}_{k_s} \end{pmatrix}, \tag{7.27}$$

where $c_1, \dots, c_s$ are the eigenvalues of $\mathbf{T}$ and the corresponding eigenspaces $V_{k_1}, \dots, V_{k_s}$ of $\mathbf{T}$ correspond to different irreducible representations $\mathbf{D}_{k_1}, \dots, \mathbf{D}_{k_s}$ of G. Thus the degeneracies $k_1, \dots, k_s$ of the eigenvalues of $\mathbf{T}$ are dimensions of irreducible representations of G. It can happen that two or more eigenvalues c_i coincide, thus implying that $\mathbf{T}$ is proportional to the identity on the carrier spaces of *reducible* representations of G (direct sum of the irreducible representations corresponding to the same eigenvalues). In these cases we say that there is an *accidental degeneracy*.[6]

We shall show in Chap. 9 how Schur's lemma allows to deduce important information on the degeneracy of the energy levels of a quantum mechanical system from the knowledge of its symmetries.

---

[6] An accidental degeneracy may hint towards the existence of a larger group G' acting on the space V_n and containing G, whose action on V_n still commutes with $\mathbf{T}$, and of which the eigenspaces of $\mathbf{T}$ define now irreducible representations. In Chap. 9 we shall consider an important application of Schur's Lemma to quantum mechanics, in which T is the Hamiltonian operator $\hat{H}$ and G the symmetry group of the system, whose unitary action on the space of states commutes with $\hat{H}$. In this case the eigenvalues c_i corresponding to irreducible representations of G are the energy levels of the system and the presence of an accidental degeneracy hints towards the existence of extra hidden symmetries (not previously recognized) which enlarge G to a group G'.

7.3 Infinitesimal Transformations and Lie Algebras

In the following we shall be mainly concerned with *infinitesimal transformations of a group G*, generalizing the definition, given in Sect. 4.5.1 for the rotation group, of the Lie algebra of infinitesimal generators. As we shall show shortly, the knowledge of the structure of the group in an *infinitesimal neighborhood* of the unit element (i.e. of the algebra of its infinitesimal generators), is sufficient to reconstruct, at least *locally*, the structure of the group itself.[7] In order to show this let us expand, as we did for rotations, a generic group element in a given representation $\mathbf{D}$ in Taylor series with respect to its parameters $\{\theta^r\} = (\theta^1, \ldots, \theta^q)$, assuming them to be small:

$$\mathbf{D}(g(\theta^r)) = 1 + \theta^r \left.\frac{\partial \mathbf{D}}{\partial \theta^r}\right|_{\theta^t = 0} + \mathcal{O}(|(\boldsymbol{\theta})|^2)$$
$$= 1 + \theta^r \mathbf{L}_r + \mathcal{O}(|(\boldsymbol{\theta})|^2), \tag{7.28}$$

where:

$$\mathbf{L}_r \equiv \left.\frac{\partial \mathbf{D}}{\partial \theta^r}\right|_{\theta^r = 0}$$

define the *infinitesimal generators* of $\mathbf{D}(g)$. These matrices clearly depend on the representation $\mathbf{D}$ of the group G we are using.

Just as we did in Sect. 4.5.1, let us write a generic transformation in G, defined by finite values (θ^r) of the parameters, as resulting from the iterated action of a large number N of "small" transformations with parameters $\delta\theta^r \equiv \frac{\theta^r}{N} \ll 1$:

$$\mathbf{D}(g(\theta^r)) = \mathbf{D}(g(\delta\theta^r))^N = \mathbf{D}\left(g\left(\frac{\theta^r}{N}\right)\right)^N. \tag{7.29}$$

To first order each infinitesimal transformation $\mathbf{D}(g(\delta\theta^r))$ can be written using the expansion (7.28) and neglecting second order terms in the infinitesimal parameters:

$$\mathbf{D}(g(\delta\theta^r)) \approx 1 + \delta\theta^r \mathbf{L}_r = 1 + \frac{\theta^r}{N} \mathbf{L}_r. \tag{7.30}$$

We can then write the following approximated expression:

$$\mathbf{D}(\theta^r) \approx \left[1 + \frac{\theta^r}{N} \mathbf{L}_r\right]^N.$$

The larger N the better the above approximation is. In the limit $N \to \infty$ we obtain

$$\mathbf{D}(\theta^r) = \exp(\theta^r \mathbf{L}_r), \tag{7.31}$$

---

[7]It is important to note that locally the same Lie algebra can describe Lie groups which are globally different. This is for example the case of the groups SO(3) and SU(2), see Appendix F.

the exponential of a matrix being defined by Eq. (4.129). We can summarize the above result as follows. Given an element $g(\theta) \in G$ in the neighborhood of the identity element $g = I$, that is for values of the parameters θ^r in a neighborhood of $\theta^r \equiv 0$, we may associate with it a unique matrix $\mathbf{A}(\theta^r)^i{}_j \equiv \theta^r (\mathbf{L}_r)^i{}_j$ such that

$$\mathbf{D}(\theta^r)^i{}_j = \left(e^{\mathbf{A}(\theta^r)}\right)^i{}_j = \sum_{n=0}^{\infty} \frac{1}{n!} \left[\mathbf{A}(\theta^r)^n\right]^i{}_j,$$

where we have used the short-hand notation $\mathbf{D}(\theta^r) \equiv \mathbf{D}(g(\theta^r))$. $\mathbf{A}(\theta^r)$ is referred to as the *infinitesimal generator* of the transformation $\mathbf{D}(g)$. As the parameters θ^r are varied $\mathbf{A}(\theta^r)^i_j = \theta^r (\mathbf{L}_r)^i{}_j$ describes a vector space $\mathcal{A}$ of parameters θ^r with respect to the basis of infinitesimal generators $(\mathbf{L}_r)^i{}_j$. In particular the higher order terms in the expansion (7.28) are written in terms of powers of $\mathbf{A}(\theta^r)$. For example, to second order the Taylor expansion (7.28) of $\mathbf{D}(\theta^r)$ reads:

$$\mathbf{D}(\theta^r) = 1 + \theta^r \, \mathbf{L}_r + \frac{1}{2}\theta^r \theta^s \mathbf{L}_r \, \mathbf{L}_s + \mathcal{O}(|(\theta)|^3). \tag{7.32}$$

From (7.32) we compute, to the same order, the *inverse* transformation:

$$\mathbf{D}(\theta^r)^{-1} = 1 - \theta^r \, \mathbf{L}_r + \frac{1}{2}\theta^r \theta^s \mathbf{L}_r \, \mathbf{L}_s + \mathcal{O}(|(\theta)|^3). \tag{7.33}$$

Consider the matrix representation $\mathbf{D}(\theta_1)$, $\mathbf{D}(\theta_2)$ of two group elements, $g_1 = g(\theta_1)$, $g_2 = g(\theta_2)$, where, for the sake of simplicity, we write θ for the set of n parameters $\{\theta^1, \ldots, \theta^n\}$, the lower index in θ_1, θ_2 referring to two different elements. We define the *commutator* of $\mathbf{D}(\theta_1)$, $\mathbf{D}(\theta_2)$ as the matrix $\mathbf{D}^{-1}(\theta_1)\mathbf{D}^{-1}(\theta_2)\mathbf{D}(\theta_1)\mathbf{D}(\theta_2)$. This matrix must be a representation $\mathbf{D}(\theta_3)$ of some group element $g_3 = g(\theta_3) \equiv g_1^{-1} \cdot g_2^{-1} \cdot g_1 \cdot g_2$. Using Eqs. (7.32) and (7.33) a simple computation shows that the terms *linear* in the θ parameters cancel against each other so that the expansion of the group commutator becomes

$$\mathbf{D}^{-1}(\theta_1)\mathbf{D}^{-1}(\theta_2)\mathbf{D}(\theta_1)\mathbf{D}(\theta_2) = 1 + \theta_1^r \, \theta_2^s \, [\mathbf{L}_r, \, \mathbf{L}_s] + \cdots \tag{7.34}$$

where $[\mathbf{L}_r, \, \mathbf{L}_s]$ is the algebra *commutator* defined as $\mathbf{L}_r \mathbf{L}_s - \mathbf{L}_s \mathbf{L}_r$. On the other hand from the group composition law we also have

$$\mathbf{D}(\theta_3) \equiv \mathbf{D}^{-1}(\theta_1)\mathbf{D}^{-1}(\theta_2)\mathbf{D}(\theta_1)\mathbf{D}(\theta_2) = 1 + \theta_3^m \, \mathbf{L}_m + \cdots . \tag{7.35}$$

Since Eqs. (7.34) and (7.35) must coincide, we deduce

$$\theta_1^k \theta_2^l \, [\mathbf{L}_k, \, \mathbf{L}_l] = \theta_3^m \, \mathbf{L}_m, \tag{7.36}$$

that is

$$[\mathbf{L}_k, \mathbf{L}_l] = C_{kl}{}^m \mathbf{L}_m, \tag{7.37}$$

where we have set

$$C_{kl}{}^m \theta_1^{[k} \theta_2^{l]} = \theta_3^m. \tag{7.38}$$

The set of constants $C_{kl}{}^m$ are referred to as the *structure constants of the Lie group*. From (7.37) we see that the structure constants are *antisymmetric* in their lower indices.

We can easily verify that the infinitesimal generators L_r satisfy the identity

$$[\mathbf{L}_k, [\mathbf{L}_l, \mathbf{L}_m]] + [\mathbf{L}_l, [\mathbf{L}_m, \mathbf{L}_k]] + [\mathbf{L}_m, [\mathbf{L}_k, \mathbf{L}_l]] = 0 \tag{7.39}$$

called *Jacobi identity*. As a consequence, by use of the Eqs. (7.37) and (7.39), we obtain that the structure constants must satisfy the identity

$$C_{kl}{}^n C_{mn}{}^p + C_{lm}{}^n C_{kn}{}^p + C_{mk}{}^n C_{ln}{}^p = 0, \tag{7.40}$$

or, equivalently:

$$C_{[kl}{}^n C_{m]n}{}^p = 0, \tag{7.41}$$

where the complete antisymmetrization in three indices has been defined after Eq. (5.17) of Chap. 5.

A vector space of matrices $\mathcal{A}$ which is closed under commutation, namely such that the commutator of any two of its elements is still in $\mathcal{A}$, is an example of a *Lie algebra*. Its algebraic structure is defined by the commutation relations between its basis elements, as in Eq. (7.37), i.e. by its structure constants $C_{mn}{}^p$. The Lie algebra, as we have seen, describes exhaustively the structure[8] of the abstract group G in the neighborhood of the identity of G. It follows that the structure constants $C_{rs}{}^p$, do not depend on the particular representation $\mathbf{D}$ of G.

7.4 Representation of a Group on a Field

Let us consider an n-dimensional flat space of points $\mathbf{M}_n$ and its associated vector space V_n described by the vectors $\overrightarrow{AB}$ connecting couples of points in $\mathbf{M}_n$. The space $\mathbf{M}_n$ can be the Euclidean space E_n if the metric tensor defined on it is δ_{ij} (in this case we shall be mainly interested in our three-dimensional Euclidean space E_3), or, for $n = 4$, the Minkowski space M_4 of special relativity if the metric is $\eta_{\mu\nu}$. It is useful at this point to recall the notations used in Chap. 4 for describing Cartesian coordinates

---

[8]By structure we mean the correspondence between any two elements of G and the third element representing their product.

in the various spaces: The collection of generic Cartesian coordinates on $\mathbf{M}_n$ is denoted by $\mathbf{r} = (x^i)$ while our familiar Cartesian rectangular coordinates are also denoted by $\mathbf{x} = (x, y, z)$ and the space-time coordinates of an event in Minkowski space are also collectively denoted by $x = (x^\mu) = (ct, \mathbf{x})$. Let us introduce a second p-dimensional vector space V_p and let us we consider a map $\Phi^\alpha \colon \mathbf{M}_n \to V_p$ which associates with each point $P \in \mathbf{M}_n$, labeled by Cartesian coordinates x^i, $i = 1, \ldots, n$, a vector in V_p of components $\Phi^\alpha(x^i)$, $\alpha = 1, \ldots, p$, with respect to a chosen basis,

$$\Phi^\alpha \colon \quad \forall x^i \in \mathbf{M}_n \to \Phi^\alpha(x^i) \in V_p. \tag{7.42}$$

This function is called a *field*, defined on $\mathbf{M}_n$, with values in V_p. The index α is called the internal index since it labels the *internal components* Φ^α of the field, which are degrees of freedom not directly related to its space-time propagation. An example is the index $\alpha = 1, 2$ labeling the physical polarizations of a photon. V_p is consequently called the *internal space*. If, as V_p, we take the space $V_{n^{k+l}}$ of type-(k, l) tensors, the corresponding field $\Phi^{i_1 \cdots i_k}{}_{j_1 \ldots j_l}(x^i)$ is called a *tensor field*. We have already introduced the notion of tensor fields in Chap. 4, Sect. 4.3, and illustrated their transformation properties under a change in the Cartesian coordinates (affine transformations) on $\mathbf{M}_n$. There we discussed, as an example, the case of a tensor $T^{ij}{}_k(x^i)$ which has values in the n^3-dimensional vector space $V_p = V_{n^3}$ of type-$(2, 1)$ tensors. Its transformation law is given by Eq. (4.73), its indices transforming under the homogeneous part $\mathbf{D} = (D^i{}_j)$ (element of GL(n)) of the affine transformation, according to their positions. Thinking of $T^{ij}{}_k$ as the $p = n^3$ components of a vector in V_p, they are subject to the *linear* action of the matrix $\left(\mathbf{D} \otimes \mathbf{D} \otimes \mathbf{D}^{-T}\right)^{ijs}{}_{lmk}$ defining the representation the GL(n) transformation on $(2, 1)$ tensors. This transformation property is generalized in a straightforward way to generic type-(k, l) tensor fields. If we wish to restrict to transformations preserving the Euclidean or Lorentzian metrics on E_3 or M_4, as we shall mostly do in the following, we need to restrict the homogeneous part of the affine transformation to O(3) or to O(1, 3), respectively.

There are several instances in physics of tensor fields. In particular rank $(1, 0)$ and rank $(0, 1)$ tensors are (contravariant or covariant) *vector fields*, while rank $(0, 0)$ tensors are *scalar fields*.

Let us give some examples. Well known three-dimensional vector fields are the gravity field $\mathbf{g}(x, y, z)$ in Newtonian mechanics or the electric and magnetic fields $\mathbf{E}(x, y, z, t)$ and $\mathbf{B}(x, y, z, t)$ of the Maxwell theory. More precisely they are vectors with respect to the rotation group SO(3). They are instances of maps between the Euclidean (E_3) or Minkowski space (M_4) and the Euclidean three-dimensional vector space ($V_p = V_3$).

The four-vector potential $A_\mu(x^\nu)$ is again a vector field albeit with respect to Lorentz transformations SO(1, 3). Here $\mathbf{M}_n = M_4$, and $V_p = V_4$, the space of four-vectors $\mathbf{V} = (V^\mu) = (V^0, V^1, V^2, V^3)$ associated with Minkowski space.

An example of rank $(0, 2)$ tensor field is the covariant field strength $F_{\mu\nu} = \partial_\mu A_\nu - \partial_\nu A_\mu$. Here again $\mathbf{M}_n$ is Minkowski space M_4 while V_p is the six-dimensional space of the antisymmetric tensors.

The field of temperatures in a given region of ordinary space x, y, z ($\mathbf{M}_n = E_3$) is an example of *scalar field* since V_p is the one-dimensional vector space of the real numbers $V_1 \equiv \mathbb{R}$; a scalar field is also the wave function $\Psi(x, y, z, t)$, solution to the Schroedinger equation, which associates with each point in space-time M_4 a complex number, that is an element of the two-dimensional space $V_2 \equiv \mathbb{C}$ (in this case we talk about a complex scalar field).

In all these examples the transformation group G on the tensors, is chosen to be either SO(3) or SO(1, 3) (or their affine extensions, like the Poincaré group on M_4, keeping in mind that tensor fields, just like vectors, always transform under the homogeneous part of the coordinate transformation, like the Lorentz group).

We wish now to generalize this discussion to a generic transformation group G and to a carrier space V_p supporting a generic representation space, not necessarily of vector or tensor character. Indeed besides the known cases of the electromagnetic field $A_\mu(x)$ and of its field strength $F_{\mu\nu}(x)$, which are tensor fields, when discussing the *Dirac equation* in Chap. 10, we shall be dealing with a field belonging to a representation of the Lorentz group, called *spinor representation*, which cannot be constructed in terms of tensors. This field will provide the relativistic description of particles with spin $1/2$ like the electron.

Let us denote by $\mathbf{R}(g)$ the representation of G acting on V_n, and by $\mathbf{D}(g)$ the one acting on V_p. We shall always consider V_n to be either the space of three-vectors on E_3, namely the vectors $\Delta\mathbf{x}$, or that of four-vectors on M_4, Δx^μ. If G acts as an affine group on the chosen Cartesian coordinate system on $\mathbf{M}_n$, like the Poincaré group on M_4, the representations $\mathbf{R}$ and $\mathbf{D}$ only refer to the action of the homogeneous part of G, like the Lorentz subgroup of the Poincaré group.

Let us now introduce a Cartesian coordinate system on $\mathbf{M}_n$ with origin O and basis $\{\mathbf{u}_i\}$ of V_n, and a basis $\{\omega_\alpha\}$ on V_p. Under a generic transformation $g \in G$ two vectors $\mathbf{V} = (V^i)$ in V_n and $\mathbf{W} = (W^\alpha)$ in V_p transform as follows:

$$V^i \rightarrow V'^i = R(g)^i{}_j V^j \quad \Leftrightarrow \quad \mathbf{V} \rightarrow \mathbf{V}' = \mathbf{R}(g)\,\mathbf{V}, \tag{7.43}$$

$$W^\alpha \rightarrow W'^\alpha = D(g)^\alpha{}_\beta W^\beta \quad \Leftrightarrow \quad \mathbf{W} \rightarrow \mathbf{W}' = \mathbf{D}(g)\,\mathbf{W}. \tag{7.44}$$

The transformation property of a generic field $\boldsymbol{\Phi}(\mathbf{r}) \equiv (\Phi^\alpha(\mathbf{r}))$ on $\mathbf{M}_n$ with values in V_p under a transformation G is then the direct generalization of the analogous law for tensor fields:

$$\Phi^\alpha(\mathbf{r}) \rightarrow \Phi'^\alpha(\mathbf{r}') = D^\alpha{}_\beta\, \Phi^\beta(\mathbf{r}) = D^\alpha{}_\beta\, \Phi^\beta(\mathbf{R}^{-1}(\mathbf{r}' + \mathbf{r}_0))$$

$$\Leftrightarrow \quad \boldsymbol{\Phi}(\mathbf{r}) \rightarrow \boldsymbol{\Phi}'(\mathbf{r}') = \mathbf{D}\,\boldsymbol{\Phi}(\mathbf{r}) = \mathbf{D}\,\boldsymbol{\Phi}(\mathbf{R}^{-1}(\mathbf{r}' + \mathbf{r}_0)), \tag{7.45}$$

where, for the sake of notational simplicity we have suppressed the explicit dependence of the matrices $\mathbf{D} = (D^\alpha{}_\beta)$ and $\mathbf{R}^i{}_j$ (and of the translation parameters $\mathbf{r}_0$) on the group element $g \in G$. In the above equation $\mathbf{R}$ is a 3×3 rotation matrix on $\mathbf{r} = \mathbf{x} = (x, y, z)$ if $\mathbf{M}_n = E_3$ and $G = SO(3)$, or a 4×4 Lorentz transformation matrix $(\Lambda^\mu{}_\nu)$ on $\mathbf{r} = x \equiv (x^\mu) = (ct, x, y, z)$ if $\mathbf{M}_n = M_4$ and G is the Poincaré group. In this latter case, under a generic Poincaré transformation $(\Lambda, x_0) \in G$, the space-time coordinates transform as in (4.191)[9]:

$$x' = \Lambda x - x_0 \quad \Leftrightarrow \quad x'^\mu = \Lambda^\mu{}_\nu x^\nu - x_0^\mu, \tag{7.46}$$

where $x_0 \equiv (x_0^\mu)$ parametrize the space-time translations, and Eq. (7.45) reads:

$$\Phi^\alpha(x) \rightarrow \Phi'^\alpha(x') = D^\alpha{}_\beta \, \Phi^\beta(x) = D^\alpha{}_\beta \, \Phi^\beta(\Lambda^{-1}(x' + x_0))$$
$$\Leftrightarrow \quad \Phi(x) \rightarrow \Phi'(x') = \mathbf{D}\,\Phi(x) = \mathbf{D}\,\Phi(\Lambda^{-1}(x' + x_0)), \tag{7.47}$$

where $\mathbf{D} = (D^\alpha{}_\beta) = \mathbf{D}(\Lambda)$ is the matrix implementing the Lorentz transformation on the internal space.

Besides considering groups of Cartesian coordinate transformations acting both on the space-time vectors of V_n (e.g. the Lorentz group acting on Δx^μ as part of the more general Poincaré group), and on the space V_p, we could consider *groups of internal transformations*, namely transformation groups acting only on V_p, that is on the internal degrees of freedom of the field, while the space or space-time vectors are left unchanged. In this case $\mathbf{R}$ is the trivial representation $\mathbf{1}$ and $\mathbf{r}_0 = \mathbf{0}$. Such transformations act on a field as follows:

$$\Phi^\alpha(\mathbf{r}) \rightarrow \Phi'^\alpha(\mathbf{r}) = D^\alpha{}_\beta \, \Phi^\beta(\mathbf{r}) \quad \Leftrightarrow \quad \Phi(\mathbf{r}) \rightarrow \Phi'(\mathbf{r}) = \mathbf{D}\,\Phi(\mathbf{r}). \tag{7.48}$$

An example is the group which transforms a wave function, i.e. a complex scalar field $\Phi(\mathbf{r})$, by multiplication with a phase:

$$\Phi(\mathbf{r}) \rightarrow \Phi'(\mathbf{r}) = \mathbf{D}(\varphi)\,\Phi(\mathbf{r}) = e^{i\varphi}\,\Phi(\mathbf{r}). \tag{7.49}$$

The reader can easily verify that the set consisting of phases $\mathbf{D}(\varphi) = e^{i\varphi}$ is a one-parameter abelian Lie group with respect to multiplication. It has the simple structure $\mathbf{D}(\varphi_1)\,\mathbf{D}(\varphi_2) = \mathbf{D}(\varphi_3)$, where $\varphi_3 = \varphi_1 + \varphi_2$. This group G is denoted by U(1) and called the *unitary one-dimensional group*. When illustrating in the next chapters, the relation between symmetry transformations and conserved quantities, we shall see that the internal U(1) symmetry of a system, i.e. the invariance of a system under internal U(1) transformations, is related to the conservation of a charge which, in electromagnetism, is the electric charge.

Let us now come back to the case in which G is a transformation group acting on the space-time reference frames. The simplest instance of field is the *scalar field* in which $\mathbf{D}$ is the trivial representation $\mathbf{1}$, defining a type-$(0, 0)$ tensor, with $p = 1$ that is $V_p = \mathbb{R}$ (real scalar field) or $\mathbb{C}$ (complex scalar field). A complex scalar field $\Phi(\mathbf{r})$ can be described as a couple of real scalar fields $\Phi_1(\mathbf{r})$, $\Phi_2(\mathbf{r})$, defined at each point $\mathbf{r}$ by the real and imaginary parts of $\Phi(\mathbf{r})$:

---

[9]Note that the analogous of the Poincaré group in the three dimensional Euclidean space E_3 is the known group of congruences of Euclidean geometry, acting on the space coordinates as in (4.102).

$$\Phi(\mathbf{r}) = \Phi_1(\mathbf{r}) + i\,\Phi_2(\mathbf{r}) \equiv (\Phi_1(\mathbf{r}),\ \Phi_2(\mathbf{r})). \tag{7.50}$$

The transformation law (7.45) reduces, for a scalar field, to

$$\Phi'(\mathbf{r}') = \Phi(\mathbf{r}(\mathbf{r}')) = \Phi(\mathbf{R}^{-1}(\mathbf{r}' + \mathbf{r}_0)). \tag{7.51}$$

If $\mathbf{M}_n = \mathbf{M}_4$ and G is the Poincaré group, $\mathbf{r}$ is the space-time coordinate vector (x^μ) and $\mathbf{R} = \Lambda = (\Lambda^\mu{}_\nu) \in SO(1,3)$.

In general, as discussed in Sect. 4.3, the coordinates $\mathbf{r} = (x^i)$ and $\mathbf{r}' = (x'^i)$ refer to the same point P of $\mathbf{M}_n$, therefore the numerical value of the scalar field must be the same, even if, when substituting $x^i = x^i(x'^j) = (R^{-1})^i{}_j\,(x'^j + x_0^j)$ the *functional form* changes from Φ to Φ'. Writing $\Phi'(\mathbf{r}') = \Phi(\mathbf{r})$ we are considering the transformation $\mathbf{r}' = \mathbf{R}\,\mathbf{r} - \mathbf{r}_0$ from a *passive point of view* since *space-points are considered fixed while only the coordinate frame is changed*.

However the same transformation can be also considered from a different point of view, namely as a *change in the functional form* of $\Phi(\mathbf{r})$

$$\Phi(\mathbf{r}) \to \Phi'(\mathbf{r}), \tag{7.52}$$

with $\Phi'(\mathbf{r}) = \Phi(\mathbf{R}^{-1}(\mathbf{r} - \mathbf{r}_0))$. In this case we consider the transformation as an *active* transformation, since the emphasis is on the *functional change* of Φ. The given change of coordinate in this case is thought of as due to a change of the geometric point.[10]

When considering the change in the functional form from an *active point of view* it is sometimes convenient to denote the new functional form Φ' taken by Φ as consequence of the coordinate change induced by an element $g \in G$, as the action of an operator O_g on Φ.[11] Eq. (7.51) takes the following form:

$$O_g\,\Phi(\mathbf{r}) = \Phi(\mathbf{R}^{-1}(\mathbf{r} + \mathbf{r}_0)), \tag{7.53}$$

where, as usual, $\mathbf{R} = \mathbf{R}(g)$ and $\mathbf{r}_0 = \mathbf{r}_0(g)$.

Consider, for the sake of simplicity, a group G acting in a homogeneous way on the coordinates (i.e. $\mathbf{r}_0 \equiv \mathbf{0}$) and apply in succession two transformations $g_1, g_2 \in G$, the resulting transformation corresponding to the product $g_2 \cdot g_1 \in G$. We have:

$$x^i \xrightarrow{\ g_1\ } x'^i = R(g_1)^i{}_j\,x^j \xrightarrow{\ g_2\ } x''^i = R(g_2)^i{}_j\,R(g_1)^j{}_k\,x^k = R(g_2 \cdot g_1)^i{}_k\,x^k$$

---

[10]Note that in the discussion of the vector and tensor calculus in Sect. 4.1 the emphasis was on the *passive point of view* since the reference frame was changed by the transformations. Therefore the whole of the vector and tensor calculus was developed taking this point of view. The active point, as previously mentioned, will be actually adopted in Chap. 9 when discussing the action of a group on the Hilbert space of states in quantum mechanics.

[11]Here by operator we mean a linear mapping of the vector space of square-integrable functions on $\mathbf{M}_n$ into itself, according to the definition given earlier. O_g is actually a transformation and it is therefore invertible.

or, expressing x^i in terms of x''^k

$$x^i = [\mathbf{R}(g_2 \cdot g_1)^{-1}]^i{}_k \, x''^k.$$

Actually the operators O_R give a homomorphic *realization* of the group G, where by *realization* we mean a homomorphic mapping on the function space. Indeed from

$$\Phi'(\mathbf{r}) \equiv O_g \, \Phi(\mathbf{r}) = \Phi(\mathbf{R}(g)^{-1} \mathbf{r}), \qquad (7.54)$$

using the short-hand notation $R_1 \equiv R(g_1)$ and $R_2 \equiv R(g_2)$, it follows

$$
\begin{aligned}
O_{g_2} \cdot O_{g_1} \, \Phi(\mathbf{r}) &\equiv O_{g_2} \left(O_{g_1} \, \Phi(\mathbf{r}) \right) = O_{g_2} \, \Phi'(\mathbf{r}) = \Phi'(\mathbf{R}_2^{-1} \, \mathbf{r}) \\
&= \Phi(\mathbf{R}_1^{-1} \mathbf{R}_2^{-1} \, \mathbf{r}) = \Phi((\mathbf{R}_2 \, \mathbf{R}_1)^{-1} \, \mathbf{r}) = \Phi(\mathbf{R}(g_2 \cdot g_1)^{-1} \, \mathbf{r}),
\end{aligned}
$$
$$(7.55)$$

where we have defined $\Phi'(\mathbf{r}) \equiv O_{g1}\Phi(\mathbf{r})$.
However the same result is also obtained acting on Φ with the operator $O_{g_2 \cdot g_1}$ corresponding to the group element $g_2 \cdot g_1$:

$$O_{g_2 \cdot g_1} \Phi(\mathbf{r}) = \Phi(\mathbf{R}(g_2 \cdot g_1)^{-1} \mathbf{r}).$$

Therefore we conclude that

$$O_{g_2 \cdot g_1} = O_{g_2} \cdot O_{g_1}. \qquad (7.56)$$

O is thus a *homomorphism* of G into the group of linear transformations on the space of functions $\Phi(x)$ on $\mathbf{M}_n$. It is easy to verify that O maps the unit element of G into the identity transformation I which maps a generic function $\Phi(x)$ into itself. Moreover $O_g^{-1} = O_{g^{-1}}$. The mapping $O : g \in G \ \to \ O_g$, has the same properties as a representation D. However the linear transformations O_g are not implemented by matrices, since they affect the functional form of the field they act on. For this reason O should be referred to as a realization of G on fields rather than a representation.

7.4.1 Invariance of Fields

The relation (7.53) is referred to general transformations of Cartesian coordinates (affine transformations), whose homogeneous part describes a linear transformation on V_n (i.e. belongs to the group GL(n)). This relation is actually valid also for *any (invertible) coordinate transformation* (thus including curvilinear coordinates)

$$x'^i = f^i(x^1, x^2, \ldots, x^n), \qquad (7.57)$$

where $f(x) \equiv (f^i(x))$ are differentiable functions which can be inverted to express the old coordinates $(x^1, \ldots, x^n)$ in terms of the new ones $(x'^1, \ldots, x'^n)$: $x^i = f^{-1\,i}(x')$, or simply $x = f^{-1}(x')$. Also the effect of this coordinate transformation on $\Phi(x)$ can be represented by the action of an operator O_f

$$O_f \Phi(x) = \Phi(f^{-1}(x)). \tag{7.58}$$

Only for *linear coordinate transformations* (among Cartesian coordinates) $f^i(x)$ reduces to: $x'^i = R^i_j x^j - x^i_0$. Let us now recall the definition of *invariance of a function Φ*:

If the *functional form* of Φ does not change under a coordinate transformation (7.57), $O_f \Phi(x) = \Phi(x)$ then Φ is *invariant*. From the relation (7.51) and the requirement of invariance we obtain

$$\Phi(x) = \Phi(f^{-1}(x)). \tag{7.59}$$

From the active point of view this means that even if the geometric point is changed, the functional form remains the same.[12]

As an example we may take the coordinate transformation corresponding to the rotation of the Cartesian coordinate system by an angle θ in the plane x, y, given by the general SO(2) element

$$\mathbf{r}' = f(\mathbf{r}) = \mathbf{R}(\theta)\,\mathbf{r}, \quad \mathbf{R}(\theta) = \begin{pmatrix} \cos\theta & \sin\theta \\ -\sin\theta & \cos\theta \end{pmatrix}. \tag{7.60}$$

The function

$$\Phi(x, y) = x^4 + y^2, \tag{7.61}$$

is not invariant, as can be easily verified by substitution of the coordinates in terms of the new ones. Instead the function

$$\Phi(x, y) = x^2 + y^2, \tag{7.62}$$

is invariant; indeed

$$x'^2 + y'^2 = (x \cos\theta + y \sin\theta)^2 + (-x \sin\theta + y \cos\theta)^2 = x^2 + y^2. \tag{7.63}$$

In general to verify the invariance one replaces x with $f^{-1}(x)$ and checks if the same function is obtained or not.

So far we have been considering the action of a group of transformations on a *scalar field*. In the general case where the representation acts on a field $\Phi^\alpha(x^i)$ which is not a scalar, but has internal components transforming in a given representation $\mathbf{D}$ of G, the transformation law is given by Eq. (7.45). Also in this more general case it

---

[12]Equivalently, from the passive point of view, invariance means that a change in the coordinate frame does not change the functional form.

is useful to describe the effect of a transformation $g \in G$ in terms of an operator O_g acting on the field:

$$g \in G: \quad \Phi^\alpha(\mathbf{r}) \xrightarrow{\;g\;} \Phi'^\alpha(\mathbf{r}') = O_g \Phi^\alpha(\mathbf{r}') = D^\alpha{}_\beta \, \Phi^\beta(\mathbf{R}^{-1}(\mathbf{r}' + \mathbf{r}_0)), \quad (7.64)$$

where, as usual, we have suppressed the explicit dependence on g of $\mathbf{D}$, $\mathbf{R}$ and $\mathbf{r}_0$. Also in this case the operators O_g give a homomorphic image of the group transformations, their action on V_p being given in terms of the matrices D of the representation of G.[13] Indeed, using the matrix notation and restricting to homogeneous coordinate transformations, for any two given elements g_1, g_2 of G we have:

$$O_{g_2} O_{g_1} \Phi(\mathbf{r}) = O_{g_2} \Phi'(\mathbf{r}) = \mathbf{D}_2 \Phi'(\mathbf{R}_2^{-1} \, \mathbf{r}) = \mathbf{D}_2 \mathbf{D}_1 \Phi(\mathbf{R}_1^{-1} \mathbf{R}_2^{-1} \mathbf{r}), \quad (7.66)$$

where, as usual we have used the short-hand notation: $\mathbf{R}_1 = \mathbf{R}(g_1)$, $\mathbf{R}_2 = \mathbf{R}(g_2)$, $\mathbf{D}_1 = \mathbf{D}(g_1)$, $\mathbf{D}_2 = \mathbf{D}(g_2)$. On the other hand applying the operator corresponding to $g_2 \cdot g_1$ we also have

$$O_{g_2 \cdot g_1} \Phi(\mathbf{r}) = \mathbf{D}(g_2 \cdot g_1) \Phi(\mathbf{R}(g_2 \cdot g_1)^{-1} \mathbf{r}), \quad (7.67)$$

Comparing Eqs. (7.66) and (7.67) and taking into account that $\mathbf{D}$ and $\mathbf{R}$ are representations of G

$$\mathbf{D}(g_2 \cdot g_1) = \mathbf{D}(g_2)\,\mathbf{D}(g_1), \quad \mathbf{R}(g_2 \cdot g_1) = \mathbf{R}(g_2)\,\mathbf{R}(g_1) \quad (7.68)$$

we find

$$O_{g_2 \cdot g_1} = O_{g_2} O_{g_1}. \quad (7.69)$$

Just as in the scalar field case, the homomorphism O between elements $g \in G$ and operators O_g defines a realization of G on the field $\Phi(r)$. The reader can easily extend the above proof to groups G acting as non-homogeneous linear coordinate transformations: $\mathbf{r}_0(g) \neq \mathbf{0}$.[14] The concept of invariance given for scalar functions can be easily extended to functions transforming in a non trivial representation $\mathbf{D}$ of G. We say that the field $\Phi^\alpha(x)$ is *invariant* under the action of G if

---

[13] Note that also in this more general case we may consider the given transformation from an *active point of view*, by redefining in the two sides of last equality in (7.64) $\mathbf{r}' \to \mathbf{r}$:

$$\Phi^\alpha(\mathbf{r}) \xrightarrow{\;g\;} \Phi'^\alpha(\mathbf{r}) = D^\alpha{}_\beta \Phi^\beta(\mathbf{R}^{-1}(\mathbf{r} + \mathbf{r}_0)). \quad (7.65)$$

[14] If the transformation of the field is due to a *general coordinate transformation* the matrix $\mathbf{D}$, which is constant for linear transformations, will be given by the corresponding Jacobian at the point x. For example, if a vector field $v^i(x)$ is invariant under $x \to f(x)$, then

$$v'^i(x') = v^i(x') = J^i{}_j(x)\, v^j(x),$$

where $x = f^{-1}(x')$ and $J^i{}_j(x) = \frac{\partial f^i}{\partial x^j}(x)$.

$$\Phi'^{\alpha}(\mathbf{r}) = \Phi^{\alpha}(\mathbf{r}) = D^{\alpha}{}_{\beta}\Phi^{\beta}(\mathbf{R}^{-1}(\mathbf{r} + \mathbf{r}_0)), \qquad (7.70)$$

which, for tensor fields acted on by $GL(n)$, reduces to (4.74) and its generalization.

7.4.2 Infinitesimal Transformations on Fields

In this subsection we consider a Lie group G of parameters (θ^r) and describe, just as we did for a generic matrix representation, the action of the operator O_g, corresponding to an element $g \in G$ in a neighborhood of the identity, in terms of infinitesimal generators $\mathcal{L}_r$ as follows:

$$O_g = e^{\theta^r \mathcal{L}_r}, \qquad (7.71)$$

where $\mathcal{L}_r$ are operators acting on the basis $\Phi^{\alpha}(x^i)$ both *linearly*, that is as matrices on the internal index α and as *differential operators* with respect to the dependence on the coordinates x^i. The presence of a differential operator in $\mathcal{L}_r$ is due to the fact that the operators O_g, which provide a homomorphic image of the group G, act simultaneously on the linear (internal) vector space V_p, spanned by the vector-components $\Phi^{\alpha}(x^i)$, as well as on the functional dependence of the field on the coordinates x^i. Therefore the infinitesimal operators $\mathcal{L}_r$ contain, besides the matrix algebra operators acting on the field components $\Phi^{\alpha}(x^i)$, also infinitesimal differential operators acting on the functional space. The proof that a generic O_g, in a neighborhood of the identity operator I, can be expressed as the exponential of a Lie algebra element $\theta^r \mathcal{L}_r$ is analogous to the one given for matrix representations and thus we are not going to repeat it here. In fact $(\mathcal{L}_r)$ represent a basis for the Lie algebra of the generators of G in the realization O. We are interested instead in deriving the general expression of the operators $\mathcal{L}_r$.

Consider infinitesimal transformations defined by infinitesimal parameters $\delta\theta^r \ll 1$. Expanding the exponential in Eq. (7.71) we can write O_g to first order in $\delta\theta^r$ as

$$O_g = I + \delta\theta^r \mathcal{L}_r.$$

It follows that

$$O_g \Phi^{\alpha}(\mathbf{r}) \cong \Phi^{\alpha}(\mathbf{r}) + \delta\theta^r \mathcal{L}_r \Phi^{\alpha}(\mathbf{r}) = \Phi^{\alpha}(\mathbf{r}) + \delta\Phi^{\alpha}(\mathbf{r}), \qquad (7.72)$$

where we have expressed the infinitesimal local variation of the field as given by

$$\delta\Phi^{\alpha}(\mathbf{r}) = \delta\theta^r \mathcal{L}_r \Phi^{\alpha}(\mathbf{r}). \qquad (7.73)$$

In order to determine the action of $\mathcal{L}_r$ on $\Phi^{\alpha}(x^i)$ we begin by writing the infinitesimal form of $\mathbf{R}(g)$ and $\mathbf{D}(g)$, supposing at first the action of G on the Cartesian coordinates $\mathbf{r} = (x^i)$ to be homogeneous:

$$R(g)^i{}_j \simeq \delta^i_j + \delta\theta^r \, (L_r)^i{}_j,$$
$$D(g)^\alpha{}_\beta \simeq \delta^\alpha_\beta + \delta\theta^r (L_r)^\alpha{}_\beta, \qquad\qquad (7.74)$$

where $(L_r)^i{}_j$ and $(L_r)^\alpha{}_\beta$ are the matrices describing the Lie algebra generators in the two representations (recall that $i, j = 1, \ldots, n$ label the coordinates on the space $\mathbf{M}_n$ while $\alpha, \beta = 1, \ldots, p$ are indices of the representation space V_p). It follows that

$$x'^i = R(g)^i{}_j \, x^j \simeq x^i + \delta x^i \quad \Rightarrow \quad \delta x^i = \delta\theta^r \, (L_r)^i{}_j \, x^j,$$

and

$$D^\alpha{}_\beta(g) \, \Phi^\beta(\mathbf{r}) \simeq \delta^\alpha_\beta + \delta\theta^r (L_r)^\alpha{}_\beta \, \Phi^\beta(\mathbf{r}).$$

Let us work out both sides of the finite relation

$$\Phi'^\alpha(\mathbf{r}') = D(g)^\alpha{}_\beta \Phi^\beta(\mathbf{r}), \qquad\qquad (7.75)$$

to first order in the infinitesimal $\delta\theta^r$ parameters. On the left hand side we have

$$\Phi'^\alpha(x'^i) \simeq \Phi'^\alpha(x^i + \delta x^i) = \Phi'^\alpha(x^i) + \frac{\partial \Phi^\alpha}{\partial x^i}(x^i) \, \delta x^i, \qquad\qquad (7.76)$$

where, expanding $\Phi^\alpha(x^i + \delta x^i)$ we only kept first order terms in δx^i (since $\delta x^i = O(\delta\theta^r)$) and, for the same reason, we have replaced $\frac{\partial}{\partial x^i}\Phi'^\alpha(x^i)$ with $\frac{\partial}{\partial x^i}\Phi^\alpha(x^i)$ in the derivative term.

On the other hand, using the *infinitesimal generators* defined in Eq. (7.28), the right hand side of (7.75) reads, to first order in $\delta\theta^r$:

$$D(g)^\alpha{}_\beta \, \Phi^\beta \simeq [\delta^\alpha_\beta + \delta\theta^r (L_r)^\alpha{}_\beta] \, \Phi^\beta(x^i). \qquad\qquad (7.77)$$

From (7.72), (7.76) and (7.77) we find

$$\Phi'^\alpha(\mathbf{r}) - \Phi^\alpha(\mathbf{r}) \equiv \delta\Phi^\alpha(\mathbf{r}) = \delta\theta^r \mathcal{L}_r \, \Phi^\alpha(\mathbf{r}) = \left[\delta\theta^r (L_r)^\alpha{}_\beta - \delta^\alpha_\beta \, \delta x^i \frac{\partial}{\partial x^i}\right] \Phi^\beta(\mathbf{r})$$

$$= \delta\theta^r \left[(L_r)^\alpha{}_\beta - \delta^\alpha_\beta (L_r)^i{}_j \, x^j \frac{\partial}{\partial x^i}\right] \Phi^\beta(\mathbf{r}). \qquad\qquad (7.78)$$

Since this equality must hold for each component $\delta\theta^r$, $r = 1, 2, \ldots, q$ we finally find the action of the infinitesimal operator $\mathcal{L}_r$ on $\Phi^\alpha(x^i)$:

$$\mathcal{L}_r \Phi^\alpha(\mathbf{r}) = \left[(L_r)^\alpha{}_\beta - \delta^\alpha_\beta (L_r)^i{}_j \, x^j \frac{\partial}{\partial x^i}\right] \Phi^\beta(\mathbf{r}). \qquad\qquad (7.79)$$

In conclusion the action of the generators of the Lie algebra $\mathcal{A}$ of G on the component-functions $\Phi^\alpha(\mathbf{r})$, is given by the following operator:

$$\mathcal{L}_r \equiv (L_r)^\alpha{}_\beta - \delta^\alpha_\beta\,(L_r)^i{}_j\,x^j\,\frac{\partial}{\partial x^i}, \tag{7.80}$$

where the first term acts linearly on the vector space of the representation labeled by the index α, while the second term is a differential operator acting on the dependence of the field on the coordinates. Note that $(L_r)^\alpha{}_\beta$ and $(L_r)^i{}_j$ are the *same infinitesimal generator* albeit in *different representations:* the p-dimensional representation on the space V_p and the n-dimensional representation in the space of the coordinates.

The most interesting case for us is, of course, that in which G implements the coordinate transformations corresponding to the relativistic invariance of a theory: The Poincaré group on Minkowski space-time $\mathbf{M}_n = M_4$. Let first G be the Lorentz group SO(1, 3). We have

$$x'^\mu = \Lambda^\mu{}_\nu x^\nu, \tag{7.81}$$

where, according to our general conventions, the coordinate indices $i, j, \ldots$ have been renamed $\mu, \nu \ldots$. Furthermore the infinitesimal parameters $\delta\theta^r$ will be written as $\delta\theta^{\rho\sigma}$, the infinitesimal Lorentz parameters in (4.171). According to Eqs. (4.166), (4.169), (4.170), the infinitesimal transformation is given by (the homogeneous part of) Eq. (4.196)

$$\delta x^\mu = \frac{1}{2}\,\delta\theta_{\rho\sigma}\,(L^{\rho\sigma})^\mu{}_\nu\,x^\nu = \delta\theta^\mu{}_\nu\,x^\nu. \tag{7.82}$$

Inserting in Eq. (7.78) the general infinitesimal transformation of the field under the Lorentz group takes the following form:

$$\begin{aligned}
\delta\Phi^\alpha(x) &= \frac{1}{2}\,\delta\theta^{\rho\sigma}\,\mathcal{L}_{\rho\sigma}\Phi^\alpha(x)\\
&= \frac{1}{2}\,\delta\theta^{\rho\sigma}\left[(L_{\rho\sigma})^\alpha{}_\beta\,\Phi^\beta + (x_\rho\partial_\sigma - x_\sigma\partial_\rho)\Phi^\alpha\right],
\end{aligned} \tag{7.83}$$

from which we deduce the expression of the infinitesimal Lorentz generators $\mathcal{L}_{\rho\sigma}$ as differential operators acting on fields:

$$\mathcal{L}_{\rho\sigma} = \left[(L_{\rho\sigma})^\alpha{}_\beta + \delta^\alpha_\beta\,(x_\rho\partial_\sigma - x_\sigma\partial_\rho)\right]. \tag{7.84}$$

So far we have mainly been considering groups, as the Lorentz one, acting on coordinates as in Eq. (7.81) that is in a linear and *homogeneous* way. We know, however, that the most general relativistic theory is invariant under the *Poincaré group* whose action on the coordinates, see Eq. (7.46), is linear but not homogeneous since it contains the subgroup of *space-time translations*, see Sect. 4.72. Let us restrict ourselves to the subgroup of *constant* translations on the x^μ coordinates

$$x^\mu \rightarrow x'^\mu = x^\mu - x_0^\mu$$

and, more specifically, to *infinitesimal* translations, $x_0^\mu = \epsilon^\mu \ll 1$.

Since constant translations do not affect the components of relative position vectors, they have a trivial action on V_n. Moreover they do not affect the internal components of a field Φ^α as well. Therefore

$$\Phi'^\alpha(x^\mu - \epsilon^\mu) = \Phi^\alpha(x^\mu) \rightarrow \Phi'^\alpha(x^\mu) - \epsilon^\nu \frac{\partial \Phi^\alpha}{\partial x^\nu}(x^\mu) = \Phi^\alpha(x^\mu), \qquad (7.85)$$

that is

$$\delta \Phi^\alpha \equiv \Phi'^\alpha(x^\mu) - \Phi^\alpha(x^\mu) = \epsilon^\nu \frac{\partial}{\partial x^\nu} \Phi^\alpha(x^\mu) = \epsilon^\nu \, \mathcal{P}_\nu \Phi^\alpha(x^\mu). \qquad (7.86)$$

Thus a basis of *infinitesimal generators of four-dimensional translations* is given by

$$\mathcal{P}_\mu = \frac{\partial}{\partial x^\mu}. \qquad (7.87)$$

This is the representation of the infinitesimal generators of translations on the fields. In Sect. 4.7.2 we gave a matrix representation $\mathbf{P}_\mu$ of the same generators. There we have proven that the Lie subalgebra of translations is *abelian*, as it is apparent also from this new realization $\mathcal{P}_\mu$ of its generators, since

$$\left[\frac{\partial}{\partial x^\mu}, \frac{\partial}{\partial x^\nu} \right] = 0.$$

This of course agrees with the Lie algebra of the *Poincaré* group worked out in Chap. 4.

Putting together (7.83) and (7.86) we find the following result:

Under an infinitesimal transformation of the Poincaré group (7.46) the classical field Φ^α transforms as follows:

$$\delta \Phi^\alpha(x) = \left(\frac{1}{2} \delta \theta^{\rho\sigma} \, \mathcal{L}_{\rho\sigma} + \epsilon^\mu \, \mathcal{P}_\mu \right) \Phi^\alpha(x)$$

$$= \frac{1}{2} \delta \theta^{\rho\sigma} \left[(L_{\rho\sigma})^\alpha{}_\beta \Phi^\beta(x) + (x_\rho \partial_\sigma - x_\sigma \partial_\rho) \Phi^\alpha(x) \right] + \epsilon^\mu \frac{\partial}{\partial x^\mu} \Phi^\alpha(x). \qquad (7.88)$$

In particular, for a scalar field, the term $(L_{\rho\sigma})^\alpha{}_\beta \Phi^\beta$ is absent, and we find:

$$\delta \phi(x) = \frac{1}{2} \delta \theta^{\rho\sigma} (x_\rho \partial_\sigma - x_\sigma \partial_\rho) \phi(x) + \epsilon^\mu \frac{\partial}{\partial x^\mu} \phi(x). \qquad (7.89)$$

A finite Poincaré transformation (Λ, x_0), Λ being defined by finite parameters $\theta^{\mu\nu}$, can be written in terms of the action of an operator $O_{(\Lambda, x_0)}$ defined by exponentiating, see Eq. (4.197), the infinitesimal generators in (7.88):

$$\Phi'^{\alpha}(x) = O_{(\Lambda, x_0)}\, \Phi^{\alpha}(x) = D^{\alpha}{}_{\beta}(\Lambda)\, \Phi^{\alpha}(\Lambda^{-1}\,(x + x_0)), \qquad (7.90)$$

where

$$O_{(\Lambda, x_0)} = e^{x_0^{\mu} \mathcal{P}_{\mu}}\, e^{\frac{1}{2} \theta^{\rho\sigma}\, \mathcal{L}_{\rho\sigma}}, \quad D(\Lambda)^{\alpha}{}_{\beta} = \left(e^{\frac{1}{2} \theta^{\rho\sigma}\, L_{\rho\sigma}}\right)^{\alpha}{}_{\beta}. \qquad (7.91)$$

We close this subsection by observing that since the structure constants do not depend of the representation, the infinitesimal generators close the same algebra irrespective of the representation. Thus we have

$$
\begin{aligned}
&\text{representation}\, D: && [L_r, L_s]^{\alpha}_{\beta} = C_{rs}{}^{P}\, (L_p)^{\alpha}_{\beta}, \\
&\text{representation}\, R: && [L_r, L_s]^{i}_{j} = C_{rs}{}^{P}\, (L_p)^{i}_{j}, \\
&\text{realization}\, O: && [\mathcal{L}_r, \mathcal{L}_s] = C_{rs}{}^{P}\, \mathcal{L}_p,
\end{aligned}
\qquad (7.92)
$$

as can be easily verified in general. In the case of the Poincaré group, we can verify, using the explicit expression for the infinitesimal Lorentz and translation generators $\mathcal{L}_{\mu\nu}$, $\mathcal{P}_{\mu}$, the following commutation relations:

$$\left[\mathcal{L}^{\mu\nu}, \mathcal{L}^{\rho\sigma}\right] = \eta^{\nu\rho}\, \mathcal{L}^{\mu\sigma} + \eta^{\mu\sigma}\, \mathcal{L}^{\nu\rho} - \eta^{\mu\rho}\, \mathcal{L}^{\nu\sigma} - \eta^{\nu\sigma}\, \mathcal{L}^{\mu\rho}, \qquad (7.93)$$

$$\left[\mathcal{L}^{\mu\nu}, \mathcal{P}_{\rho}\right] = \mathcal{P}^{\mu}\, \delta^{\nu}_{\rho} - \mathcal{P}^{\nu}\, \delta^{\mu}_{\rho}, \qquad (7.94)$$

$$\left[\mathcal{P}_{\mu}, \mathcal{P}_{\nu}\right] = 0. \qquad (7.95)$$

which share the same structure constants with those in Eq. (4.201).

7.4.3 Application to Non-Relativistic Quantum Mechanics

Let us apply the previous considerations to the non-relativistic Schrödinger wave function $\psi(\mathbf{x}, t)$, $\mathbf{x} = (x, y, z) \in E_3$, describing the state of a particle at a time t. Since we consider now only transformations in the Euclidean space $\mathbf{M}_n = E_3$, we shall neglect the dependence of the wave function on time. Let us consider an infinitesimal rotation $\mathbf{R} \in SO(3)$

$$x'^{i} = R^{i}{}_{j}\, x^{j}, \quad (i, j, k = 1, 2, 3),$$

where, for small angles $\delta\theta^{r}$, the rotation matrix is given by Eq. (7.74)

$$R^{i}{}_{j} \simeq \delta^{i}_{j} + \delta\theta^{k}\, (L_k)^{i}{}_{j}; \quad \delta\theta^{k} \simeq 0.$$

The infinitesimal generators L_1, L_2, L_3 are represented by the matrices given in Eqs. (4.116) and (4.119). Since the wave function ψ is a (complex) *scalar field*, the action of $SO(3)$ on the (internal) space $\mathbb{C}$ of complex values of ψ^{15} is trivial:

$$D^\alpha{}_\beta = \delta^\alpha{}_\beta \Rightarrow (L_r)^\alpha{}_\beta = 0.$$

Thus from (7.78) we find

$$\delta\psi(\mathbf{x}) = \delta\theta^k \, \mathcal{L}_k \, \psi(\mathbf{x}) = -\delta\theta^r \, (L_r)^i{}_j \, x^j \, \frac{\partial}{\partial x^i} \psi(\mathbf{x}),$$

where, according to (4.120)

$$
\begin{aligned}
\mathcal{L}_1 &= -x^3 \frac{\partial}{\partial x^2} + x^2 \frac{\partial}{\partial x^3}, \\
\mathcal{L}_2 &= -x^1 \frac{\partial}{\partial x^3} + x^3 \frac{\partial}{\partial x^1}, \\
\mathcal{L}_3 &= -x^2 \frac{\partial}{\partial x^1} + x^1 \frac{\partial}{\partial x^2},
\end{aligned}
\tag{7.96}
$$

are the *differential operators* representing the action of $SO(3)$ on the wave function $\psi(\mathbf{x})$. They can be rewritten in a more compact form as follows:

$$\mathcal{L}_i = \epsilon_{ijk} \, x^j \, \frac{\partial}{\partial x^k}. \tag{7.97}$$

Let us now consider the action of an infinitesimal *three-dimensional translation*

$$\mathbf{x}' = \mathbf{x} - \boldsymbol{\epsilon},$$

where $\boldsymbol{\epsilon} = (\epsilon^i)$, $\epsilon^i \ll 1$. We have

$$\psi'(\mathbf{x}') = \psi'(\mathbf{x} - \boldsymbol{\epsilon}) = \psi(\mathbf{x}) \quad \Rightarrow \quad \psi'(\mathbf{x}) - \epsilon^i \frac{\partial \psi(\mathbf{r})}{\partial x^i} = \psi(\mathbf{x}), \tag{7.98}$$

that is, according to Eq. (7.86)

$$\delta\psi(\mathbf{x}) = \psi'(\mathbf{x}) - \psi(\mathbf{x}) = \epsilon^i \, \mathcal{P}_i \psi(\mathbf{x}) \quad \Rightarrow \quad \mathcal{P}_i = \frac{\partial}{\partial x^i}.$$

In conclusion, infinitesimal rotations and translations on the wave function $\psi(\mathbf{r})$ are represented by the differential operators in (7.97) and (7.99), respectively. Form the physical point of view the operators $\mathcal{L}_i$ and $\mathcal{P}_i$ are proportional to the *quantum mechanical* operators $\hat{M}_i$, $\hat{p}_i$ associated with the angular momentum (see Eq. (4.131))

---

[15]Recall that, although complex numbers span a two-dimensional vector space, their components are inert under the SO(3) group.

and the linear momentum, respectively. In order to have *hermitian operators* with the right physical dimensions one defines

$$\hat{p}_i = -i\hbar \mathcal{P}_i = -i\hbar \frac{\partial}{\partial x^i} \quad \widehat{M}_i = -i\hbar \mathcal{L}_i. \tag{7.99}$$

The identification of the above operators with the aforementioned physical quantities will be motivated in detail in the next chapters, when we will be dealing with symmetries and conservation laws in quantum mechanics. Note that the $\mathbf{M}_i$ matrices in Eq. (4.131) and the operators $\hat{M}_i$ are different realizations of the same physical quantity, i.e. the components of the orbital angular momentum. The physical interpretation of the operators $\hat{p}_i$ and $\hat{M}_i$ is consistent with the fact that, writing in the expression for $\hat{M}_i$ the partial derivatives in terms of the momentum operator we find

$$\widehat{M}_i = -i\hbar \mathcal{L}_i = \epsilon_{ijk}\, x^j \left(-i\hbar \frac{\partial}{\partial x^k}\right) = \epsilon_{ijk}\, x^j\, \hat{p}_k, \tag{7.100}$$

or, simply

$$\widehat{\mathbf{M}} \equiv \mathbf{x} \times \hat{\mathbf{p}},$$

where $\widehat{\mathbf{M}} = (\widehat{M}_1, \widehat{M}_2, \widehat{M}_3)$ and $\mathbf{p} = (\hat{p}_1, \hat{p}_2, \hat{p}_3)$.[16]

7.4.4 References

For further reading see Refs. [2, 5, 14].

---

[16]We shall use the convention of denoting by a hatted symbol $\widehat{\mathscr{O}}$ the quantum mechanical operator acting on wave functions, associated with the observable $\mathscr{O}$. Occasionally, for the sake of notational simplicity, the hat will be omitted, provided the operator nature of the quantity be manifest from the context.

Chapter 8
Lagrangian and Hamiltonian Formalism

In this chapter we give a short account of the Lagrangian and Hamiltonian formulation of classical non-relativistic and relativistic theories. For pedagogical reasons we first address the case of systems of particles, described by a finite number of degrees of freedom. Afterwards, starting from Sect. 8.5, we extend the formalism to *fields*, that is to dynamical quantities described by functions of the points in space. Their consideration implies the study of dynamical systems carrying a *continuous infinity* of canonical coordinates, labeled by the three spatial coordinates.

8.1 Dynamical System with a Finite Number of Degrees of Freedom

8.1.1 The Action Principle

Let us consider a mechanical system consisting of an arbitrary number of point-like particles. We recall that the number of coordinates necessary to determine the configuration of the system at a given instant, defines the number of its *degrees of freedom*. These coordinates are not necessarily the Cartesian ones, but are *parameters* chosen in such a way as to characterize in the simplest way the properties of the system. They are referred to as *generalized coordinates or Lagrangian coordinates,* usually denoted by $q_i(t)$, $i = 1 \ldots n$, where n is the number of degrees of freedom. The space parameterized by the Lagrangian coordinates is the *configuration space*. Each point P in this space, of coordinates $P(t) \equiv (q_i(t))$, $(i = 1 \ldots n)$, defines the *configuration* of the system, that is the position of all the particles at a given instant. During the time evolution of the dynamical system the point P will therefore describe a trajectory in the configuration space.

© Springer International Publishing Switzerland 2016
R. D'Auria and M. Trigiante, *From Special Relativity to Feynman Diagrams*,
UNITEXT for Physics, DOI 10.1007/978-3-319-22014-7_8

The mechanical properties of the system are encoded in a *Lagrangian*, that is a function of the Lagrangian coordinates $q_i(t)$, their time derivatives $\dot{q}_i(t)$ and time t:

$$L = L(q_i(t), \dot{q}_i(t)), t), \qquad q = (q_1, q_2, \ldots, q_n).$$

Given the Lagrangian, the time evolution of the system is then derived from *Hamilton's principle of stationary action*.

Let us define the *action* S of the system as the integral of the Lagrangian along some curve γ in configuration space between two points corresponding to the configurations of the system at the instants t_1, t_2[1]:

$$S[q; t_1, t_2] = \int\limits_{(\gamma)\, t_1}^{t_2} L(q(t), \dot{q}(t), t)\, dt. \tag{8.1}$$

Notice that while L depends on the values of q_i and $\dot{q}_i$ at a given time t, S depends on the functions q_i, namely on all the values $q_i(t)$, with $t_1 \leq t \leq t_2$, defining a path γ in the configuration space. Thus, for fixed t_1, t_2, S is said to be a *functional* of q_i.

Hamilton's *principle of stationary (or least) action* states that *among all the possible paths γ connecting the two points $q(t_1) \equiv (q_i(t_1))$ and $q(t_2) \equiv (q_i(t_2))$ in the configuration space, the actual path described by the system during its time evolution between the instants t_1 and t_2 is given by the curve γ corresponding to an extremum of S. This extremum is found by performing a small deformation of γ keeping its end-points fixed, that is by performing arbitrary variations $\delta q_i(t)$ of the coordinates at any instant t, obeying the condition*

$$\delta q(t_1) = \delta q(t_2) = 0, \tag{8.2}$$

and by requiring $S[q; t_1, t_2]$ to be stationary with respect to such a variation.

In formulae, the actual path γ of the dynamical system in configuration space is found by solving the variational problem[2]:

$$\delta S = \int\limits_{t_1\, (\gamma)}^{t_2} \left(\frac{\partial L}{\partial q_i} \delta q_i + \frac{\partial L}{\partial \dot{q}_i} \delta \dot{q}_i \right) dt = 0, \qquad \delta q(t_1) = \delta q(t_2) = 0. \tag{8.3}$$

Note that by $\delta q_i(t)$ we denote the infinitesimal *local* change of the Lagrangian coordinates, namely $\delta q_i \simeq q_i'(t) - q_i(t)$.

---

[1] Here and in the following we shall often use the shorter notation $q(t)$ for the set of the coordinates $\{q_i\} = (q_1, \ldots, q_n)$ and similarly for their time derivatives, $\dot{q} = \{\dot{q}_i\} = (\dot{q}_1, \ldots, \dot{q}_n)$.

[2] To avoid clumsiness in the following formulae we shall often adopt the Einstein convention of summing over repeated indices $i, j, \ldots$ of the Lagrangian coordinates, even though these indices are in general just suffixes with no tensorial property.

To find the trajectory by extremizing of the action $\mathcal{S}[\gamma;\ t_1,\ t_2]$ one observes that, being $\delta q_i(t)$ a variation at a fixed instant t, the variation symbol δ commutes with the time derivative:

$$\delta\frac{d}{dt}q_i = \frac{d}{dt}\delta q_i.$$

Integrating by parts the second term in the integrand of (8.3), we obtain:

$$\delta\mathcal{S} = \int_{t_1}^{t_2} \left[\left(\frac{\partial L}{\partial q_i}\delta q_i + \left(\frac{\partial L}{\partial \dot{q}_i}\right)\frac{d}{dt}\delta q_i\right)\right] dt \tag{8.4}$$

$$= \int_{t_1}^{t_2} \left(\frac{\partial L}{\partial q_i} - \frac{d}{dt}\frac{\partial L}{\partial \dot{q}_i}\right)\delta q_i\, dt = 0, \tag{8.5}$$

where we have used the fact that the total derivative term $\frac{d}{dt}\left(\frac{\partial L}{\partial \dot{q}_i}\delta q_i\right)$ gives a vanishing contribution by virtue of condition (8.2):

$$\int_{t_1}^{t_2} \frac{d}{dt}\left(\frac{\partial L}{\partial \dot{q}_i}\delta q_i\right) = \left(\frac{\partial L}{\partial \dot{q}_i}\delta q_i\right)\Big|_{t_1}^{t_2} = 0. \tag{8.6}$$

Equation (8.4) has to hold for *arbitrary* variations δq_i; this implies that the integrand must vanish identically. We thus obtain:

$$\frac{\partial L}{\partial q_i} - \frac{d}{dt}\frac{\partial L}{\partial \dot{q}_i} = 0 \qquad \forall i = 1, 2, \ldots, n. \tag{8.7}$$

Equations (8.7) are the *Euler-Lagrange equation* of the system under study. They are a system of differential equations whose solution for given boundary conditions determines the time evolution of the system.

We note that the Lagrangian $L(q(t), \dot{q}(t)), t)$ is not uniquely defined; adding to it the total derivative $\frac{df}{dt}$ of an arbitrary function $f(q, t)$, does not affect the Euler-Lagrange equations.[3] Indeed, if we let:

$$L(q, \dot{q}, t) \rightarrow L'(q, \dot{q}, t) = L(q, \dot{q}, t) + \frac{d}{dt}f,$$

so that $\mathcal{S} = \int dt\, L \longrightarrow \mathcal{S}' = \int dt\, L'$, performing the variation $q_i(t) \rightarrow q_i(t) + \delta q_i(t)$, with $\delta q_i(t_1) = \delta q_i(t_2) = 0$, we obtain:

---

[3]Note that the function f can depend on q_i and t only, $f = f(q, t)$, in order for df/dt not to depend on derivatives of q_i of order higher than one.

$$\delta S' = \delta S + \int_{t_1}^{t_2} \frac{d}{dt} \delta f \, dt = \delta S + \delta f \mid_{t_1}^{t_2}.$$

However, by virtue of condition (8.2), $\delta f = \frac{\partial f}{\partial q_i} \delta q_i$, computed at the initial and final instants, is zero, and we conclude that, being $\delta S = \delta S'$, the same Euler-Lagrangian equations are obtained.[4]

In order to determine the general form of the Lagrangian of a mechanical system let us first work out the Lagrangian of a free particle in the non-relativistic case.

Anticipating our discussion on the symmetries of a system, we observe that this Lagrangian cannot explicitly depend either on the position vector $\mathbf{x}$ or on time t, since the classical theory is based on the assumption of *homogeneity of space and time*, as discussed at the end of Sect. 1.1.1. Moreover, it cannot depend on the direction of the velocity *vector* $\mathbf{v}$, because of the *isotropy of space* (there is no preferred direction). The Lagrangian must then be a function of the modulus v of $\mathbf{v}$ only: $L = L(v^2)$. From the equations of motion it follows

$$\frac{d}{dt} \frac{\partial L}{\partial v^i} = 0, \quad \forall i = 1, 2, 3,$$

so that $\mathbf{v} = const.$, that is we recover the *principle of inertia*.

As we shall discuss more systematically in the sequel, a transformation is a *symmetry* of a system if this leaves the Lagrangian invariant modulo an additional total derivative. If we now perform a Galilean transformation with an infinitesimal velocity ϵ, requiring it to be a symmetry, the Lagrangian can vary at most by a total derivative in order to describe the same inertial motion. Now, an infinitesimal Galilean transformation applied to L gives:

$$L(v'^2) = L\left[v^2 + 2\mathbf{v} \cdot \epsilon + O(\epsilon)^2\right] = L(v^2) + \frac{\partial L}{\partial v^2} 2\mathbf{v} \cdot \epsilon \qquad (8.10)$$

---

[4] We can directly show that the presence of an additional total derivative does not affect the equations of motion by showing that df/dt identically satisfies Eq. (8.7). This is readily done by observing that

$$\frac{df}{dt}(q, \dot{q}, t) = \dot{q}_i \frac{\partial f}{\partial q_i} + \frac{\partial f}{\partial t},$$

so that

$$\frac{\partial}{\partial q_i} \frac{df}{dt} = \dot{q}_j \frac{\partial f}{\partial q_i \partial q_j} + \frac{\partial}{\partial q_i} \frac{\partial f}{\partial t}. \qquad (8.8)$$

On the other hand we have

$$\frac{d}{dt}\left(\frac{\partial}{\partial \dot{q}_i} \frac{df}{dt}\right) = \frac{d}{dt}\left(\frac{\partial f}{\partial q_i}\right) = \dot{q}_j \frac{\partial f}{\partial q_i \partial q_j} + \frac{\partial}{\partial t} \frac{\partial f}{\partial q_i}. \qquad (8.9)$$

Subtracting side by side Eq. (8.8) from Eq. (8.9) and using the property that the partial derivations with respect to q_i and t commute, we conclude that the Euler-Lagrange equations are identically satisfied by the total derivative.

The term $\frac{\partial L}{\partial v^2} 2\mathbf{v} \cdot \boldsymbol{\epsilon}$ will be a total derivative if $\frac{\partial L}{\partial v^2}$ is independent of v, namely it is a constant α. It follows that:

$$L = \frac{1}{2} m v^2,$$ (8.11)

where we have set $\alpha = m/2$. The right hand side of Eq. (8.11), defines the *kinetic energy* of our system of particles.

Assuming that for a system of N non-interacting (i.e. free) particles the Lagrangian be additive, we have[5]:

$$L = \sum_{k=1}^{N} \frac{1}{2} m_{(k)} v_{(k)}^2 = T,$$ (8.12)

where we have denoted the total kinetic energy of the system by T.

Let us now consider an isolated system of N *interacting particles*: the Lagrangian is obtained by adding to the free Lagrangian an appropriate function of the coordinates, that we will denote by $-U$:

$$L = T - U(\mathbf{x}_{(1)}, \ldots, \mathbf{x}_{(N)}).$$ (8.13)

where T in the kinetic energy defined by the sum in Eq. (8.12) and the function U defines the *potential energy* of the system. Using the Cartesian coordinates of the particles as Lagrangian coordinates, the equations of motion (8.7) read, in this case:

$$\frac{\partial L}{\partial x_{(k)}^i} - \frac{d}{dt} \frac{\partial L}{\partial \dot{x}_{(k)}^i} = 0,$$ (8.14)

where $x_{(k)}^i$ ($i = 1, 2, 3$) denotes the ith coordinate of the kth particle, ($k = 1, \ldots, N$).[6] Using the Lagrangian (8.13), the Euler-Lagrange equations give:

$$m_{(k)} \frac{d\dot{x}_{(k)}^i}{dt} = -\frac{\partial U}{\partial x_{(k)}^i}.$$ (8.15)

Identifying the *force* acting on the kth particle with the right hand side of Eq. (8.15)

$$F_{(k)}^i = -\frac{\partial U}{\partial x_{(k)}^i},$$ (8.16)

we retrieve the Newton equation.

---

[5]This last property gives a physical meaning to our definition of inertial mass m. Indeed, the multiplication by a constant does not change the equations of motion, but it is equivalent to a change of the mass unit in the Lagrangian (8.12). However all the mass ratios, having a physical meaning, are unchanged.

[6]The number of degrees of freedom of the system is $n = 3N$.

Finally we consider the case of a *non-isolated* system A interacting with a system B, whose dynamics is known. For greater generality we use the generalized coordinates $q_\ell, \ell = 1, \ldots, n$.

To determine the Lagrangian of the system A, we consider the Lagrangian of the system $A + B$, and use for the coordinates of B their explicit time dependence. We then start with a Lagrangian of the form

$$L = T(q_A, \dot{q}_A) + T(q_B, \dot{q}_B) - U(q_A, q_B), \tag{8.17}$$

and replace the q_B's by their explicit dependence on time $q_B(t)$. We can then neglect the term $T(q_B, \dot{q}_B)$, which, being an explicitly known function of time, can always be written as a total derivative. The Lagrangian of A becomes:

$$L_A = T(q_A, \dot{q}_A) - U(q_A, q_B(t)). \tag{8.18}$$

This means that if the system is not isolated the Lagrangian is written as in the case of an isolated system, the only difference being that now the potential energy is an *explicit function of time* through $q_B(t)$.

8.1.2 Lagrangian of a Relativistic Particle

We may easily determine the Lagrangian of a free relativistic particle, by requiring L to be *invariant* under the group of Lorentz transformations (we shall explain in the following the concept of invariance in a more systematic way). This ensures covariance of the equations of motion so that the inertial motion will be maintained in any reference frame.

The simplest relativistic *invariant* quantity under Lorentz transformations, is the *proper time* τ defined by

$$d\tau^2 = dt^2 - \frac{|d\mathbf{x}|^2}{c^2} = dt^2 \left(1 - \frac{|\mathbf{v}|^2}{c^2}\right).$$

It is natural to expect the relativistic action for a free particle to be proportional to its proper time, namely:

$$S = \alpha \int_{t_1}^{t_2} d\tau = \alpha \int_{t_1}^{t_2} \frac{d\tau}{dt} dt = \alpha \int_{t_1}^{t_2} \sqrt{1 - \frac{v^2}{c^2}} dt, \tag{8.19}$$

where $v \equiv |\mathbf{v}|$. The corresponding Lagrangian reads:

$$L(\mathbf{v}) = \alpha \frac{d\tau}{dt} = \alpha \sqrt{1 - \frac{v^2}{c^2}}, \tag{8.20}$$

modulo a total time derivative. The constant α can be fixed by requiring that in the non-relativistic limit, $\frac{v}{c} \ll 1$, the Lagrangian (8.20) reduces to the form $L = \frac{1}{2} mv^2$. Expanding the square root to order $O(v^2/c^2)$ we find:

$$L(\mathbf{v}) \simeq \alpha - \frac{\alpha}{2} \frac{v^2}{c^2} = \frac{m}{2} v^2 + \text{const.}$$

Neglecting the inessential additive constant, we can then identify: $\alpha = -mc^2$. The relativistic free particle Lagrangian is thus given by:

$$L(\mathbf{v}) = -mc^2 \sqrt{1 - \frac{v^2}{c^2}}, \tag{8.21}$$

and for a system of N non-interacting particles, we have

$$L(\mathbf{v_i}) = \sum_{k=1}^{N} \left(-m_{(k)} c^2 \sqrt{1 - \frac{v_{(k)}^2}{c^2}} \right). \tag{8.22}$$

8.2 Conservation Laws

In this section we show that, if the Lagrangian of a system of particles is *invariant* under a group of transformations, then the dynamical system enjoys a set of *conservation laws*. In general we shall refer to the invariance property of a Lagrangian with respect to a group of transformations G as a *symmetry* under this group.

We first show that if a Lagrangian is *invariant* under time translations, $t \to t + \delta t$, then *energy is conserved*.

For simplicity, we assume that the invariance under time translations is due to the fact that the Lagrangian does not explicitly depend on t, namely, $\frac{\partial L}{\partial t} = 0$.[7] Then we may write:

$$\frac{dL}{dt} = \frac{\partial L}{\partial q_i} \dot{q}_i + \frac{\partial L}{\partial \dot{q}_i} \ddot{q}_i.$$

We now use the equations of motion (8.7) and obtain:

$$\frac{dL}{dt} = \frac{d}{dt} \frac{\partial L}{\partial \dot{q}_i} \dot{q}_i + \frac{\partial L}{\partial \dot{q}_i} \ddot{q}_i = \frac{d}{dt} \left(\frac{\partial L}{\partial \dot{q}_i} \dot{q}_i \right),$$

that is:

$$\frac{dH}{dt} = 0, \tag{8.23}$$

---

[7]The proof in a more general case is given in the following subsection.

where we have defined:

$$H = -L + \frac{\partial L}{\partial \dot{q}_i}\dot{q}_i. \tag{8.24}$$

We conclude that the quantity $H(q, \dot{q}) = -L + \frac{\partial L}{\partial \dot{q}_i}\dot{q}_i$ is conserved.

It is easy to recognize that H is the *energy* of the system. To show this in a general way, let us consider a system of particles interacting with a potential energy $U(q_1, \ldots, q_n)$:

$$L = T(q, \dot{q}) - U(q).$$

Here $T(q, \dot{q})$ is the kinetic energy which, in Cartesian coordinates, reads:

$$T = \frac{1}{2}\sum_{k=1}^{N}\sum_{i=1}^{3} m_{(k)}\dot{x}^i_{(k)}\dot{x}^i_{(k)}. \tag{8.25}$$

Let us now switch to the generalized (or Lagrangian) coordinates q_j, writing[8]:

$$x^i_{(k)} = f^i_{(k)}(q_1, \ldots, q_n); \qquad \dot{x}^i_{(k)} = \frac{\partial f^i_{(k)}}{\partial q_j}\dot{q}_j. \tag{8.26}$$

In terms of the Lagrangian coordinates the kinetic energy takes the form:

$$T = \sum_{i,j=1}^{n} a_{ij}(q)\dot{q}_i\dot{q}_j. \tag{8.27}$$

where we have set

$$a_{\ell j}(q) = \sum_{k=1}^{N}\sum_{i=1}^{3} m_{(k)}\frac{\partial f^i_{(k)}}{\partial q_\ell}\frac{\partial f^i_{(k)}}{\partial q_j}, \quad \ell, j = 1, \ldots, n.$$

This shows that the kinetic energy is *homogeneous of degree two* in the Lagrangian velocities $\dot{q}_i$. Applying the Euler theorem for *homogeneous functions*, we find:

$$\sum_{\ell=1}^{n} \frac{\partial T}{\partial \dot{q}_\ell}\dot{q}_\ell = 2T. \tag{8.28}$$

---

[8]With an abuse of notation, we shall use the same Latin indices $i, j, k, \ldots$ to label the three-dimensional Euclidean coordinates x^i and the generalized coordinates q^i, though the reader should bear in mind that in the latter case they run over the total number n of degrees of freedom of the system. Moreover the index k, when written within brackets, is also used to label the particle in the system. The meaning of these indices will be clear form the context.

Moreover:
$$\frac{\partial L}{\partial \dot{q}_\ell} = \frac{\partial T}{\partial \dot{q}_\ell}.$$

It follows:

$$H = \sum_{\ell=1}^{n} \frac{\partial L}{\partial \dot{q}_\ell} \dot{q}_\ell - L = \sum_{\ell=1}^{n} \frac{\partial T}{\partial \dot{q}_\ell} \dot{q}_\ell - L = T + U.$$

However $T + U$ is by definition the energy of the system and therefore our statement is proven.[9]

Let us now consider a system of particles whose Lagrangian $L(\mathbf{x}_{(k)}, \dot{\mathbf{x}}_{(k)})$, in Cartesian coordinates, is *invariant under translations* of the coordinates $x^i_{(k)}$; we show that the *total momentum is conserved*.

Indeed, under a *constant* translation, the position vector of the kth particle transforms as follows:

$$\mathbf{x}_{(k)} \rightarrow \mathbf{x}'_{(k)} = \mathbf{x}_{(k)} - \boldsymbol{\epsilon}; \qquad \boldsymbol{\epsilon} = \text{const.} \tag{8.29}$$

where $\boldsymbol{\epsilon}$ is a *constant vector*. In particular we have $\dot{\mathbf{x}}'_{(k)} = \dot{\mathbf{x}}_{(k)}$. *Invariance* of L, amounts to requiring it to have the same functional form in the old and in the new variables, so that:

$$L(x, \dot{x}) = L(x', \dot{x}') \Leftrightarrow \delta L = L(x', \dot{x}') - L(x, \dot{x}) = -\sum_{(k)} \frac{\partial L}{\partial x^i_{(k)}} \epsilon^i = 0. \tag{8.30}$$

Being ϵ^i arbitrary parameters, we conclude that

$$\sum_{(k)} \frac{\partial L}{\partial x^i_{(k)}} = 0, \tag{8.31}$$

Using Eqs. (8.7), the previous equation takes the following form:

$$\frac{d}{dt} \sum_{(k)} \frac{\partial L}{\partial \dot{x}^i_{(k)}} = 0. \tag{8.32}$$

On the other hand, in Cartesian coordinates, one has:

$$\frac{\partial L}{\partial \dot{x}^i_{(k)}} = \frac{\partial T}{\partial \dot{x}^i_{(k)}} = m_i \dot{x}^i_{(k)} = p^i_{(k)}, \tag{8.33}$$

---

[9]Note that since we have assumed $\frac{\partial L}{\partial t} = 0$ we have $\dot{U} = 0$, meaning that our system is isolated.

so that Eq. (8.32) becomes:

$$\frac{d}{dt}\sum_{(k)}\mathbf{P}_{(k)} = \frac{d}{dt}\mathbf{P} = 0, \tag{8.34}$$

where we have denoted by $\mathbf{P}$ the *total momentum* of the isolated system. It follows that:

The invariance of the Lagrangian under spatial translations implies the conservation of the total momentum.

It is easy to see that if L is invariant only with respect to translations along some directions (parametrized by the certain components of ϵ), only the corresponding components of the total momentum will be conserved. For example if we have an external force with $F_z = 0$ then the Lagrangian will be invariant only with respect to translations along the z-direction, $\epsilon = (0, 0, \epsilon_z)$ and we reach the conclusion that only P_z is conserved.

Let us note here that invariance under the time and space translations are unrelated in the classical theory, the former being related to time shifts and the latter to space translations of the general Galilean group (1.15). In a relativistic theory, instead, both invariances are part of *the invariance of the Lagrangian under the subgroup of four-dimensional translations of the Poincaré group*, namely

$$x^\mu \to x^\mu - \epsilon^\mu.$$

Therefore, energy and momentum conservation, which may hold separately in the classical theory are strictly related in relativistic mechanics, as discussed in Chap. 2, and we may speak of the conservation of the total four-momentum $p^\mu = (E/c, \mathbf{p})$.

Finally let us assume that the Lagrangian of the system is invariant under *spatial rotations*. In Cartesian coordinates, an infinitesimal rotation changes the kth position vector as follows (see Eq. (4.123)):

$$\mathbf{x}_{(k)} \to \mathbf{x}'_{(k)} = \mathbf{x}_{(k)} - \delta\boldsymbol{\theta} \times \mathbf{x}_{(k)}, \tag{8.35}$$

where $\delta\boldsymbol{\theta}$ is a *constant* infinitesimal vector whose direction coincides with the rotation axis and whose modulus is given by the infinitesimal rotation angle $\delta\theta$.

Requiring invariance amounts to setting $L(\mathbf{x}, \dot{\mathbf{x}}) = L(\mathbf{x}', \dot{\mathbf{x}}')$, that is:

$$0 = \delta L = \sum_{(k)}\left(\frac{\partial L}{\partial x^i_{(k)}}\delta x^i_{(k)} + \frac{\partial L}{\partial \dot{x}^i_{(k)}}\delta \dot{x}^i_{(k)}\right)$$

Using the Euler-Lagrange equations, we have

$$\delta L = -\sum_{(k)}\left(\frac{\partial L}{\partial x^i_{(k)}}\epsilon_{ij\ell}\delta\theta^j x^\ell_{(k)} + \frac{\partial L}{\partial \dot{x}^i_{(k)}}\epsilon_{ij\ell}\delta\theta^j \dot{x}^\ell_{(k)}\right)$$

$$= -\sum_{(k)} \left(\frac{d}{dt}\frac{\partial L}{\partial \dot{x}^i_{(k)}} \epsilon_{ij\ell}\delta\theta^j x^\ell_{(k)} + \frac{\partial L}{\partial \dot{x}^i_{(k)}} \epsilon_{ij\ell}\delta\theta^j \dot{x}^\ell_{(k)} \right)$$

$$= -\sum_{(k)} \frac{d}{dt}\left(\frac{\partial L}{\partial \dot{x}^i_{(k)}} \epsilon_{ij\ell}\delta\theta^j x^\ell_{(k)} \right).$$

It follows that

$$\frac{d}{dt}\sum_{(k)} \delta\theta^j \left(\epsilon_{ij\ell}x^\ell_{(k)}p^i_{(k)} \right) = \delta\theta \cdot \frac{d}{dt}\left(\sum_{(k)} \mathbf{x}_{(k)} \times \mathbf{p}_{(k)} \right) = 0.$$

Since the $\delta\theta^i$ are independent parameters, we obtain

$$\frac{d\mathbf{M}_{tot}}{dt} = 0$$

where

$$\mathbf{M}_{tot} = \sum_{(k)} \mathbf{x}_{(k)} \times \mathbf{p}_{(k)},$$

is the total angular momentum. Therefore the *invariance of a Lagrangian under rotations implies the conservation of the total angular momentum*. As in the previous case, if we have invariance only under a rotation about the ith-axis $(X, Y$ or $Z)$, parametrized by the ith component $\delta\theta^i$ of the infinitesimal rotation vector, only the corresponding component of $M_{tot,i}$ will be conserved.

8.2.1 The Noether Theorem for a System of Particles

The three conservation laws described in the previous subsection are associated with the invariance properties of the Lagrangian under space and time transformations. That means that the isotropy and homogeneity of space and time are not spoiled by interactions.

Actually, these conservation laws are just some of the implications of a general theorem, the *Noether theorem*, which we shall discuss in the present subsection. It essentially states that a conserved quantity is associated with each invariance of the *action*.

Let us first define our setting: We consider a system with a finite number of degrees of freedom, which, in the reference frame S, is described in terms of a Lagrangian $L(q, \dot{q}, t)$ function of the generalized coordinates $q_i, (i = 1, \ldots, n)$, their time derivatives $\dot{q}_i$ and time t.

Let the Lagrangian coordinates and time be subject to an arbitrary transformation of the form:

$$t' = t'(t), \qquad q'_i(t') = q'_i(q, t),$$ (8.36)

the only restriction being that Eq. (8.36) be invertible. Such transformations of the Lagrangian coordinates are often referred to as *point transformations*. In a different reference frame S', where the coordinates t', q'_i are used, the Lagrangian of the system will be given by a different function $L'(q', \dot{q}', t')$ of the new set of coordinates q'_i, $(i = 1, \ldots, n)$, $\dot{q}'_i$ and of time t'.

We now observe that the *action* S of a dynamical system is a *scalar* under the transformations (8.36), so that, taking into account the discussion made in Chap. 4, the two actions S and S' written in terms of their respective coordinates and times, are related by the condition:

$$S'[q'_i; t'_1, t'_2] = S[q_i; t_1, t_2].$$ (8.37)

where

$$S[q_i; t_1, t_2] = \int_{t_1}^{t_2} dt \, L(q(t), \dot{q}(t), t),$$ (8.38)

in the reference frame S and

$$S'[q'_i; t'_1, t'_2] = \int_{t'_1}^{t'_2} dt' \, L'(q'(t'), \dot{q}'(t'), t'),$$ (8.39)

in the reference frame S'. Similarly also L transforms as a scalar quantity, so that L and L' in the two RF's are related by:

$$L'(q'(t'), \dot{q}'(t'), t') = L(q(t), \dot{q}(t), t).$$ (8.40)

The new equations of motion in S' are clearly derived in the same way from the new Lagrangian:

$$\frac{\partial L'}{\partial q'_i} - \frac{d}{dt'} \frac{\partial L'}{\partial \dot{q}'_i} = 0 \qquad \forall i = 1, 2, \ldots, n.$$ (8.41)

However, the functional dependence of $L'(q'(t), \dot{q}'(t), t')$ on its arguments $q'(t)$, $\dot{q}'(t), t'$ is in general different from that of $L(q(t), \dot{q}(t), t)$ on $q(t), \dot{q}(t), t$. Similarly the actions S and S' are different functionals of (q_i) and (q'_i), respectively. It follows that the equations of motion derived from them will in general have a *different form*.

A transformation of the kind (8.36) is a *symmetry* of the system, namely the system is *invariant* under (8.36), if the equations of motion, as a system of differential equations, have the *same form* in the new and the old variables $q'_i(t')$ and $q_i(t)$.

In light of our discussion in Sect. 8.1.1 we can easily convince ourselves that the Euler-Lagrange equations in the two reference frames, described by different generalized coordinates and times, will have the same form provided the functional dependence of L' and L on their respective arguments is the same, modulo an additional total derivative which does not affect the equations of motion. Using the general relation (8.40), this amounts to saying that

$$L(q'(t'), \dot{q}'(t'), t') = L(q(t), \dot{q}(t), t) + \frac{df}{dt}. \tag{8.42}$$

At the level of the action the above property can be stated as follows:

$$S[q_i'; t_1', t_2'] = \int_{t_1'}^{t_2'} dt' L(q', \dot{q}', t') - \int_{t_1}^{t_2} dt L(q, \dot{q}, t) = S[q_i; t_1, t_2], \tag{8.43}$$

where we have ignored the total derivative since it yields equivalent actions.

Summarizing we have seen that a transformation of the kind (8.36) is a symmetry of the system if it leaves the action *invariant*, that is if the actions S and S' exhibit the *same functional dependence* on the paths described by q_i and q_i':

$$S[q_i'; t_1', t_2'] = S[q_i; t_1, t_2], \tag{8.44}$$

or, equivalently, if

$$\delta S \equiv \int_{t_1'}^{t_2'} dt' \, L(q', \dot{q}', t') - \int_{t_1}^{t_2} dt \, L(q, \dot{q}, t) = 0. \tag{8.45}$$

After these preliminaries we may state the *Noether theorem* as follows:

If the action of a dynamical system is invariant under a continuous group of (non singular) transformations of the generalized coordinates and time, of the form $q_i' = q_i'(q, t)$, $t' = t'(t)$, and if the equations of motion are satisfied, then the quantity:

$$Q \equiv \sum_i \frac{\partial L}{\partial \dot{q}^i} \delta q^i + L \delta t, \tag{8.46}$$

is conserved.

We stress that the variations $\delta q_i = q_i'(t) - q_i(t)$ corresponding to infinitesimal *local* transformations of the form (8.36), *are not arbitrary* as those used in the discussion of the Hamilton action principle, but correspond to the subclass of transformations leaving the action invariant.

Let us now start from the invariance property (8.45) to derive the conserved quantities.

We set

$$\delta S = \int_{t_1'}^{t_2'} dt' \, L(q'(t'), \dot{q}'(t'), t') - \int_{t_1}^{t_2} dt \, L(q(t), \dot{q}(t), t)$$

$$= \int_{t_1'}^{t_2'} dt \, L(q'(t), \dot{q}'(t), t) - \int_{t_1}^{t_2} dt \, L(q, \dot{q}, t), \tag{8.47}$$

where, on the right hand side of the above equation, we have used the fact that t' is an integration variable. We now decompose the integration over (t_2', t_1') as follows: Setting $t_1' = t_1 + \delta t_1$, $t_2' = t_2 + \delta t_2$, we can write

$$\int_{t_1'}^{t_2'} = \int_{t_1}^{t_2} + \int_{t_2}^{t_2+\delta t_2} - \int_{t_1}^{t_1+\delta t_1}, \tag{8.48}$$

so that the first integral of Eq. (8.47) can be written as follows:

$$\int_{t_1'}^{t_2'} dt \, L\left(q'(t), \dot{q}'(t), t\right) \equiv \int_{t_1}^{t_2} dt \, L\left(q'(t), \dot{q}'(t), t\right) + \delta t_2 \, L\left(q(t_2), \dot{q}(t_2), t_2\right)$$

$$- \delta t_1 \, L\left(q(t_1), \dot{q}(t_1), t_1\right). \tag{8.49}$$

In deriving (8.49) we have replaced, in the last two terms

$$L\left(q'(t), \dot{q}'(t), t\right) \equiv L\left(q(t) + \delta q, \dot{q}(t) + \delta \dot{q}(t), t\right), \tag{8.50}$$

with $L\left(q(t), \dot{q}(t), t\right)$ since their difference is infinitesimal (of order $O(\delta q)$) and therefore its product with the infinitesimal quantities δt_2, δt_1 is of higher order.

Next we substitute Eq. (8.49) into Eq. (8.47), obtaining:

$$\delta S = \int_{t_1}^{t_2} \left[L\left(q'(t), \dot{q}'(t), t\right) - L\left(q(t), \dot{q}(t), t\right) \right] dt + L\left(q(t), \dot{q}(t), t\right) \delta t \big|_{t_1}^{t_2}.$$

$$\tag{8.51}$$

The integral on the right hand side can be now expanded as follows:

$$\int_{t_1}^{t_2} \left[L\left(q'(t), \dot{q}'(t), t\right) - L\left(q(t), \dot{q}(t), t\right)\right] dt$$

$$= \int_{t_1}^{t_2} \left(\frac{\partial L}{\partial q^i} \delta q^i + \frac{\partial L}{\partial \dot{q}^i} \delta \dot{q}^i \right) dt$$

$$= \int_{t_1}^{t_2} \left(\left[\frac{\partial L}{\partial q^i} - \frac{d}{dt}\frac{\partial L}{\partial \dot{q}^i} \right] \delta q^i + \frac{d}{dt}\left(\frac{\partial L}{\partial \dot{q}^i} \delta q^i \right) \right) dt, \tag{8.52}$$

where we have integrated by parts $\frac{\partial L}{\partial \dot{q}^i} \delta \dot{q}^i = \frac{\partial L}{\partial \dot{q}^i} \frac{d}{dt} \delta q^i$. Upon substituting Eq. (8.52) into (8.51), we find:

$$\delta S = \int_{t_1}^{t_2} \left[\frac{\partial L}{\partial q^i} - \frac{d}{dt}\frac{\partial L}{\partial \dot{q}^i} \right] \delta q^i \, dt + \int_{t_1}^{t_2} \frac{d}{dt}\left(\frac{\partial L}{\partial \dot{q}^i} \delta q^i + L\delta t \right) dt \tag{8.53}$$

where we used the obvious equality:

$$L\delta t |_{t_1}^{t_2} = \int_{t_1}^{t_2} dt \frac{d}{dt}(L\delta t). \tag{8.54}$$

If the equations of motion are satisfied, the first integral on the right hand side of Eq. (8.53) vanishes and Eq. (8.53) becomes:

$$\delta S = \int_{t_1}^{t_2} \frac{d}{dt}\left(\frac{\partial L}{\partial \dot{q}^i} \delta q^i + L\delta t \right) dt. \tag{8.55}$$

From this it follows that, *if the action is invariant,* $\delta S = 0$, *the quantity*

$$Q(t) = \sum_i \frac{\partial L}{\partial \dot{q}^i} \delta q^i + L\delta t, \tag{8.56}$$

is conserved, which is the content of the Noether theorem. Indeed from (8.55), taking into account the arbitrariness of t_2, t_1, we have:

$$Q(t_2) = Q(t_1), \quad \forall t_1, t_2 \Leftrightarrow \frac{dQ}{dt} = 0. \tag{8.57}$$

Quite generally the transformations leaving the action invariant form a *group* with g parameters θ^r $(r = 1, 2, \ldots, g)$, so that we may write

$$\delta q^i = \delta \theta^r f_r^i(q, t).$$

For instance, rotations in space close the three-parameter group $SO(3)$, the parameters being the three angles. Factorizing the linear dependence of Q on the infinitesimal parameters $\delta \theta^r$, $Q = \delta \theta^r Q_r$, and being the parameters θ^r independent, we end up with g conserved charges associated with the variations δq_i ($g + 1$ if we include also the time translations).

The conserved quantities Q_r are referred to as *Noether charges*.

As an application, let us derive the conservation laws of energy, momentum and angular momentum directly from the Noether theorem.

Suppose the Lagrangian is invariant under time translations:

$$t' = t - dt \;\Rightarrow\; \delta t = -dt, \tag{8.58}$$

Then the coordinates used in the two RF's only differ by the infinitesimal time delay, so that $q_i'(t') = q_i(t)$. From this relation we deduce the infinitesimal relation between the two coordinate systems

$$q'^i(t - dt) = q^i(t) \;\Rightarrow\; q'^i(t) - \dot{q}^i(t)\, dt = q^i(t)$$
$$\Rightarrow \delta q^i(t) \equiv q'^i(t) - q^i(t) = \dot{q}^i(t)\, dt. \tag{8.59}$$

Correspondingly we find:

$$Q = Q_t \delta t = \left(\frac{\partial L}{\partial \dot{q}^i} \dot{q}^i - L \right) \delta t = H\, \delta t. \tag{8.60}$$

We see that invariance under time translations implies that we have *one* conserved charge corresponding to the energy: $H \equiv Q$ (Note that the time translation depends on *one* parameter, δt).

We have thus generalized our proof of the energy conservation given earlier since the invariance under time translations is satisfied if, in particular, the Lagrangian does not explicitly depend on time, $\frac{\partial L}{\partial t} = 0$, as was assumed in the previous subsection.

If the Lagrangian is invariant under space translations and rotations the value of the conserved charges Q_i is also readily computed using the variations δq_k given, in Cartesian coordinates, by Eqs. (8.29) and (8.35). In the case of constant space translations, after renaming $\delta q^i \equiv -\epsilon^i$, following the same steps as for the energy conservation, we find:

$$Q = Q_i \epsilon^i = -\sum_k \frac{\partial L}{\partial \dot{x}_{(k)}^i} \epsilon^i \equiv -\sum_k \mathbf{p}_{(k)} \cdot \boldsymbol{\epsilon} = -\mathbf{P} \cdot \boldsymbol{\epsilon} \tag{8.61}$$

implying conservation of the total momentum. The Noether charges Q_t, Q_i in Eqs. (8.60) and (8.61) can be grouped in the four-vector $\left(\frac{Q_t}{c}, Q_i\right) = \left(\frac{H}{c}, -P_i\right) = P_\mu = \eta_{\mu\nu} P^\nu$ which is the total energy-momentum four-vector with the index lowered with the metric. Analogously, setting $\delta q^i = -\epsilon^{ilm} \delta\theta^l x^m$, invariance under spatial rotations gives

$$Q = -\sum_k \frac{\partial L}{\partial \dot{x}^i_{(k)}} \epsilon_{ilm} \delta\theta^l x^m_{(k)} \equiv -\delta\boldsymbol{\theta} \cdot \sum_k \mathbf{x}_{(k)} \times \mathbf{p}_{(k)} = -\delta\boldsymbol{\theta} \cdot \mathbf{M}, \qquad (8.62)$$

implying the conservation of the total angular momentum.[10]

8.3 The Hamiltonian Formalism

In this section we give a short review of the Hamiltonian formalism for the description of mechanical systems with a finite number of degrees of freedom.

The *Hamiltonian formulation* of mechanics can be obtained from the Lagrangian one by introducing the *canonical momenta* p_i, conjugate to the Lagrangian coordinates, defined as:

$$p_i = \frac{\partial L}{\partial \dot{q}_i}(q, \dot{q}, t). \qquad (8.63)$$

Barring degeneracies, Eqs. (8.63) can be solved with respect to the Lagrangian velocities, obtaining: $\dot{q}_i = \dot{q}_i(p, q)$.

Next one defines the *Hamiltonian* of the dynamical system as:

$$H(p, q, t) \equiv \sum p_i \dot{q}_i(p, q) - L(q, \dot{q}(p, q), t). \qquad (8.64)$$

Comparing the definitions (8.24) and (8.64), with see that the physical meaning of $H(p, q)$ is that of the *energy of the system* $H = H(\dot{q}, q)$ and indeed, by an abuse of notation, they have been denoted by the same symbol. However it must be kept in mind that while in the Lagrangian formalism H is a function of $q, \dot{q}$, in the Hamiltonian formalism $H = H(p, q)$, so that their functional definition is different.

Let us now see how the equations of motion are derived in the Hamiltonian formalism.

By differentiating both sides of Eq. (8.64) one finds (using Einstein convention for summation over repeated indices):

---

[10]As noted in the previous discussion invariance under transformations parametrized by just some components of the vector parameters ϵ and $\delta\theta$, implies the conservation of the corresponding components of the vector quantities $\mathbf{P}$ and $\mathbf{M}$.

$$dH = dp_i \dot{q}_i + p_i \frac{\partial \dot{q}_i}{\partial p_j} dp_j + p_i \frac{\partial \dot{q}_i}{\partial q_j} dq_j - \frac{\partial L}{\partial q_i} dq_i$$

$$- \frac{\partial L}{\partial \dot{q}_i} \frac{\partial \dot{q}_i}{\partial p_j} dp_j - \frac{\partial L}{\partial \dot{q}_i} \frac{\partial \dot{q}_i}{\partial q_j} dq_j - \frac{\partial L}{\partial t} dt$$

$$= dp_i \left[\dot{q}_i + \left(p_j - \frac{\partial L}{\partial \dot{q}_j} \right) \frac{\partial \dot{q}_j}{\partial p_i} \right] + dq_i \left[-\frac{\partial L}{\partial q_i} + \left(p_j - \frac{\partial L}{\partial \dot{q}_j} \right) \frac{\partial \dot{q}_j}{\partial q_i} \right] - dt \frac{\partial L}{\partial t}.$$

Upon using Eq. (8.63) and the Euler-Lagrangian equations (8.7), one obtains:

$$dH = \sum dp_i \, \dot{q}_i - \sum dq_i \, \dot{p}_i - dt \frac{\partial L}{\partial t}$$

$$= \frac{\partial H}{\partial p_i} dp_i + \frac{\partial H}{\partial q_i} dq_i + \frac{\partial H}{\partial t} dt, \tag{8.65}$$

where the last equality represents the general expression of the total derivative of H. From the above equation we conclude that

$$\dot{q}_i = \frac{\partial H}{\partial p_i} \; ; \quad \dot{p}_i = -\frac{\partial H}{\partial q_i}, \tag{8.66}$$

$$-\frac{\partial L}{\partial t} = \frac{\partial H}{\partial t}. \tag{8.67}$$

Equation (8.66) represent a first order system of differential equations for the Lagrangian coordinates q_i and their conjugate momenta p_i, which is referred to as the *Hamilton equations of motion*. The physical content of the Hamilton equations and of the Euler-Lagrange equations is of course the same, however each formalism gives different insight into the properties of the mechanical system.

Considering q_i and p_i as coordinates of a $2n$-dimensional space, called the *phase space*, the state of a mechanical system is completely determined at each instant t by a point in this space labeled by the $2n$ coordinates $q_1, \ldots, q_n$ and $p_1, \ldots, p_n$. The time evolution of the system will then be described by a trajectory in the phase space. The Lagrangian variables q_i and their conjugate momenta p_i are referred to as *canonical coordinates*.

Let us note that the Hamilton equations (8.66) can be also obtained from the *action principle* $\delta S = 0$. Indeed, in terms of the p_i and q_i variables, the action takes the form:

$$S = \int_{t_1}^{t_2} \left(\sum p_i \dot{q}_i - H(p, q) \right) dt. \tag{8.68}$$

If we require stationarity of the action with respect to *arbitrary* variations δp_i and δq_i, with the constraint that they vanish at the end points of the time interval, $\delta p_i(t_1) = \delta q_i(t_1) = \delta p_i(t_2) = \delta q_i(t_2) = 0$, one obtains:

$$0 = \delta \mathcal{S} = \delta \left[\int_{t_1}^{t_2} \left(\sum p_i \dot{q}_i - H(p, q) \right) dt \right]$$

$$= \int_{t_1}^{t_2} \sum \left[\delta p_i \dot{q}_i - \dot{p}_i \delta q_i + \frac{d}{dt}(p_i \delta q_i) - \frac{\partial H}{\partial q_i} \delta q_i - \frac{\partial H}{\partial p_i} \delta p_i \right] dt$$

$$= \int_{t_1}^{t_2} dt \sum \left[\delta p_i \left(\dot{q}_i - \frac{\partial H}{\partial p_i} \right) - \delta q_i \left(\dot{p}_i + \frac{\partial H}{\partial q_i} \right) \right] + \sum (p_i \delta q_i)|_{t_1}^{t_2},$$

where the last term is zero. Being δp_i and δq_i arbitrary, Eq. (8.66) are retrieved.

Let us now consider a dynamical variable $f(q, p, t)$, function of $p_i(t)$, $q_i(t)$ and carrying in general also an *explicit* dependence on t. Computing its time derivative we find

$$\frac{d}{dt} f(p, q, t) = \sum_i \left(\frac{\partial f}{\partial q_i} \dot{q}_i + \frac{\partial f}{\partial p_i} \dot{p}_i \right) + \frac{\partial f}{\partial t}$$

$$= \sum_i \left(\frac{\partial f}{\partial q_i} \frac{\partial H}{\partial p_i} - \frac{\partial f}{\partial p_i} \frac{\partial H}{\partial q_i} \right) + \frac{\partial f}{\partial t} = \frac{\partial f}{\partial t} + \{f, H\}. \quad (8.69)$$

where

$$\{f, H\} \equiv \sum_i \left(\frac{\partial f}{\partial q_i} \frac{\partial H}{\partial p_i} - \frac{\partial f}{\partial p_i} \frac{\partial H}{\partial q_i} \right), \quad (8.70)$$

defines the *Poisson brackets* of f with H. From Eq. (8.69) it follows that the dynamical variable $f(q, p, t)$ is a *constant of motion* if

$$\frac{\partial f}{\partial t} + \{f, H\} = 0. \quad (8.71)$$

In particular, (8.69) implies that *if f does not explicitly depend on time*, $\frac{\partial f}{\partial t} = 0$, then *$f$ is a constant of the motion if and only if*

$$\{f, H\} = 0. \quad (8.72)$$

The definition of the Poisson brackets can be extended to any pair of dynamical variables $f(p, q)$ and $g(p, q)$. We define *Poisson brackets of f and g*, denoted by the symbol $\{f, g\}$, the following quantity:

$$\{f, g\} = \sum_i \left(\frac{\partial f}{\partial q_i} \frac{\partial g}{\partial p_i} - \frac{\partial f}{\partial p_i} \frac{\partial g}{\partial q_i} \right) = -\{g, f\}. \quad (8.73)$$

From the definition it follows that the Poisson bracket is *antisymmetric* in the exchange of its two entries. In particular $\{f, f\} = 0$.

Moreover, given three dynamical variables f, g, h, the following *Jacobi identity* holds (see Appendix D):

$$\{f, \{g, h\}\} + \{g, \{h, f\}\} + \{h, \{f, g\}\} = 0. \tag{8.74}$$

Of particular relevance are the Poisson brackets between the Lagrangian coordinates and the conjugate momenta, which are readily found to be:

$$\{q_i, p_j\} = \delta_{ij}, \tag{8.75}$$

$$\{q_i, q_j\} = \{p_i, p_j\} = 0. \tag{8.76}$$

It is important to observe that when the action S is evaluated along an *actual trajectory*, defining the evolution of the system in phase space, we can regard S as a function of the upper limit of the integral; from Eq. (8.68) it follows that the increment of the action between the instants t and $t + dt$, is given by:

$$dS = p_i dq_i - H(p, q) dt,$$

that is:

$$\frac{\partial S}{\partial q_i} = p_i; \qquad \frac{\partial S}{\partial t} = -H. \tag{8.77}$$

Using Cartesian coordinates for a single particle, Eq. (8.77) can be written in a Lorentz covariant way as follows:

$$\frac{\partial S}{\partial x^\mu} = -\eta_{\mu\nu} p^\nu, \tag{8.78}$$

where $p^\nu = \left(\frac{H}{c}, \mathbf{p}\right)$ and we have considered for simplicity a single particle.

As a simple application of the Hamilton formalism we compute the relativistic Hamiltonian of a free particle.

Using as Lagrangian coordinates the Cartesian ones, from the relativistic Lagrangian (8.21) we compute the conjugate momentum:

$$p^i = \frac{\partial L}{\partial v_i} = \frac{m}{\sqrt{1 - \frac{v^2}{c^2}}} v^i = m(v) v^i, \quad (i = 1, 2, 3),$$

which coincides with the relativistic momentum. The relativistic Hamiltonian is then computed from Eq. (8.64):

$$H = \mathbf{p} \cdot \mathbf{v} - L = \frac{mv^2}{\sqrt{1 - \frac{v^2}{c^2}}} + mc^2 \sqrt{1 - \frac{v^2}{c^2}} \tag{8.79}$$

$$= \frac{mc^2}{\sqrt{1 - \frac{v^2}{c^2}}} = m(v)c^2, \tag{8.80}$$

which coincides with the relativistic expression of the energy.[11]

8.4 Canonical Transformations and Conserved Quantities

In this section we describe the *canonical transformations* of the Hamiltonian formalism. This will allow us to give a new interpretation of the conserved quantities as *generators* of those "canonical transformations" which are symmetries, namely invariances, of the dynamical system.

We observe that the *point transformations* of the q-variables used in the Lagrangian formalism for the discussion of the Noether theorem, if symmetries, do not change the general form of either the Euler-Lagrange equations, or the Hamilton equations of motion.

In the Hamiltonian formalism, however, we have as independent variables not only the canonical coordinates q_i, but also their conjugate momenta p_i playing the role of additional coordinates. The space parametrized by the $2n$ coordinates p_i, q_i was called *phase space*. Taken together these $2n$ canonical coordinates admit a much larger class of transformations. We may indeed consider arbitrary non-singular transformations on the $2n$ canonical variables q_i, p_i:

$$\begin{aligned} p_i &\rightarrow P_i = P_i(p, q, t), \\ q_i &\rightarrow Q_i = Q_i(p, q, t), \end{aligned} \tag{8.81}$$

where, for the sake of clarity, we have denoted the new variables obtained after the transformation by Q_i, P_i.[12]

We then define *canonical transformations* the subgroup of transformations leading to canonical variables Q_i, P_i satisfying a system of Hamilton equations of the same form as in (8.66) though characterized by a different Hamiltonian function H'[13]:

$$\dot{Q}_i = \frac{\partial H'}{\partial P_i}; \qquad \dot{P}_i = -\frac{\partial H'}{\partial Q_i} \tag{8.82}$$

---

[11]From the Hamiltonian point of view we must substitute $v^2 = \frac{c^2 |\mathbf{p}|^2}{c^2 m^2 + |\mathbf{p}|^2}$.

[12]Note that the transformations (8.81) form a group, the group of coordinate transformations in *phase space*.

[13]We do not consider in this case transformation of the time variable.

Note that in general $H'(P, Q, t) \neq H(P, Q, t)$, that is we do not require these transformations to leave the *functional form of the Hamiltonian H invariant*. Therefore
in general, the new canonical equations will have a different form compared to those
in the old variables p_i, q_i.

To derive the conditions under which a general transformation (8.81) is *canonical*,
we observe that the new variables P_i, Q_i will satisfy Eq. (8.82) if and only if these
equations can be derived by the *stationary action principle*, as it is the case for the old
variables p_i, q_i. Therefore, under arbitrary variations of the canonical coordinates,
we must have:

$$\delta S = \delta \int \sum_i (p_i dq_i - H dt) = 0 \leftrightarrow \delta S' = \delta \int \sum_i (P_i dQ_i - H'(Q_i, P_i, t)dt) = 0.$$

This can only happen if the integrands of $\circ S$ and $\circ S'$ differ by the *total differential
of a function F*:

$$dF + \sum_i P_i dQ_i - H'dt = \sum_i p_i dq_i - H dt. \tag{8.83}$$

If this equation is satisfied then the general transformation (8.81) is *canonical* and
the function F is called *the generating function of the canonical transformation*.

Equation (8.83) implies a set of equations defining F in terms of q_i, Q_i and t:

$$\frac{\partial F}{\partial q_i} = p_i; \quad \frac{\partial F}{\partial Q_i} = -P_i; \quad \frac{\partial F}{\partial t} = H' - H. \tag{8.84}$$

From Eq. (8.84), we deduce that the generating function F can be regarded as a
function of the old and new Lagrangian coordinates q_i, Q_i.

We may, however, construct a new generating function $\Phi(q_i, P_i)$ depending on
the old coordinates q_i and the new momenta P_i. This can be obtained from the
generating function F by the following *Legendre transformation*:

$$\Phi(q, P, t) = F + \sum_i P_i Q_i. \tag{8.85}$$

Substituting indeed F in terms of Φ in Eq. (8.83), we obtain:

$$d\Phi = \sum_i Q_i dP_i + \sum_i p_i dq_i + (H' - H)dt,$$

that is:

$$\frac{\partial \Phi}{\partial P_i} = Q_i; \quad \frac{\partial \Phi}{\partial q_i} = p_i; \quad \frac{\partial \Phi}{\partial t} = H' - H. \tag{8.86}$$

Other Legendre transformations can be defined in order to obtain generating functions of any couple of old and new canonical coordinates. For our purposes however we shall only use the generating function (8.85). The reason is the following. Choosing $\Phi(q, P) = \sum_i q_i P_i$, Eq. (8.86) gives:

$$Q_i = q_i; \quad P_i = p_i; \quad H = H', \tag{8.87}$$

that is the corresponding canonical transformation is *the identity transformation*.

This allows us to generate *infinitesimal canonical transformations* considering transformations differing by an infinitesimal amount from the identity one (8.87):

$$\Phi(q, P, t) = \sum q_i P_i - \delta\theta^r G_r(q, P, t) \qquad \delta\theta^r \ll 1, \tag{8.88}$$

where $\delta\theta^r$ ($r = 1, ..., g$) are the infinitesimal parameters of the canonical transformation. For this kind of transformations, Eq. (8.86) takes the following form:

$$Q_i = q_i - \delta\theta^r \frac{\partial G_r}{\partial P_i}, \tag{8.89}$$

$$p_i = P_i - \delta\theta^r \frac{\partial G_r}{\partial q_i}, \tag{8.90}$$

$$H' - H = -\delta\theta^r \frac{\partial G_r}{\partial t}. \tag{8.91}$$

On the other hand, on the right hand side of Eq. (8.89) we may replace $\frac{\partial G_r}{\partial P_i}$ with $\frac{\partial G_r}{\partial p_i}$ since their difference, multiplied by the infinitesimal $\delta\theta^r$, is of higher order and can thus be neglected. The previous equations, using the definition of Poisson brackets, become:

$$\delta q_i = Q_i - q_i = -\delta\theta^r \frac{\partial G_r}{\partial p_i} = -\delta\theta_r\{q_i, G_r\} \tag{8.92}$$

$$\delta p_i = P_i - p_i = \delta\theta^r \frac{\partial G_r}{\partial q_i} = -\delta\theta^r\{p_i, G_r\}; \tag{8.93}$$

$$\delta H = H' - H = -\delta\theta^r \frac{\partial G_r}{\partial t}. \tag{8.94}$$

Accordingly, the quantity $\delta\theta^r G_r$ is called *infinitesimal generator of the canonical transformation* and the G_r's build a basis of generators.

Let us now consider a dynamical variable function of P_i, q_i and let us compute its transformation under an infinitesimal canonical transformation:

$$\delta f = f(P + \delta P, q + \delta q) - f(P, q) = \frac{\partial f}{\partial P_i} \delta p_i + \frac{\partial f}{\partial q_i} \delta q_i$$

$$= -\delta \theta^r \left(\frac{\partial f}{\partial q_i} \frac{\partial G_r}{\partial p_i} - \frac{\partial f}{\partial p_i} \frac{\partial G_r}{\partial q_i} \right) = -\delta \theta^r \{f, G_r\}, \tag{8.95}$$

where, by the same token as above, we have approximated $\frac{\partial G_r}{\partial P_i}$ by $\frac{\partial G_r}{\partial p_i}$ and $\frac{\partial f}{\partial P_i}$ by $\frac{\partial f}{\partial p_i}$ since they are multiplied by an infinitesimal quantity. It is easy to verify that under the infinitesimal canonical transformation, using (8.92), (8.93) and the Jacobi identity (8.74), we have

$$\delta \{q_i, p_j\} = 0. \tag{8.96}$$

That means that the fundamental canonical brackets between the Lagrangian coordinates and conjugate momenta are left invariant under an *infinitesimal* canonical transformation and therefore also by *finite* ones.

It is important to observe that *the time evolution of a dynamical system, i.e. the correspondence between the canonical variables computed at a time t and those evaluated at a later time $t' > t$, can be considered as a particular canonical transformation whose infinitesimal generator is the Hamiltonian.*

Let us indeed consider the change of the canonical coordinates when the time is increased from t to $t + dt$:

$$q_i(t) \rightarrow q_i'(t) = q_i(t + dt) \simeq q_i(t) + dt \dot{q}_i(t),$$
$$p_i(t) \rightarrow p_i'(t) = p_i(t + dt) \simeq p_i(t) + dt \dot{p}_i(t).$$

It is easy to show that the infinitesimal generator of this transformation is H. If we indeed identify

$$\Phi(q, P) = q_i P_i + dt \, H(p, q, t),$$

and use Eqs. (8.92), (8.93), upon identifying $G = -H$ and $\delta \theta = dt$, we find

$$\delta q_i = -dt \frac{\partial G}{\partial p_i} = dt \frac{\partial H}{\partial p_i} = dt \, \dot{q}_i,$$

$$\delta p_i = dt \frac{\partial G}{\partial q_i} = -dt \frac{\partial H}{\partial q_i} = dt \, \dot{p}_i,$$

where we have used the Hamilton equations (8.66). We may therefore state that *The Hamiltonian is the infinitesimal generator of the time translations.* In other words H generates the time evolution of the dynamical system.

We further note that if we compute the Poisson brackets of the canonical variables with the Hamiltonian we find:

$$\{q_i, H\} = \frac{\partial H}{\partial p_i} \; ; \; \{p_i, H\} = -\frac{\partial H}{\partial q_i}, \tag{8.97}$$

so that the Hamilton equations of motion (8.66) can be also written as follows:

$$\dot{q}_i = \{q_i, H\} \; ; \quad \dot{p}_i = \{p_i, H\}. \tag{8.98}$$

8.4.1 Conservation Laws in the Hamiltonian Formalism

In the Lagrangian formalism the conservation laws were derived by requiring the *symmetry* transformations on the Lagrangian coordinates to leave the *functional form* of the Lagrangian invariant, modulo an additional total derivative. Applying this requirement to translations in space and time, and to rotations, we derived the conservation laws for the linear momentum, energy and angular momentum.

Let us apply the same argument of *invariance* to the Hamiltonian of a dynamical system: A canonical transformation is an *invariance* of the system if it leaves the Hamilton equations of motion invariant in form, and this is the case if the functional dependence of the Hamiltonian on the old and new canonical variables is the same[14]

$$H'(p', q', t) = H(p', q', t). \tag{8.99}$$

If we consider infinitesimal canonical transformations (8.89), (8.90), (8.91), p', q' differ from p, q by infinitesimals δp, δq, so that Eq. (8.99) amounts to requiring:

$$\delta H = H(p', q') - H(p, q) = -\delta\theta^r \{H, G_r\} = -\delta\theta^r \frac{\partial G_r}{\partial t}, \tag{8.100}$$

where we have used Eqs. (8.94) and (8.95). From Eqs. (8.69) and (8.100), being $\delta\theta^r$ arbitrary, we conclude that

$$\frac{dG_r}{dt} = \frac{\partial G_r}{\partial t} + \{G_r, H\} = 0,$$

namely that G_r are constants of motion. In particular we see that *the infinitesimal generators G_r of the canonical transformations in the Hamiltonian formalism correspond to the Noether charges Q_r of the Lagrangian formalism.*

As an example, we want to retrieve once again the three conservation laws of linear momentum, angular momentum and energy, in the Hamiltonian formalism. Let us start implementing the condition of invariance of a system of n particles under space translations. The (Cartesian) coordinates and momenta of the particles are denoted, as usual, by $\mathbf{x}_{(k)}$ and $\mathbf{p}_{(k)}$, ($k = 1, ..., N$), respectively. Let us perform the infinitesimal translations

$$\mathbf{x}_{(k)} \rightarrow \mathbf{x}'_{(k)} = \mathbf{x}_{(k)} - \boldsymbol{\epsilon}; \quad |\boldsymbol{\epsilon}| \ll 1$$

---

[14]In this subsection we use the notations p'_i, q'_i instead of P_i, Q_i.

$$\mathbf{P}_{(k)} \to \mathbf{P}'_{(k)} = \mathbf{P}_{(k)},$$

which are supposed to leave the action invariant. To compute the infinitesimal generator of a translation on the kth particle we use Eqs. (8.92) and (8.93):

$$\delta x^i_{(k)} = -\epsilon^j \frac{\partial G_j}{\partial p^i_{(k)}} = -\epsilon^i; \qquad \delta p^i_{(k)} = \epsilon^j \frac{\partial G_j}{\partial x^i_{(k)}} = 0.$$

From the above equations it follows that

$$\frac{\partial G_j}{\partial x^i_{(k)}} = 0; \quad \frac{\partial G_j}{\partial p^i_{(k)}} = \delta^i_j \Rightarrow G_j(\mathbf{x}_{(k)}, \mathbf{P}_{(k)}) = \sum_k p^j_{(k)} = P^j_{(tot)},$$

where $P^j_{(tot)}$ is the jth component of the total linear momentum.

If the Hamiltonian is invariant under space-translations then the total linear momentum $\mathbf{P}_{tot}$ has vanishing Poisson bracket with the Hamiltonian, which in turn implies that it is conserved:

$$\frac{d\mathbf{P}_{(tot)}}{dt} = 0.$$

By the same token we deduce the conservation of the total angular momentum. Indeed under an infinitesimal rotation we have:

$$\mathbf{x}_{(k)} \to \mathbf{x}'_{(k)} = \mathbf{x}_{(k)} - \delta\boldsymbol{\theta} \times \mathbf{x}_{(k)},$$
$$\mathbf{P}_{(k)} \to \mathbf{P}'_{(k)} = \mathbf{P}_{(k)} - \delta\boldsymbol{\theta} \times \mathbf{P}_{(k)},$$

from which it follows that

$$\delta x^i_{(k)} = -\delta\theta^r \frac{\partial G_r}{\partial p^i_{(k)}} = -\delta\theta^r \epsilon_{irj} x^j_{(k)} = -\left(\delta\boldsymbol{\theta} \times \mathbf{x}_{(k)}\right)^i, \qquad (8.101)$$

$$\delta p^i_{(k)} = \delta\theta^r \frac{\partial G_r}{\partial x^i_{(k)}} = -\epsilon_{irj}\delta\theta^r p^j_{(k)} = -\left(\delta\boldsymbol{\theta} \times \mathbf{P}_{(k)}\right)^i. \qquad (8.102)$$

From (8.101) and (8.102) we obtain:

$$G_i = \sum_k \epsilon_{ijr} x^j_{(k)} p^r_{(k)} = M_{i\,(tot)},$$

where $M_{i\,(tot)}$ is the ith component of the total angular momentum. Therefore, if the system is invariant under rotations, the total angular momentum $\mathbf{M}_{(tot)}$ commutes with the Hamiltonian, implying that it is conserved:

$$\frac{d}{dt}\mathbf{M}_{(tot)} = 0.$$

Finally we note that if the Hamiltonian does not explicitly depend on t, so that we have invariance under time translations,

$$\frac{\partial H}{\partial t} = 0,$$

then, since $\{H, H\} = 0$, from (8.72) it follows:

$$\frac{dH}{dt} = 0,$$

that is the *energy conservation in the canonical formalism*. In fact, as we have seen, the Hamiltonian is the infinitesimal generator of time translations, which is a symmetry if neither the Lagrangian, nor the Hamiltonian explicitly depend on time.

8.5 Lagrangian and Hamiltonian Formalism in Field Theories

Our discussion has been confined so far to mechanical systems with a *finite* number of degrees of freedom, $q_1(t), \ldots, q_n(t)$.

This, however, has been propaedeutic to our principal objective, namely the description of *continuous* systems, hereafter called *fields*. A well known example of field is the *electromagnetic field* whose description is given in terms of the four-potential $A_\mu(\mathbf{x}, t)$; that means that, at any instant t, its configuration is defined by associating with each component μ the value of $A_\mu(\mathbf{x}, t)$ *at each point* $\mathbf{x}$ *in space*.

In this case we have a *continuous infinity of canonical coordinates* $q_i(t) = A_\mu(\mathbf{x}, t)$, labeled by the three coordinates $\mathbf{x}$ for the space-point and the index μ.[15] Other examples of fields are the continuous matter fields like fluids, elastic media, etc.

Quite generally we may view a continuous system as the limit of a mechanical system described by a finite number of degrees of freedom q_i (*discrete system*), by letting i become the continuous index $\mathbf{x}$. As a consequence every sum Σ_i over the discrete label i will be replaced by an integration on $d^3\mathbf{x}$ over a spatial domain V, usually the whole three-dimensional space[16]:

$$\sum_i \rightarrow \int_V d^3\mathbf{x}.$$

---

[15]More precisely, since $\mathbf{x} \equiv (x^1, x^2, x^3)$, we have a triple infinity of Lagrangian coordinates $q_i(t)$ for each value of the index $\mu = 0, 1, 2, 3$. The three components of $\mathbf{x}$ and the index μ play the role of the index i of the discrete case.

[16]Actually in our treatment of a discrete number of degrees of freedom, we have often omitted the symbol Σ when there were repeated indices.

242	8 Lagrangian and Hamiltonian Formalism

In the following we shall consider fields $\varphi^\alpha(\mathbf{x}, t)$ carrying an (internal) index α, where α labels the components of a *"vector $\varphi \equiv (\varphi^\alpha)$"* on which a representation of a group G acts. If α has just one value, it will be omitted and we speak of a *scalar field*. In *relativistic field theories*, the group G will often be the Lorentz group $O(1, 3)$ so that the index will label a basis of the carrier of a representation of the Lorentz group.[17] For example, in the case of the electromagnetic field, the role of α is played by the index μ pertaining to the four-dimensional fundamental representation of $SO(1, 3)$.

8.5.1 Functional Derivative

When we think of fields as a continuous limit of discrete systems, the corresponding Lagrangian obtained in the limit, $L(\varphi^\alpha, \partial_t \varphi^\alpha, t)$, will depend, at a certain instant t, on the values of the fields $\varphi^\alpha(\mathbf{x}, t)$ and $\partial_t \varphi^\alpha(\mathbf{x}, t)$ *at every point in the domain V of the three-dimensional space.* We say in this case that the Lagrangian is a *functional* of $\varphi^\alpha(\mathbf{x}, t)$ and $\partial_t \varphi^\alpha(\mathbf{x}, t)$, viewed as functions of $\mathbf{x}$. It will be convenient in the following to denote by $\varphi^\alpha(t)$ the function $\varphi^\alpha(\mathbf{x}, t)$ of the point $\mathbf{x}$ in space at a given time t, and by $\dot{\varphi}^\alpha(t)$ its time derivative $\dot{\varphi}^\alpha(\mathbf{x}, t) \equiv \partial_t \varphi^\alpha(\mathbf{x}, t)$. We shall presently explore some property of functionals.

Let us consider a functional $F[\varphi]$, and perform an independent variation of $\varphi(\mathbf{x})$, at each space point $\mathbf{x}$. The corresponding variation of $F[\varphi]$ will be:

$$\delta F[\varphi] \equiv F[\varphi + \delta\varphi] - F[\varphi] = \int \frac{\delta F[\varphi]}{\delta\varphi(\mathbf{x})} \, \delta\varphi(\mathbf{x}) \, d^3\mathbf{x} \qquad (8.103)$$

where, *by definition*, $\frac{\delta F[\varphi]}{\delta\varphi(\mathbf{x})}$ is the *functional derivative* of $F[\varphi]$ with respect to φ at the point $\mathbf{x}$. Here we have suppressed the possible dependence on time of φ and of the functional F either explicitly or through φ: $\varphi = \varphi(\mathbf{x}, t)$, $F = F[\varphi(t), t]$.

From its definition it is easy to verify that the functional derivation enjoys the same properties as the ordinary one, namely it is a linear operator, vanishes on constants and satisfies the Leibnitz rule.

When the functional depends on more than a single function, its definition can be extended correspondingly, as for ordinary derivatives. Of particular relevance for us is the additional dependence of F on the time derivative $\partial_t \varphi(\mathbf{x}, t)$ of $\varphi(\mathbf{x}, t)$. Moreover we may consider a set of fields φ^α labeled by the index α pertaining to a given representation of a group G. This is the case of the Lagrangian $F = L(\varphi^\alpha(t), \dot{\varphi}^\alpha(t), t)$, where we recall once again that, in writing $\varphi(t)$, $\dot{\varphi}(t)$ among the arguments of the Lagrangian, we mean that L depends on the values $\varphi(\mathbf{x}, t)$, $\dot{\varphi}(\mathbf{x}, t)$ of these fields in *every point* $\mathbf{x}$ in space at a given time t. Applying the definition

---

[17]Somewhat improperly, by the word representation people often refer to the carrier space V_p of a representation. We shall also do this to simplify the exposition and thus talk about a basis of a representation when referring to a basis of the corresponding carrier space.

(8.103) to the two functions $\varphi^\alpha(t)$ and $\dot\varphi^\alpha(t)$ we have:

$$\delta L(\varphi^\alpha(t), \dot\varphi^\alpha(t), t) = \int d^3x \left[\frac{\delta L}{\delta \varphi^\alpha(\mathbf{x}, t)} \delta \varphi^\alpha(\mathbf{x}, t) + \frac{\delta L}{\delta \dot\varphi^\alpha(\mathbf{x}, t)} \delta \dot\varphi^\alpha(\mathbf{x}, t) \right].$$

(8.104)

Note that the Lagrangian depends on t either through φ^α and $\dot\varphi^\alpha$ or explicitly. The Lagrangian, as a *functional* with respect to the space-dependence of the fields, can be thought of as the continuous limit of a function of infinitely many discrete variables:

$$L(\varphi x_i(t), \dot\varphi_i(t), t) \overset{i \to \mathbf{x}}{\longrightarrow} L(\varphi(\mathbf{x}, t), \dot\varphi(\mathbf{x}, t)).$$

Here and in the following we shall often omit the index α if not essential to our considerations. Correspondingly, we can show that the *functional derivative* defined above can be thought of as a suitable continuous limit of the ordinary derivative with respect to a discrete set of degrees of freedom q_i, described by a Lagrangian $L(q_i, \dot q_i)$. Let us indeed regard the values of $\varphi(\mathbf{x}, t)$ at each point $\mathbf{x}$ as independent canonical coordinates. To deal with a continuous infinity of canonical coordinates, we divide the 3-space into tiny cells of volume δV^i. Let $\varphi_i(t)$ be the mean value of $\varphi(\mathbf{x}, t)$ inside the ith cell and $L(t) = L(\varphi_i(t), \dot\varphi_i(t), t)$ be the Lagrangian, depending on the values $\varphi_i(t)$, $\dot\varphi_i(t)$ of the field and its time derivative in every cell. The variation $\delta L(\varphi_i, \dot\varphi_i)$ can be written as:

$$\delta L(\varphi_i(t), \dot\varphi_i(t), t) = \sum_i \left(\frac{\partial L}{\partial \varphi^i} \delta \varphi_i + \frac{\partial L}{\partial \dot\varphi^i} \delta \dot\varphi_i \right)$$

$$= \sum_i \frac{1}{\delta V^i} \left(\frac{\partial L}{\partial \varphi^i} \delta \varphi_i + \frac{\partial L}{\partial \dot\varphi^i} \delta \dot\varphi \right) \delta V^i, \qquad (8.105)$$

If we compare this expression with Eq. (8.104), in the continuum limit one can make the following identification:

$$\frac{\delta L}{\delta \varphi(\mathbf{x}, t)} \equiv \lim_{\delta V^i \to 0} \frac{1}{\delta V^i} \frac{\partial L}{\partial \varphi^i},$$

$$\frac{\delta L}{\delta \dot\varphi(\mathbf{x}, t)} \equiv \lim_{\delta V^i \to 0} \frac{1}{\delta V^i} \frac{\partial L}{\partial(\dot\varphi^i)}, \qquad (8.106)$$

where $\mathbf{x}$ is in the ith cell. In the limit $\delta V_i \to 0$ we can set $\delta V_i \equiv d^3x$. Thus the functional derivative $\delta L(t)/\delta \varphi(\mathbf{x}, t)$ is essentially proportional to the derivative of L with respect to the value of φ at the point $\mathbf{x}$. Since in the discretized notation the action principle leads to the equations of motion:

$$\frac{\partial L(t)}{\partial \varphi_i} - \partial_t \frac{\partial L(t)}{\partial \dot\varphi_i(t)} = 0 \qquad (8.107)$$

in the continuum limit the Euler-Lagrange equations become:

$$\frac{\delta L}{\delta \varphi^{\alpha}(\mathbf{x}, t)} - \partial_t \frac{\delta L}{\delta \dot{\varphi}^{\alpha}(\mathbf{x}, t)} = 0. \tag{8.108}$$

where we have reintroduced the index α of the general case.

In the discretized notation we shall assume the Lagrangian L, which depends on the values of the fields and their time derivatives in every cell, to be the sum of quantities $\mathcal{L}_i$ defined in each cell: $\mathcal{L}_i$ depends on the values of the field $\varphi_i^{\alpha}(t)$, its gradient $\nabla \varphi_i^{\alpha}$ and its time derivative $\dot{\varphi}_i^{\alpha}(t)$ in the ith cell only:

$$L(\varphi_i^{\alpha}(t), \dot{\varphi}_i^{\alpha}(t), t) = \sum_i \mathcal{L}_i(\varphi_i^{\alpha}(t), \nabla \varphi_i^{\alpha}(t), \dot{\varphi}_i^{\alpha}(t), t). \tag{8.109}$$

Multiplying and dividing the right hand side by δV_i and taking the continuum limit $\delta V_i \to d^3\mathbf{x}$, the above equality becomes

$$L(\varphi^{\alpha}(t), \dot{\varphi}^{\alpha}(t), t) = \int_V d^3\mathbf{x} \, \mathcal{L}(\varphi^{\alpha}(x), \nabla \varphi^{\alpha}(x), \dot{\varphi}^{\alpha}(x); \mathbf{x}, t), \tag{8.110}$$

where $x \equiv (x^{\mu}) = (ct, \mathbf{x})$ and we have defined the *Lagrangian density* $\mathcal{L}$ as:

$$\mathcal{L}(\varphi^{\alpha}(x), \nabla \varphi^{\alpha}(x), \dot{\varphi}^{\alpha}(x); \mathbf{x}, t) \equiv \lim_{\delta V_i \to 0} \frac{1}{\delta V_i} \mathcal{L}_i(\varphi_i^{\alpha}(t), \nabla \varphi_i^{\alpha}(t), \dot{\varphi}_i^{\alpha}(t), t).$$

Just as $\mathcal{L}_i$ depends, at a time t, on the dynamical variables referred to the ith cell only, $\mathcal{L}$ is a *local* quantity in Minkowski space in that it depends on both $\mathbf{x}$ and t. We note the appearance in $\mathcal{L}(x)$ of the space derivatives $\nabla \varphi^{\alpha}(\mathbf{x}, t)$. This follows from the fact that in order to have an *action* which is a *scalar* under Lorentz transformations, $\mathcal{L}$ itself must be a Lorentz scalar. Since Lorentz transformations will in general shuffle time and space derivatives, $\mathcal{L}$ should then depend on all of them. The action, in terms of the Lagrangian density, will read

$$\mathcal{S}[\varphi^{\alpha}; t_1, t_2] = \int_{t_1}^{t_2} dt \, L(t) = \int dt \, d^3\mathbf{x} \, \mathcal{L}(x) = \frac{1}{c} \int_{D_4} d^4x \, \mathcal{L}(x) \tag{8.111}$$

where D_4 is a space-time domain: An event $x \equiv (x^{\mu})$ in D_4 occurs at a time t between t_1 and t_2 and at a point $\mathbf{x}$ in the volume V. In formulas we will write $D_4 \equiv [t_1, t_2] \times V \subset M_4$. Since $\mathcal{S}$ does not depend only on the time interval $[t_1, t_2]$ but also on the volume V in which the values of the fields and their derivatives are considered, we will write $\mathcal{S} \equiv \mathcal{S}[\varphi^{\alpha}; D_4]$. The boundary of D_4, to be denoted by ∂D_4, consists of all the events occurring either at $t = t_1$ or at $t = t_2$, and of events occurring at a generic $t \in [t_1, t_2]$ in a point $\mathbf{x}$ belonging to the surface S_V which encloses the volume

$V: \mathbf{x} \in S_V \equiv \partial V$. The measure of integration $d^4x \equiv dx^0 dx^1 dx^2 dx^3 = c\, dt\, d^3\mathbf{x}$ is invariant under Lorentz transformations $\mathbf{\Lambda} = (\Lambda^\mu{}_\nu)$, since the absolute value $|\det(\mathbf{\Lambda})|$ of the determinant of the corresponding Jacobian matrix $\mathbf{\Lambda}$, is equal to one:

$$x^\mu \longrightarrow x'^\mu = \Lambda^\mu{}_\nu x^\nu \Rightarrow d^4x \longrightarrow d^4x' = |\det(\mathbf{\Lambda})|\, d^4x = d^4x. \quad (8.112)$$

It follows that in order to have a *scalar* Lagrangian density, $\mathcal{L}$ must have the same dependence on $\nabla \varphi^\alpha(\mathbf{x}, t)$ as on $\dot{\varphi}^\alpha(\mathbf{x}, t)$, that is it must actually depend on the four-vector $\partial_\mu \varphi^\alpha(\mathbf{x}, t)$. Moreover, being a scalar, it must depend on the fields and their derivatives $\partial_\mu \varphi^\alpha(\mathbf{x}, t)$ only through invariants constructed out of them. For the same reason it cannot depend on t only, but, in general, on all the space-time coordinates x^μ.

Let us now consider arbitrary infinitesimal variations of the field $\varphi^\alpha(x)$ which vanish at the boundary ∂D_4 of D_4: $\delta \varphi^\alpha(x) \equiv 0$ if $x \in \partial D_4$. The corresponding variation of L can be computed by using Eq. (8.110):

$$\delta L = \int d^3\mathbf{x} \left[\frac{\partial \mathcal{L}(\mathbf{x}, t)}{\partial \varphi^\alpha(\mathbf{x}, t)} \delta \varphi^\alpha(\mathbf{x}, t) + \frac{\partial \mathcal{L}(\mathbf{x}, t)}{\partial \partial_i \varphi^\alpha(\mathbf{x}, t)} \delta \partial_i \varphi^\alpha(\mathbf{x}, t) \right.$$
$$\left. + \frac{\partial \mathcal{L}(\mathbf{x}, t)}{\partial (\dot{\varphi}^\alpha(\mathbf{x}, t))} \delta \dot{\varphi}^\alpha(\mathbf{x}, t) \right]$$
$$= \int d^3\mathbf{x} \left\{ \left[\frac{\partial \mathcal{L}(\mathbf{x}, t)}{\partial \varphi^\alpha(\mathbf{x}, t)} - \partial_i \frac{\partial \mathcal{L}(\mathbf{x}, t)}{\partial \partial_i \varphi^\alpha(\mathbf{x}, t)} \right] \delta \varphi^\alpha(\mathbf{x}, t) + \frac{\partial \mathcal{L}(\mathbf{x}, t)}{\partial \dot{\varphi}^\alpha(\mathbf{x}, t)} \delta \dot{\varphi}^\alpha(\mathbf{x}, t) \right\}, \quad (8.113)$$

where we have written $\nabla \equiv (\partial_i)_{i=1,2,3}$, used the property that $\delta \partial_i \varphi^\alpha = \partial_i \delta \varphi^\alpha$ and integrated the second term within the integral by parts, dropping the surface term, being $\delta \varphi^\alpha(x) = 0$ for $\mathbf{x} \in S_V \equiv \partial V$.

Taking into account that the quantity inside the curly brackets defines the functional derivative of L, by comparison with Eq. (8.108) we find:

$$\frac{\delta L}{\delta \varphi^\alpha(x)} = \left[\frac{\partial \mathcal{L}(x)}{\partial \varphi^\alpha(x)} - \partial_i \frac{\partial \mathcal{L}(x)}{\partial \partial_i \varphi^\alpha(x)} \right],$$
$$\frac{\delta L}{\delta \dot{\varphi}^\alpha(x)} = \frac{\partial \mathcal{L}(x)}{\partial \dot{\varphi}^\alpha(x)}. \quad (8.114)$$

It is important to note that, using the *Lagrangian density* instead of the Lagrangian, the derivatives of $\mathcal{L}(\mathbf{x}, t)$ with respect to the fields in $(\mathbf{x}, t)$ are now *the usual partial derivatives*, since they are computed at a particular point $\mathbf{x}$. Using the equalities (8.114) the Euler-Lagrange equations (8.108) take the following form:

$$\partial_t \frac{\partial \mathcal{L}}{\partial (\partial_t \varphi^\alpha)} = \frac{\partial \mathcal{L}}{\partial \varphi^\alpha} - \partial_i \frac{\partial \mathcal{L}}{\partial (\partial_i \varphi^\alpha)}, \quad (8.115)$$

or, using a Lorentz covariant notation:

$$\frac{\partial \mathcal{L}}{\partial \varphi^\alpha} - \partial^\mu \left(\frac{\partial \mathcal{L}}{\partial (\partial_\mu \varphi^\alpha)} \right) = 0. \tag{8.116}$$

8.5.2 The Hamilton Principle of Stationary Action

In the previous paragraph the equations of motion for fields have been derived using the definition of functional derivative and performing the continuous limit of the Euler-Lagrange equations for a discrete system.

Actually Eqs. (8.116), can also be derived directly from the Hamilton principle of stationary action, considering the action $\mathcal{S}$ as a functional of the fields φ^α and depending on the space-time domain D_4 on which they are defined:

$$\mathcal{S}[\varphi^\alpha; D_4] = \frac{1}{c} \int_{D_4} d^4 x \, \mathcal{L}(\varphi^\alpha, \partial_\mu \varphi^\alpha, x^\mu). \tag{8.117}$$

Here $d^4 x \equiv dx^0 d^3 \mathbf{x} = c \, dt d^3 \mathbf{x}$ is the volume element in the Minkowski space M_4, and the integration domain D_4 was defined as $[t_1, t_2] \times V \subset M_4$.

We can now generalize the *Hamilton principle of stationary action* to systems described by fields, namely systems exhibiting a continuous infinity of degrees of freedom. It states that:

The time evolution of the field configuration describing the system is obtained by extremizing the action with respect to arbitrary variations of the fields $\delta \varphi^\alpha$ which vanish at the boundary ∂D_4 of the space-time domain D_4.

More precisely, we require the action $\mathcal{S}$ to be stationary with respect to $\delta \varphi^\alpha$, that is to satisfy

$$\delta \mathcal{S} = 0,$$

under arbitrary variations of φ^α at each point $\mathbf{x}$ and at each instant t:

$$\varphi^\alpha(x) \rightarrow \varphi^\alpha(x) + \delta \varphi^\alpha(x),$$

provided:

$$\delta \varphi^\alpha(x) = 0 \qquad \forall x^\mu \in \partial D_4. \tag{8.118}$$

Let us apply this principle to the action (8.117). We have:

$$\delta \mathcal{S} = \frac{1}{c} \int_{D_4} d^4 x \left(\frac{\partial \mathcal{L}}{\partial \varphi^\alpha} \delta \varphi^\alpha + \frac{\partial \mathcal{L}}{\partial (\partial_\mu \varphi^\alpha)} \delta (\partial_\mu \varphi^\alpha) \right). \tag{8.119}$$

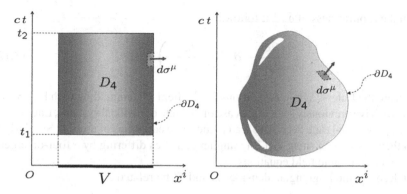

Fig. 8.1 Space-time domain D_4: of the form $[t_1, t_2] \times V$ (*left*), of generic form (*right*)

Now use the property $\delta(\partial_\mu \varphi^\alpha) = \partial_\mu(\delta \varphi^\alpha)$, and integrate by parts the second term in the integral:

$$
\int_{D_4} d^4x \frac{\partial \mathcal{L}}{\partial(\partial_\mu \varphi^\alpha)} \partial_\mu \delta \varphi^\alpha = \int_{D_4} d^4x \, \partial_\mu \left(\frac{\partial \mathcal{L}}{\partial(\partial_\mu \varphi^\alpha)} \delta \varphi^\alpha \right) - \int_{D_4} d^4x \, \partial_\mu \left(\frac{\partial \mathcal{L}}{\partial(\partial_\mu \varphi^\alpha)} \right) \delta \varphi^\alpha
$$

$$
= \int_{\partial D_4} d\sigma_\mu \left(\frac{\partial \mathcal{L}}{\partial(\partial_\mu \varphi^\alpha)} \right) \delta \varphi^\alpha - \int_{D_4} d^4x \, \partial_\mu \left(\frac{\partial \mathcal{L}}{\partial(\partial_\mu \varphi^\alpha)} \right) \delta \varphi^\alpha
$$

$$
= - \int_{D_4} d^4x \, \partial_\mu \left(\frac{\partial \mathcal{L}}{\partial(\partial_\mu \varphi^\alpha)} \right) \delta \varphi^\alpha,
$$

where we have applied the four-dimensional version of the divergence theorem by expressing the integral of a four-divergence over D_4 as an integral (boundary integral) of the four-vector $\frac{\partial \mathcal{L}}{\partial(\partial_\mu \varphi^\alpha)} \delta \varphi^\alpha$ over the three-dimensional domain ∂D_4 which encloses D_4. We have used the notation $d\sigma_\mu \equiv d^3\sigma \, n_\mu$, $d^3\sigma$ being an element of ∂D_4 to which the unit norm vector n_μ is normal, see Fig. 8.1. As for the last equality we have used Eq. (8.118) which implies the vanishing of the boundary integral.[18] Thus the partial integration finally gives:

$$
\delta S = \frac{1}{c} \int_{D_4} d^4x \left[\frac{\partial \mathcal{L}}{\partial \varphi^\alpha} - \partial_\mu \left(\frac{\partial \mathcal{L}}{\partial(\partial_\mu \varphi^\alpha)} \right) \right] \delta \varphi^\alpha. \tag{8.120}
$$

---

[18]This is true if the boundary ∂D_4 does not extend to spatial infinity; when the integration domain D_4 fills the whole space, we must require that the fields and their derivatives fall off sufficiently fast at infinity, or we may also use periodic boundary conditions. In any case the integration on an infinite domain can always be taken initially on a finite domain, and, after removing the boundary term, the integration domain can be extended to infinity.

From the arbitrariness of $\delta\varphi^\alpha$ it follows:

$$\frac{\partial \mathcal{L}}{\partial \varphi^\alpha} - \partial_\mu \left(\frac{\partial \mathcal{L}}{\partial(\partial_\mu \varphi^\alpha)} \right) = 0 \tag{8.121}$$

which are the Euler-Lagrange equations for the field φ^α, coinciding with Eq. (8.116).

As we have previously seen in the case of discrete dynamical systems, Lagrangians differing by a total time derivative lead to the same equations of motion. Similarly for field theories we can show that Lagrangian densities differing by a four-divergence $\partial_\mu f^\mu$ yield the same field equations.

Indeed, let the Lagrangian densities $\mathcal{L}$ and $\mathcal{L}'$ be related as

$$\mathcal{L}'(\varphi^\alpha(x), \partial_\mu \varphi^\alpha(x), x) = \mathcal{L}(\varphi^\alpha(x), \partial_\mu \varphi^\alpha(x), x) + \partial_\mu f^\mu,$$

where $f^\mu = f^\mu(\varphi^\alpha(x), x)$, then the two actions differ by a boundary integral:

$$\begin{aligned}
S' &= \frac{1}{c} \int_{D_4} d^4x \mathcal{L}'(\varphi^\alpha, \partial_\mu \varphi^\alpha) = \frac{1}{c} \int_{D_4} d^4x \mathcal{L}(\varphi^\alpha, \partial_\mu \varphi^\alpha) \\
&+ \frac{1}{c} \int_{D_4} d^4x \partial_\mu f^\mu = S + \frac{1}{c} \int_{\partial D_4} d\sigma_\mu f^\mu.
\end{aligned} \tag{8.122}$$

We therefore have:

$$\delta S' = \delta S + \frac{1}{c} \int_{\partial D_4} d\sigma_\mu \delta f^\mu = \delta S,$$

since $\delta f^\mu = \frac{\partial f^\mu}{\partial \varphi^\alpha(x)} \delta\varphi^\alpha(x) = 0$ on the boundary ∂D_4.

8.6 The Action of the Electromagnetic Field

As an application of our general discussion, we construct the action of the electromagnetic field in interaction with charges and currents and show that the stationary action principle gives the covariant form of the Maxwell equations discussed in Chap. 5. To this end we shall be guided by the symmetry principle. As it will be shown in detail in Sect. 8.7, the invariance of the equations of motion under space-time (i.e. Poincaré) transformations or under general field transformations is guaranteed if the Lagrangian density, as a function of the fields, their derivatives and the space-time coordinates, is invariant in form, up to a total divergence, see Eq. (8.150). As far as space- time translations are concerned, this is the case if $\mathcal{L}$ does not explicitly depend on x^μ. Covariance with respect to Lorentz transformations further requires $\mathcal{L}$ to be invariant as a function of the fields and their derivatives, namely to be a *Lorentz scalar* as a function of space-time.

The construction of the action for the electromagnetic field is relatively simple once we observe that:

- For the field $A_\mu(x)$ describing the electromagnetic field the generic index α coincides with the covariant index $\mu = 0, 1, 2, 3$ of the fundamental representation of the Lorentz group;
- The equations of motion (the Maxwell equations) are invariant under the *gauge transformations*:

$$A_\mu \to A_\mu + \partial_\mu \varphi.$$

This is guaranteed if the Lagrangian density is invariant under the same transformations, since the action would then be invariant. In the absence of charges and currents, the action should be constructed out of the gauge invariant quantity $F_{\mu\nu}$;
- The Lagrangian density must be a *scalar* under Lorentz transformations;
- In order for the equations of motion to be second-order differential equations, $\mathcal{L}$ must at most be quadratic in the derivatives of $A_\mu(x)$, that is quadratic in $F_{\mu\nu}$.

To construct Lorentz scalars which are quadratic in $F_{\mu\nu}$ we may use the invariant tensors $\eta_{\mu\nu}$, $\epsilon_{\mu\nu\rho\sigma}$ of the Lorentz group SO(1, 3).[19] It can be easily seen that the most general Lagrangian density satisfying the previous requirements has the following form:

$$\mathcal{L}(A_\mu, \partial_\mu A_\nu) = a\, F_{\mu\nu} F^{\mu\nu} + b\, \epsilon_{\mu\nu\rho\sigma} F^{\mu\nu} F^{\rho\sigma}, \tag{8.123}$$

where $F^{\mu\nu} = \eta^{\mu\rho}\eta^{\nu\sigma} F_{\rho\sigma}$ and a and b are numerical constants. On the other hand, the second term of Eq. (8.123) is the four-dimensional divergence of a four-vector so that it does not contribute to the equations of motion. Indeed:

$$\begin{aligned}
\epsilon_{\mu\nu\rho\sigma} F^{\mu\nu} F^{\rho\sigma} &= 2\epsilon_{\mu\nu\rho\sigma} \partial^\mu A^\nu F^{\rho\sigma} \\
&= \partial^\mu \left(2\epsilon_{\mu\nu\rho\sigma} A^\nu F^{\rho\sigma}\right) - 2\epsilon_{\mu\nu\rho\sigma} A^\nu \partial^\mu F^{\rho\sigma} \\
&= \partial^\mu \left(2\epsilon_{\mu\nu\rho\sigma} A^\nu F^{\rho\sigma}\right) - 2\epsilon_{\mu\nu\rho\sigma} A^\nu \partial^{[\mu} F^{\rho\sigma]} \\
&= \partial_\mu f^\mu,
\end{aligned}$$

where we have set: $f^\mu = 2\epsilon^{\mu\nu\rho\sigma} A_\nu F_{\rho\sigma}$ and use has been made of the identity: $\partial^{[\mu} F^{\rho\sigma]} = 0$.

Therefore the Lagrangian density reduces, up to a four-dimensional divergence to the single term:

$$\mathcal{L}_{em} = a\, F_{\mu\nu} F^{\mu\nu}.$$

The value of the constant a is fixed in such a way that the Lagrangian contains the positive definite (density of) "kinetic term" $1/(2c^2)\, \partial_t A_i \partial_t A_i$ with a conventional

---

[19]Recall that the latter tensor $\epsilon_{\mu\nu\rho\sigma}$ is not invariant under Lorentz transformations which are in O(1, 3) but not in SO(1, 3), namely which have determinant -1. Examples of these are the parity transformation Λ_P, or time reversal Λ_T.

factor $1/2$ which is remnant of the one appearing in the definition (8.25) of the kinetic energy.[20] Expanding $F_{\mu\nu}F^{\mu\nu} = (\partial_\mu A_\nu - \partial_\nu A_\mu)(\partial^\mu A^\nu - \partial^\nu A^\mu)$ one easily finds $a = -\frac{1}{4}$.

In the presence of charges and currents, the interaction with the source $J^\mu(x)$ requires adding an interaction term $\mathcal{L}_{int}$ to the pure electromagnetic Lagrangian. The simplest interaction is described by the Lorentz scalar term:

$$\mathcal{L}_{int} = b\, A_\mu J^\mu. \tag{8.124}$$

This term seems, however, to violate the gauge invariance of the total Lagrangian, since a gauge transformation on A_μ implies a correspondent change on the Lagrangian density:

$$\delta A_\mu = \partial_\mu \varphi \Rightarrow \delta_{(gauge)}\mathcal{L}_{int} = (\partial_\mu \varphi)\, J^\mu.$$

On the other hand, by partial integration, $\delta\mathcal{L}_{int}$ can be transformed as follows:

$$\delta\mathcal{L}_{int} = \partial_\mu(\varphi J^\mu) - \varphi\, \partial_\mu J^\mu.$$

The first term is a total four-divergence, not contributing to the equations of motion and thus can be neglected; the second term is zero *if and only if* $\partial_\mu J^\mu = 0$, that is if the continuity equation expressing the conservation of the electric charge holds. We have thus found the following important result:

Requiring gauge invariance of the action of the electromagnetic field interacting with a current, implies the conservation of the electric charge.

In conclusion, the action describing the electromagnetic field coupled to charges and currents is given by:

$$S = \frac{1}{c}\int_{M_4} d^4x \left(-\frac{1}{4}F_{\mu\nu}F^{\mu\nu} + bA_\mu J^\mu\right), \tag{8.125}$$

where the (four)-current $J^\mu(x)$ has the following general form (see Chap. 5)[21]:

$$J^\mu(x) = \frac{1}{c}\sum_k e_k \frac{dx^\mu_{(k)}}{dt}\delta^3(\mathbf{x} - \mathbf{x}_{(k)}(t)). \tag{8.126}$$

We may now apply the principle of stationary action to compute the equations of motion. Recalling the form Eq. (8.121) of the Euler-Lagrange equations for fields,

---

[20]Note that the kinetic term for A_0 is absent because of the antisymmetry of $F_{\mu\nu}$.

[21]Note that we are describing the interaction of the electromagnetic field, possessing infinite degrees of freedom, with a system of N charged particles, having $3N$ degrees of freedom represented by the N coordinate vectors $\mathbf{x}_{(k)}(t)$, $(k = i, \ldots, N)$. The Dirac delta function *formally* converts the $3N$ degrees of freedom of $\mathbf{x}_{(k)}(t)$ *into the infinite degrees of freedom associated to* $\mathbf{x}$.

we have:

$$\frac{\partial \mathcal{L}}{\partial A_\mu} - \partial_\rho \left(\frac{\partial \mathcal{L}}{\partial(\partial_\rho A_\mu)} \right) = 0. \tag{8.127}$$

The first term of Eq. (8.127) is easily computed and gives:

$$\frac{\partial \mathcal{L}}{\partial A_\mu} = b J^\mu(x).$$

As far as the second term is concerned, only the pure electromagnetic part $-1/4 F_{\mu\nu} F^{\mu\nu}$ contributes to the variation, yielding:

$$\frac{\partial(F_{\rho\sigma} F^{\rho\sigma})}{\partial(\partial_\mu A_\nu)} = 2 \left[\frac{\partial F_{\rho\sigma}}{\partial(\partial_\mu A_\nu)} \right] F^{\rho\sigma} = 4 \frac{\partial(\partial_\rho A_\sigma)}{\partial(\partial_\mu A_\nu)} F^{\rho\sigma} = \delta^\mu_\rho \delta^\nu_\sigma F^{\rho\sigma} = 4 F^{\mu\nu}(x).$$

Putting these results together, Eq. (8.127) becomes:

$$\partial_\mu F^{\mu\nu}(x) + b J^\nu(x) = 0. \tag{8.128}$$

Finally the constant b is fixed by requiring Eq. (8.128) to be identical to the Maxwell equation[22]:

$$\partial_\mu F^{\mu\nu} = -J^\nu,$$

and this fixes b to be 1. The final expression of the Lagrangian density therefore is:

$$\mathcal{L} = \mathcal{L}_{em} + \mathcal{L}_{int} = -\frac{1}{4} F_{\mu\nu} F^{\mu\nu} + A_\mu J^\mu. \tag{8.129}$$

In order to give a complete description of the charged particles in interaction with the electromagnetic field, we must add to $\mathcal{L}$ (8.129) the Lagrangian density $\mathcal{L}_{part}$ associated with system of particles.

Let us consider for the sake of simplicity the case of N particles of charges e_k and masses m_k, $k = 1, \ldots, N$. The total action will have the following form[23]:

$$\mathcal{S}_{tot} = \mathcal{S}_{em} + \mathcal{S}_{int} + \mathcal{S}_{part}, \tag{8.130}$$

where:

$$\mathcal{S}_{em}[\partial_\mu A_\nu] = \frac{1}{c} \int d^4x \left(-\frac{1}{4} F_{\mu\nu} F^{\mu\nu} \right),$$

---

[22]Note that this condition just fixes the charge normalization.

[23]The index k given to $\mathbf{x}_{(k)}$ in the following formulae has the function of indicating that the coordinate vector $\mathbf{x}_{(k)}(t)$ is a *dynamical variable*, and not the labeling of the space points, as is the case for $\mathbf{x}$.

$$\mathcal{S}_{part}[\dot{\mathbf{x}}_{(k)}] = \sum_{k=1}^{N} (-m_k c^2) \int dt \left(1 - \frac{v_{(k)}^2}{c^2}\right)^{\frac{1}{2}},$$

$$\mathcal{S}_{int}[A_\mu(x), \mathbf{x}_{(k)}, \dot{\mathbf{x}}_{(k)}] = \frac{1}{c} \int d^4x\, A_\mu(\mathbf{x}, t) J^\mu(\mathbf{x}, t)$$

$$= \frac{1}{c} \int dt \sum_k \int d^3\mathbf{x} \left[\frac{e_k}{c} A_\mu(x) \delta^{(3)}(\mathbf{x} - \mathbf{x}_{(k)}) \frac{dx_{(k)}^\mu}{dt}\right]$$

$$= \frac{1}{c} \sum_k \int dt \frac{e_k}{c} A_\mu(\mathbf{x}_k, t) \frac{dx_{(k)}^\mu}{dt}, \qquad (8.131)$$

where in deriving the expression of $\mathcal{S}_{int}$[24] we have used the explicit form of the four-current given in (8.126).

$$L_{int} = \int d^3\mathbf{x}\, A_\mu(\mathbf{x}, t) J^\mu(\mathbf{x}, t) = \sum_k \frac{e_k}{c} A_\mu(\mathbf{x}_{(k)}, t) \frac{dx_{(k)}^\mu}{dt}$$

$$= \sum_k \left(e_k A_0(\mathbf{x}_{(k)}, t) + \frac{e_k}{c} A_i(\mathbf{x}_{(k)}, t) v_{(k)}^i\right). \qquad (8.132)$$

We recall that $\mathbf{x}$ are labels of the points in space, while $\mathbf{x}_{(k)}(t)$ are the particle coordinates, that is dynamical variables, as stressed in the footnote.

We now observe that since $\mathcal{S}_{em}$ does not contain the variables $x_{(k)}^i$, we may compute the equation of motion of the kth charged particle by varying only $\hat{L} = L_{part} + L_{int}$:

$$\hat{L} = L_{part} + L_{int} = -\sum_k \left(m_k c^2 \sqrt{1 - \frac{v_{(k)}^2}{c^2}} + e_k A_0(\mathbf{x}_{(k)}, t) + \frac{e_k}{c} A_i(\mathbf{x}_{(k)}, t) v_{(k)}^i\right).$$

For the sake of simplicity in the following we neglect the index (k) of the particle. The first term of the Euler-Lagrange equations:

$$\frac{\partial \hat{L}}{\partial x^i} - \frac{d}{dt} \frac{\partial \hat{L}}{\partial v^i} = 0, \qquad (8.133)$$

---

[24]Note that also $\mathcal{S}_{part}$ can be written as a four-dimensional integral:

$$\mathcal{S}_{int} = -\sum_k m_k c \int d^4x\, \delta^{(3)}(\mathbf{x} - \mathbf{x}_{(k)}) \left(1 - \frac{1}{c^2}\left(\frac{d\mathbf{x}}{dt}\right)^2\right).$$

reads:

$$\frac{\partial \hat{L}}{\partial x^i} = \frac{\partial L_{int}}{\partial x^i} = e \frac{\partial A_0}{\partial x^i} + \frac{e}{c} \left(\frac{\partial A_j}{\partial x^i} \right) v^j. \tag{8.134}$$

The second term contains the time derivative of the *canonical momentum* p^i *conjugate to* x^i, namely:

$$p^i = \frac{\partial \hat{L}}{\partial v^i} = \frac{\partial (L_{par} + L_{int})}{\partial v^i} = m(v)v^i + \frac{e}{c} A_i. \tag{8.135}$$

We see that in the presence of the electromagnetic field *the canonical conjugate momentum is different from the momentum* $p^i_{(0)} = m(v)v^i$ *of a free particle.*[25] In fact we have the following relation:

$$p^i = p^i_{(0)} + \frac{e}{c} A_i. \tag{8.136}$$

Taking into account (8.133), (8.135) and (8.136), the equation of motion of the charged particle becomes:

$$\frac{d}{dt} \left(p^i_{(0)} + \frac{e}{c} A_i \right) - e \partial_i A_0 - \frac{e}{c} \partial_i A_j v^j = 0. \tag{8.137}$$

We now recall that $A_0 = -V$, where V is the electrostatic potential. Moreover, since

$$\frac{d A_i}{dt} = \frac{\partial A_i}{\partial x^j} \frac{dx^j}{dt} + c \frac{\partial A_i}{\partial x^0},$$

and $E^i = F_{i0} = \partial_i A_0 - \partial_0 A_i$, Eq. (8.137) becomes:

$$\frac{dp^i_{(0)}}{dt} = e E^i - \frac{e}{c} \left(\partial_j A_i - \partial_i A_j \right) v^j$$

$$= e E^i - \frac{e}{c} F_{ji} v^j = e E^i + \frac{e}{c} \epsilon_{ijk} v^j B_k$$

$$= e \left(E^i + \frac{1}{c} (\mathbf{v} \times \mathbf{B})^i \right).$$

Thus we have retrieved from the variational principle the well known equation of motion of a charged particle subject to electric and magnetic fields, since the right hand side is by definition the Lorentz force.

---

[25] Here and in the following we use the subscript 0 to denote the usual free-particle momentum $p^i_{(0)} = m(v)v^i$ and the symbol p^i for the momentum canonically conjugated to x^i.

8.6.1 The Hamiltonian for an Interacting Charge

As we have computed the Lagrangian $L_{int} + L_{par}$ for a charged particle, we pause for a moment with our treatment of the Lagrangian formalism in field theories and compute the Hamiltonian of a charge interacting with the electromagnetic field.
From the definition (8.64) we find[26]:

$$H(\mathbf{p}, \mathbf{x}) = \mathbf{p} \cdot \mathbf{v} - L_{int} - L_{par} = \mathbf{p} \cdot \mathbf{v} - e A_0 - \frac{e}{c} \mathbf{A} \cdot \mathbf{v} + \frac{m^2 c^2}{m(v)},$$

where we have used the relation

$$-L_{part} = mc^2 \sqrt{1 - \frac{v^2}{c^2}} = \frac{m^2 c^2}{m(v)}.$$

It follows:

$$H(\mathbf{p}, \mathbf{x}) = \left(\mathbf{p} - \frac{e}{c} \mathbf{A}\right) \cdot \mathbf{v} + \frac{m^2 c^2}{m(v)} + e V(\mathbf{x}). \tag{8.138}$$

We now use Eq. (8.135) to express v^i in terms of p^i:

$$\mathbf{v} = \frac{\left(\mathbf{p} - \frac{e}{c} \mathbf{A}\right)}{m(v)} = \frac{\mathbf{p}_{(0)}}{m(v)}.$$

Taking into account the relativistic relations:

$$E^2 = |\mathbf{p}_{(0)}|^2 c^2 + m^2 c^4; \qquad m(v) = E/c^2, \tag{8.139}$$

where E is the energy of the free particle, we can write:

$$H(\mathbf{p}, \mathbf{x}) = \frac{E^2}{m(v)c^2} + e V = c\sqrt{m^2 c^2 + \left|\mathbf{p} - \frac{e}{c} \mathbf{A}\right|^2} + e V(\mathbf{x}). \tag{8.140}$$

From the above equation we find:

$$(H + e A_0)^2 - c^2 \sum_{i=1}^{3} \left(p^i - \frac{e}{c} A_i\right)^2 = m^2 c^4. \tag{8.141}$$

---

[26]Recall that the vector $\mathbf{A} \equiv (A_i)$ is the spatial part of the four-vector $A_\mu \equiv (A_0, \mathbf{A})$, so that $A^\mu \equiv (A_0, -\mathbf{A})$. On the other hand $\mathbf{p}$ is the spatial component of $p^\mu \equiv (p^0, \mathbf{p})$, so that $p_\mu \equiv (p^0, -\mathbf{p})$.

Next we use the property $A^0 = A_0$, $A^i = -A_i$ to put (8.141) in relativistic invariant form:

$$\left(p^\mu + \frac{e}{c}A^\mu\right)\left(p_\mu + \frac{e}{c}A_\mu\right) = m^2c^2. \tag{8.142}$$

where we have set $\frac{H}{c} = p^0$.

Note that Eq. (8.142) can be obtained from the relativistic relation $p^\mu_{(0)}p_{(0)\mu} = m^2c^2$ of a free particle through the substitution:

$$p^\mu_{(0)} \rightarrow p^\mu + \frac{e}{c}A^\mu, \tag{8.143}$$

in agreement with Eq. (8.136). This substitution gives the correct coupling between the electromagnetic field and the charged particle and is usually referred to as *minimal coupling*.

8.7 Symmetry and the Noether Theorem

In this section we explore the connection between symmetry transformation and conservation laws in field theory.

We consider a relativistic theory described by an action of the following form:

$$S\left[\varphi^\alpha, D_4\right] = \frac{1}{c}\int_{D_4} d^4x \mathcal{L}(\varphi^\alpha, \partial_\mu\varphi^\alpha, x^\mu). \tag{8.144}$$

where $\mathcal{L}(\varphi^\alpha, \partial_\mu\varphi^\alpha, x)$ is the Lagrangian density.

We consider a generic transformation of the coordinates x^μ and of the fields φ^α:

$$\begin{aligned}
x^\mu \in D_4 &\rightarrow x'^\mu = x'^\mu(x) \in D'_4, \\
\varphi^\alpha &\rightarrow \varphi'^\alpha = \varphi'^\alpha(\varphi^\alpha, x), \\
\partial_\mu\varphi^\alpha &\rightarrow \partial'_\mu\varphi'^\alpha = \partial'_\mu\varphi'^\alpha(\varphi^\alpha, \partial_\mu\varphi^\alpha, x).
\end{aligned} \tag{8.145}$$

where $\partial'_\mu = \frac{\partial}{\partial x'^\mu}$. A transformation on space-time coordinates will in general deform the domain D_4, which we had originally taken to be a direct product of a time interval and a space volume V, into a region D'_4 with a different shape.

As already discussed in the case of a discrete set of degrees of freedom the actual value of the action computed on a generic four-dimensional domain D_4 does not depend on the set of fields and coordinates we use, since it is a *scalar*; in other words:

$$S'\left[\varphi'^\alpha; D'_4\right] = S\left[\varphi^\alpha; D_4\right], \tag{8.146}$$

or, more explicitly

$$\frac{1}{c} \int_{D_4'} d^4x' \mathcal{L}'(\varphi'^\alpha(x'), \partial_\mu' \varphi'^\alpha(x'), x') = \frac{1}{c} \int_{D_4} d^4x \mathcal{L}(\varphi^\alpha(x), \partial_\mu \varphi^\alpha(x), x), \quad (8.147)$$

where the transformed Lagrangian density $\mathcal{L}'$ in S' is given by:

$$\mathcal{L}'(\varphi'^\alpha, \partial_\mu' \varphi'^\alpha, x') = \mathcal{L}(\varphi^\alpha, \partial_\mu \varphi^\alpha, x), \quad (8.148)$$

the transformed fields and coordinates being related to the old ones by Eq. (8.145). However, as we have already emphasized in the case of a discrete system, the fact the action is a scalar does not imply that the Euler-Lagrange equations derived from S ed S' have the same *form*. The latter property holds only when the transformations (8.145) correspond to an *invariance (or symmetry)* of the system. This is the case when the action is invariant, namely when:

$$S\left[\varphi'^\alpha; D_4'\right] = S\left[\varphi^\alpha; D_4\right]. \quad (8.149)$$

Note that Eq. (8.149) implies that the Lagrangian $\mathcal{L}$ is *invariant* under the transformations (8.145) only up to the four-divergence of an arbitrary four-vector f^μ, which, as we know, does not change the equations of motion:

$$\mathcal{L}(\varphi'^\alpha(x'), \partial_\mu' \varphi'^\alpha(x'), x') = \mathcal{L}(\varphi^\alpha(x), \partial_\mu \varphi^\alpha(x), x) + \partial_\mu f^\mu, \quad (8.150)$$

where $f^\mu = f^\mu(\varphi^\alpha(x), x)$.[27]

In the sequel we shall consider transformations differing by an infinitesimal amount from the identity, to which they are connected with continuity. We write these transformations in the following form:

$$x'^\mu = x^\mu + \delta x^\mu,$$
$$\varphi'^\alpha(x) = \varphi^\alpha(x) + \delta \varphi^\alpha(x), \quad (8.151)$$

where δx^μ and $\delta \varphi^\alpha(x)$ are infinitesimals and, just as we did in Chap. 7, we define the *local* variation of the field as the difference $\delta \varphi^\alpha(x) \equiv \varphi'^\alpha(x) - \varphi^\alpha(x)$ between the transformed and the original fields evaluated in the same values of the coordinates $x = (x^\mu)$, see for instance Eq. (7.72). The invariance of the action under infinitesimal transformations is expressed by the equation

---

[27] We note that the invariance of the action means that two configurations $[\varphi^\alpha(x), x^\mu \in D_4]$ and $[\varphi'^\alpha(x'), x'^\mu \in D_4']$ related by the transformation (8.145) are solutions to the same partial differential equations.

$$c\delta S = \int_{D_4'} d^4x' \mathcal{L}(\varphi'^\alpha(x'), \partial'_\mu \varphi'^\alpha(x'), x') - \int_{D_4} d^4x \mathcal{L}(\varphi^\alpha(x), \partial_\mu \varphi^\alpha(x), x) = 0,$$

(8.152)

where, for the time being, we do not consider the contribution of a four-divergence $\partial_\mu f^\mu$ since it leads to equivalent actions.[28] The *Noether theorem* states that:

If the action of a physical system described by fields is invariant under a group of continuous global transformations, it is possible to associate with each parameter θ_r of the transformation group a four-current J_r^μ obeying the continuity equation $\partial_\mu J_r^\mu = 0$, and, correspondingly, a conserved charge Q_r, where

$$Q_r = \int d^3x \, J_r^0.$$

(8.153)

Here by *global transformations* we mean transformations whose parameters do not depend on the space-time coordinates x^μ.

The proof of the theorem requires working out the consequences of Eq. (8.152) along the same lines as for the proof of the analogous theorem for systems with a finite number of degrees of freedom. For the sake of clarity we shall give, at each step of the proof, the reference to the corresponding formulae of Sect. 8.2.1.

We begin by observing that since x' is an integration variable, we may rewrite δS as follows (cfr. (8.47)):

$$c\delta S = \int_{D_4'} d^4x \mathcal{L}(\varphi'^\alpha(x), \partial_\mu \varphi'^\alpha(x), x) - \int_{D_4} d^4x \mathcal{L}(\varphi^\alpha(x), \partial_\mu \varphi^\alpha(x), x). \quad (8.154)$$

The integration domains of the two integrals of Eq. (8.154) are D_4' and D_4 respectively. In the discrete case we had $[t_1', t_2']$ and $[t_1, t_2]$ instead of D_4' and D_4. It is then convenient to write the first integral over D_4' as the sum of an integral over D_4 and an integral over the "difference" $D_4' - D_4$ between the two domains:

$$\int_{D_4'} = \int_{D_4} + \int_{D_4' - D_4} .$$

(8.155)

The domain $D_4' - D_4$, see Fig. 8.2, can be decomposed in infinitesimal four-dimensional *hypercubes* having as basis the three-dimensional elementary volume $d^3\sigma$ on the boundary hypersurface ∂D_4 and height given by the elementary shift δx^μ of a point on $d^3\sigma$ due to the transformation (8.145). We also define $d\sigma^\mu \equiv n^\mu d^3\sigma$ as explained after Eq. (8.122).

---

[28]This freedom will be taken into account when discussing the energy momentum tensor in the next section.

Fig. 8.2 Space-time
domains D_4 and D_4'

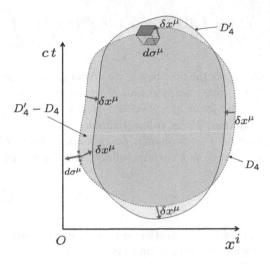

Thus we may write an elementary volume in $D_4' - D_4$ as follows:

$$d^4x = d\sigma_\mu \delta x^\mu,$$

so that the first integral on the right hand side of (8.154) reads

$$\int_{D_4'} d^4x(\cdots) = \int_{D_4} d^4x(\cdots) + \int_{D_4'-D_4} (\cdots)d^4x = \int_{D_4} d^4x(\cdots)$$

$$+ \int_{\partial D_4} d\sigma_\mu \delta x^\mu(\cdots). \tag{8.156}$$

A comparison with the analogous decomposition made in the discrete case, (8.48), reveals that ∂D_4 plays the role of the boundary of the interval $t_1 - t_2$ (consisting in that case of just two points) and δx^μ generalizes δt.

We may now insert this decomposition in Eq. (8.154) obtaining (see Eq. (8.51)):

$$c\delta S = \int_{D_4} d^4x \mathcal{L}(\varphi'^\alpha(x), \partial_\mu \varphi'^\alpha(x), x) - \int_{D_4} d^4x \mathcal{L}(\varphi^\alpha(x), \partial_\mu \varphi^\alpha(x), x)$$

$$+ \int_{\partial D_4} d\sigma_\mu \delta x^\mu \mathcal{L}(\varphi^\alpha(x), \partial_\mu \varphi^\alpha(x), x), \tag{8.157}$$

where, in the last integral, we have replaced $\mathcal{L}(\varphi'^\alpha(x), \partial_\mu \varphi'^\alpha(x), x)$ with $\mathcal{L}(\varphi^\alpha(x), \partial_\mu \varphi^\alpha(x), x)$, since their difference, being multiplied by δx^μ would have been an infinitesimal of higher order (see the analogous equation (8.49)).

On the other hand the difference between the first two integrals can be written as follows (see (8.52)):

$$\int_{D_4} d^4x \left[\mathcal{L}(\varphi'^\alpha(x), \partial_\mu \varphi'^\alpha(x), x) - \mathcal{L}(\varphi^\alpha(x), \partial_\mu \varphi^\alpha(x), x) \right]$$

$$= \int_{D_4} d^4x \left[\frac{\partial \mathcal{L}}{\partial \varphi^\alpha} \delta \varphi^\alpha + \frac{\partial \mathcal{L}}{\partial(\partial_\mu \varphi^\alpha)} \delta \partial_\mu \varphi^\alpha \right] \tag{8.158}$$

$$= \int_{D_4} d^4x \left[\frac{\partial \mathcal{L}}{\partial \varphi^\alpha} - \partial_\mu \frac{\partial \mathcal{L}}{\partial(\partial_\mu \varphi^\alpha)} \right] \delta \varphi^\alpha(x) + \int_{D_4} d^4x \partial_\mu \left(\frac{\partial \mathcal{L}}{\partial(\partial_\mu \varphi^\alpha)} \delta \varphi^\alpha \right),$$

where, as usual, we have applied the property

$$\delta(\partial_\mu \varphi^\alpha) = \partial_\mu \delta \varphi^\alpha.$$

Finally we substitute Eqs. (8.157) and (8.158) into Eq. (8.152) obtaining, for the variation of the action (see (8.53)):

$$c\delta S = \int_{D_4} d^4x \left[\frac{\partial \mathcal{L}}{\partial \varphi^\alpha} - \partial_\mu \frac{\partial \mathcal{L}}{\partial(\partial_\mu \varphi^\alpha)} \right] \delta \varphi^\alpha(x) + \int_{D_4} d^4x \partial_\mu \left(\frac{\partial \mathcal{L}}{\partial(\partial_\mu \varphi^\alpha)} \delta \varphi^\alpha \right)$$

$$+ \int_{\partial D_4} d\sigma_\mu \delta x^\mu \mathcal{L}(\varphi^\alpha(x), \partial_\mu \varphi^\alpha(x), x). \tag{8.159}$$

If the Euler-Lagrange equations (8.121) are satisfied, the first integral in Eq. (8.159) vanishes; moreover the last integral can be written as an integral on ∂D_4 by use of the four- dimensional Gauss theorem (or divergence theorem) in reverse:

$$\int_{\partial D_4} d\sigma_\mu \delta x^\mu \mathcal{L} = \int_{D_4} d^4x \partial_\mu (\delta x^\mu \mathcal{L}). \tag{8.160}$$

We have thus obtained:

$$\delta S = \frac{1}{c} \int_{D_4} d^4x \partial_\mu \left[\frac{\partial \mathcal{L}}{\partial(\partial_\mu \varphi^\alpha)} \delta \varphi^\alpha + \delta x^\mu \mathcal{L} \right]. \tag{8.161}$$

The above equation gives the desired result: It states that when $\delta S = 0$, the integral in Eq. (8.161) is zero. Taking into account that the integration domain is arbitrary, we must have:

$$\partial_\mu J^\mu = 0, \tag{8.162}$$

where:

$$J^\mu = \frac{\partial \mathcal{L}}{\partial(\partial_\mu \varphi^\alpha)} \delta\varphi^\alpha + \delta x^\mu \mathcal{L}. \tag{8.163}$$

In terms of the infinitesimal, global parameters $\delta\theta^r$, $r = 1, \ldots, g$ of the continuous transformation group G, the infinitesimal variations $\delta\varphi^\alpha$ and δx^μ can be written as:

$$\delta\varphi^\alpha = \delta\theta^r \Phi_r^\alpha; \qquad \delta x^\mu = \delta\theta^r X_r^\mu. \tag{8.164}$$

where Φ_r^α and X_r^μ are, in general, functions of the fields φ^α and coordinates x^μ. Thus we may write:

$$J^\mu = \delta\theta^r J_r^\mu,$$

where

$$J_r^\mu = \left(\frac{\partial \mathcal{L}}{\partial(\partial_\mu \varphi^\alpha)} \Phi_r^\alpha + X_r^\mu \mathcal{L} \right). \tag{8.165}$$

Taking into account that the $\delta\theta^r$ are independent, constant parameters, we can state that we have a set of g conserved currents $\partial_\mu J_r^\mu = 0$. To each conserved current J_r^μ there corresponds a conserved charge Q_r:

$$Q_r = \int_{\mathbb{R}^3} d^3\mathbf{x} J_r^0, \tag{8.166}$$

where we take as V the entire three-dimensional space $\mathbb{R}^3$. Indeed:

$$\frac{dQ_r}{dt} = c \int_{\mathbb{R}^3} d^3\mathbf{x} \frac{\partial}{\partial x^0} J_r^0 = -\int_{\mathbb{R}^3} d^3\mathbf{x} \frac{\partial}{\partial x^i} J_r^i = -\int_{S_\infty} d^2\sigma \sum_{i=1}^{3} J_r^i n^i = 0.$$

where the last surface integral is zero being evaluated at infinity where the currents are supposed to vanish.

8.8 Space-Time Symmetries

As already stressed in the first Chapter of this book, in order to satisfy the principle of relativity a physical theory must fulfil the requirement of *invariance under the Poincaré group*. The latter was discussed in detail in Chap. 4 and contains, as subgroups, the Lorentz group and the four-dimensional translation group. Invariance of a theory, describing an isolated system of fields, under Poincaré transformations implies that its predictions cannot depend on a particular direction or on a specific space-time region in which we observe the system, consistently with our assumption of *homogeneity and isotropy* of Minkowski space.

The Noether theorem allows us to derive conservation laws as a consequence of this invariance. Let us first work out the conserved charges associated with the invariance of the theory under *space-time translations*:

$$x^\mu \to x'^\mu = x^\mu - \epsilon^\mu \longrightarrow \delta x^\mu = -\epsilon^\mu, \tag{8.167}$$

$$\varphi'^\alpha(x - \epsilon) = \varphi^\alpha(x) \Rightarrow \delta\varphi^\alpha(x) = \varphi'^\alpha(x) - \varphi^\alpha(x) = \frac{\partial\varphi^\alpha(x)}{\partial x^\mu}\epsilon^\mu. \tag{8.168}$$

Comparing this with the general formula (8.164) we can identify the index r with the space-time index ν, the parameters $\delta\theta^r$ with ϵ^μ so that

$$\Phi_r^\alpha = \Phi_\nu^\alpha = \partial_\mu\varphi^\alpha\delta_\nu^\mu; \qquad X_r^\mu = X_\nu^\mu = -\delta_\nu^\mu.$$

Requiring invariance of the action under the transformations (8.167)–(8.168), and inserting the values of δx^μ and $\delta\varphi^\alpha$ in the general expression of the current (8.163) we obtain:

$$J^\mu = J_\rho^\mu\epsilon^\rho = \left(\frac{\partial\mathcal{L}}{\partial(\partial_\mu\varphi^\alpha)}\partial_\rho\varphi^\alpha - \delta_\rho^\mu\mathcal{L}\right)\epsilon^\rho \equiv c\,\epsilon^\rho\,T^\mu{}_\rho \tag{8.169}$$

where we have introduced the *energy-momentum*[29] tensor $T_{\mu|\rho}$:

$$T_{\mu|\rho} \equiv \frac{1}{c}\left[\frac{\partial\mathcal{L}}{\partial(\partial^\mu\varphi^\alpha)}\partial_\rho\varphi^\alpha - \eta_{\mu\rho}\mathcal{L}\right], \tag{8.170}$$

so that we have the general conservation law

$$\partial_\mu T^\mu{}_{|\nu} = 0. \tag{8.171}$$

We note that both the indices of $T_{\mu|\nu}$ are Lorentz indices, but we have separated them by a bar since the first index is the index of the four-current while the second index is the index r labeling the parameters. This being understood, in the following we suppress the bar between the two indices of $T_{\mu|\nu}$.

The four Noether charges associated with the space-time translations are obtained by integration of $J_\mu^0 \equiv cT^0{}_\mu$ over the whole three-dimensional space:

$$Q_\mu = c\int d^3\mathbf{x}\, T_{0\mu} \doteq c\,P_\mu, \tag{8.172}$$

and, from the Noether-conservation law $\partial_\nu T^{\nu\mu} = 0$, we obtain in the usual way that:

$$\frac{d}{dt}P^\mu = 0.$$

---

[29]The alternative name of *stress-energy tensor* is also used.

To understand the physical meaning of the energy-momentum tensor and of the conserved four-vector

$$P^\mu \equiv \int_V d^3\mathbf{x}\,\mathscr{P}^\mu \equiv \int_V d^3\mathbf{x}\,T^{0\mu}, \tag{8.173}$$

where we have defined $\mathscr{P}^\mu \equiv T^{0\mu}$, we recall that in the case of systems with a finite number of degrees of freedom, the conserved four charges associated with space-time translations are the components of the four-momentum. It is natural then to interpret P^μ as the *total conserved four-momentum* associated with the continuous system under consideration, described by the fields $\varphi^\alpha(x)$.

As a consequence of this the tensor $T_{\mu\nu}$ can be thought of as describing the *density of energy and momentum and their currents in space and time*. In particular, $\mathscr{P}^\mu \equiv T^{0\mu}$ represents the *spatial density of the four-momentum*. We conclude that the *four conserved charges Q_μ/c associated with the space-time translations, which are an invariance of an isolated system, are the components of the total four-momentum.*

Let us now consider the further six conserved charges associated with the invariance with respect to Lorentz transformations.

Under such a transformation, the fields φ^α will transform according to the $SO(1,3)$ representation, labeled by the index α, which they belong to; its infinitesimal form has being given in Eq. (7.83), namely:

$$\delta\varphi^\alpha = \frac{1}{2}\delta\theta^{\rho\sigma}\left[(L_{\rho\sigma})^\alpha_{\ \beta}\varphi^\beta + (x_\rho\partial_\sigma - x_\sigma\partial_\rho)\varphi^\alpha\right]. \tag{8.174}$$

If the action is invariant under the Lorentz group, substitution of the variations (8.174) and (7.82) into Eq. (8.163) gives the following conserved current:

$$J_\mu = -\frac{c}{2}\delta\theta^{\rho\sigma}\mathcal{M}_{\mu|\rho\sigma}, \tag{8.175}$$

where we have introduced the tensor:

$$\mathcal{M}_{\mu|\rho\sigma} = -\frac{1}{c}\left[\frac{\partial\mathcal{L}}{\partial(\partial^\mu\varphi^\alpha)}\left((L_{\rho\sigma})^\alpha_{\ \beta}\varphi^\beta + (x_\rho\partial_\sigma - x_\sigma\partial_\rho)\varphi^\alpha\right)\right.$$
$$\left. + \left(x_\sigma\eta_{\mu\rho} - x_\rho\eta_{\mu\sigma}\right)\mathcal{L}\right], \tag{8.176}$$

and used the identification of the index r with the antisymmetric couple of indices $(\mu\nu)$ labeling the Lorentz generators, so that

$$X^\mu_r \equiv X^\mu_{\rho\sigma} = \delta^\mu_\rho x_\sigma - \delta^\mu_\sigma x_\rho,$$
$$\Phi^\alpha_r \equiv \Phi^\alpha_{\rho\sigma} = (L_{\rho\sigma})^\alpha_{\ \beta}\varphi^\beta + (x_\rho\partial_\sigma - x_\sigma\partial_\rho)\varphi^\alpha. \tag{8.177}$$

Comparing (8.176) with the definition of the *energy-momentum tensor* $T_{\mu\rho}$, Eq. (8.170), little insight reveals that the two terms proportional to x_μ within square brackets in the former can be expressed in terms of $T_{\mu\rho}$ as $x_\sigma T_{\mu\rho} - x_\rho T_{\mu\sigma}$. Therefore the conserved current $\mathcal{M}_{\mu|\rho\sigma}$ takes the simpler form:

$$\mathcal{M}_{\mu|\rho\sigma} = -\left[\frac{1}{c}\frac{\partial\mathcal{L}}{\partial(\partial^\mu\varphi^\alpha)}(L_{\rho\sigma})^\alpha{}_\beta\varphi^\beta + (x_\rho T_{\mu\sigma} - x_\sigma T_{\mu\rho})\right]. \qquad (8.178)$$

Being this current associated with Lorentz transformations which are always a symmetry of a relativistic theory, the Noether theorem implies:

$$\partial_\mu\mathcal{M}^\mu{}_{\rho\sigma} = 0. \qquad (8.179)$$

Using the explicit form (8.178) in the conservation law (8.179) together with (8.171), we derive the following two equations:

$$\partial^\mu H_{\mu\rho\sigma} + T_{\rho\sigma} - T_{\sigma\rho} = 0, \qquad (8.180)$$
$$\partial^\mu T_{\mu\nu} = 0. \qquad (8.181)$$

where we have set

$$H_{\mu\rho\sigma} \equiv \frac{1}{c}\frac{\partial\mathcal{L}}{\partial(\partial^\mu\varphi^\alpha)}(L_{\rho\sigma})^\alpha{}_\beta\varphi^\beta. \qquad (8.182)$$

We have encountered instances of the energy-momentum tensor earlier in this book: On the right hand side of the Einstein equation (3.86) as the source of space-time curvature and in Chap. 5 in relation to a system of charged particles and to the electromagnetic field, see Eqs. (5.61) and (5.71), respectively. In all these cases it described the matter-energy distribution in space and its currents and was introduced as a *symmetric matrix*. This is not the case for the quantity $T^{\mu\nu}$ defined in Eq. (8.170), which, in general, *is not a symmetric tensor*. In fact its antisymmetric part is related in equation (8.180) to the divergence $\partial^\mu H_{\mu\rho\sigma}$ of $H_{\mu\rho\sigma}$, which can be non-vanishing whenever the field φ carries a Lorentz-representation index α, related to its internal degrees of freedom. As we show below, it is possible to define a symmetric tensor $\Theta^{\mu\nu}$ having the same physical content as $T^{\mu\nu}$. This is the energy-momentum tensor we have introduced in our earlier discussions (Eqs. (3.86), (5.61) and (5.71)), as we shall show at the end of this section for the case of the electromagnetic field. In terms of the symmetric energy-momentum tensor $\Theta^{\mu\nu}$, as we illustrate shortly, the conserved currents associated with the Lorentz generators will also acquire a simpler expression which is analogous to that of angular momentum in terms of the position vector and the linear momentum.

We first notice that the definition (8.170) does not determine the energy-momentum tensor uniquely. If we indeed redefine $T^{\mu\nu}$ as:

$$T^{\mu\nu} \rightarrow T^{\mu\nu} + \partial_\rho U^{\nu\mu\rho} \qquad U^{\nu\mu\rho} = -U^{\nu\rho\mu} \qquad (8.183)$$

it still satisfies the conservation law, since $\partial_\mu \partial_\rho U^{\nu\mu\rho} \equiv 0$, due to the antisymmetry of $U^{\nu\mu\rho}$ in its last two indices. This possibility is related to the freedom we have of adding to the Lagrangian a four-divergence $\partial_\mu f^\mu$. Although we had neglected such freedom when proving the Noether theorem, one can exploit it to obtain a *symmetric energy-momentum tensor*.[30]

To show this, let us perform the following redefinitions:

$$\Theta_{\mu\nu} = T_{\mu\nu} + \partial^\lambda U_{\nu\mu\lambda}; \qquad U_{\nu\mu\lambda} = -U_{\nu\lambda\mu}, \qquad (8.184)$$

$$\hat{\mathcal{M}}_{\mu|\rho\sigma} = \mathcal{M}_{\mu\rho\sigma} - \partial^\lambda \left(x_\rho U_{\sigma\mu\lambda} - x_\sigma U_{\rho\mu\lambda} \right). \qquad (8.185)$$

where $\Theta_{\mu\nu}$ is the new energy momentum tensor. As already remarked these redefinitions do not spoil the conservation law associated with the energy-momentum tensor, since, due to the antisymmetry of $U_{\mu\nu\lambda}$ in the last two indices, we still have $\partial^\mu \Theta_{\mu\nu} = 0$. Moreover, by the same token, it is easily shown that $\hat{\mathcal{M}}_{\mu|\rho\sigma}$ is still conserved, i.e. $\partial^\mu \hat{\mathcal{M}}_{\mu|\rho\sigma} = 0$, since taking into account (8.179) the additional term $\partial^\lambda \left(x_\rho U_{\sigma\mu\lambda} - x_\sigma U_{\rho\mu\lambda} \right)$ is divergenceless by virtue of the antisymmetry of $U_{\sigma\mu\lambda}$ in its last two indices:

$$\partial^\mu \partial^\lambda \left(x_\rho U_{\sigma\mu\lambda} - x_\sigma U_{\rho\mu\lambda} \right) = 0. \qquad (8.186)$$

Let us now show that $\hat{\mathcal{M}}_{\mu\rho\sigma}$ can be written in the simpler form:

$$\hat{\mathcal{M}}_{\mu|\rho\sigma} = -x_\rho \Theta_{\mu\sigma} + x_\sigma \Theta_{\mu\rho}, \qquad (8.187)$$

by a suitable choice of $U_{\sigma\mu\lambda}$. If we prove this, then, from the conservation of the current $\hat{\mathcal{M}}_{\mu|\rho\sigma}$ we have

$$0 = \partial^\mu \hat{\mathcal{M}}_{\mu\rho\sigma} = -\delta^\mu_\rho \Theta_{\mu\sigma} + \delta^\mu_\sigma \Theta_{\mu\rho} = -\Theta_{\rho\sigma} + \Theta_{\sigma\rho}, \qquad (8.188)$$

which implies that $\Theta_{\mu\nu}$ is symmetric. To prove Eq. (8.187) we first write the explicit form of $\hat{\mathcal{M}}_{\mu\rho\sigma}$ by expressing $T_{\mu\nu}$ in Eq. (8.178) in terms of $\Theta_{\mu\nu}$ and use the following identity:

$$-x_\rho \partial^\lambda U_{\sigma\mu\lambda} + x_\sigma \partial^\lambda U_{\rho\mu\lambda} = -\partial^\lambda \left(x_\rho U_{\sigma\mu\lambda} - x_\sigma U_{\rho\mu\lambda} \right) - U_{\rho\mu\sigma} + U_{\sigma\mu\rho}.$$

The four-divergence on the right hand side cancels against the opposite term in (8.185) and we end up with:

$$\hat{\mathcal{M}}_{\mu|\rho\sigma} = -H_{\mu\rho\sigma} - x_\rho \Theta_{\mu\sigma} + x_\sigma \Theta_{\mu\rho} - U_{\sigma\mu\rho} + U_{\rho\mu\sigma}. \qquad (8.189)$$

---

[30]We shall illustrate an application of this mechanism to the case of the electromagnetic field at the end of this section.

Thus in order for $\hat{\mathcal{M}}$ to have the form (8.187) we need to find a tensor $U_{\nu\mu\lambda}$ satisfying the following condition:

$$U_{\rho\mu\sigma} - U_{\sigma\mu\rho} = \frac{1}{c}\frac{\partial \mathcal{L}}{\partial \partial^\mu \varphi^\alpha}(L_{\rho\sigma})^\alpha{}_\beta \varphi^\beta \equiv H_{\mu\rho\sigma}. \tag{8.190}$$

The solution is[31]

$$U_{\mu\rho\sigma} = \frac{1}{2}\left[H_{\mu\rho\sigma} - H_{\sigma\mu\rho} - H_{\rho\sigma\mu}\right]. \tag{8.191}$$

Let us now discuss the physical meaning of the conservation law (8.179), by computing the conserved "charges" $Q_{\rho\sigma}$ associated with the 0-component of the current $\mathcal{M}_{\mu|\rho\sigma}$ (8.188). Let us rename $Q_{\rho\sigma} \to J_{\rho\sigma}$, since, as we shall presently see, they are related to the angular momentum. Then, integrating over the whole space $V = \mathbb{R}^3$:

$$J_{\rho\sigma} = \int_V d^3\mathbf{x}\, M_{0\rho\sigma} = -\int_V d^3\mathbf{x}\left[\frac{\partial \mathcal{L}}{\partial \dot{\varphi}^\alpha}(L_{\rho\sigma})^\alpha{}_\beta \varphi^\beta + (x_\rho T_{0\sigma} - x_\sigma T_{0\rho})\right], \tag{8.192}$$

are the conserved charged associated with Lorentz invariance:

$$\frac{d}{dt}J^{\rho\sigma} = 0. \tag{8.193}$$

In particular for spatial indices $(\mu\nu) = (ij)$ we find:

$$J_{ij} = -\int_V d^3\mathbf{x}\left[\frac{\partial \mathcal{L}}{\partial \dot{\varphi}^\alpha}(L_{ij})^\alpha{}_\beta \varphi^\beta + (x_i \mathscr{P}_j - x_j \mathscr{P}_i)\right] = -\epsilon_{ijk}J^k, \tag{8.194}$$

where $\mathscr{P}^i$ is the momentum density.

Let us first consider the case of a *scalar field* φ which, by definition, does not have internal components transforming under Lorentz transformations, so that the first term of Eq. (8.194) is absent. The second term in the integrand of Eq. (8.194) is easily recognized as the *density of orbital angular momentum*. Therefore $J_{ij} \equiv -\epsilon_{ijk}M^k$ is the conserved orbital angular momentum, which, for a scalar field, coincides with the total angular momentum.

If, however, we have a field φ^α transforming, through the index α, in a non-trivial representation of the Lorentz group, the first term in Eq. (8.178) is not zero; it is clear that it should also describe an angular momentum which must then refer to the

---

[31] The solution (8.191) can be obtained by writing, besides Eq. (8.190), two analogous equations obtained by cyclic permutation of the indices $\rho\mu\sigma$. Subtracting the last two equations from the first and using the antisymmetry property (8.183) we find (8.191).

intrinsic degrees of freedom of the field.[32] In fact the first term describes the *intrinsic angular momentum or spin* of the field.

In general if the field is not spinless the conservation law implies that only the sum of the orbital angular momentum and of the spin, that is only the *total angular momentum* is conserved.

Note that so far we have been discussing the conservation of the three charges J_{ij} associated with the invariance under three dimensional rotations and corresponding to the components of the total angular momentum. It is interesting to understand the meaning of the other three conservation laws (8.179) associated with the invariance under Lorentz boosts, that is with the components J^{0i} of the $J^{\mu\nu}$ charges. Restricting for the sake of simplicity to the case of a scalar field, we have from (8.192) and (8.193), setting $(L_{ij})^\alpha{}_\beta = 0$,

$$\frac{d}{dt}\left[x^0 \int d^3\mathbf{x} T^{0i} - \int x^i T^{00} d^3\mathbf{x}\right] = 0. \tag{8.195}$$

Taking into account the conservation of P^i, defined by the first integral, we obtain:

$$c P^i = \frac{d}{dt}\int x^i T^{00} d^3\mathbf{x}. \tag{8.196}$$

On the other hand since $c T^{00}$ represents the energy density, we have $T^{00} d^3\mathbf{x} = dE/c = cdm$, where E is the total energy related to the total mass by the familiar relation $E = mc^2$. It then follows:

$$\mathbf{P} = \frac{d}{dt}\int \mathbf{x} dm. \tag{8.197}$$

In words: *The conservation law associated with the Lorentz boosts implies that the relativistic center of mass moves at constant velocity.*

Let us end this section by proving that, in the case of a free electromagnetic field, the tensor $\Theta^{\mu\nu}$ coincides with the symmetric energy-momentum tensor $T_{em}^{\mu\nu}$ defined in (5.71). To this end let us compute, by using Eq. (8.170), the Noether currents $T_{\mu\nu}$ associated with the invariance of the action $\mathcal{S}_{em}$ of a free electromagnetic field under space-time translations. From Eq. (8.129), with $J_\mu = 0$, and Eq. (8.170) we find:

$$\begin{aligned} T_{\mu\rho} &= \frac{1}{c}\left(\frac{\partial \mathcal{L}_{em}}{\partial(\partial^\mu A_\sigma)}\partial_\rho A_\sigma - \eta_{\mu\rho}\mathcal{L}_{em}\right) \\ &= \frac{1}{c}\left(-\partial_\mu A_\sigma \partial_\rho A^\sigma + \partial_\sigma A_\mu \partial_\rho A^\sigma + \frac{1}{4}\eta_{\mu\rho} F_{\delta\xi}F^{\delta\xi}\right). \end{aligned} \tag{8.198}$$

---

[32]Recall from Chap. 4 that, since $L_{\rho\sigma}$ are Lorentz generators, $L_i = -\epsilon_{ijk}L_{jk}/2$ are generators of the rotation group.

Notice that this tensor is not symmetric: $T_{\mu\rho} \neq T_{\rho\mu}$. Next we evaluate $H_{\mu\nu\rho}$, defined in (8.182), using the explicit form (4.170) of the infinitesimal Lorentz generators on the internal index μ of the field A_μ:

$$H_{\mu\rho\sigma} = \frac{1}{c} \left(-A_\sigma \partial_\mu A_\rho + A_\rho \partial_\mu A_\sigma + A_\sigma \partial_\rho A_\mu - A_\rho \partial_\sigma A_\mu \right). \tag{8.199}$$

From Eq. (8.191) we can now compute the tensor $U_{\rho\mu\sigma}$:

$$U_{\rho\mu\sigma} = \frac{1}{c} \left(A_\rho \partial_\mu A_\sigma - A_\rho \partial_\sigma A_\mu \right) = -U_{\rho\sigma\mu}. \tag{8.200}$$

Finally, using (8.184) we evaluate $\Theta_{\mu\nu}$ to be

$$\Theta_{\mu\nu} = T_{\mu\nu} + \partial^\lambda U_{\nu\mu\lambda} = -\frac{1}{c} \left(F_{\mu\sigma} F_\nu{}^\sigma - \frac{1}{4} \eta_{\mu\nu} F_{\rho\sigma} F^{\rho\sigma} \right). \tag{8.201}$$

As expected the symmetric tensor $\Theta^{\mu\nu}$ coincides with the definition of the energy-momentum tensor $T_{em}^{\mu\nu}$ for the electromagnetic field, given in (5.71).

8.8.1 Internal Symmetries

The symmetries and the associated conserved charges discussed in the previous section are space-time symmetries, namely symmetries associated with translations and Lorentz transformations under which, in a relativistic theory, the action is invariant.

We now want to give an example of a symmetry which does not involve changes in the space-time coordinates x^μ, but that is rather implemented by transformations acting on the internal index α of a field $\varphi^\alpha(x)$. In this case the index α labels the basis of a representation of the corresponding symmetry group G. Such symmetries are called *internal symmetries* and G is the internal symmetry group:

$$x^\mu \to x'^\mu = x^\mu \implies \delta x^\mu = 0,$$
$$\varphi^\alpha(x) \to \varphi'^\alpha(x) = \varphi^\alpha(x) + \delta\varphi^\alpha(x), \tag{8.202}$$

where

$$\delta\varphi^\alpha = \delta\theta^r (L_r)^\alpha{}_\beta \varphi^\beta \qquad \delta\theta^r \ll 1.$$

From Eq. (8.163) the conserved currents have the simpler form:

$$\delta\theta^r J_r^\mu = \frac{\partial \mathcal{L}}{\partial(\partial_\mu \varphi^\alpha)} \delta\varphi^\alpha. \tag{8.203}$$

The simplest, albeit important, example is the case in which we have *two* real scalar fields φ_1, φ_2, or, equivalently, a complex scalar field φ, φ^*, the two descriptions being related by:

$$\varphi_1 = \frac{1}{\sqrt{2}}(\varphi + \varphi^*) \; ; \quad \varphi_2 = -\frac{i}{\sqrt{2}}(\varphi - \varphi^*),$$

and a Lagrangian density of the following form:

$$\mathcal{L} = c^2 \left(\partial_\mu \varphi^* \partial^\mu \varphi - \frac{m^2 c^2}{\hbar^2} \varphi^* \varphi \right). \tag{8.204}$$

The Euler-Lagrangian equations are:

$$\hbar^2 \partial_\mu \partial^\mu \varphi + m^2 c^2 \varphi = 0 \tag{8.205}$$

and its complex conjugate.

As will be shown in the next chapter this equation is the natural *relativistic* extension of the Schrödinger equation for a particle of mass m and wave function φ. It is referred to as the *Klein-Gordon equation*, and the Lagrangian (8.204) is the *Klein-Gordon Lagrangian density*.

We observe that the Lagrangian density $\mathcal{L}$ of Eq. (8.204) is invariant under the following transformation:

$$\varphi(x) \rightarrow \varphi'(x) = e^{-i\alpha} \varphi(x) \tag{8.206}$$

where α is a constant parameter.

In the real basis, the transformation belongs to the group SO(2):

$$\begin{pmatrix} \varphi_1' \\ \varphi_2' \end{pmatrix} = \begin{pmatrix} \cos \alpha & \sin \alpha \\ -\sin \alpha & \cos \alpha \end{pmatrix} \begin{pmatrix} \varphi_1 \\ \varphi_2 \end{pmatrix}. \tag{8.207}$$

In the complex basis the transformation (8.206) defines a one-parameter Lie group of unitary transformations denoted by U(1), which is isomorphic to, i.e. has the same structure as, SO(2).

The infinitesimal version of Eq. (8.206) is:

$$\varphi(x) \rightarrow \varphi'(x) \simeq \varphi(x) - i\alpha\varphi(x) \Rightarrow \delta\varphi(x) = -i\alpha\varphi(x); \; \delta\varphi^* = i\alpha\varphi^*.$$

Using a suitable multiplicative coefficient to normalize the conserved current J^μ to the dimension of the electric current, we obtain from (8.203):

$$\alpha J^\mu = \frac{e}{c\hbar} \left[\frac{\partial \mathcal{L}}{\partial(\partial_\mu \varphi)} \delta\varphi + \frac{\partial \mathcal{L}}{\partial(\partial_\mu \varphi^*)} \delta\varphi^* \right] = -i\frac{ec}{\hbar} \left[\varphi \partial^\mu \varphi^* - \varphi^* \partial^\mu \varphi \right] \alpha. \tag{8.208}$$

Let us verify the conservation law $\partial_\mu J^\mu = 0$ explicitly:

$$i\frac{\hbar}{ec}\partial_\mu J^\mu = \partial_\mu \varphi \partial^\mu \varphi^* + \varphi \partial_\mu \partial^\mu \varphi^* - \partial_\mu \varphi^* \partial^\mu \varphi - \varphi^* \partial_\mu \partial^\mu \varphi$$

$$= \frac{m^2 c^2}{\hbar^2} \varphi \varphi^* - \frac{m^2 c^2}{\hbar^2} \varphi \varphi^* = 0$$

where we have used the equation of motion (8.205). We shall see in Chap. 10 that the conserved charge:

$$Q = \int d^3x J^0 = i\frac{e}{\hbar} \int d^3\mathbf{x}(\varphi^* \partial_t \varphi - \varphi \partial_t \varphi^*) \tag{8.209}$$

can be identified with *electric charge* of a scalar field φ interacting with the electromagnetic field.

Let us note that if the field were real, $\varphi(x) = \varphi^*(x)$, that is if we had just one field, there would be no invariance of the Lagrangian and the charge Q would be zero. As it will be shown in the sequel, this is a general feature: when a field is interpreted as the wave function of a particle, a *real* field describes a neutral particle, as it happens for the photon field $A_\mu(x) = A_\mu^*(x)$, while fields associated with charged particles are intrinsically *complex*.

8.9 Hamiltonian Formalism in Field Theory

In the previous section we have described systems with a continuum of degrees of freedom using the Lagrangian formalism. We want now to discuss the dynamics of such systems using the Hamiltonian formalism.

The most direct way to derive the Hamiltonian description of field dynamics is to use the limiting procedure discussed in Sect. 8.5.1 for the Lagrangian formalism.

Consider a theory describing a field $\varphi(x)$ (let us suppress the internal index α for the time being). Just as we did in Sect. 8.5.1, we divide the 3-dimensional domain V in which we study the system, into tiny cells of volume δV^i, defining the Lagrangian coordinates $\varphi_i(t)$ as the mean value of $\varphi(\mathbf{x}, t)$ within the ith cell. We thus have a discrete dynamical system and define the momenta p_i conjugate to φ_i as

$$p_i = \frac{\partial L(t)}{\partial \dot{\varphi}_i(t)}. \tag{8.210}$$

The Hamiltonian of the system is given by:

$$H = \sum_i p_i \dot{\varphi}_i - L, \tag{8.211}$$

with equations of motion:

$$\dot{\varphi}_i = \frac{\partial H}{\partial p_i}; \qquad \dot{p}_i = -\frac{\partial H}{\partial \varphi_i}. \tag{8.212}$$

Recall now from the discussion in Sect. 8.5.1 that, in the continuum limit (δV_i infinitesimal)

$$p_i(t) = \frac{\partial L}{\partial \dot{\varphi}_i(t)} = \delta V_i \frac{\delta L}{\delta \dot{\varphi}(\mathbf{x}, t)} = \delta V_i \frac{\partial \mathcal{L}(\mathbf{x}, t)}{\partial \dot{\varphi}(\mathbf{x}, t)} = \delta V_i \, \pi(\mathbf{x}, t), \tag{8.213}$$

where $\mathbf{x} \in \delta V_i$ and we have defined the field $\pi(\mathbf{x}, t)$ as:

$$\pi(\mathbf{x}, t) \equiv \frac{\partial \mathcal{L}(\mathbf{x}, t)}{\partial \dot{\varphi}(\mathbf{x}, t)}, \tag{8.214}$$

so that $p_i(t)$ represents the mean value of $\pi(\mathbf{x}, t)$ within the ith cell δV_i (multiplied by δV_i). The field $\pi(x)$ is the *momentum conjugate to the* $\varphi(x)$. Expressing $p_i(t)$ in terms of $\pi(x)$ through Eq. (8.213), upon identifying in the continuum limit $\delta V_i = d^3\mathbf{x}$, we may write the Hamiltonian (8.211) as:

$$H = \int_V [\pi(x)\dot{\varphi}(x) - \mathcal{L}(x)]d^3\mathbf{x}, \tag{8.215}$$

where we have used the definition (8.110) of Lagrangian density. The integrand in the above equation

$$\mathcal{H} = \pi(x)\dot{\varphi}(x) - \mathcal{L}(x), \tag{8.216}$$

defines the *Hamiltonian density*. Using the notion of functional derivative, the Hamilton equations of motion can be derived in a way analogous to Eq. (8.106):

$$\frac{\delta H(t)}{\delta \varphi(\mathbf{x}, t)} = \lim_{\delta V^i \to 0} \frac{1}{\delta V_i} \frac{\partial H(t)}{\partial \varphi_i(t)},$$

$$\frac{\delta H(t)}{\delta \pi(\mathbf{x}, t)} = \lim_{\delta V^i \to 0} \frac{1}{\delta V_i} \frac{\partial H(t)}{\partial \pi_i(t)}, \tag{8.217}$$

and combining Eqs. (8.106), (8.217) with Eqs. (8.212) and (8.213) we obtain:

$$\dot{\pi}(x) = -\frac{\delta H(t)}{\delta \varphi(x)} = -\frac{\partial \mathcal{H}(x)}{\partial \varphi(x)}, \tag{8.218}$$

$$\dot{\varphi}(x) = \frac{\delta H(t)}{\delta \pi(\mathbf{x}, t)} = \frac{\partial \mathcal{H}(x)}{\partial \pi(x)}. \tag{8.219}$$

Using the same limiting procedure one can see that the Poisson brackets of two functionals $F[\varphi, \pi]$, $G[\varphi, \pi]$ is defined as:

$$\{F, G\} = \int\limits_V \left(\frac{\delta F}{\delta\varphi(x)} \frac{\delta G}{\delta\pi(x)} - \frac{\delta F}{\delta\pi(x)} \frac{\delta G}{\delta\varphi(x)} \right) d^3x, \qquad (8.220)$$

so that, the time derivative of F gives:

$$\dot{F}(t) = \frac{\partial F}{\partial t} + \int\limits_V \left(\frac{\delta F}{\delta\varphi(x)} \dot{\varphi}(x) + \frac{\delta F}{\delta\pi(x)} \dot{\pi}(x) \right) d^3x \qquad (8.221)$$

$$= \frac{\partial F}{\partial t} + \int\limits_V \left(\frac{\delta F}{\delta\varphi(x)} \frac{\delta H}{\delta\pi(x)} - \frac{\delta F}{\delta\pi(x)} \frac{\delta H}{\delta\varphi(x)} \right) d^3x$$

$$= \frac{\partial F}{\partial t} + \{F, H\},$$

where we have used Eqs. (8.218 and 8.219).

In particular, if F does not have an explicit dependence on time:

$$\dot{F}(t) = \{F, H\}.$$

In this case the dynamical variable F is conserved *if and only if its Poisson bracket with the Hamiltonian vanishes.*

Writing:

$$\varphi(\mathbf{x}, t) = \int\limits_V \delta^3(\mathbf{x} - \mathbf{x}')\varphi(\mathbf{x}', t) d^3x$$

$$\pi(\mathbf{x}, t) = \int\limits_V \delta^3(\mathbf{x} - \mathbf{x}')\pi(\mathbf{x}', t) d^3x$$

from the definition of *functional derivative* we have:

$$\frac{\delta\varphi(\mathbf{x}, t)}{\delta\varphi(\mathbf{x}', t)} = \frac{\delta\pi(\mathbf{x}, t)}{\delta\pi(\mathbf{x}', t)} = \delta^3(\mathbf{x} - \mathbf{x}'). \qquad (8.222)$$

Applying this relations we find:

$$\{\varphi(\mathbf{x}, t), H\} = \frac{\delta H}{\delta\pi(\mathbf{x}, t)} \qquad (8.223)$$

$$\{\pi(\mathbf{x}, t), H\} = -\frac{\delta H}{\delta\varphi(\mathbf{x}, t)}, \qquad (8.224)$$

and using the Hamilton equations (8.218–8.219):

$$\dot{\varphi}(\mathbf{x}, t) = \{\varphi(\mathbf{x}, t), H\}, \tag{8.225}$$

$$\dot{\pi}(\mathbf{x}, t) = \{\pi(\mathbf{x}, t), H\}. \tag{8.226}$$

From Eq. (8.222) we also derive the fundamental relations:

$$\{\varphi(\mathbf{x}, t), \pi(\mathbf{x}', t)\} = \delta^3(\mathbf{x} - \mathbf{x}'), \tag{8.227}$$

$$\{\varphi(\mathbf{x}, t), \varphi(\mathbf{x}', t)\} = \{\pi(\mathbf{x}, t), \pi(\mathbf{x}', t)\} = 0. \tag{8.228}$$

In order to simplify notation, we have developed the Hamilton formalism using just one field. If we have several fields in some non trivial representation of a group G, we need an additional index α. The extension of the previous formalism to several fields is, however, straightforward. For example the momenta conjugate to the fields are defined as:

$$\pi_\alpha(x) \equiv \frac{\partial \mathcal{L}(x)}{\partial \dot{\varphi}^\alpha(x)}. \tag{8.229}$$

Similarly, in defining the Poisson brackets, we need, besides the integration on the $\mathbf{x}$ variable, also a sum over the index α:

$$\{F, G\} = \sum_\alpha \int_V \left(\frac{\delta F}{\delta \varphi^\alpha(x)} \frac{\delta G}{\delta \pi_\alpha(x)} - \frac{\delta F}{\delta \pi_\alpha(x)} \frac{\delta G}{\delta \varphi^\alpha(x)} \right) d^3\mathbf{x}. \tag{8.230}$$

Furthermore, the relations (8.227) and (8.228) generalize as follows:

$$\{\varphi^\alpha(\mathbf{x}, t), \pi_\beta(\mathbf{x}', t)\} = \delta^\alpha_\beta \delta^3(\mathbf{x} - \mathbf{x}'), \tag{8.231}$$

$$\{\varphi^\alpha(\mathbf{x}, t), \varphi^\beta(\mathbf{x}', t)\} = \{\pi_\alpha(\mathbf{x}, t), \pi_\beta(\mathbf{x}', t)\} = 0. \tag{8.232}$$

An important case is that of two real scalar fields φ^1, φ^2 which, as shown in Sect. 8.8.1, is equivalent to a single complex scalar field and its complex conjugate. In this case, using the real notation we have indices $\alpha, \beta = 1, 2$. If however, as we shall mostly do in the next chapters, we use the complex scalar fields $\varphi(\mathbf{x}, t), \varphi^*(\mathbf{x}, t)$, then the Poisson brackets (8.231) become

$$\{\varphi(\mathbf{x}, t), \pi(\mathbf{y}, t)\} = \delta^3(\mathbf{x} - \mathbf{y}), \tag{8.233}$$

$$\{\varphi^*(\mathbf{x}, t), \pi^*(\mathbf{y}, t)\} = \delta^3(\mathbf{x} - \mathbf{y}), \tag{8.234}$$

all the other Poisson brackets being zero.

8.9.1 Symmetry Generators in Field Theories

We have seen in Sect. 8.4.1 that the infinitesimal generators of continuous canonical transformations $\delta\theta^r G_r(t)$ *generate* transformations $\delta q_i, \delta p_i$ leaving the Hamilton equations in the standard form (8.67)–(8.218). Moreover, when the Hamiltonian is left *invariant*, $H' = H$, for each parameter θ^r associated with the continuous symmetry group G, the infinitesimal generator $G_r(t)$ provides a constant of motion which coincides with the "charge" given by Noether theorem.

The same of course applies to continuous theories described by fields, namely, the *generators of canonical symmetry transformations* of a field theory are precisely the conserved Noether charges. Therefore, in analogy with Eqs. (8.92) and (8.93), we may write:

$$\delta\varphi^\alpha(\mathbf{x}, t) = -\{\varphi^\alpha(\mathbf{x}, t), G(t)\} \tag{8.235}$$

$$\delta\pi_\alpha(\mathbf{x}, t) = -\{\pi_\alpha(\mathbf{x}, t), G(t)\}. \tag{8.236}$$

where $G(t) \equiv \delta\theta^r G_r(t)$. When the Hamiltonian is left invariant it coincides, aside from an overall sign, with the charge $Q(t) \equiv \delta\theta^r Q_r(t)$ of the Noether theorem.

In the case of Poincaré transformations given by space-time translations and Lorentz transformations, let us show that the infinitesimal generator has the following form:

$$G(t) = -\epsilon^\mu P_\mu(t) + \frac{1}{2}\delta\theta^{\mu\nu} J_{\mu\nu}(t), \tag{8.237}$$

where the explicit expression of the generators is obtained from Eqs. (8.170) and (8.192) identifying $\frac{\partial \mathcal{L}}{\partial \dot\varphi^\alpha} \equiv \pi_\alpha$:

$$P^\rho = \int \left(\pi_\alpha(\mathbf{x}, t)\partial^\rho\varphi^\alpha(\mathbf{x}, t) - \eta^{0\rho}\mathcal{L}(x)\right)\delta^3\mathbf{x}, \tag{8.238}$$

$$J_{\rho\sigma} = -\left[\int \left(\pi_\alpha(\mathbf{x}, t)(L_{\rho\sigma})^\alpha{}_\beta\varphi^\beta(\mathbf{x}, t) + (x_\rho\mathscr{P}_\sigma - x_\sigma\mathscr{P}_\rho)\right)\right]d^3\mathbf{x}. \tag{8.239}$$

Let us first consider the case of space-time translations, that is we take $G(t) = -\epsilon^\mu P_\mu$. Taking into account the fundamental Poisson brackets (8.231) and the general formulae (8.235), (8.236), we obtain:

$$\delta\varphi^\alpha = -\{\varphi^\alpha(\mathbf{x}, t), [-\epsilon_\rho P^\rho(t)]\} = \epsilon_\rho\frac{\delta P^\rho(t)}{\delta\pi_\alpha}(\mathbf{x}, t) = \epsilon_\rho\partial^\rho\varphi^\alpha(\mathbf{x}, t), \tag{8.240}$$

so that, for time or space translations we find, respectively:

$$\delta\varphi^\alpha = \delta t \{\varphi^\alpha(\mathbf{x}, t), H(t)\} = \delta t \, \partial_t\varphi^\alpha, \tag{8.241}$$

$$\delta\varphi^\alpha = \epsilon^i \{\varphi^\alpha(\mathbf{x}, t), P_i(t)\} = \epsilon^i\partial_i\varphi^\alpha = \epsilon \cdot \nabla\varphi^\alpha, \tag{8.242}$$

where $\epsilon \equiv (\epsilon^i)$.

For infinitesimal canonical transformations generated by $J_{\mu\nu}$ we find

$$
\delta\varphi^\alpha = -\frac{\delta\theta^{\rho\sigma}}{2}\{\varphi^\alpha(\mathbf{x}, t), J_{\rho\sigma}(t)\} = -\frac{\delta\theta^{\rho\sigma}}{2}\frac{\delta J_{\rho\sigma}}{\delta\pi_\alpha}(\mathbf{x}, t)
$$

$$
= \frac{\delta\theta^{\mu\nu}}{2}\left[(L_{\mu\nu})^\alpha{}_\beta\varphi^\beta + (x_\mu\partial_\nu - x_\nu\partial_\mu)\varphi^\alpha\right]. \tag{8.243}
$$

Let us now compute the infinitesimal change of the Hamiltonian:

$$
\delta H = H(\varphi', \pi') - H(\varphi, \pi) = \sum_\alpha \int \left(\frac{\delta H}{\delta\varphi^\alpha}\delta\varphi^\alpha + \frac{\delta H}{\delta\pi_\alpha}\delta\pi_\alpha\right) d^3\mathbf{x}
$$

$$
= -\sum_\alpha \int \left(\frac{\delta H}{\delta\varphi^\alpha}\frac{\delta G}{\delta\pi_\alpha} - \frac{\delta H}{\delta\pi_\alpha}\frac{\delta G}{\delta\varphi^\alpha}\right) d^3\mathbf{x}
$$

$$
= -\{H, G(t)\}. \tag{8.244}
$$

If the transformations are a symmetry of the Hamiltonian, $\delta H = \{G, H\} = -\frac{\partial G}{\partial t}$, see Eqs. (8.99) and (8.100), we recover the result that $G(t)$ *is a conserved quantity*:

$$
\frac{dG}{dt} = \{G(t), H\} + \frac{\partial G}{\partial t} = 0, \tag{8.245}
$$

As an example let us consider Lorentz boosts for which $\delta H \neq 0$ since it transforms as the 0-component of the four vector P^μ; infinitesimally we have:

$$
\delta P^0 \equiv \frac{1}{c}\delta H = -\theta^{0i}\{H, J_{0i}\}. \tag{8.246}
$$

On the other hand

$$
\delta P^0 = \delta\theta^{0\mu}P_\mu = \delta\theta^{0i}P_i,
$$

so that combining the two expressions of δP^0 we find: $\{H, J_{0i}\} = -cP_i$. Now if we consider the component $0i$ of Eq. (8.239) we see that when the Lorentz index $\rho = 0$, it carries an explicit time dependence in the second term, namely $-\int d^3\mathbf{x}\,(x_0\mathscr{P}_i - x_i\mathscr{P}_0)$. It follows:

$$
\frac{dJ_{0i}}{dt} = -\{H, J_{0i}\} + \frac{\partial J_{0i}}{\partial t} = cP_i - c\frac{d}{dt}\int d^3\mathbf{x}t\,\mathscr{P}_i = cP_i - cP_i = 0, \tag{8.247}
$$

and therefore J_{0i} is also conserved, in agreement with the Noether theorem.

8.9.2 References

For further reading see Refs. [6], [8] (Vol. 1).

Chapter 9
Quantum Mechanics Formalism

9.1 Introduction

In this chapter we give a concise review of the quantum mechanics formalism from a perspective which generalizes the ordinary Schroedinger formulation. In this way we may reconsider the Schroedinger approach to quantum mechanics from a more geometrical and group-theoretical point of view and show the close relationship between the classical Hamiltonian theory and quantum mechanics. Moreover the formalism developed in this chapter will be useful for an appropriate exposition of the relativistic wave equations in Chap. 10 and the field quantization approach in Chap. 11.

9.2 Wave Functions, Quantum States and Linear Operators

In elementary courses in quantum mechanics the state of a system is described by a *wave function* $\psi_\alpha(\xi, t)$ where the variables ξ denote the set of the coordinates which the wave function depends on and the suffix α refers to a set of (discrete) physical quantities, or quantum numbers, which, together with ξ, define the state of the system. In the Schroedinger approach the variables ξ comprise the space coordinates $\mathbf{x} = (x, y, z)$ while, if *spin* is present, the variable α labels the corresponding polarization state. In this case $\psi_\alpha(\mathbf{x}; t)$ is referred to as the wave function in the *coordinate representation*. In this section we wish to adopt the Dirac formalism which allows a quantum description of a system that is independent of its explicit *coordinate representation*. Since throughout this section we refer to states at a particular instant t, the time coordinate will not be indicated explicitly.

We recall that the essential difference between quantum theory and classical mechanics resides in the different characterization of the concept of *state* of a physical system. According to a more general point of view than the wave function

© Springer International Publishing Switzerland 2016
R. D'Auria and M. Trigiante, *From Special Relativity to Feynman Diagrams*,
UNITEXT for Physics, DOI 10.1007/978-3-319-22014-7_9

description, any quantum state can be characterized, independently of the particular representation, by a complex vector in an abstract *finite or infinite* dimensional complex vector space hereafter denoted by $V^{(c)}$. The vector nature of quantum states is in agreement with the *superposition principle* of quantum mechanics, implying that any linear combination of quantum states is again a quantum state. In this chapter we shall be dealing with *single particle* states. As we presently show the wave function description of the quantum state will then appear as *the set of components of the state vector along a particular basis*.

For the sake of clarity let us first consider the particular case of a finite dimensional space $V^{(c)}$ endowed with a hermitian scalar product. To recall the Dirac formalism and to fix the conventions, let us briefly sketch out the defining properties of $V^{(c)}$.

We introduce an n-dimensional *complex vector space* $V_n^{(c)}$ (see Chap. 7 for a formal introduction to the concept of complex vector space), whose elements are called *kets*, on which the observable dynamical quantities act as linear operators. Using the *ket* notation an element $\mathbf{a} \in V_n^{(c)}$ is denoted by the symbol $|a\rangle$ and a basis $\{\mathbf{u}_i\}$ of $V_n^{(c)}$ by $\{|u_i\rangle\}(i = 1, \ldots, n)$.

We define on $V_n^{(c)}$ a *hermitian scalar (or inner) product* associating with each pair of elements $|a\rangle, |b\rangle \in V_n^{(c)}$ a complex number that is denoted by $\langle a|b\rangle$,

$$|a\rangle, |b\rangle \in V_n^{(c)} \quad \rightarrow \quad \langle a|b\rangle \in \mathbb{C},$$

with the following properties:

$$\langle a|b\rangle = \langle b|a\rangle^*, \tag{9.1}$$

$$\langle a|a\rangle \geq 0 \ ; \ \langle a|a\rangle = 0 \Rightarrow |a\rangle = 0, \tag{9.2}$$

$$\langle a| (\alpha |b\rangle + \beta |c\rangle) = \alpha \langle a|b\rangle + \beta \langle a|c\rangle, \quad \forall \alpha, \beta \in \mathbb{C}. \tag{9.3}$$

Two vectors are said to be orthogonal if

$$\langle a|b\rangle = 0.$$

We may thus associate, by means of the scalar product, with each $|a\rangle$ a *dual* vector "bra" $\langle a|$ defining a linear correspondence from $V_n^{(c)}$ to $\mathbb{C}$

$$\langle a| : \ |b\rangle \in V_n^{(c)} \rightarrow \langle a|b\rangle \in \mathbb{C}. \tag{9.4}$$

From the properties of the scalar product it follows that the bra corresponding to the ket $\alpha|b\rangle + \beta|c\rangle$ is $\langle b|\alpha^* + \langle c|\beta^*$. The squared norm $\|a\|^2$ of a state $|a\rangle$ is the quantity $\langle a|a\rangle$, which is strictly positive if $|a\rangle$ is non-zero. The *distance* d of two elements $|a\rangle, |b\rangle$ is then defined as

$$d(a, b) = \sqrt{((\langle a| - \langle b|) (|a\rangle - |b\rangle))}.$$

A state in quantum mechanics is associated with a vector of $V_n^{(c)}$ modulo multiplication by a complex number, that is *parallel* vectors $|a\rangle$ and $\alpha|a\rangle$, $\alpha \in \mathbb{C}$ define the same quantum state:

$$\text{quantum state} \leftrightarrow \{|a\rangle\} \equiv \{\alpha|a\rangle \mid \alpha \in \mathbb{C}\}. \tag{9.5}$$

In general we shall choose to describe states by unit norm vectors $|a\rangle$: $\|a\|^2 = \langle a|a\rangle = 1$. This of course fixes the vector associated with a given state modulo an arbitrary phase factor. In next section we shall comment on a convenient choice of such factors.

Recalling the definition given in Sect. 7.2, a *linear operator* $\hat{F}$ on the vector space $V_n^{(c)}$ is defined as a (not necessarily invertible) mapping from $V_n^{(c)}$ into itself:

$$\hat{F}: \quad |v\rangle \in V_n^{(c)} \rightarrow |Fv\rangle \equiv \hat{F}|v\rangle \in V_n^{(c)}, \tag{9.6}$$

satisfying the linearity condition (7.3) which, in our new notations, reads:

$$\hat{F}(\alpha|v\rangle + \beta|w\rangle) = \alpha\hat{F}|v\rangle + \beta\hat{F}|w\rangle, \quad \forall \alpha, \beta \in \mathbb{C}. \tag{9.7}$$

Linear transformations are invertible operators on $V_n^{(c)}$. Of particular physical relevance in quantum mechanics is the notion of *expectation value* $\langle\hat{F}\rangle$ of an operator $\hat{F}$ on a state $|a\rangle$:

$$\langle\hat{F}\rangle \equiv \frac{\langle a|\hat{F}|a\rangle}{\langle a|a\rangle}.$$

Let $|u_i\rangle$, $i = 1, \ldots, n$, be a basis of ket vectors in $V_n^{(c)}$, and let $\langle u_i|$ be the dual basis of bra vectors. The basis $|u_i\rangle$ is said to be orthonormal if

$$\langle u_i|u_j\rangle = \delta_j^i. \tag{9.8}$$

With respect to this basis $\hat{F}$ can be represented by a matrix $\mathbf{F} \equiv (F^i{}_j)$, see Eq. (7.6) and footnote 3 of Chap. 7:

$$|u_i\rangle \xrightarrow{\hat{F}} |Fu_i\rangle \equiv \hat{F}|u_i\rangle = F^j{}_i|u_j\rangle, \tag{9.9}$$

that is, using (9.8)

$$F^i{}_j = \langle u_i|\hat{F}|u_j\rangle.$$

Clearly if $\hat{F}$ is not invertible, and thus is not a transformation, the matrix $\mathbf{F}$ is singular and the vectors $|Fu_i\rangle$ do not form a new basis.

The identity operator $\hat{I}$ on $V_n^{(c)}$ can be written in the form

$$\hat{I} = \sum_{i=1}^{n} |u_i\rangle\langle u_i|, \tag{9.10}$$

since it can be easily verified, using the orthonormality of the basis, that $\hat{I}|u_i\rangle = |u_i\rangle$, for all $i = 1, \ldots, n$. The corresponding matrix representation is the $n \times n$ identity matrix $\mathbf{1} \equiv (\delta_j^i)$.

In quantum mechanics there are two classes of operators which play a special role: The *hermitian and the unitary operators*. Both of them can be characterized by their properties with respect to their *hermitian conjugate* operators.

The *hermitian conjugate* $\hat{F}^\dagger$ of $\hat{F}$, is defined as the operator such that

$$\forall\ |a\rangle, |b\rangle \in V_n^{(c)}, \quad \langle a|\hat{F}|b\rangle = \langle b|\hat{F}^\dagger|a\rangle^*,$$

or, equivalently, $\langle a|Fb\rangle = \langle F^\dagger a|b\rangle$. This definition implies that $\langle F\,b| \equiv \langle b|\hat{F}^\dagger$ is the bra of $\hat{F}|b\rangle$. $\hat{F}$ is a *hermitian operator* iff $\hat{F} = \hat{F}^\dagger$, or, equivalently

$$F^i_j = \langle u_i|\hat{F}|u_j\rangle = \langle u_j|\hat{F}^\dagger|u_i\rangle^* = \langle u_j|\hat{F}|u_i\rangle^* = (F^j_i)^*,$$

that is, the matrix representing it coincides with the conjugate of its transposed (hermitian conjugate): $\mathbf{F} = \mathbf{F}^\dagger \equiv (\mathbf{F}^T)^*$, and is therefore a *hermitian matrix*. In other words an operator is hermitian if and only if its matrix representation with respect to an orthonormal basis is hermitian. From this it clearly follows that the expectation value of a hermitian operator on any state is a real number since $\langle a|\hat{F}|a\rangle = \langle a|\hat{F}^\dagger|a\rangle = \langle a|\hat{F}|a\rangle^*$.

A *unitary* operator U is defined by the condition

$$UU^\dagger = U^\dagger U = \hat{I}. \tag{9.11}$$

From the above definition we derive the corresponding unitarity property of the matrix $\mathbf{U} = (U^i_j)$ representing U:

$$\delta_j^i = \langle u_i|u_j\rangle = \langle u_i|U^\dagger U|u_j\rangle = \sum_{k=1}^{n}\langle u_i|U^\dagger|u_k\rangle\langle u_k|U|u_j\rangle = \sum_{k=1}^{n}(U^k_i)^* U^k_j,$$

i.e. $\mathbf{U}^\dagger\,\mathbf{U} = \mathbf{1}$. While a hermitian operator is not necessarily invertible, a unitary one is, being the corresponding matrix $\mathbf{U}$ non-singular: $|\det\mathbf{U}|^2 = 1$. Unitary operators are therefore linear transformations and their geometrical and physical meaning will be discussed in the next section. It can be shown, from the properties of the corresponding matrix representations, that both hermitian and unitary operators are diagonalizable, namely admit n linearly independent eigenvectors, and, moreover, have the following important properties:

(a) *The eigenvalues of hermitian operators are real;*
(b) *The eigenvalues of unitary operators have unit complex modulus;*
(c) *The eigenvectors $|\lambda_1\rangle$, $|\lambda_2\rangle$ corresponding to two different eigenvalues λ_1, λ_2 are orthogonal.*

Having defined the relevant mathematical objects, let us recall their relation to our physical world within quantum mechanics and in particular the role of hermitian operators. In quantum mechanics physical observables $\mathcal{O}$, like energy, momentum, position etc. are represented by hermitian operators $\widehat{\mathcal{O}}$ acting on states and their expectation value on a given state is defined as $|a\rangle$

$$\langle \mathcal{O} \rangle \equiv \frac{\langle a | \widehat{\mathcal{O}} | a \rangle}{\|a\|^2}. \tag{9.12}$$

This quantity has the following interpretation: If infinitely many identical systems are prepared in a state $|a\rangle$, a measurement of the observable $\mathcal{O}$ on them will give a statistical distribution of results about an average value $\langle \mathcal{O} \rangle$. In other words it can be interpreted as the most likely value that a measurement of the observable $\mathcal{O}$ on $|a\rangle$ would give. The hermitian property of $\widehat{\mathcal{O}}$, i.e. $\widehat{\mathcal{O}}^\dagger = \widehat{\mathcal{O}}$, then guarantees that its expectation value in (9.12) be a *real number*, as a measurable quantity should be. The eigenvalues λ_i of $\widehat{\mathcal{O}}$ represent all possible values that the actual measurement of $\mathcal{O}$ can give on the system and the corresponding eigenvectors $|\lambda_i\rangle$ describe states characterized by the values λ_i of $\mathcal{O}$.

Given two states $|a_1\rangle$, $|a_2\rangle$, the quantity

$$P(|a_1\rangle, |a_2\rangle) \equiv \frac{|\langle a_2 | a_1 \rangle|^2}{\|a_2\|^2 \|a_2\|^2}, \tag{9.13}$$

represents the probability of finding, upon measurement, a system which was initially prepared in the state $|a_1\rangle$, in the state $|a_2\rangle$, characterized, for instance, by a definite value of a quantity we are measuring. $P(|a_1\rangle, |a_2\rangle)$ is also called *transition probability* from the state $|a_1\rangle$ to $|a_2\rangle$. For instance $P(|\lambda_i\rangle, |a\rangle)$ represents the probability that the measurement of an observable $\mathcal{O}$ on the state $|a\rangle$ yield the value λ_i. Note that neither of the two measurable quantities (9.12) and (9.13) depends on the normalization of the state vectors. Such normalization, as anticipated earlier, is unphysical and can be fixed at convenience.

We also recall the concept of a *complete set of commuting observables*, as a maximal set of observables represented by commuting operators. It has a considerable importance in quantum mechanics, since by measuring the value of such a set of observables on the system, the state vector is uniquely determined: It is the common eigenvector associated with the eigenvalues of the corresponding hermitian operators.

So far the dimension of the vector space $V^{(c)}$ was taken to be finite. Let us now suppose the number of dimensions to be *infinite* as it happens in quantum mechanics when the eigenvalues of operators representing physical observables form an infinite but *discrete* set. Then, if every Cauchy sequence of vectors converges to an element of the space, we say that our complex infinite-dimensional vector space is a *Hilbert space*.[1]

We introduce an *orthonormal basis* in the Hilbert space given by the eigenfunctions of a hermitian operator $\hat{F}$, which we suppose to have a discrete spectrum of eigenvalues $\{F_i\}, n = 1, \dots \infty$. We may expand the state vector $|a\rangle$ of the Hilbert space along the orthonormal eigenvectors $\{|F_i\rangle\}$, labeled by the eigenvalues F_i of $\hat{F}$

$$|a\rangle = \sum_{i=1}^{\infty} a_i \, |F_i\rangle. \tag{9.14}$$

By definition the *wave function* describing the state $|a\rangle$ in the F-representation is the totality of the infinite coefficients of the expansion, namely the components a_i of the state vector along the eigenvectors $\{|F_i\rangle\}$. Since we are using an orthonormal basis, each component a_i can be written as the scalar product between the bra $\langle F_i|$ and the ket $|a\rangle$.[2]

$$a_i = \langle F_i|a\rangle. \tag{9.15}$$

Actually the Hilbert space does not cover the description of all the possible quantum states of a physical system. Indeed when the eigenstates of a hermitian operator belong to a *continuous spectrum* of eigenvalues (or to a discrete set of values followed by a continuous one), it is necessary to enlarge the Hilbert space to include *generalized functions*, like the Dirac delta function, and we may thus have *non-normalizable* wave-functions.

In this case we must allow for the dimensions of the vector space to be labeled by *continuous variables* and, correspondingly, the sum in (9.14) to be replaced by an integral over the continuous set of eigenvalues (or by an integral *and* a sum over the discrete part of the spectrum).

This is the case, for instance, of the coordinate operator $\hat{F} = \hat{\mathbf{x}} \equiv (\hat{x}, \hat{y}, \hat{z})$, the momentum operator $\hat{F} = \hat{\mathbf{p}} \equiv (\hat{p}_x, \hat{p}_y, \hat{p}_z)$, as well as the energy operator for certain systems. As far as the coordinate or momentum operators are concerned, the integral should be computed over the corresponding eigenvalues $F = \mathbf{x} = (x, y, z)$ or $\mathbf{p} = (p_x, p_y, p_z)$ and the wave function $\langle F|a\rangle$ becomes a *continuous* function of F: $\psi_{(a)}(F)$.

---

[1] We recall that a Cauchy sequence is any sequence of elements ϕ_n such that $\lim_{m,n \to \infty} d(\phi_n, \phi_m) = 0$. In particular, the finite dimensional space $V_n^{(c)}$ treated so far is trivially a *Hilbert space*.

[2] Indeed the expansion (9.14) is quite analogous to the expansion of an ordinary vector $\mathbf{v}$ along a orthonormal basis $\mathbf{u}_i$ in a finite dimensional space

$$\mathbf{v} = \sum_i v^i \mathbf{u}_i = \sum_i \mathbf{u}_i(\mathbf{u}_i \cdot \mathbf{v})$$

and the "wave function" representation $\{\langle F|v\rangle\}$ of $\mathbf{v}$ corresponds to the representation of the vector in terms of its components along the chosen basis: $\mathbf{v} \equiv \{v^i\}$.

For quantum states defined in $V^{(c)}$ the *coordinate representation* is defined by taking $\hat{F} \equiv \hat{\mathbf{x}}$ so that the expansion (9.14) takes the form

$$|a\rangle = \int_V d^3\mathbf{x} \, \langle \mathbf{x}|a\rangle |\mathbf{x}\rangle, \tag{9.16}$$

where $d^3\mathbf{x} \equiv dx \, dy \, dz$ and each eigenvector $|\mathbf{x}\rangle$ describes a single particle localized at the point $\mathbf{x} = (x, y, z)$ in space. It is defined by the equation $\hat{\mathbf{x}}|\mathbf{x}\rangle = \mathbf{x}\,|\mathbf{x}\rangle$. The volume V of integration can be finite or infinite, that is coinciding with the whole space $\mathbb{R}^3$.

In this framework, the wave function $\psi_{(a)}(\mathbf{x})$ of the Schrödinger's theory, describing the state $|a\rangle$, is the continuous set of the components of the ket $|a\rangle$ along the eigenvectors of the position operator $\hat{\mathbf{x}}$: $\psi_{(a)}(\mathbf{x}) = \langle \mathbf{x}|a\rangle$. Multiplying both sides of Eq. (9.16) by $\langle \mathbf{x}'|$ we find

$$\langle \mathbf{x}'|a\rangle = \psi_a(\mathbf{x}') = \int d^3\mathbf{x} \, \psi_a(\mathbf{x})\langle \mathbf{x}'|\mathbf{x}\rangle. \tag{9.17}$$

We see that for consistency we must set

$$\langle \mathbf{x}'|\mathbf{x}\rangle = \delta^3(\mathbf{x}' - \mathbf{x}). \tag{9.18}$$

The above normalization equation can be interpreted as the definition of the wave function $\psi_{\mathbf{x}}(\mathbf{x}')$ describing the ket $|\mathbf{x}\rangle$ in the coordinate representation. Such eigenfunction is no ordinary function, but belongs to the class of *generalized or improper functions*. The reader can then easily verify that the identity operator $\hat{I}$ can be expressed in this basis as follows: $\hat{I} = \int d^3\mathbf{x} \, |\mathbf{x}\rangle\langle \mathbf{x}|$, which generalizes Eq. (9.10) to a basis labeled by a triplet of continuously varying variables (i.e. (x, y, z)). Restoring for the moment the explicit dependence of the quantum state $|a, t\rangle$ on time, the (time-dependent) wave function is defined as

$$\psi_{(a)}(\mathbf{x}, t) = \langle \mathbf{x}|a, t\rangle. \tag{9.19}$$

Note that since $d^3\mathbf{x}$ has dimension L^3, in order for the Eq. (9.16) to be consistent the state $|\mathbf{x}\rangle$ has to be dimensionful, of dimension $L^{-\frac{3}{2}}$. This is in agreement with the normalization (9.18).

There is a one-to-one correspondence between states and wave-functions which satisfies the property that a linear combination of states corresponds to the same linear combination of the wave functions representing them (the space of wave function is said to be *isomorphic* to $V^{(c)}$). In particular we can write a hermitian scalar product of wave functions which reproduces the inner product between states:

$$\langle b|a\rangle = \langle b|\hat{I}|a\rangle = \int_V d^3\mathbf{x} \, \langle b|\mathbf{x}\rangle\langle \mathbf{x}|a\rangle = \int_V d^3\mathbf{x} \, \psi_{(b)}(\mathbf{x})^* \psi_{(a)}(\mathbf{x}), \tag{9.20}$$

so that we can write the squared norm $\|a\|^2$ of a state as

$$\|a\|^2 = \langle a|a \rangle = \int d^3\mathbf{x}\, |\psi_{(a)}(\mathbf{x})|^2. \tag{9.21}$$

We conclude that states with finite norm (i.e. normalizable) correspond to *square integrable* wave functions, belonging to the Hilbert space $\mathbf{L}^2(V)$.

Let us recall, for completeness, the probabilistic interpretation of a wave function $\psi(\mathbf{x}, t)$, normalized to one, in quantum mechanics:

The quantity $|\psi(\mathbf{x}, t)|^2\, dV$ measures the probability of finding the particle within an infinitesimal volume dV about $\mathbf{x}$ at a time t.

To complete the correspondence between abstract states and their wave function representation, we observe that operators acting on states correspond to *differential operators* acting on the corresponding wave functions:

$$|b\rangle \equiv \hat{O}|a\rangle \;\;\Leftrightarrow\;\; \psi_{(b)}(\mathbf{x}) = \hat{O}(\mathbf{x}, \nabla)\, \psi_{(a)}(\mathbf{x}), \tag{9.22}$$

where $\hat{O}(\mathbf{x}, \nabla)$ is a *local differential operator*. For example, as we shall review in Sect. 9.3.1, the momentum operator $\hat{\mathbf{p}}$ is implemented on wave functions by the operator $-i\hbar\nabla$. Observables quantities are represented by differential operators which are hermitian with respect to the inner product (9.20). For the time being, we shall denote abstract operators and their differential representation on wave functions by the same symbol. Eigenstates $|\lambda_i\rangle$ of an observable $\hat{O}$ are represented by eigenfunctions $\psi_i(\mathbf{x})$ of the corresponding differential operator, solution to a differential equation:

$$\hat{O}|\lambda_i\rangle = \lambda_i\,|\lambda_i\rangle \;\;\Leftrightarrow\;\; \hat{O}(\mathbf{x}, \nabla)\psi_i(\mathbf{x}) = \lambda_i\,\psi_i(\mathbf{x}). \tag{9.23}$$

The eigenstates of $\hat{\mathbf{p}}$ are then represented by the functions $\psi_{\mathbf{p}}(\mathbf{x}) \propto e^{\frac{i}{\hbar}\mathbf{p}\cdot\mathbf{x}}$.

It is apparent from our analysis thus far, that our space $V^{(c)}$ also contains states with no finite norm, whose wave functions are therefore not in $L^2(V)$. Simple examples are given by eigenstates of the $\hat{\mathbf{x}}$ or of the $\hat{\mathbf{p}}$ operators, represented, respectively, by delta functions and by $e^{\frac{i}{\hbar}\mathbf{p}\cdot\mathbf{x}}$. The norm of the latter is indeed infinite if the space V is infinite: $\int_V d^3\mathbf{x}|e^{\frac{i}{\hbar}\mathbf{p}\cdot\mathbf{x}}|^2 = \int_V d^3\mathbf{x} = \infty$.[3] Although the physical (probabilistic) interpretation of non-normalizable wave functions is more problematic, as we saw

---

[3] As we shall see in Sect. 9.3.1, when dealing with free one-particle states, we can avoid the use of non-normalizable wave functions, generalized functions Dirac delta-functions etc., by quantizing the physical system in a box. In this case, instead of considering the whole $\mathbb{R}^3$ as the domain of integration, we take a large box of finite volume V, so that the functions which were not $L^2(-\infty; +\infty)$-integrable become now $L^2(V)$-integrable. In this way we may always restrict ourselves to considering the Hilbert space of functions defined over a finite volume.

for the case of the position eigenstates, they are useful to express wave functions which are $L^2(V)$.[4]

Let us emphasize here the different roles played in non-relativistic quantum mechanics by the space and time variables $\mathbf{x}$, t. Just as in classical mechanics, the former are dynamical variables while the latter is a parameter. By this we do not mean that the argument $\mathbf{x}$ in $\psi(\mathbf{x}, t)$ should be intended as the position of the particle at the time t, since we adopt for the probability distribution in space the analogue of the Eulerian point of view in describing the velocity distribution of a fluid in fluid-dynamics.

If we have a system of N non-interacting particles, the corresponding space of quantum states is the tensor product of the spaces describing the quantum states of each particle (see Chap. 4, Sect. 4.2). We can therefore consider as a basis of the N-particle states the vectors:

$$|\mathbf{x}_1\rangle|\mathbf{x}_2\rangle \ldots |\mathbf{x}_N\rangle \equiv |\mathbf{x}_1\rangle \otimes |\mathbf{x}_2\rangle \otimes \cdots \otimes |\mathbf{x}_N\rangle,$$

also denoted by $|\mathbf{x}_1, \mathbf{x}_2, \ldots, \mathbf{x}_N\rangle$, describing the particles located in $\mathbf{x}_1, \mathbf{x}_2, \ldots, \mathbf{x}_N$. The corresponding representation of a state $|a, t\rangle$ is described by the wave function:

$$\psi(\mathbf{x}_1, \mathbf{x}_2, \ldots, \mathbf{x}_N, t) \equiv \langle\mathbf{x}_1|\langle\mathbf{x}_2| \ldots \langle\mathbf{x}_N|a, t\rangle.$$

Let us come back now to a single particle system.

Similarly to what we have done when defining the coordinate representation, we can choose to describe a state $|a\rangle$ in the *momentum representation* by expanding it in a basis of eigenvectors $|\mathbf{p}\rangle$ of the momentum operator:

$$|a\rangle = \text{const.} \times \int d^3\mathbf{p}\,\langle\mathbf{p}|a\rangle|\mathbf{p}\rangle = \int d^3\mathbf{p}\,\tilde{\psi}_{(a)}(\mathbf{p})\,|\mathbf{p}\rangle, \qquad (9.24)$$

where $\hat{\mathbf{p}}|\mathbf{p}\rangle = \mathbf{p}\,|\mathbf{p}\rangle$, and $\tilde{\psi}_{(a)}(\mathbf{p})$ is the wave function in the momentum representation. The proportionality factor after the first equality in Eq. (9.24) depends on the normalization of the momentum eigenstates, which will be defined in Sect. 9.3.1.

As previously pointed out, the state of a system can be completely characterized in terms of a complete set of observables. Therefore a single particle state can be identified not just by a certain position $\mathbf{x}$ (or momentum $\mathbf{p}$), but also by its *spin* state, since the spin operator $\hat{\mathbf{S}}$ commutes with $\hat{\mathbf{x}}$ (and $\hat{\mathbf{p}}$). Let us label the spin state of a

---

[4]This is not an uncommon feature. For example in the Fourier integral transform

$$f(x) = \frac{1}{\sqrt{2\pi}} \int dp\, F(p)\, e^{ipx},$$

if $f(x) \subset L^2(-\infty; +\infty)$, so does its Fourier transform. However the basis functions $\frac{1}{\sqrt{2\pi}} e^{ipx}$ are not in $L^2(-\infty; +\infty)$ since $|\frac{1}{\sqrt{2\pi}} e^{ipx}|^2 = \frac{1}{2\pi}$.

particle by a discrete index α (representing for instance the eigenvalues of $\hat{\mathbf{S}}^2$ and $\hat{S}_z$). We can then take as a basis of the Hilbert space either $\{|\mathbf{x},\ \alpha\rangle\}$ or $\{|\mathbf{p},\ \alpha\rangle\}$. In the former case, normalizing the basis elements as follows

$$\langle \mathbf{x},\ \alpha|\mathbf{x}',\ \alpha'\rangle = \delta^3(\mathbf{x} - \mathbf{x}')\,\delta_{\alpha\alpha'}, \tag{9.25}$$

Equation (9.16) generalizes to

$$|a\rangle = \int d^3\mathbf{x} \sum_\alpha |\mathbf{x},\ \alpha\rangle\langle \mathbf{x},\ \alpha|\,a\rangle = \int d^3\mathbf{x} \sum_\alpha \Phi^\alpha_{(a)}(\mathbf{x})\,|\mathbf{x},\ \alpha\rangle, \tag{9.26}$$

the wave function being defined by $\Phi^\alpha_{(a)}(\mathbf{x}) \equiv \langle \mathbf{x},\ \alpha|\ a\rangle$. Restoring the explicit dependence on time the above definition reads

$$\Phi^\alpha_{(a)}(\mathbf{x}, t) = \langle \mathbf{x},\ \alpha|a, t\rangle. \tag{9.27}$$

We stress that the wave function $\Phi_{(a)}(\mathbf{x})$ is a c-number field, that is a *classical field*.[5] As an example, the electromagnetic potential in the Coulomb gauge $\mathbf{A}(x) = \epsilon_\mathbf{k}e^{i(\mathbf{k}\cdot\mathbf{x}-\omega t)}$ can be thought of as the wave function[6] describing a photon in the state $|a\rangle = |\hbar\mathbf{k}, \alpha\rangle$, where $\mathbf{p} = \hbar\mathbf{k}$ is the momentum and, recalling that in the Coulomb gauge $\epsilon_\mathbf{k} \cdot \mathbf{k} = 0$ (see Chap. 6), the index $\alpha = 1, 2$ labels the two physical polarizations in the plane orthogonal to $\mathbf{k}$.

We may of course describe the state in other representations. If we take, for example, the complete set of the eigenfunctions of the Hamiltonian operator, possessing a discrete spectrum of eigenvalues E_n and eigenstates $|E_n\rangle$

$$|a\rangle = \sum_{E_n} |E_n\rangle\langle E_n|a\rangle,$$

the set $\langle E_n|a\rangle$ will now represent the wave function of the same state in the *energy representation*. Its relation to the wave function in the *coordinate representation* is given by

$$\langle \mathbf{x}|\,a\rangle = \sum_n \langle \mathbf{x}|E_n\rangle\langle E_n|a\rangle, \tag{9.28}$$

$\langle \mathbf{x}|E_n\rangle$ being the eigenfunctions of the Hamiltonian.

---

[5]As we shall see in Chap. 11 a consistent interpretation of a quantum relativistic theory requires that the interpretation of $\Phi_{(a)}(\mathbf{x})$ as a quantum mechanical wave function must be abandoned and that the classical field be promoted to a quantum mechanical operator.

[6]Clearly a normalization factor should be used when giving this interpretation since A_μ has dimension $(Newton)^{\frac{1}{2}}$ and not $(length)^{\frac{3}{2}}$, as a wave.

9.3 Unitary Operators

As pointed out in Sect. 7.2, when describing transformations in three-dimensional Euclidean space or in Minkowski space, we have adopted the so-called *passive* point of view, that is we have assumed that transformations (rotations, translations, Lorentz transformations etc.) act on the reference frame $\{O, \mathbf{u}_i\} \rightarrow \{O', \mathbf{u}_i'\}$ while points and vectors are fixed. That means that the geometrical meaning of vectors and points is not altered by a transformation, only their description in terms of coordinates or components undergoes a change.

This same point of view was adopted in Chap. 7 for the description of the transformation of a field under a change in coordinates. There, writing $\phi'(x') = \phi(x)$ we were considering the transformation $x' = f(x)$ from a *passive* point of view. However we pointed out that the transformation $\phi(x) \rightarrow \phi'(x)$ with $\phi'(x) = \phi(f^{-1}(x))$ could also be considered from an *active* point of view, thereby putting the emphasis on the *functional change* of ϕ.

In the following transformations on the Hilbert space of states, namely on the ket vectors $|v\rangle$, will be mainly considered for the time being from the *active* point of view. That means that we will describe a linear transformation U on a ket-vector as acting on the vector itself, while the basis with respect to which it is described is kept fixed:

$$|v\rangle \rightarrow |v'\rangle = |Uv\rangle \equiv U|v\rangle,$$

where U is a linear transformation. Here both $|v'\rangle$ and $|v\rangle$ are then represented in components with respect to the *same basis* $|\mathbf{u}_i\rangle$ defined by the simultaneous eigenstates of a complete system of observables.[7] This means that the effect of a space-time coordinate transformation is described at the level of quantum states by means of the action of an operator mapping the original state vector of the system into a new one.

To motivate this consider, for example, a particle that, with respect to a reference frame S has definite momentum $\mathbf{p}$ and thus is in a state $|\mathbf{p}\rangle$. In a different frame S', obtained by a rotation $\mathbf{R}$ of the first, the same particle will be described as having a momentum $\mathbf{p}' = \mathbf{R}\,\mathbf{p}$, that is as being in the new quantum state $|\mathbf{p}'\rangle$, different from the original one $|\mathbf{p}\rangle$. The effect of the transformation is then to change the quantum state of the system and thus is naturally described on the space of states from an active point of view: The new state results from the action of an operator U on the old one: $|\mathbf{p}'\rangle = U\,|\mathbf{p}\rangle$. Such active description of a transformation is referred to as the Schroedinger representation. We shall also consider the Heisenberg representation in which operators rather than states are affected by a transformation, and which realize the passive description of transformations on a quantum system.

Let us choose an orthonormal basis $\{|\mathbf{u}_i\rangle\}$ for the Hilbert space of states (e.g. the eigenstates of the momentum operator). The action of an operator U from the active point of view was described in Eqs. (7.6), (7.7) and, in the new notations, in Eq. (9.9). In the chosen basis the transformation U is represented by a non-

---

[7]We warn the reader that, in the case of space-time transformations, the orthonormal basis $|\mathbf{u}_i\rangle$ has nothing to do with the space, or space-time, reference frame which undergoes the transformation.

singular matrix $\mathbf{U} = (U^i{}_j)$. If we write the original state $|v\rangle$ and the transformed one $|v'\rangle \equiv |U\,v\rangle \equiv U\,|v\rangle$ in components with respect to the same basis $\{|\mathbf{u}_i\rangle\}$

$$|v\rangle = v^i\,|\mathbf{u}_i\rangle, \quad |v'\rangle = v'^i\,|\mathbf{u}_i\rangle, \tag{9.29}$$

the old an new components are related by the action of $\mathbf{U}$: $v'^i = U^i{}_j\,v^j$. If the basis elements form a denumerable infinity, then $\mathbf{U}$ has infinitely many rows and columns. If the basis elements form a non-denumerable infinity, as it is the case of the coordinate (or momentum) representation, the action of U is more conveniently expressed in terms of a differential operator on wave functions. We may consider transformations belonging to a group G, like the Lorentz transformations. In this case U provides a representation of G on the space of states (which is more appropriately called *realization* if the transformations are realized in terms of differential operators on wave functions):

$$g \in G: \quad |a\rangle \in V^{(c)} \xrightarrow{\;g\;} |a'\rangle = |g\,a\rangle = U(g)|a\rangle,$$
$$\forall g_1,\, g_2 \in G: \quad U(g_1 \cdot g_2) = U(g_1) \cdot U(g_2). \tag{9.30}$$

The transformation $U(g)$ on states, associated with a coordinate transformation g, must be defined in such a way that the expectation value $\langle \mathcal{O}\rangle$ of any observable quantity, like the position vector $\mathbf{x}$ or the linear momentum $\mathbf{p}$, transform accordingly under g. For instance we must have that the expectation value of the position operator $\hat{\mathbf{x}} = (\hat{x}^i)$ on a particle state $|a\rangle$ transforms under a rotation $R \in SO(3)$ as the position vector $\mathbf{x}$ of a classical particle, namely as follows:

$$\langle x^i\rangle \equiv \langle a|\hat{x}^i|a\rangle \xrightarrow{\;R\;} \langle x^i\rangle' = \langle R\,a|\hat{x}^i|R\,a\rangle = R^i{}_j\,\langle x^i\rangle. \tag{9.31}$$

We recall that in quantum mechanics given two states $|b\rangle$ and $|a\rangle$, the probability of transition from $|a\rangle$ to the state $|b\rangle$ is given by (or it is proportional to) $|\langle b|a\rangle|^2$, if the state is normalizable (or not normalizable), see Eq. (9.13). Since this probability is a measurable quantity, it must have the same value in every reference frame. It follows that the action of a generic element g of the transformation group G, represented on states by the operator U, must leave $|\langle a|b\rangle|^2$ invariant for any pair of kets $|a\rangle$ and $|b\rangle$. In formulae, if

$$\forall g \in G: \quad \begin{cases} |a\rangle \\ |b\rangle \end{cases} \xrightarrow{\;g\;} \begin{cases} |a'\rangle = |g\,a\rangle = U(g)|a\rangle \\ |b'\rangle = |g\,b\rangle = U(g)|b\rangle, \end{cases} \tag{9.32}$$

is the action of $g \in G$ on the given kets, we require that:

$$|\langle a|b\rangle|^2 = |\langle a'|b'\rangle|^2 = |\langle a|U(g)^\dagger U(g)|b\rangle|^2.$$

A theorem by Wigner, which we are not going to prove, states that it is possible to fix the multiplicative phases in the definition of the (unit norm) state vectors in such a way that that $U(g)$ is either *unitary*

$$\langle a|U(g)^\dagger U(g)|b\rangle = \langle a|b\rangle \quad \Rightarrow \quad U(g)^\dagger = U(g)^{-1},$$

or *antiunitary*[8]

$$\langle a|U(g)^\dagger U(g)|b\rangle = \langle a|b\rangle^*.$$

We shall show in the next chapter that the discrete transformation $t \to -t$ called *time reversal* is an example of antiunitary operator. Clearly not all transformations of a group G can be realized as antiunitary operators, since, as the reader can easily verify, the product of two antiunitary transformations is unitary.

If $U(g)$ is unitary for any $g \in G$, we say that U defines a *unitary representation* of G on $V^{(c)}$. In the following we restrict our discussion to the *unitary* representations only:

$$\forall g \in G \ : \ U(g)^\dagger U(g) = U(g)U(g)^\dagger = \hat{I}.$$

According to our analysis of Lie algebras in Chap. 7, the structure of a Lie group G in a neighborhood of the identity element $U(g_0) = \hat{I}$ is captured by the Lie algebra $\mathcal{A}$ of infinitesimal generators, so that a generic element $U(g)$ can be expressed as the exponential of an element of $\mathcal{A}$:

$$U(g) = e^{\frac{i}{\hbar}\theta^r \hat{G}_r}, \tag{9.34}$$

where (θ^r) are the parameters defining the element g of G and $\hat{G}_r$ is a basis of $\mathcal{A}$ and consists of operators in the quantum-space of states $V^{(c)}$. Note that, with respect to the notation used in Chap. 7, the infinitesimal generators here are rescaled by a factor $i/\hbar$. As usual infinitesimal transformations, parametrized by $\delta\theta^r \ll 1$, can be expressed by truncating the exponential to first order in the parameters:

$$U(g) \approx \hat{I} + \frac{i}{\hbar}\delta\theta^r \, \hat{G}_r. \tag{9.35}$$

Writing the unitary condition to first order in the infinitesimal parameters $\delta\theta^r$ we find the following property of the infinitesimal operators:

---

[8]An antiunitary operator $\hat{A}$ does not fit the definition of linear operators given in Eq. (9.7). In fact $\hat{A}$ is an example of an *antilinear* operator defined by the property

$$\hat{A}(\alpha|v\rangle + \beta|w\rangle) = \alpha^* \hat{A}|v\rangle + \beta^* \hat{A}|w\rangle, \quad \forall \alpha, \beta \in \mathbb{C}. \tag{9.33}$$

$$U(g)^\dagger U(g) = U(g) U(g)^\dagger = \hat{I} \Leftrightarrow \left(\hat{I} - \frac{i}{\hbar} \delta\theta^r \hat{G}_r^\dagger \right) \left(\hat{I} + \frac{i}{\hbar} \delta\theta^r \hat{G}_r \right)$$

$$\simeq \hat{I} - \frac{i}{\hbar} \delta\theta^r (\hat{G}_r^\dagger - \hat{G}_r) = \hat{I} \quad \Leftrightarrow \quad \hat{G}_r^\dagger = \hat{G}_r,$$

namely we find that the infinitesimal generators $\hat{G}_r$, defined in Eq. (9.34), are *hermitian*. The hermiticity condition allows us to associate each G_r with *observable* quantities that is with operators whose eigenvalues are real and therefore interpretable as the result of a measurement of a physical quantity.[9] In Chaps. 4 and 7, see Sect. 7.3, the exponential representation of a finite transformation in (9.34) was motivated by the fact that any finite transformation, in a neighborhood of the identity element, can be realized by iterating infinitely many infinitesimal transformations. If $\hat{F} = \theta^r \hat{G}_r / \hbar$ is a finite element of $\mathcal{A}$, iterating a large number $n \gg 1$ of times the infinitesimal transformation generated by the infinitesimal element $\frac{i}{n} \hat{F}$, in the limit $n \to \infty$, we generate a finite group element

$$U = \lim_{n \to \infty} \left(1 + \frac{i}{n} \hat{F} \right)^n = \hat{I} + i \hat{F} + \frac{i^2}{2!} \hat{F}^2 + \cdots = e^{i \hat{F}}. \tag{9.36}$$

By suitably choosing $\hat{F}$ we can reach, through the *exponential map* (9.36), any element of G in a finite neighborhood of the identity.

So far we have described the effect of transformations (e.g. of coordinate transformations closing a group G) on the quantum description of a system in terms of the action of unitary operators on the state vectors. Such description defines the so called *Schrödinger picture (or representation)*, in which the state of a system belongs to a unitary representation of the transformation group G. As explained above, the condition defining such unitary action is that the expectation value $\langle \hat{O} \rangle$ of an observable $\hat{O}$ on a state $|a\rangle$ transform under a change in the RF as the corresponding classical quantity $\mathcal{O}$:

$$\langle \mathcal{O} \rangle \equiv \langle a | \hat{O} | a \rangle \xrightarrow{U} \langle \mathcal{O} \rangle' = \langle a' | \hat{O} | a' \rangle = \langle a | U^\dagger \hat{O} U | a \rangle. \tag{9.37}$$

We can adopt an alternative description, called the *Heisenberg representation*, in which transformations affect the operators $\hat{O}$ associated with observables, leaving state vectors unchanged. Since in both representations the effect of a transformation on the expectation value $\langle \mathcal{O} \rangle$ of an observable should be the same, we deduce the following transformation rules in the Heisenberg picture:

$$\hat{O} \xrightarrow{U} \hat{O}' = U^\dagger \hat{O} U; \quad |a\rangle \in V^{(c)} \xrightarrow{U} |a\rangle. \tag{9.38}$$

If U represents a Lie group of transformations G, we can consider the effect on $\hat{O}$ of an infinitesimal transformation defined by parameters $\delta\theta^r \ll 1$:

---

[9]Note that the imaginary unit i in Eq (9.34) has been inserted in order to deal with hermitian generators.

$$\hat{O}' = \left(\hat{I} - \frac{i}{\hbar} \delta\theta^r \hat{G}_r\right) \hat{O} \left(\hat{I} + \frac{i}{\hbar} \delta\theta^r \hat{G}_r\right) = \hat{O} + \frac{i}{\hbar} \delta\theta^r [\hat{O}, \hat{G}_r] = \hat{O} + \delta\hat{O},$$

where we have neglected second order terms in $\delta\theta$ and used the property $\hat{G}_r^\dagger = \hat{G}_r$. We deduce that

$$\delta\hat{O} = \hat{O}' - \hat{O} = \frac{i}{\hbar} \delta\theta^r [\hat{O}, \hat{G}_r]. \tag{9.39}$$

Compare now Eq. (9.39) with Eq. (8.95) describing the infinitesimal transformation property of the corresponding observable $\mathcal{O}$ in the classical theory. We observe that the former can be obtained from the latter by *replacing the Poisson brackets of the classical theory with the commutator of the quantum theory*:

$$\{\cdot, \cdot\} \rightarrow -\frac{i}{\hbar}[\cdot, \cdot], \tag{9.40}$$

and the *classical* observable $\mathcal{O}(p, q)$ and G_r with their *quantum* counterparts $\hat{O}$, $\hat{G}_r$. Taking into account that in the classical theory $G_r(p, q)$ are the infinitesimal generators of *canonical transformations*, we conclude that *canonical transformations are implemented in the quantum theory by unitary operators* U. This was to be expected since just as Poisson brackets in the classical theory were invariant under canonical transformations, commutators between quantum operators are invariant under *unitary* transformations (9.38), as it can be easily verified.

9.3.1 Application to Non-Relativistic Quantum Theory

Let us apply the above considerations in the context of non-relativistic quantum mechanics.

According to Eq. (9.40) the Heisenberg commutation conditions can be deduced from the Poisson brackets (8.75) of the fundamental quantum canonical variables. We have

$$\left[\hat{x}^i, \hat{p}_j\right] = i\hbar\, \delta^i_j\, \hat{I}. \tag{9.41}$$

Using these commutation relations we can introduce the operators $\hat{G}_r$ corresponding to the infinitesimal generators of rotations, space and time translations in the quantum theory. Let us recall from Chap. 7 that the angular momentum $\mathbf{M} = (M_i)$, the linear momentum $\mathbf{p} = (p_i)$ and the Hamiltonian H are the infinitesimal generators of rotations, spatial and time translations. The corresponding quantum hermitian operators will generate the same transformations implemented on quantum states. Promoting the classical dynamical variables to quantum operators in the Hilbert space, we then obtain the following infinitesimal generators:

$$\hat{G}_r : \begin{cases} \hat{M}_i \ (i=1,2,3) & \begin{array}{l} \text{angular momentum operator} \\ \hat{\mathbf{M}} = \hat{\mathbf{x}} \times \hat{\mathbf{p}} \text{ generating SO(3)} \\ \text{rotations.} \end{array} \\[2em] \hat{p}_i \ \ i=1,2,3 & \begin{array}{l} \text{momentum operator } \hat{\mathbf{p}}, \text{ generating} \\ \text{space translations} \end{array} . \\[2em] \hat{H} & \begin{array}{l} \text{Hamiltonian operator, generating} \\ \text{time-evolution.} \end{array} \end{cases}$$

They are operators corresponding to physical observables and generate transformations on the system which can be described as an action either on the quantum states (Schroedinger representation) or on the hermitian operators corresponding to classical observables (Heisenberg representation). Let us first describe in some detail the action of these operators on other dynamical variables.

The operators $\hat{\mathbf{p}}$ generate translations in $\mathbf{x}$:

$$\mathbf{x} \to \mathbf{x}' = \mathbf{x} - \boldsymbol{\epsilon}; \quad \delta\mathbf{x} = \mathbf{x}' - \mathbf{x} = -\boldsymbol{\epsilon}, \tag{9.42}$$

the corresponding finite unitary transformation being:

$$U(\epsilon) = e^{\frac{i}{\hbar}\hat{\mathbf{p}}\cdot\boldsymbol{\epsilon}}. \tag{9.43}$$

If we take an infinitesimal displacement $\boldsymbol{\epsilon}$ the variation of $\hat{\mathbf{x}}$ is

$$\delta\hat{x}^i = \frac{i}{\hbar}\epsilon^j \left[\hat{x}^i, \hat{p}_j\right] = -\epsilon^i \hat{I}.$$

reproducing Eq. (9.42) on the operator $\hat{\mathbf{x}}$. For finite transformations we then have:

$$\hat{\mathbf{x}}' = U(\epsilon)^\dagger \hat{\mathbf{x}} U(\epsilon) = \hat{\mathbf{x}} - \boldsymbol{\epsilon}\hat{I}. \tag{9.44}$$

It is now straightforward to check that the expectation value $\langle\mathbf{x}\rangle$ of $\hat{\mathbf{x}}$ on a state $|a\rangle$ (relative to a frame S) has the right transformation property

$$\langle\mathbf{x}\rangle \equiv \langle a|\hat{\mathbf{x}}|a\rangle \xrightarrow{U(\epsilon)} \langle\mathbf{x}\rangle' = \langle a'|\hat{\mathbf{x}}|a'\rangle = \langle a|U(\epsilon)^\dagger \hat{\mathbf{x}} U(\epsilon)|a\rangle$$
$$= \langle a|(\hat{\mathbf{x}} - \boldsymbol{\epsilon}\hat{I})|a\rangle = \langle\mathbf{x}\rangle - \boldsymbol{\epsilon}, \tag{9.45}$$

where $|a'\rangle = U(\epsilon)|a\rangle = e^{\frac{i}{\hbar}\hat{\mathbf{p}}\cdot\boldsymbol{\epsilon}}|a\rangle$ is the state of the particle as observed in the frame S' translated with respect to S. Similarly it can be easily shown that if $|\mathbf{x}\rangle$ is the eigenstate of $\hat{\mathbf{x}}$ corresponding to the eigenvalue $\mathbf{x}$, $U(\epsilon)|\mathbf{x}\rangle$ is the eigenstate of $\hat{\mathbf{x}}$ corresponding to the eigenvalue $\mathbf{x} - \boldsymbol{\epsilon}$. To this end let us apply $\hat{\mathbf{x}}$ to the transformed vector $U(\epsilon)|\mathbf{x}\rangle$ (we suppress, for the sake of simplicity, the argument of U):

$$\hat{\mathbf{x}} U|\mathbf{x}\rangle = U (U^\dagger \hat{\mathbf{x}} U)|\mathbf{x}\rangle = U \left(\hat{\mathbf{x}} - \boldsymbol{\epsilon}\hat{I}\right)|\mathbf{x}\rangle = (\mathbf{x} - \boldsymbol{\epsilon}) U|\mathbf{x}\rangle, \tag{9.46}$$

where we have used Eq. (9.44). From the above derivation we conclude that, modulo a proportionality factor, we can make the following identification:

$$U(\epsilon)|\mathbf{x}\rangle = |\mathbf{x} - \epsilon\rangle. \tag{9.47}$$

Applying instead $U(\epsilon)$ to the eigenvector $|\mathbf{p}\rangle$ of $\hat{\mathbf{p}}$, corresponding to an eigenvalue $\mathbf{p}$, its effect amounts to a multiplication by a phase: $U(\epsilon)|\mathbf{p}\rangle = e^{\frac{i}{\hbar}\hat{\mathbf{p}}\cdot\epsilon}|\mathbf{p}\rangle = e^{\frac{i}{\hbar}\mathbf{p}\cdot\epsilon}|\mathbf{p}\rangle$. Let us now use this property and Eq. (9.47) to write the wave function $\psi_{\mathbf{p}}(\mathbf{x})$ associated with an eigenstate of the momentum operator:

$$\psi_{\mathbf{p}}(\mathbf{x}) = \langle\mathbf{x}|\mathbf{p}\rangle = \langle\mathbf{x} = \mathbf{0}|U(\mathbf{x})|\mathbf{p}\rangle = \langle\mathbf{0}|\mathbf{p}\rangle\, e^{\frac{i}{\hbar}\mathbf{p}\cdot\mathbf{x}}, \tag{9.48}$$

where we have written $|\mathbf{x}\rangle = U(-\mathbf{x})\,|\mathbf{x} = \mathbf{0}\rangle$ and used the property $U(-\mathbf{x})^{\dagger} = U(\mathbf{x})$. We see that $\psi_{\mathbf{p}}(\mathbf{x}) \propto e^{\frac{i}{\hbar}\mathbf{p}\cdot\mathbf{x}}$, which has an infinite norm as observed in Sect. 9.2 after Eq. (9.21). Physically this descends from the fact that a particle with definite momentum is completely delocalized in space, as implied by Heisenberg's uncertainty principle.

We can now write the relation between the coordinate and the momentum representations, see Eq. (9.24)

$$\psi_{(a)}(\mathbf{x}) = \langle\mathbf{x}|a\rangle = \text{const.} \times \int d^3\mathbf{p}\,\langle\mathbf{x}|\mathbf{p}\rangle\langle\mathbf{p}|a\rangle = \int d^3\mathbf{p}\,\tilde{\psi}_{(a)}(\mathbf{p})\, e^{\frac{i}{\hbar}\mathbf{p}\cdot\mathbf{x}}, \tag{9.49}$$

where we have absorbed normalization factors like $\langle\mathbf{x} = \mathbf{0}|\mathbf{p}\rangle$ in the definition of $\tilde{\psi}_{(a)}(\mathbf{p})$. We see that $\tilde{\psi}_{(a)}(\mathbf{p})$ is the familiar Fourier transform of $\psi_{(a)}(\mathbf{x})$. Particles which are localized at each time within a finite region of space of size Δx are described by wave packets $\psi(\mathbf{x})$, whose Fourier transform $\tilde{\psi}(\mathbf{p})$ is peaked on some average value $\bar{\mathbf{p}}$ of the linear momentum and has a width of size Δp, related to Δx by Heisenberg's uncertainty principle: $\Delta x \Delta p \gtrsim \hbar$. The probabilistic interpretation of such a wave function is that the particle it describes is localized within a volume $(\Delta x)^3$ and moves with a momentum $\mathbf{p}$ which is undetermined within a region $(\Delta p)^3$ about $\bar{\mathbf{p}}$.

As mentioned in Sect. 9.2 it is possible to extend the space of square-integrable wave functions corresponding to normalizable states to include functions like $e^{\frac{i}{\hbar}\mathbf{p}\cdot\mathbf{x}}$.

We can indeed avoid the problem of dealing with non-normalizable states if we quantize the particle in a box, just as we did for the photon in Chap. 5: We consider the particle as propagating inside a parallelepiped of sides L_A, L_B, L_C along the three directions X, Y, Z and volume $V = L_A L_B L_C$. We then impose periodic boundary conditions on the wave function, as a consequence of which the eigenvalues of $\hat{\mathbf{p}}$ are quantized:

$$\mathbf{p} = (p_x, p_y, p_z) = 2\pi\hbar\left(\frac{n_x}{L_A}, \frac{n_y}{L_B}, \frac{n_z}{L_C}\right), \quad n_x, n_y, n_z \in \mathbb{Z}, \tag{9.50}$$

and the corresponding eigenstates are normalizable to one:

$$\langle \mathbf{p} | \mathbf{p}' \rangle = \delta_{\mathbf{p}, \mathbf{p}'}. \tag{9.51}$$

Writing the identity operator as $\hat{I} = \sum_{\mathbf{p}} |\mathbf{p}\rangle\langle\mathbf{p}|$ we rederive the relation (9.49) between the coordinate and momentum representation of a state in the form of a Fourier series

$$\psi_{(a)}(\mathbf{x}) = \sum_{\mathbf{p}} \tilde{\psi}_{(a)}(\mathbf{p})\, e^{\frac{i}{\hbar}\mathbf{p}\cdot\mathbf{x}}. \tag{9.52}$$

In the infinite volume limit $L_A, L_B, L_C \rightarrow \infty$, see discussion in Sect. 5.6.2 of Chap. 5 and set $\mathbf{k} = \mathbf{p}/\hbar$, we recover a continuous momentum spectrum and the sum over the discrete momentum values becomes an integral through the replacement:

$$\sum_{\mathbf{p}} \rightarrow \int \frac{d^3\mathbf{p}}{(2\pi\hbar)^3}\, V. \tag{9.53}$$

This limit amounts to requiring that V be much larger than the size $(\Delta x)^3$ of the wave packet describing the particle. In the large volume limit the normalization condition (9.51) becomes

$$\langle \mathbf{p} | \mathbf{p}' \rangle = \frac{(2\pi\hbar)^3}{V}\, \delta^3(\mathbf{p} - \mathbf{p}'), \tag{9.54}$$

where we have used the prescription

$$\delta_{\mathbf{p},\mathbf{p}'} \longrightarrow \frac{(2\pi\hbar)^3}{V}\, \delta^3(\mathbf{p} - \mathbf{p}'), \tag{9.55}$$

which follows from Eq. (5.126) of Chap. 5 upon replacing $\mathbf{k}$ with $\mathbf{p}/\hbar$.[10]

Using Eqs. (9.54) and (9.53), the identity operator can be written as

$$\hat{I} = \sum_{\mathbf{p}} |\mathbf{p}\rangle\langle\mathbf{p}| \rightarrow \int \frac{d^3\mathbf{p}}{(2\pi\hbar)^3}\, V\, |\mathbf{p}\rangle\langle\mathbf{p}|. \tag{9.56}$$

The *one-particle volume* V is a normalization factor which should ultimately drop off the expression of observable quantities, as it will be shown in Chap. 12, when computing transition probabilities and cross sections of interaction processes.

---

[10] We have also used the property of delta-functions $\delta(cx) = \delta(x)/c$, so that $\delta^3(\mathbf{k} - \mathbf{k}') = \hbar^3\, \delta^3(\mathbf{p} - \mathbf{p}')$.

Let us now describe the effect of a spatial translation (9.42) on the wave function $\psi(\mathbf{x}) \equiv \psi_{(a)}(\mathbf{x})$ of a particle which is in state $|a\rangle$ with respect to the frame $\mathcal{S}$. An observer in the translated frame $\mathcal{S}'$ will observe the particle in the state $|a'\rangle = U(\epsilon)\,|a\rangle$ and describe it through the following wave function:

$$\psi'(\mathbf{x}') \equiv \psi_{(a')}(\mathbf{x}') = \langle \mathbf{x}'|a'\rangle = \langle \mathbf{x}'|e^{\frac{i}{\hbar}\hat{\mathbf{p}}\cdot\epsilon}|a\rangle = \langle \mathbf{x}'+\epsilon|a\rangle = \psi(\mathbf{x}'+\epsilon).$$

We thus find the correct transformation property of the wave function given in infinitesimal form in equation (7.98).

Writing $\psi'(\mathbf{x})$ as resulting from the action of a differential operator $O_\epsilon = e^{\frac{i}{\hbar}\hat{\mathbf{p}}\cdot\epsilon}$ on $\psi(\mathbf{x})$: $\psi'(\mathbf{x}) = \psi(\mathbf{x}+\epsilon) = e^{\frac{i}{\hbar}\hat{\mathbf{p}}\cdot\epsilon}\,\psi(\mathbf{x})$ and expanding the expression for infinitesimal shift parameters $\epsilon^i \ll 1$, along the lines of Sect. (7.4.3), we derive the form of $\hat{p}_i$ as differential operators on wave functions:

$$\hat{p}_i = -i\hbar\frac{\partial}{\partial x^i} \quad \Leftrightarrow \quad \hat{\mathbf{p}} = -i\hbar\,\nabla. \tag{9.57}$$

As we have seen in Sect. 4.5.1, ordinary rotations are described by the following transformations:

$$g \in SO(3) \rightarrow U(\boldsymbol{\theta}) = e^{\frac{i}{\hbar}\theta^i\hat{M}_i}, \quad i = 1, 2, 3,$$

where $\boldsymbol{\theta} \equiv (\theta^i)$ and the $\hat{M}_i$ operators satisfy the commutation rules (4.132)

$$\left[\hat{M}_i,\ \hat{M}_j\right] = i\hbar\,\epsilon_{ijk}\,\hat{M}_k.$$

If we take as $\hat{O}$ the same operators $\hat{M}_i$ and compute their variation (9.39), with $\hat{G}_i = \hat{M}_i$ we find

$$\delta\hat{M}_i = \frac{i}{\hbar}\,\delta\theta^j\left[\hat{M}_i,\ \hat{M}_j\right] = -\epsilon_{ijk}\,\delta\theta^j\,\hat{M}^k,$$

that is

$$\delta\hat{\mathbf{M}} = -\delta\boldsymbol{\theta}\times\hat{\mathbf{M}}.$$

This means that $\hat{\mathbf{M}}$ transforms under rotations as a three-dimensional vector. As far as the effect of rotations on the position and momentum operators is concerned, we may further verify that

$$[\hat{M}_i, \hat{x}^j] = i\hbar\,\epsilon_{ijk}\,\hat{x}^k; \quad [\hat{M}_i, \hat{p}_j] = i\hbar\,\epsilon_{ijk}\,\hat{p}_k,$$

implying that $\hat{\mathbf{x}}$ and $\hat{\mathbf{p}}$ transform under rotations as the vectors they represent

$$\delta\hat{\mathbf{x}} = -\delta\boldsymbol{\theta}\times\hat{\mathbf{x}}, \quad \delta\hat{\mathbf{p}} = -\delta\boldsymbol{\theta}\times\hat{\mathbf{p}}.$$

Under finite transformations, recalling our discussion of the rotation group and its algebra given in Sect. 4.5.1 (see specifically equation (4.123)), we have

$$\hat{M}^i \to \hat{M}'^i = U(\boldsymbol{\theta})^\dagger \, \hat{M}^i \, U(\boldsymbol{\theta}) = R(\boldsymbol{\theta})^i{}_j \, \hat{M}^j, \tag{9.58}$$

$$\hat{x}^i \to \hat{x}'^i = U(\boldsymbol{\theta})^\dagger \hat{x}^i U(\boldsymbol{\theta}) = R(\boldsymbol{\theta})^i{}_j \hat{x}^j, \tag{9.59}$$

$$\hat{p}^i \to \hat{p}'^i = U(\boldsymbol{\theta})^\dagger \hat{p}^i U(\boldsymbol{\theta}) = R(\boldsymbol{\theta})^i{}_j \hat{p}^j, \tag{9.60}$$

which imply the correct transformation rules under rotations of the corresponding expectation values.

Let us now investigate the action of a rotation on the physical states represented by kets. We take, for the sake of definiteness, as a basis of the Hilbert space either the eigenstates $|\mathbf{x}\rangle$ of the operator $\hat{\mathbf{x}}$ or the eigenstates $|\mathbf{p}\rangle$ of $\hat{\mathbf{p}}$, defined in Sect. 9.2. Consider, for instance, the action of a rotation $U(\boldsymbol{\theta})$ on $|\mathbf{p}\rangle$: Applying the operator $\hat{\mathbf{p}}$ to the transformed vector $|\mathbf{p}'\rangle = U(\boldsymbol{\theta})\,|\mathbf{p}\rangle$ we have

$$\hat{\mathbf{p}} \, U(\boldsymbol{\theta})|\mathbf{p}\rangle = U(\boldsymbol{\theta})U(\boldsymbol{\theta})^\dagger \hat{\mathbf{p}} \, U(\boldsymbol{\theta}) \, |\mathbf{p}\rangle = U(\boldsymbol{\theta}) \, \hat{\mathbf{p}}' \, |\mathbf{p}\rangle = U(\boldsymbol{\theta}) \, \left(\mathbf{R}(\boldsymbol{\theta})\hat{\mathbf{p}}\right) \, |\mathbf{p}\rangle$$
$$= (\mathbf{R}(\boldsymbol{\theta})\mathbf{p})U(\boldsymbol{\theta})|\mathbf{p}\rangle.$$

It follows that $U(\boldsymbol{\theta})|\mathbf{p}\rangle$ is eigenstate of $\hat{\mathbf{p}}$ corresponding to the eigenvalue $\mathbf{Rp} \equiv (R^i{}_j \, p^j)$. Therefore, neglecting a possible normalization coefficient

$$U(\boldsymbol{\theta})|\mathbf{p}\rangle = |\mathbf{R}(\boldsymbol{\theta}) \, \mathbf{p}\rangle.$$

In an analogous way we may show

$$U(\boldsymbol{\theta})|\mathbf{x}\rangle = |\mathbf{R}(\boldsymbol{\theta}) \, \mathbf{x}\rangle.$$

The transformation property of a wave function under rotations is readily derived: Let $|a\rangle$ and $|a'\rangle = U(\boldsymbol{\theta})\,|a\rangle$ be the states of a same (spin-less) particle in $\mathcal{S}$ and in the rotated frame $\mathcal{S}'$, $\psi(\mathbf{x})$ and $\psi'(\mathbf{x}')$ the corresponding wave functions. We have:

$$\psi'(\mathbf{x}') = \langle \mathbf{x}'|a'\rangle = \langle \mathbf{x}'|U(\boldsymbol{\theta})\,|a\rangle = \langle \mathbf{R}(\boldsymbol{\theta})^{-1}\,\mathbf{x}'|a\rangle = \psi(\mathbf{R}(\boldsymbol{\theta})^{-1}\,\mathbf{x}') = \psi(\mathbf{x}).$$

Writing $\psi'(\mathbf{x}) = O_\theta \, \psi(\mathbf{x}) = e^{\frac{i}{\hbar}\hat{\mathbf{M}}\cdot\boldsymbol{\theta}}\psi(\mathbf{x})$ and expanding for small angles $\theta^i \ll 1$ we find the explicit expression (7.99) for the angular momentum components as differential operators on wave functions.

We note however that, writing the effect of rotations on a state just by means of the action of the (orbital) angular momentum operator $\hat{\mathbf{M}}$ is correct only if the particle does not carry spin degrees of freedom as it has been discussed in Chap. 8. If this is not the case we may think of the rotation as acting simultaneously on the coordinates by means of the $\hat{M}_i$ generators and on the spin degrees of freedom by means of the generators $\hat{S}_i$. That means that the infinitesimal generator of the rotations is given by the *total angular momentum* operator $\hat{\mathbf{J}}$,

$$\hat{\mathbf{J}} = \hat{\mathbf{M}} + \hat{\mathbf{S}}. \tag{9.61}$$

The effect of a finite rotation $g(\theta) \in SO(3)$ is

$$U(\boldsymbol{\theta}) \,|\mathbf{x}, \alpha\rangle = D^{\beta}{}_{\alpha} |\mathbf{R}\,\mathbf{x}, \beta\rangle, \tag{9.62}$$

where the explicit dependence on θ was suppressed and

$$U(\boldsymbol{\theta}) = e^{\frac{i}{\hbar}\hat{\mathbf{J}}\cdot\boldsymbol{\theta}} \tag{9.63}$$

$$D(\boldsymbol{\theta})^{\beta}{}_{\alpha} = \left(e^{\frac{i}{\hbar}\hat{\mathbf{S}}\cdot\boldsymbol{\theta}}\right)^{\beta}{}_{\alpha} \quad \text{acts on the spin-components,} \tag{9.64}$$

$$R(\boldsymbol{\theta})^{j}{}_{i} = \left(e^{\frac{i}{\hbar}\hat{\mathbf{M}}\cdot\boldsymbol{\theta}}\right)^{j}{}_{i} \quad \text{acts on the space-components.} \tag{9.65}$$

Similarly, for the momentum eigenstates $|\mathbf{p}, \alpha\rangle$, we find

$$U|\mathbf{p}, \alpha\rangle = D^{\beta}{}_{\alpha} |\mathbf{R}\,\mathbf{p}, \beta\rangle.$$

Consider now a local, scalar differential operator $\hat{A}(\mathbf{x})$ (here we suppress, for the sake of notational simplicity, the obvious dependence of $\hat{A}$ on the partial derivatives: $\hat{A}(\mathbf{x}) \equiv \hat{A}(\mathbf{x}, \nabla)$), acting on wave functions and depending on $\mathbf{x}$ and on partial derivatives with respect to the coordinates (by scalar we mean representing an observable which does not transform under spatial rotations). An example of $\hat{A}(\mathbf{x})$ is the Hamiltonian operator $\hat{H}(\hat{\mathbf{p}}, \mathbf{x}) = \hat{H}(-i\hbar\nabla, \mathbf{x})$ in the coordinate representation. Let us illustrate how $\hat{A}(\mathbf{x})$ transforms under a coordinate transformation $f : \mathbf{x} \to \mathbf{x}' = f(\mathbf{x})$, which can be a rotation, a translation, or a general congruence. Let the transformation be implemented on wave functions by the operator O_f:

$$O_f \psi(\mathbf{x}) = \psi(f^{-1}(\mathbf{x})) \quad \Leftrightarrow \quad O_f^{-1}\psi(\mathbf{x}) = \psi(f(\mathbf{x})). \tag{9.66}$$

If f is a rotation, O_f is the transformation O_θ defined above, while $\mathbf{x}' = f(\mathbf{x}) = \mathbf{R}(\theta)\mathbf{x}$ and $f^{-1}(\mathbf{x}) = \mathbf{R}(\theta)^{-1}\mathbf{x}$. Let $\Phi(\mathbf{x})$ denote the result of the action of $\hat{A}(\mathbf{x})$ on $\psi(\mathbf{x})$:

$$\hat{A}(\mathbf{x})\psi(\mathbf{x}) = \Phi(\mathbf{x}), \tag{9.67}$$

and let us act on both sides by O_f^{-1}:

$$O_f^{-1}\hat{A}(\mathbf{x})\psi(\mathbf{x}) = O_f^{-1}\Phi(\mathbf{x}) = \Phi(f(\mathbf{x})) = \hat{A}(f(\mathbf{x}))\psi(f(\mathbf{x})), \tag{9.68}$$

where it is understood that $\hat{A}(f(\mathbf{x})) = \hat{A}(\mathbf{x}')$ is the operator $\hat{A}$ in which also the partial derivatives are computed with respect to the new coordinates x'^i: $\hat{A}(\mathbf{x}') \equiv \hat{A}(\mathbf{x}', \nabla')$. On the other hand we have:

$$O_f^{-1}\hat{A}(\mathbf{x})\psi(\mathbf{x}) = O_f^{-1}\hat{A}(\mathbf{x})O_f\,O_f^{-1}\psi(\mathbf{x}) = O_f^{-1}\hat{A}(\mathbf{x})O_f\psi(f(\mathbf{x}))$$
$$= \hat{A}'(\mathbf{x})\,\psi(f(\mathbf{x})). \tag{9.69}$$

Comparing (9.68) with (9.69), being $\psi(\mathbf{x})$ a generic function, we deduce the transformation property of the local differential operator $\hat{A}(\mathbf{x})$ under a coordinate transformation:

$$\hat{A}'(\mathbf{x}) = O_f^{-1}\hat{A}(\mathbf{x})O_f = \hat{A}(f(\mathbf{x})). \tag{9.70}$$

The above equation defines the transformation property of a *scalar operator* $\hat{A}(\mathbf{x})$. It expresses Eq. (9.38) in the coordinate representation. The scalar operator is *invariant* under f iff $\hat{A}'(\mathbf{x}) = \hat{A}(\mathbf{x})$, namely if

$$O_f^{-1}\hat{A}(\mathbf{x})O_f = \hat{A}(f(\mathbf{x})) = \hat{A}(\mathbf{x}), \tag{9.71}$$

or, equivalently:

$$[\hat{A}(\mathbf{x}),\ O_f] \equiv \hat{A}(\mathbf{x})\,O_f - O_f\,\hat{A}(\mathbf{x}) = 0. \tag{9.72}$$

We conclude that *a local differential (scalar) operator is invariant under a coordinate transformation f if it commutes with O_f.*

9.3.2 The Time Evolution Operator

In non-relativistic quantum mechanics space and time are treated on a different footing. So far we have considered quantum states and their transformations under unitary operators *at a fixed time t*, and we have shown that they play in quantum mechanics the same role as canonical transformations in the classical theory. In classical mechanics *time-evolution*, namely the correspondence between the state of a system at a given instant and its evolved at a later time, is a canonical transformation generated by the Hamiltonian of the system. In quantum mechanics, however, in order to describe the time-evolution of a system, we must find an operator on $V^{(c)}$ that connects the states of a system at two generic instants, say $|a, t\rangle$ and $|a, t_0\rangle$. From the superposition principle in quantum mechanics it follows that if at t_0

$$|a, t_0\rangle = \alpha_1\,|b, t_0\rangle + \alpha_2\,|c, t_0\rangle,$$

the same superposition must hold at any other time t:

$$|a, t\rangle = \alpha_1\,|b, t\rangle + \alpha_2\,|c, t\rangle.$$

This implies that the mapping U between $|a, t_0\rangle$ and $|a, t\rangle$

$$|a, t\rangle = U(t, t_0)\, |a, t_0\rangle \qquad (9.73)$$

must be a linear operator. Requiring also the conservation of the norm of a state during its time-evolution (conservation of probability), we must have

$$\langle a, t|a, t\rangle = \langle a, t_0|a, t_0\rangle \ \Rightarrow\ \langle a, t_0|U^\dagger U|a, t_0\rangle = \langle a, t_0|a, t_0\rangle,$$

implying

$$U^\dagger U = \hat{I},$$

that is, the time-evolution operator U has to be *unitary*. Moreover if $U(t, t_0)$ maps $|a, t_0\rangle$ into $|a, t\rangle$, $U(t, t_0)^{-1} = U(t, t_0)^\dagger$ maps $|a, t\rangle$ into $|a, t_0\rangle$, so that $U(t, t_0)^\dagger = U(t_0, t)$. We finally require U to satisfy the condition $U(t_0, t_0) = \hat{I}$.

In order to determine the time-evolution of $|a, t\rangle$ we compute the change of $|a, t\rangle$ under an infinitesimal change in the parameter t. We have

$$\left.\frac{\partial|a, t\rangle}{\partial t}\right|_{t=t_0} = \lim_{t \to t_0} \frac{|a, t\rangle - |a, t_0\rangle}{t - t_0} = \left\{\lim_{t \to t_0} \frac{U - \hat{I}}{t - t_0}\right\} |a, t_0\rangle. \qquad (9.74)$$

Let us denote the limit of the operator inside the curly brackets by $-i\hat{H}/\hbar$; we can then write, at a generic time t, the differential equation

$$\hat{H}|a, t\rangle = i\hbar\frac{\partial}{\partial t}|a, t\rangle. \qquad (9.75)$$

The operator $\hat{H}$ is the infinitesimal generator of time-evolution and, in analogy with classical mechanics, is identified with the *quantum Hamiltonian*. If we substitute Eq. (9.73) in (9.74) we obtain an equation for the evolution operator:

$$i\hbar\frac{dU(t, t_0)}{dt} = \hat{H}U(t, t_0) \Leftrightarrow i\hbar\frac{dU(t, t_0)^\dagger}{dt} = -U(t, t_0)^\dagger\, \hat{H}, \qquad (9.76)$$

where we have used the hermiticity property of $\hat{H}$: $\hat{H}^\dagger = \hat{H}$. If the Hamiltonian is time-independent, as it is the case for a free particle, we can easily write the formal solution to the above equation with the initial condition $U(t_0, t_0) = \hat{I}$:

$$U(t, t_0) = U(t - t_0) = e^{-\frac{i}{\hbar}\hat{H}(t-t_0)}. \qquad (9.77)$$

The equation for the wave function

$$\psi(\mathbf{x}, t) = \langle\mathbf{x}|a, t\rangle,$$

is obtained by scalar multiplication of both sides of (9.75) by the bra $\langle \mathbf{x}|$. Taking into account Eq. (9.19) we obtain

$$\hat{H}\psi(\mathbf{x}, t) = i\hbar \frac{\partial}{\partial t}\psi(\mathbf{x}, t), \tag{9.78}$$

that is the *Schrödinger equation*, where $\hat{H}$ is now the Hamiltonian operator realized as a differential operator on wave functions. For a free particle

$$\hat{H} = \frac{|\hat{\mathbf{p}}|^2}{2m} = -\frac{\hbar^2}{2m}\nabla^2,$$

and eq. (9.78) reads:

$$-\frac{\hbar^2}{2m}\nabla^2\psi(\mathbf{x}, t) = i\hbar \frac{\partial}{\partial t}\psi(\mathbf{x}, t), \tag{9.79}$$

where $\nabla^2 \equiv \nabla \cdot \nabla = \sum_{i=1}^{3}\partial_i^2$. Note that in this formulation the dynamical variables described by hermitian operators are not evolving in time, that is they are time-independent, while states, or equivalently wave-functions, are time-dependent. Thinking of time-evolution as of a particular kind of transformation, we have previously referred to such description as the *Schroedinger picture*.

In the *Heisenberg picture* on the other hand, transformations (including time-evolution) act on operators while states stay inert. In this representation therefore states are time-independent while operators $\hat{O}(t)$ representing observables evolve in time. To see how, let us specialize Eq. (9.38) to the time-evolution and apply it to an observable $\hat{O}(t)$:

$$\hat{O}(t) = U(t - t_0)^\dagger \hat{O}(t_0) U(t - t_0) = e^{-\frac{i}{\hbar}\hat{H}(t_0 - t)} \hat{O}(t_0) e^{-\frac{i}{\hbar}\hat{H}(t - t_0)}, \tag{9.80}$$

where we have used the property $U(t - t_0)^\dagger = U(t_0 - t)$. Clearly at $t = t_0$, being $U(t_0, t_0) = \hat{I}$, any Heisenberg dynamical variable, as well as the state of the system, is the same as the corresponding one in the Schroedinger picture. To find the equation of motion for the operator $\hat{O}(t)$ we differentiate both sides of (9.80) with respect to t:

$$\frac{d}{dt}\hat{O}(t) = \left(\frac{d}{dt}U(t - t_0)^\dagger\right)\hat{O}(t_0)U(t - t_0) + U(t - t_0)^\dagger\hat{O}(t_0)\frac{d}{dt}U(t - t_0)$$

$$= \frac{i}{\hbar}\left(\hat{H}U(t - t_0)^\dagger\hat{O}(t_0)U(t - t_0) - U(t - t_0)^\dagger\hat{O}(t_0)U(t - t_0)\hat{H}\right),$$

where we have used (9.76). Using Eq. (9.80) again we find

$$\frac{d}{dt}\hat{O}(t) = \frac{i}{\hbar}[\hat{H}, \hat{O}(t)], \tag{9.81}$$

which is referred to as the *quantum Hamilton equations of motion*.

Let us compare this equation with the Hamilton equations of motion of the classical theory, Eq. (8.98). We see that the time-evolution of a dynamical variable in quantum mechanics can be obtained from the classical formula Eq. (8.98) by replacing the Poisson bracket between the classical observable quantities with the commutator between the corresponding quantum operators, according to the prescription (9.40).

We give another example of this procedure by examining the condition under which a quantum dynamical variable is conserved. In the classical case this happens when the Hamiltonian of the system is invariant under the action of a group of transformations G.

Quantum mechanically the transformation of the Hamiltonian operator $\hat{H}$ under the transformations $U(g)$ of G reads

$$\forall g \in G \; : \quad \hat{H}' = U(g)^\dagger \, \hat{H} \, U(g).$$

The infinitesimal form of the above transformation is given by Eq. (9.39) with $\hat{O} = \hat{H}$:

$$\delta \hat{H} = \hat{H}' - \hat{H} = \frac{i}{\hbar} \delta\theta^r \, [\hat{H}, \hat{G}_r], \quad \forall \theta^r, \tag{9.82}$$

where $\hat{G}_r$ denote the infinitesimal generators of G. As in the classical case, in quantum mechanics the group G is a symmetry or an invariance of the theory if under the action of G-transformations, the Hamiltonian is left invariant (here we assume $\hat{G}_r$ not to explicitly depend on time):

$$\delta \hat{H} = 0 \; \Rightarrow \; \forall r \; : \quad [\hat{H}, \hat{G}_r] = 0.$$

On the other hand from Eq. (9.81), using the invariance condition, we obtain

$$\frac{d}{dt} \hat{G}_r(t) = \frac{i}{\hbar} [\hat{H}, \hat{G}_r] = 0,$$

that is the generators $\hat{G}_r$ of G are *conserved*. Equation (9.82) amounts to saying that, a system is invariant with respect to the transformations in G if and only if the Hamiltonian operator commutes with all the infinitesimal generators of G. Using the exponential representation of a finite time-evolution operator $U(t - t_0)$ and of a finite G-transformation $U(g)$, this property implies that for any $g \in G$ and t, t_0: $U(t - t_0)U(g) = U(g)U(t - t_0)$, that is the result of a time-evolution and of a G-transformation (e.g. a change in the RF) does not depend on the order in which the two are effected on the system.

Let us now mention an important application of Schur's Lemma, see Sect. 7.2, to quantum mechanics. Let G be a symmetry group of a quantum mechanical system. We know, from our previous discussion, that the Hamiltonian operator $\hat{H}$ commutes with the action U of G on the Hilbert space $V^{(c)}$. Its matrix representation on the states will then have the form (7.27), where $c_1, \ldots, c_s$ (s may be infinite!) are the energy levels $E_1, \ldots, E_s$ of the system, and $k_1, \ldots k_s$ their degeneracies. This means that the k_ℓ states $|E_\ell\rangle$ of the system corresponding to a given energy level E_ℓ, define a subspace of $V^{(c)}$ on which an irreducible representation $\mathbf{D}_{k_\ell}$ of the symmetry group G acts. We can easily show this by writing the Schroedinger equation for a state $|E_\ell\rangle$:

$$\hat{H}|E_\ell\rangle = E_\ell |E_\ell\rangle. \tag{9.83}$$

Consider a generic symmetry transformation $g \in G$ and the transformed state $|E_\ell\rangle' = U(g)|E_\ell\rangle$. This state corresponds to the same energy level E_ℓ as the original one, since

$$\hat{H}|E_\ell\rangle' = \hat{H}U(g)|E_\ell\rangle = U(g)\hat{H}|E_\ell\rangle = E_\ell\, U(g)|E_\ell\rangle = E_\ell|E_\ell\rangle', \tag{9.84}$$

where we have used the property that $\hat{H}$ commutes with $U(g)$. Since the above property holds for any $g \in G$, the eigenspace of the Hamiltonian operator corresponding to a given energy level supports a representation of the symmetry group G.

In the generic case, in which there is no accidental degeneracy, $\mathbf{D}_{k_1}, \dots, \mathbf{D}_{k_s}$ are irreducible representations of G. If $\mathbf{D}_{k_i}$ is not irreducible, this may indicate that there exists a larger symmetry group G' containing G, whose action on V_{k_i} is irreducible. In other words *accidental degeneracies may signal the existence of a larger symmetry* of the system. As an example consider the Hydrogen atom which is a system consisting of an electron and a proton, with charges $\pm|e|$. The classical Hamiltonian of the system reads:

$$H(\mathbf{p}, \mathbf{x}) = \frac{|\mathbf{p}|^2}{2m_e} - \frac{e^2}{4\pi |\mathbf{x}|}, \tag{9.85}$$

and is manifestly invariant under rotations $H(\mathbf{p}, \mathbf{x}) = H(\mathbf{p}', \mathbf{x}')$, where $\mathbf{x}' = \mathbf{R}(\theta)\,\mathbf{x}$, $\mathbf{p}' = \mathbf{R}(\theta)\,\mathbf{p}$, since it only depends on the norms of the two vectors. In quantum mechanics, the Hamiltonian operator in the coordinate representation $\hat{H}(\hat{\mathbf{p}}, \mathbf{x}) = \hat{H}(-i\hbar \nabla, \mathbf{x})$ reads

$$\hat{H}(-i\hbar \nabla, \mathbf{x}) = -\frac{\hbar^2}{2m} \nabla^2 - \frac{e^2}{4\pi |\mathbf{x}|}. \tag{9.86}$$

It shares the same symmetry as its classical counterpart: If O_θ is the differential operator defined in Sect. 9.3.1, which implements a rotation on wave functions ($\psi(\mathbf{x}) \to O_\theta \psi(\mathbf{x}) = \psi(\mathbf{R}(\theta)^{-1}\mathbf{x})$), then, applying Eq. (9.70) to $\hat{H}$ we find:

$$\hat{H}'(-i\hbar \nabla, \mathbf{x}) = O_\theta^{-1}\hat{H}(-i\hbar \nabla, \mathbf{x})O_\theta = \hat{H}(-i\hbar \nabla', \mathbf{x}') = \hat{H}(-i\hbar \nabla, \mathbf{x}),$$

namely the Hamiltonian operator is invariant under rotations. Thus by Schur's lemma we expect the wave functions corresponding to a given energy level to define a basis of a representation of SO(3). This is actually the case, although such representation is completely reducible. In other words, there is an accidental degeneracy which can be explained by the existence of a larger symmetry group of the system, which contains, besides the rotation group SO(3) generated by the angular momentum $\hat{\mathbf{M}}$, a further hidden symmetry SO(3)$'$, commuting with the first one, generated by the so called *Laplace–Runge–Lenz vector*. We say that the symmetry group is actually $G = SO(3) \times SO(3)'$.

9.4 Towards a Relativistically Covariant Description

Consider the effect on states of a space-time translation. Suppose a same particle is observed from two different frames $\mathcal{S}$, $\mathcal{S}'$ whose Cartesian rectangular coordinates coincide at all times. The only difference is that the chronometers in the two systems were not set to start at the same time but measure two times, t, t' respectively, related by $t = t' + \epsilon$. The state $|a', t'\rangle$ observed from $\mathcal{S}'$ must coincide with the state $|a, t\rangle$ measured from $\mathcal{S}$ at the same time, so we can write:

$$|a', t'\rangle = |a, t\rangle = |a, t' + \epsilon\rangle = e^{-\frac{i}{\hbar}\hat{H}\epsilon}|a, t'\rangle. \qquad (9.87)$$

Suppose now the two spatial coordinate systems are related by a rigid translation, so that the coordinate vectors of the particle in the two RF's are related as follows: $\mathbf{x} = \mathbf{x}' + \boldsymbol{\epsilon}$. The relation between the two quantum descriptions of the particle becomes:

$$|a', t'\rangle = e^{\frac{i}{\hbar}\hat{\mathbf{p}}\cdot\boldsymbol{\epsilon}}|a, t' + \epsilon\rangle = e^{\frac{i}{\hbar}\hat{\mathbf{p}}\cdot\boldsymbol{\epsilon}}\, e^{-\frac{i}{\hbar}\hat{H}\epsilon}|a, t'\rangle.$$

We see that the effect of the coordinate transformation is implemented on the state by the unitary transformation

$$U(\epsilon^\mu) \equiv e^{\frac{i}{\hbar}\hat{\mathbf{p}}\cdot\boldsymbol{\epsilon}}\, e^{-\frac{i}{\hbar}\hat{H}\epsilon} = e^{-\frac{i}{\hbar}\hat{P}^\mu \epsilon_\mu}, \qquad (9.88)$$

where we have defined the *four-momentum* operator $\hat{P}^\mu \equiv (\frac{1}{c}\hat{H}, \hat{\mathbf{p}})$ and the four-vector $(\epsilon^\mu) \equiv (c\,\epsilon, \boldsymbol{\epsilon})$. Note that in this derivation we have used the property that, for a free particle, $\hat{H}$ and $\hat{\mathbf{p}}$ commute.

Consider now the wave function description of the particle state in the two RF's. Using the definition (9.27) we find:

$$\begin{aligned}\Phi^\alpha_{(a')}(\mathbf{x}', t') &\equiv \langle \mathbf{x}', \alpha|a', t'\rangle = \langle \mathbf{x}', \alpha|e^{\frac{i}{\hbar}\hat{\mathbf{p}}\cdot\boldsymbol{\epsilon}}|a, t' + \epsilon\rangle = \langle \mathbf{x}' + \boldsymbol{\epsilon}, \alpha|a, t' + \epsilon\rangle \\ &= \Phi^\alpha_{(a)}(\mathbf{x}' + \boldsymbol{\epsilon}, t' + \epsilon) = O\,\Phi^\alpha_{(a)}(\mathbf{x}', t'),\end{aligned} \qquad (9.89)$$

which is the correct transformation property under space-time translations of a field $\Phi^\alpha_{(a)}(x^\mu) \equiv \Phi^\alpha_{(a)}(\mathbf{x}, t)$ on Minkowski space. Note that, from an active point of view, the above transformation can be written as the effect on $\Phi^\alpha_{(a)}(x^\mu)$ of a differential operator O which implements a space-time translation on wave functions:

$$\Phi^\alpha_{(a')}(x^\mu) \equiv O\,\Phi^\alpha_{(a)}(x^\mu) = \Phi^\alpha_{(a)}(x^\mu + \epsilon^\mu), \qquad (9.90)$$

where we have just renamed x'^{μ} in (9.89) by x^{μ}. Writing $O = e^{-\frac{i}{\hbar}\hat{P}^{\mu}\epsilon_{\mu}}$, we can find the explicit realization of the four-momentum operator on wave functions from the infinitesimal form of Eq. (9.90) ($\epsilon^{\mu} \ll 1$). The derivation is analogous to that of Eq. (7.86) and yields the following identification

$$\hat{P}^{\mu} = i\hbar \eta^{\mu\nu} \frac{\partial}{\partial x^{\nu}}. \tag{9.91}$$

If we use the short-hand notation $\partial_{\mu} \equiv \frac{\partial}{\partial x^{\mu}}$ and $\partial^{\mu} \equiv \eta^{\mu\nu}\partial_{\nu}$, we can simply write $\hat{P}^{\mu} = i\hbar \partial^{\mu}$.

To achieve a relativistically covariant description of states, we need to define on them Lorentz (and in general Poincaré) transformations, that is they should have a definite transformation property under Poincaré transformations.[11] To this end let us define the following vectors:

$$|x, \alpha\rangle = |(x^{\mu}), \alpha\rangle = |(ct, \mathbf{x}), \alpha\rangle \equiv e^{\frac{i}{\hbar}\hat{H}t}|\mathbf{x}, \alpha\rangle = e^{\frac{i}{\hbar}\hat{P}\cdot x}|(x^{\mu} = 0), \alpha\rangle. \tag{9.92}$$

The wave function corresponding to a given state would then read:

$$\Phi^{\alpha}_{(a)}(x^{\mu}) = \langle \mathbf{x}, \alpha | a, t\rangle = \langle \mathbf{x}, \alpha | e^{-\frac{i}{\hbar}\hat{H}t}|a\rangle = \langle x, \alpha | a\rangle, \tag{9.93}$$

where $|a\rangle \equiv |a, t = 0\rangle$. So far we have just performed redefinitions. Let us now define the action of Poincaré transformations on these states. In analogy with the transformation property of states under rotations (9.62) and space translations (9.47) in the non-relativistic theory, we try to define the action of a Poincaré transformation (Λ, x_0) on the basis $|x, \alpha\rangle$ by means of a unitary operator $U(\Lambda, x_0)$:

$$|x, \alpha\rangle \xrightarrow{(\Lambda, x_0)} U(\Lambda, x_0)|x, \alpha\rangle \equiv D^{\beta}{}_{\alpha}|x', \beta\rangle = D^{\beta}{}_{\alpha}|\Lambda x - x_0, \beta\rangle, \tag{9.94}$$

where we have used the general transformation law (7.46). Consistency requires for these states a normalization condition which generalizes Eq. (9.25):

$$\langle x, \alpha | x', \beta\rangle = \delta^4(x' - x)\delta^{\alpha}_{\beta} = \frac{1}{c}\delta(t' - t)\delta^3(\mathbf{x}' - \mathbf{x})\delta^{\alpha}_{\beta}. \tag{9.95}$$

The index α labels internal degrees of freedom which can be made to freely vary by means of Poincaré transformations *at fixed point* x^{μ} in M_4. We may convince ourselves that the largest group of transformations which leaves x^{μ} fixed is the full Lorentz group.[12] Consider the state corresponding to the origin of the RF $x^{\mu} \equiv 0$

---

[11] In fact we shall characterize a single particle state as belonging to an irreducible representation of the Poincaré group.

[12] This statement seems to be at odds with what we have learned from our earlier discussion about Lorentz transformations: Under a Lorentz transformation a generic position four-vector x^{μ} transforms into a different one $x'^{\mu} = \Lambda^{\mu}{}_{\nu} x^{\nu}$, and the only four vector which is left invariant is the null

and act on it by means of a Lorentz transformation:

$$|0, \alpha\rangle \xrightarrow{\Lambda} U(\Lambda)|0, \alpha\rangle \equiv D^\beta{}_\alpha |0, \beta\rangle. \tag{9.96}$$

By virtue of homogeneity of space-time, whatever statement about the point $x^\mu = 0$ equally applies to any other point x^μ. We conclude that the matrix $\mathbf{D} \equiv (D^\beta{}_\alpha) = \mathbf{D}(\Lambda)$ acting on the internal index α *is a representation of the Lorentz group*. Assuming $\{|x, \alpha\rangle\}$ to be a basis of the Hilbert space we define the coordinate representation of a state by expanding it in this basis:

$$|a\rangle = \int d^4x \sum_\alpha \Phi^\alpha(x) |x, \alpha\rangle. \tag{9.97}$$

Acting on $|a\rangle$ by means of $U(\Lambda, x_0)$ we deduce the transformation property of the coefficients $\Phi^\alpha(x)$:

$$\begin{aligned}|a'\rangle = U(\Lambda, x_0)|a\rangle &= \int d^4x \, \Phi^\alpha(x) \, U(\Lambda, x_0)|x, \alpha\rangle \\ &= \int d^4x \, \Phi^\alpha(x) \, D^\beta{}_\alpha |\Lambda x - x_0, \beta\rangle = \int d^4x' \, \Phi^\alpha(x) \, D^\beta{}_\alpha |x', \beta\rangle \\ &= \int d^4x' \, \Phi'^\beta(x') \, |x', \beta\rangle,\end{aligned} \tag{9.98}$$

where $x' = \Lambda x - x_0$ and we have used the invariance of the elementary space-time volume under Poincaré transformations: $d^4x = d^4x'$. We conclude that

$$\Phi'^\alpha(x') = D^\alpha{}_\beta \, \Phi^\beta(x). \tag{9.99}$$

We have thus retrieved the general transformation property (7.47) of a relativistic field under Poincaré transformations.

(Footnote 12 continued)
one $(x^\mu) \equiv \mathbf{0} = (0, 0, 0, 0)$ defining the origin O of the RF. For a given Lorentz transformation Λ in SO(1, 3) and a point P described by $x \equiv (x^\mu)$, we can define the Poincaré transformation $\Lambda_x \equiv (\mathbf{1}, -x)(\Lambda, 0)(\mathbf{1}, x)$, see Sect. 4.7.2 of Chap. 4 for the notation, which consists in a first translation $(\mathbf{1}, x)$ mapping P into the origin O $(x \to \mathbf{0})$, then a Lorentz transformation Λ which leaves O invariant $(\mathbf{0} \to \mathbf{0})$, followed by a second translation which brings O back into P $(\mathbf{0} \to x)$. By construction Λ_x, which is not pure Lorentz since it contains translations, leaves x invariant. The transformations Λ_x, corresponding to $\Lambda \in$ SO(1, 3), close a group *which has the same structure as the Lorentz group*, though being implemented by different transformations: The correspondence $\Lambda \leftrightarrow \Lambda_x$ for a given x is one-to-one, and, moreover, if $\Lambda \Lambda' = \Lambda''$ then $\Lambda_x \Lambda'_x = \Lambda''_x$. The two groups are said to be *isomorphic*. Transformation groups sharing the same structure represent the same symmetry. We shall denote the group consisting of the Λ_x transformation by SO(1, 3)$_x$. It can be regarded as the *copy* of the Lorentz group, depending on the point x, which leaves x invariant.

As we did in Sect. 7.4.2, we can describe the effect of a Poincaré transformation (Λ, x_0) on $\Phi^\alpha(x)$ in terms of the active action of an operator $O_{(\Lambda, x_0)}$, as in Eq. (7.90):

$$\Phi^\alpha(x) \xrightarrow{(\Lambda, x_0)} \Phi'^\alpha(x') = O_{(\Lambda, x_0)}\, \Phi^\alpha(x'). \qquad (9.100)$$

We then write $O_{(\Lambda, x_0)}$ as the exponential of infinitesimal differential operators, as in Eq. (7.91):

$$O_{(\Lambda, x_0)} = e^{-\frac{i}{\hbar} x_0^\mu \hat{P}_\mu}\, e^{\frac{i}{2\hbar} \theta^{\rho\sigma} \hat{J}_{\rho\sigma}}, \quad D(\Lambda)^\alpha{}_\beta = \left(e^{\frac{i}{2\hbar} \theta^{\rho\sigma} \Sigma_{\rho\sigma}}\right)^\alpha{}_\beta, \qquad (9.101)$$

where, with respect to the notations used in Sect. 7.4.2 we have defined $\hat{P}_\mu = i\hbar\, \mathcal{P}_\mu$ and

$$\hat{J}_{\rho\sigma} = -i\hbar\, \mathcal{L}_{\rho\sigma} = \hat{M}_{\rho\sigma} + \Sigma_{\rho\sigma},$$
$$\hat{M}^{\rho\sigma} = -i\hbar\,(x^\rho \partial^\sigma - x^\sigma \partial^\rho) = -x^\rho \hat{P}^\sigma + x^\sigma \hat{P}^\rho, \quad (\Sigma_{\rho\sigma})^\alpha{}_\beta = -i\hbar\,(L_{\rho\sigma})^\alpha{}_\beta. \qquad (9.102)$$

The commutation relations (7.95) in these new generators read:

$$\left[\hat{J}^{\mu\nu}, \hat{J}^{\rho\sigma}\right] = -i\hbar\,\left(\eta^{\nu\rho}\, \hat{J}^{\mu\sigma} + \eta^{\mu\sigma}\, \hat{J}^{\nu\rho} - \eta^{\mu\rho}\, \hat{J}^{\nu\sigma} - \eta^{\nu\sigma}\, \hat{J}^{\mu\rho}\right), \qquad (9.103)$$

$$\left[\hat{J}^{\mu\nu}, \hat{P}_\rho\right] = -i\hbar\,\left(\hat{P}^\mu \delta^\nu_\rho - \hat{P}^\nu \delta^\mu_\rho\right); \quad \left[\hat{P}_\mu, \hat{P}_\nu\right] = 0. \qquad (9.104)$$

The differential operator $\hat{M}_{\rho\sigma}$ realizes the action of the Lorentz generators on the coordinate dependence of the field, while the matrix $\Sigma_{\rho\sigma}$ defines the corresponding action on the internal components. Clearly, since these two operators act on different degrees of freedom, i.e. different components, they commute:

$$[\hat{M}_{\rho\sigma}, \Sigma_{\mu\nu}] = 0. \qquad (9.105)$$

For the same reason $\Sigma_{\mu\nu}$ commutes with the four-momentum operator $\hat{P}_\mu$. Both $\hat{M}_{\rho\sigma}$ and $\Sigma_{\mu\nu}$ satisfy the same commutation relations (9.103) as $\hat{J}_{\rho\sigma}$, being generators of different Lorentz representations.

Recalling that the angular momentum operator $\hat{\mathbf{J}} = (\hat{J}_i)$ generates the rotation subgroup of SO(1, 3), its components are related to the Lorentz generators as in Eq. (4.176):

$$\hat{J}_i = -\frac{1}{2}\, \epsilon_{ijk}\, \hat{J}^{jk} = \epsilon_{ijk}\, x^i\, \hat{p}^j - \frac{1}{2}\, \epsilon_{ijk}\, \Sigma^{jk} = \hat{M}_i + \hat{S}_i, \qquad (9.106)$$

where we have used the general expression (9.61). From the above equation we deduce the expression of the spin component operators $\hat{S}_i$, which implement the effect of a rotation on the internal components, in terms of the Lorentz generators: $\hat{S}_i = -\frac{1}{2}\, \epsilon_{ijk}\, \Sigma^{jk}$. We shall give a more intrinsic definition of spin in the next section, when describing relativistic states in the momentum representation.

Aside from the operators $\hat{P}^\mu$ which generate the space-time translations and which are associated with the components of the four-momentum p^μ, also the Lorentz generators $\hat{J}^{\mu\nu}$ define, through their eigenvalues, observables $J^{\mu\nu}$ transforming under Lorentz transformations as components of a rank-2 antisymmetric tensor.

The wave function $\Phi^\alpha(x)$ describes then a free particle with a given spin. We have not yet imposed, however, the condition that the particle has a certain mass m, namely that its momentum satisfies the *mass-shell condition*: $p^2 = m^2 c^2$. Such condition will be implemented on $\Phi^\alpha(x)$ by a differential operator obtained upon replacing, in the *mass-shell* equation $p^2 - m^2 c^2 = 0$, the four-momentum with the corresponding operator $\hat{P}^\mu$. We end up with the following, manifestly Lorentz-invariant, differential equation:

$$\left(\hat{P}^\mu \hat{P}_\mu - m^2 c^2\right) \Phi^\alpha(x) = 0 \quad \Leftrightarrow \quad \left(\partial^\mu \partial_\mu + \frac{m^2 c^2}{\hbar^2}\right) \Phi^\alpha(x) = 0. \quad (9.107)$$

We shall examine, in the next chapter, the solutions to the above equation and their physical interpretation.

To make contact with our previous non-relativistic discussion, we have introduced the basis of vectors $|x, \alpha\rangle$ and used it to define the relativistic wave-function. There are however problems in defining such states within a relativistic framework. Let us mention some of them:

- The state $|x, \alpha\rangle$ would describe a particle located in $\mathbf{x}$ at a time t. Due to the possibility, in relativistic processes, of creating new particles, provided the energy involved is large enough, and in the light of Heisenberg's uncertainty principle, there is a physical obstruction in determining the position of a particle at a given time with indefinite precision: The smaller the distances we wish to probe, in order to locate a particle with a sufficiently high precision, the larger the momentum and thus the energy we need to transfer to the particle and, if the energy transferred is large enough to produce one or more particles identical to the original one, we may end up with a system of virtually undistinguishable particles, thus making our initial position measurement meaningless. This is also related to the problem with interpreting the relativistic field $\Phi^\alpha(x)$ as the wave function associated with a given single-particle state, like we did in the non-relativistic theory. We shall comment on this in some more detail in the introduction to next chapter;
- The normalization (9.95), which guarantees that all the states of the basis have a positive norm, is not Lorentz-invariant. Indeed, while $\delta^4(x - x')$ is Lorentz-invariant, $\delta_{\alpha, \beta}$ would be invariant only if the representation $\mathbf{D}$ were *unitary*. There is however a property in group theory which states that *unitary representations of the Lorentz group can only be infinite dimensional* (like the one acting on the infinitely many independent quantum states of a particle). Being $\mathbf{D}$ finite-dimensional, it cannot be unitary, namely $\mathbf{D}^\dagger \mathbf{D} \neq 1$. As an example, suppose the representation $\mathbf{D}$ is the fundamental (defining) representation of the Lorentz group, that is $\mathbf{D}(\Lambda) = \Lambda = (\Lambda^\mu{}_\nu)$. If these matrices were unitary, being real, they would be orthogonal. We have learned, however, that Λ are *pseudo-orthogonal* matrices,

namely they leave the Minkowski metric $\eta_{\mu\nu}$, rather than the Euclidean one $\delta_{\mu\nu}$, invariant. In other words $\Lambda^T \Lambda \neq 1$.[13]

These are some of the reasons why the coordinate basis $\{|x, \alpha\rangle\}$ is ill defined in a relativistic context. We find it however pedagogically useful to use such states in order to introduce the main objects and notations of relativistic field theory starting from non-relativistic quantum mechanics. From now on the main object associated with a given particle will be the relativistic field $\Phi^\alpha(x)$.

9.4.1 The Momentum Representation

Since relativistic effects pose in principle no obstruction in determining the momentum and energy of a particle with indefinite precision, the basis of eigenvectors $|p, r\rangle = |(p^\mu), r\rangle$ of the four-momentum operator:

$$\hat{P}^\mu |p, r\rangle = p^\mu |p, r\rangle, \tag{9.108}$$

is the most appropriate in order to describe the single-particle relativistic quantum states and thus the action on them of the Poincaré group. It is important at this point to stress the differences between the coordinate and the momentum bases, besides the problems mentioned above with defining the former. The internal index r of the state $|p, r\rangle$ labels the degrees of freedom which can be made to freely vary, using Lorentz transformations, *keeping p^μ fixed*. In the coordinate basis we could act on α with the full Lorentz group, generated by $\Sigma_{\mu\nu}$, while keeping x^μ fixed. This was related to the fact that we can move everywhere in space-time by means of space-time translations, and in particular we could move to the origin of the RF, whose invariance under Lorentz transformations is manifest, see Eq. (9.96). The set of Lorentz transformations keeping p^μ fixed would coincide with the full Lorentz group, as for x^μ, only if we were able to Poincaré-transform a generic p^μ into $p^\mu = 0$, corresponding to the absence of a particle (the *vacuum state*). This is clearly not possible, since the momentum four-vector is inert under space-time translations and only transforms under Lorentz transformations, which, however, cannot alter the Lorentz-invariant rest-mass $m^2 = p^2/c^2$ of the particle (if we work

---

[13] A problem related to the non-unitarity of $\mathbf{D}$ is the fact that if we defined $\Phi^\alpha(x) = \langle x, \alpha|a\rangle$, as we did in the non-relativistic theory, it would no longer have the correct transformation property (9.99) under Poincaré transformations. For spin $1/2$ and 1 particles, we can however define a real symmetric matrix $\gamma = (\gamma_{\alpha\beta})$ squaring to the identity $\gamma^2 = 1$, such that $\gamma \mathbf{D}(\Lambda)^\dagger \gamma = \mathbf{D}(\Lambda)^{-1}$: For spin 1 particles $\mathbf{D}$ is the fundamental representation, namely $\mathbf{D}(\Lambda) = \Lambda$, and $\gamma = \eta$, while for spin $1/2$ particles, as it will be shown in next chapter, $\mathbf{D}$ is the spinorial representation and $\gamma = \gamma^0$. We can use this matrix to define $\Phi^\alpha(x) \equiv \langle x, \beta|a\rangle \gamma^{\alpha\beta}$. We shall however, for notational convenience, still write $\Phi^\alpha(x) = \langle x, \alpha|a\rangle$, keeping though this subtlety in mind. We can also use the γ matrix to define a Lorentz-invariant normalization for the $|x, \alpha\rangle$ states: $\langle x, \alpha|x', \beta\rangle = \delta^4(x' - x)\gamma_{\alpha\beta}$. Such normalization is however problematic since γ is not positive definite and thus some of these states would have negative norm!

with *proper Lorentz transformations*, also the sign of p^0 is invariant). The subgroup of the Poincaré group which leaves a covariant object unchanged is called the *little group $G^{(0)}$* of the object. The little group of x^μ is thus the full Lorentz group $O(1, 3)$ (see comment in footnote 12), while the little group of p^μ is a proper subgroup of the Lorentz group. From this it follows that, while α is an index of a representation of the Lorentz group, the index r of $|p, r\rangle$ will label a representation of the little group $G_p^{(0)}$ of p^μ. Clearly the matrix representation of $G_p^{(0)}$ will depend upon p^μ, since it consists of matrices $\Lambda_p^{(0)} = (\Lambda_p^{(0)\,\mu}{}_\nu)$ such that:

$$\Lambda_p^{(0)} \in G_p^{(0)} \quad : \quad \Lambda_p^{(0)\,\mu}{}_\nu p^\nu = p^\mu, \tag{9.109}$$

its structure however only depends on m^2. We can indeed, for a given m^2, evaluate $G^{(0)} = G_{\bar{p}}^{(0)}$ in a RF $\mathcal{S}_0$ in which $p^\mu = \bar{p}^\mu$ is simplest, namely has the largest number of vanishing components (*standard four-momentum*). Any other four-momentum vector $p = (p^\mu)$ with the *same* value of m^2 will be related to $\bar{p} = (\bar{p}^\mu)$ by a Lorentz boost Λ_p: $p = \Lambda_p \bar{p}$. The explicit matrix form of Λ_p was given in (4.190) of Chap. 4. The little group $G^{(0)}$ can be taken as the definition of the *spin* of a particle: It represents the residual symmetry once we keep its four-momentum fixed at some representative value $\bar{p}$.

Let us now see how the structure of the little group depends on m^2 (for a more rigorous discussion of this issue see section E.2 of Appendix E). If $m^2 > 0$, we can define a rest-frame for the particle, in which $\mathbf{p} = \mathbf{0}$ and thus choose $\bar{p} = (mc, \mathbf{0})$ as the standard four-momentum. The little group $G^{(0)}$ clearly contains the group of rotations in three dimensions, since a particle at rest is a system with spherical symmetry. Its generators consist in the components of the total angular momentum $\hat{\mathbf{J}}$, which coincide with the spin components $\hat{\mathbf{S}}$ since, in the rest frame, the orbital part is zero $\hat{\mathbf{M}}|\bar{p}, r\rangle = 0$. We recover the definition of the spin of a massive particle as its angular momentum when it is at rest. In this case the index r spans a representation of the spin group SU(2) (the group SU(2), i.e. the group of 2×2 unitary matrices with determinant 1, has the same local structure as SO(3), though it admits more representations, like the double-valued representation pertaining to particles with spin $1/2$. The spin group is therefore SU(2) rather than SO(3), see Appendix F for a definition of the group SU(2) and its relation to the rotation group). If $m^2 = 0$, we cannot define a rest-frame for the particle. The best that we can do is to go to a RF $\mathcal{S}_0$ in which the X-axis coincides with its direction of motion and thus choose $\bar{p} = (E, E, 0, 0)/c$. Clearly in this RF we have symmetry under rotations about the X-axis, which is generated by the *helicity* operator defined in Sect. 6.3, $\hat{\Gamma} = \frac{\hat{\mathbf{J}} \cdot \bar{\mathbf{p}}}{|\bar{\mathbf{p}}|}$, where $\bar{\mathbf{p}} = (E, 0, 0)/c$. In this case the generators of $G^{(0)}$ consist of other two generators, which we impose to vanish on the state, since they would generate infinitely many internal degrees of freedom, see Appendix E. Helicity therefore provides the definition of the spin for massless particles.

Consider the action of the spin group $G^{(0)}$ on a state $|\bar{p}, r\rangle$. Since it does not affect $\bar{p}$, it will only act on the internal index r:

$$\Lambda^{(0)} \in G^{(0)} : \quad U(\Lambda^{(0)}) |\bar{p}, r\rangle = \mathcal{R}^{r'}_{r} |\bar{p}, r'\rangle, \tag{9.110}$$

where the matrix $\mathcal{R} = (\mathcal{R}^{r'}_{r}) = \mathcal{R}(\Lambda^{(0)})$ represents the action of the $G^{(0)}$-element $\Lambda^{(0)}$ in the representation pertaining to the spin of the particle. Note that $\mathcal{R}$ is always defined to be *unitary*. This is possible since, in contrast to the full Lorentz group, $G^{(0)}$ consists of rotations only (it is SU(2) for massive and effectively SO(2) for massless particles) and thus admits finite dimensional unitary representations.

How does a generic transformation $\Lambda \in$ SO(1, 3) act on a state $|p, r\rangle$? The infinite-dimensional, unitary representation of the Lorentz group acting on the states $|p, r\rangle$ is constructed starting from the finite-dimensional unitary representation $\mathcal{R}$ of $G^{(0)}$ acting on the particle states in the RF S_0 as in Eq. (9.110). The particle state $|p, r\rangle$ in a RF in which the momentum is p is defined by acting on $|\bar{p}, r\rangle$ by means of the boost Λ_p relating $\bar{p}$ to p: $p = \Lambda_p \bar{p}$

$$|p, r\rangle \equiv U(\Lambda_p) |\bar{p}, r\rangle. \tag{9.111}$$

This suffices to define the action of a Lorentz transformation Λ on a generic state $|p, r\rangle$. As it is shown in detail in Appendix E, the action of the unitary operator $U(\Lambda)$ which realizes Λ on states reads:

$$U(\Lambda) |p, r\rangle = \mathcal{R}^{r'}_{r} |\Lambda p, r'\rangle, \tag{9.112}$$

where now the rotation matrix $\mathcal{R}$ in $G^{(0)}$ depends on both Λ and p: $\mathcal{R} = \mathcal{R}(\Lambda, p)$. If Λ is a simple boost, the corresponding rotation $\mathcal{R}(\Lambda, p)$ is called *Wigner rotation*. The method of constructing the unitary infinite-representation of the Lorentz group on the states $|p, r\rangle$ starting from the finite-dimensional representation of the little group is called method of *induced representations*, see Appendix E for a more detailed discussion.

Clearly the effect of a translation on $|p, r\rangle$ is trivial since it amounts to multiplying the state by a phase $e^{\frac{i}{\hbar} P \cdot x_0}$.

Single-particle states are characterized by *irreducible representations* of the Poincaré group. Consequently we require the representation $\mathcal{R}$ of the spin group $G^{(0)}$ to be irreducible, as motivated in Appendix E. On these states the mass-shell condition has not been imposed yet. Just as we did for the eigenstates $|\mathbf{p}, \alpha\rangle$ in Sect. 9.3.1, we can derive the wave function description $\Phi^{\alpha}_{p,r}(x)$ of the states $|p, r\rangle$ by projecting them on the basis $|x, \alpha\rangle$ (see footnote 13 for subtleties with this projection) and writing each vector $|x, \alpha\rangle$ as $e^{\frac{i}{\hbar} \hat{P} \cdot x} |x = 0, \alpha\rangle$, where the point $x = (x^\mu) = 0$ is the origin of the coordinate system:

$$\Phi^{\alpha}_{p,r}(x) = \langle x, \alpha | p, r\rangle = \langle x = 0, \alpha | e^{-\frac{i}{\hbar} \hat{P} \cdot x} | p, r\rangle = c_p \, u^{\alpha}(p, r) \, e^{-\frac{i}{\hbar} p \cdot x}, \tag{9.113}$$

where we have defined $c_p u^\alpha(p, r) \equiv \langle x = 0, \alpha | p, r \rangle$, c_p being a Lorentz-invariant normalization factor: $c_{\Lambda p} = c_p$. Note that $\Phi^\alpha_{p,r}(x)$, for different $p = (p^\mu)$, define a complete set of eigenfunctions of the four-momentum operator (9.91):

$$\hat{P}^\mu \, \Phi^\beta_{p,r}(x) = i\hbar \eta^{\mu\nu} \, \partial_\nu \Phi^\beta_{p,r}(x) = p^\mu \, \Phi^\beta_{p,r}(x). \tag{9.114}$$

For a given particle, the components p^μ are not independent but constrained by the mass-shell condition $p^2 = m^2 c^2$. From now on we describe a single particle state in terms of the simultaneous eigenstates $|\mathbf{p}, r\rangle$ of the linear momentum operator $\hat{\mathbf{p}}$ and of the Hamiltonian $\hat{H}$ whose energy eigenvalue E is fixed by the mass-shell condition $E = E_\mathbf{p} \equiv \sqrt{m^2 c^4 - |\mathbf{p}|^2 c^2} > 0$:

$$|\mathbf{p}, r\rangle \equiv |p, r\rangle_{p^0 = \frac{E_\mathbf{p}}{c}}. \tag{9.115}$$

Such states were defined in Sect. 9.3.1 and normalized as in Eq. (9.54). Their wave function representation is given by Eq. (9.113) in which p^0 is now fixed by the mass-shell condition to $E_\mathbf{p}/c$. The normalization condition (9.54), using Eqs. (9.113) and (9.20), reads:

$$\langle \mathbf{p}, r | \mathbf{p}', s \rangle = \int d^3\mathbf{x} e^{\frac{i}{\hbar}(p-p')\cdot x} \, c_p^* c_{p'} \sum_\alpha u^\alpha(p, r)^* \, u^\alpha(p', s)$$

$$= (2\pi\hbar)^3 \, \delta^3(\mathbf{p} - \mathbf{p}') \, |c_p|^2 \, u(p, r)^\dagger u(p, s) = \frac{(2\pi\hbar)^3}{V} \, \delta^3(\mathbf{p} - \mathbf{p}') \, \delta_{rs},$$

where we have defined the vector $u(p, r) \equiv (u^\alpha(p, r))$. The above equation implies for the vectors $u(p, r)$ the following normalization:

$$u(p, r)^\dagger u(p, s) = \frac{1}{|c_p|^2 \, V} \, \delta_{rs}. \tag{9.116}$$

Comparing Eqs. (9.100) and (9.112) we find for $\Phi^\alpha_{p,r}(x)$ the following transformation property under a Lorentz transformation:

$$O_\Lambda \, \Phi^\alpha_{p,r}(x) = D(\Lambda)^\alpha_{\ \beta} \, \Phi^\beta_{p,r}(\Lambda^{-1} x) = c_p \, D(\Lambda)^\alpha_{\ \beta} \, u^\alpha(p, r) \, e^{-\frac{i}{\hbar} p \cdot (\Lambda^{-1} x)}$$

$$= c_p \, D(\Lambda)^\alpha_{\ \beta} \, u^\alpha(p, r) \, e^{-\frac{i}{\hbar}(\Lambda p) \cdot x} = \mathcal{R}(\Lambda, p)^{r'}_{\ r} \, \Phi^\alpha_{\Lambda p, r}(x)$$

$$= c_p \, \mathcal{R}(\Lambda, p)^{r'}_{\ r} \, u^\alpha(\Lambda p, r') \, e^{-\frac{i}{\hbar}(\Lambda p) \cdot x}, \tag{9.117}$$

where we have used the Lorentz invariance of c_p. From the above equation which we deduce:

$$D(\Lambda)^\alpha_{\ \beta} \, u^\beta(p, r) = \mathcal{R}(\Lambda, p)^{r'}_{\ r} \, u^\alpha(\Lambda p, r'). \tag{9.118}$$

The vectors $u(p, r)$ for a scalar particle are proportional to 1, being the corresponding Lorentz and spin representations, spanned by the indices α and r, trivial. For a spin

1/2 particle, like the electron or the positron, as we shall see in the next chapter, they will be of two kinds: One denoted by the same symbol $u(p, r)$, the other by $v(p, r)$. In this case $r = 1, 2$ labels the two spin states $S_z = \pm \hbar/2$, while α labels the four components corresponding to the spinorial representation of the Lorentz group. As far as the photon is concerned, the role of $u^\alpha(p, r)$ will be played by the polarization vector $\varepsilon^r_\mu(p)$, where $\alpha \equiv \mu$ labels the Lorentz representation $\mathbf{D}(\Lambda) = \Lambda$ of the potential four-vector, while $r = 1, 2$ label the two transverse polarizations.

As remarked earlier, the state $|a\rangle$ of a particle is in general described by a wave packet $\Phi^\alpha(x)$ propagating in space and can be represented in terms of its momentum representation by using Eqs. (9.93) and (9.113):

$$\Phi^\alpha(x) = \langle x, \alpha | a \rangle = \int \frac{d^3\mathbf{p}}{(2\pi\hbar)^3} \, V \sum_\beta \langle x, \alpha | \mathbf{p}, \beta \rangle \langle \mathbf{p}, \beta | a \rangle$$

$$= \int \frac{d^3\mathbf{p}}{(2\pi\hbar)^3} \, V \, c_p \sum_r u(p, r)^\alpha \tilde{\Phi}(\mathbf{p}, r) \, e^{-\frac{i}{\hbar} p \cdot x}$$

$$= \int \frac{d^3\mathbf{p}}{(2\pi\hbar)^3} \, V \, c_p \sum_r u(p, r)^\alpha \tilde{\Phi}(\mathbf{p}, r) \, e^{-\frac{i}{\hbar}(Et - \mathbf{p} \cdot \mathbf{x})}, \qquad (9.119)$$

where, we have defined $\tilde{\Phi}^\alpha(\mathbf{p}) \equiv \langle \mathbf{p}, \alpha | a \rangle$. A wave packet is thus expressed as a superposition of plane waves $e^{-\frac{i}{\hbar}(Et - \mathbf{p} \cdot \mathbf{x})}$ with angular frequency $\omega = E/\hbar$ and wave number $\mathbf{k} = \mathbf{p}/\hbar$. It propagates in space at a speed which is the *group velocity* of the wave, and which is given by the well known formula $v = \frac{d\omega}{d|\mathbf{k}|}$. In terms of the energy and linear momentum, using the relativistic relation between $E = E_\mathbf{p}$ and $|\mathbf{p}|$, we find:

$$v = \frac{d\omega}{d|\mathbf{k}|} = \frac{dE}{d|\mathbf{p}|} = \frac{|\mathbf{p}| \, c^2}{E}, \qquad (9.120)$$

which is indeed the expression of the velocity of a free particle in special relativity.

9.4.2 Particles and Irreducible Representations of the Poincaré Group

An elementary particle state is characterized as transforming in an irreducible representation of the Poincaré group. Such representations are uniquely defined by the mass m and the spin s of the particle which, as we are going to show below, are indeed invariants of the group, namely if we change inertial RF their values are unaffected. Therefore all states belonging to the base space of an irreducible representation of the Poincaré group, being quantum descriptions of a same particle from different inertial RF's, share the same values of m and s. To show that these observables are Poincaré invariant, we first need to express them in terms of Lorentz-invariant quantities. On a single particle state the action of the Poincaré generators $\hat{J}_{\rho\sigma}$, $\hat{P}_\mu$ is defined. With respect to the Lorentz group, these are a rank-2 antisymmetric tensor and a four-vector, respectively. Using the Lorentz-invariant tensor $\epsilon_{\mu\nu\rho\sigma}$ we can define a second four-vector, besides $\hat{P}_\mu$:

$$\hat{W}_\mu \equiv -\frac{1}{2}\,\epsilon_{\mu\nu\rho\sigma}\,\hat{J}^{\nu\rho}\hat{P}^\sigma, \qquad (9.121)$$

which is called the *Pauli-Lubanski* four-vector. We first notice that the component $\hat{M}_{\nu\rho}$ of $\hat{J}_{\nu\rho}$ does not contribute to $\hat{W}_\mu$ since

$$-\frac{1}{2}\,\epsilon_{\mu\nu\rho\sigma}\,\hat{M}^{\nu\rho}\hat{P}^\sigma = -\hbar^2\,\epsilon_{\mu\nu\rho\sigma}x^\nu\,\partial^\rho\partial^\sigma = 0, \qquad (9.122)$$

where we have used the property that the ϵ-tensor is totally antisymmetric in its four indices and that two partial derivatives commute: $\partial^\rho\partial^\sigma = \partial^\sigma\partial^\rho$. By the same token one can show that $[\hat{W}_\mu, \hat{P}_\nu] = 0$, which implies that $\hat{W}_\mu$, just as $\hat{P}_\mu$, is invariant under space-time translations. Clearly the mass m of a particle is Lorentz-invariant since it is the eigenvalue of the Lorentz-invariant operator $\frac{1}{c^2}\hat{P}_\mu\hat{P}^\mu$.

Consider now a particle with mass $m \neq 0$. In its rest frame $\mathcal{S}_0$ its linear momentum vanishes, $p_i = 0$, that is its state is annihilated by the operators $\hat{p}^i$, while the eigenvalue of the time-component $\hat{P}^0$ corresponds to the rest energy mc. In this frame we can replace the components of the operator $\hat{P}^\mu$ in the expression of $\hat{W}_\mu$ by their eigenvalues $\bar{p} = (mc, 0, 0, 0)$. The only non-vanishing components of $\hat{W}_\mu$ in this RF are the space ones:

$$\hat{W}_i = -\frac{mc}{2}\,\epsilon_{ijk0}\,\Sigma^{jk} = \frac{mc}{2}\,\epsilon_{ijk}\,\Sigma^{jk} = -mc\,\hat{S}_i\,, \quad \hat{W}_0 = 0, \qquad (9.123)$$

where we have used the convention that $\epsilon_{ijk0} = -\epsilon_{0ijk} = -\epsilon_{ijk}$, namely that $\epsilon_{0123} = +1$, and the definition of the spin-component operators, see Eqs. (9.101)–(9.106). The squared norm $\hat{W}_\mu\hat{W}^\mu$ of $\hat{W}_\mu$ is Lorentz-invariant and, on the states in the rest frame, reads

$$\hat{W}_\mu\hat{W}^\mu = -\sum_{i=1}^{3}\hat{W}^i\hat{W}^i = -m^2c^2\,|\hat{\mathbf{S}}|^2 = -\hbar^2\,m^2c^2\,s(s+1), \qquad (9.124)$$

where we have replaced $|\hat{\mathbf{S}}|^2$ by its eigenvalue $\hbar^2\,s(s+1)$ defining the spin s of the particle. Being $\hat{W}_\mu\hat{W}^\mu$ Lorentz-invariant, its eigenvalue will not change if we switch to a generic RF. We then conclude that the mass m and the spin s of an elementary particle are Poincaré invariant quantities. Using the four-vectors $\hat{P}^\mu$, $\hat{W}^\mu$, we could in principle build a third Lorentz-invariant operator $\hat{P} \cdot \hat{W} \equiv \hat{P}^\mu\hat{W}_\mu$. Such operator is however null, being proportional to $\epsilon_{\mu\nu\rho\sigma}\,\Sigma^{\mu\nu}\,\partial^\rho\partial^\sigma = 0$. Since there are no other independent invariant, commuting with the previous ones, which can be constructed out of the eigenvalues of the Poincaré generators, we conclude that *a single particle state is completely defined by the values of the mass m and the spin s*.[14]

Consider now the case of a massless particle. The standard four-momentum vector can be chosen to be $\bar{p} = (E, E, 0, 0)/c = (E/c, \bar{\mathbf{p}})$. Let us compute the four vector $\hat{W}_\mu$ on the states $|\bar{p}, r\rangle$:

$$\hat{W}_\mu = -\frac{|\bar{\mathbf{p}}|}{2} \left(\epsilon_{\mu\nu\rho 0} + \epsilon_{\mu\nu\rho 1} \right) \hat{J}^{\nu\rho}. \tag{9.125}$$

In components:

$$\hat{W}_0 = -\frac{|\bar{\mathbf{p}}|}{2} \epsilon_{01ij} \hat{J}^{ij} = |\bar{\mathbf{p}}| \hat{J}^1; \quad \hat{W}_1 = \frac{|\bar{\mathbf{p}}|}{2} \epsilon_{01ij} \hat{J}^{ij} = -|\bar{\mathbf{p}}| \hat{J}^1,$$

$$\hat{W}_a = |\bar{\mathbf{p}}| \epsilon_{ab} (\hat{J}^{0b} - \hat{J}^{1b}); \quad a, b = 2, 3. \tag{9.126}$$

As proven in the Appendix E, in order to have finitely many spin states, we need the two operators $\hat{N}^a \equiv \hat{J}^{0a} - \hat{J}^{1a}$ to vanish on the states $|\bar{p}, r\rangle$, so that $\hat{W}_a = 0$ and we can effectively write:

$$\hat{W}_\mu = \hat{J}^1 \, \bar{p}_\mu = \hat{\Gamma} \, \bar{p}_\mu, \tag{9.127}$$

where we have defined the helicity operator as

$$\hat{\Gamma} \equiv \frac{\hat{\mathbf{J}} \cdot \bar{\mathbf{p}}}{|\bar{\mathbf{p}}|} = \hat{J}^1. \tag{9.128}$$

In going from $\mathcal{S}_0$ to any other frame $\mathcal{S}$, $\bar{p}_\mu$ and $\hat{W}_\mu$ are four vectors transforming by the same Lorentz transformation, so that, in $\mathcal{S}$, $\hat{W}_\mu = p_\mu \, \hat{\Gamma}$. We conclude that $\hat{\Gamma}$ is a Lorentz-invariant operator. The condition that the single particle state transform in an irreducible representation of the little group further implies that there can be just two helicity states:

$$\hat{\Gamma} |\hat{p}, \pm s\rangle = \pm \hbar s |\hat{p}, \pm s\rangle, \tag{9.129}$$

The state of a single massless particle is completely defined by the value if its helicity, which is a Poincaré invariant quantity.[15]

---

[14]If we consider proper Lorentz transformations ($\Lambda^0{}_0 \geq 0$, $\det\Lambda = 1$), the sign of p^0, eigenvalue of $\hat{P}^0$, is invariant as well.

[15]Here we are restricting to proper Lorentz transformations. The parity transformation $\Lambda_P : p^0 \to p^0$, $\mathbf{p} \to -\mathbf{p}$ reverses the sign of Γ.

9.5 A Note on Lorentz-invariant Normalizations

In this note we show that the normalization that we have adopted for single particle states is Lorentz-invariant. To this end let us consider a particle of mass m and let S_0 denote its rest frame in which $\bar{\mathbf{p}} = \mathbf{0}$ and $\bar{p}^0 = mc$. S_0 then moves, relative to a given RF S, at the corresponding velocity $\mathbf{v}$ of the particle. The relation between the four-momenta $\bar{p}$ and p of the particle in S_0 and S, respectively, is given by the Lorentz boost Λ_p, using Eq. (4.190). If we write $\bar{p} = \Lambda_p^{-1} p$, expressing the transformation matrix in terms of $\mathbf{v}$ we have:

$$\bar{p}^0 = \gamma(v)\left(p^0 - \frac{\mathbf{v} \cdot \mathbf{p}}{c}\right) = \gamma(v)\left(1 - \frac{v^2}{c^2}\right)p^0 = \frac{1}{\gamma}p^0,$$

$$\bar{p}^i = p^i + (\gamma - 1)\frac{\mathbf{v} \cdot \mathbf{p}}{c^2}v^i - \gamma\frac{\mathbf{v}}{c}p^0, \tag{9.130}$$

where we have used the relation $\mathbf{p} = p^0 \mathbf{v}/c$ and $v^2 \equiv |\mathbf{v}|^2$ (here, as usual, upper and lower indices for three-dimensional vectors are the same: $v^i = v_i$, $p^i = p_i$). If we perturb the rest state of the particle in S_0 by an infinitesimal velocity, but keeping the relative motion between the two frames unchanged, the momentum p relative to S will vary by an infinitesimal amount $p \rightarrow p + dp$. We can relate the infinitesimal variation of $\bar{\mathbf{p}}$ to that of $\mathbf{p}$ by computing the Jacobian matrix $\mathcal{J}_p{}^i{}_j = \frac{\partial \bar{p}^i}{\partial p^j}$:

$$d\bar{p}^i = \frac{\partial \bar{p}^i}{\partial p^j}dp^j = J_p{}^i{}_j dp^j. \tag{9.131}$$

This Jacobian is computed from the transformation law (9.130) by taking into account that p^0 is not independent of $\mathbf{p}$, being $p^0 = \sqrt{|\mathbf{p}|^2 + m^2 c^2}$. Using the property:

$$\frac{\partial p^0}{\partial p^i} = \frac{p^i}{p^0} = \frac{v^i}{c}, \tag{9.132}$$

from Eq. (9.130) we find:

$$\mathcal{J}_p{}^i{}_j = \frac{\partial \bar{p}^i}{\partial p^j} = \delta^i_j + (\gamma - 1)\frac{v^i v_j}{v^2} - \gamma\frac{v^i}{c}\frac{\partial p^0}{\partial p^j}$$

$$= \delta^i_j + \left(\frac{1}{\gamma} - 1\right)\frac{v^i v_j}{v^2}. \tag{9.133}$$

The reader can easily verify that the matrix $\mathcal{J}_p$ has one eigenvalue $1/\gamma$ corresponding to the eigenvector $\mathbf{v}$, and two eigenvalues 1, corresponding to the two vectors perpendicular to $\mathbf{v}$. The determinant of this matrix, being the product of its eigenvalues, is therefore:

$$\det(\mathcal{J}_p) = \frac{1}{\gamma}. \tag{9.134}$$

We can now compute the transformation property of an infinitesimal volume in momentum space when moving from $\mathcal{S}$ to $\mathcal{S}_0$:

$$d^3\bar{\mathbf{p}} \equiv d\bar{p}_x d\bar{p}_y d\bar{p}_z = |\det(\mathcal{J}_p)| d^3\mathbf{p} = \frac{1}{\gamma} d^3\mathbf{p}. \tag{9.135}$$

Note from Eq. (9.130) that the same relation holds for the energies: $\bar{p}^0 = mc = p^0/\gamma$. We conclude that:

$$\frac{d^3\mathbf{p}}{p^0} = \frac{d^3\bar{\mathbf{p}}}{\bar{p}^0}, \tag{9.136}$$

namely that $d^3\mathbf{p}/p^0$ is Lorentz-invariant.

Let us now consider the relation between the position vectors of the particle in the two frames:

$$\bar{x}^0 = \gamma(v) \left(x^0 - \frac{\mathbf{v} \cdot \mathbf{x}}{c} \right),$$
$$\bar{x}^i = x^i + (\gamma - 1) \frac{\mathbf{v} \cdot \mathbf{x}}{c^2} v^i - \gamma \frac{\mathbf{v}}{c} x^0. \tag{9.137}$$

The above equations allow us to compute the relation between the measures $dV_0 = d^3\bar{\mathbf{x}}$ and $dV = d^3\mathbf{x}$ which are infinitesimal cubic volumes in $\mathcal{S}_0$ and $\mathcal{S}$, respectively. The two quantities are related by the Jacobian matrix $\mathcal{J}_x{}^i{}_j \equiv \frac{\partial \bar{x}^i}{\partial x^j}$ which can be computed from (9.137). We need however to observe that, in contrast to the case of the four- momentum, where the energy p^0 depends on the remaining space components, the four components of the position vector are independent and thus $\partial x^0/\partial x^i = 0$. The Jacobian matrix $\mathcal{J}_x{}^i{}_j$ is then easily computed to be:

$$\mathcal{J}_x{}^i{}_j = \delta_j^i + (\gamma - 1) \frac{v^i v_j}{v^2}. \tag{9.138}$$

The eigenvalues of the above matrix are γ (eigenvector $\mathbf{v}$) and twice 1 (eigenvectors perpendicular to the velocity), so that $\det(\mathcal{J}_x) = \gamma$ and

$$dV_0 = d^3\bar{\mathbf{x}} = |\det(\mathcal{J}_x)| dV = \gamma \, dV. \tag{9.139}$$

The same relation holds for a finite volume: $V_0 = \int dV_0 = \gamma \int dV = \gamma V$. This result was also obtained in Chap. 1 as a consequence of the contraction of lengths along the direction of relative motion.

We conclude that the following quantities are Lorentz-invariant:

$$d^3\mathbf{p} \, V = d^3\bar{\mathbf{p}} \, V_0 \; ; \; E V = mc^2 \, V_0. \tag{9.140}$$

From equation (9.139) it also follows the transformation property of the *density of particles*. Indeed the particle density is computed in a given RF $\mathcal{S}$ as the ratio between the number of particles N contained in a volume V and V: $\rho \equiv N/V$. Since N does not depend on the RF, ρ transforms under a Lorentz transformation as $1/V$, namely, if $\rho_0 = N/V_0$ is the density in $\mathcal{S}_0$, we have:

$$\rho = \gamma \rho_0. \tag{9.141}$$

Consider now the transformation property of a δ^3-function computed in the difference between two momenta $\mathbf{p}$ and $\mathbf{q}$, which can be shown to be:

$$\delta^3(\bar{\mathbf{p}} - \bar{\mathbf{q}}) = \frac{1}{|\det(\mathcal{J}_p)|}\, \delta^3(\mathbf{p} - \mathbf{q}) = \gamma\, \delta^3(\mathbf{p} - \mathbf{q}). \tag{9.142}$$

From the above property we conclude that $\frac{1}{V}\,\delta^3(\mathbf{p} - \mathbf{q})$, which is the normalization that we have chosen for the momentum eigenstates, is *Lorentz-invariant* as well.

Since the value of the product $E\,V$ does not depend on the RF, we can fix for the volume, as measured in $\mathcal{S}_0$, an arbitrary value V_0 and write in a generic RF $\mathcal{S}$:

$$2E\,V = 2mc^2 V_0 \;\Rightarrow\; V = \frac{c_0}{2E}, \tag{9.143}$$

where we have defined $c_0 \equiv 2mc^2 V_0$. Since the definition of c_0 is referred to a specific RF $\mathcal{S}_0$, it is Lorentz-invariant by construction. In all formulas we can then replace the normalization volume V by $1/(2E)$ times this Lorentz-invariant normalization factor c_0 which however finally drops off the expression of any measurable quantity, as we shall show in the last chapter when evaluating transition probabilities and cross sections for interaction processes.

The above conclusions also apply to massless particles, although in this case a rest frame $\mathcal{S}_0$ cannot be defined.

9.5.1 References

For further reading see Refs. [4, 8] (Vols. 3, 4), [3].

Chapter 10
Relativistic Wave Equations

10.1 The Relativistic Wave Equation

In the previous chapter we have recalled the basic notions of non-relativistic quantum mechanics. We have seen that, in the Schroedinger representation, the physical state of a free particle of mass m is described by a wave function $\psi(\mathbf{x}, t)$ which is itself a *classical field* having a probabilistic interpretation. For a single free particle this function is solution to the Schroedinger equation (9.79). A system of N interacting particles will be described by a wave function $\psi(\mathbf{x}_1, \mathbf{x}_2, \ldots, \mathbf{x}_N; t)$ whose squared modulus represents the probability density of finding the particles at the points $\mathbf{x}_1, \mathbf{x}_2, \ldots, \mathbf{x}_N$ at the time t. In this description *the number N of particles is always constant* that is it cannot vary during the interaction.

Note that the conservation of the number of particles is related to the conservation of mass in a non-relativistic theory: The sum of the rest masses of the particles, and in fact the identity of each particle, cannot change during the interaction. A change in this number would imply a variation in the sum of the corresponding rest masses.

Strictly related to this property of the Schroedinger equation is the fact that the total probability is conserved in time. Let us recall the argument in the case of a single particle.

The normalization of $\psi(\mathbf{x}, t)$ is fixed by requiring that the probability of finding the particle anywhere in space at any time t be one:

$$\int_V d^3\mathbf{x}\, |\psi(\mathbf{x}, t)|^2 = 1,$$

where $V = \mathbb{R}^3$ representing the whole space. This total probability should not depend on time, and indeed, by using Schroedinger's equation and Gauss' law we find:

© Springer International Publishing Switzerland 2016
R. D'Auria and M. Trigiante, *From Special Relativity to Feynman Diagrams*,
UNITEXT for Physics, DOI 10.1007/978-3-319-22014-7_10

$$\frac{d}{dt}\int_V d^3\mathbf{x}\,|\psi(\mathbf{x},t)|^2 = \int_V d^3\mathbf{x}\,\left(\psi^*\frac{\partial}{\partial t}\psi + \psi\frac{\partial}{\partial t}\psi^*\right)$$

$$= \frac{i\hbar}{2m}\int_V d^3\mathbf{x}\,\left[(\nabla^2\psi)\psi^* - \psi\nabla^2\psi^*\right]$$

$$= \frac{i\hbar}{2m}\int_V d^3\mathbf{x}\,\nabla\cdot\left(\psi^*\nabla\psi - \psi\nabla\psi^*\right)$$

$$= \frac{i\hbar}{2m}\int_{S_\infty} dS\,\mathbf{n}\cdot(\psi^*\nabla\psi - \psi\nabla\psi^*) = 0, \qquad (10.1)$$

$\mathbf{n}$ being the unit vector orthogonal to dS and S_∞ is the surface at infinity which ideally encloses the whole space V. The last integral over S_∞ in the above equation vanishes since both ψ and $\nabla\psi$ vanish sufficiently fast at infinity. Thus *the total probability is conserved in time*.

Equation (10.1) can also be neatly expressed, in a local form, as a *continuity equation*:

$$\partial_t\rho + \nabla\cdot\mathbf{j} = 0, \quad \rho \equiv |\psi(\mathbf{x},t)|^2, \quad \mathbf{j} \equiv \frac{-i\hbar}{2m}(\psi^*\nabla\psi - \psi\nabla\psi^*), \qquad (10.2)$$

which, as we have seen, holds by virtue of Schroedinger's equation.

Can the above properties still be valid in a relativistic theory? Let us give some physical arguments about why the very concept of wave function looses its meaning in the context of a relativistic theory. As emphasized in Chap. 9, in non-relativistic quantum mechanics $\mathbf{x}$ and t play different roles, the former being a dynamical variable as opposed to the latter.

Furthermore we know that one of the most characteristic features of elementary particles is their possibility of generation, annihilation, and reciprocal transformation as a consequence of their interaction. Photons can be generated by electrons in motion within atoms, neutrinos are emitted in β-decays, a neutral pion, a composite particle of a quark and an anti-quark, can decay and produce two photons, a fast electron moving close to a nucleus can produce photons which in turn may transform in electron-positron pairs, and so on.

That means that in phenomena arising from high energy particle interactions, *the number of particles is no longer conserved*.

Consequently some concepts of the non-relativistic formulation of quantum mechanics must be consistently revised.

First of all, we must give up the possibility of localizing in space and time a particle with absolute precision, which was instead allowed in the non-relativistic theory. Indeed if in a relativistic theory we were to localize a particle within a domain of linear dimension Δx less than $\hbar/2mc$, by virtue of the Heisenberg uncertainty principle $\Delta x\,\Delta p_x \geq \hbar/2$, the measuring instrument should exchange with the particle a momentum $\Delta p_x \geq mc$, carried for example by a photon. Such a photon of momentum Δp_x, would carry an energy $\Delta E = c \, \Delta p_x \geq mc^2$ which is greater than or equal to the rest energy of the particle. This would be in principle sufficient to create a particle (or better a couple particle-antiparticle, as we shall see) of rest mass m which may be virtually undistinguishable from the original one.

It is therefore impossible to localize a particle in a region whose linear size is of the order of the Compton wavelength $\hbar/mc$. In the case of photons, having $m = 0$ and $v = c$, the notion of position of the particle simply does not exist.

The existence of a minimal uncertainty $\Delta x \approx \hbar/mc$ in the position of a particle also implies a basic uncertainty in time, since from the inequality $\Delta t \Delta E \geq \frac{\hbar}{2}$ and the condition $\Delta E \leq mc^2 \approx \frac{\hbar c}{\Delta x}$ deduced above, it follows that $\Delta t \gtrsim \frac{\hbar}{\Delta E} \gtrsim \frac{\Delta x}{c} \approx \hbar/mc^2$ (note that in the non-relativistic theory $c = \infty$ so that Δt can be zero). As far as the uncertainty in the momentum of a particle is concerned, we note that from $\Delta x \lesssim c \, \Delta t$ it follows that $\Delta p \gtrsim \frac{\hbar}{c \Delta t}$, that is the uncertainty in the momentum p_x can be made as small as we wish ($\Delta p_x \to 0$) just by waiting for a sufficiently long time ($\Delta t \to \infty$). This can certainly be done for *free particles*. Localizing a particle in space and time with indefinite precision is thus conceptually not possible within a relativistic context and the interpretation of $\rho \equiv |\psi(\mathbf{x}, t)|^2$ as the probability density of finding a particle in $\mathbf{x}$ at a time t should be substantially reconsidered. By the same token, we can conclude that, using the momentum representation $\tilde{\psi}(\mathbf{p})$ of the wave function instead, we can consistently define a *probability density in the momentum space* as $|\tilde{\psi}(\mathbf{p})|^2$.

The argument given above relies on the possibility, in high energy processes, for particles to be created and destroyed. This fact, as anticipated earlier, is at odds with the Schroedinger's formulation of quantum mechanics, which is based on the notion of single particle state, or, in general of multi-particle states with a fixed number of particles. Such description is no longer appropriate in a relativistic theory.

In order to have a more quantitative understanding of this state of affairs let us go back to the quantum description of the electromagnetic field given in Chap. 6.

We have seen that in the Coulomb gauge ($A^0 = 0$, $\nabla \cdot \mathbf{A} = 0$), the *classical field* $\mathbf{A}(\mathbf{x}, t)$ satisfies the Maxwell equation:

$$\Box \mathbf{A} = \frac{1}{c^2} \frac{\partial^2}{\partial t^2} \mathbf{A} - \nabla^2 \mathbf{A} = 0. \qquad (10.3)$$

Suppose that we do not quantize the field as we did in Chap. 6, but consider the Maxwell equation as the wave equation for the classical field $\mathbf{A}(\mathbf{x}, t)$, just as the Schroedinger equation is the wave equation of the classical field $\psi(\mathbf{x}, t)$. We may ask whether a solution $\mathbf{A}(\mathbf{x}, t)$ to Maxwell's equations can be consistently given the same probabilistic interpretation as a solution $\psi(\mathbf{x}, t)$ to the Schroedinger equation. In other words, does the quantity $|A_\mu(\mathbf{x}, t)|^2 d^3\mathbf{x}$ make sense as probability of finding a photon with a given polarization in a small neighborhood $d^3\mathbf{x}$ of a point $\mathbf{x}$ at a time t?

To answer this question we consider the Fourier expansion of the classical field $\mathbf{A}(\mathbf{x}, t)$ given in (6.16):

$$\mathbf{A}(\mathbf{x}, t) = \sum_{k_1 k_2 k_3} \left(\boldsymbol{\epsilon}_{\mathbf{k}} \, e^{-i \, k \cdot x} + \boldsymbol{\epsilon}_{\mathbf{k}}^* \, e^{i \, k \cdot x} \right), \tag{10.4}$$

where $\boldsymbol{\epsilon}_{\mathbf{k}}$ can be written as in Eq. (6.43)

$$\boldsymbol{\epsilon}_{\mathbf{k}} = c \sqrt{\frac{\hbar}{2\omega_k V}} \sum_{\alpha=1}^{2} a_{\mathbf{k},\alpha} \, \mathbf{u}_{\mathbf{k},\alpha}, \tag{10.5}$$

but with the operators $a, a^\dagger$ replaced by numbers a, a^* since we want to consider $\mathbf{A}(\mathbf{x}, t)$ as a *classical* field. If Maxwell's propagation equation could be regarded as a quantum wave equation, then, according to ordinary quantum mechanics, the (complex) component of the Fourier expansion of $\mathbf{A}(\mathbf{x}, t)$

$$\mathbf{A}_{\mathbf{k},\alpha}(x) \equiv c \sqrt{\frac{\hbar}{2\omega_k V}} \, a_{\mathbf{k},\alpha} \mathbf{u}_{\mathbf{k},\alpha} e^{-ik \cdot x},$$

can be given the interpretation of *eigenstate* of the four momentum operator $\hat{P}_\mu$, describing a free particle with polarization $\mathbf{u}_{\mathbf{k},\alpha}$, energy $E = \hbar\omega$ and momentum $\mathbf{p} = \hbar\mathbf{k}$, respectively and satisfying the relation $E/c = |\mathbf{p}|$. This would imply that $\mathbf{A}_{\mathbf{k},\alpha}(x)$ represents the wave function of a *photon* with definite values of energy and momentum. Consequently it would seem reasonable to identify the four potential $A_\mu(\mathbf{x}, t)$ as the *photon wave function* expanded in a set of eigenstates, so that the Maxwell equation for the vector potential would be the natural relativistic generalization of the non-relativistic Schroedinger's equation.

We note however that, while the Schroedinger's equation is of *first order in the time derivatives*, the Maxwell equation, being relativistic and therefore Lorentz-invariant, contains the operator $\square \equiv 1/c^2\partial_t^2 - \nabla^2$ which is of *second order both in time and in spatial coordinates*. This makes a great difference as far as the conservation of probability is concerned since the proof (10.1) of the continuity equation (10.2) makes use of the Schroedinger equation (9.78). More specifically such proof strongly relies on the fact that the Schroedinger equation is of first order in the time derivative and of second order in the spatial ones.

The fact that Maxwell's propagation equation, involves *second order derivatives* with respect to time, makes it impossible to derive a continuity equation for the "would be" probability density $\rho \equiv |\mathbf{A}(x)|^2$: $\partial_t\rho + \nabla \cdot \mathbf{j} \neq 0$. Indeed the first order time derivatives are actually Cauchy data of the Maxwell propagation equation. As a consequence *the quantity ρ cannot be interpreted as a probability density, since the total probability of finding a photon in the whole space would not be conserved.*

On the other hand, as we have illustrated when discussing the quantization of the electromagnetic field, these difficulties are circumvented if we quantize the infinite set of canonical variables associated with $A_\mu(\mathbf{x}, t)$ by the usual prescription of converting Poisson brackets into commutators. This is effected by converting the coefficients $a_{\mathbf{k},\alpha}$, $a^*_{\mathbf{k},\alpha}$, defined in Eq. (10.5), and thus each Fourier component $\epsilon_\mathbf{k}$, into operators through the general procedure introduced in Chap. 6 under the name of *second quantization*.[1] In this new framework the classical field $\mathbf{A}_\mu(\mathbf{x}, t)$ becomes a *quantum field*, that is an operator, and the quantum states of the electromagnetic field are described in the *occupation number representation* by the multi-photon state $|\{N_{\mathbf{k},\alpha}\}\rangle$, characterized by $N_{\mathbf{k},\alpha}$ photons in each single-particle state $(\mathbf{k}, \alpha)$.

We may therefore expect the same considerations to apply, as we shall see, also to *free* particles of spin 0 and 1/2, for which a consistent relativistic description can be achieved by a *quantum field theory* in which particles are seen as quantized excitations of a field, in the same way as photons were defined as quantum excitations of the electromagnetic field.

Notwithstanding the difficulties of interpretation mentioned above, it is however our purpose to give in this chapter a treatment of the *classical wave equations* for spin 0 and 1/2 particles in some detail for two reasons: First we want to give a precise quantitative discussion of how inconsistencies show up when trying to interpret the relativistic fields as wave functions of one-particle states, thus tracing back the historical development of relativistic quantum theories. Second, the formal development of these equations will allow us to assemble those formulae which we shall need in the next chapter where the "second quantization" of the spin 0 and 1/2 fields, besides the Lorentz-covariant quantization of the electromagnetic field, will be developed, that is the classical fields will be treated as dynamic variables and, as such, promoted to quantum operators. As shown for the electromagnetic case, the second quantization procedure allows to describe the system in terms of states which differ in the number of particles they describe and thus provides an ideal framework in which to analyze relativistic processes involving the creation and destruction of particles, namely in which the number and the identities of the interacting particles are not conserved. This will be dealt with in Chap. 12, where a relativistically covariant, perturbative description of fields in interaction will be developed for the electromagnetic field in interaction with a Dirac field. This analysis provides however a paradigm for the description of all the other fundamental interactions among elementary particles.

10.2 The Klein-Gordon Equation

Let us consider a relativistic field theory describing a *classical* field $\Phi^\alpha(x^\mu)$. Such field is defined by its transformation property (7.47) under a generic Poincaré transformation $(\Lambda, \mathbf{x}_0)$ (7.46):

---

[1] The name second quantization is somewhat improper since, just as in the first quantization, only dynamical quantities are promoted to operators acting on states. However, while in the first quantization these quantities include the position and the momentum of a particle, in this new framework, the dynamical quantities to be quantized are fields, the space-coordinates being just a labels.

$$(\mathbf{\Lambda}, x_0) : \quad x^\mu \to x'^\mu = \Lambda^\mu{}_\nu x^\nu - x_0^\mu,$$

$$\Phi^\alpha(x) \to \Phi'^\alpha(x') = D^\alpha{}_\beta \, \Phi^\beta(x) = D^\alpha{}_\beta \, \Phi^\beta(\mathbf{\Lambda}^{-1}(x' + x_0)),$$

where $\mathbf{D} = (D^\alpha{}_\beta) = \mathbf{D}(\mathbf{\Lambda})$ represents the action of the Lorentz transformation $\mathbf{\Lambda}$ on the *internal degrees of freedom* of the field, labeled by α and defining a representation of the Lorentz group SO(1, 3). In Chap. 7 and in Chap. 9, see Eq. (9.101), the action of a Poincaré transformation on $\Phi^\alpha(x)$ was described in terms of the infinitesimal generators $\hat{J}_{\mu\nu}$ associated with the Lorentz part, and $\hat{P}_\mu$ generating space-time translations. The latter provide the operator representation, in a relativistic quantum theory, of the four-momentum of a particle:

$$\hat{P}^\mu \equiv \left(\frac{1}{c}\,\hat{H},\,\hat{\mathbf{p}}\right) = i\hbar\,\eta^{\mu\nu}\,\partial_\nu. \tag{10.6}$$

The identification of the Hamiltonian operator, function of the particle position and the momentum operator, with the generator of time evolution $i\hbar\,\partial_t$ is expressed by the Schroedinger equation (9.78), and describes the dynamics of the system. For a free particle this equation has the form (9.79), which is clearly not Lorentz covariant, since it is obtained from the non-relativistic relation $E = |\mathbf{p}|^2/2m$ upon replacing

$$\mathbf{p} \to \hat{\mathbf{p}} = -i\hbar\,\nabla, \quad E \to \hat{H} = i\hbar\,\partial_t. \tag{10.7}$$

In seeking for the simplest Lorentz-covariant generalization of the Schrödinger equation describing a free particle, we should start from the mass-shell condition of relativistic mechanics which relates the linear momentum and the energy with the rest mass of the particle

$$\mathbf{p}^2 + m^2 c^2 = \frac{E^2}{c^2} \longleftrightarrow p^\mu p_\mu - m^2 c^2 = 0. \tag{10.8}$$

Implementing the same canonical prescription (10.7) on Φ^α we end up with Eq. (9.107) of the previous chapter, which can be written in the following compact form:

$$\left(\Box + \frac{m^2 c^2}{\hbar^2}\right)\Phi^\alpha(x) = 0. \tag{10.9}$$

By construction the above equation represents a manifestly *Lorentz-invariant* generalization of the Schroedinger equation[2] and is referred to as the *Klein-Gordon equation*.

We note that *this equation should hold for particles of any spin*, that is for any representation of the Lorentz group carried by the index α. For example, in the case of the electromagnetic field, setting $\phi^\alpha(x) \equiv A_\mu(x)$ and $m = 0$ we obtain

$$\Box A_\mu(x) = 0, \tag{10.10}$$

---

[2]Extension of the invariance to the full Poincaré group is obvious.

that is the Maxwell propagation equation for the electromagnetic four-potential describing particles of spin 1, in the Lorentz gauge. We shall see in the sequel that also the wave functions associated with spin-1/2 particles satisfy the Klein Gordon equation.

In the rest of this section we shall treat exclusively the case of spin-0 fields, that is fields that are *scalar* under Lorentz transformations. We shall consider a *complex scalar field*, ϕ, or equivalently two real scalar fields (see Chap. 7, Sect. 7.4).

In this case the equation of motion (10.9) can be derived from the Hamilton principle of stationary action, starting from the following Lagrangian density (8.204):

$$\mathcal{L} = c^2 \left(\partial_\mu \phi^* \partial^\mu \phi - \frac{m^2 c^2}{\hbar^2} \phi^* \phi \right). \tag{10.11}$$

Indeed in this case the Euler-Lagrange equations

$$\frac{\partial \mathcal{L}}{\partial \phi(x)^*} - \partial_\mu \left(\frac{\partial \mathcal{L}}{\partial \partial_\mu \phi(x)^*} \right) = 0; \quad \frac{\partial \mathcal{L}}{\partial \phi(x)} - \partial_\mu \left(\frac{\partial \mathcal{L}}{\partial \partial_\mu \phi(x)} \right) = 0,$$

give:

$$\left(\Box + \frac{m^2 c^2}{\hbar^2} \right) \phi(x) = 0. \tag{10.12}$$

together with its complex conjugate.

As a complete set of solutions we can take the plane waves (Eq. 9.113)

$$\Phi_p(x) \propto e^{-\frac{i}{\hbar} p^\mu x_\mu}, \tag{10.13}$$

with wave number $\mathbf{k} = \mathbf{p}/\hbar$ and angular frequency $\omega = E/\hbar$. These are the eigenfunctions of the operator $\hat{P}^\mu$ which describe the wave functions of particles with definite value of energy E and momentum $\mathbf{p}$, see Chap. 9. Substituting the exponentials (10.13) in Eq. (10.12) we find

$$\frac{E^2}{c^2} - |\mathbf{p}|^2 = m^2 c^2, \tag{10.14}$$

or

$$E = \pm E_\mathbf{p} = \pm \sqrt{|\mathbf{p}|^2 c^2 + m^2 c^4}. \tag{10.15}$$

We see that solutions exist for both positive and *negative* values of the energy corresponding to the exponentials:

$$e^{-\frac{i}{\hbar}(E_\mathbf{p} t - \mathbf{p}\cdot\mathbf{x})}; \quad e^{\frac{i}{\hbar}(E_\mathbf{p} t + \mathbf{p}\cdot\mathbf{x})}. \tag{10.16}$$

Strictly speaking this is not a problem as long as we consider only *free fields*. Indeed
the conservation of energy would forbid transition between positive and negative
energy solutions and a positive energy state will remain so. Therefore we could regard
as physical only those solutions corresponding to positive energy $E > 0$. However
the very notion of free particle is far from reality since real particles interact with
each other, usually in scattering processes. During an interaction transitions between
quantum states are induced, according to perturbation theory. Therefore we cannot
neglect the existence of negative energy states. For example, a particle with energy
$E = +E_\mathbf{p}$ could decay into a particle of energy $E = -E_\mathbf{p}$, through the emission
of a photon of energy $2E_\mathbf{p}$. Moreover the existence of negative energies is in some
sense contradictory since, as shown in the following, from a field theoretical point
of view, the Hamiltonian of the theory is *positive definite*.[3]

Thus, the existence of negative energy solutions is a true problem when trying to
achieve a relativistic generalization of the Schroedinger equation.[4]

A second problem arises when trying to give a probabilistic interpretation to the
wave function $\psi(\mathbf{x}, t)$. As we have anticipated in the introduction with each solution
to the Schroedinger equation we can associate a *positive* probability $\rho = |\psi(\mathbf{x}, t)|^2$,
and a current density $\mathbf{j} = \frac{i\hbar}{2m}(\psi \nabla \psi^* - \psi^* \nabla \psi)$ satisfying the *continuity equation*
(10.2), which assures that the total probability is conserved.

We can attempt to follow the same route for the Klein-Gordon equation, and
associate with its solution a conserved current, i.e. a current j^μ for which we can
write a continuity equation in the form $\partial_\mu j^\mu = 0$. Although this can be done, as
we are going to illustrate below, the conserved quantity associated with j^μ cannot
be consistently identified with a total probability. To construct j^μ let us multiply
Eq. (10.9) by ϕ^*

$$\phi^* \left(\square + \frac{m^2 c^2}{\hbar^2}\right) \phi = 0,$$

and subtract the complex conjugate expression. We obtain:

$$\phi^* \left(\square + \frac{m^2 c^2}{\hbar^2}\right) \phi - \phi \left(\square + \frac{m^2 c^2}{\hbar^2}\right) \phi^* = 0,$$

which can be written as a conservation law:

$$\partial_\mu j^\mu(x) = 0, \tag{10.17}$$

where[5]

$$j^\mu = i \left(\phi^* \partial^\mu \phi - \partial^\mu \phi^* \phi\right). \tag{10.18}$$

---

[3]Furthermore, erasing the negative energy solutions would spoil the completeness of the eigenstates
of $\hat{P}^\mu$ and the expansion in plane waves would be no longer correct.

[4]A possible interpretation of the negative-energy states as 'holes' in the sea of positive-energy ones
was originally proposed by Dirac. For further reading on this we refer the reader to the references
at the end of the chapter.

[5]The factor i has been inserted in order to have a *real* current.

Note however that $j^0 = \frac{i}{c}(\phi^*\dot{\phi} - \phi\dot{\phi}^*)$ *is not positive definite* and thus cannot be identified with a probability density. In fact this current has a different physical interpretation. If we define

$$J^\mu = \frac{ce}{\hbar} j^\mu = \frac{i\,ce}{\hbar}\left(\phi^*\partial^\mu\phi - \partial^\mu\phi^*\phi\right), \qquad (10.19)$$

we recognize this as the *conserved current* in Eq. (8.208), associated with the invariance of the Lagrangian (10.11) under the symmetry transformation (8.206). The corresponding conserved Noether charge was given by Eq. (8.209), namely:

$$Q = \int d^3\mathbf{x}\, J^0 = i\frac{e}{\hbar}\int d^3\mathbf{x}(\phi^*\partial_t\phi - \phi\partial_t\phi^*), \qquad (10.20)$$

and was interpreted in Chap. 8 as the charge carried by a complex field.[6]

Notwithstanding the above difficulties we shall develop in the following all the properties of the Klein-Gordon equation since they will be very useful in the second quantized version of the scalar field theory.

Let us now write down the *most general* solution to the Klein-Gordon equation. It can be written in a form in which relativistic invariance is manifest:

$$\phi(x) = \frac{1}{(2\pi\hbar)^3}\int d^4p\,\tilde{\phi}(p)\,\delta(p^2 - m^2c^2)e^{-\frac{i}{\hbar}p\cdot x} \qquad (10.21)$$

where $d^4p = dp^0\,d^3\mathbf{p}$. Let us comment on this formula. We have first solved Eq. (10.12), as we did for Maxwell's equation in the vacuum (5.96), in a finite size box of volume V, see Sect. 5.6, so that the momenta of the solutions have *discrete* values $\mathbf{p} = \hbar\mathbf{k} = \hbar\left(\frac{2\pi n_1}{L_A}, \frac{2\pi n_2}{L_B}, \frac{2\pi n_3}{L_C}\right)$ as a consequence of the *periodic boundary conditions* on the box. We have then considered the large volume limit $V \to \infty$, see Sect. 5.6.2, in which the components of the linear momentum become continuous variables and the discrete sum over $\mathbf{p}$ is replaced by a triple integral, according to the prescription Eq. (5.122):

$$\sum_{\mathbf{p}} \to \frac{V}{(2\pi\hbar)^3}\int d^3\mathbf{p}. \qquad (10.22)$$

This explains the factor $1/(2\pi\hbar)^3$ in Eq. (10.21) while the normalization volume V has been absorbed in the definition of $\tilde{\phi}(p)$.

Secondly, the Dirac delta function $\delta(p^2 - m^2c^2)$ makes the integrand non-zero only for $p^0 = \frac{E}{c} = \pm\frac{E_\mathbf{p}}{c}$, thus implementing condition (10.15). Indeed, applying

---

[6]Actually this "charge" can be any conserved quantum number associated with invariance under U(1) transformations, like baryon or lepton number etc. However we will always refer to the *electric charge*.

the Klein-Gordon operator to Eq. (10.21) and using the property $x \, \delta(x) = 0$ we find:

$$\left(\Box + \frac{m^2 c^2}{\hbar^2}\right) \phi(x) \propto \int d^4 p \, \tilde{\phi}(p) \, (-p^2 + m^2 c^2) \delta(p^2 - m^2 c^2) e^{-\frac{i}{\hbar} p \cdot x} = 0,$$

(10.23)

that is the Klein-Gordon equation is satisfied by the expression (10.21).

The representation (10.21) of the general solution of the Klein-Gordon equation has the advantage of being explicitly Lorentz-invariant, but it is not very manageable. A more convenient representation is found by eliminating the constraint implemented by the delta function. This can be done by integrating over p^0 so that only the integration on $d^3 \mathbf{p}$ remains.

For this purpose we recall the following property of the Dirac delta function:

Given a function $f(x)$ with a certain number n of *simple* zeros, $f(x_i) = 0$, x_i, $(i = 1, \ldots, n)$, then

$$\delta(f(x)) = \sum_{i=1}^{n} \frac{1}{|f'(x_i)|} \delta(x - x_i).$$

(10.24)

We apply this formula to the function $f(E) = p^2 - m^2 c^2 = \frac{E^2}{c^2} - |\mathbf{p}|^2 - m^2 c^2$. It has two simple zeros corresponding to $E = \pm E_\mathbf{p}$. Taking into account that

$$|f'(\pm E_\mathbf{p})| = \frac{2}{c^2} E_\mathbf{p},$$

(10.25)

the derivative being computed with respect to E, and using (10.24), we find:

$$\delta(p^2 - m^2 c^2) = \frac{c^2}{2E_\mathbf{p}} \left(\delta(E - E_\mathbf{p}) + \delta(E + E_\mathbf{p}) \right).$$

(10.26)

Substituting this expression in Eq. (10.21) one obtains:

$$\phi(x) = \frac{c}{(2\pi\hbar)^3} \int d^3\mathbf{p} \int \frac{dE}{2E_\mathbf{p}} \tilde{\phi}(p) \left(\delta(E - E_\mathbf{p}) + \delta(E + E_\mathbf{p}) \right) e^{-\frac{i}{\hbar} p \cdot x}$$

$$= \frac{c}{(2\pi\hbar)^3} \int \frac{d^3\mathbf{p}}{2E_\mathbf{p}} \left(\tilde{\phi}(E_\mathbf{p}, \mathbf{p}) e^{-\frac{i}{\hbar}(E_p t - \mathbf{p} \cdot \mathbf{x})} \right.$$

$$\left. + \tilde{\phi}(-E_\mathbf{p}, -\mathbf{p}) e^{-\frac{i}{\hbar}(-E_p t - (-\mathbf{p}) \cdot \mathbf{x})} \right).$$

(10.27)

Note that in the second term of the integrand we have replaced the integration variable $\mathbf{p}$ with $-\mathbf{p}$; such change is immaterial since the integration in $d^3\mathbf{p}$ runs over all the directions of $\mathbf{p}$. This replacement however allows us to rewrite the argument of the exponential $e^{-\frac{i}{\hbar}(-E_p t - (-\mathbf{p})\cdot\mathbf{x})}$ as $\frac{i}{\hbar}$ times the product of the four-vectors $p^\mu = \left(\frac{1}{c} E_\mathbf{p}, \mathbf{p}\right)$ and x^μ:

$$e^{-\frac{i}{\hbar}(-E_p t - (-\mathbf{p})\cdot\mathbf{x})} = e^{\frac{i}{\hbar}(E_p t - \mathbf{p}\cdot\mathbf{x})} = e^{\frac{i}{\hbar}p\cdot x}. \tag{10.28}$$

Thus Eq. (10.27) takes the final form:

$$\phi(x) = \frac{c}{(2\pi\hbar)^3} \int \frac{d^3\mathbf{p}}{2E_p} \left[\tilde{\phi}_+(p)\, e^{-\frac{i}{\hbar}p\cdot x} + \tilde{\phi}_-(p)\, e^{\frac{i}{\hbar}p\cdot x}\right]$$

$$= \frac{1}{(2\pi\hbar)^3} \int \frac{d^3\mathbf{p}}{2p^0} \left[\tilde{\phi}_+(p)\, e^{-\frac{i}{\hbar}p\cdot x} + \tilde{\phi}_-(p)\, e^{\frac{i}{\hbar}p\cdot x}\right], \tag{10.29}$$

where $p^0 \equiv E_\mathbf{p}/c$ and we have defined

$$\tilde{\phi}_+(p) \equiv \tilde{\phi}(E_\mathbf{p}, \mathbf{p}); \qquad \tilde{\phi}_-(p) \equiv \tilde{\phi}(-E_\mathbf{p}, -\mathbf{p}). \tag{10.30}$$

They represent the Fourier transforms of the positive and negative energy solutions.

It is important to note that in the particular case of a *real* field $\phi(x)$, $\phi(x) = \phi^*(x)$, the two Fourier coefficients would be related by complex conjugation, $\tilde{\phi}_+^* = \tilde{\phi}_-$. Instead in the present case of a complex scalar field there is no relation between them. We also note, by comparing Eqs. (10.21) and (10.29), that *the quantity $\frac{d^3\mathbf{p}}{2E_\mathbf{p}}$ is Lorentz-invariant* (see also Sect. 9.5).

In summary Eq. (10.29) represents the most general solution of the Klein-Gordon equation for a complex scalar field $\phi(x)$, given in terms of both positive and negative energy solutions. Moreover Eq. (10.29), though not manifestly, is *Lorentz-invariant* since it has been derived from (10.21).

For future purpose it is interesting to compute the conserved charge (10.20) in terms of the Fourier coefficients (10.30).

To this end let us first compute the Fourier integral form of $\dot{\phi}(x)$ from (10.29):

$$\dot{\phi}(x) = -ic \int \frac{d^3\mathbf{p}}{2(2\pi\hbar)^3\,\hbar} \left[\tilde{\phi}_+(p)e^{-\frac{i}{\hbar}p\cdot x} - \tilde{\phi}_-(p)e^{\frac{i}{\hbar}p\cdot x}\right]. \tag{10.31}$$

Inserting the general solution (10.29) and (10.31) in the left hand side of the following equation:

$$\frac{\hbar}{e} Q = i \int d^3\mathbf{x}\,(\phi^*\dot{\phi}) + \text{c.c.}$$

we find a number of terms involving two momentum and one volume integrals. The integral in $d^3\mathbf{x}$ can be performed over the exponentials and yields delta functions according to the property:

$$\int d^3 \mathbf{x} e^{\pm \frac{i}{\hbar} (\mathbf{p} - \mathbf{p}') \cdot \mathbf{x}} = (2\pi\hbar)^3 \delta^3 (\mathbf{p} - \mathbf{p}'). \qquad (10.32)$$

Let us consider each term separately. The terms containing the products $\phi_+\phi_+$ give the following contribution:

$$\frac{c}{(2\pi\hbar)^6} \int d^3 \mathbf{x} \int \frac{d^3 \mathbf{p}}{4\,\hbar\,p_0} \int d^3 \mathbf{q} \left[\tilde{\phi}_+^*(p) \tilde{\phi}_+(q) e^{+\frac{i}{\hbar}((p_0 - q_0)x_0 - (\mathbf{p} - \mathbf{q})) \cdot \mathbf{x}} + \text{c.c.} \right]$$

$$= \frac{c}{(2\pi\hbar)^3} \int \frac{d^3 \mathbf{p}}{4\,\hbar\,p_0} \int d^3 \mathbf{q} \left[\tilde{\phi}_+^*(p) \tilde{\phi}_+(q) e^{+\frac{i}{\hbar}(p_0 - q_0)x_0} \delta^3 (\mathbf{p} - \mathbf{q}) + \text{c.c.} \right]$$

$$= \frac{c}{(2\pi\hbar)^3} \int \frac{d^3 \mathbf{p}}{4\,\hbar\,p_0} \left[\tilde{\phi}_+^*(p) \tilde{\phi}_+(p) + \text{c.c.} \right] = \frac{c}{(2\pi\hbar)^3 \hbar} \int \frac{d^3 \mathbf{p}}{2\,p_0} \tilde{\phi}_+^*(p) \tilde{\phi}_+(p).$$

where we have used the fact that if $\mathbf{p} = \mathbf{q}$, then $E_{\mathbf{p}} = E_{\mathbf{q}}$. Similarly, for the $\phi_-\phi_-$ terms we find:

$$-\frac{c}{(2\pi\hbar)^6} \int d^3 \mathbf{x} \int \frac{d^3 \mathbf{p}}{4\hbar p_0} \int d^3 \mathbf{q}\, \tilde{\phi}_-^*(p) \tilde{\phi}_-(q) e^{-\frac{i}{\hbar}((p_0 - q_0)x_0 - (\mathbf{p} - \mathbf{q})) \cdot \mathbf{x}} + \text{c.c.}$$

$$= -\frac{c}{(2\pi\hbar)^3} \int \frac{d^3 \mathbf{p}}{2\hbar p_0} \tilde{\phi}_-^*(p) \tilde{\phi}_-(p).$$

Finally the terms containing the mixed products $\phi_+\phi_-$ give a vanishing contribution:

$$\frac{c}{(2\pi\hbar)^6} \int d^3 \mathbf{x} \int \frac{d^3 \mathbf{p}}{4\hbar\,p_0} \int d^3 \mathbf{q} \left[\tilde{\phi}_-^*(p) \tilde{\phi}_+(q) e^{-\frac{i}{\hbar}((p_0 + q_0)x_0 - (\mathbf{p} + \mathbf{q})) \cdot \mathbf{x}} - \right.$$

$$\left. - \tilde{\phi}_+^*(p) \tilde{\phi}_-(q) e^{\frac{i}{\hbar}((p_0 + q_0)x_0 - (\mathbf{p} + \mathbf{q})) \cdot \mathbf{x}} \right] + \text{c.c.}$$

$$= \frac{c}{(2\pi\hbar)^3} \int \frac{d^3 \mathbf{p}}{4\hbar\,p_0} \left[\tilde{\phi}_-^*(p_0, \mathbf{p}) \tilde{\phi}_+(p_0, -\mathbf{p})\, e^{-\frac{2i}{\hbar} p_0 x_0} - \right.$$

$$\left. - \tilde{\phi}_+^*(p_0, -\mathbf{p}) \tilde{\phi}_-(p_0, \mathbf{p})\, e^{\frac{2i}{\hbar} p_0 x_0} \right] + \text{c.c.} = 0.$$

The last equality is due to the fact that the expression within brackets, being the difference between two complex conjugate terms, is purely imaginary and therefore, when adding to it its own complex conjugate, we obtain zero.

The final result is therefore:

$$Q = \frac{1}{(2\pi\hbar)^3} \frac{ec}{\hbar^2} \int \frac{d^3 \mathbf{p}}{2\,p_0} \left[\tilde{\phi}_+^*(p) \tilde{\phi}_+(p) - \tilde{\phi}_-^*(p) \tilde{\phi}_-(p) \right] \qquad (10.33)$$

confirming the fact that Q is not a positive definite quantity.

In the introduction we have pointed out that the difficulties in giving a probabilistic interpretation to wave functions satisfying a relativistic equation is ultimately related to the fact that in the relativistic processes the number and identities of the particles involved is not conserved. We also know, however, that in any experiment performed so far, the *electric charge is always conserved*. We may therefore argue that the conserved quantity Q should be interpreted as the *total charge* and J^0 as the charge density. Furthermore, from (10.33), it follows that solutions with positive and negative energy have *opposite charge*. This will have a consistent physical interpretation only when, in next chapter, we shall pursue the second quantization program and promote the field $\phi(x)$ to a quantum operator acting on multi-particle states. The quantity Q will be reinterpreted as the *charge operator* and the positive and negative energy solutions will describe the creation and destruction on a state of positive energy solutions associated with *particles* and *antiparticles* having opposite charge.

Note that a real field has charge $Q \equiv 0$, since $\phi_- = \phi_+^*$, so that it must describe a *neutral particle* coinciding with its own antiparticle. This is the case, for example, of the electromagnetic field.

10.2.1 Coupling of the Complex Scalar Field $\phi(x)$ to the Electromagnetic Field

We show in this section that the charge Q introduced in the previous section can be given the interpretation of *electric charge* carried by the particle whose wave function is described by a *complex scalar field*. To this end, we observe that the presence of electric charge can only be ascertained by letting the particle interact with an electromagnetic field. In other words the interpretation of the quantity $J^\mu = (\rho, \mathbf{j}/c)$ as the *electric four-current* can be justified only by studying the interaction of $\phi(x)$ with the electromagnetic field $A_\mu(x)$.

We have seen in Chap. 8 that the passage from the Hamilton function of a free particle to the Hamilton function of a particle interacting with the electromagnetic field can be effected by the *minimal coupling substitution* $p^\mu \rightarrow p^\mu + \frac{e}{c} A^\mu$.

Using the analogy with the classical case, the quantum equation describing the interaction of a complex scalar with the electromagnetic field A_μ can therefore be derived through the substitution:

$$\hat{P}^\mu \rightarrow \hat{P}^\mu + \frac{e}{c} A^\mu = i\hbar \, \partial^\mu + \frac{e}{c} A^\mu, \tag{10.34}$$

into the mass-shell condition (9.107) of Chap. 9:

$$\left[\left(\hat{P}^\mu + \frac{e}{c} A^\mu \right) \left(\hat{P}_\mu + \frac{e}{c} A_\mu \right) - m^2 c^2 \right] \phi = 0,$$

thus obtaining, using (10.34), the new field equation:

$$\left[\left(\partial^\mu - i\frac{e}{\hbar c}A^\mu\right)\left(\partial_\mu - i\frac{e}{\hbar c}A_\mu\right) + \frac{m^2 c^2}{\hbar^2}\right]\phi(x) = 0. \tag{10.35}$$

Defining the *covariant derivative* D_μ as:

$$D_\mu = \partial_\mu - \frac{ie}{\hbar c}A^\mu, \tag{10.36}$$

Equation (10.35) becomes:

$$\left(D^\mu D_\mu + \frac{m^2 c^2}{\hbar^2}\right)\phi(x) = 0. \tag{10.37}$$

which can be derived from the Lagrangian density

$$\mathcal{L} = c^2\left[(D_\mu\phi)^* D^\mu\phi - \frac{m^2 c^2}{\hbar^2}|\phi|^2\right]. \tag{10.38}$$

We observe that the equation of motion (10.35) is *not invariant under the gauge transformation* $A_\mu(x) \to A_\mu(x) + \partial_\mu\varphi(x)$. However, it can be easily checked that gauge invariance can be restored if we extend the gauge transformation also to the complex scalar field as follows:

$$\phi(x) \to \phi'(x) = \phi(x)\, e^{i\frac{e}{\hbar c}\varphi(x)}. \tag{10.39}$$

Note that in the particular case of a constant gauge parameter $\varphi(x) = $ const. we have no transformation of the gauge field and we retrieve the invariance under the U(1)-transformation (8.206) with a constant parameter $\alpha = -e\varphi/(\hbar c)$, also called *global-U(1)* transformation, which implies the conservation of the electric charge. The name of covariant derivative given to (10.36) stems from the fact that under the combined transformations

$$\begin{aligned}
\phi'(x) &= e^{i\frac{e}{\hbar c}\varphi(x)}\,\phi(x),\\
A'_\mu(x) &= A_\mu(x) + \partial_\mu\varphi(x),
\end{aligned} \tag{10.40}$$

we have

$$D^\mu\phi \to e^{i\frac{e}{\hbar c}\varphi(x)}\,D^\mu\phi,$$

that is $D^\mu\phi$ transforms exactly as ϕ. It follows that the Lagrangian density (10.38), being a sum of moduli squared, is *invariant* under (10.40). Equation (10.40) define the so called *local-U(1)* transformations, since they involve a U(1) transformation of the complex scalar field with a local, i.e. space-time dependent, parameter.

We may now read off the Lagrangian density describing the interaction of A_μ with ϕ by expanding the right hand side of (10.38) up to terms which are linear in the electric charge e and comparing these with the general form of the coupling between the electromagnetic field ad an electric current, given in Eq. (8.125) of Chap. 8. We find:

$$\mathcal{L} = \mathcal{L}_0 + \mathcal{L}_{int},$$

with

$$\mathcal{L}_0 = c^2 \left(\partial_\mu \phi^* \partial^\mu \phi - \frac{m^2 c^2}{\hbar^2} \phi^* \phi \right) \quad \text{and} \quad \mathcal{L}_{int} = A_\mu J^\mu,$$

the four-current J^μ being given by Eq. (10.19). As shown in Chap. 8 *this current is conserved* $\partial_\mu J^\mu = 0$ as a consequence of the invariance of the Lagrangian under the global-U(1) transformation (8.206)

$$\phi(x) \longrightarrow \phi'(x) = e^{-i\alpha} \phi(x),$$

with a *constant parameter* α. This justifies our previous guess that

$$Q = \int d^3 x J^0(\mathbf{x}, t),$$

is the conserved electric charge carried by the field ϕ.

We have learned that the interaction of a charged scalar field with the electromagnetic one is described by a Lagrangian (10.38) which is invariant under local-U(1) transformations. This guarantees that the minimal coupling between A_μ and ϕ docs not spoil the gauge invariance associated with the vector potential. The Lagrangian (10.38) is obtained from the one in Eq. (10.35), describing the free scalar field, through the substitution: $\partial_\mu \to D_\mu$.

10.3 The Hamiltonian Formalism for the Free Scalar Field

The Klein-Gordon equation can be cast into a Hamiltonian form following the procedure discussed in Chap. 8. Rewriting the Lagrangian density (10.12) as:

$$\mathcal{L} = \dot{\phi}^* \dot{\phi} - c^2 \nabla \phi^* \cdot \nabla \phi - \frac{m^2 c^4}{\hbar^2} \phi^* \phi, \tag{10.41}$$

the Hamiltonian H and the Hamiltonian density $\mathcal{H}$ then reads:

$$H = \int d^3 \mathbf{x} \, \mathcal{H}(\phi, \phi^*, \pi, \pi^*), \tag{10.42}$$

$$\mathcal{H} = \pi \dot{\phi} + \pi^* \dot{\phi}^* - \mathcal{L} = 2\dot{\phi}^* \dot{\phi} - \mathcal{L}, \tag{10.43}$$

where

$$\pi(\mathbf{x}, t) = \frac{\partial \mathcal{L}}{\partial \dot{\phi}(\mathbf{x}, t)} = \dot{\phi}^*(\mathbf{x}, t), \tag{10.44}$$

$$\pi^*(\mathbf{x}, t) = \frac{\partial \mathcal{L}}{\partial \dot{\phi}(\mathbf{x}, t)^*} = \dot{\phi}(\mathbf{x}, t). \tag{10.45}$$

Substituting these values into (10.43) we obtain:

$$H = \int d^3 \mathbf{x} \mathcal{H}; \quad \mathcal{H} = \pi \pi^* + c^2 \nabla \phi^* \cdot \nabla \phi + \frac{m^2 c^4}{\hbar^2} \phi^* \phi, \tag{10.46}$$

showing that the Hamiltonian density, and hence the Hamiltonian, are *positive definite*.

The Hamilton equations of motion are:

$$\dot{\pi} = -\frac{\delta H}{\delta \phi}; \quad \dot{\phi} = \frac{\delta H}{\delta \pi} \tag{10.47}$$

$$\dot{\pi}^* = -\frac{\delta H}{\delta \phi^*}; \quad \dot{\phi}^* = \frac{\delta H}{\delta \pi^*}. \tag{10.48}$$

The equation for the conjugate momentum density π^* gives the propagation equation:

$$\dot{\pi}^* = -\frac{\delta H}{\delta \phi^*} \implies \frac{\partial^2 \phi}{\partial t^2} = c^2 \nabla^2 \phi - \frac{m^2 c^4}{\hbar^2} \phi,$$

where, in computing the functional derivative, we have integrated the term $\nabla \delta \phi^* \cdot \nabla \phi$ by parts and we neglected the total divergence since it gives a vanishing contribution when integrated over the whole space. Thus we have retrieved the Klein-Gordon equation (10.12) in the Hamiltonian formalism.

Since the Hamiltonian density has the physical meaning of an *energy density* it could have been computed alternatively, in the Lagrangian formalism, in terms of the canonical energy-momentum tensor associated with the Lagrangian density (10.12). Indeed, from the definition (8.170), and taking into account that we have two independent fields ϕ and ϕ^*, we compute the energy-momentum tensor to be:

$$T_{\mu\nu} = \frac{1}{c} \left[\frac{\partial \mathcal{L}}{\partial(\partial^\mu \phi)} \partial_\nu \phi + \frac{\partial \mathcal{L}}{\partial(\partial^\mu \phi^*)} \partial_\nu \phi^* - \eta_{\mu\nu} \mathcal{L} \right], \tag{10.49}$$

where

$$\frac{\partial \mathcal{L}}{\partial(\partial^\mu \phi)} = c^2 \partial_\mu \phi^*; \quad \frac{\partial \mathcal{L}}{\partial \partial^\mu \phi^*} = c^2 \partial_\mu \phi.$$

Substituting in Eq. (10.49) we find:

$$T_{\mu\nu} = c \left(\partial_\mu \phi^* \partial_\nu \phi + \partial_\nu \phi^* \partial_\mu \phi \right) - \eta_{\mu\nu} \frac{\mathcal{L}}{c}. \tag{10.50}$$

In particular we may verify the identity between energy density $c\, T_{00}$ and Hamiltonian density:

$$T_{00} = \frac{1}{c}(2\dot\phi^*\dot\phi - \mathcal{L}) = \frac{1}{c}\left(\dot\phi^*\dot\phi + c^2 \nabla\phi^* \cdot \nabla\phi + \frac{m^2 c^4}{\hbar^2}|\phi|^2 \right) = \frac{\mathcal{H}}{c}.$$

that is

$$H = c \int d^3\mathbf{x}\, T_{00} = \int d^3\mathbf{x} \left(\pi\pi^* + c^2 \nabla\phi^* \cdot \nabla\phi + \frac{m^2 c^4}{\hbar^2}|\phi|^2 \right). \tag{10.51}$$

As far as the momentum of the field is concerned we find

$$P^i = \int d^3\mathbf{x} \left(\dot\phi^* \partial^i \phi + \dot\phi\, \partial^i \phi^* \right) \quad \Rightarrow \quad \mathbf{P} = -\int d^3\mathbf{x} \left(\pi\, \nabla\phi + \pi^*\, \nabla\phi^* \right). \tag{10.52}$$

10.4 The Dirac Equation

In the previous sections we have focussed our attention on a scalar field, whose distinctive property is the absence of internal degrees of freedom since it belongs to a trivial representation of the Lorentz group. This means that its intrinsic angular momentum, namely its *spin*, is zero.

We have also studied, both at the classical level and in a second quantized setting, the electromagnetic field which, as a four-vector, transforms in the fundamental representation of the Lorentz group. Its internal degrees of freedom are described by the two transverse components of the polarization vector. At the end of Chap. 6 we have associated with the photon a unit spin: $s = 1$ (in units of $\hbar$). As explained there, by this we really mean that the photon *helicity* is $\Gamma = 1$.

Our final purpose is to give an elementary account of the quantum description of electromagnetic interactions. The most important electromagnetic interaction at low energy is the one between matter and radiation. Since the elementary building blocks of matter are electrons and quarks, which have half-integer spin ($s = 1/2$), such processes will involve the interaction between photons and spin 1/2 particles. It is therefore important to complete our analysis of classical fields by including the *fermion* fields, that is fields associated with spin 1/2 particles.

In this section and in the sequel we discuss the relativistic equation describing particles of spin 1/2, known as *the Dirac equation*.

10.4.1 The Wave Equation for Spin 1/2 Particles

Historically Dirac discovered his equation while attempting to construct a relativistic equation which, unlike Klein-Gordon equation, would allow for a consistent interpretation of the modulus squared of the wave function as a probability density. As we shall see in the following, this requirement can be satisfied if, unlike in the Klein-Gordon case, the equation is of *first order in the time derivative*. On the other hand, the requirement of relativistic invariance implies that the equation ought to be of *first order in the space derivatives* as well. The resulting equation will be shown to describe particles of spin $s = \frac{1}{2}$.

Let $\psi^\alpha(x)$ be the classical field representing the wave function. The most general form for a first order wave equation is the following:

$$i\hbar \frac{\partial \psi}{\partial t} = (-ic\hbar\,\alpha^i\,\partial_i + \beta\,mc^2)\,\psi = \hat{H}\,\psi. \tag{10.53}$$

In writing Eq. (10.53) we have used a matrix notation suppressing the index α of $\psi^\alpha(x)$ and the indices of the matrices α^i, β acting on ψ^α namely $\alpha^i = (\alpha^i)^\alpha{}_\beta$, $\beta = (\beta)^\alpha{}_\beta$.

In order to determine the matrices α^i, β we require the solutions to Eq. (10.53) to satisfy the following properties:

(i) $\psi^\alpha(x)$ must satisfy the Klein-Gordon equation for a free particle which implements the mass-shell condition:

$$E^2 - |\mathbf{p}|^2 c^2 = m^2 c^4;$$

(ii) It must be possible to construct a conserved current in terms of ψ^α whose 0-component is *positive definite* and which thus can be interpreted as a probability density;

(iii) Equation (10.53) must be Lorentz covariant. This would imply Poincaré invariance.

To satisfy the first requirement we apply the operator $i\hbar \frac{\partial}{\partial t}$ to both sides of Eq. (10.53) obtaining:

$$-\hbar^2 \frac{\partial^2 \psi}{\partial t^2} = (-ic\,\hbar\,\alpha^i\,\partial_i + \beta\,mc^2)(-ic\hbar\,\alpha^j\,\partial_j + \beta\,mc^2)\psi, \tag{10.54}$$

where, because of the symmetry of $\partial_i \partial_j$, the term $\alpha^i\,\alpha^j\,\partial_i \partial_j$ can be rewritten as

$$\alpha^i\,\alpha^j\,\partial_i \partial_j = \frac{1}{2}(\alpha^i\,\alpha^j + \alpha^j\,\alpha^i)\,\partial_i \partial_j.$$

If we now require α^i and β to be *anticommuting* matrices, namely to satisfy:

$$\left\{\alpha^i, \alpha^j\right\} \equiv \alpha^i \alpha^j + \alpha^j \alpha^i = 2\,\delta^{ij}\,\mathbf{1}; \quad \left\{\alpha^i, \beta\right\} = 0, \tag{10.55}$$

and furthermore to square to the identity matrix:

$$\beta^2 = (\alpha^i)^2 = \mathbf{1} \quad (\text{no summation over } i), \tag{10.56}$$

then Eq. (10.54) becomes:

$$-\hbar^2 \frac{\partial^2 \psi^\alpha}{\partial t^2} = \left(-c^2\,\hbar^2\,\nabla^2 + m^2 c^4\right) \psi^\alpha, \tag{10.57}$$

which is the Klein-Gordon equation

$$\left(\Box + \frac{m^2 c^2}{\hbar^2}\right) \psi^\alpha = 0, \tag{10.58}$$

where the differential operator is applied to each component of ψ.

Therefore, given a set of four matrices satisfying (10.55) and (10.56), Eq. (10.53) implies the Klein-Gordon equation, for each component of ψ^α, consistently with our first requirement. Equation (10.53) is called the Dirac equation. We still need to explicitly construct the matrices α^i, β and to show that requirements (ii) and (iii) are also satisfied. In order to discuss Lorentz covariance of the Dirac equation, it is convenient to introduce a new set of matrices

$$\gamma^0 \equiv \beta; \quad \gamma^i \equiv \beta\,\alpha^i, \tag{10.59}$$

in terms of which conditions (10.55) and (10.56) can be recast in the following compact form

$$\left\{\gamma^\mu, \gamma^\nu\right\} = 2\eta^{\mu\nu}\,\mathbf{1}, \tag{10.60}$$

where, as usual, $i, j = 1, 2, 3$ and $\mu, \nu = 0, 1, 2, 3$. In terms of the matrices γ^μ Eq. (10.53) takes the following simpler form[7]:

$$\left(i\hbar\gamma^\mu\partial_\mu - mc\,\mathbf{1}\right) \psi(x) = 0. \tag{10.61}$$

It can be shown that the minimum dimension for a set of matrices γ^μ satisfying Eq. (10.60) is 4. Therefore the simplest choice is to make the internal index α run

---

[7]For the sake of simplicity, we shall often omit the identity matrix when writing combinations of spinorial matrices. We shall for instance write the Dirac equation in the simpler form $\left(i\hbar\gamma^\mu\partial_\mu - mc\right) \psi(x) = 0$.

over four values so that

$$\psi^\alpha(x) = \begin{pmatrix} \psi^1(x) \\ \psi^2(x) \\ \psi^3(x) \\ \psi^4(x) \end{pmatrix} \tag{10.62}$$

belongs to a four-dimensional representation of the Lorentz group.

It must be noted that although the Lorentz group representation $S(\Lambda)$ acting on the "vector" ψ has the same dimension as the defining representation $\Lambda = (\Lambda^\mu_\nu)$, *the two representations are different*. In our case ψ^α is called a *spinor*, or *Dirac field*, and correspondingly the matrix S^α_β belongs to the *spinor representation* of the Lorentz group (see next section).[8] This representation will be shown in Sect. 10.4.4 to describe a spin 1/2 particle. This seems to be in contradiction with the fact that ψ has four components, corresponding to its four internal degrees of freedom, which are twice as many as the spin states $s_z = \pm \frac{\hbar}{2}$ of a spin $\frac{1}{2}$ particle. We shall also prove that if we want to extend the invariance from *proper Lorentz transformation* SO(1,3) to transformations in O(1, 3) which include *parity*, that is including reflections of the three coordinate axes, all the four components of ψ are needed. It is convenient to introduce an explicit representation of the γ-matrices (10.60), called *standard or Pauli representation*, satisfying (10.60):

$$\gamma^0 = \begin{pmatrix} 1_2 & 0 \\ 0 & -1_2 \end{pmatrix}; \quad \gamma^i = \begin{pmatrix} 0 & \sigma^i \\ -\sigma^i & 0 \end{pmatrix}, \quad (i = 1, 2, 3) \tag{10.63}$$

where each entry is understood as a 2×2 matrix

$$0 \equiv \begin{pmatrix} 0 & 0 \\ 0 & 0 \end{pmatrix}; \quad 1_2 = \begin{pmatrix} 1 & 0 \\ 0 & 1 \end{pmatrix}.$$

The σ^i matrices are the Pauli matrices of the non-relativistic theory, defined as:

$$\sigma^1 = \begin{pmatrix} 0 & 1 \\ 1 & 0 \end{pmatrix}; \quad \sigma^2 = \begin{pmatrix} 0 & -i \\ i & 0 \end{pmatrix}; \quad \sigma^3 = \begin{pmatrix} 1 & 0 \\ 0 & -1 \end{pmatrix}. \tag{10.64}$$

We recall that they are hermitian and satisfy the relation:

$$\sigma^i \sigma^j = \delta^{ij} 1_2 + i \, \epsilon^{ijk} \sigma^k, \tag{10.65}$$

which implies

$$\mathrm{Tr}(\sigma^i \sigma^j) = 2 \delta^{ij}; \quad \{\sigma_i, \sigma_j\} = 2 \delta_{ij} 1_2; \quad [\sigma^i, \sigma^j] = 2i \, \epsilon_{ijk} \sigma^k. \tag{10.66}$$

---

[8] As mentioned in Chap. 7 the spinor representation cannot be obtained in terms of tensor representations of the Lorentz group.

The matrices α^i, β read:

$$\alpha^i = \begin{pmatrix} 0 & \sigma^i \\ \sigma^i & 0 \end{pmatrix}; \quad \beta = \begin{pmatrix} 1_2 & 0 \\ 0 & -1_2 \end{pmatrix}. \tag{10.67}$$

Using the representation (10.63), the Dirac equation can be written as a coupled system of two equations in the upper and lower components of the Dirac spinor $\psi^\alpha(x)$. Indeed, writing

$$\psi^\alpha(x) = \begin{pmatrix} \varphi(x) \\ \chi(x) \end{pmatrix}; \quad \varphi(x) = \begin{pmatrix} \varphi^1 \\ \varphi^2 \end{pmatrix}; \quad \chi(x) = \begin{pmatrix} \chi^1 \\ \chi^2 \end{pmatrix}, \tag{10.68}$$

where $\varphi(x)$, e $\chi(x)$ are two-component spinors, the Dirac equation (10.61) becomes

$$\left\{ i\hbar c \left[\begin{pmatrix} 1_2 & 0 \\ 0 & -1_2 \end{pmatrix} \frac{\partial}{\partial x^0} + \begin{pmatrix} 0 & \sigma^i \\ -\sigma^i & 0 \end{pmatrix} \frac{\partial}{\partial x^i} \right] - mc^2 \begin{pmatrix} 1_2 & 0 \\ 0 & 1_2 \end{pmatrix} \right\} \begin{pmatrix} \varphi \\ \chi \end{pmatrix} = 0. \tag{10.69}$$

The matrix equation (10.69) is equivalent to the following system of coupled equations:

$$i\hbar \frac{\partial}{\partial t} \varphi = -i\,\hbar c\,\boldsymbol{\sigma} \cdot \boldsymbol{\nabla} \chi + mc^2 \varphi, \tag{10.70}$$

$$i\hbar \frac{\partial}{\partial t} \chi = -i\hbar c \boldsymbol{\sigma} \cdot \boldsymbol{\nabla} \varphi - mc^2 \chi, \tag{10.71}$$

where $\boldsymbol{\sigma} \equiv (\sigma^i)$ denotes the vector whose components are the three Pauli matrices. The two-component spinors φ and χ are called *large* and *small* components of the Dirac four-component spinor, since, as we now show, in the non-relativistic limit, χ becomes negligible with respect to φ.

To show this we first redefine the Dirac field as follows:

$$\psi = \psi' e^{-i \frac{mc^2}{\hbar} t}, \tag{10.72}$$

so that Eq. (10.61) takes the following form:

$$\left(i\hbar \frac{\partial}{\partial t} + mc^2 \right) \psi' = \left[c\,\alpha^i\,(-i\hbar\partial_i) + \beta\,mc^2 \right] \psi'.$$

The rescaled spinor ψ' is of particular use when evaluating the non-relativistic limit, since it is defined by "subtracting" from the time evolution of ψ the part due to its rest energy, so that its time evolution is generated by the kinetic energy operator only: $\hat{H} - mc^2\,\hat{I}$. In other words $i\,\hbar\,\partial_t\psi'$ is of the order of the kinetic energy times ψ' and, in the non-relativistic limit, it is negligible compared to $mc^2 \psi'$. Next we decompose the field ψ' as in Eq. (10.68) and, using Eq. (10.67), we find:

$$i\hbar\frac{\partial}{\partial t}\varphi = c\,\boldsymbol{\sigma}\cdot\hat{\mathbf{p}}\,\chi, \tag{10.73}$$

$$\left(i\hbar\frac{\partial}{\partial t} + 2mc^2\right)\chi = c\,\boldsymbol{\sigma}\cdot\hat{\mathbf{p}}\,\varphi, \tag{10.74}$$

where we have omitted the prime symbols in the new φ and χ. In the non-relativistic approximation we only keep on the left hand side of the second equation the term $2\,mc^2\,\chi$, so that

$$\chi = \frac{1}{2mc}\,\boldsymbol{\sigma}\cdot\hat{\mathbf{p}}\,\varphi. \tag{10.75}$$

Substituting this expression in the equation for φ we obtain:

$$i\hbar\frac{\partial}{\partial t}\varphi = \frac{1}{2m}\,\hat{\mathbf{p}}^2\varphi = -\frac{\hbar^2}{2m}\,\nabla^2\varphi, \tag{10.76}$$

where we have used the identity:

$$(\hat{\mathbf{p}}\cdot\boldsymbol{\sigma})(\hat{\mathbf{p}}\cdot\boldsymbol{\sigma}) = |\hat{\mathbf{p}}|^2 = -\hbar^2\,\nabla^2, \tag{10.77}$$

which is an immediate consequence of the properties (10.65) of the Pauli matrices.

Equation (10.76) tells us that in the non-relativistic limit the Dirac equation reduces to the familiar Schroedinger equation for the two component spinor wave function φ. Moreover, from Eq. (10.75), we realize that the lower components χ of the Dirac spinor are of subleading order $O(\frac{1}{c})$ with respect to the upper ones φ and therefore vanish in the non-relativistic limit $c \to \infty$. This justifies our referring to them as the *small* and *large* components of ψ, respectively. We also note that in the present non-relativistic approximation, taking into account that the small components χ can be neglected, the probability density $\psi^\dagger\psi = \varphi^\dagger\varphi + \chi^\dagger\chi$ reduces to $\varphi^\dagger\varphi$ as it must be the case for the Schroedinger equation.

10.4.2 Conservation of Probability

We now show that property (ii) of Sect. 10.4.1 is satisfied by the solutions to the Dirac equation, namely that it is possible to construct a conserved probability in terms of the spinor ψ^α. Let us take the hermitian conjugate of the Dirac equation (10.61)

$$-i\,\hbar\,\partial_\mu\psi^\dagger\,\gamma^{\mu\dagger} - mc\,\psi^\dagger = 0. \tag{10.78}$$

We need now the following property of the γ^μ-matrices (10.63) which can be easily verified:

$$\gamma^0 \gamma^{\mu\dagger} = \gamma^\mu \gamma^0. \tag{10.79}$$

Multiplying both sides of Eq. (10.78) from the right by the matrix γ^0 and defining the *Dirac conjugate* $\bar{\psi}$ of ψ as

$$\bar{\psi}(x) = \psi^\dagger(x)\,\gamma^0,$$

we find:

$$-i\,\hbar\,\partial_\mu \bar{\psi}\,\gamma^\mu - mc\,\bar{\psi} = 0,$$

where we have used Eq. (10.79). Thus the field $\bar{\psi}(x)$ satisfies the equation:

$$\bar{\psi}(x)(i\,\hbar\,\overleftarrow{\partial}_\mu \gamma^\mu + mc) = 0, \tag{10.80}$$

where, by convention

$$\bar{\psi}\,\overleftarrow{\partial}_\mu \equiv \partial_\mu \bar{\psi}.$$

Next we define the following *current*:

$$J^\mu = \bar{\psi}\gamma^\mu\psi, \tag{10.81}$$

and *assume* that J^μ transforms as a four-vector. This property will be proven to hold in the next Section. Using the Dirac equation we can now easily show that $\partial_\mu J^\mu = 0$, that is J^μ *is a conserved current*:

$$\partial_\mu J^\mu = (\partial_\mu \bar{\psi})\gamma^\mu\psi + \bar{\psi}\gamma^\mu \partial_\mu\psi = \bar{\psi}\,\overleftarrow{\partial}_\mu \gamma^\mu\psi + \bar{\psi}\gamma^\mu \partial_\mu\psi$$
$$= i\,\frac{mc}{\hbar}\,\bar{\psi}\psi - i\,\frac{mc}{\hbar}\bar{\psi}\psi = 0. \tag{10.82}$$

Note that the 0-component $\rho = J^0 = \psi^\dagger\psi$ of this current is *positive definite*. If we normalize ψ so as to have dimension $[L^{-3/2}]$, then ρ has the dimensions of an inverse volume and therefore it can be consistently given the interpretation of a probability density, the total probability being conserved by virtue of Eq. (10.82). The second requirement (ii) is therefore satisfied.

10.4.3 Covariance of the Dirac Equation

We finally check that Dirac equation is *covariant* under Lorentz transformations, so that also the third requirement of Sect. 10.4.1 is satisfied.

Lorentz covariance of the Dirac equation means that if in a given reference frame (10.61) holds, then in any new reference frame, related to the former one by a Lorentz transformation, the same equation should hold, although in the transformed variables.

Let us write down the Dirac equation in two frames S' and S related by a Lorentz (or in general a Poincaré) transformation:

$$\left(i\,\hbar\gamma^{\mu}\,\partial'_{\mu} - mc\right)\psi'(x') = 0, \tag{10.83}$$

$$\left(i\,\hbar\gamma^{\mu}\partial_{\mu} - mc\right)\psi(x) = 0, \tag{10.84}$$

where $\partial'_{\mu} = \frac{\partial}{\partial x'_{\mu}}$ and $x'^{\mu} = \Lambda^{\mu}{}_{\nu}\,x^{\nu}$.

We must require Eq. (10.83) to hold in the new frame S' if Eq. (10.84) holds in the original frame S.

As explained after Eq. (10.62) we denote by $S \equiv (S^{\alpha}{}_{\beta}) = S(\boldsymbol{\Lambda})$ the spinor representation of the Lorentz transformation acting on $\psi(x)$. A Poincaré transformation on $\psi^{\alpha}(x)$ is then described as follows:

$$\psi'^{\alpha}(x') = S^{\alpha}{}_{\beta}\,\psi^{\beta}(x), \tag{10.85}$$

where, as usual, $x' = \boldsymbol{\Lambda}\,x - x_0$. We use a *matrix notation* for the spinor representation while we write explicit indices for the defining representation $\Lambda^{\mu}{}_{\nu}$ of the Lorentz group. Since :

$$\frac{\partial}{\partial x'^{\mu}} = \frac{\partial x^{\nu}}{\partial x'^{\mu}}\frac{\partial}{\partial x^{\nu}} = (\Lambda^{-1})^{\nu}{}_{\mu}\partial_{\nu},$$

we have:

$$\left(i\hbar\,\gamma^{\mu}\partial'_{\mu} - mc\right)\psi'(x') = \left(i\hbar\,\gamma^{\mu}(\Lambda^{-1})^{\nu}{}_{\mu}\partial_{\nu} - mc\right)S\,\psi(x) = 0. \tag{10.86}$$

Multiplying both sides from the left by S^{-1} we find:

$$\left[i\hbar\,(\Lambda^{-1})^{\nu}{}_{\mu}\left(S^{-1}\,\gamma^{\mu}\,S\right)\partial_{\nu} - mc\right]\psi(x) = 0. \tag{10.87}$$

We see that in order to obtain covariance, we must require

$$(\Lambda^{-1})^{\nu}{}_{\mu}\,S^{-1}\gamma^{\mu}S = \gamma^{\nu} \;\;\Rightarrow\;\; S^{-1}\gamma^{\nu}\,S = \Lambda^{\nu}{}_{\mu}\,\gamma^{\mu}. \tag{10.88}$$

In that case Eq. (10.87) becomes:

$$\left(i\hbar\gamma^{\nu}\partial_{\nu} - mc\right)\psi(x) = 0, \tag{10.89}$$

that is we retrieve (10.84). In the next section we shall explicitly construct the transformation S satisfying condition (10.88). We then conclude that *Dirac equation is covariant under Lorentz (Poincaré) transformations.*

We may now check that the current $J^\mu = \bar{\psi}\gamma^\mu\psi$ introduced in the previous section transforms as a four-vector. From Eq. (10.85) we have, suppressing spinor indices

$$\bar{\psi}'(x') = \overline{S\,\psi(x)} = \psi^\dagger(x)S^\dagger\gamma^0, \tag{10.90}$$

so that

$$\bar{\psi}'(x')\gamma^\mu\psi'(x') = \psi^\dagger(x)\gamma^0(\gamma^0 S^\dagger\gamma^0)\gamma^\mu S\psi(x) = \bar{\psi}(\gamma^0 S^\dagger\gamma^0)\gamma^\mu S\psi, \tag{10.91}$$

where we have used the property $(\gamma^0)^2 = 1$. As we are going to prove in the next subsection, the following relation holds:

$$\gamma^0 S^\dagger\gamma^0 = S^{-1}. \tag{10.92}$$

In this case, using (10.88), Eq. (10.91) becomes

$$\bar{\psi}'(x')\gamma^\mu\psi'(x') = \bar{\psi}(x)\,S^{-1}\gamma^\mu S\,\psi = \Lambda^\mu{}_\nu\,\bar{\psi}(x)\gamma^\nu\psi(x), \tag{10.93}$$

which shows that *the current J^μ transforms as a four-vector.*

10.4.4 Infinitesimal Generators and Angular Momentum

To find the explicit form of the spinor matrix $S(\Lambda)$ we require it to induce the transformation of the γ-matrices given by Eq. (10.88). Actually it is sufficient to perform the computation in the case of infinitesimal Lorentz transformations.

We can write the Poincaré-transformed spinor $\psi'(x')$ in (10.85) as resulting from the action of a differential operator $O_{(\Lambda,x_0)}$, defined in Eq. (9.101):

$$\psi'^\alpha(x') = O_{(\Lambda,x_0)}\,\psi^\alpha(x') = S^\alpha{}_\beta\,\psi^\beta(x), \tag{10.94}$$

The generators $\hat{J}^{\rho\sigma}$ of $O_{(\Lambda,x_0)}$ are expressed, see Eq. (9.102), as the sum of differential operators $\hat{M}^{\rho\sigma}$ acting on the functional form of the field, and matrices $\Sigma^{\rho\sigma}$ acting on the internal index α (which coincide with $(-i\hbar)$ times the matrices $(L^{\rho\sigma})^\alpha{}_\beta$ in Eq. (7.83)). These latter are the Lorentz generators in the spinor representation:

$$S(\Lambda) = e^{\frac{i}{2\hbar}\theta_{\rho\sigma}\Sigma^{\rho\sigma}}, \tag{10.95}$$

and satisfy the commutation relations (9.103):

$$\left[\Sigma^{\mu\nu}, \Sigma^{\rho\sigma}\right] = -i\hbar\left(\eta^{\nu\rho}\Sigma^{\mu\sigma} + \eta^{\mu\sigma}\Sigma^{\nu\rho} - \eta^{\mu\rho}\Sigma^{\nu\sigma} - \eta^{\nu\sigma}\Sigma^{\mu\rho}\right). \tag{10.96}$$

We can construct such matrices in terms of the γ^μ ones as follows:

$$\Sigma^{\mu\nu} = -\frac{\hbar}{2}\sigma^{\mu\nu}, \tag{10.97}$$

where the $\sigma^{\mu\nu}$ matrices are defined as:

$$\sigma^{\mu\nu} \equiv \frac{i}{2}[\gamma^\mu, \gamma^\nu] = -\sigma^{\nu\mu}. \tag{10.98}$$

Using the properties (10.60) of the γ^μ-matrices, the reader can verify that $\Sigma^{\mu\nu}$ defined in (10.97) satisfy the relations (10.96). The expression of an infinitesimal Lorentz transformation on $\psi(x)$ follows from Eq. (7.83), with the identification $(L^{\rho\sigma})^\alpha{}_\beta = \frac{i}{\hbar}(\Sigma^{\rho\sigma})^\alpha{}_\beta = -\frac{i}{2}(\sigma^{\rho\sigma})^\alpha{}_\beta$:

$$\begin{aligned}
\delta\psi(x) &= \frac{i}{2\hbar}\,\delta\theta_{\rho\sigma}\,\hat{J}^{\rho\sigma}\,\psi(x) \\
&= \frac{1}{2}\,\delta\theta_{\rho\sigma}\left[-\frac{i}{2}\sigma^{\rho\sigma} + x^\rho\partial^\sigma - x^\sigma\partial^\rho\right]\psi(x),
\end{aligned} \tag{10.99}$$

where we have adopted the matrix notation for the spinor indices and used the identification:

$$\hat{J}_{\rho\sigma} = \hat{M}_{\rho\sigma} + \Sigma_{\rho\sigma} = -i\hbar\,(x_\rho\partial_\sigma - x_\sigma\partial_\rho) - \frac{\hbar}{2}\sigma_{\rho\sigma}. \tag{10.100}$$

To verify that the matrices $\Sigma^{\rho\sigma}$ defined in (10.97) generate the correct transformation property Eq. (10.88) of the γ^μ matrices, let us verify Eq. (10.88) for infinitesimal Lorentz transformations:

$$\begin{aligned}
\Lambda^\mu{}_\nu &\approx \delta^\mu_\nu + \frac{1}{2}\,\delta\theta_{\rho\sigma}\,(L^{\rho\sigma})^\mu{}_\nu = \delta^\mu_\nu + \delta\theta^\mu{}_\nu, \\
S(\Lambda) &\approx 1 - \frac{i}{4}\,\delta\theta_{\rho\sigma}\,\sigma^{\rho\sigma},
\end{aligned} \tag{10.101}$$

where we have used the matrix form (4.170) of the Lorentz generators $\mathbf{L}^{\rho\sigma} = [(L^{\rho\sigma})^\mu{}_\nu]$ in the fundamental representation: $(L^{\rho\sigma})^\mu{}_\nu = \eta^{\rho\mu}\delta^\sigma_\nu - \eta^{\sigma\mu}\delta^\rho_\nu$. Equation (10.88) reads to lowest order in $\delta\theta$:

$$\left(1 + \frac{i}{4}\delta\theta_{\rho\sigma}\,\sigma^{\rho\sigma}\right)\gamma^\mu\left(1 - \frac{i}{4}\delta\theta_{\rho\sigma}\,\sigma^{\rho\sigma}\right) = \gamma^\mu + \frac{1}{2}\delta\theta_{\rho\sigma}\,(L^{\rho\sigma})^\mu{}_\nu\,\gamma^\nu.$$

The above equation implies:

$$\frac{i}{2}[\sigma^{\rho\sigma}, \gamma^\mu] = (L^{\rho\sigma})^\mu{}_\nu\,\gamma^\nu = \eta^{\rho\mu}\gamma^\sigma - \eta^{\sigma\mu}\gamma^\rho, \tag{10.102}$$

which can be verified using the properties of the γ^μ-matrices. Having checked Eq. (10.88) for infinitesimal transformations, the equality extends to finite transformations as well, since the latter can be expressed as a sequence of infinitely many infinitesimal transformations.

As far as Eq. (10.92) is concerned, from the definition (10.97) we can easily prove the following property:

$$
\gamma^0 \, (\Sigma^{\rho\sigma})^\dagger \, \gamma^0 = \frac{i\hbar}{4} \, \gamma^0 \, [\gamma^\rho, \, \gamma^\sigma]^\dagger \, \gamma^0 = \frac{i\hbar}{4} \, \gamma^0 \, [(\gamma^\sigma)^\dagger, \, (\gamma^\rho)^\dagger] \, \gamma^0
$$

$$
= \frac{i\hbar}{4} \, [\gamma^0 (\gamma^\sigma)^\dagger \gamma^0, \, \gamma^0 (\gamma^\rho)^\dagger \gamma^0] = -\frac{i\hbar}{4} \, [\gamma^\rho, \, \gamma^\sigma] = \Sigma^{\rho\sigma}.
$$

Let us now compute the left hand side of Eq. (10.92) by writing the series expansion of the exponential and use the above property of $\Sigma^{\mu\nu}$:

$$
\gamma^0 \, S^\dagger \, \gamma^0 = \gamma^0 \left[\sum_{n=0}^{\infty} \frac{1}{n!} \left(-\frac{i}{2\hbar} \, \theta_{\rho\sigma} \, \Sigma^{\rho\sigma \, \dagger} \right)^n \right] \gamma^0 = \sum_{n=0}^{\infty} \frac{1}{n!} \left(-\frac{i}{2\hbar} \, \theta_{\rho\sigma} \, \gamma^0 \, \Sigma^{\rho\sigma \, \dagger} \, \gamma^0 \right)^n
$$

$$
= \exp\left(-\frac{i}{2\hbar} \, \theta_{\rho\sigma} \, \gamma^0 \, \Sigma^{\rho\sigma \, \dagger} \, \gamma^0 \right) = \exp\left(-\frac{i}{2\hbar} \, \theta_{\rho\sigma} \, \Sigma^{\rho\sigma} \right) = S^{-1}. \quad (10.103)
$$

This proves Eq. (10.92).

In terms of the generators $\hat{J}^{\rho\sigma}$ of the Lorentz group we can define the angular momentum operator $\hat{\mathbf{J}} = (\hat{J}_i)$ as in Eq. (9.106) of last chapter:

$$
\hat{J}_i = -\frac{1}{2} \, \epsilon_{ijk} \, \hat{J}^{jk} = \hat{M}_i + \Sigma_i,
$$

$$
\hat{M}_i = \epsilon_{ijk} \, \hat{x}^i \, \hat{p}^j; \quad \Sigma_i = -\frac{1}{2} \, \epsilon_{ijk} \, \Sigma^{jk}, \quad (10.104)
$$

where, as usual $\hat{\mathbf{M}} = (\hat{M}_i)$ denotes the orbital angular momentum, while we have denoted by $\mathbf{\Sigma} = (\Sigma_i)$ the spin operators acting as matrices on the internal spinor components. Let us compute the latter using the definition (10.97) of $\Sigma^{\mu\nu}$:

$$
\Sigma_i = -\frac{1}{2} \, \epsilon_{ijk} \, \Sigma^{jk} = \frac{\hbar}{4} \, \epsilon_{ijk} \, \sigma^{jk} = \frac{\hbar}{2} \begin{pmatrix} \sigma^i & \mathbf{0} \\ \mathbf{0} & \sigma^i \end{pmatrix}, \quad (10.105)
$$

The above expression is easily derived from the definition of σ^{ij} and the explicit form of the γ^μ-matrices:

$$
\sigma^{ij} = \frac{i}{2} \, [\gamma^i, \, \gamma^j] = -\frac{i}{2} \begin{pmatrix} [\sigma^i, \, \sigma^j] & \mathbf{0} \\ \mathbf{0} & [\sigma^i, \, \sigma^j] \end{pmatrix} = \epsilon^{ijk} \begin{pmatrix} \sigma^k & \mathbf{0} \\ \mathbf{0} & \sigma^k \end{pmatrix},
$$

where we have used the properties (10.66) of the Pauli matrices and the relation $\epsilon_{ijk}\,\epsilon^{jk\ell} = 2\,\delta_i^\ell$. For a massive fermion, like the electron, $\boldsymbol{\Sigma} = (\Sigma_i)$ generate the spin group $G^{(0)} = SU(2)$, see Sect. 9.4.1, which is the little group of the four-momentum in the rest frame $\mathcal{S}_0$ in which $p = \bar{p} = (mc,\ \mathbf{0})$. In Sect. 9.4.2 we have shown that $|\boldsymbol{\Sigma}|^2 = -\hat{W}_\mu \hat{W}^\mu/(m^2 c^2)$, i.e. the spin of the particle, is a Poincaré invariant quantity. In our case, using (10.105), we have:

$$|\boldsymbol{\Sigma}|^2 = \hbar^2\, s(s+1)\,\mathbf{1} = \frac{3}{4}\,\hbar^2\,\mathbf{1}, \qquad (10.106)$$

from which we deduce that the particle has spin $s = 1/2$, namely that the states $|p, r\rangle$ belong to the two-dimensional representation of $SU(2)$, labeled by r. The matrix $\mathcal{R}(\Lambda, p)$ in Eq. (9.112) is thus an $SU(2)$ transformation generated by the matrices $\mathbf{s}_i \equiv \hbar\,\sigma_i/2$, see Appendix F:

$$\mathcal{R}(\Lambda, p) = \exp\left(\frac{i}{\hbar}\,\theta^i\,\mathbf{s}_i\right), \qquad (10.107)$$

where, if Λ were a rotation, θ^i would coincide with the rotation angles, and thus be independent of p, whereas if Λ were a boost, θ^i would depend on p and on the boost parameters.

Note that, in the spinorial representation of the Lorentz group, which acts on the index α of $\psi^\alpha(x)$, a generic rotation with angles θ^i is generated by the matrices Σ_i in Eq. (10.105) and has the form:

$$S(\Lambda_R) = e^{\frac{i}{\hbar}\theta^i\,\Sigma_i} = \begin{pmatrix} e^{\frac{i}{\hbar}\theta^i\,\mathbf{s}_i} & 0 \\ 0 & e^{\frac{i}{\hbar}\theta^i\,\mathbf{s}_i} \end{pmatrix} = \begin{pmatrix} \mathbf{S}(\theta) & 0 \\ 0 & \mathbf{S}(\theta) \end{pmatrix}, \quad (10.108)$$

$$\mathbf{S}(\theta) \equiv e^{\frac{i}{\hbar}\theta^i\,\mathbf{s}_i} = \cos\left(\frac{\theta}{2}\right) + i\,\boldsymbol{\sigma}\cdot\hat{\boldsymbol{\theta}}\,\sin\left(\frac{\theta}{2}\right), \qquad (10.109)$$

where $\boldsymbol{\theta} \equiv (\theta^i)$, $\theta \equiv |\boldsymbol{\theta}|$ and $\hat{\boldsymbol{\theta}} \equiv \boldsymbol{\theta}/\theta$. Equation (10.109) is readily obtained along the same lines as in the derivation of the 4×4 matrix representation of a Lorentz boost in Chap. 4.

Equation (10.108) shows that, with respect to the spin group $SU(2)$, the spinorial representation is completely reducible into two two-dimensional representations acting on the small and large components of the spinor, respectively. Moreover we see that a rotation by an angle θ of the reference frame about an axis, amounts to a rotation by an angle $\theta/2$ of a spinor.

If the particle is massless, $\mathcal{R}$ is an $SO(2)$ rotation generated by the helicity operator Γ in the frame in which the momentum is the standard one $p = \bar{p}$. Choosing[9]

---

[9]Note that, with respect to the last chapter, we have changed our convention for the standard momentum of a massless particle. Clearly the discussion in Chap. 9 equally applies to this new choice, upon replacing direction 1 with direction 3.

$\bar{p} = (E, 0, 0, E)/c$, $\hat{\Gamma} = \Sigma_3$ we have

$$\mathcal{R}(\Lambda, p) = \exp\left(\frac{i}{\hbar}\theta s_3\right). \tag{10.110}$$

Finally we may verify that the spin Σ *does not commute with the Hamiltonian*, i.e. it is not a conserved quantity. Indeed, let us recall the expression of the Hamiltonian given in Eq. (10.53), namely

$$\hat{H} = -ic\hbar\alpha^i\partial_i + \beta mc^2 = c\alpha^i\,\hat{p}_i + \beta mc^2 = \begin{pmatrix} mc^2 & c\hat{\mathbf{p}}\cdot\boldsymbol{\sigma} \\ c\hat{\mathbf{p}}\cdot\boldsymbol{\sigma} & -mc^2 \end{pmatrix},$$

where we have used the explicit matrix representation (10.67) of α^i, β. Using for Σ the expression (10.105) we find:

$$[\hat{H}, \Sigma^k] = ic\hbar\begin{pmatrix} 0 & \epsilon_{kij}\sigma^i\,\hat{p}^j \\ \epsilon_{kij}\sigma^i\,\hat{p}^j & 0 \end{pmatrix} = ic\hbar\epsilon_{kij}\,\alpha^i\,\hat{p}^j \neq 0. \tag{10.111}$$

We see that, considering the third component Σ_3, the commutator does not vanish, except in the special case $p^1 = p^2 = 0$, $p^3 \neq 0$. In general the component of Σ along the direction of motion, which is the helicity Γ, is conserved. This is easily proven by computing $[\hat{H}, \Sigma\cdot\hat{\mathbf{p}}] = [\hat{H}, \Sigma_i\,\hat{p}^i]$ and using the property that $\hat{H}$ commutes with $\hat{p}^i$, so that, in virtue of Eq. (10.111), $[\hat{H}, \Sigma_i\,\hat{p}^i] = [\hat{H}, \Sigma_i]\hat{p}^i = 0$.

Similarly also the orbital angular momentum is not conserved since, if we compute $[\hat{H}, \hat{M}_k]$ and use the commutation relation $[\hat{x}^i, \hat{p}_j] = i\hbar\delta^i_j$, we find:

$$[\hat{H}, \hat{M}_k] = \epsilon_{kij}[\hat{H}, \hat{x}^i]\,\hat{p}^j = c\,\epsilon_{kij}\,\alpha^\ell\,[\hat{p}_\ell, \hat{x}^i]\,\hat{p}^j = -ic\hbar\epsilon_{kij}\,\alpha^i\,\hat{p}^j.$$

Summing the above equation with Eq. (10.111) we find:

$$[\hat{H}, \hat{J}_k] = [\hat{H}, \hat{M}_k + \Sigma_k] = -ic\hbar\epsilon_{kij}\,\alpha^i\,\hat{p}^j + ic\hbar\epsilon_{kij}\,\alpha^i\,\hat{p}^j = 0,$$

namely that *the total angular momentum* $\mathbf{J} = \mathbf{M} + \Sigma$ *is conserved*.

So far we have been considering the action of the *rotation subgroup* of the Lorentz group on spinors. On the other hand we have learned in Chap. 4 that a generic proper Lorentz transformation can be written as the product of a boost and a rotation:

$$S(\Lambda) = S(\Lambda_B)\,S(\Lambda_R). \tag{10.112}$$

Let us consider now the boost part. Lorentz boosts are generated, in the fundamental representation, by the matrices $\mathbf{K}_i$ defined in Sect. 4.7.1 of Chap. 4. To find the representation of these generators on the spinors, let us expand a generic Lorentz generator in the spinor representation:

$$\frac{i}{2\hbar} \theta_{\mu\nu} \Sigma^{\mu\nu} = \frac{i}{\hbar} \theta_{0i} \Sigma^{0i} + \frac{i}{\hbar} \theta_i \Sigma^i = \Lambda_i K^i + \frac{i}{\hbar} \theta_i \Sigma^i, \quad (10.113)$$

where, as usual, $\theta_i = -\epsilon_{ijk} \theta^{jk}/2$ while $\Lambda_i \equiv \theta_{0i}$. The boost generators $K^i = i \Sigma^{0i}/\hbar$ read:

$$K^i = \frac{1}{2} \gamma^0 \gamma^i = \frac{1}{2} \alpha^i. \quad (10.114)$$

A boost transformation is thus implemented on a spinor by the following matrix

$$S(\Lambda_B) = e^{\frac{i}{\hbar} \Lambda_i \Sigma^{0i}} = e^{\lambda_i K^i}. \quad (10.115)$$

The above matrix can be evaluated by noting that $(\lambda_i K^i)^2 = -\lambda_i \lambda_j \gamma^i \gamma^j/4 = \lambda^2/4$, where $\lambda = |\boldsymbol{\lambda}|$ and we have used the anticommutation properties of the γ^i-matrices. By using this property and defining the unit vector $\hat{\lambda}^i = \lambda^i/\lambda$ the expansion of the exponential on the right hand side of Eq. (10.115) boils down to:

$$S(\Lambda_B) = \cosh\left(\frac{\lambda}{2}\right) 1 + \sinh\left(\frac{\lambda}{2}\right) \hat{\lambda}^i \alpha_i. \quad (10.116)$$

From the identifications $\cosh(\lambda) = \gamma(v)$, $\sinh(\lambda) = \gamma(v) v/c$, $\hat{\boldsymbol{\lambda}} = (\hat{\lambda}_i) = \mathbf{v}/v$, see Sect. 4.7.1 of Chap. 4, we derive:

$$\cosh\left(\frac{\lambda}{2}\right) = \sqrt{\frac{\gamma(v) + 1}{2}}; \quad \sinh\left(\frac{\lambda}{2}\right) = \sqrt{\frac{\gamma(v) - 1}{2}},$$

$$S(\Lambda_B) = \sqrt{\frac{\gamma(v) + 1}{2}} 1 + \sqrt{\frac{\gamma(v) - 1}{2}} \frac{v^i}{v} \alpha_i. \quad (10.117)$$

It is useful to express the boost Λ_p which connects the rest frame $\mathcal{S}_0$ of a massive particle to a generic one in which $p = (p^\mu) = (E_{\mathbf{p}}/c, \mathbf{p})$. In this case we can write $\gamma(v) = E/(mc^2)$, $\mathbf{v}/c = \mathbf{p}c/E_{\mathbf{p}}$ and Eq. (10.117), after some algebra, becomes:

$$S(\Lambda_p) = \frac{1}{\sqrt{2m (mc^2 + E_{\mathbf{p}})}} (p_\mu \gamma^\mu + mc \gamma^0) \gamma^0$$

$$= \frac{1}{\sqrt{2m (mc^2 + E_{\mathbf{p}})}} \begin{pmatrix} (p^0 + mc) 1_2 & \mathbf{p} \cdot \boldsymbol{\sigma} \\ \mathbf{p} \cdot \boldsymbol{\sigma} & (p^0 + mc) 1_2 \end{pmatrix}. \quad (10.118)$$

10.5 Lagrangian and Hamiltonian Formalism

The field equations of the Dirac field can be derived from the Lagrangian density:

$$\mathcal{L} = i\frac{\hbar c}{2}\left(\bar{\psi}(x)\gamma^\mu\partial_\mu\psi(x) - \partial_\mu\bar{\psi}(x)\gamma^\mu\psi(x)\right) - mc^2\bar{\psi}(x)\psi(x). \qquad (10.119)$$

Indeed, since

$$\frac{\partial\mathcal{L}}{\partial\partial_\mu\bar{\psi}(x)} = -i\frac{\hbar c}{2}\gamma^\mu\psi(x),$$

we find

$$\frac{\partial\mathcal{L}}{\partial\bar{\psi}(x)} - \partial_\mu\left(\frac{\partial\mathcal{L}}{\partial\partial_\mu\bar{\psi}(x)}\right) = 0 \;\Leftrightarrow\; \left(i\hbar\gamma^\mu\partial_\mu - mc\,\mathbf{1}\right)\psi(x) = 0, \qquad (10.120)$$

that is, the Dirac equation.

In an analogous way we find the equation for the Dirac conjugate spinor $\bar{\psi}(x)$:

$$\frac{\partial\mathcal{L}}{\partial\psi}(x) - \partial_\mu\left(\frac{\partial\mathcal{L}}{\partial\partial_\mu\psi(x)}\right) = 0 \;\Leftrightarrow\; \bar{\psi}(x)\left(i\hbar\gamma^\mu\overleftarrow{\partial}_\mu + mc\,\mathbf{1}\right) = 0. \qquad (10.121)$$

The Lagrangian density has, in addition to Lorentz invariance, a further invariance under the phase transformation

$$\psi(x) \longrightarrow \psi'(x) = e^{-i\alpha}\,\psi(x),\; \bar{\psi}(x) \longrightarrow \bar{\psi}'(x) = e^{i\alpha}\,\bar{\psi}(x). \qquad (10.122)$$

α being a constant parameter. In Sect. 10.2.1, we have referred to analogous transformations on a complex scalar field as *global U(1)* transformations, the term global refers to the property of α of being constant. This is indeed the same invariance exhibited by the Klein-Gordon Lagrangian of a complex scalar field and leads to conservation of a charge according to Noether theorem. The reader can easily verify that the conserved Noether current associated with the symmetry (10.122) is the quantity J^μ defined in (10.81), which will be shown, just as for the complex scalar field, to be proportional to the electric four-current.

Let us compute the energy-momentum tensor

$$\begin{aligned}
T^{\nu\mu} &= \frac{1}{c}\left[\frac{\partial\mathcal{L}}{\partial\partial_\nu\psi(x)}\partial^\mu\psi(x) + \partial^\mu\bar{\psi}(x)\frac{\partial\mathcal{L}}{\partial\partial_\nu\bar{\psi}(x)} - \eta^{\mu\nu}\mathcal{L}\right] \\
&= \frac{1}{c}\left[i\frac{\hbar c}{2}\left(\bar{\psi}\gamma^\nu\partial^\mu\psi - \partial^\mu\bar{\psi}\gamma^\nu\psi\right) - \eta^{\mu\nu}\mathcal{L}\right].
\end{aligned} \qquad (10.123)$$

We observe that the Lagrangian density is zero on spinors satisfying the Dirac equation. We may therefore write

$$T^{\nu\mu} = i\,\frac{\hbar}{2}\left(\bar{\psi}\gamma^{\nu}\partial^{\mu}\psi - \partial^{\mu}\bar{\psi}\gamma^{\nu}\psi\right). \tag{10.124}$$

This tensor is not symmetric. We can however verify that the divergences of $T^{\mu\nu}$ with respect to both indices vanish:

$$\partial_{\mu}T^{\nu\mu} = \partial_{\mu}T^{\mu\nu} = 0, \tag{10.125}$$

The latter equality is a consequence of the Noether theorem, being μ the index of the conserved current. As for the former, it is easily proven as follows:

$$\partial_{\mu}T^{\nu\mu} = i\,\frac{\hbar}{2}\left(\partial_{\mu}\bar{\psi}\gamma^{\nu}\partial^{\mu}\psi + \bar{\psi}\gamma^{\nu}\Box\psi - \Box\bar{\psi}\gamma^{\nu}\psi - \partial_{\mu}\bar{\psi}\gamma^{\nu}\partial^{\mu}\psi\right) = 0,$$

where we have used the Klein-Gordon equation for ψ and $\bar{\psi}$. Using property (10.125) we can define a symmetric energy momentum-tensor $\Theta^{\mu\nu}$ simply as the symmetric part of $T^{\mu\nu}$:

$$\Theta^{\mu\nu} = \frac{1}{2}\left(T^{\mu\nu} + T^{\nu\mu}\right), \tag{10.126}$$

since (10.125) guarantee that $\partial_{\mu}\Theta^{\mu\nu} = 0$. The four-momentum of the spinor field

$$P^{\mu} = \int_{V} d^{3}\mathbf{x}\,T^{0\mu},$$

has the following form

$$P^{\mu} = i\,\frac{\hbar}{2}\int_{V} d^{3}\mathbf{x}\left(\bar{\psi}\gamma^{0}\partial^{\mu}\psi - \partial^{\mu}\bar{\psi}\gamma^{0}\psi\right), \tag{10.127}$$

while the Noether energy of the field $H = cp^{0}$ reads

$$H = i\,\frac{\hbar}{2}\int_{V} d^{3}\mathbf{x}\left(\psi^{\dagger}\dot{\psi} - \dot{\psi}^{\dagger}\,\psi\right). \tag{10.128}$$

Using the Dirac equation and integrating by parts, we can easily prove that the right hand side is the sum of two equal terms:

$$i\hbar \int_V d^3\mathbf{x}\, \dot{\bar{\psi}}\gamma^0\psi = -i\hbar c \int_V d^3\mathbf{x}\, \partial_i \bar{\psi}\gamma^i \psi - mc^2 \int_V d^3\mathbf{x}\, \bar{\psi}\psi$$

$$= i\hbar c \int_V d^3\mathbf{x}\, \bar{\psi}\gamma^i \partial_i \psi - mc^2 \int_V d^3\mathbf{x}\, \bar{\psi}\psi = -i\hbar \int_V d^3\mathbf{x}\, \bar{\psi}\gamma^0 \dot{\psi},$$

so that the energy can also be written in the following simpler form:

$$H = i\hbar \int_V d^3\mathbf{x}\, \psi^\dagger \dot{\psi}. \tag{10.129}$$

This energy will be shown below to coincide with the Hamiltonian of the field. For this reason we describe it by the symbol H.

Let us now compute the conjugate momenta of the Hamiltonian formalism:

$$\pi(x) = \frac{\partial \mathcal{L}(x)}{\partial \dot{\psi}(x)} = i\frac{\hbar}{2}\psi^\dagger(x), \tag{10.130}$$

$$\pi^\dagger(x) = \frac{\partial \mathcal{L}(x)}{\partial \dot{\psi}^\dagger(x)} = -i\frac{\hbar}{2}\psi(x). \tag{10.131}$$

We note that from these equations it follows that the canonical variables π, ψ, $\pi^\dagger$, $\psi^\dagger$ *are not independent*: $\pi^\dagger \propto \psi$, $\pi \propto \psi^\dagger$. In view of the quantization of the Dirac field, we need to deal with independent canonical variables. It is useful, in this respect, to redefine the Lagrangian density in the following way:

$$\mathcal{L} = i\hbar c\, \bar{\psi}(x)\gamma^\mu \partial_\mu \psi(x) - mc^2 \bar{\psi}(x)\psi(x). \tag{10.132}$$

The reader can easily verify that the above expression differs from the previous definition (10.119) by a divergence. We then define, as the only independent variables, the components of $\psi(x)$, so that the corresponding conjugate momenta read

$$\pi(x) = \frac{\partial \mathcal{L}(x)}{\partial \dot{\psi}} = i\hbar \psi^\dagger(x). \tag{10.133}$$

From the canonical Poisson brackets (8.231) and (8.232) and the above expression of $\pi(x)$, we find:

$$\left\{ \psi^\alpha(\mathbf{x}, t), \psi^\dagger_\beta(\mathbf{y}, t) \right\} = -\frac{i}{\hbar}\delta^3(\mathbf{x} - \mathbf{y})\delta^\alpha_\beta, \tag{10.134}$$

$$\left\{ \psi^\alpha(\mathbf{x}, t), \psi^\beta(\mathbf{y}, t) \right\} = \left\{ \psi^\dagger_\alpha(\mathbf{x}, t), \psi^\dagger_\beta(\mathbf{y}, t) \right\} = 0. \tag{10.135}$$

It is convenient to rewrite the energy H in Eq. (10.129) using Dirac equation (10.53):

$$H = i\hbar \int_V d^3\mathbf{x}\, \psi^\dagger\, \dot\psi = \int_V d^3\mathbf{x}\psi^\dagger \left[-i\,\hbar c\, \alpha^i\, \partial_i + mc^2\, \beta \right] \psi. \tag{10.136}$$

The reader can verify that the energy density in the above formula coincides with the Hamiltonian density, which has the form:

$$\mathcal{H} = \pi_\alpha\, \dot\psi^\alpha - \mathcal{L}. \tag{10.137}$$

We can also verify that the Hamilton equation

$$\dot\pi^\dagger(x) = -\frac{\delta H}{\delta\psi^\dagger(x)} = -\left[-i\hbar c\, \alpha^i\, \partial_i + mc^2\, \beta \right] \psi, \tag{10.138}$$

coincides with the Dirac equation

$$i\hbar\dot\psi = (-i\hbar c\alpha^i\, \partial_i + mc^2\, \beta)\, \psi.$$

10.6 Plane Wave Solutions to the Dirac Equation

We now examine solutions to the Dirac equation having definite values of energy and momentum. A spinor field with *definite four-momentum* $p = (p^\mu)$ and spin r, must have the general plane-wave form given in (9.113):

$$\psi_{p,r}(x) = c_p\, w(p,r)\, e^{\frac{i}{\hbar}(\mathbf{p}\cdot\mathbf{x} - Et)} = c_p\, w(p,r)\, e^{-\frac{i}{\hbar} p\cdot x}, \tag{10.139}$$

where $w(p,r)$ is a four-component Dirac spinor and c_p a Lorentz-invariant normalization factor, to be fixed later. Inserting (10.139) into Eq. (10.61), and using the short-hand notation $\not{p} \equiv \gamma^\mu\, p_\mu$, we find that the generic spinor $w(p)$ satisfies the equation

$$(\not{p} - mc)\, w(p,r) = 0, \tag{10.140}$$

where $p^\mu = (\frac{E}{c}, \mathbf{p})$. If we decompose $w(p,r)$ into two-dimensional spinors as in Eq. (10.68) and use the representation (10.63) of the γ- matrices, Eq. (10.140) becomes:

$$\begin{pmatrix} \frac{E}{c} - mc & -\sigma \cdot \mathbf{p} \\ \sigma \cdot \mathbf{p} & -\frac{E}{c} - mc \end{pmatrix} \begin{pmatrix} \varphi \\ \chi \end{pmatrix} = 0. \tag{10.141}$$

We have shown that each component of $\psi(x)$ is in particular solution to the Klein-Gordon equation (10.58) which implements the mass-shell condition. This can be also verified by multiplying Eq. (10.140) to the left by the matrix $(\not{p} + mc)$:

$$(\not{p} + mc)(\not{p} - mc)w(p, r) = (\not{p}^2 + mc\,\not{p} - mc\,\not{p} - m^2c^2)\,w(p, r) = 0.$$

Using the anti-commutation properties of the γ^μ-matrices we find

$$\not{p}^2 = \gamma^\mu\gamma^\nu\,p_\mu p_\nu = \frac{1}{2}\left(\gamma^\mu\gamma^\nu + \gamma^\nu\gamma^\mu\right)p_\mu p_\nu = \eta^{\mu\nu}\,p_\mu p_\nu = p^2, \quad (10.142)$$

which implies

$$(\not{p} + mc)(\not{p} - mc)w(p, r) = (p^2 - m^2c^2)\,w(p, r) = 0, \qquad (10.143)$$

namely the mass-shell condition.

As noticed earlier, the Klein-Gordon equation contains negative energy solutions besides the positive energy ones:

$$\frac{E^2}{c^2} = \mathbf{p}^2 + m^2c^2 \;\Rightarrow\; E = \pm E_\mathbf{p} = \pm\sqrt{|\mathbf{p}|^2 c^2 + m^2 c^4}. \qquad (10.144)$$

The problem of interpreting such solutions, as already mentioned in the case of the complex scalar field, will be resolved by the *field quantization* which associates them with operators creating *antiparticles*.

We write the solutions with $E = \pm E_\mathbf{p}$ in the following form:

$$\psi_{\mathbf{p},r}^{(+)}(x) \equiv c_p\, w((E_\mathbf{p}/c, \mathbf{p}), r)\, e^{\frac{i}{\hbar}(\mathbf{p}\cdot\mathbf{x} - E_\mathbf{p} t)} = c_p\, u(p, r)\, e^{-\frac{i}{\hbar}p\cdot x},$$

$$\psi_{\mathbf{p},r}^{(-)}(x) \equiv c_p\, w((-E_\mathbf{p}/c, \mathbf{p}), r)\, e^{\frac{i}{\hbar}(\mathbf{p}\cdot\mathbf{x} + E_\mathbf{p} t)} = c_p\, v((E_\mathbf{p}/c, -\mathbf{p}), r)\, e^{\frac{i}{\hbar}(\mathbf{p}\cdot\mathbf{x} + E_\mathbf{p} t)},$$

where we have defined $u(p, r) \equiv w((\frac{E_\mathbf{p}}{c}, \mathbf{p}), r)$, $v((\frac{E_\mathbf{p}}{c}, -\mathbf{p}), r) \equiv w((-\frac{E_\mathbf{p}}{c}, \mathbf{p}), r)$. We shall choose the normalization factor c_p to be: $c_p \equiv \sqrt{\frac{mc^2}{E_\mathbf{p} V}}$. Note that the exponent in the definition of $\psi_{\mathbf{p},r}^{(-)}$ acquires a Lorentz-invariant form if we switch $\mathbf{p}$ into $-\mathbf{p}$. We can then write:

$$\psi_{\mathbf{p},r}^{(+)}(x) \equiv \sqrt{\frac{mc^2}{E_\mathbf{p} V}}\, u(p, r)\, e^{-\frac{i}{\hbar}p\cdot x}, \qquad (10.145)$$

$$\psi_{-\mathbf{p},r}^{(-)}(x) \equiv \sqrt{\frac{mc^2}{E_\mathbf{p} V}}\, v(p, r)\, e^{\frac{i}{\hbar}p\cdot x}. \qquad (10.146)$$

In the above solution we have defined $p = (p^\mu) = (\frac{E_\mathbf{p}}{c}, \mathbf{p})$ so that (10.146) describes a negative-energy state with momentum $-p$, $v(p, r) \equiv w(-p, r)$.

The general solution to the Dirac equation will be expanded in both kinds of solutions, and have the following form:

$$\psi(x) = \int \frac{d^3\mathbf{p}}{(2\pi\hbar)^3} \, V \sum_{r=1}^{2} \left(c(\mathbf{p}, r) \, \psi_{\mathbf{p},r}^{(+)}(x) + d(-\mathbf{p}, r)^* \, \psi_{\mathbf{p},r}^{(-)}(x) \right),$$

where c, d are complex numbers representing the components of $\psi(x)$ relative to the complete set of solutions $\psi_{\mathbf{p},r}^{(\pm)}(x)$. By changing $\mathbf{p}$ into $-\mathbf{p}$ in the integral of the second term on the right hand side, we have:

$$\psi(x) = \int \frac{d^3\mathbf{p}}{(2\pi\hbar)^3} \, V \sum_{r=1}^{2} \left(c(\mathbf{p}, r) \, \psi_{\mathbf{p},r}^{(+)}(x) + d(\mathbf{p}, r)^* \, \psi_{-\mathbf{p},r}^{(-)}(x) \right)$$

$$= \int \frac{d^3\mathbf{p}}{(2\pi\hbar)^3} \sqrt{\frac{mc^2 \, V}{E_\mathbf{p}}} \sum_{r=1}^{2} \left(c(\mathbf{p}, r) \, u(p, r) \, e^{-\frac{i}{\hbar} p \cdot x} + d(\mathbf{p}, r)^* \, v(p, r) \, e^{\frac{i}{\hbar} p \cdot x} \right).$$

$$(10.147)$$

We need now to explicitly construct the spinors $u(p, r)$, $v(p, r)$. Being $u(p, r) = w(p, r)$ and $v(p, r) = w(-p, r)$, the equation for $u(p, r)$ is the same as Eq. (10.140), while the one for $v(p, r)$ is obtained from (10.140) by replacing $p \rightarrow -p$:

$$(\not{p} - mc) \, u(p, r) = 0; \quad (\not{p} + mc) \, v(p, r) = 0. \qquad (10.148)$$

The Lorentz covariance of the above equations implies that $S(\Lambda)u(p, r)$ and $S(\Lambda)v(p, r)$ must be a combination of $u(\Lambda p, s)$ and $v(\Lambda p, s)$,[10] with coefficients given by the rotation $\mathcal{R}(\Lambda, p)^s{}_r$ of Eq. (10.107), or (10.110) for massless particles, according to our discussion in Sect. 9.4.1:

$$S(\Lambda) \, u(p, r) = \mathcal{R}(\Lambda, p)^{r'}{}_r \, u(\Lambda p, r')$$

$$S(\Lambda) \, v(p, r) = \mathcal{R}(\Lambda, p)^{r'}{}_r \, v(\Lambda p, r'). \qquad (10.149)$$

These are nothing but the transformation properties derived in Eq. (9.118). In the frame $\mathcal{S}_0$ in which the momentum p is the standard one $\bar{p}$, $u(\bar{p}, r)$ and $v(\bar{p}, r)$ transform covariantly under the action of the spin group. Let us construct them in this frame and then extend their definition to a generic one.

---

[10]This can be easily ascertained by multiplying both Eq. (10.148) to the left by $S(\Lambda)$. We find that $S(\Lambda)u(p, r)$ and $S(\Lambda)v(p, r)$ satisfy the following equations: $(S(\Lambda) \not{p} S(\Lambda)^{-1} - mc)$ $S(\Lambda)u(p, r) = 0$ and $(S(\Lambda) \not{p} S(\Lambda)^{-1} + mc) S(\Lambda)v(p, r) = 0$. Next we use property (10.88) and invariance of the Lorentzian scalar product $\gamma \cdot p \equiv \gamma^\mu p_\mu = \not{p}$ to write $S(\Lambda) \not{p} S(\Lambda)^{-1} = \not{p}' = \gamma^\mu p'_\nu$, where $p' = \Lambda p$. Thus the transformed spinors satisfy Eq. (10.148) with the transformed momentum p', and consequently, should be a combination of $u(p', s)$ and $v(p', s)$, respectively.

Consider a massive particle, $m \neq 0$, and let us first examine the solutions of the coupled system (10.141) in the rest frame S_0, where $\mathbf{p} = \mathbf{0}$, namely $\bar{p} = (mc, \mathbf{0})$. Equation (10.141) becomes:

$$\left(E - mc^2 \right) \varphi = 0; \quad \left(E + mc^2 \right) \chi = 0. \tag{10.150}$$

Then we have either

$$E = E_{\mathbf{p}=0} = mc^2; \quad \varphi \neq 0, \ \chi = 0,$$

or

$$E = -E_{\mathbf{p}=0} = -mc^2; \quad \varphi = 0, \ \chi \neq 0.$$

The non zero spinors in the two cases can be chosen arbitrarily. We choose them to be eigenvectors of σ^3:

$$\varphi_1 = \begin{pmatrix} 1 \\ 0 \end{pmatrix}; \quad \varphi_2 = \begin{pmatrix} 0 \\ 1 \end{pmatrix}. \tag{10.151}$$

In S_0 we can then write the positive and negative energy solutions in the momentum representation as

$$u(0, r) \equiv u(\bar{p}, r) = \begin{pmatrix} \varphi_r \\ 0 \end{pmatrix}; \quad v(0, r) \equiv v(\bar{p}, r) = \begin{pmatrix} 0 \\ \varphi_r \end{pmatrix} \quad r = 1, 2, \tag{10.152}$$

where $0 = \begin{pmatrix} 0 \\ 0 \end{pmatrix}$. Since the φ_r are eigenstates of σ^3, the *rest frame* solutions $u(0, r)$ and $v(0, r)$ are eigenstates of the operator:

$$\Sigma^3 = \begin{pmatrix} \frac{\hbar}{2} \sigma^3 & 0 \\ 0 & \frac{\hbar}{2} \sigma^3 \end{pmatrix}, \tag{10.153}$$

corresponding to the eigenvalues $\pm \hbar/2$. Once the solutions in the rest frame are given we may construct the solutions $u(p, r)$ and $v(p, r)$ of the Dirac equation in a generic frame S where $\mathbf{p} \neq 0$ as follows:

$$u(p, r) = \frac{\not{p} + mc}{\sqrt{2m \left(mc^2 + E_{\mathbf{p}} \right)}} u(0, r), \tag{10.154}$$

$$v(p, r) = \frac{-\not{p} + mc}{\sqrt{2m \left(mc^2 + E_{\mathbf{p}} \right)}} v(0, r). \tag{10.155}$$

The denominators appearing in Eqs. (10.154) and (10.155) are normalization factors determined in such a way that the spinors $u(p, r)$, $v(p, r)$ obey simple normalization conditions (see Eqs. (10.168) and (10.169) of the next section).

It is straightforward to show that $u(p, r)$ and $v(p, r)$ satisfy Eq. (10.148) by using the properties

$$(\not{p} + mc)(\not{p} - mc) = (\not{p} - mc)(\not{p} + mc)$$
$$= p^2 - m^2c^2 + mc\,\not{p} - mc\,\not{p} = p^2 - m^2c^2 = 0, \quad (10.156)$$

which descend from Eq. (10.142). Using the representation (10.63) of the γ-matrices and the explicit form of $\not{p}$, we obtain $u(p, r)$ and $v(p, r)$ in components:

$$u(p, r) = \begin{pmatrix} \sqrt{\dfrac{E_{\mathbf{p}} + mc^2}{2mc^2}}\,\varphi_r \\[2ex] \dfrac{\mathbf{p} \cdot \boldsymbol{\sigma}}{\sqrt{2m(E_{\mathbf{p}} + mc^2)}}\,\varphi_r \end{pmatrix} \; ; \; v(p, r) = \begin{pmatrix} \dfrac{\mathbf{p} \cdot \boldsymbol{\sigma}}{\sqrt{2m(E_{\mathbf{p}} + mc^2)}}\,\varphi_r \\[2ex] \sqrt{\dfrac{E_{\mathbf{p}} + mc^2}{2mc^2}}\,\varphi_r \end{pmatrix}.$$
$$(10.157)$$

Let us show that the above vectors transform as in Eq. (10.149) with respect to rotations Λ_R:

$$S(\Lambda_R)u(p, r) = e^{\frac{i}{\hbar}\theta^i\,\Sigma_i}\,u(p, r) = \begin{pmatrix} \sqrt{\dfrac{E_{\mathbf{p}} + mc^2}{2mc^2}}\,S(\theta^i)\,\varphi_r \\[2ex] \dfrac{S(\theta^i)\,\mathbf{p} \cdot \boldsymbol{\sigma}}{\sqrt{2m(E_{\mathbf{p}} + mc^2)}}\varphi_r \end{pmatrix}$$

$$= \begin{pmatrix} \sqrt{\dfrac{E_{\mathbf{p}} + mc^2}{2mc^2}}\,\varphi_r' \\[2ex] \dfrac{S(\theta^i)\,\mathbf{p} \cdot \boldsymbol{\sigma}\,S(\theta^i)^{-1}}{\sqrt{2m(E_{\mathbf{p}} + mc^2)}}\varphi_r' \end{pmatrix}, \quad (10.158)$$

where:

$$\varphi_r' \equiv S(\theta^i)\,\varphi_r = S(\theta^i)^s{}_r\,\varphi_s = \mathcal{R}^s{}_r\,\varphi_s. \quad (10.159)$$

Let us now use the property of the Pauli matrices to transform under conjugation by an SU(2) matrix $S(\boldsymbol{\theta})$, $\boldsymbol{\theta} \equiv (\theta^i)$, as the components of a three-dimensional vector $\boldsymbol{\sigma} \equiv (\sigma_i)$ under a corresponding rotation $\mathbf{R}(\boldsymbol{\theta})$, see Appendix (F):

$$\mathbf{S}(\boldsymbol{\theta})^{-1}\sigma_i\,\mathbf{S}(\boldsymbol{\theta}) = R(\boldsymbol{\theta})_i{}^j\,\sigma_j \;\Rightarrow\; \mathbf{S}(\boldsymbol{\theta})\sigma_i\,\mathbf{S}(\boldsymbol{\theta})^{-1} = R(\boldsymbol{\theta})^{-1}{}_i{}^j\,\sigma_j. \quad (10.160)$$

We can then write:

$$\mathbf{S}(\theta)\,\mathbf{p}\cdot\boldsymbol{\sigma}\mathbf{S}(\theta)^{-1} = \mathbf{p}\cdot\left(\mathbf{R}(\theta)^{-1}\,\boldsymbol{\sigma}\right) = \mathbf{p}'\cdot\boldsymbol{\sigma}, \qquad (10.161)$$

where $\mathbf{p}' \equiv \mathbf{R}(\theta)\,\mathbf{p}$. Since $\Lambda_R\, p = (p^0, \mathbf{p}')$, we conclude that

$$S(\Lambda_R)u(p,r) = \mathcal{R}^s{}_r \left(\begin{array}{c} \sqrt{\dfrac{E_\mathbf{p}+mc^2}{2mc^2}}\,\varphi_s \\[2mm] \dfrac{\mathbf{p}'\cdot\boldsymbol{\sigma}}{\sqrt{2m(E_\mathbf{p}+mc^2)}}\varphi_s \end{array}\right) = \mathcal{R}^s{}_r\, u(\Lambda_R\, p, s). \quad (10.162)$$

A similar derivation can be done for $v(p,r)$. If Λ is a boost, of the form $\Lambda = \exp(\frac{i}{\hbar}\omega_{0i}\, J^{0i})$, the corresponding representation on the spinors reads $S(\Lambda) = \exp(\frac{i}{\hbar}\omega_{0i}\, \Sigma^{0i})$. The resulting SU(2) rotation $\mathcal{R}(\Lambda, p)$, which we are not going to derive, is the Wigner rotation.

We note that $u(p, r)$ and $v(p, r)$ are not eigenstates of the third component of the spin operator Σ^3 (10.153) except in the special case of $p^1 = p^2 = 0$, $p^3 \neq 0$. This can be explained in light of the discussion done in Sect. 9.4.1 about little groups. The solutions $u(p, r)$ and $v(p, r)$, for a fixed p, transform as doublets with respect to the little group of the momentum p, which we have denoted by $G_p^{(0)}$: The action of $G_p^{(0)}$ on the solutions $u(p, r)$ and $v(p, r)$, according to Eq. (10.149), does not affect their dependence on p, and only amounts to an SU(2)-transformation on the index r. This group is related to the little group $G^{(0)} = SU(2)$ of $\bar{p} = (mc, \mathbf{0})$, generated by the Σ^i matrices as follows: $G_p^{(0)} = \Lambda_p \cdot SU(2) \cdot \Lambda_p^{-1}$. This means that its generators are $\Sigma_i' = S(\Lambda_p)\,\Sigma_i\, S(\Lambda_p)^{-1}$. If instead we act on $u(p, r)$ and $v(p, r)$ by means of a $G^{(0)} = SU(2)$-transformation, it will affect the dependence of these fields on p, mapping it into $p' = (p^0, \mathbf{R}\,\mathbf{p})$. Therefore, if $u(\bar{p}, r)$ and $v(\bar{p}, r)$ are eigenvectors of Σ_3, $u(p, r)$ and $v(p, r)$ will be eigenvectors of Σ_3'.

In Sect. 9.4.1 of last chapter, a general method was applied to the construction of the single-particle quantum states $|p, r\rangle$ acted on by a unitary irreducible representation of the Lorentz group. The method consisted in first constructing the states of the particle $|\bar{p}, r\rangle$ in some special frame S_0 in which the momentum of the particle is the standard one $\bar{p}$, and on which an irreducible representation $\mathcal{R}$ of the little group $G^{(0)}$ of $\bar{p}$ acts ($\bar{p} = (mc, \mathbf{0})$ and $G^{(0)} = SU(2)$ for massive particles, while $\bar{p} = (E, E, 0, 0)/c$ and $G^{(0)}$ is effectively SO(2) for massless particles). A generic state $|p, r\rangle$ is then constructed by acting on $|\bar{p}, r\rangle$ by means of $U(\Lambda_p)$, see Eq. (9.111), that is the representative on the quantum states of the simple Lorentz boost Λ_p connecting $\bar{p}$ to p: $p = \Lambda_p\,\bar{p}$. This suffices to define the representative $U(\Lambda)$ of a generic Lorentz transformation, see Eq. (9.112).

In this section we have applied this prescription to the construction of both the positive and negative energy eigenstates of the momentum operators. The role of $|p, r\rangle$ is now played by the spinors $u(p, r)$, $v(p, r)$, and that of $U(\Lambda)$ by the matrix $S(\Lambda)$, as it follows by comparing Eq. (10.149) with Eq. (9.112). It is instructive at this point to show that the expressions for $u(p, r)$, $v(p, r)$ given in (10.154) or, equivalently, (10.157), for massive fermions, could have been obtained from the corresponding spinors $u(0, r)$, $v(0, r)$ in S_0 using the prescription (9.111), namely by acting on them through the Lorentz boost $S(\Lambda_p)$:

$$u(p, r) = S(\Lambda_p) u(0, r); \quad v(p, r) = S(\Lambda_p) v(0, r). \tag{10.163}$$

This is readily proven using the matrix form (10.118) of $S(\Lambda_p)$ derived in Sect. 10.4.4 and the definition of $u(0, r)$, $v(0, r)$ in Eq. (10.152). The matrix product on the right hand side of Eq. (10.163) should then be compared with the matrix form of $u(p, r)$, $v(p, r)$ in (10.157).

10.6.1 Useful Properties of the $u(p, r)$ and $v(p, r)$ Spinors

In the following we shall prove some properties of the spinors $u(p, r)$ and $v(p, r)$ describing solutions with definite four-momentum.

• Let us compute the Dirac conjugates of $u(p, r)$ e $v(p, r)$:

$$\bar{u}(p, r) = u^\dagger(p, r)\gamma^0 = u^\dagger(0, r)\frac{\not{p}^\dagger + mc}{\sqrt{2m(E_\mathbf{p} + mc^2)}}\gamma^0$$

$$= u^\dagger(0, r)\gamma^0\gamma^0\frac{\not{p}^\dagger + mc}{\sqrt{2m(E_\mathbf{p} + mc^2)}}\gamma^0$$

$$= \bar{u}(0, r)\frac{\not{p} + mc}{\sqrt{2m(E_\mathbf{p} + mc^2)}}. \tag{10.164}$$

In an analogous way one finds:

$$\bar{v}(p, r) = \bar{v}(0, r)\frac{-\not{p} + mc}{\sqrt{2m(E_\mathbf{p} + mc^2)}}. \tag{10.165}$$

Recalling the property (10.156), from (10.164) and (10.165) we obtain the equations of motion obeyed by the Dirac spinors $\bar{u}(p, r)$ e $\bar{v}(p, r)$:

$$\bar{u}(p, r)(\not{p} - mc) = 0,$$
$$\bar{v}(p, r)(\not{p} + mc) = 0. \tag{10.166}$$

• Next we use the relations:

$$(\not{p} + mc)^2 = 2mc(\not{p} + mc),$$
$$(\not{p} - mc)^2 = 2mc(-\not{p} + mc), \tag{10.167}$$

which follow from Eq. (10.142) and the mass-shell condition $p^2 = m^2c^2$, to compute $\bar{u}(p, r)u(p, r')$:

$$\bar{u}(p, r)u(p, r') = \frac{2mc}{2m(E_{\mathbf{p}} + mc^2)}\bar{u}(\mathbf{0}, r)(\not{p} + mc)u(\mathbf{0}, r')$$

$$= \frac{c}{E_{\mathbf{p}} + mc^2}(\varphi_r, 0, 0)(\not{p} + mc)\begin{pmatrix} \varphi_{r'} \\ 0 \\ 0 \end{pmatrix}$$

$$= \varphi_r \cdot \varphi_{r'} = \delta_{rr'}, \tag{10.168}$$

With analogous computations one also finds:

$$\bar{v}(p, r)v(p, r') = \frac{c}{E_{\mathbf{p}} + mc^2}\bar{v}(\mathbf{0}, r)(-\not{p} + mc)v(\mathbf{0}, r')$$

$$= \frac{c}{E_{\mathbf{p}} + mc^2}(0, 0, -\varphi_r)(-\not{p} + mc)\begin{pmatrix} 0 \\ 0 \\ \varphi_{r'} \end{pmatrix}$$

$$= -\delta_{rr'}, \tag{10.169}$$

and moreover

$$\bar{u}(p, r)v(p, r') \propto \bar{u}(\mathbf{0}, r)(\not{p} + mc)(-\not{p} + mc)v(\mathbf{0}, r')$$

$$= 0 = \bar{v}(p, r)u(p, r'). \tag{10.170}$$

Summarizing, we have obtained the relations

$$\bar{u}(p, r)u(p, r') = \delta_{rr'}, = -\bar{v}(p, r)v(p, r'),$$

$$\bar{u}(p, r)v(p, r') = 0. \tag{10.171}$$

• Next we show that:

$$u^{\dagger}(p, r)u(p, r') = \frac{E_{\mathbf{p}}}{mc^2}\delta_{rr'} \geq 0, \tag{10.172}$$

$$v^{\dagger}(p, r)v(p, r') = \frac{E_{\mathbf{p}}}{mc^2}\delta_{rr'} \geq 0. \tag{10.173}$$

Indeed, using the Dirac equation $\not{p}u = mc\,u$, and $\bar{u}\,\not{p} = mc\,\bar{u}$, we find

$$u^{\dagger}(p, r)u(p, r') = \bar{u}(p, r)\gamma^0 u(p, r') = \bar{u}(p, r)\frac{m\gamma^0 + m\gamma^0}{2m}u(p, r')$$

$$= \bar{u}(p, r)\frac{\not{p}\gamma^0 + \gamma^0\not{p}}{2mc}u(p, r').$$

Using now the property

$$\not{p}\gamma^0 + \gamma^0 \not{p} = \{\not{p}, \gamma^0\} = p_\mu \{\gamma^\mu, \gamma^0\} = 2\eta^{\mu 0} p_\mu = \frac{2E_\mathbf{p}}{c},$$

the last term, can be rewritten as follows:

$$\frac{E_\mathbf{p}}{mc^2}\,\bar{u}(p,r)u(p,r') = \frac{E_\mathbf{p}}{mc^2}\,\delta_{rr'},$$

so that Eq. (10.172) is retrieved. Equation (10.173) is obtained in an analogous way.

We conclude that $u^\dagger(p,r)u(p,r')$ and $v^\dagger(p,r)v(p,r')$ *are not Lorentz-invariant quantities*, since they transform as $E_\mathbf{p}$, that is as the time component of a four-vector. This agrees with the previous result that $J^\mu = \bar{\psi}\gamma^\mu \psi$ is a four-vector whose time component is $J^0 = \psi^\dagger \psi > 0$.

We can also prove the following orthogonality condition:

$$u(\mathbf{p},r)^\dagger v(-\mathbf{p},s) = 0, \tag{10.174}$$

where we have used the short-hand notation $u(\mathbf{p},r) \equiv u((E_\mathbf{p}/c, \mathbf{p}),r)$, $v(\mathbf{p},r) \equiv v((E_\mathbf{p}/c, \mathbf{p}),r)$. To prove the above equation we use the Dirac equation for $v(-\mathbf{p},s)$: $\not{p}'v(-\mathbf{p},s) = -mc\,v(-\mathbf{p},s)$, where $p' \equiv (E_\mathbf{p}/c, -\mathbf{p})$. We can then write:

$$u(\mathbf{p},r)^\dagger v(-\mathbf{p},s) = \bar{u}(\mathbf{p},r)\,\gamma^0\,v(-\mathbf{p},s) = \frac{1}{2mc}\bar{u}(\mathbf{p},r)\,(\not{p}\gamma^0 - \gamma^0\,\not{p}')\,v(-\mathbf{p},s)$$

$$= \frac{1}{2mc}\bar{u}(\mathbf{p},r)\,(p_i\gamma^i\,\gamma^0 + \gamma^0\,\gamma^i\,p_i)\,v(-\mathbf{p},s) = 0. \tag{10.175}$$

From property (10.174) it also follows that *positive and negative energy states are represented by mutually orthogonal spinors if they have the same momentum:*

$$\left[\psi_\mathbf{p}^{(+)}(x)\right]^\dagger \psi_\mathbf{p}^{(-)}(x) = 0. \tag{10.176}$$

Recalling from Eqs. (10.145) and (10.146) that

$$\psi_{\mathbf{p},r}^{(+)}(x) = c_p\,u(\mathbf{p},r)\,e^{-\frac{i}{\hbar}(E_\mathbf{p}t - \mathbf{p}\cdot\mathbf{x})}; \qquad \psi_{\mathbf{p},r}^{(-)}(x) = c_p\,v(-\mathbf{p},r)\,e^{\frac{i}{\hbar}(E_\mathbf{p}t + \mathbf{p}\cdot\mathbf{x})},$$

from the orthogonality condition (10.174) it indeed follows that

$$\psi_{\mathbf{p},r}^{(+)}(x)^\dagger \psi_{\mathbf{p},s}^{(-)}(x) = |c_p|^2\,u^\dagger(\mathbf{p},r)\,v(-\mathbf{p},s)\,e^{\frac{2i}{\hbar}E_\mathbf{p}t} = 0. \tag{10.177}$$

Having fixed the normalization factor c_p in Eq. (10.139) to be $\sqrt{\frac{mc^2}{V E_p}}$, we now observe that Eqs. (10.172) and (10.173) represent the right normalization (9.116) of the u and v vectors in order for the corresponding positive and negative energy solutions $\psi_{p,r}^{(\pm)}(x)$ to be normalized as in (9.54):

$$\left(\psi_{p,r}^{(\pm)}, \psi_{p',r'}^{(\pm)} \right) = \int d^3x\, \psi_{p,r}^{(\pm)}(x)^\dagger\, \psi_{p',r'}^{(\pm)}(x) = \frac{(2\pi\hbar)^3}{V}\, \delta^3(\mathbf{p} - \mathbf{p}')\, \delta_{rr'},$$

as the reader can easily verify.[11] Similarly, using the orthogonality condition (10.174), which applies to the above expression only when $\mathbf{p} = \mathbf{p}'$, we can show that positive and negative energy solutions are mutually orthogonal:

$$\left(\psi_{p,r}^{(+)}, \psi_{p',r'}^{(-)} \right) = \int d^3x\, |c_p|^2\, u(\mathbf{p}, r)^\dagger\, v(-\mathbf{p}', r')\, e^{\frac{i}{\hbar}(E_p + E_p')t}\, e^{-\frac{i}{\hbar}(\mathbf{p} - \mathbf{p}') \cdot \mathbf{x}}$$

$$\times \propto (2\pi\hbar)^3\, \delta^3(\mathbf{p} - \mathbf{p}')\, e^{\frac{2i}{\hbar} E_p t}\, u(\mathbf{p}, r)^\dagger\, v(-\mathbf{p}, r') = 0.$$

- Finally we define *projection operators* $\Lambda_+(\pm p)$ on the positive and negative energy solutions:

$$\Lambda_+(p)^\alpha{}_\beta \equiv \sum_{r=1}^{2} u(p, r)^\alpha \bar{u}(p, r)_\beta, \tag{10.178}$$

$$\Lambda_-(p)^\alpha{}_\beta \equiv -\sum_{r=1}^{2} v(p, r)^\alpha \bar{v}(p, r)_\beta. \tag{10.179}$$

Using the formulae (10.171) we see $\Lambda_\pm(p)$ are indeed projection operators:

$$\Lambda_+(p)\, u(p, r) = u(p, r); \quad \Lambda_+(p)\, v(p, r) = 0, \tag{10.180}$$

$$\Lambda_-(p) u(p, r) = 0; \quad \Lambda_-(p) v(p, r) = v(p, r). \tag{10.181}$$

The explicit form of $\Lambda_\pm$ is immediately derived from Eq. (10.167) since they express the fact that $\not{p} \pm mc$ are proportional to projection operators. Thus we have:

$$\Lambda_+(p) = \frac{1}{2mc}(\not{p} + mc), \tag{10.182}$$

$$\Lambda_-(p) = -\frac{1}{2mc}(\not{p} - mc). \tag{10.183}$$

---

[11] In the above derivation the time-dependent exponential $e^{\frac{i}{\hbar}(p^0 - p'^0)x^0}$ equals one since the equality $\mathbf{p} = \mathbf{p}'$ implemented by the delta-function implies $p^0 = p'^0$.

10.6.2 Charge Conjugation

We show the existence of an operator in the Dirac relativistic theory which transforms *positive energy* solutions into *negative energy* solutions, and viceversa. One can prove on general grounds that there exists a matrix in spinor space, called the *charge-conjugation matrix*, with the following properties

$$C^{-1}\gamma_\mu C = -\gamma_\mu^T; \quad C^T = -C; \quad C^\dagger = C^T = C^{-1}. \tag{10.184}$$

In the standard representation we may identify the C matrix as

$$C = i\gamma^2\gamma^0 = \begin{pmatrix} 0 & -i\sigma^2 \\ -i\sigma^2 & 0 \end{pmatrix}. \tag{10.185}$$

Given a Dirac field $\psi(x)$, we define its *charge conjugate* spinor $\psi^c(x)$ as follows:

$$\psi^c(x) \equiv C\,\bar\psi^T(x). \tag{10.186}$$

The operation which maps $\psi(x)$ into its charge conjugate $\psi^c(x)$ is called *charge-conjugation*. Let us show that charge conjugation is a correspondence between positive and negative energy solutions.

To this end let us consider the positive energy plane wave described by the spinor $u(p, r)$. Its Dirac conjugate $\bar u$ will satisfy the following equation:

$$\bar u(p)\,(\not p - mc) = 0.$$

By transposition we have

$$\left(\gamma_\mu^T p^\mu - mc\right)\bar u^T(p) = 0$$

If we now multiply the above equation to the left by the C matrix and use properties (10.184) we obtain

$$(\not p + mc)\,C\bar u^T(p) = 0, \tag{10.187}$$

which shows that charge-conjugate spinor $u^c(p) = C\bar u^T(p)$ satisfies the second of Eq. (10.148) and should therefore coincide with a spinor $v(p)$ defining the negative energy solution $\psi_{-\mathbf p}^{(-)}$ with opposite momentum $-\mathbf p$. Besides changing the value of the momentum, charge-conjugation also reverses the spin orientation. Going, for the sake of simplicity, to the rest frame, where a positive energy solution with spin projection $\hbar/2$ along a given direction, is described by

$$u(\mathbf 0, r = 1) = (1, 0, 0, 0)^T,$$

(see Eq. (10.152)), we find for the charge conjugate spinor $u^c \equiv C\gamma^0 u^*$ (note that $\gamma^{0T} = \gamma^0$)

$$u^c(0, r) = C\gamma^0 u^*(0, r = 1) = (0, 0, 0, 1)^T = v(0, r = 2),$$

that is a negative energy spinor with spin projection $-\hbar/2$. In general the reader can verify that

$$u^c(0, r) = \epsilon_{rs} v(0, s), \tag{10.188}$$

where summation over $s = 1, 2$ is understood, and (ϵ_{rs}) is the matrix $i\sigma_2$: $\epsilon_{11} = \epsilon_{22} = 0$, $\epsilon_{12} = -\epsilon_{21} = 1$.

Let us now evaluate $u^c(p, r)$ using the explicit form of $u(p, r)$ given in Eq. (10.154):

$$u^c(p, r) = C\gamma^0 u(p, r)^* = C\gamma^0 \frac{\displaystyle \not{p}^* + mc}{\sqrt{2m(mc^2 + E_{\mathbf{p}})}} u(0, r)^*$$

$$= C\frac{\not{p}^T + mc}{\sqrt{2m(mc^2 + E_{\mathbf{p}})}} \gamma^0 u(0, r)^* = \frac{-\not{p} + mc}{\sqrt{2m(mc^2 + E_{\mathbf{p}})}} u^c(0, r)$$

$$= \epsilon_{rs} \frac{-\not{p} + mc}{\sqrt{2m(mc^2 + E_{\mathbf{p}})}} v(0, s) = \epsilon_{rs} v(p, s). \tag{10.189}$$

In the above derivation we have used the properties $C\not{p}^T C^{-1} = -\not{p}$ and $\gamma^0 \not{p}^* = \not{p}^T \gamma^0$.

We shall see in the next chapter that, upon quantizing the Dirac field, negative energy solutions $\psi^{(-)}_{-\mathbf{p},r}$ with momentum $-\mathbf{p}$ and a certain spin component (up or down relative to a given direction) are reinterpreted as creation operators of *antiparticles* with positive energy, momentum $\mathbf{p}$ and opposite spin component. Thus the charge conjugation operation can be viewed as the operation which interchanges particles with antiparticles with the same momentum and spin. As far as the electric charge is concerned we need to describe the coupling of a *charge conjugate spinor* to an external electromagnetic field as it was done for the scalar field. This will be discussed in Sect. 10.7. We anticipate that the electric charge of a charge conjugate spinor describing an antiparticle is *opposite* to that of the corresponding particle.

10.6.3 Spin Projectors

In Sect. 10.6.1 we have labeled the spin states of the massive solutions to the Dirac equation by the eigenvalues, in the rest frame, of Σ_3: $u(0, r)$, $v(0, r)$, for $r = 1, 2$ correspond to the eigenvalues $+\hbar/2$ and $-\hbar/2$ of Σ_3. This amounts to choosing the two-component vectors φ_r to correspond to the eigenvalues $+1$ and -1 of σ_3. We could have chosen $u(0, r)$, $v(0, r)$ to be eigenvectors of the spin-component $\Sigma \cdot \mathbf{n}$ along a generic direction $\mathbf{n}$ in space, $|\mathbf{n}| = 1$. The corresponding eigenvalues would still be $\pm \hbar/2$. Clearly, for generic $\mathbf{n}$, $\Sigma \cdot \mathbf{n}$ is not conserved, namely it does not commute with the Hamiltonian, as proven in Sect. 10.4.4. This is not the case if $\mathbf{n} = \mathbf{p}/|\mathbf{p}|$, in which case the corresponding component of the spin vector defines the helicity $\Gamma = \Sigma \cdot \mathbf{p}/|\mathbf{p}|$ which is indeed conserved (see discussion after Eq. (10.111)).

We now ask whether it is possible to give a *covariant* meaning to the value of the spin orientation along a direction $\mathbf{n}$. We wish in other words to define a Lorentz-invariant operator O_n which reduces to $\Sigma \cdot \mathbf{n}$ in the rest frame, namely such that, if in $\mathcal{S}_0$:

$$(\Sigma \cdot \mathbf{n}) \, u(0, r) = \varepsilon_r \frac{\hbar}{2} u(0, r); \quad (\Sigma \cdot \mathbf{n}) \, v(0, r) = \varepsilon_r \frac{\hbar}{2} v(0, r), \quad (10.190)$$

where $\varepsilon_1 = 1$, $\varepsilon_2 = -1$, in a generic frame $\mathcal{S}$:

$$O_n \, u(p, r) = \varepsilon_r \frac{\hbar}{2} u(p, r); \quad O_n \, v(p, r) = \varepsilon_r \frac{\hbar}{2} v(p, r). \quad (10.191)$$

Clearly, using Eq. (10.163), we must have:

$$O_n = S(\Lambda_p) \, (\Sigma \cdot \mathbf{n}) S(\Lambda_p)^{-1} = \Sigma' \cdot \mathbf{n}, \quad (10.192)$$

where Σ'_i are the generators of the little group $G_p^{(0)} \equiv SU(2)_p$ of p.

We shall however compute O_n in a simpler way, using the *Pauli-Lubanski* four-vector $\hat{W}_\mu$ introduced in Sect. 9.4.2, which, on spinor solutions with definite momentum p^μ, acts by means of the following matrices:

$$W_\mu \equiv -\frac{1}{2} \epsilon_{\mu\nu\rho\sigma} \Sigma^{\nu\rho} p^\sigma, \quad (10.193)$$

It is useful to write it in a simpler way by introducing the matrix γ^5 (see Appendix G):

$$\gamma^5 = i\gamma^0 \gamma^1 \gamma^2 \gamma^3 = \frac{i}{4!} \epsilon_{\mu\nu\rho\sigma} \gamma^\mu \gamma^\nu \gamma^\rho \gamma^\sigma = \begin{pmatrix} 0 & 1_2 \\ 1_2 & 0 \end{pmatrix}. \quad (10.194)$$

Note that γ^5 anticommutes with all the γ^μ-matrices and thus commutes with the Lorentz generators $\Sigma^{\mu\nu}$ which contain products of two γ^μ-matrices. From this we conclude that γ^5 commutes with a generic Lorentz transformation $S(\Lambda)$, since it commutes with its infinitesimal generator.

Using the γ^5 matrix the Pauli-Lubanski four-vector (10.193) takes the simpler form:

$$W_\mu = -\frac{1}{2}\epsilon_{\mu\nu\rho\sigma}\left(-\frac{\hbar}{2}\sigma^{\nu\rho}\right)p^\sigma = \frac{\hbar}{4}\epsilon_{\mu\nu\rho\sigma}\sigma^{\nu\rho}p^\sigma$$

$$= \frac{i\hbar}{2}\gamma^5\sigma_{\mu\nu}\,p^\nu = -i\gamma^5\Sigma_{\mu\nu}p^\nu, \tag{10.195}$$

where we have used the identity

$$\gamma^5\sigma_{\mu\sigma} = -\frac{i}{2}\epsilon_{\mu\sigma\nu\rho}\sigma^{\nu\rho},$$

given in Appendix G, which can be verified by direct computation, starting from the definition of γ^5. Using the Lorentz transformation properties (10.88) of the γ^μ-matrices, and the invariance of the $\epsilon_{\mu\nu\rho\sigma}$-tensor under proper transformations, we can easily verify that W^μ transforms like the γ^μ-matrices:

$$S(\Lambda)\,W^\mu\,S(\Lambda)^{-1} = \Lambda^{-1\mu}{}_\nu\,W^\nu. \tag{10.196}$$

Let us now introduce the four-vector $n^\mu(\mathbf{p}) = (n^0(\mathbf{p}), \mathbf{n}(\mathbf{p}))$ having the following properties:

$$\begin{cases} n^2 = n^\mu n_\mu = -1, \\ n_\mu p^\mu = 0. \end{cases} \tag{10.197}$$

In the rest frame, $\mathbf{p} = 0$ and $E = mc^2 \neq 0$, the previous relations yield:

$$n_\mu p^\mu = n^0 E = 0 \;\Rightarrow\; n^0 = 0,$$

$$n^2 = (n^0)^2 - |\mathbf{n}|^2 = -1 \;\Rightarrow\; |\mathbf{n}| = 1, \tag{10.198}$$

that is $n_\mu(\mathbf{p} = 0) = (0, \mathbf{n})$. We may now compute the scalar quantity $n^\mu W_\mu$:

$$n^\mu W_\mu = \frac{i\hbar}{2}\gamma^5\sigma_{\mu\nu}\,n^\mu\,p^\nu = -\frac{\hbar}{4}\gamma^5(\gamma_\mu\gamma_\nu - \gamma_\nu\gamma_\mu)n^\mu p^\nu$$

$$= -\frac{\hbar}{4}\gamma^5(2\gamma_\mu\gamma_\nu - 2\eta_{\mu\nu})\,n^\mu p^\nu = -\frac{\hbar}{2}\gamma^5\gamma_\mu n^\mu \slashed{p}. \tag{10.199}$$

where the property $n \cdot p = 0$ has been used. In the rest frame $\mathbf{p} = 0$, $n^\mu W_\mu$ becomes:

$$(n \cdot W)(\mathbf{p} = 0) = \frac{\hbar}{2}\gamma^5\,(n^i\gamma^i)\,p^0\gamma^0 = -\frac{\hbar}{2}mc\,\gamma^5\,\gamma^0\,\gamma^i\,n^i = -\frac{\hbar}{2}mc\,\gamma^5\,\alpha^i\,n^i$$

$$= -mc\,\Sigma \cdot \mathbf{n}, \tag{10.200}$$

where we have used the property

$$\Sigma^i = \frac{\hbar}{2}\gamma^5 \alpha^i, \tag{10.201}$$

which can be verified using Eqs. (10.67), (10.105) and (10.194). Thus we have found a Lorentz scalar quantity that in the rest frame reduces to $\mathbf{n} \cdot \mathbf{\Sigma}$:

$$O_n \equiv -\frac{1}{mc} n^\mu W_\mu \xrightarrow{\mathbf{p}=0} O_n^{(0)} = \mathbf{n} \cdot \mathbf{\Sigma}. \tag{10.202}$$

In the particular case of n pointing along the z-axis, $n = n_z = (0, 0, 0, 1)$, from Eq. (10.105) we find

$$O_{n_z}^{(0)} = -\frac{1}{mc} n^\mu W_\mu \bigg|_{\mathbf{p}=0} = \begin{pmatrix} \frac{\hbar}{2}\sigma_3 & 0 \\ 0 & \frac{\hbar}{2}\sigma_3 \end{pmatrix} = \Sigma_3. \tag{10.203}$$

Clearly, using the transformation property (10.196) of W^μ and the Lorentz invariance of the expression of O_n, in a generic frame $\mathcal{S}$ we find

$$O_n = -\frac{1}{mc} n^\mu W_\mu = S(\Lambda_p) O_n^{(0)} S(\Lambda_p)^{-1}, \tag{10.204}$$

that is if $u(\mathbf{0}, r)$, $v(\mathbf{0}, r)$ are eigenvectors on $\mathbf{\Sigma} \cdot \mathbf{n}$, $u(p, r)$, $v(p, r)$ are eigenvectors on O_n corresponding to the same eigenvalues, which is the content of Eqs. (10.190) and (10.191).

We can define projectors $\mathcal{P}_r$ on eigenstates of O_n corresponding to the eigenvalues $\varepsilon_r \hbar/2 = \pm\hbar/2$:

$$\mathcal{P}_r \equiv \frac{1}{2}\left(1 + \varepsilon_r \frac{2}{\hbar} O_n\right) = \frac{1}{2}\left(1 + \varepsilon_r \frac{1}{mc}\gamma^5 \not{n}\not{p}\right). \tag{10.205}$$

In the rest frame the above projector reads:

$$\mathcal{P}_r^{(0)} \equiv \frac{1}{2}\left(1 + \varepsilon_r \gamma^5 n^i \alpha_i\right) = \begin{pmatrix} 1_2 + \varepsilon_r \, \mathbf{n} \cdot \boldsymbol{\sigma} & 0 \\ 0 & 1_2 + \varepsilon_r \, \mathbf{n} \cdot \boldsymbol{\sigma} \end{pmatrix}. \tag{10.206}$$

The matrices $\mathcal{P}_r$ project on both positive and negative energy solutions with the same spin component along $\mathbf{n}$. Let us now define two operators $\Lambda_{+,r}$, $\Lambda_{-,r}$ projecting on positive and negative solutions with a given spin component r, respectively:

$$\Lambda_{+,r} \, u(p, s) = \delta_{rs} \, u(p, s); \quad \Lambda_{+,r} \, v(p, s) = 0,$$
$$\Lambda_{-,r} \, u(p, s) = 0; \quad \Lambda_{-,r} \, v(p, s) = \delta_{rs} \, v(p, s). \tag{10.207}$$

They have the following general form:

$$(\Lambda_{+,r})^\alpha{}_\beta = u^\alpha(p,r)\,\bar{u}_\beta(p,r); \quad (\Lambda_{-,r})^\alpha{}_\beta = -v^\alpha(p,r)\,\bar{v}_\beta(p,r), \quad (10.208)$$

as it follows from the orthogonality properties (10.168) and (10.169). To find the explicit expression of these matrices in terms of p and n, we notice that they are obtained by multiplying to the right and to the left the projectors $\mathcal{P}_r$ on the spin state r by the projectors $\Lambda_\pm$ on the positive and negative energy states:

$$\Lambda_{\pm,r} = \Lambda_\pm \, \mathcal{P}_r \, \Lambda_\pm = \Lambda_\pm \frac{1}{2}\left(1 \pm \varepsilon_r \, \gamma^5 \slashed{n}\right) = \pm\frac{1}{4\,mc}\left(\slashed{p} \pm mc\right)\left(1 \pm \varepsilon_r \, \gamma^5 \slashed{n}\right),$$

where we have used the property:

$$\left(1 + \varepsilon_r \frac{1}{mc} \gamma^5 \slashed{n}\slashed{p}\right)(\slashed{p} \pm mc) = (\slashed{p} \pm mc)(1 \pm \varepsilon_r \, \gamma^5 \slashed{n}), \quad (10.209)$$

which can be easily verified using the fact that $\slashed{p}$ and $\slashed{n}$ anticommute: $\slashed{n}\slashed{p} = -\slashed{p}\slashed{n}$.

10.7 Dirac Equation in an External Electromagnetic Field

We shall now study the coupling of the Dirac field to the electromagnetic field A_μ.

To this end, as we did for the complex scalar field in Sect. 10.2.1, we apply the *minimal coupling* prescription, namely we substitute in the free Dirac equation

$$p^\mu \to p^\mu + \frac{e}{c}\,A^\mu, \quad (10.210)$$

that is, in terms of the quantum operator

$$i\hbar\partial^\mu \to i\hbar\partial^\mu + \frac{e}{c}\,A^\mu. \quad (10.211)$$

In the convention which we adopt throughout the book, the electron has charge $e = -|e| < 0$.

The coupled Dirac equation takes the following form:

$$\left[(i\hbar\partial_\mu + \frac{e}{c}\,A_\mu)\gamma^\mu - mc\right]\psi(x) = 0. \quad (10.212)$$

Using the *covariant derivative* introduced in Eqs. (10.36) and (10.212) becomes

$$\left[i\hbar\,\gamma^\mu D_\mu - mc\right]\psi(x) = 0. \quad (10.213)$$

Just as in the case of the complex scalar field, the resulting equation is not invariant under gauge transformations

$$A_\mu(x) \to A_\mu(x) + \partial_\mu \varphi(x), \tag{10.214}$$

unless we also apply to the Dirac wave function the following simultaneous phase transformation

$$\psi(x) \to \psi(x) e^{\frac{ie}{\hbar c} \varphi(x)}. \tag{10.215}$$

In connection with the discussion of the meaning of the charge-conjugation operation, it is instructive to see how the Dirac equation in the presence of an external electromagnetic field transforms under charge-conjugation. The equation for the charge-conjugate spinor $\psi^c = C \overline{\psi}^T = C \gamma^0 \psi^*$ is easily derived from (10.212) and reads:

$$\left((i\hbar \partial_\mu - \frac{e}{c} A_\mu) \gamma^\mu - mc \right) \psi^c(x) = 0. \tag{10.216}$$

We see that ψ and ψ^c describe particles with opposite charge. This justifies the statement given at the end of Sect. 10.6.2 that antiparticles have opposite charge with respect to the corresponding particles.[12]

Let us now recast Eq. (10.212) in a Hamiltonian form. Solving with respect to the time derivative, we have:

$$i\hbar \frac{\partial \psi}{\partial t} = \left[-c \left(i\hbar \partial_i + \frac{e}{c} A_i \right) \alpha^i + \beta mc^2 - eA^0 \right] \psi = \hat{H} \psi, \tag{10.218}$$

where $H = H_{free} + H_{int}$, H_{free} being given by Eq. (10.53) and $H_{int} = -e (A_0 + A_i \alpha^i)$. In order to study the physical implications of the minimal coupling it is convenient to study its non-relativistic limit.

We proceed as in Sect. 10.4.1. We first redefine the Dirac field as in Eq. (10.72), so that the Dirac equation (10.218) takes the following form:

$$\left(i\hbar \frac{\partial}{\partial t} + mc^2 \right) \psi' = \left[-c \left(i\hbar \partial_i + \frac{e}{c} A_i \right) \alpha^i + \beta mc^2 - eA^0 \right] \psi'.$$

Next we decompose the field ψ' as in Eq. (10.70) (omitting prime symbols on φ and χ) and find:

$$\left(i\hbar \frac{\partial}{\partial t} + eA^0 \right) \varphi = c \sigma \cdot \left(\hat{\mathbf{p}} - \frac{e}{c} \mathbf{A} \right) \chi, \tag{10.219}$$

---

[12]We also observe that the Dirac equation is invariant under the transformations

$$\psi \to \psi^c, \quad A_\mu \to -A_\mu. \tag{10.217}$$

$$\left(i\hbar\frac{\partial}{\partial t} + eA^0 + 2mc^2\right)\chi = c\,\boldsymbol{\sigma}\cdot\left(\hat{\mathbf{p}} - \frac{e}{c}\mathbf{A}\right)\varphi. \tag{10.220}$$

As explained earlier, in the non-relativistic limit, we only keep on the left hand side of the second equation the term $2mc^2\,\chi$, since the rest energy mc^2 of the particle is much larger than the kinetic and potential energies, so that

$$\chi = \frac{1}{2mc}\boldsymbol{\sigma}\cdot\left(\hat{\mathbf{p}} - \frac{e}{c}\mathbf{A}\right)\varphi,$$

so that only the *large* upper component φ remains.

Substituting the expression for χ into the first of Eq. (10.219) we obtain:

$$\left(i\hbar\frac{\partial}{\partial t} + e\,A^0\right)\varphi = \frac{1}{2m}\left[\boldsymbol{\sigma}\cdot\left(\hat{\mathbf{p}} - \frac{e}{c}\mathbf{A}\right)\right]^2\varphi. \tag{10.221}$$

To evaluate the right hand side we note that given two vectors $\mathbf{a}, \mathbf{b}$ the following identity holds as a consequence of the Pauli matrix algebra:

$$(\mathbf{a}\cdot\boldsymbol{\sigma})(\mathbf{b}\cdot\boldsymbol{\sigma}) = \mathbf{a}\cdot\mathbf{b} + i\boldsymbol{\sigma}\cdot(\mathbf{a}\times\mathbf{b}).$$

In our case

$$\mathbf{a} = \mathbf{b} = \left(\hat{\mathbf{p}} - \frac{e}{c}\mathbf{A}\right) = -\left(i\hbar\nabla + \frac{e}{c}\mathbf{A}\right),$$

but the wedge product does not vanish, since ∇ and $\mathbf{A}$ do not commute. We find:

$$\left(\hat{\mathbf{p}} - \frac{e}{c}\mathbf{A}\right)\times\left(\hat{\mathbf{p}} - \frac{e}{c}\mathbf{A}\right)\varphi = i\,\frac{e\hbar}{c}\left(-\mathbf{A}\times\nabla + \nabla\times\mathbf{A}\right)\varphi + i\,\frac{e\hbar}{c}\mathbf{A}\times\nabla\varphi$$

$$= i\,\frac{e\hbar}{c}\mathbf{B}\,\varphi. \tag{10.222}$$

Substituting in (10.221) we finally obtain:

$$i\hbar\frac{\partial\varphi}{\partial t} = \left[\frac{1}{2m}\,|i\hbar\,\nabla + \frac{e}{c}\mathbf{A}|^2 + e\,V - \frac{e}{mc}\mathbf{s}\cdot\mathbf{B}\right]\varphi \equiv \hat{H}\varphi, \tag{10.223}$$

where we have defined, as usual, $\mathbf{s} \equiv \hbar\,\boldsymbol{\sigma}/2$, and written A_0 as $-V$, V being the electric potential. Equation (10.223) is called the *Pauli equation*. It differs from the Schroedinger equation of an electron interacting with the electromagnetic field by the presence in the Hamiltonian of the interaction term:

$$H_{magn} = -\frac{e}{mc}\mathbf{s}\cdot\mathbf{B} = -\boldsymbol{\mu}_s\cdot\mathbf{B}, \tag{10.224}$$

which has the form of the potential energy of a magnetic dipole in an external magnetic field with:

$$\boldsymbol{\mu}_s = \frac{e}{mc}\,\mathbf{s} = g\,\frac{e}{2mc}\,\mathbf{s}, \tag{10.225}$$

representing the electron *intrinsic magnetic moment*. The factor $g = 2$ is called the *g-factor* and the gyromagnetic ratio associated with the spin, defined as $|\boldsymbol{\mu}|/|\mathbf{s}|$, is $g|e|/(2mc)$. Recall that the magnetic moment associated with the orbital motion of a charge e reads

$$\boldsymbol{\mu}_{orbit} = \frac{e}{2mc}\,\mathbf{M}, \tag{10.226}$$

$\mathbf{M}$ being the orbital angular momentum. The gyromagnetic ratio $|\boldsymbol{\mu}|/|\mathbf{s}| = |e|/(mc)$ is twice *the one associated with the orbital angular momentum*. This result was found by Dirac in 1928.[13]

Let us write the Lagrangian density for a fermion with charge e, coupled to the electromagnetic field:

$$\mathcal{L} = \bar{\psi}(x)\left(i\hbar c\;\slashed{D} - mc^2\right)\psi(x). \tag{10.227}$$

The reader can easily verify that the above Lagrangian yields Eq. (10.212), or, equivalently (10.213). Just as we did for the scalar field, we can write $\mathcal{L}$ as the sum of a part describing the free fermion, plus an interaction term $\mathcal{L}_I$, describing the coupling to the electromagnetic field:

$$\begin{aligned}
\mathcal{L} &= \mathcal{L}_0 + \mathcal{L}_I,\\
\mathcal{L}_0 &= \bar{\psi}(x)\left(i\hbar c\;\slashed{\partial} - mc^2\right)\psi(x),\\
\mathcal{L}_I &= A_\mu(x)\,J^\mu(x) = e\,A_\mu(x)\bar{\psi}(x)\gamma^\mu\psi(x),
\end{aligned} \tag{10.228}$$

where we have defined the electric current four-vector J^μ as:

$$J^\mu(x) \equiv e\,j^\mu(x) = e\bar{\psi}(x)\gamma^\mu\psi(x). \tag{10.229}$$

In Sect. 10.4.2 we have shown that, by virtue of the Dirac equation, J^μ is a conserved current, namely that it is divergenceless: $\partial_\mu J^\mu = 0$. As pointed out earlier, the electric four-current defined above, is the conserved Noether current associated with the global U(1) invariance (10.122).

---

[13] We recall that the Zeeman effect can only be explained if $g = 2$. We see that this value is correctly predicted by the Dirac relativistic equation in the non-relativistic limit.

10.8 Parity Transformation and Bilinear Forms

It is important to observe that the standard representation of the γ-matrices given in Sect. 10.4.1 is by no means unique. Any other representation preserving the basic anticommutation rules works exactly the same way. It is only a matter of convenience to use one or another. In particular the expression (10.97) of the Lorentz generators $\Sigma^{\mu\nu}$ in terms of γ^μ-matrices is representation-independent.

In this section we introduce a different representation, called the *Weyl representation*, defined as follows:

$$\gamma^0 = \begin{pmatrix} 0 & 1_2 \\ 1_2 & 0 \end{pmatrix}; \quad \gamma^i = \begin{pmatrix} 0 & -\sigma^i \\ \sigma^i & 0 \end{pmatrix}; \quad i = 1, 2, 3. \tag{10.230}$$

It is immediate to verify that the basic anticommutation rules (10.60) are satisfied. Defining

$$\sigma^\mu = (1_2, -\sigma^i); \quad \bar{\sigma}^\mu = (1_2, \sigma^i), \tag{10.231}$$

Equation (10.230) can be given the compact form

$$\gamma^\mu = \begin{pmatrix} 0 & \sigma^\mu \\ \bar{\sigma}^\mu & 0 \end{pmatrix}. \tag{10.232}$$

The standard (Pauli) and the Weyl representations are related by a unitary change of basis:

$$\gamma^\mu_{\text{Pauli}} = U^\dagger \gamma^\mu_{\text{Weyl}} U.$$

Decomposing as usual the spinor ψ into two-dimensional spinors ξ e ζ:

$$\psi = \begin{pmatrix} \xi \\ \zeta \end{pmatrix}, \tag{10.233}$$

one can show that, in the Weyl representation, the *proper* Lorentz transformations act separately on the two spinors, without mixing them. As we are going to show below, this means that the four-dimensional spinor representation, *irreducible* with respect to the full Lorentz group $O(1, 3)$ becomes *reducible* into two two-dimensional representations under the subgroup of the *proper* Lorentz group $SO(1, 3)$.

To show this we observe that since infinitesimal transformations in the spinor representation of the Lorentz group are, by definition, connected with continuity to the identity, they ought to have unit determinant, and therefore they can only belong to the subgroup of proper Lorentz transformations $SO(1, 3)$.

We can compute, in the Weyl basis, the matrix form of the $\Sigma^{\mu\nu}$ generators:

$$\Sigma^{\mu\nu} = -\frac{\hbar}{2}\sigma^{\mu\nu} = -\frac{i\hbar}{4}\left[\gamma^\mu, \gamma^\nu\right] \tag{10.234}$$

$$= -\frac{i\hbar}{4} \begin{pmatrix} \sigma^\mu \bar{\sigma}^\nu - \sigma^\nu \bar{\sigma}^\mu & 0 \\ 0 & \bar{\sigma}^\mu \sigma^\nu - \bar{\sigma}^\nu \sigma^\mu \end{pmatrix}, \qquad (10.235)$$

and restricting $\mu\nu$ to space indices we have

$$\Sigma_i = -\frac{1}{2}\epsilon_{ijk}\Sigma^{jk} = \frac{\hbar}{2} \begin{pmatrix} \sigma_i & 0 \\ 0 & \sigma_i \end{pmatrix}. \qquad (10.236)$$

The generators Σ_i of rotations $S(\mathbf{\Lambda}_R)$ have the same form as in the Pauli representation. The corresponding finite transformation will therefore be implemented on spinors by the same matrix $S(\mathbf{\Lambda}_R)$ in Eq. (10.108).

Moreover from Eq. (10.100) the spinor representation of the infinitesimal *boost* generators J^{0i}, are also given in terms of a block diagonal matrix

$$\Sigma^{0i} = -i\hbar\, K_i = -i\hbar\frac{\alpha^i}{2} = -\frac{i\hbar}{2} \begin{pmatrix} \sigma_i & 0 \\ 0 & -\sigma_i \end{pmatrix}. \qquad (10.237)$$

It follows that if we use the decomposition (10.233) a *proper* Lorentz transformation can never mix the upper and lower components of the Dirac spinor ψ. The explicit finite form of the *proper* Lorentz transformations in the spinor representation can be found by exponentiation of the generators, following the method explained in Chap. 7.

A generic proper Lorentz transformation can be written as the product of a rotation and a boost transformation, as in Eq. (10.112). The rotation part was given in Eq. (10.108), while the boost part $S(\mathbf{\Lambda}_B)$ was given in Eq. (10.116) in terms of the matrices α^i, whose matrix representation now, in the Weyl basis, is different. One finds that under $\mathbf{\Lambda}_R$ and $\mathbf{\Lambda}_B$ the two two-dimensional spinors ξ, ζ transform as follows:

$$\xi \xrightarrow{\Lambda_R} \left[\cos\frac{\theta}{2} + i\,\sigma\cdot\hat{\theta}\sin\frac{\theta}{2}\right]\xi; \quad \zeta \xrightarrow{\Lambda_R} \left[\cos\frac{\theta}{2} + i\,\sigma\cdot\hat{\theta}\sin\frac{\theta}{2}\right]\zeta,$$

$$\xi \xrightarrow{\Lambda_B} \left[\cosh\frac{\Lambda}{2} + \sigma\cdot\hat{\lambda}\sinh\frac{\lambda}{2}\right]\xi; \quad \zeta \xrightarrow{\Lambda_B} \left[\cosh\frac{\lambda}{2} - \sigma\cdot\hat{\lambda}\sinh\frac{\lambda}{2}\right]\zeta,$$

where $\theta \equiv |\theta|$; $\lambda \equiv |\lambda|$; $\hat{\lambda} = \frac{\lambda}{|\lambda|}$; $\hat{\theta} = \frac{\theta}{|\theta|}$.

The above results refer to proper Lorentz transformations, that is they exclude transformations with negative determinant: $\det \mathbf{\Lambda} = -1$. Let us now consider Lorentz transformations with $\det \mathbf{\Lambda} = -1$. Keeping $\Lambda^0_0 > 0$, the typical transformation with $\det \mathbf{\Lambda} = -1$ is the *parity transformation* $\mathbf{\Lambda}_P \in O(1,3)$ defined by the following *improper* Lorentz matrix:

$$(\Lambda_P)^\mu{}_\nu = \begin{pmatrix} 1 & 0 & 0 & 0 \\ 0 & -1 & 0 & 0 \\ 0 & 0 & -1 & 0 \\ 0 & 0 & 0 & -1 \end{pmatrix}. \qquad (10.238)$$

On the space-time coordinates x^μ it acts as follows :

$$x^0 \rightarrow x^0; \quad \mathbf{x} \rightarrow -\mathbf{x}, \tag{10.239}$$

that is it corresponds to a change of the orientation of the three coordinate axes.

We now show that Λ_P acts on spinors as follows:

$$S(\Lambda_P) = \eta_P \, \gamma^0, \tag{10.240}$$

where $\eta_P = \pm 1$.

We may indeed verify that

$$S(\Lambda_P)^{-1} \gamma^\mu S(\Lambda_P) = \Lambda_P{}^\mu{}_\nu \, \gamma^\nu.$$

which generalizes the general formula (10.88) to the parity transformation. The above property is readily proven, using Eq. (10.240):

$$\begin{aligned}
S(\Lambda_P)^{-1} \gamma^0 S(\Lambda_P) &= \gamma^0 = \Lambda_P{}^0{}_0 \, \gamma^0, \\
S(\Lambda_P)^{-1} \gamma^i S(\Lambda_P) &= -\gamma^i = \Lambda_P{}^i{}_j \, \gamma^j.
\end{aligned} \tag{10.241}$$

The action of a parity transformation on a Dirac field $\psi(x)$ is therefore:

$$\psi(x) \xrightarrow{P} \eta_P \, \gamma^0 \, \psi(x^0, -\mathbf{x}). \tag{10.242}$$

If we take into account that in the Weyl representation the γ-matrices are given by Eq. (10.230) and are off-diagonal, we see that the parity transformation Λ_P transforms ξ and ζ into one another:

$$\begin{cases} \xi \rightarrow \eta_P \, \zeta, \\ \zeta \rightarrow \eta_P \, \xi. \end{cases} \tag{10.243}$$

This result shows that while for *proper* Lorentz transformations the representation of the Lorentz group is reducible since it acts separately on the two spinor components, if we consider the full the Lorentz group, including also improper transformations like parity, the representation becomes irreducible and we are bound to use four-dimensional spinors.

Let us now write the Dirac equation in this new basis. On momentum eigenstates $w(p) \, e^{-\frac{i}{\hbar} p \cdot x}$ it reads:

$$(\not{p} - mc) \, w(p) = 0 \;\; \Rightarrow \;\; \begin{cases} (p^0 - \mathbf{p} \cdot \boldsymbol{\sigma}) \, \xi = mc \, \zeta, \\ (p^0 + \mathbf{p} \cdot \boldsymbol{\sigma}) \, \zeta = mc \, \xi, \end{cases} \tag{10.244}$$

where we have written $w = (\xi, \zeta)$. For massless spinors $m = 0$ the above equations decouple:

$$(p^0 - \mathbf{p} \cdot \boldsymbol{\sigma})\,\xi = 0; \quad (p^0 + \mathbf{p} \cdot \boldsymbol{\sigma})\,\zeta = 0, \tag{10.245}$$

which will have solutions for $p^0 > 0$ and $p^0 < 0$. The above equations fix the helicity Γ of the solution which, as we know, is a conserved quantity and labels the internal degrees of freedom of a massless particle.[14] On the two bi-dimensional spinors ξ, ζ, the helicity is indeed $\Gamma = \hbar\,\mathbf{p} \cdot \boldsymbol{\sigma}/(2|\mathbf{p}|) = \hbar\,\mathbf{p} \cdot \boldsymbol{\sigma}/(2\,p^0)$: It is positive for negative energy solutions ζ and positive energy solutions ξ, while it is negative for positive energy solutions ζ and negative energy solutions ξ.

In nature there are three spin 1/2 particles, called neutrinos and denoted by ν_e, ν_μ, ν_τ, which, until recently, were believed to be massless.

In next chapter we shall be dealing with the other improper Lorentz transformation besides parity, which is *time-reversal*.

10.8.1 Bilinear Forms

Let us now consider the matrix γ^5, introduced in Eq. (10.194). Its explicit form in the Weyl representation is

$$\gamma^5 = i\gamma^0\gamma^1\gamma^2\gamma^3 = \frac{i}{4!}\epsilon_{\mu\nu\rho\sigma}\gamma^\mu\gamma^\nu\gamma^\rho\gamma^\sigma = \begin{pmatrix} 1_2 & 0 \\ 0 & -1_2 \end{pmatrix}. \tag{10.246}$$

Let us investigate the transformation properties of γ^5 under a general Lorentz transformation:

$$\begin{aligned}
S(\Lambda)^{-1}\gamma^5 S(\Lambda) &= \frac{i}{4!}\epsilon_{\mu\nu\rho\sigma}S^{-1}\gamma^\mu S\, S^{-1}\gamma^\nu S\, S^{-1}\gamma^\rho S\, S^{-1}\gamma^\sigma S \\
&= \frac{i}{4!}\epsilon_{\mu\nu\rho\sigma}\Lambda^\mu{}_{\mu'}\Lambda^\nu{}_{\nu'}\Lambda^\rho{}_{\rho'}\Lambda^\sigma{}_{\sigma'}\gamma^{\mu'}\gamma^{\nu'}\gamma^{\rho'}\gamma^{\sigma'} \\
&= \det(\Lambda)\frac{i}{4}\epsilon_{\mu\nu\rho\sigma}\gamma^\mu\gamma^\nu\gamma^\rho\gamma^\sigma \\
&= \det(\Lambda)\,\gamma^5. \tag{10.247}
\end{aligned}$$

In particular under a *parity transformation*, being $\det \Lambda_P = -1$, we have:

$$S(\Lambda_P)^{-1}\gamma^5 S(\Lambda_P) = -\gamma^5, \tag{10.248}$$

---

[14]Recall that helicity is invariant under proper Lorentz transformations and labels irreducible representations of $SO(1, 3)$ with $m = 0$.

that is, it transforms as a *pseudoscalar*. By the same token we can show that:

$$S(\Lambda)^{-1}\gamma^5\gamma^\mu S(\Lambda) = \det(\Lambda)\,\Lambda^\mu{}_\nu\,(\gamma^5\gamma^\nu). \tag{10.249}$$

so that $\gamma^5\gamma^\mu$ transforms as an *pseudo-vector*, that is as an ordinary vector under proper Lorentz transformations, and with an additional minus sign under parity.

Defining

$$\gamma^{\mu\nu} \equiv \frac{1}{2}[\gamma^\mu, \gamma^\nu],$$

we verify that $\gamma^{\mu\nu}$ transforms an *antisymmetric tensor of rank two*:

$$S(\Lambda)^{-1}\gamma^{\mu\nu}S(\Lambda) = \frac{1}{2}\left[S^{-1}\gamma^\mu S,\, S^{-1}\gamma^\nu S\right] = \Lambda^\mu{}_\rho\Lambda^\nu{}_\sigma\gamma^{\rho\sigma}, \tag{10.250}$$

while $\gamma_5\gamma^{\mu\nu}$ transforms like a pseudo- (or axial-) tensor, that is with an additional minus sign under parity as it follows from Eq. (10.248):

$$S(\Lambda)^{-1}\gamma_5\gamma^{\mu\nu}S(\Lambda) = \det(\Lambda)\Lambda^\mu{}_\rho\Lambda^\nu{}_\sigma\gamma^5\gamma^{\mu\nu}. \tag{10.251}$$

These properties allow us to construct bilinear forms in the spinor fields ψ which have definite transformation under the full Lorentz group.

Indeed if we consider a general bilinear form of the type:

$$\bar\psi(x)\gamma^{\mu_1\cdots\mu_k}\psi(x), \tag{10.252}$$

as shown in Appendix G the independent bilinears are:

$$\bar\psi(x)\psi(x);\ \ \bar\psi(x)\gamma^\mu\psi(x);\ \ \bar\psi(x)\gamma^{\mu\nu}\psi(x);\ \ \bar\psi(x)\gamma^5\psi(x);$$
$$\bar\psi(x)\gamma^5\gamma^\mu\psi(x). \tag{10.253}$$

To exhibit their transformation properties we perform the transformation

$$\psi'(x') = S\psi(x) \to \overline{\psi}'(x') = \overline{S\psi(x)} = \psi^\dagger(x)S^\dagger\gamma^0, \tag{10.254}$$

and use the relation (10.92) of Sect. 10.3.3, namely

$$\gamma^0 S^\dagger\gamma^0 = S^{-1}. \tag{10.255}$$

Using Eqs. (10.247) and (10.248) it is easy to show that $\bar\psi(x)\psi(x)$ is a scalar field while $\bar\psi(x)\gamma^5\psi(x)$ is a pseudoscalar, i.e. under parity they transform as follows:

$$\bar\psi(x)\psi(x) \to \bar\psi'(x')\psi'(x');\ \ \bar\psi(x)\gamma^5\psi(x) \to -\bar\psi'(x')\gamma^5\psi'(x'). \tag{10.256}$$

By the same token, and using Eqs. (10.250) and (10.251) as well, we find analogous transformation properties for the remaining fermion bilinears. The result is summarized in the following table:

bilinear	P-transformed	kind
$\bar{\psi}(x)\psi(x)$	$\bar{\psi}(x_P)\psi(x_P)$	scalar field
$\bar{\psi}(x)\gamma^5\psi(x)$	$-\bar{\psi}(x_P)\gamma^5\psi(x_P)$	pseudo-scalar field
$\bar{\psi}(x)\gamma^\mu\psi(x)$	$\eta_{\mu\mu}\,\bar{\psi}(x_P)\gamma^\mu\psi(x_P)$	vector field
$\bar{\psi}(x)\gamma^5\gamma^\mu\psi(x)$	$-\eta_{\mu\mu}\,\bar{\psi}(x_P)\gamma^5\gamma^\mu\psi(x_P)$	axial-vector field
$\bar{\psi}(x)\gamma^{\mu\nu}\psi(x)$	$\eta_{\mu\mu}\,\eta_{\nu\nu}\,\bar{\psi}(x_P)\gamma^{\mu\nu}\psi(x_P)$	(antisymmetric) tensor field

where, in the second column, there is no summation over the μ and ν indices, and $x_P \equiv (x_P^\mu) = (ct, -\mathbf{x})$.

10.8.2 References

For further reading see Refs. [3], [8, vol. 4], [9, 13].

Chapter 11
Quantization of Boson and Fermion Fields

11.1 Introduction

In the previous chapter we have examined the relativistic wave equations for spin 0 and spin 1/2 particles. The corresponding fields $\phi(x)$ and $\psi^\alpha(x)$ were *classical* in the same sense that the Schroedinger wave function $\psi(\mathbf{x}, t)$ is a classical field. In contrast to the non-relativistic Schroedinger construction, we have seen that requiring *relativistic invariance* of the quantum theory, that is invariance under Poincaré transformations, unavoidably leads to serious difficulties when trying to interpret the field as representing the physical state of the system: It implies the appearance of a non-conserved probability density and, most of all, the appearance of negative energy states. Note that the latter difficulty is in some sense contradictory because if we just consider the field aspect of the wave equations, the field energy, expressed in terms of the canonical energy momentum tensor, is positive.

As we have anticipated in the previous chapter and shall show in the present one, the key to a consistent quantization procedure is provided by *the quantization of a free electromagnetic field* discussed as an example in Chap. 6, where the would-be wave function represented by the classical field $A_\mu(\mathbf{x}, t)$ was interpreted as a quantum "mechanical" system with infinite degrees of freedom described by a system of infinitely many decoupled harmonic oscillators. Within the Hamiltonian formulation of the theory, the infinite dynamical variables associated with the degrees of freedom of $A_\mu(x)$ were quantized according to the same prescription used for systems with a finite number of degrees of freedom, namely *trading dynamical variables with operators whose commutation rules are determined by the Heisenberg prescription*[1]:

$$\{A, B\}_{P.B.} = -\frac{i}{\hbar} [A, B]. \tag{11.1}$$

---

[1] In this chapter we denote the Poisson brackets by the symbol $\{,\}_{P.B.}$ since we want to reserve the symbol $\{A, B\}$ to the *anticommutator* of quantum operators, $\{A, B\} = AB + BA$.

© Springer International Publishing Switzerland 2016

R. D'Auria and M. Trigiante, *From Special Relativity to Feynman Diagrams*,

UNITEXT for Physics, DOI 10.1007/978-3-319-22014-7_11

Actually when we try to extend this procedure to free fields other than the electromagnetic one, we shall find that in order to ensure the *positivity of energy* we have to treat fields with *integer* and *half-integer* spin on a different footing, respectively. Integer spin fields, also called *bosonic fields*, like the electromagnetic field discussed Chaps. 5 and 6, or the Klein-Gordon field, will be quantized by a straightforward extension of the canonical Heisenberg quantization method already used for the electromagnetic field, namely trading classical Poisson brackets with *commutators* according to Eq. (11.1). Fields with half-integer spin, like the Dirac field, will instead require a quantization procedure based on *anticommutators* rather than *commutators*. Only in this case we can obtain a consistent description of the quantized fermion field in which the energy is positive definite. Besides the self-consistency of the procedure, it will turn out that the different quantization rules for relativistic bosonic and fermionic fields give a natural explanation of the *connection between spin and statistics* namely the *Pauli principle* for spin 1/2 particles, which, in a non-relativistic theory, must be introduced as an independent assumption. Actually, while the quantization of bosonic fields using commutators yields a consistent theory, the same prescription applied to fermionic fields will be seen to violate the microcausality of the theory which is a fundamental requirement of the relativity principle.

11.2 Quantization of the Klein Gordon Field

In the previous chapters we have been dealing with classical fields of different spin: bosonic with spin 0 and 1 (massless) and fermionic with spin 1/2. These fields are distinguished, at the classical level, by their different transformation properties under the Lorentz group. A bosonic field $\phi^\alpha(x)$ sits in a tensor representation of the Lorentz group. For example, while the scalar field is a Lorentz singlet, the electromagnetic field $A_\mu(x)$ transforms in the defining representation of the Lorentz group. A fermionic field, like the spin 1/2 Dirac field, transforms instead in the spinor representation (or for higher half-integer spins in higher spinor representations).

In Chap. 8 we have given the fundamental Poisson brackets between the classical field $\Phi^\alpha(x)$ and its conjugate momentum density $\pi_\alpha(x)$. We have also pointed out that, in a quantum theory, the dynamical variables $\Phi^\alpha(x)$ and their conjugate momenta $\pi_\alpha(x)$ are promoted to *linear operators* $\hat{\Phi}^\alpha(x)$, $\hat{\pi}_\alpha(x)$ acting on the Hilbert space of the physical states. As mentioned in the introduction their commutation properties depend on their being bosonic or fermionic. For every boson field $\phi^\alpha(x)$ the quantization procedure is effected using the canonical Heisenberg *equal time commutation rules* through the prescription (11.1). A bosonic quantum field theory will thus be characterized by the following commutators between the *field operators*:

$$\left[\hat{\phi}^\alpha(\mathbf{x}, t), \hat{\pi}_\beta(\mathbf{y}, t)\right] = i\hbar\, \delta^\alpha_\beta \delta^3(\mathbf{x} - \mathbf{y}),$$

$$\left[\hat{\phi}^\alpha(\mathbf{x}, t), \hat{\phi}^\beta(\mathbf{y}, t)\right] = \left[\hat{\pi}_\alpha(\mathbf{x}, t), \hat{\pi}_\beta(\mathbf{y}, t)\right] = 0. \tag{11.2}$$

Taking into account that the complex conjugation of a classical dynamical variable must be replaced by the hermitian conjugate of the corresponding quantum operator, we can also write the hermitian conjugate counterparts of (11.2):

$$\left[\hat{\phi}^{\dagger \alpha}(\mathbf{x}, t), \hat{\pi}_\beta^\dagger(\mathbf{y}, t) \right] = i\hbar \, \delta_\beta^\alpha \delta^3(\mathbf{x} - \mathbf{y}),$$

$$\left[\hat{\phi}^{\dagger \alpha}(\mathbf{x}, t), \hat{\phi}^{\dagger \beta}(\mathbf{y}, t) \right] = \left[\hat{\pi}_\alpha^\dagger(\mathbf{x}, t), \hat{\pi}_\beta^\dagger(\mathbf{y}, t) \right] = 0. \tag{11.3}$$

Note that the classical relation (8.214) now becomes

$$\hat{\pi}_\alpha(\mathbf{x}, t) = \frac{\partial}{\partial t} \hat{\phi}_\alpha^\dagger(\mathbf{x}, t); \quad \pi^{\dagger \alpha}(\mathbf{x}, t) = \frac{\partial}{\partial t} \hat{\phi}^\alpha(\mathbf{x}, t). \tag{11.4}$$

The same replacement (11.1) implies that the classical Hamilton equations, given in terms of the Poisson brackets in (8.227) and (8.228), at the quantum level become

$$i\hbar \frac{\partial}{\partial t} \hat{\phi}^\alpha(\mathbf{x}, t) = \left[\hat{\phi}^\alpha(\mathbf{x}, t), \hat{H} \right],$$

$$i\hbar \frac{\partial}{\partial t} \hat{\pi}_\alpha(\mathbf{x}, t) = \left[\hat{\pi}_\alpha(\mathbf{x}, t), \hat{H} \right], \tag{11.5}$$

where the Hamiltonian operator is obtained from the classical expression (8.215), (8.216) by promoting the field variables to quantum operators. We note that this replacement implies time evolution in the quantum system to be described in the *Heisenberg picture* since the classical dynamical variables are time dependent. Thus the quantum state of the system is time independent.

In this section we restrict our discussion to the dynamics of a *free complex scalar field*, which, as discussed in Sects. 8.8.1 and 8.9, is equivalent to *two* real scalar fields. Since by definition a scalar field sits in the trivial representation of the Lorentz group, it corresponds to a spin-0 field, carrying no representation indices. Its classical description is given in terms of the Lagrangian (10.11) from which the classical Klein-Gordon equation (10.12) is derived. In that case, following (11.1), the Poisson brackets (8.232) become the equal-time commutators (11.2) with no indices α, β:

$$\left[\hat{\phi}(\mathbf{x}, t), \hat{\pi}(\mathbf{y}, t) \right] = i\hbar \delta^3(\mathbf{x} - \mathbf{y}),$$

$$\left[\hat{\phi}^\dagger(\mathbf{x}, t), \hat{\pi}^\dagger(\mathbf{y}, t) \right] = i\hbar \delta^3(\mathbf{x} - \mathbf{y}), \tag{11.6}$$

all the other possible commutators being zero.

To derive the quantum equations of motion we first need to compute the Hamiltonian operator. Recall that the classical Hamiltonian density is given by Eq. (10.46), that we rewrite here for convenience:

$$\mathcal{H} = \pi \pi^* + c^2 \nabla \phi^* \nabla \phi + \frac{m^2 c^4}{\hbar^2} \phi^* \phi. \tag{11.7}$$

Caution is however required when trading the classical fields in the above expression by their quantum counterparts $\hat{\varphi}(x)$, $\hat{\varphi}^\dagger(x)$, $\hat{\pi}(x)$, $\hat{\pi}^\dagger(x)$, since operators appearing in products in (11.7), being computed at the *same space-time point* $x^\mu = y^\mu$, do not in general commute (see Sect. 11.4 below). This implies that a certain order must be chosen. The convention we shall use will be shown in the sequel to lead to consistent results in the development of the theory. It consists in the following substitutions:

$$\text{(classical fields)} \qquad \text{(field operators)},$$

$$\pi^*(x)\pi(x) \quad \rightarrow \quad \hat{\pi}(x)\hat{\pi}^\dagger(x),$$

$$\phi^*(x)\phi(x) \quad \rightarrow \quad \hat{\phi}^\dagger(x)\hat{\phi}(x), \tag{11.8}$$

$$\nabla\phi^*(x) \cdot \nabla\phi(x) \rightarrow \nabla\hat{\phi}^\dagger(x) \cdot \nabla\hat{\phi}(x).$$

The resulting Hamiltonian operator reads:

$$\hat{H} = \int d^3\mathbf{x}\left[\hat{\pi}(x)\hat{\pi}^\dagger(x) + c^2\nabla\hat{\phi}^\dagger(x) \cdot \nabla\hat{\phi}(x) + \frac{m^2c^4}{\hbar^2}\hat{\phi}^\dagger(x)\hat{\phi}(x)\right]. \tag{11.9}$$

Let us now use this Hamiltonian in the quantum Hamilton equations (11.5)

$$i\hbar\frac{\partial}{\partial t}\hat{\phi}(\mathbf{x}, t) = \left[\hat{\phi}(\mathbf{x}, t), \hat{H}\right]; \quad i\hbar\frac{\partial}{\partial t}\hat{\pi}(\mathbf{x}, t) = \left[\hat{\pi}(\mathbf{x}, t), \hat{H}\right], \tag{11.10}$$

and show that it reproduces the quantum version of the classical Klein-Gordon equation.

Applying Eq. (11.6) to the first of (11.10) we find

$$i\hbar\frac{\partial}{\partial t}\hat{\phi}(\mathbf{x}, t) = \left[\hat{\phi}(\mathbf{x}, t), \hat{H}(t)\right] = \int d^3\mathbf{y}\left[\hat{\phi}(\mathbf{x}, t), \hat{\pi}(\mathbf{y}, t)\right]\hat{\pi}^\dagger(\mathbf{y}, t)$$

$$= i\hbar\hat{\pi}^\dagger(\mathbf{x}, t), \tag{11.11}$$

which is the expression (11.4) of the conjugate momentum operators. The same computations applied to the last of (11.10) (or better to its hermitian conjugate), yields

$$i\hbar\frac{\partial}{\partial t}\hat{\pi}^\dagger(\mathbf{x}, t) = c^2\int d^3\mathbf{y}\left[\pi^\dagger(\mathbf{x}, t), \left(\frac{\partial\hat{\phi}^\dagger}{\partial y^i}\frac{\partial\hat{\phi}}{\partial y^i}\right)(\mathbf{y}, t)\right]$$

$$+ \frac{m^2c^2}{\hbar^2}\int d^3\mathbf{y}\left[\hat{\pi}^\dagger(\mathbf{x}, t), \hat{\phi}^\dagger(\mathbf{y}, t)\right]\hat{\phi}(\mathbf{y}, t)$$

$$= -c^2 \int d^3y \left[\hat{\pi}^\dagger(\mathbf{x}, t), \hat{\phi}^\dagger(\mathbf{y}, t) \right] \nabla^2 \hat{\phi}(\mathbf{y}, t)$$

$$- i \frac{m^2 c^2}{\hbar} \hat{\phi}(\mathbf{x}, t)$$

$$= i\hbar \left(c^2 \nabla^2 \hat{\phi} - \frac{m^2 c^2}{\hbar^2} \hat{\phi} \right) (\mathbf{x}, t). \tag{11.12}$$

Substituting in the left hand side the value of $\pi^\dagger$ given by Eq. (11.11) we obtain

$$\frac{1}{c^2} \frac{\partial^2 \hat{\phi}}{\partial t^2} - \nabla^2 \hat{\phi} - \frac{m^2 c^2}{\hbar^2} \hat{\phi} = 0, \tag{11.13}$$

so that *the quantum field operator obeys the same Klein-Gordon equation as its classical counterpart.*

In an analogous way we reproduce the quantum version of Eq. (10.52)

$$\hat{\mathbf{P}}(t) = - \int d^3y \left[\hat{\pi}(y) \nabla \hat{\phi}(y) + \nabla \hat{\phi}^\dagger(y) \hat{\pi}^\dagger(y) \right], \tag{11.14}$$

where $y = (ct, \mathbf{y})$, which yields the right transformation property of the field operator under infinitesimal space-translations (see Eq. (9.39))

$$\delta_\epsilon \hat{\phi}(\mathbf{x}, t) = \frac{i}{\hbar} \left[\hat{\phi}(\mathbf{x}, t), \epsilon \cdot \hat{\mathbf{P}}(t) \right] = -\frac{i}{\hbar} \int d^3y \left[\hat{\phi}(\mathbf{x}, t), \hat{\pi}(\mathbf{y}, t) \right] \epsilon \cdot \nabla \hat{\phi}(\mathbf{y}, t)$$

$$= \epsilon \cdot \nabla \hat{\phi}(\mathbf{x}, t). \tag{11.15}$$

We can thus also write:

$$- i\hbar \nabla \hat{\phi}(\mathbf{x}, t) = \left[\hat{\phi}(\mathbf{x}, t), \hat{\mathbf{P}}(t) \right]. \tag{11.16}$$

Recalling that $\hat{\mathbf{P}}$ is the three-dimensional counterpart of the four-momentum $\hat{P}^\mu = (\hat{H}/c, \hat{\mathbf{P}})$ of the field, (11.11) and (11.16) can be written in a Lorentz covariant form as

$$\left[\hat{\phi}(\mathbf{x}, t), \hat{P}^\mu(t) \right] = i\hbar \partial^\mu \hat{\phi}(\mathbf{x}, t) = i\hbar \eta^{\mu\nu} \partial_\nu \hat{\phi}(\mathbf{x}, t). \tag{11.17}$$

Solving the quantum Klein-Gordon theory means to explicitly construct the Hilbert space of states $V^{(c)}$ and the dynamical variables $\hat{\phi}^\alpha, \hat{\pi}_\alpha, \hat{H}$ acting as operators on it, such that the commutation relations (11.6), and the equations of motion (11.10) are satisfied. In the *free* field case we are now considering, this can be achieved by constructing the quantum states of the system in terms of states describing a definite number of particles with given momenta, that is in terms of *simultaneous eigenstates* of the *occupation number operator*. This representation is called the *Fock space representation*, or *occupation number representation*, and was in fact used for the quantization of the free spin 1 electromagnetic field given in Chap. 6.

Let us shortly recall our procedure in that case: We started representing the free field as a collection of infinitely many decoupled harmonic oscillators, one for each plane wave, i.e. for each wave number vector $\mathbf{k}$ and polarization. The quantization of the field was effected by quantizing each constituent harmonic oscillator: A complete set of quantum states was constructed as the product of the various quantum oscillator states, each characterized by an integer number, the occupation number, representing the corresponding oscillation mode. In this picture a single particle state, that is a photon state with energy $\hbar\omega$, linear momentum $\hbar\mathbf{k}$ and a given polarization i (helicity), was associated with each plane wave, i.e. harmonic oscillator, and the occupation number of the corresponding quantum state represented the number of photons with those physical properties. A quantum field state in this representation is completely defined by specifying the occupation number of each oscillator state, that is the number of photons with a given four-momentum $\hbar k^\mu$ and helicity i. Such states differ in the number of particles they describe, each photon representing a quantum of field-excitation.

The same procedure will be applied in the present chapter to the quantization of scalar and fermion fields. The key ingredient for this construction, namely the representation of the field as a collection of decoupled harmonic oscillators, is guaranteed by the fact that all free fields satisfy the Klein-Gordon equation and can thus be expanded in plane waves.

For interacting fields instead *no closed solution to the problem of quantization is known* in general and we have to resort to a perturbative approach, to be developed in the next chapter.[2]

As classical and quantum equations of motion are formally *identical*, we can expand the quantum field $\hat{\phi}(x)$ in plane waves with positive and negative angular frequencies, that is in a complete set of eigenfunctions with definite four-momentum defined in (10.13) and (10.16).

We replace in the classical expansion of (10.29) the c-number coefficients of the exponentials by operator coefficients as follows:

$$\tilde{\phi}_+(\mathbf{p}) \rightarrow \hat{\phi}_+(\mathbf{p}) = \frac{\hbar}{c}\sqrt{2E_\mathbf{p}V}\, a(\mathbf{p}),$$

$$\tilde{\phi}_-(\mathbf{p}) \rightarrow \hat{\phi}_-(\mathbf{p}) = \frac{\hbar}{c}\sqrt{2E_\mathbf{p}V}\, b^\dagger(\mathbf{p}), \qquad (11.18)$$

---

[2]Actually we shall only consider the quantum description of the interaction between the electromagnetic field and a Dirac field.

where the normalization has been chosen such that $a(\mathbf{p})$ and $b^\dagger(\mathbf{p})$ are dimensionless operators. Therefore we have the following expansions:

$$\hat{\phi}(x) = \int \frac{d^3\mathbf{p}}{(2\pi\hbar)^3} \frac{\hbar\sqrt{V}}{\sqrt{2E_\mathbf{p}}} \left(a(\mathbf{p})e^{-\frac{i}{\hbar}P\cdot x} + b^\dagger(\mathbf{p})e^{\frac{i}{\hbar}P\cdot x}\right), \tag{11.19}$$

$$\hat{\phi}(x)^\dagger = \int \frac{d^3\mathbf{p}}{(2\pi\hbar)^3} \frac{\hbar\sqrt{V}}{\sqrt{2E_\mathbf{p}}} \left(a^\dagger(\mathbf{p})e^{\frac{i}{\hbar}P\cdot x} + b(\mathbf{p})e^{-\frac{i}{\hbar}P\cdot x}\right), \tag{11.20}$$

$$\hat{\pi}(x) = \frac{\partial}{\partial t}\hat{\phi}^\dagger = i\int \frac{d^3\mathbf{p}}{(2\pi\hbar)^3} \sqrt{\frac{V E_\mathbf{p}}{2}} \left[a^\dagger(\mathbf{p})e^{\frac{i}{\hbar}P\cdot x} - b(\mathbf{p})e^{-\frac{i}{\hbar}P\cdot x}\right], \tag{11.21}$$

$$\hat{\pi}(x)^\dagger = \frac{\partial}{\partial t}\hat{\phi} = -i\int \frac{d^3\mathbf{p}}{(2\pi\hbar)^3} \sqrt{\frac{V E_\mathbf{p}}{2}} \left[a(\mathbf{p})e^{-\frac{i}{\hbar}P\cdot x} - b^\dagger(\mathbf{p})e^{\frac{i}{\hbar}P\cdot x}\right]. \tag{11.22}$$

Note that had we considered an *hermitian field* $\hat{\phi}^\dagger(x) = \hat{\phi}(x)$, corresponding to a real classical field, *hermiticity* would have identified $b^\dagger \equiv a^\dagger$. For a non-hermitian field instead *the a and b operators are independent*.[3]

Equations (11.19) and/or (11.20) can be inverted to compute the operators $a(\mathbf{p})$, $b(\mathbf{p})$ and their hermitian conjugates in terms of $\hat{\phi}$ and $\hat{\phi}^\dagger$. To this end let us define the following function:

$$f_\mathbf{p} = \frac{1}{\sqrt{2E_\mathbf{p}V}} e^{-\frac{i}{\hbar}P\cdot x},$$

and prove that

$$a(\mathbf{q}) = i\int d^3\mathbf{x}\left[f_\mathbf{q}^*(x)\partial_t\hat{\phi}(x) - \hat{\phi}(x)\partial_t f_\mathbf{q}^*(x)\right]$$

$$= i\int d^3\mathbf{x}\left[f_\mathbf{q}^*(x)\,\hat{\pi}^\dagger(x) - \hat{\phi}(x)\partial_t f_\mathbf{q}^*(x)\right], \tag{11.23}$$

$$a^\dagger(\mathbf{q}) = -i\int d^3\mathbf{x}\left[f_\mathbf{q}(x)\partial_t\hat{\phi}^\dagger(x) - \hat{\phi}^\dagger(x)\partial_t f_\mathbf{q}(x)\right] \tag{11.24}$$

$$= -i\int d^3\mathbf{x}\left[f_\mathbf{q}(x)\hat{\pi}(x) - \hat{\phi}^\dagger(x)\partial_t f_\mathbf{q}(x)\right]. \tag{11.25}$$

Let us recall here some useful properties which we shall extensively use in the following:

$$\int d^3\mathbf{x}\, e^{\pm\frac{i}{\hbar}\mathbf{p}\cdot\mathbf{x}} = (2\pi\hbar)^3\,\delta^3(\mathbf{p}); \quad f(\mathbf{p}) = \int d^3\mathbf{q}\,\delta^3(\mathbf{p} - \mathbf{q})f(\mathbf{q}),$$

$$E_\mathbf{p} \equiv \sqrt{m^2c^4 + |\mathbf{p}|^2c^2} = E_{-\mathbf{p}}. \tag{11.26}$$

---

[3] We have used a similar argument after Eq. (10.30), in the classical case.

Consider the first of (11.23) and let us rewrite the first term on the right hand side using for $\hat{\phi}$ the expansion (11.19):

$$i \int d^3x \, f_{\mathbf{q}}^*(x) \partial_t \hat{\phi}(x) = \int d^3x \int \frac{d^3p}{(2\pi\hbar)^3} \sqrt{\frac{E_{\mathbf{p}}}{E_{\mathbf{q}}}} \left[\frac{a(\mathbf{p})}{2} e^{-\frac{i}{\hbar}(p-q)\cdot x} - \frac{b^\dagger(\mathbf{p})}{2} e^{\frac{i}{\hbar}(p+q)\cdot x} \right]$$

$$= \frac{a(\mathbf{q})}{2} - \frac{b^\dagger(-\mathbf{q})}{2} e^{\frac{2i}{\hbar} E_{\mathbf{q}} t}, \tag{11.27}$$

where we have used the fact that $\mathbf{p} = \mathbf{q}$ implies $E_{\mathbf{p}} = E_{\mathbf{q}}$. By the same token we prove that:

$$- i \int d^3x \, \hat{\phi}(x) \partial_t f_{\mathbf{q}}^*(x) = \frac{a(\mathbf{q})}{2} + \frac{b^\dagger(-\mathbf{q})}{2} e^{\frac{2i}{\hbar} E_{\mathbf{q}} t}. \tag{11.28}$$

Summing (11.27) and (11.28) the terms with $b^\dagger(-\mathbf{q})$ drop out and we find the first of (11.23). We can prove similar formulas for b and $b^\dagger$:

$$b(\mathbf{q}) = i \int d^3x \left[f_{\mathbf{q}}^*(x) \partial_t \hat{\phi}^\dagger(x) - \hat{\phi}^\dagger(x) \partial_t f_{\mathbf{q}}^*(x) \right]$$

$$= i \int d^3x \left[f_{\mathbf{q}}^*(x) \, \hat{\pi}(x) - \hat{\phi}^\dagger(x) \partial_t f_{\mathbf{q}}^*(x) \right], \tag{11.29}$$

$$b^\dagger(\mathbf{q}) = -i \int d^3x \left[f_{\mathbf{q}}(x) \partial_t \hat{\phi}(x) - \hat{\phi}(x) \partial_t f_{\mathbf{q}}(x) \right]$$

$$= -i \int d^3x \left[f_{\mathbf{q}}(x) \hat{\pi}^\dagger(x) - \hat{\phi}(x) \partial_t f_{\mathbf{q}}(x) \right], \tag{11.30}$$

by showing that the following properties hold:

$$i \int d^3x \, f_{\mathbf{q}}^*(x) \partial_t \hat{\phi}^\dagger(x) = -\frac{a^\dagger(-\mathbf{q})}{2} e^{\frac{2i}{\hbar} E_{\mathbf{q}} t} + \frac{b(\mathbf{q})}{2},$$

$$-i \int d^3x \, \hat{\phi}^\dagger(x) \partial_t f_{\mathbf{q}}^*(x) = \frac{a^\dagger(-\mathbf{q})}{2} e^{\frac{2i}{\hbar} E_{\mathbf{q}} t} + \frac{b(\mathbf{q})}{2}. \tag{11.31}$$

From (11.23) and (11.29), and from the commutation relations (11.6), we may now compute the commutators among $a(\mathbf{q}), a^\dagger(\mathbf{q}), b(\mathbf{q}), b^\dagger(\mathbf{q})$

$$[a(\mathbf{p}), a^\dagger(\mathbf{q})]$$

$$= \int d^3x d^3y \left[f_{\mathbf{p}}^*(x) \hat{\pi}^\dagger(x) - \hat{\phi}(x) \partial_t f_{\mathbf{p}}^*(x), \, f_{\mathbf{q}}(y) \hat{\pi}(y) - \hat{\phi}^\dagger(y) \partial_t f_{\mathbf{q}}^*(y) \right]$$

$$= \int d^3x d^3y \left(f_{\mathbf{p}}^*(x) \partial_t f_{\mathbf{q}}(y) \, [\hat{\phi}^\dagger(y), \hat{\pi}^\dagger(x)] - \partial_t f_{\mathbf{p}}^*(x) f_{\mathbf{q}}(y) \, [\hat{\phi}(x), \hat{\pi}(y)] \right)$$

$$= i\hbar \int d^3x d^3y \left(f_{\mathbf{p}}^*(x) \partial_t f_{\mathbf{q}}(y) - \partial_t f_{\mathbf{p}}^*(x) f_{\mathbf{q}}(y) \right) \delta^3(\mathbf{x} - \mathbf{y})$$

$$= i\hbar \int d^3\mathbf{x} \left(f_\mathbf{p}^*(x)\partial_t f_\mathbf{q}(x) - \partial_t f_\mathbf{p}^*(x) f_\mathbf{q}(x) \right)$$

$$= \frac{i\hbar}{2V} \left(-\frac{i}{\hbar}\sqrt{\frac{E_\mathbf{q}}{E_\mathbf{p}}} - \frac{i}{\hbar}\sqrt{\frac{E_\mathbf{p}}{E_\mathbf{q}}} \right) (2\pi\hbar)^3 \delta^3(\mathbf{p}-\mathbf{q}) = \frac{(2\pi\hbar)^3}{V}\delta^3(\mathbf{p}-\mathbf{q}).$$

Analogous computations give the complete set of commutation relations:

$$\left[a(\mathbf{p}), a(\mathbf{q})^\dagger \right] = (2\pi\hbar)^3\delta^3(\mathbf{p}-\mathbf{q})\cdot\frac{1}{V},$$

$$\left[b(\mathbf{p}), b(\mathbf{q})^\dagger \right] = (2\pi\hbar)^3\delta^3(\mathbf{p}-\mathbf{q})\frac{1}{V}, \tag{11.32}$$

all the other possible commutation relations being zero. The reverse is also true: Given $\hat{\phi}(x)$, $\hat{\phi}^\dagger(x)$ expressed in terms of operators $a, b, a^\dagger, b^\dagger$, satisfying the relations (11.32), the canonical commutation rules (11.2) are satisfied. Let us check the first of (11.2). Using the expansions (11.19)–(11.22) and assuming the commutation rules (11.32), we find

$$\left[\hat{\phi}(\mathbf{x}, t), \hat{\pi}(\mathbf{y}, t) \right] = \int \frac{d^3\mathbf{p}}{(2\pi\hbar)^3} \int \frac{d^3\mathbf{q}}{(2\pi\hbar)^3} \frac{\hbar^2 V}{2\sqrt{E_\mathbf{p} E_\mathbf{q}}}$$

$$\times \frac{i}{\hbar} E_\mathbf{q} \left[[a(\mathbf{p}), a(\mathbf{q})^\dagger]e^{-\frac{i}{\hbar}(E_\mathbf{p}-E_\mathbf{q})t}e^{\frac{i}{\hbar}(\mathbf{p}\cdot\mathbf{x}-\mathbf{q}\cdot\mathbf{y})} \right.$$

$$\left. - [b(\mathbf{p})^\dagger, b(\mathbf{q})]e^{\frac{i}{\hbar}(E_\mathbf{p}-E_\mathbf{q})t}e^{-\frac{i}{\hbar}(\mathbf{p}\cdot\mathbf{x}-\mathbf{q}\cdot\mathbf{y})} \right]$$

$$= \int \frac{d^3\mathbf{p}}{(2\pi\hbar)^3} \frac{\hbar^2}{2E_\mathbf{p}} \left[\frac{i}{\hbar} E_\mathbf{p}e^{\frac{i}{\hbar}\mathbf{p}(\mathbf{x}-\mathbf{y})} + \frac{i}{\hbar} E_\mathbf{p}e^{-\frac{i}{\hbar}\mathbf{p}(\mathbf{x}-\mathbf{y})} \right]$$

$$= i\hbar\delta^3(\mathbf{x}-\mathbf{y}).$$

With analogous computations one verifies the other commutation rules in (11.2).

For a finite volume V, the components of the linear momentum $\mathbf{p}$ have discrete values, see Eq. (9.50) of Chap. 9, and the integral in $d^3\mathbf{p}$ becomes a sum, according to the identification (9.53). In particular, using the prescription (9.55), the commutation relations (11.32) for discrete momenta simplify to:

$$\left[a(\mathbf{p}), a(\mathbf{q})^\dagger \right] = \delta_{\mathbf{p},\mathbf{q}}; \quad \left[b(\mathbf{p}), b(\mathbf{q})^\dagger \right] = \delta_{\mathbf{p},\mathbf{q}}, \tag{11.33}$$

all other commutators being zero.[4]

Let us now express, in the large volume limit, the Hamiltonian operator (11.9) in terms of the operators $a, a^\dagger$ and $b, b^\dagger$. Upon using the expansions (11.19)–(11.22),

---

[4]In the discrete notation we shall often use the following symbols $a_\mathbf{p} \equiv a(\mathbf{p})$, $b_\mathbf{p} \equiv b(\mathbf{p})$.

the first term on the right hand side of (11.9) gives:

$$
\int d^3\mathbf{x}\,\hat{\pi}(x)\hat{\pi}^\dagger(x)
$$

$$
= \int d^3\mathbf{x} \int \frac{d^3\mathbf{p}}{(2\pi\hbar)^3} \frac{d^3\mathbf{q}}{(2\pi\hbar)^3} \frac{V}{2}\sqrt{E_\mathbf{p}E_\mathbf{q}}
$$

$$
\times \left[a^\dagger(\mathbf{p})a(\mathbf{q})e^{\frac{i}{\hbar}((E_\mathbf{p}-E_\mathbf{q})t-(\mathbf{p}-\mathbf{q})\cdot\mathbf{x})} + b(\mathbf{p})b^\dagger(\mathbf{q})e^{-\frac{i}{\hbar}((E_\mathbf{p}-E_\mathbf{q})t-(\mathbf{p}-\mathbf{q})\cdot\mathbf{x})} \right.
$$

$$
\left. - a^\dagger(\mathbf{p})b^\dagger(\mathbf{q})e^{\frac{i}{\hbar}((E_\mathbf{p}+E_\mathbf{q})t-\frac{i}{\hbar}(\mathbf{p}+\mathbf{q})\cdot\mathbf{x})} - b(\mathbf{p})a(\mathbf{q})e^{-\frac{i}{\hbar}((E_\mathbf{p}+E_\mathbf{q})t-(\mathbf{p}+\mathbf{q})\cdot\mathbf{x})} \right]
$$

$$
= \int \frac{d^3\mathbf{p}}{(2\pi\hbar)^3}\frac{V}{2}E_\mathbf{p}\left[a^\dagger(\mathbf{p})a(\mathbf{p}) + b(\mathbf{p})b^\dagger(\mathbf{p}) - \left(a^\dagger(\mathbf{p})b^\dagger(-\mathbf{p})e^{\frac{2i}{\hbar}E_\mathbf{p}t} \right. \right.
$$

$$
\left. \left. + b(\mathbf{p})a(-\mathbf{p})e^{-\frac{2i}{\hbar}E_\mathbf{p}t} \right) \right], \tag{11.34}
$$

where we have used the properties (11.26). With an analogous calculation, and using the following expansion

$$
\nabla\hat{\phi}(x) = i\int \frac{d^3\mathbf{p}}{(2\pi\hbar)^3}\sqrt{\frac{V}{2E_\mathbf{p}}}\,\mathbf{p}\left(a(\mathbf{p})\,e^{-\frac{i}{\hbar}\,p\cdot x} - b^\dagger(\mathbf{p})\,e^{\frac{i}{\hbar}\,p\cdot x} \right), \tag{11.35}
$$

the second term on the right hand side of (11.9) reads:

$$
\int d^3\mathbf{x}\nabla\hat{\phi}(x)^\dagger\cdot\nabla\hat{\phi}(x)
$$

$$
= \int d^3\mathbf{x} \int \frac{d^3\mathbf{p}}{(2\pi\hbar)^3} \frac{d^3\mathbf{q}}{(2\pi\hbar)^3} \frac{V}{2\sqrt{E_\mathbf{p}E_\mathbf{q}}}\,\mathbf{p}\cdot\mathbf{q}
$$

$$
\times \left[a^\dagger(\mathbf{p})a(\mathbf{q})e^{\frac{i}{\hbar}((E_\mathbf{p}-E_\mathbf{q})t-(\mathbf{p}-\mathbf{q})\cdot\mathbf{x})} + b(\mathbf{p})b^\dagger(\mathbf{q})e^{-\frac{i}{\hbar}((E_\mathbf{p}-E_\mathbf{q})t-(\mathbf{p}-\mathbf{q})\cdot\mathbf{x})} \right.
$$

$$
\left. - a^\dagger(\mathbf{p})b^\dagger(\mathbf{q})e^{\frac{i}{\hbar}((E_\mathbf{p}+E_\mathbf{q})t-\frac{i}{\hbar}(\mathbf{p}+\mathbf{q})\cdot\mathbf{x})} - b(\mathbf{p})a(\mathbf{q})e^{-\frac{i}{\hbar}((E_\mathbf{p}+E_\mathbf{q})t-(\mathbf{p}+\mathbf{q})\cdot\mathbf{x})} \right]
$$

$$
= \int \frac{d^3\mathbf{p}}{(2\pi\hbar)^3}\frac{V}{2E_\mathbf{p}}|\mathbf{p}|^2\left[a^\dagger(\mathbf{p})a(\mathbf{p}) + b(\mathbf{p})b^\dagger(\mathbf{p}) + a^\dagger(\mathbf{p})b^\dagger(-\mathbf{p})e^{\frac{2i}{\hbar}E_\mathbf{p}t} \right.
$$

$$
\left. + b(\mathbf{p})a(-\mathbf{p})e^{-\frac{2i}{\hbar}E_\mathbf{p}t} \right], \tag{11.36}
$$

while the third term has the following expansion

$$
\int d^3\mathbf{x}\hat{\phi}^\dagger(x)\hat{\phi}(x) = \int \frac{d^3\mathbf{p}}{(2\pi\hbar)^3}\frac{\hbar^2 V}{2E_\mathbf{p}}\left[a^\dagger(\mathbf{p})a(\mathbf{p}) + b(\mathbf{p})b^\dagger(\mathbf{p}) \right.
$$

$$
\left. + \left(a^\dagger(\mathbf{p})b^\dagger(-\mathbf{p})e^{\frac{2i}{\hbar}E_\mathbf{p}t} + b(\mathbf{p})a(-\mathbf{p})e^{-\frac{2i}{\hbar}E_\mathbf{p}t} \right) \right]. \tag{11.37}
$$

Summing up the three results, we finally obtain

$$\int d^3\mathbf{x}\left[\hat{\pi}(x)\hat{\pi}^\dagger(x) + c^2\nabla\hat{\phi}^\dagger(x)\cdot\nabla\hat{\phi}(x) + \frac{m^2c^4}{\hbar^2}\hat{\phi}^\dagger(x)\hat{\phi}(x)\right]$$

$$= \int\frac{d^3\mathbf{p}}{(2\pi\hbar)^3}\frac{V}{2E_\mathbf{p}}\left[(E_\mathbf{p}^2 + c^2|\mathbf{p}|^2 + m^2c^4)(a^\dagger(\mathbf{p})a(\mathbf{p}) + b(\mathbf{p})b^\dagger(\mathbf{p}))\right.$$

$$\left. + (-E_\mathbf{p}^2 + c^2|\mathbf{p}|^2 + m^2c^4)\left(a^\dagger(\mathbf{p})b^\dagger(-\mathbf{p})e^{\frac{2i}{\hbar}E_\mathbf{p}t} + b(\mathbf{p})a(-\mathbf{p})e^{-\frac{2i}{\hbar}E_\mathbf{p}t}\right)\right]$$

$$= \int\frac{d^3\mathbf{p}}{(2\pi\hbar)^3}V E_\mathbf{p}(a^\dagger(\mathbf{p})a(\mathbf{p}) + b(\mathbf{p})b^\dagger(\mathbf{p})). \tag{11.38}$$

where we have used the definition of $E_\mathbf{p}$ in (11.26).

The Hamiltonian operator has therefore the following form,

$$\hat{H} = \int\frac{d^3\mathbf{p}}{(2\pi\hbar)^3}V E_\mathbf{p}(a^\dagger(\mathbf{p})a(\mathbf{p}) + b(\mathbf{p})b^\dagger(\mathbf{p}))$$

$$= \int\frac{d^3\mathbf{p}}{(2\pi\hbar)^3}V E_\mathbf{p}\left[a^\dagger(\mathbf{p})a(\mathbf{p}) + b^\dagger(\mathbf{p})b(\mathbf{p}) + \frac{(2\pi\hbar)^3}{V}\delta^3(\mathbf{p} - \mathbf{p})\right]. \tag{11.39}$$

The Dirac delta function appearing in the last term of the right hand side is an infinite constant devoid of physical significance since it associates with the vacuum an infinite energy. This is apparent if we consider the particle in a finite-size box, with volume V. The momentum becomes discretized and Eq. (11.39) will have then the form:

$$\hat{H} = \sum_\mathbf{p} E_\mathbf{p}\left[a^\dagger(\mathbf{p})a(\mathbf{p}) + b^\dagger(\mathbf{p})b(\mathbf{p}) + 1\right]. \tag{11.40}$$

The vacuum energy part would read $\sum_\mathbf{p} E_\mathbf{p} = \infty$. This inessential infinite constant can be formally eliminated in the same way as we did for the electromagnetic field in Chap. 6, that is by introducing the *normal ordering prescription* when computing physical quantities. Let us recall the definition of "normal ordering" : :. An operator product is *normal ordered* if all the creation operators stand to the left of all destruction operators. For instance:

$$: a(\mathbf{p})a^\dagger(\mathbf{p}) : = : a^\dagger(\mathbf{p})a(\mathbf{p}) := a^\dagger(\mathbf{p})a(\mathbf{p}),$$

$$: b(\mathbf{p})b^\dagger(\mathbf{p}) : = : b^\dagger(\mathbf{p})b(\mathbf{p}) := b^\dagger(\mathbf{p})b(\mathbf{p}). \tag{11.41}$$

With the normal order prescription the Hamiltonian (11.9) is replaced by

$$\hat{H} = \int d^3\mathbf{x} : \left[\hat{\pi}\hat{\pi}^\dagger + c^2\nabla\hat{\phi}^\dagger\cdot\nabla\hat{\phi} + \frac{m^2c^4}{\hbar^2}\hat{\phi}^\dagger\hat{\phi}\right] : \tag{11.42}$$

As a consequence Eq. (11.39) takes the following form:

$$\hat{H} = \int \frac{d^3\mathbf{p}}{(2\pi\hbar)^3} V E_\mathbf{p} : \left[a^\dagger(\mathbf{p})a(\mathbf{p}) + b(\mathbf{p})b^\dagger(\mathbf{p})) \right] :$$
$$= \int \frac{d^3\mathbf{p}}{(2\pi\hbar)^3} V E_\mathbf{p} \left[a^\dagger(\mathbf{p})a(\mathbf{p}) + b^\dagger(\mathbf{p})b(\mathbf{p}) \right], \qquad (11.43)$$

where no infinite constant appears. For finite volume V the normal ordered Hamiltonian reads

$$\hat{H} = \sum_\mathbf{p} E_\mathbf{p} : \left[a^\dagger(\mathbf{p})a(\mathbf{p}) + b(\mathbf{p})b^\dagger(\mathbf{p}) \right] :$$
$$= \sum_\mathbf{p} E_\mathbf{p} \left[a^\dagger(\mathbf{p})a(\mathbf{p}) + b^\dagger(\mathbf{p})b(\mathbf{p}) \right]. \qquad (11.44)$$

It is instructive to compare the above expression with the corresponding one (6.62) found in Chap. 6 for the electromagnetic field. Identifying $\hbar\omega_k$ with the energy $E_\mathbf{p}$ of a photon of momentum $\mathbf{p} = \hbar\mathbf{k}$, we recognize that the two expressions for the energy are quite similar. The only differences between (11.44) and (6.62) consist, on the one hand, in the absence in the former of the polarization index, as it must be the case for a spinless field, (recall that the electromagnetic field has spin 1 and therefore has a polarization index related to the helicity of the photon); on the other hand we have the presence, on the right hand side of (11.44), of additional operators $b, b^\dagger$, which, as will be shown in the following, are always present for a *charged field*. They are not present in the electromagnetic field due to the hermiticity of $\hat{A}_\mu(x)$.

We can proceed in the same way to evaluate the total quantum momentum of the field $\hat{P}^i$, given in (11.14), in terms of the operators $a(\mathbf{p})$, $b(\mathbf{p})$ and their hermitian conjugates. Using (11.35), (11.21) and (11.22) we find:

$$\hat{\mathbf{P}} = \int \frac{d^3\mathbf{p}}{(2\pi\hbar)^3} \frac{V}{2}\mathbf{p} \left[\left(a^\dagger(\mathbf{p})a(\mathbf{p}) + b(\mathbf{p})b^\dagger(\mathbf{p}) \right) \right.$$
$$\left. + a^\dagger(\mathbf{p})b^\dagger(-\mathbf{p}) e^{\frac{2i}{\hbar}E_\mathbf{p}t} - b(-\mathbf{p})a(\mathbf{p}) e^{-\frac{2i}{\hbar}E_\mathbf{p}t} \right] + h.c.,$$

where $h.c.$ denotes the hermitian conjugate terms. In the last term on the right hand side of the above equation we have performed a change in the integration variable, namely $\mathbf{p} \to -\mathbf{p}$, gaining a minus sign. Note that the last two terms in the integral sum up to an anti-hermitian operator, which cancels against its hermitian conjugate. The first two terms instead are hermitian, so that we end up with:

$$\hat{\mathbf{P}} = \int \frac{d^3\mathbf{p}}{(2\pi\hbar)^3} V\mathbf{p} \left(a^\dagger(\mathbf{p})a(\mathbf{p}) + b(\mathbf{p})b^\dagger(\mathbf{p}) \right)$$
$$= \sum_\mathbf{p} \mathbf{p} \left(a^\dagger(\mathbf{p})a(\mathbf{p}) + b(\mathbf{p})b^\dagger(\mathbf{p}) \right), \qquad (11.45)$$

where the last equality refers to the case of a finite volume V and discrete momenta. Note that in this case no normal ordering is necessary since, when writing $bb^\dagger$ in terms of $b^\dagger b$ in (11.45) we have

$$\sum_{\mathbf{p}} \mathbf{p}\, b(\mathbf{p}) b^\dagger(\mathbf{p}) = \sum_{\mathbf{p}} \mathbf{p}\, \left(b^\dagger(\mathbf{p}) b(\mathbf{p}) + 1\right) = \sum_{\mathbf{p}} \mathbf{p}\, b^\dagger(\mathbf{p}) b(\mathbf{p}), \qquad (11.46)$$

due to the cancellation of $\mathbf{p}$ against $-\mathbf{p}$ when summing the constant term over all possible values of $\mathbf{p}$.

Putting together the results obtained for $\hat{H}$ and $\hat{P}^i$, we may define the four-momentum quantum operator

$$\hat{P}^\mu = \int \frac{d^3\mathbf{p}}{(2\pi\hbar)^3} V\, p^\mu \left(a^\dagger(\mathbf{p}) a(\mathbf{p}) + b^\dagger(\mathbf{p}) b(\mathbf{p})\right). \qquad (11.47)$$

So far we have defined the quantum operator associated with a Klein-Gordon field. We still need to define the Hilbert space of quantum states on which such operator acts. This will allow us to give a particle interpretation of our results.

Our discussion so far paralleled the one for the electromagnetic field in Chap. 6. When we wrote the field operators $\hat{\phi}(x)$, $\hat{\phi}^\dagger(x)$ in terms of the a, b operators and of their hermitian conjugates satisfying the commutation relations (11.32) (or (11.33)), we have described the quantum system as a collection of infinitely many decoupled quantum harmonic oscillators of two kinds: The "(a)" and the "(b)" oscillators, associated with the positive and negative energy solutions to the Klein-Gordon equation. Each value of $\mathbf{p}$ defines a corresponding oscillator of type (a) and (b), the operators $a(\mathbf{p})$, $a^\dagger(\mathbf{p})$ and $b(\mathbf{p})$, $b^\dagger(\mathbf{p})$ being the corresponding destruction and creation operators, respectively. For the two kinds of oscillators we define the (hermitian) *number operators*:

$$\hat{N}_{\mathbf{p}}^{(a)} = a^\dagger(\mathbf{p}) a(\mathbf{p}); \quad \hat{N}_{\mathbf{p}}^{(b)} = b^\dagger(\mathbf{p}) b(\mathbf{p}). \qquad (11.48)$$

We see that both the energy (11.43) and the momentum (11.45) are expressed as infinite sums over such operators. In particular the Hamiltonian operator $\hat{H}$ is the sum over the Hamiltonian operators $\hat{H}_{\mathbf{p}}^{(a)}$, $\hat{H}_{\mathbf{p}}^{(b)}$ of the various oscillators (we use here, for the sake of simplicity, the finite volume notation):

$$\hat{H} = \sum_{\mathbf{p}} \left(\hat{H}_{\mathbf{p}}^{(a)} + \hat{H}_{\mathbf{p}}^{(b)}\right), \qquad (11.49)$$

$$\hat{H}_{\mathbf{p}}^{(a)} \equiv E_{\mathbf{p}}\, \hat{N}_{\mathbf{p}}^{(a)} = E_{\mathbf{p}}\, a^\dagger(\mathbf{p}) a(\mathbf{p}); \quad \hat{H}_{\mathbf{p}}^{(b)} \equiv E_{\mathbf{p}}\, \hat{N}_{\mathbf{p}}^{(b)} = E_{\mathbf{p}}\, b^\dagger(\mathbf{p}) b(\mathbf{p}).$$

Since these harmonic oscillators correspond to independent, decoupled degrees of freedom of the scalar field, operators associated with different oscillators commute, as it is apparent from (11.32). In particular the hermitian operators $\hat{N}_{\mathbf{p}}^{(a)}$, $\hat{N}_{\mathbf{p}}^{(b)}$ form a commuting system[5] and thus can be diagonalized simultaneously. As a consequence of this the quantum states of the field can be expressed as products of the infinite states pertaining to the constituent quantum oscillators, each constructed as an eigenstate of the corresponding number operator.

Recall indeed, from elementary quantum mechanics, that the states of a, say, type (a), quantum oscillator, corresponding to a momentum $\mathbf{p}$, have the form $|N_\mathbf{p}\rangle^{(a)}$, and are eigenstates of $\hat{N}_\mathbf{p}^{(a)}$:

$$\hat{N}_\mathbf{p}^{(a)}|N_\mathbf{p}\rangle^{(a)} = N_\mathbf{p}|N_\mathbf{p}\rangle^{(a)}, \tag{11.51}$$

the energy of such state being $N_\mathbf{p} E_\mathbf{p}$. The action on it of $a(\mathbf{p})$ or $a^\dagger(\mathbf{p})$, lowers or raises $N_\mathbf{p}$ by one unit, respectively. In other words they destroy or create quanta of energy $E_\mathbf{p}$. This follows from the commutation relations:

$$[\hat{N}_\mathbf{p}^{(a)}, a_\mathbf{p}^\dagger] = a_\mathbf{p}^\dagger; \quad [\hat{N}_\mathbf{p}^{(a)}, a_\mathbf{p}] = -a_\mathbf{p}, \tag{11.52}$$

from which we find

$$\hat{N}_\mathbf{p}^{(a)} a_\mathbf{p}^\dagger|N_\mathbf{p}\rangle^{(a)} = a_\mathbf{p}^\dagger \hat{N}_\mathbf{p}^{(a)}|N_\mathbf{p}\rangle^{(a)} + a_\mathbf{p}^\dagger|N_\mathbf{p}\rangle^{(a)} = (N_\mathbf{p} + 1)|N_\mathbf{p}\rangle^{(a)},$$
$$\hat{N}_\mathbf{p}^{(a)} a_\mathbf{p}|N_\mathbf{p}\rangle^{(a)} = a_\mathbf{p} \hat{N}_\mathbf{p}^{(a)}|N_\mathbf{p}\rangle^{(a)} - a_\mathbf{p}|N_\mathbf{p}\rangle^{(a)} = (N_\mathbf{p} - 1)|N_\mathbf{p}\rangle^{(a)},$$

that is the states $a_\mathbf{p}^\dagger|N_\mathbf{p}\rangle^{(a)}$ and $a_\mathbf{p}|N_\mathbf{p}\rangle^{(a)}$ correspond to the eigenvalues $N_\mathbf{p} + 1$ and $N_\mathbf{p} - 1$ of $\hat{N}_\mathbf{p}^{(a)}$, respectively.

Requiring $N_\mathbf{p}$, as well as the energy, to be non-negative, the sequence $N_\mathbf{p}-1$, $N_\mathbf{p} - 2, \ldots$ must terminate with zero, corresponding to ground state $|0\rangle^{(a)}$ for which $a_\mathbf{p}|0\rangle^{(a)} = 0$, so that

$$\hat{N}_\mathbf{p}^{(a)}|0\rangle^{(a)} = 0.$$

The eigenvalues $N_\mathbf{p}$ are then non-negative integers ($N_\mathbf{p} = 0, 1, 2, \ldots$), also called *occupation numbers*, and the corresponding eigenstates are constructed by applying the creation operator $a^\dagger(\mathbf{p})$ to $|0\rangle^{(a)}$ $N_\mathbf{p}$-times:

$$|N_\mathbf{p}\rangle^{(a)} = \frac{1}{\sqrt{(N_\mathbf{p})!}} a_\mathbf{p}^\dagger a_\mathbf{p}^\dagger \cdots a_\mathbf{p}^\dagger |0\rangle^{(a)},$$

---

[5]The fact that $[\hat{N}^{(a)}, \hat{N}^{(b)}] = 0$ immediately follows from the property that a and b commute, as stressed after (11.33). Consider now the number operators corresponding to oscillators of a given kind, say (a):

$$[\hat{N}_\mathbf{p}^{(a)}, \hat{N}_{\mathbf{p}'}^{(a)}] = a^\dagger(\mathbf{p})[a(\mathbf{p}), a^\dagger(\mathbf{p}')]a(\mathbf{p}') + a^\dagger(\mathbf{p}')[a^\dagger(\mathbf{p}), a(\mathbf{p}')]a(\mathbf{p})$$

$$= \left(a^\dagger(\mathbf{p})a(\mathbf{p}') - a^\dagger(\mathbf{p}')a(\mathbf{p})\right) \delta_{\mathbf{p}\mathbf{p}'} = 0, \tag{11.50}$$

the same result obviously holds for the operators $\hat{N}^{(b)}$.

where the denominator is fixed by normalizing the state to one. What we have said for the type-(a) oscillator equally applies to the type-(b) ones. The ground states of the two kinds of oscillators will be denoted by $|0\rangle^{(a)}$ and $|0\rangle^{(b)}$ respectively. We can summarize the construction of single-oscillator states, for a given momentum $\mathbf{p}$, as follows

Type a-oscillators $\qquad\qquad\qquad$; Type b-oscillators

$$\hat{N}_{\mathbf{p}}^{(a)}|N_{\mathbf{p}}\rangle^{(a)} = N_{\mathbf{p}}|N_{\mathbf{p}}\rangle^{(a)} \qquad ; \ \hat{N}_{\mathbf{p}}^{(b)}|N_{\mathbf{p}}\rangle^{(b)} = N_{\mathbf{p}}|N_{\mathbf{p}}\rangle^{(b)}$$

$$a_{\mathbf{p}}|0\rangle^{(a)} = 0 \qquad\qquad\qquad ; \ b_{\mathbf{p}}|0\rangle^{(b)} = 0$$

$$|N_{\mathbf{p}}\rangle^{(a)} = \frac{(a_{\mathbf{p}}^{\dagger})^{N_{\mathbf{p}}}}{\sqrt{(N_{\mathbf{p}})!}}|0\rangle^{(a)} \qquad ; \ |N_{\mathbf{p}}\rangle = \frac{(b_{\mathbf{p}}^{\dagger})^{N_{\mathbf{p}}}}{\sqrt{(N_{\mathbf{p}})!}}|0\rangle^{(b)}$$

$$a_{\mathbf{p}}^{\dagger}|N_{\mathbf{p}}\rangle^{(a)} = \sqrt{N_{\mathbf{p}} + 1}|N_{\mathbf{p}} + 1\rangle^{(a)} \ ; \ b^{\dagger}(\mathbf{p})|N_{\mathbf{p}}\rangle^{(b)} = \sqrt{N_{\mathbf{p}} + 1}\,|N_{\mathbf{p}} + 1\rangle^{(b)}$$

$$a_{\mathbf{p}}|N_{\mathbf{p}}\rangle^{(a)} = \sqrt{N_{\mathbf{p}}}|N_{\mathbf{p}} - 1\rangle^{(a)} \qquad ; \ b_{\mathbf{p}}|N_{\mathbf{p}}\rangle^{(b)} = \sqrt{N_{\mathbf{p}}}\,|N_{\mathbf{p}} - 1\rangle^{(b)}$$

We may now construct the Hilbert space of quantum field states, labeled by the eigenvalues of the number operators. The states $|\{N\}\rangle^{(a)}$ of the system of type-(a) oscillators are constructed as tensor products of the single-oscillator states $|N_{\mathbf{p}}\rangle^{(a)}$ over all possible values $\mathbf{p}_1, \mathbf{p}_2, \ldots$, of $\mathbf{p}$:

$$\begin{aligned}
|\{N\}\rangle^{(a)} &\equiv |N_{\mathbf{p}_1}, N_{\mathbf{p}_2}, \ldots\rangle^{(a)} = |N_{\mathbf{p}_1}\rangle^{(a)}|N_{\mathbf{p}_2}\rangle^{(a)} \cdots \\
&= \left[\frac{(a^{\dagger}(\mathbf{p}_1))^{N_{\mathbf{p}_1}}(a^{\dagger}(\mathbf{p}_2))^{N_{\mathbf{p}_2}} \cdots}{\sqrt{(N_{\mathbf{p}_1})!\,(N_{\mathbf{p}_2})! \cdots}}\right]|0, 0, \ldots, 0\rangle^{(a)},
\end{aligned} \qquad (11.53)$$

where $|0, 0, \ldots, 0\rangle^{(a)}$ denotes the product over all $\mathbf{p}$ of the ground states $|0\rangle^{(a)}$ associated with each type-(a) oscillator. By the same token we construct a complete set of states from the system of type-(b) oscillators $|\{N\}\rangle^{(b)}$. The full Hilbert space of states of the quantum field will be the product of the Hilbert spaces associated with each type of oscillators, and will therefore be generated by the following complete set of vectors:

$$|\{N\}; \{N'\}\rangle \equiv |\{N\}\rangle^{(a)} \otimes |\{N'\}\rangle^{(b)}, \qquad (11.54)$$

Each of the above states are constructed by repeatedly applying $a^{\dagger}$ and $b^{\dagger}$ operators to the "vacuum" state:

$$|0\rangle \equiv |0, 0, \ldots, 0\rangle^{(a)} \otimes |0, 0, \ldots, 0\rangle^{(b)}, \qquad (11.55)$$

For instance

$$a^\dagger(\mathbf{p})|0\rangle = |0,\ldots 0, 1, 0, \ldots 0\rangle^{(a)} \otimes |0, \ldots 0\rangle^{(b)},$$
$$b^\dagger(\mathbf{p})|0\rangle = |0, \ldots 0\rangle^{(a)} \otimes |0, \ldots 0, 1, 0, \ldots 0\rangle^{(b)},$$

where the position of the entry 1 corresponds to type-(a), respectively (b), oscillator state labeled by the momentum $\mathbf{p}$. The states $|\{N^{(a)}\}; \{N^{(b)}\}\rangle$ are, by construction, eigenstates of all the number operators $\hat{N}_\mathbf{p}^{(a)}$, $\hat{N}_\mathbf{p}^{(b)}$ and the Hilbert space they generate is called Fock space.

Recall now the expression of the momentum operator $\hat{\mathbf{P}}$ of the field in the continuous as well as in the discrete (i.e. finite volume) notations (here and in the following we denote by an arrow the change to the finite-volume notation)

$$\hat{\mathbf{P}} = \int \frac{d^3\mathbf{p}}{(2\pi\hbar)^3} V \mathbf{p} \, (\hat{N}_\mathbf{p}^{(a)} + \hat{N}_\mathbf{p}^{(b)}) \to \sum_\mathbf{p} \mathbf{p} \, (\hat{N}_\mathbf{p}^{(a)} + \hat{N}_\mathbf{p}^{(b)}), \tag{11.56}$$

which completes, with the Hamiltonian operator $\hat{H}$ in (11.49), the four momentum operator:

$$\hat{P}^\mu = \int \frac{d^3\mathbf{p}}{(2\pi\hbar)^3} V p^\mu (\hat{N}_\mathbf{p}^{(a)} + \hat{N}_\mathbf{p}^{(a)}) \to \sum_\mathbf{p} p^\mu (\hat{N}_\mathbf{p}^{(a)} + \hat{N}_\mathbf{p}^{(a)}). \tag{11.57}$$

Just as we did for the quantized electromagnetic field, the quantum field states are interpreted as describing a *multiparticle* system: Each type-(a) and type-(b) oscillator defines a single particle state with definite momentum $\mathbf{p}$ and the occupation number $N_\mathbf{p}$ is interpreted as the number of particles in that state. This time however the quantized excitations of the field are described in terms of *two kinds of particles*, according to the type of oscillator. Conventionally those describing excitations of type-(a) and type-(b) oscillators are referred to as *particles* and *antiparticles* respectively. For instance the state $|\{N\}; \{N'\}\rangle$ describes $N_{\mathbf{p}_1}$ particles and $N'_{\mathbf{p}_1}$ antiparticles with momentum $\mathbf{p}_1$; $N_{\mathbf{p}_2}$ particles and $N'_{\mathbf{p}_2}$ antiparticles with momentum $\mathbf{p}_2$, and so on.

With this interpretation the quantum Hamiltonian and momentum operators are simply understood as the sum of the energies and momenta of the particles and antiparticles in the system, each carrying a quantum of energy $E_\mathbf{p}$ and of momentum $\mathbf{p}$. Every single-particle (antiparticle) state contributes to the energy and momentum of the total field state $|\{N\}; \{N'\}\rangle$ an amount $N_\mathbf{p} E_\mathbf{p}$ and $N_\mathbf{p} \mathbf{p}$ ($N'_\mathbf{p} E_\mathbf{p}$ and $N'_\mathbf{p} \mathbf{p}$), respectively, proportional to the corresponding occupation number.

Therefore when this number varies by a unit, the total energy and momentum of the state vary by $E_\mathbf{p}$ and $\mathbf{p}$, respectively. It is important to note that even if antiparticles are associated with negative energy solutions to the *classical* Klein-Gordon equation, they contribute a *positive energy* $E_\mathbf{p}$ to the Hamiltonian, that is *antiparticles are positive energy particles*. Let us observe in this respect that the photon, associated with the excitations of the electromagnetic field, coincides with its own antiparticle, since in that case, as often pointed out, the field $\hat{A}_\mu(x)$ is hermitian, thus implying $a = b$.

We conclude that the operators $a^\dagger(\mathbf{p})$ and $b^\dagger(\mathbf{p})$ create a particle and an antiparticle with momentum $\mathbf{p}$, respectively, while $a(\mathbf{p})$ and $b(\mathbf{p})$ destroy them. In an analogous way, denoting by $\hat{\phi}_+(x)$ and $\hat{\phi}_-(x)$ the positive and negative energy components of the field operator $\hat{\phi}(x)$ in (11.19):

$$\hat{\phi}(x) = \hat{\phi}_+(x) + \hat{\phi}_-(x),$$

$$\hat{\phi}_+(x) = \int \frac{d^3\mathbf{p}}{(2\pi\hbar)^3} \frac{\hbar\sqrt{V}}{\sqrt{2E_\mathbf{p}}} a(\mathbf{p})e^{-\frac{i}{\hbar}p\cdot x} \rightarrow \sum_\mathbf{p} \frac{\hbar}{\sqrt{2E_\mathbf{p}V}} a(\mathbf{p})e^{-\frac{i}{\hbar}p\cdot x},$$

$$\hat{\phi}_-(x) = \int \frac{d^3\mathbf{p}}{(2\pi\hbar)^3} \frac{\hbar\sqrt{V}}{\sqrt{2E_\mathbf{p}}} b^\dagger(\mathbf{p})e^{\frac{i}{\hbar}p\cdot x} \rightarrow \sum_\mathbf{p} \frac{\hbar}{\sqrt{2E_\mathbf{p}V}} b^\dagger(\mathbf{p})e^{\frac{i}{\hbar}p\cdot x},$$

$$(11.58)$$

the former destroys a particle at the space-time point $x \equiv (x^\mu)$ (since it contains $a(\mathbf{p})$) while the latter creates an antiparticle at x (since it contains $b^\dagger(\mathbf{p})$). The reverse is true for $\hat{\phi}_-^\dagger(x)$, $\hat{\phi}_+^\dagger(x)$, defined as the negative and positive frequency components of $\hat{\phi}^\dagger(x)$, respectively:

$$\hat{\phi}_-^\dagger(x) = \int \frac{d^3\mathbf{p}}{(2\pi\hbar)^3} \frac{\hbar\sqrt{V}}{\sqrt{2E_\mathbf{p}}} a^\dagger(\mathbf{p})e^{\frac{i}{\hbar}p\cdot x},$$

$$\hat{\phi}_+^\dagger(x) = \int \frac{d^3\mathbf{p}}{(2\pi\hbar)^3} \frac{\hbar\sqrt{V}}{\sqrt{2E_\mathbf{p}}} b(\mathbf{p})e^{-\frac{i}{\hbar}p\cdot x}. \tag{11.59}$$

It is implicit from the above discussion that we are working in the Heisenberg picture in which operators, like $\hat{\phi}(\mathbf{x}, t)$ depend on time while states are constant. This is necessary in order to have a relativistically covariant framework, see Sect. 6.2 of Chap. 6.

The Fock space formalism is particularly suited for providing a *multiparticle description* of a quantum relativistic *free* field theory. It is however interesting to write down the familiar non-relativistic *wave function* of a system of particles, using the $\mathbf{x}$-representation instead of the Fock representation. We define a state describing n particles located at the points $\mathbf{x}_1, \dots, \mathbf{x}_n$ at a time t as

$$|\mathbf{x}_1, \dots, \mathbf{x}_n; t\rangle \equiv \hat{\phi}_-^\dagger(\mathbf{x}_1, t) \dots \hat{\phi}_-^\dagger(\mathbf{x}_n, t)|0\rangle, \tag{11.60}$$

where the effect of $\hat{\phi}_-^\dagger(\mathbf{x}_i, t)$ is that of creating a particle in $\mathbf{x}_i$ at the time t. On the other hand a generic n particle state in the Fock-representation is defined as

$$|N_1, N_2, \dots\rangle^{(a)} = \frac{(a_1^\dagger)^{N_1}(a_2^\dagger)^{N_2}\dots}{(N_1!N_2!\dots)^{\frac{1}{2}}}|0\rangle^{(a)}, \tag{11.61}$$

where $N_1 + N_2 + \cdots = n$ and we have used the short-hand notation $N_1 \equiv N_{\mathbf{p}_1}$, $N_2 \equiv N_{\mathbf{p}_2}$, and so on. The wave function $\phi^{(n)}_{N_1,N_2,\ldots}(\mathbf{x}_1 \ldots \mathbf{x}_n, t)$ realizing the coordinate representation of the state (11.61) therefore reads

$$\phi^{(n)}_{N_1,N_2,\ldots}(\mathbf{x}_1, \ldots, \mathbf{x}_n, t) = \langle \mathbf{x}_1 \ldots \mathbf{x}_n; t | N_1, N_2, \ldots \rangle^{(a)}. \qquad (11.62)$$

By the same token we construct the coordinate representation of a multi-antiparticle state $|N'_1, N'_2, \ldots \rangle^{(b)}$ or of a generic particle-antiparticle state $|\{N\}; \{N'\}\rangle$. We conclude from this that the multi-particle wave function (describing both particles and antiparticles) is *completely symmetric with respect to the exchange of the particles* (antiparticles) since the operators $\hat{\phi}^\dagger_-(\mathbf{x}_i, t)$ and $\hat{\phi}^\dagger_-(\mathbf{x}_j, t)$ ($\hat{\phi}_-(\mathbf{x}_i, t)$ and $\hat{\phi}_-(\mathbf{x}_j, t)$) commute. In other words: *Spin-zero particles obey the Bose-Einstein statistics.* This result, which obviously also holds for the photon field, can be shown to be valid for all particles of integer spin.

11.2.1 Electric Charge and Its Conservation

We have seen that a complex scalar field, being equivalent to two real fields, has extra (internal) degrees of freedom which are related to the existence of antiparticles. We now show that these extra degrees of freedom are connected with the presence of a *charge* carried by the field. We recall that in the classical Hamilton formulation the current and the charge associated with the Klein-Gordon field are given by Eqs. (8.208) and (8.209) of Chap. 8, see also (10.19) and (10.20) of Chap. 10. At the quantum level they become the following operators:

$$\hat{j}^\mu = -i\frac{ec}{\hbar} : \left[\partial^\mu \hat{\phi}^\dagger(x) \hat{\phi}(x) - \hat{\phi}^\dagger(x) \partial^\mu \hat{\phi}(x) \right] :, \qquad (11.63)$$

and

$$\hat{Q} = i\frac{e}{\hbar} \int d^3x : \left(\hat{\phi}^\dagger(x) \hat{\pi}^\dagger(x) - h.c. \right) :, \qquad (11.64)$$

where $h.c.$ denotes, as usual, the hermitian conjugate of the preceding terms. The explicit computation of $\hat{Q}$ is quite similar to that of Q in Eq. (10.33) of Chap. 10. If we compute the first term of (11.64) we find

$$i\frac{e}{\hbar} \int d^3x \hat{\phi}^\dagger(x) \hat{\pi}^\dagger(x)$$

$$= i\frac{e}{\hbar} \int d^3x \int \frac{d^3\mathbf{p}}{(2\pi\hbar)^3} \frac{d^3\mathbf{p}}{(2\pi\hbar)^3} (-i\hbar) \frac{V}{2} \sqrt{\frac{E_\mathbf{q}}{E_\mathbf{p}}}$$

$$\times \left[a^\dagger(\mathbf{p})a(\mathbf{q}) e^{\frac{i}{\hbar}(E_\mathbf{p}-E_\mathbf{q})t - \frac{i}{\hbar}(\mathbf{p}-\mathbf{q})\cdot\mathbf{x}} - b(\mathbf{q})b^\dagger(\mathbf{q}) e^{\frac{i}{\hbar}(E_\mathbf{p}-E_\mathbf{q})t - \frac{i}{\hbar}(\mathbf{p}-\mathbf{q})\cdot\mathbf{x}} \right.$$

$$- a^\dagger(\mathbf{p})b^\dagger(\mathbf{q})e^{\frac{i}{\hbar}(E_\mathbf{p}+E_\mathbf{q})t-\frac{i}{\hbar}(\mathbf{p}+\mathbf{q})\cdot\mathbf{x}} + b(\mathbf{p})a(\mathbf{q})e^{-\frac{i}{\hbar}(E_\mathbf{p}+E_\mathbf{q})t+\frac{i}{\hbar}(\mathbf{p}+\mathbf{q})\cdot\mathbf{x}}\Bigg]$$

$$= e \int \frac{d^3\mathbf{p}}{(2\pi\hbar)^3} \frac{V}{2} \left[a^\dagger(\mathbf{p})a(\mathbf{p}) - b(\mathbf{p})b^\dagger(\mathbf{p}) - a^\dagger(\mathbf{p})b^\dagger(-\mathbf{p})e^{\frac{2i}{\hbar}E_\mathbf{p}t} \right.$$
$$\left. + b(-\mathbf{p})a(\mathbf{p})e^{-\frac{2i}{\hbar}E_\mathbf{q}t} \right],$$

where, in the last integral we have changed $\mathbf{p} \to -\mathbf{p}$. If we sum the above expression with its hermitian conjugate, the quantities containing $a^\dagger a$ and $b^\dagger b$, being hermitian, will sum up. The terms containing ba and $a^\dagger b^\dagger$, on the other hand, add up to a antihermitian operator which cancels against its hermitian conjugate. We then obtain:

$$\hat{Q} = e \int \frac{d^3\mathbf{p}}{(2\pi\hbar)^3} V : \left[a(\mathbf{p})^\dagger a(\mathbf{p}) - b(\mathbf{p})b(\mathbf{p})^\dagger \right] :$$
$$= e \int \frac{d^3\mathbf{p}}{(2\pi\hbar)^3} V (a(\mathbf{p})^\dagger a(\mathbf{p}) - b(\mathbf{p})^\dagger b(\mathbf{p})), \tag{11.65}$$

or, equivalently, in terms of the number operators $\hat{N}_\mathbf{p}^{(a)}$, $\hat{N}_\mathbf{p}^{(b)}$,

$$\hat{Q} = e \int \frac{d^3\mathbf{p}}{(2\pi\hbar)^3} V \left(\hat{N}_\mathbf{p}^{(a)} - \hat{N}_\mathbf{p}^{(b)} \right), \tag{11.66}$$

Using the finite-volume notation, the charge operator has the following simple form:

$$\hat{Q} = \sum_\mathbf{p} e(\hat{N}_\mathbf{p}^{(a)} - \hat{N}_\mathbf{p}^{(b)}). \tag{11.67}$$

This formula shows that if particles have charge e, antiparticles have opposite charge $-e$. We conclude that, as anticipated in Sect. 10.6.2, *antiparticles have the same mass as the corresponding particles but opposite charge.*

We have learned that, in the classical Klein-Gordon theory, the charge Q is conserved. This was related, in the Hamiltonian framework, to the fact that it generates phase transformations which leave the Hamiltonian invariant. We show that the same properties hold in the quantum theory.

From the classical treatment we expect the generator $\hat{G}$ of a global phase transformation to be related to $\hat{Q}$ as follows:

$$\hat{G}(t) = -\frac{\hbar}{e} \hat{Q}(t), \tag{11.68}$$

so that the corresponding unitary transformation reads:

$$U(\alpha) = e^{\frac{i}{\hbar}\alpha\hat{G}(t)} = e^{-\frac{i}{e}\alpha\hat{Q}(t)}, \tag{11.69}$$

where α is the parameter introduced in Eq. (8.206).

To show that the above operator implements a phase transformation on $\hat{\phi}(x)$, let us transform the latter by means of $U(\alpha)$, computed *at the same time*. For $\alpha \ll 1$ we have

$$\hat{\phi}'(\mathbf{x}, t) = U^\dagger(\alpha)\hat{\phi}(\mathbf{x}, t)U(\alpha) \approx \hat{\phi}(\mathbf{x}, t) - i\frac{\alpha}{e}\left[\hat{\phi}(\mathbf{x}, t), \hat{Q}(t)\right] = \hat{\phi}(x) + \delta\hat{\phi}(x),$$

$$\delta\hat{\phi}(\mathbf{x}, t) = -i\frac{\alpha}{e}\left[\hat{\phi}(\mathbf{x}, t), \hat{Q}(t)\right]. \tag{11.70}$$

On the other hand, from the explicit form of $\hat{Q}$ in Eq. (11.64) and the canonical commutation relations between $\hat{\phi}$ and $\hat{\pi}$, we also have

$$\left[\hat{\phi}(\mathbf{x}, t), \hat{Q}(t)\right] = -\frac{ie}{\hbar}\int d^3\mathbf{y}\left[\hat{\phi}(\mathbf{x}, t), \hat{\pi}(\mathbf{y}, t)\right]\hat{\phi}(\mathbf{y}, t) = e\hat{\phi}(\mathbf{x}, t), \tag{11.71}$$

from which it follows that

$$\delta\hat{\phi}(\mathbf{x}, t) = -i\alpha\hat{\phi}(\mathbf{x}, t), \tag{11.72}$$

as in the classical case. Furthermore we can easily verify that the transformation (11.72) leaves the Hamiltonian (11.42) invariant:

$$\delta\hat{H} = [\hat{Q}, \hat{H}] = 0. \tag{11.73}$$

Combining this result with the quantum equation

$$\frac{d\hat{Q}}{dt} = -\frac{i}{\hbar}[\hat{Q}, \hat{H}],$$

we find that the charge operator is conserved.

11.3 Transformation Under the Poincaré Group

We recall that, in the Heisenberg picture, any operator on the Hilbert space of states transforms according to Eq. (9.38) of Chap. 9. In particular the action of a Poincaré transformation (Λ, x_0) on a *scalar field operator* $\hat{\phi}(x)$ reads

$$\hat{\phi}(x) = \hat{\phi}'(x') = U^\dagger(\Lambda, x_0)\hat{\phi}(x')U(\Lambda, x_0) = O_{(\Lambda, x_0)}\hat{\phi}(x'), \tag{11.74}$$

where, as usual, $x' = \Lambda x - x_0$. We can indeed easily verify that the above transformation law for the *quantum field* is in agreement with the transformation law of the *classical field*, namely of the *wave function*. Recall from (11.60) and (11.62), the general relation between a multi-particle state and the corresponding coordinate representation. For a single particle state $|s\rangle$ in the Fock space, the corresponding wave function $\phi_{(s)}(x)$ is expressed as (see Eq. (9.19)):

$$\phi_{(s)}(x) = \langle 0|\hat{\phi}(x)|s\rangle. \tag{11.75}$$

If we now perform the Poincaré transformation (Λ, x_0) on the field operator, according to (11.74), we find

$$\phi_{(s)}(x) \xrightarrow{(\Lambda, x_0)} \phi'(x') = \langle 0|\hat{\phi}'(x')|s\rangle = \langle 0|U^\dagger(\Lambda, x_0)\hat{\phi}(x')U(\Lambda, x_0)|s\rangle$$
$$= \langle 0|\hat{\phi}(x)|s\rangle = \phi(x)_{(s)} = \phi(\Lambda^{-1}(x' + x_0)) = O_{(\Lambda, x_0)}\phi(x'), \tag{11.76}$$

where we have used the property, which we shall always assume to hold, that the vacuum state is invariant under the Poincaré group: $U(\Lambda, x_0)|0\rangle = |0\rangle$. Equation (11.76) is indeed the correct transformation law for a classical scalar field. It is important however to bear in mind that the unitary operator $U(\Lambda, x_0)$ acts on the Fock space of states, while $O_{(\Lambda, x_0)}$ acts on the space of functions. Clearly Eq. (11.76) defines a relation between the infinitesimal Poincaré generators of the two operators. Let us denote here by $\hat{\mathbb{J}}^{\rho\sigma}$ and $\hat{\mathbb{P}}^\mu$ the generators of $U(\Lambda, x_0)$, that is the representation of the Poincaré generators on the quantum states. We can write:

$$U(\Lambda, x_0) = e^{-\frac{i}{\hbar}\hat{\mathbb{P}}_\mu x_0^\mu} e^{\frac{i}{2\hbar}\theta_{\rho\sigma}\hat{\mathbb{J}}^{\rho\sigma}}. \tag{11.77}$$

Recalling the expression (9.101) of $O_{(\Lambda, x_0)}$ in terms of its infinitesimal generators $\hat{J}^{\rho\mu}$ and $\hat{P}^\mu$, given in (9.102), we can write an infinitesimal Poincaré transformation of the field operator as follows:

$$\delta\hat{\phi}(x) = \frac{i}{\hbar}\left[\hat{\phi}(x), \frac{1}{2}\delta\theta_{\rho\sigma}\hat{\mathbb{J}}^{\rho\sigma} - \epsilon \cdot \hat{\mathbb{P}}\right] = \frac{i}{\hbar}\left(\frac{1}{2}\delta\theta_{\rho\sigma}\hat{J}^{\rho\sigma} - \epsilon \cdot \hat{P}\right)\hat{\phi}(x),$$

where we have expanded in Eq. (11.74) both $U(\Lambda, x_0)$ and $O_{(\Lambda, x_0)}$ to first order in the infinitesimal Poincaré parameters $\delta\theta_{\rho\sigma}$, $x_0^\mu = \epsilon^\mu$. Using the explicit form of $\hat{J}^{\rho\mu}$ and $\hat{P}^\mu$ in Eq. (9.102) we find:

$$\frac{i}{\hbar}[\hat{\phi}(x), \hat{\mathbb{J}}_{\rho\sigma}] = (x_\rho\partial_\sigma - x_\sigma\partial_\rho)\hat{\phi}(x); \quad \frac{i}{\hbar}[\hat{\phi}(x), \hat{\mathbb{P}}_\mu] = -\partial_\mu\hat{\phi}(x), \tag{11.78}$$

so that the realization of $\hat{\mathbb{J}}_{\mu\nu}$ and $\hat{\mathbb{P}}^\mu$ in terms of the field operator is obtained from (8.240) and (8.243) respectively, by promoting all the classical fields to quantum operators.

Coming back to the finite unitary transformation (11.74) it is interesting to see how the creation and annihilation operators transform under a Lorentz transformations.

Let us show that the following transformation laws for $a(p)$ and $b(p)$:

$$U^\dagger(\Lambda)a(p)U(\Lambda) = a(\Lambda^{-1}p); \quad U^\dagger(\Lambda)b(p)U(\Lambda) = b(\Lambda^{-1}p). \quad (11.79)$$

induce on $\hat\phi(x)$ the corresponding transformation (11.74). Indeed we have

$$
\begin{aligned}
U^\dagger(\Lambda)\hat\phi(x')U(\Lambda) &= \int \frac{d^3p}{(2\pi\hbar)^3} \frac{\hbar\sqrt{V}}{\sqrt{2E_p}} \left[U^\dagger(\Lambda)a(p)U(\Lambda)e^{-\frac{i}{\hbar}p\cdot x'} \right.\\
&\qquad\qquad\qquad\qquad \left. + U^\dagger(\Lambda)b^\dagger(p)U(\Lambda)e^{\frac{i}{\hbar}p\cdot x'} \right] \\
&= \int \frac{d^3p}{(2\pi\hbar)^3} \frac{\hbar\sqrt{V}}{\sqrt{2E_p}} \left[a(\Lambda^{-1}p)e^{-\frac{i}{\hbar}p\cdot x'} + b^\dagger(\Lambda^{-1}p)e^{\frac{i}{\hbar}p\cdot x'} \right] \\
&= \int \frac{d^3p'}{(2\pi\hbar)^3} \frac{\hbar\sqrt{V'}}{\sqrt{2E_{p'}}} \left[a(p')e^{-\frac{i}{\hbar}(\Lambda p')\cdot x'} + b^\dagger(p')e^{\frac{i}{\hbar}(\Lambda p')\cdot x'} \right] \\
&= \int \frac{d^3p'}{(2\pi\hbar)^3} \frac{\hbar\sqrt{V'}}{\sqrt{2E_{p'}}} \left[a(p')e^{-\frac{i}{\hbar}p'\cdot(\Lambda^{-1}x')} + b^\dagger(p')e^{\frac{i}{\hbar}p'\cdot(\Lambda^{-1}x')} \right] \\
&= \hat\phi(\Lambda^{-1}x'), \quad\quad\quad\quad\quad\quad\quad\quad\quad\quad\quad (11.80)
\end{aligned}
$$

so that Eq. (11.74) is verified.[6]

11.3.1 Discrete Transformations

In the study of the Lorentz transformations, we have mostly considered the proper subgroup SO(1, 3) corresponding to transformations with unit determinant that are connected with continuity to the identity transformation. This allows us to consider their infinitesimal action on the fields.

In Chap. 4 we have also defined other Lorentz transformations. These include the *parity transformation* or *space reflection* P and the *time reversal transformation* T, whose active action of a scalar field is

$$P : \hat\phi(\mathbf{x}, t) \to \eta_P \hat\phi(-\mathbf{x}, t), \quad\quad\quad (11.81)$$
$$T : \hat\phi(\mathbf{x}, t) \to \eta_T \hat\phi(\mathbf{x}, -t). \quad\quad\quad (11.82)$$

These transformations are respectively implemented on four-vectors by the matrices Λ_P and Λ_T given in Chap. 4, with negative determinant: $\det \Lambda_P = \det \Lambda_T = -1$. The complex factors η_P, η_T can only be ± 1 since parity and time reversal are *involutive*, namely applying them twice gives the identity transformation.

---

[6]In the above derivation we have used the Lorentz invariance of the measure $d^3\mathbf{p}V$ and of $E_\mathbf{p}V$, see Sect. 9.5 of Chap. 9.

Let us first consider the parity transformation. In classical canonical mechanics, a parity transformation implies the inversion of the position vector **x** and the linear momentum of a particle, while it leaves its angular momentum (including spin), invariant:

$$P : \mathbf{x} \to -\mathbf{x}; \quad \mathbf{p} \to -\mathbf{p}; \quad \mathbf{J} \to \mathbf{J}. \tag{11.83}$$

In field theory, we note that the Klein-Gordon equation is invariant under (11.81). The sign η_P in the transformation defines the *intrinsic parity* of the field, the sign plus or minus corresponding to *scalar* or *pseudoscalar* field, respectively.[7] Since any transformation is determined by a unitary transformation in the Hilbert space of the states, let us denote by $U(P)$ the one implementing parity, so that

$$U(P)^\dagger \hat{\phi}(\mathbf{x}, t) U(P) = \eta_P \, \hat{\phi}(-\mathbf{x}, t), \tag{11.84}$$

where $\eta_P = \pm 1$ denotes the intrinsic parity of the field. Using the expansion (11.19) it is easy to see that the transformations (11.84) can be realized in terms of the oscillators $a(\mathbf{p})$, $b(\mathbf{p})$ as follows:

$$U(P)^\dagger a(\mathbf{p}) U(P) = \eta_P \, a(-\mathbf{p}); \quad U(P)^\dagger b(\mathbf{p}) U(P) = \eta_P \, b(-\mathbf{p}). \tag{11.85}$$

Let us give an explicit realization of the operator $U(P)$ in terms of $a, b, a^\dagger, b^\dagger$, depending on the intrinsic parity of the field. Consider the operator $e^{i\lambda S}$, with

$$S = \sum_\mathbf{p} \left(a_\mathbf{p}^\dagger a_{-\mathbf{p}} + b_\mathbf{p}^\dagger b_{-\mathbf{p}} \right),$$

where, for the sake of clarity, we have used a discrete notation: $a_\mathbf{p} \equiv a(\mathbf{p})$, $b_\mathbf{p} \equiv b(\mathbf{p})$. Now use the identity (see for instance [2] for a general proof)

$$e^{i\lambda S} O e^{-i\lambda S} = O + i\lambda [S, O] + \frac{i^2 \lambda^2}{2!} [S, [S, O]] + \cdots . \tag{11.86}$$

Since

$$[S, a_\mathbf{p}] = -a_{-\mathbf{p}} \qquad [S, [S, a_\mathbf{p}]] = a_\mathbf{p},$$

we find

$$e^{i\lambda S} a_\mathbf{p} e^{-i\lambda S} = a_\mathbf{p} \cos \lambda - i a_{-\mathbf{p}} \sin \lambda,$$

---

[7]The intrinsic parity can only be fixed by experiment involving interactions, so that it is meaningful only when specified relative to other particles.

and the same relation for $b_{\mathbf{p}}$. Setting $\lambda = \eta_P \pi/2$, we get rid of the term in $a_{\mathbf{p}}$, obtaining

$$e^{i\frac{\pi}{2}\eta_P S} a_{\mathbf{p}} e^{-i\frac{\pi}{2}\eta_P S} = -i\eta_P \, a_{-\mathbf{p}}, \tag{11.87}$$

$$e^{i\frac{\pi}{2}\eta_P S} b_{\mathbf{p}} e^{-i\frac{\pi}{2}\eta_P S} = -i\eta_P \, b_{-\mathbf{p}}. \tag{11.88}$$

This is close to (11.85), but not yet correct. To get the exact result we further multiply $e^{i\lambda S}$ by the operator $e^{i\lambda' S'}$, defined in such a way that

$$e^{i\lambda' S'} a_{\mathbf{p}} e^{-i\lambda' S'} = i a_{\mathbf{p}}, \tag{11.89}$$

$$e^{i\lambda' S'} b_{\mathbf{p}} e^{-i\lambda' S'} = i b_{\mathbf{p}}. \tag{11.90}$$

This is achieved by taking

$$S' = \sum_{\mathbf{p}} \left(a_{\mathbf{p}}^{\dagger} a_{\mathbf{p}} + b_{\mathbf{p}}^{\dagger} b_{\mathbf{p}} \right),$$

and $\lambda' = -\pi/2$. The reader can show that $[S, S'] = 0$.

Combining these results and defining

$$U(P) \equiv e^{i\frac{\pi}{2} S'} e^{-i\eta_P \frac{\pi}{2} S} = \exp i\frac{\pi}{2} (S' - \eta_P S)$$

$$= \exp i\frac{\pi}{2} \sum_{\mathbf{p}} \left(a_{\mathbf{p}}^{\dagger} a_{\mathbf{p}} + b_{\mathbf{p}}^{\dagger} b_{\mathbf{p}} - \eta_P \, a_{\mathbf{p}}^{\dagger} a_{-\mathbf{p}} - \eta_P \, b_{\mathbf{p}}^{\dagger} b_{-\mathbf{p}} \right). \tag{11.91}$$

Note that $U(P)$ is indeed a unitary operator satisfying

$$U(P)|0\rangle = |0\rangle,$$

as can be easily seen by expanding the exponentials. Thus the *vacuum state has even parity*. Moreover considering the momentum operator (11.45) we see that

$$U(P)^{\dagger} \hat{\mathbf{P}} U(P) = -\hat{\mathbf{P}},$$

consistently with the fact that the eigenvalues of the physical momentum are ordinary vectors under a space reflection.

On the other hand $U(P)$ *commutes with the Hamiltonian*, implying the conservation of the parity operator: $[U(P), \hat{H}] = 0$.[8]

On the quantum field $\hat{\phi}(\mathbf{x}, t)$ we can also define a transformation with no analogue in the non-relativistic quantum theory: the *charge conjugation* C. It corresponds to

_____

[8] Since the parity transformation is involutive, $U(P)^2 = \hat{I}$, its eigenvalues can only be ± 1.

exchanging particles for antiparticles, that is

$$a_\mathbf{p} \to \eta_C \, b_\mathbf{p}; \quad b_\mathbf{p} \to \eta_C \, a_\mathbf{p}, \tag{11.92}$$

or, in terms of the field operator,

$$\hat{\phi}(x) \to \eta_C \, \hat{\phi}^\dagger(x),$$

where η_C is a constant which, defining C as an involutive transformation, can be chosen to be ± 1. This operation is clearly a symmetry of the *charged* scalar theory. The construction of the unitary operator $U(C)$ implementing such transformation on the Hilbert space of the states, namely

$$U(C)^\dagger \, a_\mathbf{p} \, U(C) = \eta_C \, b_\mathbf{p},$$
$$U(C)^\dagger \, b_\mathbf{p} \, U(C) = \eta_C \, a_\mathbf{p},$$
$$U(C)^\dagger \, \hat{\phi}(x) \, U(C) = \eta_C \, \hat{\phi}^\dagger(x), \tag{11.93}$$

can be done by the same procedure used for the parity transformation. The result is

$$U(C) = \exp\left[\frac{i\pi}{2} \eta_C \sum_\mathbf{p} \left(a_\mathbf{p}^\dagger a_\mathbf{p} + b_\mathbf{p}^\dagger b_\mathbf{p} - \eta_C \, a_\mathbf{p}^\dagger b_\mathbf{p} - \eta_C \, b_\mathbf{p}^\dagger a_\mathbf{p} \right) \right]. \tag{11.94}$$

It is easily verified that $U(C)$ is unitary and satisfies $U(C)|0\rangle = |0\rangle$. Moreover, from (11.63) and (11.64) it follows

$$U(C)^\dagger \hat{J}^\mu U(C) = -\hat{J}^\mu; \quad U(C)^\dagger \hat{Q} U(C) = -\hat{Q}. \tag{11.95}$$

That means that, under charge conjugation the sign of the charge is flipped, according to our previous discussion in Sect. 11.2.1.

We finally consider *time reversal* $T: t \to -t$. In classical canonical mechanics time-reversal leaves the position $\mathbf{x}$ of a particle unchanged while it reverses its velocity

$$\mathbf{v} = d\mathbf{x}/dt \to d\mathbf{x}/d(-t) = -\mathbf{v}$$

and thus its linear momentum $\mathbf{p} \to -\mathbf{p}$, as well as the angular momentum (including spin). In summary

$$T : \mathbf{x} \to \mathbf{x}; \quad \mathbf{p} \to -\mathbf{p}; \quad \mathbf{J} \to -\mathbf{J}. \tag{11.96}$$

Time reversal is a symmetry of classical Newtonian mechanics, where force is taken to depend only on the position on the particle:

$$\frac{d^2\mathbf{x}(t)}{dt^2} = \mathbf{F}(\mathbf{x}(t)) \quad \Leftrightarrow \quad \frac{d^2\mathbf{x}(-t)}{dt^2} = \frac{d^2\mathbf{x}(-t)}{d(-t)^2} = \mathbf{F}(\mathbf{x}(-t)), \qquad (11.97)$$

that is if $\mathbf{x}(t)$ is a solution to the Newton equation, also $\mathbf{x}(-t)$ is. T, however, is not a symmetry when we consider, for instance, the action of the Lorentz force on a moving charge, which depends on the velocity of the particle, and thus in general is not invariant under it.

As far as field theory is concerned, the Klein-Gordon equation is invariant under the transformation

$$\hat{\phi}(\mathbf{x}, t) \rightarrow \pm \hat{\phi}(\mathbf{x}, -t), \qquad (11.98)$$

but the equal time commutation relations, for example

$$[\hat{\phi}(\mathbf{x}, t), \frac{\partial}{\partial t}\hat{\phi}^\dagger(\mathbf{y}, t)] = i\hbar\delta^3(\mathbf{x} - \mathbf{y}), \qquad (11.99)$$

do not exhibit this invariance unless (11.99) is accompanied by the change $i \rightarrow -i$. That means that we must include in the *time reversal operator* $U(T)$

$$U(T)^\dagger \hat{\phi}(\mathbf{x}, t)U(T) = \eta_T \hat{\phi}(\mathbf{x}, -t), \qquad (11.100)$$

the *complex conjugation* operator K. This operator is defined by the following properties[9]

$$K (\lambda_1 |a\rangle + \lambda_2|b\rangle) = \lambda_1^*|Ka\rangle + \lambda_2^*|Kb\rangle; \quad \langle a|Kb\rangle = \langle b|Ka\rangle, \quad \langle Ka|Kb\rangle = \langle b|a\rangle, \qquad (11.101)$$

where $|a\rangle$ is a generic state, λ is a c-number and the last equation expresses the fact that the norm is not affected by complex conjugation: given a generic state $|a\rangle$, we indeed have $\langle Ka|Ka\rangle = \langle a|a\rangle$. The first of (11.101) represents the property of K of being *antilinear*. We define, for a generic antilinear operator A, its hermitian conjugate $A^\dagger$ through the relation $\langle a|Ab\rangle = \langle b|A^\dagger a\rangle$, for any $|a\rangle$ and $|b\rangle$ (note the difference with respect to the definition of the analogous quantity for a linear operator S: $\langle a|Sb\rangle = \langle S^\dagger a|b\rangle$). An antilinear operator A is *antiunitary* if, and only if, for any two states: $\langle Aa|Ab\rangle = \langle b|a\rangle$ (one can show that if A preserves the norm of any vector it is antiunitary, so that antiunitarity for an antilinear operator is equivalent to *norm-preserving*). If A is antiunitary we have $A^\dagger A = AA^\dagger = \hat{I}$, since $\langle Aa|Ab\rangle = \langle b|A^\dagger Aa\rangle = \langle b|a\rangle$, for any $|a\rangle$ and $|b\rangle$. The second of (11.101) characterizes then K as a hermitian operator ($K = K^\dagger$), while the last of (11.101)

---

[9]For a formal treatment of this issue see for instance A. Messiah, *Quantum Mechanics*, Dover 1999.

expresses the fact that K is antiunitary: $K^\dagger K = K K^\dagger = \hat{I}$. The complex conjugation K, being antiunitary and hermitian, squares to the identity: $K^2 = \hat{I}$. From this property one can show that there exists an orthonormal basis of vectors $|u_n\rangle$ on which $K|u_n\rangle = |u_n\rangle$ (*real basis*). With respect to a real basis, the action of K on any vector simply amounts to *changing its components into their complex conjugates*, hence the name *complex conjugation* for K.

If $\hat{O}$ is a linear operator and A is an antilinear transformation, the following relation can be easily derived:

$$\langle a|A^\dagger \hat{O} A|b\rangle = \langle a|A^\dagger \hat{O} Ab\rangle = \langle \hat{O} Ab|Aa\rangle = \langle Ab|\hat{O}^\dagger|Aa\rangle,$$

which implies that, if we take A to be antiunitary and $\hat{O} = i\,\hat{I}$, we then have: $\langle a|A^\dagger i\, A|b\rangle = -i\,\langle Ab|Aa\rangle = -i\,\langle a|b\rangle$, for any two states $|a\rangle$ and $|b\rangle$. Symbolically this property can be expressed by the relation:

$$A^\dagger i\, A = -i.$$

where the identity operator on the right hand side is understood. We must require $U(T)$ to satisfy the same properties (11.101) as K, namely to be antiunitary, so that

$$U(T)^\dagger \,[\hat{\phi}(\mathbf{x}, t), \frac{\partial}{\partial t}\hat{\phi}^\dagger(\mathbf{y}, t)]\, U(T) = U(T)^\dagger \, i\hbar\, U(T)\delta^3(\mathbf{x} - \mathbf{y}),$$

implies

$$[\hat{\phi}(\mathbf{x}, -t), \frac{\partial}{\partial(-t)}\hat{\phi}^\dagger(\mathbf{y}, -t)] = -i\hbar\delta^3(\mathbf{x} - \mathbf{y}),$$

and the equal time commutation relations are left invariant.

Let us write the time-reversal operator as

$$U(T) = \mathcal{U} K, \tag{11.102}$$

where $\mathcal{U}$ is a unitary transformation. The operator $U(T)$ defined above is indeed antiunitary since, defining for a generic couple of states $|Ta\rangle \equiv U(T)|a\rangle$, $|Tb\rangle \equiv U(T)|b\rangle$, we have:

$$\langle Ta|Tb\rangle = \langle \mathcal{U} Ka|\mathcal{U} Kb\rangle = \langle Ka|Kb\rangle = \langle b|a\rangle. \tag{11.103}$$

Requiring it to satisfy (11.100), we get

$$\mathcal{U}^\dagger a_\mathbf{p} \mathcal{U} = \eta_T\, a_{-\mathbf{p}}; \quad \mathcal{U}^\dagger b_\mathbf{p} \mathcal{U} = \eta_T\, b_{-\mathbf{p}}, \tag{11.104}$$

$\eta_T = \pm 1$, so that we can take $\mathcal{U}$ to have the same form as $U(P)$ in Eq. (11.91), with $\eta_P \to \eta_T$. To show this, let us implement $U(T)$ defined above on the field operator

$\hat{\phi}(x)$ defined in (11.19) and written in the discrete notation:

$$U(T)^\dagger \, \hat{\phi}(x)U(T) = \sum_{\mathbf{p}} \frac{\hbar}{\sqrt{2E_{\mathbf{p}}V}} \left(\eta_T \, a_{-\mathbf{p}} \, e^{\frac{i}{\hbar} \, p \cdot x} + \eta_T \, b_{-\mathbf{p}}^\dagger \, e^{-\frac{i}{\hbar} \, p \cdot x} \right)$$

$$= \eta_T \sum_{\mathbf{p}} \frac{\hbar}{\sqrt{2E_{\mathbf{p}}V}} \left(a_{\mathbf{p}} \, e^{-\frac{i}{\hbar} \, p \cdot x_T} + b_{\mathbf{p}}^\dagger \, e^{\frac{i}{\hbar} \, p \cdot x_T} \right) = \eta_T \, \hat{\phi}(\mathbf{x}, -t),$$

where $x_T \equiv (x_T^\mu) = (-ct, \mathbf{x})$. In the above derivation we have used $K^\dagger \, e^{\pm \frac{i}{\hbar} \, p \cdot x} \, K = e^{\mp \frac{i}{\hbar} p \cdot x}$ and changed the summation variable from $\mathbf{p}$ to $-\mathbf{p}$.

We can also verify that

$$U(T)^\dagger \, \hat{\mathbf{P}} \, U(T) = -\hat{\mathbf{P}},$$

and, as far as the current operator is concerned, we also find from (11.63)

$$U(T)^\dagger \, \hat{\mathbf{j}}(\mathbf{x}, t) \, U(T) = -\hat{\mathbf{j}}(\mathbf{x}, -t); \quad U(T)^\dagger \, \hat{j}^0(\mathbf{x}, t) \, U(T) = \hat{j}^0(\mathbf{x}, -t).$$

Both these results are in agreement with our physical intuition.

11.4 Invariant Commutation Rules and Causality

Let us note that all the commutation rules among field operators considered so far are *equal-time* commutators. We now consider commutators at different times. We show that the commutator

$$D(x - y) = \frac{c}{\hbar} \left[\hat{\phi}(x), \hat{\phi}^\dagger(y) \right], \tag{11.105}$$

is a *Lorentz-invariant* function. Furthermore, if the four-dimensional distance between $x \equiv (x^\mu)$ e $y \equiv (y^\mu)$ is *space-like*, that is if $(x - y)^2 = (x^0 - y^0)^2 - |\mathbf{x} - \mathbf{y}|^2 < 0$, then *the commutator is zero*.

To show these properties we decompose $\hat{\phi}(x)$ and $\hat{\phi}^\dagger(x)$ in their positive energy and negative energy parts, according to (11.58) and (11.59). We clearly have $[\hat{\phi}_+(x), \hat{\phi}_+^\dagger(y)] = [\hat{\phi}_-(x), \hat{\phi}_-^\dagger(y)] = 0$, so that

$$D(x - y) = \frac{c}{\hbar}[\hat{\phi}_+(x), \hat{\phi}_-^\dagger(y)] + \frac{c}{\hbar}[\hat{\phi}_-(x), \hat{\phi}_+^\dagger(y)]. \tag{11.106}$$

From the commutation rules (11.32) between $a, a^\dagger$ and $b, b^\dagger$, we find

$$\frac{c}{\hbar}[\hat{\phi}_\pm(x), \hat{\phi}_\mp^\dagger(y)] = \pm D_\pm(x - y), \tag{11.107}$$

where

$$D_{\pm}(x - y) = \hbar c \int \frac{d^3\mathbf{p}}{(2\pi\hbar)^3} \frac{1}{2E_\mathbf{p}} e^{\mp \frac{i}{\hbar} p \cdot (x-y)}, \tag{11.108}$$

and therefore Eq. (11.106) becomes

$$D(x - y) = D_+(x - y) - D_-(x - y) = -2i\hbar c \int \frac{d^3\mathbf{p}}{(2\pi\hbar)^3} \frac{1}{2E_\mathbf{p}} \sin \frac{p \cdot (x - y)}{\hbar}. \tag{11.109}$$

Using then Eq. (10.26) we find that $D_{(+)}(x - y)$ and $D(x - y)$ can be written in the following *manifestly Lorentz-invariant form*

$$D_+(x - y) = \hbar \int \frac{d^4 p}{(2\pi\hbar)^3} \delta(p^2 - m^2 c^2)\theta(p^0)e^{-\frac{i}{\hbar}p \cdot (x-y)},$$

$$D_-(x - y) = \hbar \int \frac{d^4 p}{(2\pi\hbar)^3} \delta(p^2 - m^2 c^2)\theta(-p^0)e^{-\frac{i}{\hbar}p \cdot (x-y)},$$

$$D(x - y) = \hbar \int \frac{d^4 p}{(2\pi\hbar)^3} \delta(p^2 - m^2 c^2)\epsilon(p^0)e^{-\frac{i}{\hbar}p \cdot (x-y)}, \tag{11.110}$$

where

$$\theta(p^0) = \begin{cases} 1 & \text{for } p^0 > 0 \\ 0 & \text{for } p^0 < 0, \end{cases} \tag{11.111}$$

and $\epsilon(p^0) = \theta(p^0) - \theta(-p^0)$.

The relativistic invariance of the three D-functions follows from the fact that the functions $\theta(p^0), \theta(-p^0), \epsilon(p^0)$ are themselves invariant under proper Lorentz transformations since the four-vector p^μ is *non-spacelike*. Indeed the restriction $E > 0$ due to the presence of $\theta(p^0)$ on the right hand side of Eq. (11.110) implies that, when expanding $\delta(p^2 - m^2 c^2)$ according to the (10.24)–(10.26), we must only take the positive energy solution $E > 0$ of $p^2 - m^2 c^2 = 0$. This choice is Lorentz-invariant (more precisely it is invariant under proper Lorentz transformations) since p^μ is non-spacelike, as it was proven in full generality in Chap. 4: If $p^0 > 0$ in a given reference frame it will keep the same sign in any other frame.[10] The same argument holds for the functions $\theta(-p^0)$ and $\epsilon(p^0)$.

---

[10]Let us repeat here the argument given in Chap. 4 in a more compact form. Suppose $p^0 > 0$, in a frame S. In a Lorentz transformed frame S' moving at velocity $\mathbf{v}$ relative to S, we have

$$p'^0 = \gamma \left(p^0 - \frac{\mathbf{v}}{c} \cdot \mathbf{p}\right) \geq \gamma \left(p^0 - \frac{|\mathbf{v}|}{c}|\mathbf{p}|\right):$$

We conclude that $D(x - y) = D(x^0 - y^0, \mathbf{x} - \mathbf{y})$ as well as $D_\pm(x - y)$ are Lorentz-invariant functions, namely

$$D_\pm(x - y) = D_\pm(\mathbf{\Lambda} \cdot (x - y)),$$

where $\mathbf{\Lambda} \equiv (\Lambda^\mu{}_\nu)$ is a proper Lorentz transformation and $\mathbf{\Lambda} \cdot (x-y) \equiv \Lambda^\mu{}_\nu (x^\nu - y^\nu)$. Recalling our discussion in Sect. 1.5.1, we know that when $(x - y)^2 < 0$ there exists a frame where $x^0 = y^0$, in which

$$D(x - y) \equiv D(0, \mathbf{x} - \mathbf{y}) = 2i\hbar c \int \frac{d^3 \mathbf{p}}{(2\pi\hbar)^3} \frac{1}{2E_\mathbf{p}} \sin \frac{\mathbf{p} \cdot (\mathbf{x} - \mathbf{y})}{\hbar} = 0, \quad (11.112)$$

since the integrand is odd for $\mathbf{p} \to -\mathbf{p}$. This implies that $D_+(0, \mathbf{x} - \mathbf{y}) = D_-(0, \mathbf{x} - \mathbf{y})$ as can be also verified from their explicit expressions:

$$
\begin{aligned}
D_\pm(0, \mathbf{x} - \mathbf{y}) &= \hbar c \int \frac{d^3 \mathbf{p}}{(2\pi\hbar)^3} \frac{1}{2E_\mathbf{p}} e^{\mp \frac{i}{\hbar} \mathbf{p} \cdot (\mathbf{x} - \mathbf{y})} \\
&= \hbar c \int \frac{d^3 \mathbf{p}}{(2\pi\hbar)^3} \frac{1}{2E_\mathbf{p}} e^{\frac{i}{\hbar} \mathbf{p} \cdot (\mathbf{x} - \mathbf{y})}. \quad (11.113)
\end{aligned}
$$

On the other hand Lorentz invariance of the D-functions implies that *the properties* (11.112) *and* (11.113) *must hold in any any Lorentz frame*. Therefore from (11.109) and (11.105) we conclude that *if the four-dimensional distance is spacelike*, $(x - y)^2 < 0$, *the function* $D(x - y)$, *vanishes*

$$(x - y)^2 < 0 \quad \Rightarrow \quad D(x - y) = \frac{c}{\hbar} [\hat{\phi}(x), \hat{\phi}(y)^\dagger] = 0, \quad (11.114)$$

and moreover

$$D_+(x - y) = D_-(x - y). \quad (11.115)$$

This result is important in order to have a *causal theory*. The operators $\hat{\phi}(x)$ and $\hat{\phi}^\dagger(y)$ are associated with the field excitations which can be interpreted as particles at the points x^μ ed y^μ of space-time. If the two events related to the presence of the two quanta are separated by a space-like distance, $(x - y)^2 < 0$, they cannot

---

(Footnote 10 continued)

However $p^0 = \sqrt{|\mathbf{p}|^2 + m^2 c^2} \geq |\mathbf{p}|$ and therefore, since $\frac{|\mathbf{v}|}{c} < 1$

$$p^0 \geq \gamma \left(1 - \frac{|\mathbf{v}|}{c}\right) |\mathbf{p}| > 0.$$

be correlated since this would imply the presence of a signal traveling at a velocity greater than c, thus violating the *causality principle*, as explained in Chap. 1.[11]

The requirement of commutativity of two observables separated by a space-like distance is referred to as the *principle of microcausality*. It is also worth noting that this result is guaranteed by the cancellation of the contributions from $D_\pm(x - y)$ in the commutator, related in turn to the presence of positive and negative energy solutions $\hat{\phi}_+(x), \hat{\phi}_-(x)$. The very presence of these two solutions and, in particular, of the negative energy ones, so embarrassing for the classical Klein-Gordon equation, is therefore essential for the consistency of the quantum field theory.

For the sake of completeness we now show that $D_+(0, \mathbf{x} - \mathbf{y})$ is different from zero, and give its explicit expression.

$$D_+(0, \mathbf{x} - \mathbf{y}) = \hbar c \int \frac{d^3\mathbf{p}}{(2\pi\hbar)^3} \frac{1}{2E_\mathbf{p}} e^{\frac{i\mathbf{p}\cdot(\mathbf{x}-\mathbf{y})}{\hbar}}. \tag{11.116}$$

Using polar coordinates for the variable $\mathbf{p}$, we have $d^3\mathbf{p} = |\mathbf{p}|^2 \sin\theta d|\mathbf{p}|d\theta d\varphi$, so that:

$$\begin{aligned}
\frac{\hbar}{c} D_+(0, \mathbf{x} - \mathbf{y}) &= \frac{\hbar^2}{(2\pi\hbar)^3}(2\pi) \int_0^\infty \frac{d|\mathbf{p}| |\mathbf{p}|^2}{2E_\mathbf{p}} \int_{-1}^1 d(\cos\theta) e^{\frac{i}{\hbar}|\mathbf{p}||\mathbf{x}-\mathbf{y}|\cos\theta} \\
&= \frac{1}{(2\pi)^2\hbar} \int_0^\infty \frac{d|\mathbf{p}| |\mathbf{p}|^2}{2E_\mathbf{p}} \frac{\hbar \left(e^{\frac{i}{\hbar}|\mathbf{p}||\mathbf{x}-\mathbf{y}|} - e^{-\frac{i}{\hbar}|\mathbf{p}||\mathbf{x}-\mathbf{y}|}\right)}{i|\mathbf{p}||\mathbf{x}-\mathbf{y}|} \\
&= \frac{1}{(2\pi)^2} \int_0^\infty \frac{d|\mathbf{p}| |\mathbf{p}|}{E_\mathbf{p}|\mathbf{x}-\mathbf{y}|} \sin\left(\frac{|\mathbf{p}|}{\hbar}|\mathbf{x}-\mathbf{y}|\right) \\
&= \frac{1}{(2\pi)^2c} \int_0^\infty \frac{d|\mathbf{p}| |\mathbf{p}|}{\sqrt{|\mathbf{p}|^2 + m^2c^2}} \sin\left(\frac{|\mathbf{p}|}{\hbar}|\mathbf{x}-\mathbf{y}|\right) \frac{1}{|\mathbf{x}-\mathbf{y}|} \\
&= \frac{1}{(2\pi)^2} \frac{m}{|\mathbf{x}-\mathbf{y}|} K_1\left(\frac{mc}{\hbar}|\mathbf{x}-\mathbf{y}|\right).
\end{aligned} \tag{11.117}$$

---

[11] We note that the requirement of causality refers to *observables* and in general field operators are not necessarily observables. However quite generally observables in a physical system are constructed in terms of local functions of the field variables so that the requirement of causality can be expressed in terms of the fields themselves. The requirement of commuting operators for space-like separations is also called "*locality*" and, correspondingly, the quantum field theory is referred to as a "*local theory*". Locality assures that the results of two measures made at a space-like distance cannot have any influence on one another, there being no correlation between the two events.

where $K_n(z)$ are the modified Bessel functions of the second type and we have used the general formula

$$\int\limits_0^\infty dz z \frac{\sin(bz)}{\sqrt{z^2+\gamma^2}} e^{-\beta\sqrt{z^2+\gamma^2}} = \frac{b\gamma}{\sqrt{b^2+\beta^2}} K_1(\gamma\sqrt{b^2+\beta^2}). \tag{11.118}$$

In our case we have $z = |\mathbf{p}|$, $\gamma = mc$, $b = \frac{|\mathbf{x}-\mathbf{y}|}{\hbar}$, $\beta = 0$. The asymptotic behavior of $K_1(s)$ as $s \to \infty$ is:

$$K_1(s) = \sqrt{\frac{\pi}{2s}} e^{-s}\left(1 + O\left(\frac{1}{s}\right)\right) \simeq \sqrt{\frac{\pi}{2s}} e^{-s}, \tag{11.119}$$

and therefore for large space-time separation $|\mathbf{x} - \mathbf{y}| \to \infty$

$$D_+(0, \mathbf{x} - \mathbf{y}) \approx \frac{1}{(2\pi)^2}\sqrt{\frac{\pi mc}{2\hbar}} \frac{1}{|\mathbf{x}-\mathbf{y}|^{\frac{3}{2}}} e^{-\frac{mc}{\hbar}|\mathbf{x}-\mathbf{y}|}, \tag{11.120}$$

that is D_+ is sensibly different from zero only within spatial distances of the order of the Compton wave-length $\lambda = \frac{\hbar}{mc}$ of the particle.

11.4.1 Green's Functions and the Feynman Propagator

The invariant D-functions discussed in the previous paragraph are strictly related to another invariant function which plays a major role in the theory of interacting fields: the *Feynman propagator function* $D_F(x - y)$. It is defined to be the *vacuum expectation value* of the so called *time-ordered product*:[12]

$$D_F(x - y) = \frac{c}{\hbar} \langle 0|T\hat{\phi}(x)\hat{\phi}^\dagger(y)|0\rangle, \tag{11.122}$$

where

$$T\hat{\phi}(x)\hat{\phi}^\dagger(y) = \begin{cases} \hat{\phi}(x)\hat{\phi}^\dagger(y) & x^0 > y^0, \\ \hat{\phi}^\dagger(y)\hat{\phi}(x) & y^0 > x^0. \end{cases} \tag{11.123}$$

and we note that there is no ambiguity when $x^0 = y^0$ since in this case $\hat{\phi}(x)$ and $\hat{\phi}^\dagger(y)$ commute. Furthermore the fact that $\hat{\phi}(x)$ and $\hat{\phi}^\dagger(y)$ commute at space-like distances

---

[12]For a hermitian (and thus neutral) field

$$D_F(x - y) = \frac{c}{\hbar} \langle 0|T\hat{\phi}(x)\hat{\phi}(y)|0\rangle. \tag{11.121}$$

ensures that time ordering remains invariant under Lorentz transformations, and thus that the *Feynman propagator is Lorentz-invariant.*

To compute $D_F(x - y)$ we note that if $x^0 > y^0$, using Eq. (11.108) and the fact that the destruction operators annihilate the vacuum, we have

$$\langle 0|T\hat{\phi}(x)\hat{\phi}^\dagger(y)|0\rangle = \langle 0|\hat{\phi}_+(x)\hat{\phi}^\dagger_-(y)|0\rangle$$

$$= \langle 0|\left[\hat{\phi}_+(x), \hat{\phi}^\dagger_-(y)\right]|0\rangle = \frac{\hbar}{c}D_+(x - y).$$

Similarly for $y^0 > x^0$ we get

$$\langle 0|T\hat{\phi}(x)\hat{\phi}^\dagger(y)|0\rangle = \langle 0|[\hat{\phi}^\dagger_+(y), \hat{\phi}_-(x)]|0\rangle = \frac{\hbar}{c}D_-(x - y). \quad (11.124)$$

We may then write the Feynman propagator in the following compact form:

$$D_F(x - y) = \theta(x^0 - y^0)D_+(x - y) + \theta(y^0 - x^0)D_-(x - y). \quad (11.125)$$

The physical meaning of the Feynman propagator is easily understood if we observe that for $x^0 > y^0$, $D_F(x - y) = \langle 0|\hat{\phi}_+(x)\hat{\phi}^\dagger_-(y)|0\rangle$, that is $D_F(x - y)$ measures the probability amplitude that a particle be created at $\mathbf{y}$ at the instant y^0 and then destroyed at $\mathbf{x}$ at the instant x^0, while, if $y^0 > x^0$, $D_F(x - y) = \langle 0|\hat{\phi}^\dagger_+(y)\hat{\phi}_-(x)|0\rangle$ is the probability amplitude that an antiparticle be created in $\mathbf{x}$ at the time x^0 and then destroyed in $\mathbf{y}$ at the time y^0.

If we now use the explicit expression of $D_+(x - y)$ and $D_-(x - y)$ we may write

$$D_F(x - y) = \left[\theta(x^0 - y^0)D_+(x - y) + \theta(y^0 - x^0)D_-(x - y)\right]$$

$$= c\hbar \int \frac{d^3\mathbf{p}}{(2\pi\hbar)^3} \frac{1}{2E_\mathbf{p}} \left[\theta(x^0 - y^0)e^{-\frac{i}{\hbar}p\cdot(x-y)} + \theta(y^0 - x^0)e^{\frac{i}{\hbar}p\cdot(x-y)}\right].$$

$$(11.126)$$

We are now going to prove that, using the Cauchy residue theorem, we can write the above expression for the Feynman propagator in terms of a an integral in the complex p^0 plane along the path C_F in Fig. 11.1, as follows:

$$D_F(x - y) = i\hbar^2 \int \frac{d^3\mathbf{p}}{(2\pi\hbar)^3} \int_{C_F} \frac{dp^0}{2\pi\hbar} \frac{e^{-\frac{i}{\hbar}p\cdot(x-y)}}{p^2 - m^2c^2}$$

$$= i\hbar^2 \int \frac{d^3\mathbf{p}}{(2\pi\hbar)^3} \int_{C_F} \frac{dp^0}{2\pi\hbar} \frac{e^{-\frac{i}{\hbar}(p^0(x^0-y^0)-\mathbf{p}\cdot(\mathbf{x}-\mathbf{y}))}}{(p^0 - \bar{p}^0)(p^0 + \bar{p}^0)}, \quad (11.127)$$

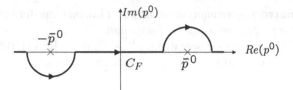

Fig. 11.1 Integration in the complex p^0 plane

$$\boxed{x^0 > y^0}$$

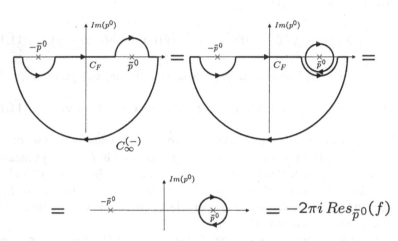

Fig. 11.2 The $x^0 > y^0$ case

where, as usual, $p \cdot (x - y) \equiv p^0(x^0 - y^0) - \mathbf{p} \cdot (\mathbf{x} - \mathbf{y})$ and $\bar{p}^0 \equiv \sqrt{|\mathbf{p}|^2 + m^2 c^2} = E_\mathbf{p}/c > 0$. To show this we observe that if $x^0 > y^0$ we can close the contour C_F in the lower p^0 half-plane, where the imaginary part of p^0 is negative, along a semi-circle $C_\infty^{(-)}$ of infinite radius, so that the integral along $C_\infty^{(-)}$ be exponentially suppressed, see Fig. 11.2. Indeed, denoting the integrand in (11.127) by $f(p^0, \mathbf{p})$:

$$f(p^0, \mathbf{p}) \equiv \frac{i\hbar}{2\pi} \frac{1}{p^2 - m^2 c^2} e^{-\frac{i}{\hbar} p \cdot (x - y)},$$

the integral of f along $C_\infty^{(-)}$ vanishes since $\lim_{\text{Im}(p^0) \to -\infty} e^{\frac{1}{\hbar} \text{Im}(p^0)(x^0 - y^0)} = 0$. We can then write:

$$\int_{C_F} f(p^0, \mathbf{p}) \, dp^0 = \int_{C_F + C_\infty^{(-)}} f(p^0, \mathbf{p}) \, dp^0 = -2\pi i \, \text{Res}_{\bar{p}^0}(f)$$

$$= -2\pi i \, [(p^0 - \bar{p}^0) \, f(p^0, \mathbf{p})]_{p^0 = \bar{p}^0} = \frac{\hbar}{2\bar{p}^0} \, e^{-\frac{i}{\hbar} p \cdot (x - y)} \Big|_{p^0 = \bar{p}^0}.$$

Therefore, for $x^0 > y^0$, the integral in (11.127) reads:

$$\int \frac{d^3\mathbf{p}}{(2\pi\hbar)^3} \int_{C_F} dp^0 \, f(p^0, \mathbf{p}) = \hbar c \int \frac{d^3\mathbf{p}}{(2\pi\hbar)^3} \frac{e^{-\frac{i}{\hbar} p \cdot (x-y)}}{2 E_\mathbf{p}} = D_F(x-y).$$

which is consistent with Eq. (11.126). If instead $y^0 > x^0$ we close the contour in the upper half-plane along the semi-circle $C_\infty^{(+)}$, see Fig. 11.3, and obtain:

$$\int_{C_F} f(p^0, \mathbf{p}) \, dp^0 = \int_{C_F + C_\infty^{(+)}} f(p^0, \mathbf{p}) \, dp^0 = 2\pi i \, Res_{-\bar{p}^0}(f)$$

$$= \frac{\hbar}{2 \bar{p}^0} e^{\frac{i}{\hbar} (\bar{p}^0 (x^0 - y^0) + \mathbf{p} \cdot (\mathbf{x} - \mathbf{y}))}.$$

Inserting the above result in (11.127) and changing the integration variable from $\mathbf{p}$ to $-\mathbf{p}$, we find

$$\int \frac{d^3\mathbf{p}}{(2\pi\hbar)^3} \int_{C_F} dp^0 \, f(p^0, \mathbf{p}) = \hbar c \int \frac{d^3\mathbf{p}}{(2\pi\hbar)^3} \frac{e^{\frac{i}{\hbar} p \cdot (x-y)}}{2 E_\mathbf{p}} = D_F(x-y),$$

which completes the proof of Eq. (11.127).

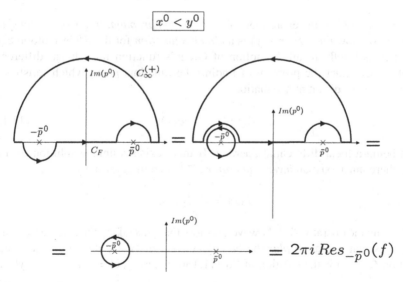

Fig. 11.3 The $y^0 > x^0$ case

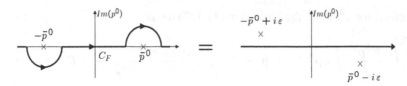

Fig. 11.4 Prescription for the Feynman propagator: Shifting the poles in the p^0 plane

Summarizing, the Feynman propagator is defined by the integral over the four-momentum space of $f(p^0, \mathbf{p})$, with the prescription that the integral over p^0 be computed along C_F. Such prescription is equivalent to integrating over the real p^0 axis and shifting at the same time the poles to $\pm(\bar{p}^0 - i\,\epsilon)$), where ϵ is an infinitesimal displacement, as shown in Fig. 11.4. The denominator $p^2 - m^2 c^2$ of $f(p^0, \mathbf{p})$, with this prescription, changes into $p^2 - m^2 c^2 + i\,\epsilon$ and Eq. (11.127) can be also written as follows

$$D_F(x - y) = i\hbar^2 \int\limits_{C_F} \frac{d^4 p}{(2\pi\hbar)^4} \frac{e^{-\frac{i}{\hbar} p\cdot(x-y)}}{p^2 - m^2 c^2}$$

$$= \int \frac{d^4 p}{(2\pi\hbar)^4} e^{-\frac{i}{\hbar} p\cdot(x-y)} D_F(p), \qquad (11.128)$$

where

$$D_F(p) \equiv \frac{i\hbar^2}{p^2 - m^2 c^2 + i\epsilon}, \qquad (11.129)$$

Equation (11.129) is the expression of the *Feynman propagator in momentum space*.

We now show that $D_F(x - y)$ is a *Green's function* for the Klein-Gordon equation. Let us briefly recall the notion of Green's function for a linear differential equation. Consider the problem of finding the function $f(x^\mu)$ which satisfies the *inhomogeneous* differential equation

$$\mathcal{L}(x) f(x) = g(x), \qquad (11.130)$$

$\mathcal{L}(x)$ being a local differential operator. If there exists a unique solution for each $g(x)$, there must exist an *inverse operator* $\mathcal{L}^{-1}$ such that, formally:

$$f(x) = \mathcal{L}^{-1} g(x).$$

Computing the operator $\mathcal{L}^{-1}$, however, is more than just taking the inverse to $\mathcal{L}$, since it denotes that operation *plus* the *boundary conditions*. By definition the Green's function $G(x, y)$ is the solution of Eq. (11.130) where $g(x) = -i\,\delta^4(x^\mu - y^\mu)$ and corresponds to $\mathcal{L}^{-1}$ *together with* the associated boundary conditions. The solution of the differential equation (11.130) is then given by the formula

$$f(x^\mu) = f_0(x^\mu) + i \int d^4y \, G(x^\mu, y^\mu) g(y^\mu). \tag{11.131}$$

where $f_0(x^\mu)$ is the general solution of the associated homogeneous equation. This is easily verified applying the operator $\mathcal{L}$ to both sides of (11.131).

Let us then consider the Klein-Gordon equation describing the interaction of a classical field $\phi(x)$ with an external source $J(x)$:

$$\left(\Box_x + \frac{m^2c^2}{\hbar^2}\right)\phi(x) = J(x). \tag{11.132}$$

Identifying $\mathcal{L}(x^\mu)$ with the operator $\Box_x + \frac{m^2c^2}{\hbar^2}$ and $g(x^\mu)$ with $J(x^\mu)$, the general solution of Eq. (11.132) can be written as

$$\phi(x) = \phi^0(x) + i \int d^4y \, D(x, y) J(y) \tag{11.133}$$

where $\phi^0(x)$ is the general solution of the homogeneous part of the Klein-Gordon equation $\left(\Box_x + \frac{m^2c^2}{\hbar^2}\right)\phi^0(x) = 0$ while the last term is a particular solution of the inhomogeneous equation. Acting with the Klein-Gordon operator on (11.128) we find

$$\left(\Box + \frac{m^2c^2}{\hbar^2}\right) D_F(x - y) = i\hbar^2 \int \frac{d^4p}{(2\pi\hbar)^4} e^{-\frac{i}{\hbar}p\cdot(x-y)} \frac{1}{\hbar^2}\left(-p^2 + m^2c^2\right) D_F(p)$$

$$= -i\delta^{(4)}(x - y), \tag{11.134}$$

so that $D_F(x - y)$ is the *Green's function* of the Klein-Gordon equation together with the boundary conditions implicit in the choice of the integration contour C_F.

Since the Green's function is not unique we may introduce two further Green's functions corresponding to different boundary conditions, defined as follows:

$$D_R(x - y) = \frac{c}{\hbar}\theta(x^0 - y^0)\left[\hat{\phi}(x), \hat{\phi}^\dagger(y)\right], \tag{11.135}$$

named *retarded Green's function* and

$$D_A(x - y) = \frac{c}{\hbar}\theta(y^0 - x^0)\left[\hat{\phi}(x), \hat{\phi}^\dagger(y)\right],$$

named *advanced Green's function*. Direct computation yields the following expression for the retarded Green's function:

$$
\begin{aligned}
D_R(x - y) &= \frac{c}{\hbar}\theta(x^0 - y^0)\left(\left[\hat{\phi}_+(x), \hat{\phi}_-^\dagger(y)\right] + \left[\hat{\phi}_-(x), \hat{\phi}_+^\dagger(y)\right]\right) \\
&= \theta(x^0 - y^0)\left(D_+(x - y) - D_-(x - y)\right) \\
&= \theta(x^0 - y^0)\hbar c \int \frac{d^3\mathbf{p}}{(2\pi\hbar)^3}\frac{1}{2E_\mathbf{p}}\left(e^{-\frac{i}{\hbar}p(x-y)} - e^{\frac{i}{\hbar}p(x-y)}\right).
\end{aligned}
$$

$$(11.136)$$

Proceeding as in the case of the Feynman propagator we may write $D_R(x - y)$ as the integral to the function $f(p^0, \mathbf{p})$ with a specific prescription

$$
D_R(x - y) = i\hbar^2 \int \frac{d^3 p}{(2\pi\hbar)^3} \int_{C_R} \frac{dp^0}{2\pi\hbar}\frac{e^{-\frac{i}{\hbar}p\cdot(x-y)}}{p^2 - m^2 c^2} = \int \frac{d^3 p}{(2\pi\hbar)^3} \int_{C_R} dp^0 f(p^0, \mathbf{p}),
$$

where C_R is the contour shown in Fig. 11.5. As shown in the figure, if $x^0 > y^0$ the contour should be closed on the lower half plane, giving two residues, which reproduce the expression (11.136), while if $x^0 < y^0$ the contour should be closed in

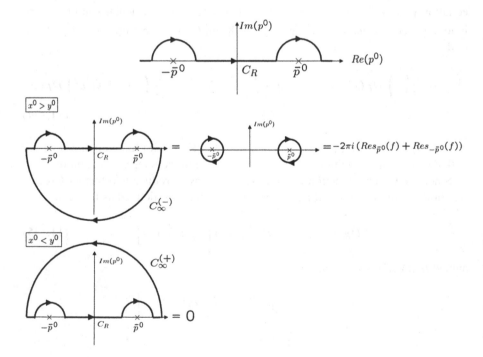

Fig. 11.5 Prescription for the retarded Green's function

Fig. 11.6 Prescription for the advanced Green's function

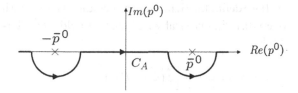

the upper half plane, yielding zero. By the same token one shows that the advanced Green's function $D_A(x - y)$ can be written as

$$D_A(x - y) = i\hbar^2 \int \frac{d^4 p}{(2\pi\hbar)^4} \frac{e^{-\frac{i}{\hbar} p(x-y)}}{p^2 - m^2 c^2},$$

where the p^0 integration now has to be done along the contour C_A of Fig. 11.6.

11.5 Quantization of the Dirac Field

The consistent quantization of the bosonic spin 0 scalar field illustrated in this chapter and of the spin 1 electromagnetic field pursued in Chap. 6 was effected by replacing the classical Poisson brackets between dynamical variables with commutators between *operators*. This quantization procedure leads, among other things, to a positive definite energy and in general to a consistent description of the quantized dynamical variables.

If we now turn to the quantization of the Dirac field which describes spin 1/2 particles and is solution to the Dirac equation, we shall show that the aforementioned prescription *does not work*. Indeed we shall shortly see that in order to have a positive definite field energy at the quantum level we shall be forced to trade the canonical Poisson brackets in the classical theory for *anticommutation rules* among operators.

To show this it is instructive to first pursue the canonical approach which trades the classical Poisson brackets (10.134), (10.135) for *commutation rules* showing that it unavoidably leads to inconsistent results.

Let us start from the classical Poisson brackets of the previous chapter, Sect. 10.5. Replacing ψ and $\pi = i\hbar\psi^\dagger$ with the field operators $\hat{\psi}(x)$ and $\hat{\pi}(x) = i\hbar\hat{\psi}^\dagger$ and implementing the replacement (11.1), we find

$$\left[\hat{\psi}^\alpha(\mathbf{x}, t), \hat{\psi}_\beta^\dagger(\mathbf{y}, t)\right] = \delta^3(\mathbf{x} - \mathbf{y})\delta_\beta^\alpha,$$

$$\left[\hat{\psi}^\alpha(\mathbf{x}, t), \hat{\psi}^\beta(\mathbf{y}, t)\right] = \left[\hat{\psi}_\alpha^\dagger(\mathbf{x}, t), \hat{\psi}_\beta^\dagger(\mathbf{y}, t)\right] = 0,$$

(11.137)

where $\hat{\psi}(\mathbf{x}, t)$ and $\hat{\psi}^\dagger(\mathbf{x}, t)$, being time-dependent operators, are described in the Heisenberg picture.

If we define the Hamiltonian operator by replacing in the expression of the classical one (10.136), classical with quantum fields, the Heisenberg equation obeyed by $\hat{\psi}$ reads:

$$
\begin{aligned}
\dot{\hat{\psi}}(\mathbf{x}, t) &= -\frac{i}{\hbar}\left[\hat{\psi}(\mathbf{x}, t), \hat{H}(t)\right] \\
&= -\frac{i}{\hbar}\int_V d^3\mathbf{y}\left\{\left[\hat{\psi}(\mathbf{x}, t), \hat{\psi}^\dagger(\mathbf{y}, t)\right](-i\hbar c\,\alpha^i\partial_i + mc^2\,\beta)\hat{\psi}(\mathbf{y}, t)\right\} \\
&= -\frac{i}{\hbar}(-i\hbar c\,\alpha^i\partial_i + mc^2\beta)\hat{\psi}(\mathbf{x}, t),
\end{aligned}
\tag{11.138}
$$

implying:

$$
i\hbar\dot{\hat{\psi}}(x) = (-i\hbar c\,\alpha^i\partial_i + mc^2\,\beta)\hat{\psi}(x),
$$

that is

$$
\left(i\hbar\gamma^\mu\partial_\mu - mc\right)\hat{\psi}(x) = 0.
\tag{11.139}
$$

As for the complex scalar field we find that *the field operators satisfy the same equations as their classical counterparts.*

Proceeding as in the classical case we expand the field $\hat{\psi}(x)$ in terms of the positive and negative energy solutions of the classical equation of motion, as in (10.147):

$$
\hat{\psi}(x) = \int\frac{d^3\mathbf{p}}{(2\pi\hbar)^3}\sqrt{\frac{mc^2 V}{E_\mathbf{p}}}\sum_{r=1}^2\left[c(\mathbf{p}, r)u(p, r)e^{-i\frac{p\cdot x}{\hbar}} + d^\dagger(\mathbf{p}, r)v(p, r)e^{i\frac{p\cdot x}{\hbar}}\right],
\tag{11.140}
$$

$$
\hat{\psi}^\dagger(x) = \int\frac{d^3\mathbf{p}}{(2\pi\hbar)^3}\sqrt{\frac{mc^2 V}{E_\mathbf{p}}}\sum_{r=1}^2\left[c^\dagger(\mathbf{p}, r)u^\dagger(p, r)e^{i\frac{p\cdot x}{\hbar}} + d(\mathbf{p}, r)v^\dagger(p, r)e^{-i\frac{p\cdot x}{\hbar}}\right],
\tag{11.141}
$$

where $c(\mathbf{p}, r)$ and $d^\dagger(\mathbf{p}, r)$ are now *operators* and the overall normalization has been chosen in order for them to be dimensionless. Just as we did for the scalar field operator, it is convenient to split $\hat{\psi}(x)$ into its positive and negative energy components, $\hat{\psi}_+(x)$ and $\hat{\psi}_-(x)$ respectively:

$$
\hat{\psi}(x) = \hat{\psi}_+(x) + \hat{\psi}_-(x),
$$

$$
\hat{\psi}_+(x) = \int\frac{d^3\mathbf{p}}{(2\pi\hbar)^3}\sqrt{\frac{mc^2 V}{E_\mathbf{p}}}\sum_{r=1}^2 c(\mathbf{p}, r)u(p, r)e^{-i\frac{p\cdot x}{\hbar}},
$$

$$
\hat{\psi}_-(x) = \int\frac{d^3\mathbf{p}}{(2\pi\hbar)^3}\sqrt{\frac{mc^2 V}{E_\mathbf{p}}}\sum_{r=1}^2 d^\dagger(\mathbf{p}, r)v(p, r)e^{i\frac{p\cdot x}{\hbar}}.
\tag{11.142}
$$

We also define $\hat{\psi}^\dagger_+(x) \equiv (\hat{\psi}_-(x))^\dagger$ and $\hat{\psi}^\dagger_-(x) \equiv (\hat{\psi}_+(x))^\dagger$, as the corresponding components of $\hat{\psi}^\dagger(x)$. Let us now determine the commutation rules obeyed by the $c, d, c^\dagger, d^\dagger$ operators. In order to do this we observe that the expansions (11.140) and (11.141) can be inverted to compute the operator coefficients $c(\mathbf{p}, r)$ and $d(\mathbf{p}, r)$ in terms of $\hat{\psi}(x)$ and $\hat{\bar{\psi}}(x)$. For example we can show that:

$$c(\mathbf{p}, r) = \sqrt{\frac{mc^2}{V E_\mathbf{p}}} \int_V d^3x \bar{u}(\mathbf{p}, r)\gamma^0 \hat{\psi}(x) e^{\frac{i}{\hbar}P\cdot x}. \tag{11.143}$$

Indeed:

$$
\begin{aligned}
c(\mathbf{p}, r) &= \sqrt{\frac{mc^2}{V E_\mathbf{p}}} \int_V d^3x \bar{u}(\mathbf{p}, r)\gamma^0 \hat{\psi}(x) e^{\frac{i}{\hbar}P\cdot x} \\
&= \int_V d^3x \int \frac{d^3q}{(2\pi\hbar)^3} \frac{mc^2}{\sqrt{E_\mathbf{p}E_\mathbf{q}}} \sum_{s=1}^2 \Big[\bar{u}(\mathbf{p}, r)\gamma^0 u(\mathbf{q}, s)c(\mathbf{q}, s)e^{-i\frac{(q-p)\cdot x}{\hbar}} \\
&\quad + \bar{u}(\mathbf{p}, r)\gamma^0 v(\mathbf{q}, s)d^\dagger(\mathbf{q}, s)e^{i\frac{(q+p)\cdot x}{\hbar}} \Big] \\
&= \int \frac{d^3q}{(2\pi\hbar)^3} \frac{mc^2}{\sqrt{E_\mathbf{p}E_\mathbf{q}}} \sum_{s=1}^2 \Big[\bar{u}(\mathbf{p}, r)\gamma^0 u(\mathbf{q}, s)c(\mathbf{q}, s)(2\pi\hbar)^3\delta^3(\mathbf{p} - \mathbf{q}) \\
&\quad + \bar{u}(\mathbf{p}, r)\gamma^0 v(\mathbf{q}, s)d^\dagger(\mathbf{q}, s)(2\pi\hbar)^3\delta^3(\mathbf{p} + \mathbf{q})e^{\frac{2i}{\hbar}E_\mathbf{p}t} \Big] \\
&= \frac{mc^2}{E_\mathbf{p}} \sum_{s=1}^2 \Big[\bar{u}(\mathbf{p}, r)\gamma^0 u(\mathbf{p}, s)c(\mathbf{p}, s) + \bar{u}(\mathbf{p}, r)\gamma^0 v(-\mathbf{p}, s)d^\dagger(-\mathbf{p}, s)e^{\frac{2i}{\hbar}E_\mathbf{p}t} \Big] \\
&= \frac{mc^2}{E_\mathbf{p}} \sum_{s=1}^2 \frac{E_\mathbf{p}}{mc^2}\delta_{rs}c(\mathbf{p}, s) = c(\mathbf{p}, r),
\end{aligned}
$$

where in the last line we used (10.172) and (10.174).

In an analogous way we can compute the operator coefficient $d(\mathbf{p}, r)$:

$$d(\mathbf{p}, r) = \sqrt{\frac{mc^2}{V E_\mathbf{p}}} \int_V d^3x \hat{\bar{\psi}}(x)\gamma^0 v(p, r) e^{\frac{i}{\hbar}P\cdot x},$$

and their hermitian conjugates

$$c^\dagger(\mathbf{p}, r) = \sqrt{\frac{mc^2}{V E_\mathbf{p}}} \int_V d^3x \hat{\bar{\psi}}(x)\gamma^0 u(p, r) e^{-\frac{i}{\hbar}P\cdot x}, \tag{11.144}$$

$$d^\dagger(\mathbf{p}, r) = \sqrt{\frac{mc^2}{VE_\mathbf{p}}} \int_V d^3x \bar{v}(\mathbf{p}, r) \gamma^0 \hat{\psi}(x) e^{-\frac{i}{\hbar} p \cdot x}.$$

It is now possible to compute the commutators among the operators $c, c^\dagger, d, d^\dagger$. We find

$$\left[c(\mathbf{p}, r), c^\dagger(\mathbf{q}, s)\right] = \frac{mc^2}{V\sqrt{E_\mathbf{p} E_\mathbf{q}}} \int_V d^3x \int_V d^3y \, u^\dagger(p, r) \left[\hat{\psi}(\mathbf{x}, t), \hat{\psi}^\dagger(\mathbf{y}, t)\right] u(q, s)$$

$$\times e^{\frac{i}{\hbar}(p \cdot x - q \cdot y)} = \frac{(2\pi\hbar)^3}{V} \delta^3(\mathbf{p} - \mathbf{q}) \delta_{rs}, \tag{11.145}$$

where we used (11.137) and (10.172). Analogous computations also give

$$\left[d(\mathbf{p}, r), d^\dagger(\mathbf{p}, s)\right] = \frac{(2\pi\hbar)^3}{V} \delta^3(\mathbf{p} - \mathbf{q}) \delta_{rs}, \tag{11.146}$$

while all the other commutators are zero

$$[c, c] = \left[c^\dagger, c^\dagger\right] = [d, d] = \left[d^\dagger, d^\dagger\right] = 0.$$

$$[c, d] = \left[c, d^\dagger\right] = 0.$$

Using (10.136) it is now easy to compute the energy $\hat{H}$

$$\hat{H} = i\hbar \int_V d^3x \hat{\psi}^\dagger \partial_t \hat{\psi} = i\hbar \int_V d^3x \int \frac{d^3\mathbf{p}}{(2\pi\hbar)^3} \sqrt{\frac{mc^2 V}{E_\mathbf{p}}} \sum_{r=1}^{2} \left[c^\dagger(\mathbf{p}, r) u^\dagger(\mathbf{p}, r) e^{\frac{ip \cdot x}{\hbar}}\right.$$

$$\left. + d(\mathbf{p}, r) v^\dagger(\mathbf{p}, r) e^{-\frac{ip \cdot x}{\hbar}}\right] \int \frac{d^3\mathbf{q}}{(2\pi\hbar)^3} \sqrt{mc^2 V E_\mathbf{q}} \left(\frac{-i}{\hbar}\right)$$

$$\times \sum_{r=1}^{2} \left[c(\mathbf{q}, s) u(\mathbf{q}, s) e^{-\frac{iq \cdot x}{\hbar}} - d^\dagger(\mathbf{q}, s) v(\mathbf{q}, s) e^{\frac{iq \cdot x}{\hbar}}\right].$$

The terms containing $c^\dagger(\mathbf{p}, r) d^\dagger(\mathbf{q}, s)$ and $d(\mathbf{p}, r) c(\mathbf{q}, s)$ give a vanishing contribution since the integration in $d\mathbf{x}$ implies $\mathbf{q} = -\mathbf{p}$ and the resulting factors are zero in virtue of Eqs. (10.174):

$$u^\dagger(\mathbf{p}, r) v(-\mathbf{p}, s) = 0; \quad v^\dagger(\mathbf{p}, r) u(-\mathbf{p}, s) = 0.$$

Using then (10.172) and (10.173) we finally find:

$$\hat{H} = \int \frac{d^3\mathbf{p}}{(2\pi\hbar)^3} V E_{\mathbf{p}} \left[c^\dagger(\mathbf{p}, r)c(\mathbf{p}, r) - d(\mathbf{p}, r)d^\dagger(\mathbf{p}, r) \right]. \tag{11.147}$$

A similar computation gives the quantum momentum operator as it follows from (10.127):

$$\hat{P}^i = -i\hbar \int_V d^3\mathbf{x} \hat{\psi}^\dagger \partial_i \hat{\psi}$$

$$= \int \frac{d^3\mathbf{p}}{(2\pi\hbar)^3} V\, p^i \left[c^\dagger(\mathbf{p}, r)c(\mathbf{p}, r) - d(\mathbf{p}, r)d^\dagger(\mathbf{p}, r) \right]. \tag{11.148}$$

where a partial integration has been used.

Finally we compute the conserved charge $\hat{Q}$ associated with the four-current $\hat{j}^\mu = e\hat{\bar{\psi}}\gamma^\mu\hat{\psi}$:

$$\hat{Q} = e \int_V d^3\mathbf{x} \hat{\psi}^\dagger(x)\hat{\psi}(x)$$

$$= e \int \frac{d^3\mathbf{p}}{(2\pi\hbar)^3} V \left[c^\dagger(\mathbf{p}, r)c(\mathbf{p}, r) + d(\mathbf{p}, r)d^\dagger(\mathbf{p}, r) \right]. \tag{11.149}$$

Let us now observe that Eq. (11.147) implies that the *energy H can have both positive and negative values, even if we normal order it by dropping the infinite ground state energy*:

$$\hat{H} = \int \frac{d^3\mathbf{p}}{(2\pi\hbar)^3} V E_{\mathbf{p}} \left[c^\dagger(\mathbf{p}, r)c(\mathbf{p}, r) - d^\dagger(\mathbf{p}, r)d(\mathbf{p}, r) \right]. \tag{11.150}$$

On the other hand, *the charge operator $\hat{Q}$ turns out to have the same sign for particles and antiparticles* (it is positive definite if $e > 0$, negative definite is $e < 0$).

Thus the negative energy problem is not solved by field quantization and, moreover, also the charge operator, being positive definite, gives an inconsistent result, owing to the experimental fact that particles and antiparticles (e.g. an electron and a positron) have opposite charges. Such conclusions are of course unacceptable.

To avoid these difficulties we replace the equal-time commutation relations (11.137) of the Dirac field with equal-time *anticommutation relations*, that is we set

$$\left\{ \hat{\psi}^\alpha(\mathbf{x}, t), \hat{\psi}^\dagger_\beta(\mathbf{y}, t) \right\} = \delta^3(\mathbf{x} - \mathbf{y})\delta^\alpha_\beta,$$

$$\left\{ \hat{\psi}^\alpha(\mathbf{x}, t), \hat{\psi}^\beta(\mathbf{y}, t) \right\} = \left\{ \hat{\psi}^\dagger_\alpha(\mathbf{x}, t), \hat{\psi}^\dagger_\beta(\mathbf{y}, t) \right\} = 0. \tag{11.151}$$

Thus for spin 1/2 fields we assume that the correspondence between the classical Poisson brackets is given by

$$\{ \ , \ \}_{P.B.} \rightarrow \ -\frac{i}{\hbar}\{ \ , \ \},$$

where $\{A, B\} = A \cdot B + B \cdot A$ is the anticommutator. The Heisenberg equation obeyed by $\hat{\psi}(\mathbf{x}, t)$ is, however, still written in terms of a *commutator*:

$$i\hbar\dot{\hat{\psi}}(\mathbf{x}, t) = \left[\hat{\psi}(\mathbf{x}, t), \hat{H}(t)\right]. \tag{11.152}$$

Indeed, if we write Eq. (11.152) explicitly

$$i\hbar\dot{\hat{\psi}}(\mathbf{x}, t) = \left[\hat{\psi}(\mathbf{x}, t), \hat{H}(t)\right]$$

$$= \left[\hat{\psi}(x), \int_V d^3\mathbf{y} \, \hat{\psi}^\dagger(\mathbf{y}, t)(-i\hbar c \, \alpha^i \partial_i + \beta mc^2) \, \hat{\psi}(\mathbf{y}, t)\right], \tag{11.153}$$

and use the identity

$$[C, A \, B] = \{A, C\} \, B - A \, \{B, C\},$$

identifying the operators in the last line of (11.153) with C, A, B respectively, we easily retrieve the Dirac equation for $\hat{\psi}(\mathbf{x}, t)$.

Along the same lines which previously led to the derivation of the commutation rules (11.145) and (11.146), we now find the following anticommutation rules among the operators $c, c^\dagger, d, d^\dagger$:

$$\left\{c(\mathbf{p}, r), c^\dagger(\mathbf{q}, s)\right\} = \frac{(2\pi\hbar)^3}{V}\delta^3(\mathbf{p} - \mathbf{q})\delta_{rs} = \left\{d(\mathbf{p}, r), d^\dagger(\mathbf{q}, s)\right\}, \tag{11.154}$$

all other anticommutators being zero. The Hamiltonian is still given by Eq. (11.147) since commutation rules were not used in deriving (11.149). Now, however, we can use the anticommutation relations (11.154) to write

$$d(\mathbf{p}, r)d^\dagger(\mathbf{p}, r) \rightarrow -d^\dagger(\mathbf{p}, r)d(\mathbf{p}, r) + \frac{(2\pi\hbar)^3}{V}\delta^3(\mathbf{0}). \tag{11.155}$$

Subtracting the infinite zero-point energy we find

$$\hat{H} = \int \frac{d^3\mathbf{p}}{(2\pi\hbar)^3}VE_\mathbf{p}\left[c^\dagger(\mathbf{p}, r)c(\mathbf{p}, r) + d^\dagger(\mathbf{p}, r)d(\mathbf{p}, r)\right]. \tag{11.156}$$

The same result is obtained using a normal ordering prescription, that is by requiring that in all physical operators the creation operators must be to the left of the destruction operators. However, since in the present case the operators are *fermionic*, that is they obey *anticommutation rules*, we must take into account an extra minus sign when swapping the position of two of them:

$$: cc^\dagger := -c^\dagger c = - : c^\dagger c :, \quad : dd^\dagger := -d^\dagger d = - : d^\dagger d : . \quad (11.157)$$

Defining all the physical quantities in terms of *normal ordered* products of field operators, we have:

$$\hat{H} = i\hbar \int_V d^3\mathbf{x} : \hat{\psi}^\dagger \partial_t \hat{\psi} := \int \frac{d^3\mathbf{p}}{(2\pi\hbar)^3} V E_\mathbf{p} \sum_{r=1}^{2} (\hat{N}_{\mathbf{p},r}^{(c)} + \hat{N}_{\mathbf{p},r}^{(d)}), \quad (11.158)$$

$$\hat{P}^i = i\hbar \int_V d^3\mathbf{x} : \hat{\psi}^\dagger \partial^i \hat{\psi} := \int \frac{d^3\mathbf{p}}{(2\pi\hbar)^3} V p^i \sum_{r=1}^{2} (\hat{N}_{\mathbf{p},r}^{(c)} + \hat{N}_{\mathbf{p},r}^{(d)}), \quad (11.159)$$

$$\hat{Q} = \int_V d^3\mathbf{x} : \hat{\psi}^\dagger \hat{\psi} := \int \frac{d^3\mathbf{p}}{(2\pi\hbar)^3} V \sum_{r=1}^{2} (\hat{N}_{\mathbf{p},r}^{(c)} - \hat{N}_{\mathbf{p},r}^{(d)}), \quad (11.160)$$

where $\hat{N}_{\mathbf{p},r}^{(c)} = c^\dagger(\mathbf{p}, r)c(\mathbf{p}, r)$ and $\hat{N}_{\mathbf{p},r}^{(d)} = d^\dagger(\mathbf{p}, r)d(\mathbf{p}, r)$.

We see that the adoption of the anticommutation rules (11.151) leads to an Hamiltonian operator which is *positive definite* while the *charge operator may assume both positive and negative values*. In conclusion, much like in the case of the complex scalar field associated with spin 0 particles, and the electromagnetic field associated with the spin 1 photons, we have found that for spin 1/2 particles the quantum field is represented as an infinite collection of two types of quantum harmonic oscillators: For each single particle state $(\mathbf{p}, r)$, there are oscillators of type "c" (associated with the classical positive energy solutions) whose excitations are interpreted as *particles* in the state $(\mathbf{p}, r)$; and oscillators of type "d" (associated with the classical negative energy solutions) whose excitations, for each state $(\mathbf{p}, r)$, are interpreted as *antiparticles* in the same state. The essential difference between the bosonic and the fermionic case is the necessity of using anticommutation rules for the quantization of the latter in order to obtain a sensible theory.

The most important implication of the anticommutation rules for the c and d operators is obtained when we construct the Fock space of states for the fermion field in an analogous way as we did for the scalar field. Indeed multiparticle states are obtained by acting on the vacuum state of the whole system by means of the creation operators $c^\dagger(\mathbf{p}, r)$ (for particles) and $d^\dagger(\mathbf{p}, r)$ (for antiparticles). We can easily convince ourselves that in this construction *two identical particles cannot occupy the same state*. If we indeed try to act *twice* on the vacuum, or on a generic state, with the same creation operator to create two identical particles in a given state, we find zero. This is a consequence of the anticommutation relations (11.151), which imply

$$(c^\dagger)^2 = \frac{1}{2}\{c^\dagger, c^\dagger\} = 0 = (d^\dagger)^2,$$

for each $(\mathbf{p}, r)$. Therefore the states of the system are of the type

$$|0\rangle; \quad |N_{\mathbf{p},r}^{(c)} = 1\rangle = c^\dagger(\mathbf{p}, r)|0\rangle; \quad |N_{\mathbf{p},r}^{(d)} = 1\rangle = d^\dagger(\mathbf{p}, r)|0\rangle.$$

It follows that the particle and antiparticle occupation numbers *for each single particle state* $(\mathbf{p}, r)$ *can only take the values 0 or 1*:

$$N_{\mathbf{p},r}^{(c)} = 0, 1 \quad ; \quad N_{\mathbf{p},r}^{(d)} = 0, 1.$$

This is indeed the content of Pauli's exclusion principle for particles of spin $\frac{1}{2}$ which states that *two spin 1/2 particles cannot exist in a same quantum state* $(\mathbf{p}, r)$.

At the end of Sect. 11.2 we have shown that spin 0 particles obey the Bose-Einstein statistics. We now show that for spin 1/2 particles the Schroedinger wave function is completely antisymmetric under the exchange of particles. Following the same steps as for spin 0 particles, the wave function $\Psi_{N_1..N_k}^{(n)}(\mathbf{x}_1 \dots \mathbf{x}_n, t)$ in the Schroendiger representation is

$$\Psi_{N_1..N_k}^{(n)}(\mathbf{x}_1, \dots, \mathbf{x}_n, t) = \langle \mathbf{x}_1 \dots \mathbf{x}_n; t | N_1, \dots, N_k \rangle, \qquad (11.161)$$

where $N_i = 0, 1$ since $(c_i^\dagger)^2 = 0$. Moreover, since

$$|\mathbf{x}_1, \dots \mathbf{x}_n; t\rangle = \hat{\psi}_-^\dagger(\mathbf{x}_1, t) \cdots \hat{\psi}_-^\dagger(\mathbf{x}_n, t)|0\rangle, \qquad (11.162)$$

and the quantum operators anticommute $\left\{ \hat{\psi}_-^\dagger(\mathbf{x}_i, t), \hat{\psi}_-^\dagger(\mathbf{x}_j, t) \right\} = 0$, the wave function $\Psi^{(n)}$ is antisymmetric in the exchange of $\mathbf{x}_i$ and $\mathbf{x}_j$. We conclude that $\Psi_{N_1...N_k}^{(n)}(\mathbf{x}_1, \dots \mathbf{x}_n; t)$ is *completely antisymmetric in the exchange of the particles positions* $\mathbf{x}_i$, thus implying that spin-1/2 particles obey the Fermi-Dirac statistics.

The connection between spin and statistics or, equivalently, between spin and the type of the commutation relations used for quantization, is one of the most significant predictions of local relativistic quantum field theory. It is specifically a *relativistic effect*, since it can be shown that in the non-relativistic Schroedinger theory, using the Fock-space representation, both quantization procedures based on commutators and anticommutators, give a consistent theory. Therefore such a connection, so essential for explaining the stability of ordinary matter in the non-relativistic domain, *is a consequence of the relativistic formulation*, that is of the principle of relativity expressed by the Lorentz invariance of physical laws.

Let us end this section by commenting on the definition of the single particle (or antiparticle) states. A single spin-1/2 particle state (describing say electron) with momentum $\mathbf{p}$ and spin component $s_z = \varepsilon_r \, \hbar/2$ ($\varepsilon_1 = +1$, $\varepsilon_2 = -1$), is obtained by acting on the vacuum state $|0\rangle$ by means of $c(\mathbf{p}, r)^\dagger$:

$$|\mathbf{p}, r\rangle^{(c)} = c(\mathbf{p}, r)^\dagger |0\rangle. \tag{11.163}$$

The normalization is the usual Lorentz-invariant one:

$$
\begin{aligned}
^{(c)}\langle \mathbf{p}, r | \mathbf{q}, s \rangle^{(c)} &= \langle 0 | c(\mathbf{p}, r) c(\mathbf{q}, s)^\dagger | 0 \rangle = \langle 0 | \{ c(\mathbf{p}, r) c(\mathbf{q}, s)^\dagger \} | 0 \rangle \\
&= \frac{(2\pi\hbar)^3}{V} \delta^3(\mathbf{p} - \mathbf{q}) \delta_{rs}.
\end{aligned}
\tag{11.164}
$$

As for the antiparticle state, recall from our discussion on charge conjugation in Chap. 10, that the component r of $d(\mathbf{p}, r)$ is associated with an antiparticle (say a positron) of opposite spin component $s_z = -\varepsilon_r \, \hbar/2$. Thus if $|\mathbf{p}, r\rangle^{(d)}$ describes an antiparticle with momentum $\mathbf{p}$ and spin component $s_z = \varepsilon_r \, \hbar/2$, we have:

$$|\mathbf{p}, r\rangle^{(d)} = \epsilon_{rs} \, d(\mathbf{p}, s)^\dagger |0\rangle, \tag{11.165}$$

where the effect of ϵ_{sr} is to reverse the spin component.

11.6 Invariant Commutation Rules for the Dirac Field

As for the Klein-Gordon field we now compute the general anticommutation rules for Dirac fields at different times. Using the decomposition (11.142) of the field operator $\hat{\psi}(x)$ into its positive and negative energy components we can write the general anticommutators among Dirac fields as follows

$$\{\hat{\psi}(x), \hat{\bar{\psi}}(y)\} = \{\hat{\psi}_+(x), \hat{\bar{\psi}}_-(x)\} + \{\hat{\psi}_-(x), \hat{\bar{\psi}}_+(y)\}, \tag{11.166}$$

where we have suppressed the spinor indices α, β.

Taking into account Eq. (11.154), the anticommutators on the right hand side give

$$\{\hat{\psi}_+(x), \hat{\bar{\psi}}_-(y)\} = \int \frac{d^3\mathbf{p}}{(2\pi\hbar)^3} \frac{mc^2}{E_\mathbf{p}} \sum_{r=1}^{2} u(\mathbf{p}, r)\bar{u}(\mathbf{p}, r) e^{-\frac{i}{\hbar} p \cdot (x-y)}, \tag{11.167}$$

$$\{\hat{\psi}_-(x), \hat{\bar{\psi}}_+(y)\} = \int \frac{d^3\mathbf{p}}{(2\pi\hbar)^3} \frac{mc^2}{E_\mathbf{p}} \sum_{r=1}^{2} v(\mathbf{p}, r)\bar{v}(\mathbf{p}, r) e^{\frac{i}{\hbar} p \cdot (x-y)}, \tag{11.168}$$

respectively. Using the spin sum given by (10.182) and (10.183) and summing the two results we obtain

$$\{\hat{\psi}(x), \hat{\bar{\psi}}(y)\} = \int \frac{d^3\mathbf{p}}{(2\pi\hbar)^3} \frac{c}{2E_\mathbf{p}} \left((\not{p}+mc)e^{-\frac{i}{\hbar}p\cdot(x-y)} + (\not{p}-mc)e^{\frac{i}{\hbar}p\cdot(x-y)} \right)$$

$$= \int \frac{d^3\mathbf{p}}{(2\pi\hbar)^3} \frac{c}{2E_\mathbf{p}} (i\hbar\not{\partial} + mc) \left[e^{-\frac{i}{\hbar}p\cdot(x-y)} - e^{\frac{i}{\hbar}p\cdot(x-y)} \right].$$

$$(11.169)$$

Comparing (11.167), (11.168) with the definitions (11.108) we find

$$\{\hat{\psi}_\pm(x), \hat{\bar{\psi}}_\mp(y)\} = \pm \left(i\not{\partial} + \frac{mc}{\hbar} \right) D_\pm(x - y) \equiv S_\pm(x - y),   (11.170)$$

so that we may rewrite (11.169) as follows:

$$\{\hat{\psi}(x), \hat{\bar{\psi}}(y)\} = S(x - y) = \left(i\not{\partial} + \frac{mc}{\hbar} \right) D(x - y),   (11.171)$$

where we have defined

$$S(x - y) = S_+(x - y) + S_-(x - y).   (11.172)$$

Note that $S(x - y)$, $S_\pm(x - y)$ all satisfy the Klein-Gordon equation. Moreover, if $(x - y)^2 < 0$, $D(x - y)$ vanishes, and so does its derivatives with respect to x, since if we increase x by an infinitesimal amount dx, $x + dx$ is still at a space-like distance from y and thus the zero value of D is unaffected. We conclude that $S(x - y)$, that is the anticommutator between two spinor-field operators, vanishes at space-like distances $(x - y)^2 < 0$. In Dirac theory the *local observables* are expressed in terms of *fermion bilinears*. It can be verified that bilinears in the Dirac fields satisfy *microcausality*. Indeed

$$[\hat{\bar{\psi}}_\alpha(x)\hat{\psi}^\beta(x), \hat{\bar{\psi}}_\gamma(y)\hat{\psi}^\delta(y)]$$

$$= \hat{\bar{\psi}}_\gamma(y)[\hat{\bar{\psi}}_\alpha(x)\hat{\psi}^\beta(x), \hat{\psi}^\delta(y)] + [\hat{\bar{\psi}}_\alpha(x)\hat{\psi}^\beta(x), \hat{\bar{\psi}}_\gamma(y)]\hat{\psi}^\delta(y)$$

$$= -\hat{\bar{\psi}}_\gamma(y)S^\delta{}_\alpha(y - x)\hat{\psi}^\beta(x) + \hat{\bar{\psi}}_\alpha(x)S^\beta{}_\gamma(x - y)\hat{\psi}^\delta(y).   (11.173)$$

which is zero if $(x - y)^2 < 0$ since $S(x - y)$ is. Therefore the commutators of bilinears for space-like separations is zero, ensuring microcausality for the Dirac theory.

11.6.1 The Feynman Propagator for Fermions

We extend the concept of *time ordering* introduced for bosonic particles in Sect. 11.4.1 to Dirac fermions. For notational convenience we shall, from now on, omit the "hat" symbol ˆ on the field operator whenever there is no possibility of confusing it with the corresponding classical quantity. We define the *Feynman propagator for spin 1/2 fields* as

$$S_F(x - y) = \langle 0|T\psi(x)\overline{\psi}(y)|0\rangle, \tag{11.174}$$

where the *time-ordered product* is

$$T\psi(x)\overline{\psi}(y) = \begin{cases} \psi(x)\overline{\psi}(y) & x^0 > y^0, \\ -\overline{\psi}(y)\psi(x) & y^0 > x^0. \end{cases} \tag{11.175}$$

Note the difference in sign when $y^0 > x^0$ with respect to the bosonic case. If $x^0 > y^0$ we have:

$$S_F(x - y) = \langle 0|(\psi_+(x) + \psi_-(x))(\overline{\psi}_+(y) + \overline{\psi}_-(y))|0\rangle$$
$$= \langle 0|\psi_+(x)\overline{\psi}_-(y)|0\rangle = \langle 0|\{\psi_+(x), \overline{\psi}_-(y)\}|0\rangle = S_+(x - y).$$

Similarly for $x^0 < y^0$ we find

$$S_F(x - y) = -\{\psi_-(x), \overline{\psi}_+(y)\} = -S_-(x - y). \tag{11.176}$$

The Feynman propagator becomes

$$S_F(x - y) = \theta(x^0 - y^0)S_+(x - y) - \theta(y^0 - x^0)S_-(x - y)$$
$$= \frac{1}{\hbar}\Big[\theta(x^0 - y^0)\,(i\hbar\partial\!\!\!/ + mc)\,D_+(x - y)$$
$$+\theta(y^0 - x^0)\,(i\hbar\partial\!\!\!/ + mc)\,D_-(x - y)\Big]. \tag{11.177}$$

Let us now move the θ factors past the differential operator $(i\hbar\partial\!\!\!/ + mc)$. Since the latter contains a derivative with respect to x^0, we will have to write:

$$\theta(x^0 - y^0)\,(i\hbar\partial\!\!\!/ + mc)\,(\cdots) = (i\hbar\partial\!\!\!/ + mc)\,[\theta(x^0 - y^0)(\cdots)]$$
$$-i\hbar\gamma^0\,[\partial_0\theta(x^0 - y^0)](\cdots).$$

Using the property of distributions[13] $\frac{d}{dz}\theta(z) = \delta(z)$, Eq. (11.177) can be recast in the following form:

$$S_F(x - y) = \frac{1}{\hbar} (i\hbar \not\partial + mc) \left[\theta(x^0 - y^0)D_+(x - y) + \theta(y^0 - x^0)D_-(x - y)\right]$$
$$-i\gamma^0\delta(x^0 - y^0)\left[D_+(x - y) - D_-(x - y)\right]. \tag{11.178}$$

The last term vanishes since

$$\delta(x^0 - y^0)\left(D_+(x - y) - D_-(x - y)\right) = \delta(x^0 - y^0)D(x - y) = 0,$$

in virtue of the microcausality condition (11.114). We end up with the following expression for the Feynman propagator:

$$S_F(x - y) = \frac{1}{\hbar} (i\hbar \not\partial + mc) \left[\theta(x^0 - y^0)D_+(x - y) + \theta(y^0 - x^0)D_-(x - y)\right]$$
$$= \frac{1}{\hbar} (i\hbar \not\partial + mc) D_F(x - y), \tag{11.179}$$

where $D_F(x - y)$ is the Feynman propagator for the spinless field, as defined in Eq. (11.125). Using now the definitions (11.128) and (11.129) we obtain the final result

$$S_F(x - y) = \int_{C_F} \frac{dp^4}{(2\pi\hbar)^4} S_F(p) e^{-\frac{i}{\hbar}p\cdot(x - y)}, \tag{11.180}$$

$$S_F(p) = i\hbar \frac{\not p + mc}{p^2 - m^2c^2 + i\epsilon}. \tag{11.181}$$

With an abuse of notation, it is common in the literature to denote the spinorial matrix $S_F(p)$ in (11.181) with the following symbol:

$$S_F(p) = \frac{i\hbar}{\not p - mc + i\epsilon}. \tag{11.182}$$

Formally multiplying both the numerator and denominator by the matrix $\not p + mc$, we retrieve the right hand side of Eq. (11.181).

---

[13]This property is easily proven on a generic test function $f(z)$: $\int_{-\infty}^{\infty} f(z)\frac{d}{dz}\theta(z)dz = -\int_{-\infty}^{\infty} f'(z)\theta(z)dz = -\int_{0}^{\infty} f'(z)dz = f(0) = \int_{-\infty}^{\infty} f(z)\delta(z)dz.$

11.6.2 Transformation Properties of the Dirac Quantum Field

We consider first the transformation properties of the Dirac field operator $\hat{\psi}^\alpha(x)$ under the space-time symmetries of the Poincaré group.[14] Equations (11.161) and (11.162) allow to define the general relation between the relativistic wave function $\psi^\alpha_{(a)}(x)$ describing the state $|a\rangle$ of a spin $1/2$ particle in the coordinate representation (configuration space) and the field operator $\hat{\psi}^\alpha(x)$:

$$\psi^\alpha_{(a)}(x) = \langle 0|\hat{\psi}^\alpha(x)|a\rangle. \tag{11.183}$$

Take, for instance, a single particle state $|a\rangle$ which is described by a wave packet, superposition of monochromatic plane-waves associated with the states $|\mathbf{p}, r\rangle$ (see Chap. 9):

$$|a\rangle = \int \frac{d^3\mathbf{p}}{(2\pi\hbar)^3} V \sum_{r=1}^{2} f(\mathbf{p}, r) |\mathbf{p}, r\rangle = \int \frac{d^3\mathbf{p}}{(2\pi\hbar)^3} V \sum_{r=1}^{2} f(\mathbf{p}, r) c(\mathbf{p}, r)^\dagger |0\rangle$$

$$= \sum_{\mathbf{p}} \sum_{r=1}^{2} f(\mathbf{p}, r) c(\mathbf{p}, r)^\dagger |0\rangle, \tag{11.184}$$

where in the last line we have used the discrete momentum notation. If we substitute in Eq. (11.183) the above expansion and use the expression (11.140) for $\hat{\psi}(x)$, we find (we keep, for the sake of simplicity, the discrete momentum notation):

$$\psi^\alpha_{(a)}(x) = \sum_{\mathbf{p}} \sum_{\mathbf{q}} \sqrt{\frac{mc^2}{E_\mathbf{p} V}} \sum_{r,s=1}^{2} f(\mathbf{q}, r) u(\mathbf{q}, s) \langle 0|c(\mathbf{p}, s) c(\mathbf{q}, r)^\dagger |0\rangle e^{-\frac{i}{\hbar} p\cdot x}$$

$$= \sum_{\mathbf{p}} \sum_{\mathbf{q}} \sqrt{\frac{mc^2}{E_\mathbf{p} V}} \sum_{r,s=1}^{2} f(\mathbf{q}, r) u(\mathbf{q}, s) \langle 0|\{c(\mathbf{p}, s), c(\mathbf{q}, r)^\dagger\}|0\rangle e^{-\frac{i}{\hbar} p\cdot x}$$

$$= \sum_{\mathbf{p}} \sqrt{\frac{mc^2}{E_\mathbf{p} V}} \sum_{r=1}^{2} f(\mathbf{q}, r) u(\mathbf{q}, r) e^{-\frac{i}{\hbar} p\cdot x}, \tag{11.185}$$

where we have used the finite-volume version of the anticommutator in (11.154): $\{c(\mathbf{p}, s), c(\mathbf{q}, r)^\dagger\} = \delta_{\mathbf{p},\mathbf{q}} \delta_{rs}$. We then retrieve for $\psi^\alpha_{(a)}(x)$ the general form of the classical positive energy solution to the Dirac equation.

In Chap. 10 we have written the general transformation property of a Dirac field $\psi(x)$ induced by a space-time symmetry transformation. A generic Poincaré transformation is implemented on the quantum states, as discussed in Chap. 9, by a unitary transformation $U(\Lambda, x_0)$. In light of the relation (11.183) we can write for the field operator $\hat{\psi}(x)$ the following transformation law:

---

[14] In this subsection we restore the "hat" symbol on the field operators.

$$\hat{\psi}^\alpha(x) \xrightarrow{(\Lambda,x_0)} \hat{\psi}'^\alpha(x') = U(\Lambda, x_0)^\dagger \hat{\psi}^\alpha(x')U(\Lambda, x_0) = S(\Lambda)^\alpha{}_\beta \hat{\psi}^\beta(x)$$

$$= O_{(\Lambda,x_0)}\hat{\psi}^\alpha(x'), \qquad (11.186)$$

where, as usual $x' = \Lambda x - x_0$. Using (11.183), it is straightforward to see that Eq. (11.186) implies (10.85). Performing indeed a Poincaré transformation on the state $|a\rangle$ ($|a\rangle \to |a'\rangle = U(\Lambda, x_0)|a\rangle$), the corresponding wave function given by (11.183) transforms as follows

$$\psi^\alpha(x) \xrightarrow{(\Lambda,x_0)} \psi'^\alpha(x') = \langle 0|\hat{\psi}^\alpha(x')U|a\rangle = \langle 0|UU^\dagger\hat{\psi}^\alpha(x')U|a\rangle$$

$$= S(\Lambda)^\alpha{}_\beta \langle 0|\hat{\psi}^\beta(x)|a\rangle = S(\Lambda)^\alpha{}_\beta \psi^\beta(x), \qquad (11.187)$$

where $\psi(x) \equiv \psi_{(a)}(x)$, $\psi'(x) \equiv \psi_{(a')}(x)$, $U \equiv U(\Lambda, x_0)$.

Just as we did for the scalar field in Sect. 11.3, we write the unitary operator $U(\Lambda, \epsilon)$ corresponding to an infinitesimal transformation in terms of its generators $\mathbb{J}^{\mu\nu}$, $\mathbb{P}^\mu$ so that, expanding both $U(\Lambda, \epsilon)$ and $O_{(\Lambda,\epsilon)}$ to first order in the Poincaré parameters $\delta\theta_{\rho\sigma}, \epsilon^\mu \ll 1$, we can express the infinitesimal variation of $\hat{\psi}(x)$ as follows:

$$\delta\hat{\psi}^\alpha(x) = \frac{i}{\hbar}[\hat{\psi}^\alpha(x), \frac{1}{2}\delta\theta_{\rho\sigma}\,\mathbb{J}^{\rho\sigma} - \epsilon \cdot \hat{\mathbb{P}}] = \frac{i}{\hbar}\left(\frac{1}{2}\delta\theta_{\rho\sigma}\,\hat{J}^{\rho\sigma} - \epsilon \cdot \hat{P}\right)\hat{\psi}^\alpha(x),$$

where, as usual, $\hat{J}^{\rho\sigma}$, $\hat{P}^\mu$ are the infinitesimal generators of $O_{(\Lambda,x_0)}$ which implements the Poincaré transformation on the internal components and the functional form of $\hat{\psi}^\alpha(x)$. Using the explicit form of $\hat{J}^{\rho\sigma}$ in (10.100) we deduce from (11.188) the following commutation relations for $\hat{\psi}(x)$:

$$\frac{i}{\hbar}[\hat{\psi}^\alpha(x), \mathbb{J}^{\rho\sigma}] = -\frac{i}{2}(\sigma^{\rho\sigma})^\alpha{}_\beta\,\hat{\psi}^\beta(x) + (x^\rho\partial^\sigma - x^\sigma\partial^\rho)\hat{\psi}^\alpha(x),$$

$$\frac{i}{\hbar}[\hat{\psi}^\alpha(x), \mathbb{P}_\mu] = -\partial_\mu\hat{\psi}^\alpha(x). \qquad (11.188)$$

The above commutators completely define the transformation properties of $\hat{\psi}$ under the action of the Poincaré group. Of course U, as well as its generators $\mathbb{J}^{\mu\nu}$, $\mathbb{P}^\mu$, act on the c and d operators in the expansion of $\hat{\psi}$. Let us define what such an action should be in order to reproduce the correct transformation property (11.186). To this end let us recall that the $u(p, r)$ and $v(p, r)$ spinors transform under a Lorentz transformation as in Eq. (10.149) of Chap. 10, where the matrix $\mathcal{R}(\Lambda, p)^r{}_s$ is a rotation in the spin-group, namely a SU(2) (for massive particles) or an SO(2) (for massless particles) transformation depending on the momentum **p** and the Lorentz transformation itself. Let us show that the transformation law (11.186) is correctly reproduced if:

$$U(\Lambda, x_0)^\dagger c(p, s)U(\Lambda, x_0) = e^{-\frac{i}{\hbar}P \cdot x_0} \mathcal{R}(\Lambda, \Lambda^{-1}p)^s{}_r \, c(\Lambda^{-1}p, r),$$
$$U(\Lambda, x_0)^\dagger d(p, s)U(\Lambda, x_0) = e^{-\frac{i}{\hbar}P \cdot x_0} [\mathcal{R}(\Lambda, \Lambda^{-1}p)^s{}_r]^* \, d(\Lambda^{-1}p, r).$$

$$(11.189)$$

Computing the hermitian conjugate of last equation we find:

$$U(\Lambda, x_0)^\dagger d^\dagger(p, s)U(\Lambda, x_0) = e^{\frac{i}{\hbar}P \cdot x_0} \mathcal{R}(\Lambda, \Lambda^{-1}p)^s{}_r \, d^\dagger(\Lambda^{-1}p, r). \quad (11.190)$$

Applying the above properties, the transformation rule for the spinor field operator reads:

$$U(\Lambda, x_0)^\dagger \hat{\psi}(x')U(\Lambda, x_0)$$

$$= \sum_{\mathbf{p},r} \sqrt{\frac{mc^2}{E_\mathbf{p}V}} \left(U^\dagger c(p, r)U \, u(p, r) \, e^{-\frac{i}{\hbar}p \cdot x'} \right.$$

$$\left. + U^\dagger d^\dagger(p, r)U \, v(p, r) \, e^{\frac{i}{\hbar}p \cdot x'} \right) = \sum_{\mathbf{p},r,s} \sqrt{\frac{mc^2}{E_\mathbf{p}V}} \mathcal{R}(\Lambda, \Lambda^{-1}p)^r{}_s$$

$$\times \left(c(\Lambda^{-1}p, s) \, u(p, r) \, e^{-\frac{i}{\hbar}p \cdot (x'+x_0)} + d^\dagger(\Lambda^{-1}p, s)v(p, r) \, e^{\frac{i}{\hbar}p \cdot (x'+x_0)} \right)$$

$$= \sum_{\mathbf{p}',r,s} \sqrt{\frac{mc^2}{E_{\mathbf{p}'}V'}} \mathcal{R}(\Lambda, p')^r{}_s$$

$$\times \left(c(p', s) \, u(\Lambda p', r) \, e^{-\frac{i}{\hbar}(\Lambda p') \cdot (x'+x_0)} + d^\dagger(p', s)v(\Lambda p', r) \, e^{\frac{i}{\hbar}(\Lambda p') \cdot (x'+x_0)} \right)$$

$$= \sum_{\mathbf{p}',s} \sqrt{\frac{mc^2}{E_{\mathbf{p}'}V'}} \left(c(p', s) \, S(\Lambda)u(p', s) \, e^{-\frac{i}{\hbar}p' \cdot x} + d^\dagger(p', s)S(\Lambda)v(p', s) \, e^{\frac{i}{\hbar}p' \cdot x} \right)$$

$$= S(\Lambda) \sum_{\mathbf{p}',s} \sqrt{\frac{mc^2}{E_{\mathbf{p}'}V'}} \left(c(p', s) \, u(p', s) \, e^{-\frac{i}{\hbar}p' \cdot x} + d^\dagger(p', s)v(p', s) \, e^{\frac{i}{\hbar}p' \cdot x} \right)$$

$$= S(\Lambda)\hat{\psi}(x), \quad (11.191)$$

where we have changed summation variable from p to $p' = \Lambda^{-1}p$ and, as usual, wrote $x = \Lambda^{-1}(x' + x_0)$. We have moreover used the transformation properties (10.149).

11.6.3 Discrete Transformations

Let us now consider the three discrete transformations corresponding to *parity P*, *charge conjugation C* and *time-reversal T* for the Dirac quantum field. In the previous chapter we have seen that for the *classical* Dirac field the space reflection corresponds to the active transformation (see Eq. (10.242)):

$$\psi(\mathbf{x}, t) \rightarrow \psi'(\mathbf{x}, t) = \eta_P \, \gamma^0 \psi(-\mathbf{x}, t), \tag{11.192}$$

with respect to which it is easily verified that the Dirac equation is invariant. For the quantized field we must seek a unitary operator $U(P)$ such that[15]

$$U(P)^\dagger \, \psi(\mathbf{x}, t) \, U(P) = \eta_P \gamma^0 \psi(-\mathbf{x}, t). \tag{11.193}$$

We can define $U(P)$ through its action on the operators $c(\mathbf{p}, r), d^\dagger(\mathbf{p}, r)$, which should reproduce (11.193). Using the properties

$$\gamma^0 u(\mathbf{p}, r) = u(-\mathbf{p}, r), \qquad \gamma^0 v(\mathbf{p}, r) = -v(-\mathbf{p}, r)$$

which can be easily derived from the explicit form of the spinors $u(p, r)$ and $v(p, r)$ given in (10.154) and (10.155), we find the operators c and $d^\dagger$ should transform under parity as follows

$$U(P)^\dagger c(\mathbf{p}, r) U(P) = \eta_P \, c(-\mathbf{p}, r), \tag{11.194}$$

$$U(P)^\dagger d^\dagger(\mathbf{p}, r) U(P) = -\eta_P \, d^\dagger(-\mathbf{p}, r). \tag{11.195}$$

The explicit form of $U(P)$ can be obtained following the same procedure as in the scalar field case. The result is

$$U(P) = e^{\frac{i\pi}{2} \sum_{\mathbf{p},r} [c^\dagger(\mathbf{p},r)c(\mathbf{p},r) - d^\dagger(\mathbf{p},r)d(\mathbf{p},r) - \eta_P \, c^\dagger(\mathbf{p},r)c(-\mathbf{p},r) - \eta_P \, d^\dagger(\mathbf{p},r)d(-\mathbf{p},r)]}.$$

In the case of *charge conjugation* one seeks a unitary operator $U(C)$ such that

$$U(C)^\dagger \, \psi \, U(C) = \eta_C \psi^c, \tag{11.196}$$

where $\psi^c = C\overline{\psi}^T$ is the charge conjugate field defined in Sect. 10.6.2, and the matrix $C = i\gamma^2\gamma^0$ satisfies (10.184). We have seen in sect. 10.6.2 that $\psi \rightarrow \psi^c$ leaves the (free) Dirac equation invariant. Moreover also the anticommutation rules are invariant under the same substitution. Indeed writing the (equal-time) anticommutation relations as

$$\{\psi^\alpha(x), \overline{\psi}(y)_\beta\} = (\gamma^0)^\alpha_{\ \beta} \delta^{(3)}(\mathbf{x} - \mathbf{y}) \tag{11.197}$$

---

[15] We suppress here and in the following the "hat" symbol for the field operator.

multiplying by $C^{\rho\beta}C^{-1}_{\alpha\sigma}$, and contracting over the repeated indices α, β, we obtain

$$\{\psi^{\alpha}(x)C^{-1}_{\alpha\sigma}, \psi^{c}_{\rho}(y)\} = (C^{-T}\gamma^{0}C^{T})_{\sigma}{}^{\rho}\,\delta^{(3)}(\mathbf{x}-\mathbf{y}).$$

We now observe that

$$\overline{\psi^{c}} = -\psi^{T}C^{-1}$$

so that, using the property (10.184) we have

$$\{(\overline{\psi^{c}})_{\sigma}(x), (\psi^{c})^{\rho}(y)\} = (\gamma^{0})^{\rho}{}_{\sigma}\delta^{(3)}(\mathbf{x}-\mathbf{y}) \qquad (11.198)$$

Exchanging $x \longleftrightarrow y$ we prove the invariance.

We observe now that the operator $\psi^{c}(x)$ has the following form:

$$\psi^{c}(x) = \sum_{\mathbf{p},r}\sqrt{\frac{mc^{2}}{E_{\mathbf{p}}V}}\left(c(p,r)^{\dagger}u^{c}(p,r)e^{\frac{i}{\hbar}p\cdot x} + d(p,r)v^{c}(p,r)e^{-\frac{i}{\hbar}p\cdot x}\right).$$

Recalling the following relations (see (10.189)):

$$u^{c}(p,r) = \epsilon_{rs}v(p,s); \quad v^{c}(p,r) = -\epsilon_{rs}\,u(p,s),$$

it is straightforward to prove that the action of $U(C)$ on the c, d operators should be:

$$U(C)^{\dagger}c(p,r)U(C) = \eta_{C}\epsilon_{rs}\,d(p,s); \quad U(C)^{\dagger}d(p,r)U(C) = -\eta_{C}\epsilon_{rs}\,c(p,s).$$

We leave to the reader the exercise of finding the explicit form of the unitary operator $U(C)$.

Let us give here the transformation properties of fermion bilinears. As mentioned in last chapter, all physical quantities associated with the Dirac field, like the conserved current J^{μ}, are expressed in terms of fermion bilinears of the form $\bar{\psi}(x)\Gamma\psi(x)$, where Γ can be $\mathbf{1}$, γ^{μ}, γ^{5}, $\gamma^{5}\gamma^{\mu}$, $\gamma^{\mu\nu}$. In the quantum theory of the free fermion field, physical quantities should be expressed in terms of normal-ordered bilinears : $\bar{\psi}(x)\Gamma\psi(x)$: in the fermion field operator $\psi(x)$. Consider the effect of charge conjugation on a generic fermion bilinear:

$$: \bar{\psi}(x)\Gamma\psi(x) : \xrightarrow{C} : \bar{\psi}^{c}(x)\Gamma\psi^{c}(x) :, \qquad (11.199)$$

where we have used the property $|\eta_{C}|^{2} = 1$. The transformed bilinear can also be written in the following form:

$$: \bar{\psi}^{c}(x)\Gamma\psi^{c}(x) := - : \psi^{T}(x)C^{-1}\Gamma\,C\,\gamma^{0}\psi^{\dagger}(x) :, \qquad (11.200)$$

where we have used the property $\bar{\psi}^c = -\psi^T C^{-1}$ and the † symbol in $\psi^\dagger$ should be intended as the hermitian conjugate of each component ψ^α as a quantum operator, and not as the transposed of the complex conjugate of the spinorial vector (ψ^α).

Now consider the following property of normal ordered products of Dirac field operators:

$$: \psi_\alpha^\dagger(x)\psi^\beta(x) : = - : \psi^\beta(x)\psi_\alpha^\dagger(x) :, \tag{11.201}$$

which can be easily proven by decomposing each field operator into its positive and negative energy components and using the definition (11.157) of normal-ordering. The transformed bilinear can then be recast in the following form:

$$: \bar{\psi}^c(x)\Gamma\psi^c(x) :=: \bar{\psi}(x)\,C\Gamma^T\,C^{-1}\psi(x) :, \tag{11.202}$$

where we have used the properties of γ^0 and of the C-matrix described in last chapter. From Eq. (11.202) we can deduce the transformation properties of the fermion bilinears, which are summarized below:

$$\begin{aligned}
\text{scalar:} \quad & : \bar{\psi}(x)\psi(x) : \xrightarrow{C} : \bar{\psi}(x)\psi(x) : & (11.203)\\
\text{pseudo-scalar:} \quad & : \bar{\psi}(x)\gamma^5\psi(x) : \xrightarrow{C} : \bar{\psi}(x)\gamma^5\psi(x) :, \\
\text{vector:} \quad & : \bar{\psi}(x)\gamma^\mu\psi(x) : \xrightarrow{C} - : \bar{\psi}(x)\gamma^\mu\psi(x) :, \\
\text{pseudo-vector:} \quad & : \bar{\psi}(x)\gamma^5\gamma^\mu\psi(x) : \xrightarrow{C} : \bar{\psi}(x)\gamma^5\gamma^\mu\psi(x) :, \\
\text{antisymmetric tensor:} \quad & : \bar{\psi}(x)\gamma^{\mu\nu}\psi(x) : \xrightarrow{C} - : \bar{\psi}(x)\gamma^{\mu\nu}\psi(x) :,
\end{aligned}$$

where we have used the property $C\gamma^5 C^{-1} = \gamma^5$, so that $C(\gamma^5\gamma^\mu)^T C^{-1} = C(\gamma^\mu)^T\gamma^5 C^{-1} = C(\gamma^\mu)^T C^{-1}\gamma^5 = \gamma^5\gamma^\mu$.

Finally we consider the *time-reversal*. From the discussion given in the Klein-Gordon case we expect that it will be represented by an *antiunitary operator* of the form

$$U(T) = \mathcal{U}\,K,$$

where K is the complex conjugation operator defined in Sect. 11.3.1. The general transformation property of the spinor field operator reads:

$$U(T)^\dagger\psi(\mathbf{x}, t)U(T) = \eta_T\,\mathbf{S}(T)\,\psi(\mathbf{x}, -t), \tag{11.204}$$

where the matrix $\mathbf{S}(T)$ implements the effect of time reversal on the spinor components of $\psi(x)$. Let us determine $\mathbf{S}(T)$.

We multiply the Dirac equation to the left by $U(T)^\dagger$ and to the right by $U(T)$. When these operators pass across the γ^μ-matrices and the i factor, their effect is to complex-conjugate them, being $U(T)$ antilinear. We find

$$\left(i\hbar(\gamma^\mu)^* \partial_\mu + mc\right) U(T)^\dagger \psi(\mathbf{x}, t) U(T) = 0.$$

Using Eq. (11.204) and multiplying the equation to the left by the spinorial matrix $\mathbf{S}(T)^{-1}$, we find:

$$\left(i\hbar \mathbf{S}(T)^{-1}(\gamma^\mu)^* \mathbf{S}(T) \frac{\partial}{\partial x^\mu} + mc\right) \psi(x_T) = 0,$$

where we have defined $x_T \equiv (x_T^\mu) = (-ct, \mathbf{x})$. Taking into account that $\frac{\partial}{\partial x_T^\mu} = -\eta^{\mu\mu} \frac{\partial}{\partial x^\mu}$ (no summation over μ), in order for the above equation to be equivalent to the Dirac equation (though in the time-reversed coordinates):

$$\left(-i\hbar\gamma^\mu \frac{\partial}{\partial x_T^\mu} + mc\right) \psi(x_T) = 0 \iff \left(-i\hbar\gamma^\mu \frac{\partial}{\partial x^\mu} + mc\right) \psi(x) = 0,$$

we must require for the matrix $\mathbf{S}(T)$ the following property:

$$\mathbf{S}(T)^{-1}(\gamma^\mu)^* \mathbf{S}(T) = \eta^{\mu\mu} \gamma^\mu \quad \text{no summation over } \mu. \tag{11.205}$$

The reader can verify that the matrix below satisfies this condition:

$$\mathbf{S}(T) = \gamma^5 C, \tag{11.206}$$

where γ^5 is the matrix introduced in Sect. 10.6.3.

We may compute the effect of time reversal on the four-current $j^\mu = \bar\psi \gamma^\mu \psi$. We have

$$\begin{aligned} U(T)^\dagger j^\mu(\mathbf{x}, t) U(T) &= U(T)^\dagger \bar\psi(x) U(T) (\gamma^\mu)^* U(T)^\dagger \psi(x) U(T) \\ &= \psi^\dagger(x_T) \mathbf{S}(T)^\dagger \gamma^{0*} \gamma^{\mu*} \mathbf{S}(T) \psi(x_T) = \bar\psi(x_T) \mathbf{S}(T)^{-1} \gamma^{\mu*} \mathbf{S}(T) \psi(x_T) \\ &= \eta^{\mu\mu} j^\mu(x_T), \end{aligned} \tag{11.207}$$

where we have used the property $\mathbf{S}(T)^\dagger \gamma^0 = -\gamma^0 \mathbf{S}(T) = \gamma^0 \mathbf{S}(T)^{-1}$, being $\mathbf{S}(T)^{-1} = -\mathbf{S}(T)$. Similarly we can verify that the action of time reversal on the other spinor bilinears reads:

$$\text{scalar:} \quad : \bar\psi(x)\psi(x) : \xrightarrow{T} : \bar\psi(x_T)\psi(x_T) : \tag{11.208}$$

$$\text{pseudo-scalar:} \quad : \bar\psi(x)\gamma^5\psi(x) : \xrightarrow{T} : \bar\psi(x_T)\gamma^5\psi(x_T) :,$$

$$\text{pseudo-vector:} \quad : \bar\psi(x)\gamma^5\gamma^\mu\psi(x) : \xrightarrow{T} \eta^{\mu\mu} : \bar\psi(x_T)\gamma^5\gamma^\mu\psi(x_T) :,$$

$$\text{antisymmetric tensor:} \quad : \bar\psi(x)\gamma^{\mu\nu}\psi(x) : \xrightarrow{T} \eta^{\mu\mu}\eta^{\nu\nu} : \bar\psi(x_T)\gamma^{\mu\nu}\psi(x_T) :,$$

To define the action of $U(T)$ on the c and d operators, we first observe, using (10.154), (10.155) and (11.205) that[16]:

$$\mathbf{S}(T)u(\mathbf{p}, r) = \epsilon_{rs}\, u(-\mathbf{p}, s)^*; \quad \mathbf{S}(T)v(\mathbf{p}, r) = \epsilon_{rs}\, v(-\mathbf{p}, s)^*, \quad (11.209)$$

where the effect of ϵ_{rs} is to flip the spin component, as time reversal should do. From the above relations we may conclude that the action of $U(T)$ on the c and d operators should be:

$$U(T)^\dagger c(\mathbf{p}, r)U(T) = \eta_T\, \epsilon_{sr}\, c(-\mathbf{p}, s); \quad U(T)^\dagger d(\mathbf{p}, r)U(T) = \eta_T\, \epsilon_{sr}\, d(-\mathbf{p}, s).$$

$$(11.210)$$

We leave the proof that (11.210) reproduce (11.204) as well as the explicit construction of the $U(T)$ operator to the reader as an exercise.

11.7 Covariant Quantization of the Electromagnetic Field

In Chap. 6 the quantization of the electromagnetic field was achieved using the *Coulomb (or radiation) gauge*

$$\nabla \cdot A = 0 = A_0.$$

This approach has the advantage that only the physical degrees of freedom of the Maxwell field, namely, for each value of the momentum $\mathbf{p}$, the two polarization states orthogonal to $\mathbf{p}$, are quantized. The Coulomb gauge, however, refers to a particular frame where $A_0(x) = 0$ so that the *Lorentz invariance* of the theory is not manifest: Moving to another RF, the corresponding Lorentz transformation would not in general preserve such condition and would switch on the time-component of the vector potential. For practical calculations it is important to have a covariant formalism. In the classical case this is achieved by choosing the manifestly covariant *Lorentz gauge* $\partial_\mu A^\mu = 0$, but, as we shall see below, this cannot be imposed as an *operatorial equation* when $A_\mu(x)$ is quantized. We must rather impose a suitable version of it on the physical states of the theory.

To understand what concretely goes wrong when we try to naively apply to the electromagnetic field, the quantization procedure that we have followed for the lower spin fields, let us observe that the Maxwell Lagrangian density (8.129) depends on the derivatives of $A_\mu(x)$ only through the field strength $F_{\mu\nu}$, which does not contain

---

[16]To prove it suffices to note that $\mathbf{S}(T)\not{p}\mathbf{S}(T)^{-1} = \gamma^{\mu*}p'_\mu$, where $(p'^\mu) \equiv (p^0, -\mathbf{p})$ and that $\mathbf{S}(T)u(\mathbf{0}, r) = \epsilon_{rs}\, u(\mathbf{0}, s)$.

the time derivative $\dot{A}_0$ of A_0. As a consequence of this, the momentum π^0, conjugate to A_0 is zero:

$$\pi^0 = \frac{\partial \mathcal{L}}{\partial \dot{A}_0} = 0. \tag{11.211}$$

This clearly poses a problem when we try to quantize the system by promoting the field components and their conjugate momenta to operators satisfying the canonical commutation relations. This problem arises from the gauge symmetry associated with the field $A_\mu(x)$, which is telling us that the number of variables we use to describe it exceeds the number of its physical degrees of freedom (which is two).

On the other hand, as pointed out above, the covariant structure of Maxwell's equations and a canonical quantization procedure which only takes into account the independent (physical) degrees of freedom, are incompatible: If we choose, for instance, to promote only the spatial components $\mathbf{A}(x)$ of $A_\mu(x)$, and their conjugate momenta, to operators, leaving $A_0(x)$ to be a classical field, we are breaking the manifest Lorentz covariance of the theory. A possible way out is to modify the Lagrangian of the electromagnetic field so as to manifestly break gauge invariance. In doing so, all the four components of $A_\mu(x)$ become independent degrees of freedom, which can thus be quantized. The alternative Lagrangian density proposed, along these lines, by *Fermi* is obtained by adding to the Maxwell's Lagrangian density (8.129) a term proportional to $(\partial_\mu A^\mu)^2$, so as to obtain:

$$\mathcal{L} = -\frac{1}{4} F_{\mu\nu} F^{\mu\nu} - \frac{1}{2} (\partial_\mu A^\mu)^2 = -\frac{1}{2} \partial_\mu A_\nu \partial^\mu A^\nu + \frac{1}{2} \partial_\mu A_\nu \partial^\nu A^\mu - \frac{1}{2} (\partial_\mu A^\mu)^2. \tag{11.212}$$

Writing

$$-\frac{1}{2} (\partial_\mu A^\mu)^2 = -\frac{1}{2} \partial_\mu A_\nu \partial^\nu A^\mu + \text{(4-divergences)}, \tag{11.213}$$

and neglecting the four-divergences, the Lagrangian density (11.212) reads:

$$\mathcal{L} = -\frac{1}{2} \partial_\mu A_\nu \partial^\mu A^\nu = -\frac{1}{2c^2} \dot{A}_\mu \dot{A}^\mu + \frac{1}{2} \nabla A_\mu \cdot \nabla A^\mu, \tag{11.214}$$

and the corresponding Euler-Lagrange equations of motion are

$$\Box A_\mu = 0. \tag{11.215}$$

We observe that, even if the equations of motion have the same form as Maxwell's equations in the Lorentz gauge, they are by no means equivalent to them since the condition $\partial_\mu A^\mu = 0$ has not been imposed. To recover the description of the field in terms of the two physical degrees of freedom we shall eventually have to impose a constraint on the physical states (see next section).

The conjugate momentum is

$$\pi^\mu = \frac{\partial \mathcal{L}}{\partial \dot{A}_\mu} = -\frac{1}{c^2}\dot{A}^\mu, \tag{11.216}$$

and the classical Hamiltonian of the field turns out to be

$$H(t) = \int d^3\mathbf{x}[\pi^\mu \dot{A}_\mu - \mathcal{L}] = -\int d^3\mathbf{x}\frac{1}{2}[c^2\pi^\mu\pi_\mu + \nabla A_\mu \cdot \nabla A^\mu]$$

$$= \int d^3\mathbf{x}\frac{1}{2c^2}\left[\sum_{i=1}^{3}\left((\dot{A}_i)^2 + c^2|\nabla A_i|^2\right) - \left((\dot{A}_0)^2 + c^2|\nabla A_0|^2\right)\right]. \tag{11.217}$$

We can now promote, according to the general quantization prescription, the components of A_μ and the corresponding conjugate momenta, to operators.[17] The equal time commutation relations will then read:

$$[A_\mu(\mathbf{x}, t), \pi^\nu(\mathbf{y}, t)] = i\hbar\,\delta_\mu^\nu\delta^3(\mathbf{x} - \mathbf{y}), \tag{11.218}$$

$$[A_\mu(\mathbf{x}, t), A_\nu(\mathbf{y}, t)] = [\pi^\mu(\mathbf{x}, t), \pi^\nu(\mathbf{y}, t)] = 0. \tag{11.219}$$

It is also a simple matter to check that the quantum equations of motion

$$\dot{A}_\mu(\mathbf{x}, t) = -\frac{i}{\hbar}[A_\mu(\mathbf{x}, t), H(t)], \tag{11.220}$$

$$\dot{\pi}^\mu(\mathbf{x}, t) = -\frac{i}{\hbar}[\pi^\mu(\mathbf{x}, t), H(t)], \tag{11.221}$$

lead to Eq. (11.215); indeed

$$\dot{A}_\mu(x) = \frac{i}{2\hbar}\left[A_\mu(x), \int d^3\mathbf{y}c^2\pi^\nu(y)\pi_\nu(y)\right] = -c^2\int d^3\mathbf{y}\delta_\mu^\nu\delta^3(\mathbf{x} - \mathbf{y})\pi_\nu(y)$$

$$= -c^2\pi_\mu(x),$$

$$\dot{\pi}^\mu(x) = \frac{i}{\hbar}\int d^3\mathbf{y}\left[\pi^\mu(x), \frac{\partial A_\nu(y)}{\partial y^i}\right]\frac{\partial A^\nu(y)}{\partial y^i}$$

$$= -\frac{i}{\hbar}\int d^3\mathbf{y}\left[\pi^\mu(x), A_\nu(y)\right]\nabla^2 A^\nu(y) = -\int d^3\mathbf{y}\delta^3(\mathbf{x} - \mathbf{y})\nabla^2 A^\mu(y)$$

$$= -\nabla^2 A^\mu(x). \tag{11.222}$$

---

[17]Here and in the remainder of this chapter we denote the quantized fields without the "hat symbol".

Comparing the two commutators we obtain

$$\frac{1}{c^2}\ddot{A}_\mu(x) = \nabla^2 A_\mu(x),$$

which coincides with (11.215).

Let us now expand the field operator in plane waves, as we did in Chap. 5, Eq. (5.123). We recall that, in the present theory, all components of the polarization vector $\epsilon_\mu(\mathbf{k})$ are in principle independent. As far as the classical field $A_\mu(x)$ in Eq. (5.123) is concerned, it is convenient to expand, for each monochromatic wave, the four vector $\epsilon_\mu(\mathbf{k})$ in a basis $\varepsilon_\mu^{(\lambda)}(\mathbf{k})$ of four independent real four-vectors:

$$\epsilon_\mu(\mathbf{k}) = c\sqrt{\frac{\hbar}{2\omega_\mathbf{k} V}} \sum_{\lambda=0}^{3} \varepsilon_\mu^{(\lambda)}(\mathbf{k}) a_\lambda(\mathbf{k}), \qquad (11.223)$$

where the factor in front of the right hand side is introduced in order for the vectors $\varepsilon_\mu^{(\lambda)}$ and the *Fourier coefficients* $a_\lambda(\mathbf{k})$, to be dimensionless. In terms of these quantities, the expansion (5.123) for the classical field reads:

$$A_\mu(x) = c \int \frac{d^3k}{(2\pi)^3} \sqrt{\frac{\hbar V}{2\omega_\mathbf{k}}} \sum_{\lambda=0}^{3} \varepsilon_\mu^{(\lambda)}(\mathbf{k}) \left[a_\lambda(\mathbf{k}) e^{-ik\cdot x} + a_\lambda(\mathbf{k})^* e^{ik\cdot x} \right],$$

$$(11.224)$$

When we consider the quantum field operator, the complex coefficients a and a^* in the above expansion become operators a and $a^\dagger$, so that we can write:

$$A_\mu(x) = c \int \frac{d^3k}{(2\pi)^3} \frac{\sqrt{\hbar V}}{\sqrt{2\omega_\mathbf{k}}} \sum_{\lambda=0}^{3} \varepsilon_\mu^{(\lambda)}(\mathbf{k}) \left[a_\lambda(\mathbf{k}) e^{-ik\cdot x} + a_\lambda^\dagger(\mathbf{k}) e^{ik\cdot x} \right],$$

$$(11.225)$$

$$\pi^\mu(x) = \frac{i}{c} \int \frac{d^3k}{(2\pi)^3} \sqrt{\frac{\hbar V \omega_\mathbf{k}}{2}} \sum_{\lambda=0}^{3} \varepsilon_\mu^{(\lambda)}(\mathbf{k}) \left[a_\lambda(\mathbf{k}) e^{-ik\cdot x} - a_\lambda^\dagger(\mathbf{k}) e^{ik\cdot x} \right].$$

$$(11.226)$$

As mentioned earlier, the polarization vectors $\epsilon_\mu^{(\lambda)}(\mathbf{k})$ are a set of four real unconstrained polarization four-vectors which will be chosen to satisfy, in the four dimensional space-time, the orthonormality condition

$$\varepsilon^{(\lambda)}(\mathbf{k}) \cdot \varepsilon^{(\lambda')}(\mathbf{k}) = \eta^{\lambda\lambda'}, \qquad (11.227)$$

and the completeness condition

$$\sum_{\lambda,\sigma=0}^{3} \varepsilon_{\mu}^{(\lambda)}(\mathbf{k})\varepsilon_{\nu}^{(\sigma)}(\mathbf{k})\,\eta_{\lambda\sigma} = \eta_{\mu\nu}. \qquad (11.228)$$

It is useful to have an explicit expression of the four polarization vectors in a given reference frame. We first observe that in three-dimensional space, for each value of the wave number vector $\mathbf{k}$, we may take as a complete set of orthonormal vectors the transverse polarization vectors $\varepsilon^{(1)}(\mathbf{k})$, $\varepsilon^{(2)}(\mathbf{k})$ and the longitudinal vector $\mathbf{n} = \frac{\mathbf{k}}{|\mathbf{k}|}$ $(\varepsilon^{(3)}(\mathbf{k}) = -\mathbf{n})$ satisfying $\varepsilon^{(r)}(\mathbf{k}) \cdot \varepsilon^{(r')}(\mathbf{k}) = \delta^{rr'}$, $r, r' = 1, 2$ and $\varepsilon^{(r)}(\mathbf{k}) \cdot \mathbf{n} = 0$ together with the completeness relation

$$\sum_{r=1}^{2} \varepsilon_{i}^{(r)}(\mathbf{k})\varepsilon_{j}^{(r)}(\mathbf{k}) + \frac{k_i k_j}{|\mathbf{k}|^2} = \delta_{ij}. \qquad (11.229)$$

This is sufficient for the description of the polarization vectors used in Chap. 6, where we worked in the Coulomb gauge in which $A_0(x) = 0$. However we can formally extend the completeness relation (11.229) to a four-dimensional setting by writing the transverse polarization vectors and the longitudinal vector $\mathbf{n}$ as follows:

$$\varepsilon_{\mu}^{(3)}(\mathbf{k}) = \begin{pmatrix} 0 \\ -\mathbf{n} \end{pmatrix} \quad \varepsilon_{\mu}^{(1,2)}(\mathbf{k}) = \begin{pmatrix} 0 \\ \varepsilon^{(1,2)}(\mathbf{k}) \end{pmatrix}. \qquad (11.230)$$

Of course the above vectors refer to a particular RF $\mathcal{S}_0$ in which the time component is zero. A Lorentz transformation will in general alter this property. In a arbitrary frame the longitudinal vector $\varepsilon_{\mu}^{(3)}(\mathbf{k})$ can be written as

$$\varepsilon_{\mu}^{(3)}(\mathbf{k}) = \frac{k_\mu - \eta_\mu(k \cdot \eta)}{(k \cdot \eta)} \qquad (11.231)$$

where $k_\mu = (k_0, -\mathbf{k})^T$, $k_0 = |\mathbf{k}|$ and we have introduced a time-like vector η_μ with unit norm $(\eta^\mu \eta_\mu = 1)$ which, in the RF $\mathcal{S}_0$, has the form

$$\eta_\mu = (1, \mathbf{0})^T, \qquad (11.232)$$

so that in this frame the expression (11.231) for $\varepsilon_{\mu}^{(3)}(\mathbf{k})$ reduces to the form (11.230). To obtain a complete set of orthonormal four-dimensional vectors in Minkowski space, we further add the time-like polarization vector

$$\varepsilon_{\mu}^{(0)}(\mathbf{k}) = \eta_\mu. \qquad (11.233)$$

It is now easily verified that the given four vectors $\varepsilon_\mu^{(\lambda)}(\mathbf{k})$, $\lambda = 0, 1, 2, 3$ satisfy the two conditions (11.227) and (11.228) corresponding to the orthonormality and completeness relations in Minkowski space.

Aside from the presence of the polarization vectors, the expansion (11.225) is quite analogous, for each of the four values of μ, to the expansion of four real scalar fields and their conjugate momenta as given in (11.19) and (11.21) (the reality of A_μ being expressed by the relation $b^\dagger = a^\dagger$). Therefore, along the same lines as in Sect. 11.2 (see the discussion from (11.23) to (11.32)) we can compute a_λ and $a_\lambda^\dagger$ in terms of $A_\mu(x)$ ad $\pi^\mu(x)$ by inverting (11.226) and (11.225). Applying the canonical commutation relations (11.218), we find for a_λ and $a_\lambda^\dagger$ the following relations

$$\left[a_\lambda(\mathbf{k}), a_\sigma^\dagger(\mathbf{k}') \right] = -\eta_{\lambda\sigma} \frac{(2\pi)^3}{V} \delta^3(\mathbf{k} - \mathbf{k}'), \tag{11.234}$$

$$[a, a] = \left[a^\dagger, a^\dagger \right] = 0. \tag{11.235}$$

The computation of the invariant commutation rules follows the same lines as in Sect. 11.4; indeed from (11.215) one finds

$$[A_\mu(x), A_\nu(y)] = -\hbar c \left(\sum_{\lambda,\lambda'=0}^{3} c_\mu^{(\lambda)}(\mathbf{k}) \epsilon_\nu^{(\lambda')}(\mathbf{k}) \eta_{\lambda\lambda'} \right) D(x - y), \tag{11.236}$$

where $D(x - y)$ is the function of Eq. (11.110) where we set $m = 0$. If we assume the polarization vectors to satisfy the completeness relation (11.228), we end up with

$$[A_\mu(x), A_\nu(y)] = -\hbar c \, \eta_{\mu\nu} D(x - y). \tag{11.237}$$

Similarly, following the steps of Sect. 11.4.1, one can evaluate the Feynman propagator

$$
\begin{aligned}
D_{F\mu\nu}(x - y) &= \frac{1}{c\hbar} \langle 0| T A_\mu(x) A_\nu(y) |0 \rangle \\
&= \frac{1}{c\hbar} \langle 0| \left[\theta(x^0 - y^0) A_\mu(x) A_\nu(y) + \theta(y^0 - x^0) A_\nu(y) A_\mu(x) \right] |0 \rangle \\
&= -\eta_{\mu\nu} D_F(x - y) = \int_{C_F} \frac{d^4 p}{(2\pi\hbar)^4} \frac{-i\hbar^2 \eta_{\mu\nu}}{p^2 + i\epsilon} e^{-i\frac{p}{\hbar}\cdot(x-y)}
\end{aligned}
\tag{11.238}
$$

where $D_F(x - y)$ is the Feynman propagator for the scalar field, computed in Sect. 11.4.1, C_F is the contour defined in Fig. 11.1, and $p^\mu = \hbar k^\mu$ is the photon four-momentum. The Feynman propagator in momentum space is therefore

$$D_{F\mu\nu}(p) = -\frac{i\hbar^2}{p^2 + i\epsilon} \eta_{\mu\nu}. \tag{11.239}$$

Note that, just as the Feynman propagator for the scalar and spinor fields, $D_{F\mu\nu}(x-y)$ is the Green's function associated with the equation of motion for $A_\mu(x)$, namely it satisfies the equation

$$\Box_x D_{F\mu\nu}(x - y) = i\eta_{\mu\nu}\,\delta^4(x - y). \tag{11.240}$$

This can be easily verified by taking the four-dimensional Fourier transform of both sides and using the fact that the Fourier transform of $\delta^4(x - y)$ is 1.

Let us mention that we could have broken the gauge invariance of the Maxwell Lagrangian by adding a term proportional to $\partial_\mu A^\mu$ with a generic coefficient, thus obtaining:

$$\begin{aligned}
\mathcal{L} &= -\frac{1}{4}\,F_{\mu\nu}F^{\mu\nu} - \frac{1}{2\alpha}\,(\partial_\mu A^\mu)^2 \\
&= -\frac{1}{2}\,\partial_\mu A_\nu \partial^\mu A^\nu + \frac{1}{2}\left(1 - \frac{1}{\alpha}\right)(\partial_\mu A^\mu)^2,
\end{aligned} \tag{11.241}$$

where we have neglected additional four-divergences and α is a generic number, which we have previously fixed to 1. Imposing the Lorentz gauge $\partial_\mu A^\mu = 0$ the above Lagrangian is equivalent to the original one. This time, however, the conjugate momenta read:

$$\pi^\mu(x) = -\frac{1}{c^2}\,\dot{A}^\mu + \frac{\eta^{\mu 0}}{c}\left(1 - \frac{1}{\alpha}\right)(\partial_\nu A^\nu). \tag{11.242}$$

The equations of motion now have the following form:

$$\left[\Box\,\delta_\mu^\rho - \left(1 - \frac{1}{\alpha}\right)\partial_\mu \partial^\rho\right] A_\rho(x) = 0. \tag{11.243}$$

The equation for the Feynman propagator changes accordingly, from (11.240) to:

$$\left[\Box\,\delta_\mu^\rho - \left(1 - \frac{1}{\alpha}\right)\partial_\mu \partial^\rho\right] D_{F\rho\nu}(x - y) = i\eta_{\mu\nu}\,\delta^4(x - y). \tag{11.244}$$

Going to the momentum representation by evaluating the Fourier transform of both sides, we find:

$$\left[p^2\,\delta_\mu^\rho - \left(1 - \frac{1}{\alpha}\right)p_\mu p^\rho\right] D_{F\rho\nu}(p) = -i\hbar^2 \eta_{\mu\nu}, \tag{11.245}$$

which is now solved by:

$$D_{F\mu\nu}(p) = -\frac{i\hbar^2}{p^2 + i\epsilon}\left(\eta_{\mu\nu} - (1 - \alpha)\frac{p_\mu p_\nu}{p^2}\right), \tag{11.246}$$

as the reader can easily verify.

11.7.1 Indefinite Metric and Subsidiary Conditions

As it stands this theory exhibits the embarrassing property that the Hilbert space contains states with *negative norm*, which is the price we have to pay for preserving manifest Lorentz covariance at the quantum level. To see this it is sufficient to observe that the commutation relations (11.234) are the usual ones only if $\lambda = \lambda' = 1, 2, 3$. When $\lambda = \lambda' = 0$, however, we have

$$\left[a_0(\mathbf{k}), a_0^\dagger(\mathbf{k}') \right] = -(2\pi)^3 \delta^{(3)}(\mathbf{k} - \mathbf{k}') \frac{1}{V}.$$

For the sake of simplicity, let us switch once again to the discrete momentum notation, pertaining to a finite volume V, and treat correspondingly the discrete variable $\mathbf{k}$ as an index, defining: $a_{\mathbf{k}\mu} \equiv a_\mu(\mathbf{k})$, $\varepsilon_{\mathbf{k}\mu}^{(\lambda)} \equiv \varepsilon_\mu^{(\lambda)}(\mathbf{k})$. We can then write

$$\left[a_{\mathbf{k}0}, a_{\mathbf{k}'0}^\dagger \right] = -\delta_{\mathbf{k}\mathbf{k}'}.$$

The appearance of the minus sign is of course related to the indefinite character of the Lorentz metric.

To see why negative norm states appear, let us compute the norm of the state $a_{\mathbf{k}0}^\dagger|0\rangle$:

$$\langle 0|a_{\mathbf{k}0} a_{\mathbf{k}0}^\dagger|0\rangle = \langle 0| \left[a_{\mathbf{k}0}, a_{\mathbf{k}0}^\dagger \right]|0\rangle = -1,$$

or more generally for a state containing $N_{\mathbf{k}}^{(0)}$ "timelike" (i.e. excitations of the $\mu = 0$ quantum oscillator) photons

$$\langle N_{\mathbf{k}}^{(0)}|N_{\mathbf{k}}^{(0)}\rangle = (-1)^{N_{\mathbf{k}}^{(0)}}. \tag{11.247}$$

Negative norm states are clearly unacceptable on physical grounds since they would lead to negative probabilities. Furthermore their existence implies that the expectation value of the quantum Hamiltonian can be negative. To show this we first observe that quantization of the Hamiltonian (11.217) can be computed exactly as in the case of the (real) scalar field leading to

$$\begin{aligned}
\hat{H} &= \sum_{\mathbf{k}} \hbar \omega_{\mathbf{k}} \left(\sum_{\lambda=1}^{3} a_{\mathbf{k}\lambda}^\dagger a_{\mathbf{k}\lambda} - a_{\mathbf{k}0}^\dagger a_{\mathbf{k}0} \right) \\
&= \sum_{\mathbf{k}} \hbar \omega_{\mathbf{k}} \, a_{\mathbf{k}\lambda}^\dagger a_{\mathbf{k}\sigma} \eta^{\lambda\sigma},
\end{aligned} \tag{11.248}$$

where, as usual, we have discarded the infinite constant corresponding to the zero point energy by the normal-ordering prescription.

We note that the minus sign appearing in the Hamiltonian does not imply a negative contribution to the energy. In fact the number operator for the time-like photons reads

$$N_{\mathbf{k}}^{(0)} = -a_{\mathbf{k}\,0}^{\dagger} a_{\mathbf{k}\,0},$$

since

$$N_{\mathbf{k}}^{(0)} a_{\mathbf{k}\,0}^{\dagger}|0\rangle = -a_{\mathbf{k}\,0}^{\dagger}[a_{\mathbf{k}\,0}, a_{\mathbf{k}\,0}^{\dagger}]|0\rangle = 1 \times a_{\mathbf{k}\,0}^{\dagger}|0\rangle,$$
$$N_{\mathbf{k}}^{(0)} (a_{\mathbf{k}\,0}^{\dagger})^{2}|0\rangle = 2 \times (a_{\mathbf{k}\,0}^{\dagger})^{2}|0\rangle, \quad etc.$$

The *expectation value* of the Hamiltonian, however, can be *negative*, since, using (11.247),

$$\langle N_{\mathbf{k}}^{(0)}|H|N_{\mathbf{k}}^{(0)}\rangle = N_{\mathbf{k}}^{(0)} \hbar \omega_{\mathbf{k}} (-1)^{N_{\mathbf{k}}^{(0)}}.$$

It is important to stress that what we have quantized so far *is not* the Maxwell theory, but rather a theory based on the Lagrangian (11.214). In the classical theory the additional degrees of freedom related to the longitudinal and timelike components were eliminated by choosing the Lorentz gauge $\partial_{\mu} A^{\mu} = 0$. We may try to implement the Lorentz gauge as an operatorial constraint on the state vectors $|s\rangle$ of the Hilbert space, through the condition

$$\partial_{\mu} A^{\mu}|s\rangle = 0. \tag{11.249}$$

This is however too strong a condition, and indeed it is clearly not satisfied by the vacuum state $|0\rangle$. Indeed, decomposing the quantum field A_{μ} into positive and negative frequency parts, we obtain

$$\partial^{\mu} A_{\mu}|0\rangle = \partial^{\mu} A_{-\mu}(x)|0\rangle = 0, \tag{11.250}$$

from which it follows

$$0 = A_{+\nu}(y) \frac{\partial}{\partial x^{\mu}} A_{-\mu}(x)|0\rangle$$
$$= \frac{\partial}{\partial x^{\mu}} \left[A_{+\nu}(y), A_{-\mu}(x) \right] |0\rangle = -\hbar c\, \eta_{\mu\nu} \frac{\partial}{\partial x^{\mu}} D_{+}(y - x)|0\rangle,$$

and since the last right hand side is not zero, condition (11.250) is inconsistent. The consistent formulation of the quantum Lorentz-gauge condition, called *subsidiary condition*, is due to Gupta and Bleuler and is given by the less stringent requirement:

$$\partial^{\mu} A_{+\mu}|s\rangle = 0, \tag{11.251}$$

which is obviously satisfied by the vacuum state. The above condition can be interpreted as defining a *physical state*. Moreover (11.251) also implies that the classical Lorentz-gauge condition is satisfied in terms of its expectation value between physical states $|s\rangle$. Indeed

$$\langle s|\partial^\mu A_\mu|s\rangle = \langle s|\partial^\mu \left(A_{-\mu} + A_{+\mu}\right)|s\rangle = \langle s|\partial^\mu A_{+\mu}|s\rangle^* + \langle s|\partial^\mu A_{+\mu}|s\rangle = 0.$$

where we have used the relation $A_{+\mu} = (A_{-\mu})^\dagger$ and the general property that $\langle s|\hat{O}^\dagger|s\rangle = \langle s|\hat{O}|s\rangle^*$, which holds for any operator $\hat{O}$.

Let us now express the subsidiary condition in terms of the operators $a^\dagger_{\mathbf{k}\,0}, a_{\mathbf{k}\,0}$, by defining the operator $L(\mathbf{k})$ as follows:

$$\partial_\mu A^\mu_+(x) = -i \sum_{\mathbf{k}} c\sqrt{\frac{\hbar}{2\omega_{\mathbf{k}} V}} L(\mathbf{k}) e^{-i\,k\cdot x} \Rightarrow L(\mathbf{k}) = k^\mu \sum_{\lambda=0}^{3} \varepsilon^{(\lambda)}_{\mathbf{k}\,\mu} a_{\mathbf{k}\,\lambda},$$

$$(11.252)$$

where we have used the expansion (11.225). Equation (11.251) can be recast in the following form

$$L(\mathbf{k})|s\rangle = 0, \quad \forall\mathbf{k}. \tag{11.253}$$

In the frame in which $k = (\kappa, 0, 0, \kappa)$ ($k^0 = k^3 = \kappa$), using as polarization vectors

$$\varepsilon^{(0)}_{\mathbf{k}\mu} = (1, 0, 0, 0); \; \varepsilon^{(1)}_{\mathbf{k}\mu} = (0, -1, 0, 0); \; \varepsilon^{(2)}_{\mathbf{k}\mu} = (0, 0, -1, 0); \; \varepsilon^{(3)}_{\mathbf{k}\mu} = (0, 0, 0, -1),$$

$L(\mathbf{k})$ becomes

$$L(\mathbf{k}) = \kappa\,(a_{\mathbf{k}\,0} - a_{\mathbf{k}\,3}), \tag{11.254}$$

Moreover from (11.253) we also find

$$\langle s|a^\dagger_{\mathbf{k}\,0} a_{\mathbf{k}\,0}|s\rangle - \langle s|a^\dagger_{\mathbf{k}\,3} a_{\mathbf{k}\,3}|s\rangle = 0, \tag{11.255}$$

that is the occupation numbers associated with the timelike and longitudinal photons coincide on a physical state, so that the total contribution from these excitations to the expectation value of the quantum Hamiltonian is zero. Thus the subsidiary condition ensures that only the physical degrees of freedom of the electromagnetic field contribute to the (expectation value of) energy. The same can be shown to hold for the four-momentum P^μ of the field.

Let us finally show that the subsidiary condition (11.253) eliminates all the negative norm states. We first prove that the operator $L^\dagger$ commutes with L[18]:

$$\left[L(\mathbf{k}), L^\dagger(\mathbf{k'}) \right] = \kappa^2 \left[a_{\mathbf{k}0} - a_{\mathbf{k}3}, a_{\mathbf{k}0}^\dagger - a_{\mathbf{k}3}^\dagger \right]$$

$$= \kappa^2 \left[a_{\mathbf{k}0}, a_{\mathbf{k}0}^\dagger \right] + \kappa^2 \left[a_{\mathbf{k}3}, a_{\mathbf{k}3}^\dagger \right] = 0. \qquad (11.257)$$

The action of the operator $L^\dagger(\mathbf{k}) = \kappa \, (a_{\mathbf{k}0}^\dagger - a_{\mathbf{k}3}^\dagger)$ on a state generates a particular admixture of timelike and longitudinal photons which we shall call *pseudophoton*. Clearly a state containing only transverse photons satisfies the subsidiary condition (11.253) and thus is physical. Given a physical state $|s\rangle$, any other state obtained acting on it any number of times by $L^\dagger$ is still physical. This is easily shown using (11.257):

$$L \, (L^\dagger)^k |s\rangle = (L^\dagger)^k \, L \, |s\rangle = 0. \qquad (11.258)$$

This is not the case if we act on a physical state, setting $a_{\mathbf{k}0} = a_0$ and $a_{\mathbf{k}3} = a_3$, by a combination of $a_0^\dagger$ and $a_3^\dagger$ which is different from $L^\dagger$, say $a_0^\dagger + a_3^\dagger$. The resulting state would not be physical since:

$$L(\mathbf{k})(a_0^\dagger + a_3^\dagger)|s\rangle = [L(\mathbf{k}), \, (a_0^\dagger + a_3^\dagger)]|s\rangle = -2\kappa \, |s\rangle \neq 0. \qquad (11.259)$$

Thus *pseudophotons are the only combinations of longitudinal and timelike photons allowed in a physical state*. Let us show that a physical state containing at least one pseudophoton is perpendicular to any other state (including itself) satisfying (11.253) and thus has zero norm. Consider a physical state $(L^\dagger)^k |s\rangle$ containing a number of pseudophotons created by $(L^\dagger)^k$, and let $|s'\rangle$ be another physical state, we have:

$$\langle s'|(L^\dagger)^k|s\rangle = (\langle s'|L^\dagger)(L^\dagger)^{k-1}|s\rangle = 0, \qquad (11.260)$$

by virtue of (the hermitian conjugate of) condition (11.253): $L|s'\rangle = 0$. We conclude that states containing at least one pseudophoton have zero norm and are orthogonal to any other physical state. Adding to a physical state an other one containing at least one pseudophoton will not alter its physical content. In fact it corresponds to a gauge transformation, see below.

---

[18]It is straightforward to prove this property in a generic RF, using the general expression of $L(\mathbf{k})$ in Eq. (11.252)

$$[L(\mathbf{k}), L(\mathbf{k'})^\dagger] = k^\mu k'^\nu \varepsilon_{\mathbf{k},\mu}^{(\lambda)} \varepsilon_{\mathbf{k'},\nu}^{(\sigma)} [a_{\mathbf{k}\lambda}, \, a_{\mathbf{k'}\sigma}^\dagger]$$

$$= -k^\mu k'^\nu \varepsilon_{\mathbf{k},\mu}^{(\lambda)} \varepsilon_{\mathbf{k'},\nu}^{(\sigma)} \eta_{\lambda\sigma} \, \delta_{\mathbf{k}\mathbf{k'}} = -k^2 \, \delta_{\mathbf{k}\mathbf{k'}} = 0, \qquad (11.256)$$

where we have used the photon mass-shell condition $k^2 = 0$ and the completeness property of the polarization vectors.

We have thus far learned that a state satisfying the subsidiary condition can only contain pseudophotons besides the transverse ones, and thus we can convince ourselves that the most general physical state $|s_{ph}\rangle$ is constructed by adding to a state $|s_0\rangle$ containing just *transverse photons* (and thus no pseudophotons) other ones containing any number of pseudophotons besides the transverse ones:

$$|s_{ph}\rangle \equiv |s_0\rangle + \sum_k f_0(\mathbf{k}) L^\dagger(\mathbf{k})|s_0'\rangle + \sum_{\mathbf{k}_1,\mathbf{k}_2} f_1(\mathbf{k}_1) L^\dagger(\mathbf{k}_1) f_2(\mathbf{k}_2) L^\dagger(\mathbf{k}_2)|s_0''\rangle + \dots,$$

(11.261)

where $|s_0\rangle$, $|s_0'\rangle$, $|s_0''\rangle$ are states containing transverse photons only. The second state on the right hand side contains one pseudophoton, the third two and so on. We can easily show that the terms in (11.261) containing pseudophotons do not affect scalar products between physical states of the form (11.261), being them orthogonal to any state satisfying the subsidiary condition, including themselves. Take indeed an other state $|\bar{s}_{ph}\rangle = |\bar{s}_0\rangle + \dots$ of the form (11.261), using (11.253) and (11.257) we find:

$$\langle \bar{s}_{ph}|s_{ph}\rangle = \langle \bar{s}_0|s_0\rangle.$$

(11.262)

This important result states that the allowed admixtures of timelike and longitudinal photons (pseudophotons) do not affect the scalar products and in particular the *norm* of the states. Thus *all the physical state vectors have positive norm.* As pointed out earlier, *the states $|s_{ph}\rangle$ and $|s_0\rangle$ are physically equivalent.* Mathematically the difference between them corresponds to the gauge freedom of the classical theory. To see this explicitly, let us write, just as we did for the scalar and Dirac field, the relation between the classical potential $A_\mu(x)$ and its quantum counterpart $\hat{A}_\mu(x)$ (we temporarily restore the "hat" symbol on the field operator). If $|s\rangle$ is a physical state describing a single photon, we can describe it in configuration space through the classical potential

$$A_\mu(x) = \langle 0|\hat{A}_\mu(x)|s\rangle.$$

(11.263)

Let us now consider the physically equivalent state, obtained by adding to $|s\rangle$ a single-pseudophoton state:

$$|s'\rangle = |s\rangle + \sum_k i f(\mathbf{k}) L(\mathbf{k})^\dagger |0\rangle.$$

(11.264)

Let us show that the classical field $A_\mu(x)$ changes accordingly by a gauge transformation:

$$A_\mu'(x) = \langle 0|\hat{A}_\mu(x)|s'\rangle = A_\mu(x) + \delta A_\mu(x),$$

$$\delta A_\mu(x) = \sum_{\mathbf{k},\mathbf{k}'} c \sqrt{\frac{\hbar}{2\omega_{\mathbf{k}'} V}} i f(\mathbf{k}) \varepsilon_{\mathbf{k}'\mu}^{(\lambda)} \langle 0|a_{\mathbf{k}'\lambda}, L(\mathbf{k})^\dagger|0\rangle e^{-ik'\cdot x}$$

$$= \sum_{\mathbf{k},\mathbf{k'}} c \sqrt{\frac{\hbar}{2\omega_{\mathbf{k'}} V}} if(\mathbf{k}) \varepsilon^{(\lambda)}_{\mathbf{k'}\mu} k^{\nu} \varepsilon^{(\sigma)}_{\mathbf{k}\nu} \langle 0|[a_{\mathbf{k'}\lambda}, a^{\dagger}_{\mathbf{k}\sigma}]|0\rangle e^{-ik'\cdot x}$$

$$= \sum_{\mathbf{k}} c \sqrt{\frac{\hbar}{2\omega_{\mathbf{k}} V}} (-ik_{\mu}) f(\mathbf{k}) e^{-ik\cdot x} = \partial_{\mu}\Lambda(x), \qquad (11.265)$$

where

$$\Lambda(x) \equiv \sum_{\mathbf{k}} c \sqrt{\frac{\hbar}{2\omega_{\mathbf{k}} V}} f(\mathbf{k}) e^{-ik\cdot x}, \qquad (11.266)$$

where we have used (11.228).

Thus adding to $|s\rangle$ a single pseudophoton state amounts to a gauge transformation on the corresponding classical field.

A physical state is therefore more appropriately described in terms of a *class* or a set of vectors which differ from one another by zero norm states (i.e. containing pseudophotons). Different choices of vectors within a same class differ by a gauge transformation and thus define the same physical object. In a consistent quantum theory of the electromagnetic field, measurable quantities should be gauge invariant. The expectation value of an observable $\hat{O}$ on a physical state should not therefore depend on the choice of vectors within the corresponding class. Take two physical states $|s_{ph}\rangle = |s_0\rangle + \ldots$ and $|\bar{s}_{ph}\rangle = |\bar{s}_0\rangle + \ldots$ of the form (11.261), and therefore in the same classes as $|s_0\rangle$ and $|\bar{s}_0\rangle$, respectively. Gauge invariance requires the following condition on the matrix element of any observable $\hat{O}$ between these two states:

$$\langle \bar{s}_{ph}|\hat{O}|s_{ph}\rangle = \langle \bar{s}_0|\hat{O}|s_0\rangle. \qquad (11.267)$$

A sufficient condition for this to hold is that the operator $\hat{O}$ commute with L and $L^{\dagger}$:

$$[\hat{O}, L] = [\hat{O}, L^{\dagger}] = 0. \qquad (11.268)$$

The above condition indeed allows us to move, in the terms containing L and $L^{\dagger}$, the $L^{\dagger}$ operators to the left and the L ones to the right, past $\hat{O}$, hitting the bra and ket physical vectors, respectively, and thus giving a zero result.

In next chapter we shall study electromagnetic interaction processes, like the Compton scattering. We will learn that the probability amplitude describing the transition between the initial and final states in the process is expressed as the matrix element of an operator, the S-matrix, between the states of the incoming and outgoing particles. Gauge invariance is then guaranteed if the S-matrix satisfies conditions (11.268).

We can choose to describe physical states through the representative vectors $|s_0\rangle$ of each class, which only contain transverse photons. This corresponds to a gauge choice. In particular single photon states with definite momentum $\mathbf{p} = \hbar\mathbf{k}$ and transverse polarization r $(r = 1, 2)$ will read:

$$|\mathbf{p}, r\rangle = a_{\mathbf{p}r}^{\dagger} |0\rangle. \tag{11.269}$$

A generic physical state will then be expressed as a superposition of the above states:

$$|s_0\rangle = \sum_{\mathbf{p}} \sum_{r=1}^{2} f(\mathbf{p}, r)|\mathbf{p}, r\rangle, \tag{11.270}$$

and the corresponding description in configuration space is:

$$A_{\mu}(x) = \langle 0|\hat{A}_{\mu}(x)|s_0\rangle = \sum_{\mathbf{p}} \frac{c\hbar}{\sqrt{2E_{\mathbf{p}}V}} \sum_{r=1}^{2} \varepsilon_{\mathbf{p}\mu}^{(r)} f(\mathbf{p}, r) e^{-\frac{i}{\hbar} p \cdot x}$$

$$\xrightarrow{V \to \infty} c\hbar \int \frac{d^3\mathbf{p}}{(2\pi\hbar)^3} \sqrt{\frac{V}{2E_{\mathbf{p}}}} \sum_{r=1}^{2} \varepsilon_{\mu}^{(r)}(\mathbf{p}) f(\mathbf{p}, r) e^{-\frac{i}{\hbar} p \cdot x}. \tag{11.271}$$

We see that the transverse polarization vectors $\varepsilon_{\mu}^{(r)}(\mathbf{p})$ play the role of the vectors $u(p, r)$ introduced in Eq. (9.113) of Chap. 9.[19] Recall, from our discussion of irreducible representations of the Poincaré group, that the states of a massless particle are characterized by a definite value of its helicity Γ. From last section of Chap. 6 we have learned that the polarization vectors with definite helicity $\pm\hbar$ are given by complex combinations of the transverse vectors: $\varepsilon_{\mu}^{(1)}(\mathbf{p}) \pm i \, \varepsilon_{\mu}^{(2)}(\mathbf{p})$.[20] If we denote by $\varepsilon_{\mu}(\mathbf{p}, r)$, $\varepsilon_{\mu}(\mathbf{p}, r)^*$ such complex vectors we can write the photon field operator in the following form:

$$\hat{A}_{\mu}(x) = c\hbar \int \frac{d^3\mathbf{p}}{(2\pi\hbar)^3} \sqrt{\frac{V}{2E_{\mathbf{p}}}} \sum_{r=1}^{2} \left(\varepsilon_{\mu}(\mathbf{p}, r) a(\mathbf{p}, r) e^{-\frac{i}{\hbar} p \cdot x} \right.$$

$$\left. + \varepsilon_{\mu}(\mathbf{p}, r)^* a(\mathbf{p}, r)^{\dagger} e^{\frac{i}{\hbar} p \cdot x} \right). \tag{11.272}$$

where $a(\mathbf{p}, r)$ are the complex combinations $a_1(\mathbf{p}) \mp i \, a_2(\mathbf{p})$.

---

[19]Note that this correspondence should take into account a normalization factor due to the fact that $A_{\mu}(x)$ does not have the dimension of a wave-function: $\langle \mathbf{x}|s_0\rangle$.

[20]In Chap. 6 the direction of motion was chosen along the X-axis so that the transverse directions were 2 and 3. Here the motion is chosen along the Z axis.

11.7.2 Poincaré Transformations and Discrete Symmetries group

Let us recall the transformation property of the classical electromagnetic field under a Poincaré transformation:

$$A_\mu(x) \xrightarrow{(\Lambda, x_0)} A'_\mu(x') = \Lambda_\mu{}^\nu A_\nu(x) = O_{(\Lambda, x_0)} A_\mu(x'), \qquad (11.273)$$

where μ and ν indices are raised and lowered using the Lorentzian metric ($\Lambda_\mu{}^\nu \equiv \eta_{\mu\rho}\Lambda^\rho{}_\sigma \eta^{\sigma\nu} = (\Lambda^{-T})_\mu{}^\nu$) and, as usual, $x' = \Lambda x - x_0$. From the relation (11.263) we deduce, just as we did for the scalar and Dirac fields, the transformation property of the field operator $\hat{A}_\mu(x)$:

$$\hat{A}_\mu(x) \xrightarrow{(\Lambda, x_0)} U^\dagger \hat{A}_\mu(x') U = \Lambda_\mu{}^\nu \hat{A}_\nu(x) = O_{(\Lambda, x_0)} \hat{A}_\mu(x'), \qquad (11.274)$$

where $U = U(\Lambda, x_0)$ is the unitary operator implementing the Poincaré transformation on the physical multi-photon states. The commutation relations between the infinitesimal generators $\mathbb{J}^{\mu\nu}$, $\mathbb{P}^\mu$ of $U(\Lambda, x_0)$ and the $\hat{A}_\nu(x)$, which characterize its transformation properties, are deduced just as we did for the lower spin fields, namely by writing (11.274) for an infinitesimal transformation and expanding it to first order in the parameters.

Let us now evaluate the action of the discrete symmetries C, P, T on the photon field. Since a photon coincides with its own antiparticle, the action of C only amounts to a multiplication by a factor $\eta_C = \pm 1$:

$$U(C)^\dagger \hat{A}_\mu(x) U(C) = \eta_C \, \hat{A}_\mu(x). \qquad (11.275)$$

We shall choose $\eta_C = -1$ for reasons we are going to illustrate below, so that *the photon is odd under charge conjugation*. As for P, T, the transformation properties read:

$$U(P)^\dagger \hat{A}_\mu(x) U(P) = \eta_P \, \Lambda_{P\mu}{}^\nu \, \hat{A}_\nu(x_P) = \eta_P \, \eta_{\mu\mu} \hat{A}_\mu(x_P), \qquad (11.276)$$

$$U(T)^\dagger \hat{A}_\mu(x) U(T) = \eta_T \, \Lambda_{T\mu}{}^\nu \, \hat{A}_\nu(x_T) = -\eta_T \, \eta_{\mu\mu} \hat{A}_\mu(x_T), \qquad (11.277)$$

with no summation over μ. In the above formulas we have defined $x_P \equiv \Lambda_P x = (ct, -\mathbf{x})$, and $x_T \equiv \Lambda_T x = (-ct, \mathbf{x})$. In order to determine action of $U(C)$, $U(P)$ and $U(T)$ on the a operators which reproduces (11.275)–(11.277), one follows the same procedure illustrated for the scalar and Dirac field, which we shall not repeat here.

11.8 Quantum Electrodynamics

In Sect. 10.7 of Chap. 10, we have studied the interaction of a charged Dirac field $\psi(x)$ (such as an electron) with the electromagnetic one $A_\mu(x)$. The description of such interaction was obtained by applying to the free Dirac equation the minimal coupling prescription (10.210). The resulting equations of motion could be derived from the Lagrangian density (10.228). If we include the electromagnetic field in the description by adding to $\mathcal{L}$ in (10.228) the term $\mathcal{L}_{e.m.}$ describing the free Maxwell field we end up with the following Lagrangian density for the system:

$$\mathcal{L}_{tot} = \mathcal{L}_0 + \mathcal{L}_I, \tag{11.278}$$

where $\mathcal{L}_0$ describes the free Dirac and electromagnetic fields:

$$\mathcal{L}_0 = \mathcal{L}_{Dirac} + \mathcal{L}_{e.m.},$$
$$\mathcal{L}_{Dirac} = \bar{\psi}(i\hbar c\,\gamma^\mu \partial_\mu - mc^2)\psi,$$
$$\mathcal{L}_{e.m.} = -\frac{1}{4}\,F_{\mu\nu}\,F^{\mu\nu}, \tag{11.279}$$

while $\mathcal{L}_I$ is the interaction term in Eq. (10.228):

$$\mathcal{L}_I = A_\mu(x)\,J^\mu(x) = e\,A_\mu(x)\bar{\psi}(x)\gamma^\mu\psi(x). \tag{11.280}$$

The classical equations of motion are readily derived from $\mathcal{L}_{tot}$ and read

$$\Box A_\mu = -e\bar{\psi}\gamma_\mu\psi, \tag{11.281}$$
$$\left(i\hbar\gamma^\mu\partial_\mu - mc\right)\psi = -\frac{e}{c}\gamma^\mu\psi\,A_\mu, \tag{11.282}$$

the latter coinciding with Eq. (10.212) of last chapter.

In this section we formulate the quantum version of this theory, known as *quantum electrodynamics*, that is the quantum theory describing the interaction between an electron (or in general a charged spin $1/2$-particle) and the electromagnetic field. To this end we describe the classical system in the Hamiltonian formalism and write the Maxwell term $\mathcal{L}_{e.m.}$ in the form (11.214). We easily realize that the conjugate momenta to the Dirac and electromagnetic fields are given by the same as in the free case, namely by (10.133) and (11.216). The Hamiltonian density reads

$$\mathcal{H} = \pi_\psi\dot{\psi} + \pi^\mu\dot{A}_\mu - \mathcal{L}_{tot} = \mathcal{H}_{Dirac} + \mathcal{H}_{e.m.} + \mathcal{H}_I = \mathcal{H}_0 + \mathcal{H}_I, \tag{11.283}$$

where

$$\mathcal{H}_{Dirac} = -i\hbar c\bar{\psi}\gamma^i\partial_i\psi + mc^2\bar{\psi}\psi, \tag{11.284}$$

$$\mathcal{H}_{e.m.} = -\frac{1}{2c^2}\dot{A}_\mu\dot{A}^\mu - \frac{1}{2}\partial_i A^\nu\partial_i A_\nu, \tag{11.285}$$

$$\mathcal{H}_I = -e\bar{\psi}\gamma^\mu\psi\, A_\mu \equiv -\mathcal{L}_I, \tag{11.286}$$

and $\mathcal{H}_0 \equiv \mathcal{H}_{Dirac} + \mathcal{H}_{e.m.}$ represents the Hamiltonian density of the free fields $\psi(x)$, $A_\mu(x)$. The quantization of the system is effected by promoting $\psi^\alpha(x)$ and $A_\mu(x)$ to operators and the Poisson brackets to commutators/anticommutators[21]:

$$[A_\mu(\mathbf{x}, t), \pi^\nu(\mathbf{y}, t)] = i\hbar\delta_\mu^\nu\delta^3(\mathbf{x} - \mathbf{y}), \tag{11.287}$$

$$[A_\mu(\mathbf{x}, t), A_\nu(\mathbf{y}, t)] = [\pi^\mu(\mathbf{x}, t), \pi^\nu(\mathbf{y}, t)] = 0, \tag{11.288}$$

$$\left\{\psi^\alpha(\mathbf{x}, t), \psi_\beta^\dagger(\mathbf{y}, t)\right\} = -\frac{i}{\hbar}\delta^3(\mathbf{x} - \mathbf{y})\delta_\beta^\alpha, \tag{11.289}$$

$$\left\{\psi^\alpha(\mathbf{x}, t), \psi^\beta(\mathbf{y}, t)\right\} = \left\{\psi_\alpha^\dagger(\mathbf{x}, t), \psi_\beta^\dagger(\mathbf{y}, t)\right\} = 0. \tag{11.290}$$

Furthermore we require that Dirac and electromagnetic field operators commute at equal time:

$$[\psi^\alpha(\mathbf{x}, t), A_\mu(\mathbf{x}', t)] = [\psi^\alpha(\mathbf{x}, t), \dot{A}_\mu(\mathbf{x}', t)] = 0. \tag{11.291}$$

The time evolution is determined by the Hamilton quantum equations

$$\dot{\psi} = -\frac{i}{\hbar}[\psi, \hat{H}]; \quad \dot{A}_\mu = -\frac{i}{\hbar}[A_\mu, \hat{H}]; \quad \dot{\pi}^\mu = -\frac{i}{\hbar}[\pi^\mu, \hat{H}]. \tag{11.292}$$

where

$$\hat{H} = \int d^4x\hat{\mathcal{H}}(x),$$

is the conserved Hamiltonian. One easily verifies that the Hamilton equations of motion are equivalent to the Euler-Lagrange equations. Equations (11.292) can be *formally* integrated to read

$$\psi(\mathbf{x}, t) = e^{\frac{i}{\hbar}\hat{H}t}\psi(\mathbf{x}, 0)e^{-\frac{i}{\hbar}\hat{H}t}, \tag{11.293}$$

$$A_\mu(\mathbf{x}, t) = e^{\frac{i}{\hbar}\hat{H}t}A_\mu(\mathbf{x}, 0)e^{-\frac{i}{\hbar}\hat{H}t}.$$

Quantizing the interacting system means defining a Hilbert space of states and quantum field operators acting on it, which satisfy the canonical commutation/anti-commutation relations, as well as the equations of motion. As the operators obey

---

[21] As often done previously, we shall omit in this section the hat symbol on field operators only.

coupled equations, we cannot expand them in terms of the free field solutions. We can of course expand the field at a certain time, say $t = 0$ exactly as in Eqs. (11.225) and (11.140), with creation and destruction operators obeying the same commutation rules as in the free case. However they are no longer eigenmodes of the Hamiltonian and hence we cannot interpret them as creation and destruction operators of single particles. Indeed (11.293) imply that those operators evolve in time as

$$c_{\mathbf{p},t} = e^{-\frac{i}{\hbar}\hat{H}t} c_{\mathbf{p},0} e^{\frac{i}{\hbar}\hat{H}t}$$

and analogously for the other operators. This means the entire apparatus of the free field theories for constructing the eigenmodes of the Hamiltonian breaks down and the exact solution of the coupled equations is unknown. Indeed interacting quantum theories are too complex to be solved exactly and we must resort to perturbative methods, to be developed in the next chapter. Let us here anticipate some concepts related to this issue. In the perturbative approach the quantum Lagrangian and Hamiltonian are written in terms of the free field operators $\psi(x)$, $A_\mu(x)$, evolving with $H_0 = \int d^3\mathbf{x}(\mathcal{H}_{Dirac} + \mathcal{H}_{e.m.})$, and acting on the Fock space of free-particle states. These are expressed as the tensor product of the electron/positron states and the photon states:

$$|\{N_{e^-}\}; \{N_{e^+}\}\rangle_0 \otimes |\{N_\gamma\}\rangle_0, \qquad (11.294)$$

where N_{e^-}, N_{e^+}, N_γ are the occupation numbers of electron, positron and photon states, the subscript "0" indicates that these states pertain to the free theory ($e = 0$). Such states are constructed, as illustrated in this chapter, by acting on a vacuum state $|0\rangle_0$ by means of creation operators. Also the free field operators are expressed in terms of creation and annihilation operators, and all terms in the quantum Lagrangian and Hamiltonian are written in normal ordered form. In particular the interaction term reads:

$$\hat{\mathcal{H}}_I = -e : \bar{\psi}\gamma^\mu\psi A_\mu :\equiv -\hat{\mathcal{L}}_I. \qquad (11.295)$$

Writing everything (fields, states, Hamiltonian etc.) in terms of the solution to the free problem, corresponding the absence of interaction ($e = 0$), is the lowest order approximation from which the perturbation analysis is developed, the perturbation parameter being the dimensionless *fine structure constant* $\alpha \equiv e^2/(4\pi\hbar c) \approx 1/137 \ll 1$. All perturbative corrections, as we shall illustrate in next chapter, are expressed in terms of a series expansion in powers of the interaction Hamiltonian:

$$\hat{H}_I \equiv \int d^3\mathbf{x}\hat{\mathcal{H}}_I, \qquad (11.296)$$

(and thus in powers of the small constant α), through the so called *S-matrix*. Let us stress that each term in this expansion is expressed in terms of free fields.

Here we wish to comment on the issue of symmetries. So far we have defined symmetry transformations on free fields. The Lagrangian and Hamiltonian density operators $\hat{\mathcal{L}}_{tot}$, $\hat{\mathcal{H}}_{tot}$ according to the above prescription, are obtained from their classical expressions in (11.278), (11.279), (11.283), (11.284), by replacing the fields by their corresponding free quantum operators, and normal ordering the resulting expression. Let g be a symmetry transformation of the free theory ($e = 0$), belonging to some symmetry group G, and let $U(g)$ be the unitary operator which realizes it on the free-particle states. The invariance property is expressed in terms of the free action operator

$$\int d^4x \hat{\mathcal{L}}_0 \xrightarrow{g} \int d^4x \hat{\mathcal{L}}'_0 = \int d^4x U(g)^\dagger \, \hat{\mathcal{L}}_0 \, U(g) = \int d^4x \hat{\mathcal{L}}_0. \quad (11.297)$$

If $U(g)$ also commutes with the interaction Hamiltonian (11.296) or, equivalently with the interaction Lagrangian, we have

$$\int d^4x \hat{\mathcal{L}}'_{tot} = \int d^4x U(g)^\dagger \, \hat{\mathcal{L}}_{tot} \, U(g) = \int d^4x \hat{\mathcal{L}}_{tot}. \quad (11.298)$$

The above property is equivalent to the statement that g is a symmetry of the classical interacting theory described by $\mathcal{L}_{tot}$.

In this case the transformation g will also commute with the S-matrix which, as mentioned above, is expressed as a series expansion in powers of $\hat{H}_I$. If this is the case for any $g \in G$, then the whole group G will be a symmetry of the full quantum theory. This is the case, however, if also the vacuum $|0\rangle$ of the interacting theory is invariant under $U(g)$, for any $g \in G$:

$$U(g)|0\rangle = |0\rangle. \quad (11.299)$$

Although we do not know $|0\rangle$ a priori, we assume it to be unique and to be invariant under the symmetries of $\hat{\mathcal{L}}_{tot}$.

Let us review these symmetries

- *Poincaré invariance.* This was our guiding principle for constructing a local relativistic field theory. It is guaranteed by the fact that $\mathcal{L}_{tot}$ is written in a manifestly Poincaré invariant form (that is it is Lorentz and translation invariant) and therefore transforms as a scalar under Poincaré transformations:

$$\hat{\mathcal{L}}'_{tot}(x') = U(\Lambda, x_0)^\dagger \, \hat{\mathcal{L}}_{tot}(x') \, U(\Lambda, x_0) = \hat{\mathcal{L}}_{tot}(x), \quad (11.300)$$

where $x' = \Lambda x - x_0$.

- *Local-*$U(1)$ *invariance.* This symmetry characterizes the present theory and is described by the local transformations in (10.214) and (10.215) of Chap. 10:

$$A_\mu(x) \to A_\mu(x) + \partial_\mu \varphi(x),$$
$$\psi(x) \to \psi(x) \, e^{\frac{ie}{\hbar c} \varphi(x)}. \quad (11.301)$$

- *Parity*. The free part of the action is parity invariant, since the Lagrangian density is written in a form which is manifestly invariant under (proper and improper) Lorentz transformations. Consider the transformation property of $\mathcal{L}_I(x)$ under parity:

$$U(P)^\dagger \hat{\mathcal{L}}_I(x)U(P) = e\, U(P)^\dagger(: \bar\psi(x)\gamma^\mu\psi(x) :)U(P)U(P)^\dagger A_\mu(x)U(P)$$
$$= e\,\eta_P \; : \bar\psi(x_P)\gamma^\mu\psi(x_P) : A_\mu(x_P), \qquad (11.302)$$

where we have used the transformation properties of the fermion current, derived in Sect. 10.8.1, and of $A_\mu(x)$. The factor η_P is the photon intrinsic parity. Choosing $\eta_P = 1$, the whole action is invariant. This is consistent with the fact that parity is conserved in all electromagnetic processes.

- *Time-reversal*. The free part of the action is invariant for the reasons explained above, while $\hat{\mathcal{L}}_I(x)$ transforms as:

$$U(T)^\dagger \hat{\mathcal{L}}_I(x)U(T) = e\, U(T)^\dagger(: \bar\psi(x)\gamma^\mu\psi(x) :)U(T)U(T)^\dagger A_\mu(x)U(T)$$
$$= -e\,\eta_T \; : \bar\psi(x_T)\gamma^\mu\psi(x_T) : A_\mu(x_T), \qquad (11.303)$$

where we have used the transformation properties (11.208) of the fermion current and those of $A_\mu(x)$. The factor η_T is related to the photon field. Choosing $\eta_T = -1$, the action is invariant. This is also consistent with experimental evidence.

- *Charge-conjugation*. The free part of the action is invariant. Indeed we have:

$$U(C)^\dagger \hat{\mathcal{L}}_0(x)U(C) = \hat{\mathcal{L}}_0(x) + \text{four-divergence}. \qquad (11.304)$$

This is apparent if we consider the Maxwell contribution to $\hat{\mathcal{L}}_0$ and the mass term of the spinor field, which is proportional to the invariant bilinear $\bar\psi\psi$, see (11.203). As far as the kinetic term for $\psi(x)$ is concerned we have:

$$U(C)^\dagger \bar\psi(x)(i\hbar c\gamma^\mu\partial_\mu)\psi(x)U(C) = -\partial_\mu\bar\psi(x)i\hbar c\gamma^\mu\psi(x)$$
$$= \bar\psi(x)(i\hbar c\gamma^\mu\partial_\mu)\psi(x) + \text{four-divergence}.$$
$$(11.305)$$

The interaction Lagrangian density $\hat{\mathcal{L}}_I(x)$ transforms as:

$$U(C)^\dagger \hat{\mathcal{L}}_I(x)U(C) = e\, U(C)^\dagger(: \bar\psi(x)\gamma^\mu\psi(x) :)U(C)U(C)^\dagger A_\mu(x)U(C)$$
$$= -e\,\eta_C \; : \bar\psi(x)\gamma^\mu\psi(x) : A_\mu(x), \qquad (11.306)$$

where we have used the transformation properties (11.203) of the fermion current and those of $A_\mu(x)$. The factor η_C is again related to the photon field and choosing $\eta_C = -1$, the action is invariant. Electromagnetic processes are indeed found to respect charge-conjugation symmetry.

452   11   Quantization of Boson and Fermion Fields

Not all the fundamental interactions in nature respect the above symmetries and, in particular, the discrete ones. The *strong (nuclear) interaction*, which is responsible for the binding force among protons and neutrons within nuclei as well as for the confinement of quarks inside protons and neutrons, respects P, C, T separately, just as electromagnetic interaction does. On the other hand the weak interaction, responsible for the beta decay of certain nuclei, is known to violate parity: The *mirror image* of certain processes do not occur with the same rate as the original ones.

There is however an important theorem, which we are not going to prove, which states that in a local field theory described by a Lorentz-invariant, normal-ordered Lagrangian density $\hat{\mathcal{L}}(x)$, the numbers η_C, η_P, η_T for each field can be chosen so that *the product CPT of the three discrete transformations C, P, T is always a symmetry*. In particular $\hat{\mathcal{L}}(x)$ transforms as follows:

$$U(T)^\dagger U(P)^\dagger U(C)^\dagger \hat{\mathcal{L}}(x) U(C) U(P) U(T) = \hat{\mathcal{L}}(-x). \qquad (11.307)$$

This is known as the *CPT theorem*. The symmetry of a theory under *CPT* implies that the image under *CPT* of a process must be as likely to occur as the process itself. The effect of a *CPT* transformation is to change particles into antiparticles, and to reverse space and time directions. The latter operation maps a process into its *inverse* with the initial and final states interchanged. Furthermore the spin components of the various particles are reversed as well.

Although we are not going deal with interactions other that the electromagnetic one, let us mention that the weak, strong and electromagnetic interactions are described by a unified local, Lorentz-invariant field theory known as *standard model*. The weak interaction is found to violate parity, charge conjugation and also, to a smaller extent, the combination *CP*. Assuming it to be correctly described by the standard model, by the above theorem we expect the combination *CPT* to be preserved in the weak interaction phenomenology.

11.8.1 References

For further reading see Refs. [3], [8, vol. 4], [9, 13].

Chapter 12
Fields in Interaction

12.1 Interaction Processes

So far we have restricted our analysis to free bosonic and fermionic fields. In this case we were able to canonically quantize them, by associating with the fields and their conjugate momenta operators acting on a Hilbert space of states and satisfying the canonical commutation (or anti-commutation) relations and the equations of motion. This was possible since free fields can be represented as collections of infinitely many decoupled harmonic oscillators, each associated with a given one-particle state. Quantizing them amounted to quantizing each oscillator,[1] whose elementary excitation is now interpreted as an elementary particle in the corresponding state. This defines the correspondence between particles and fields, such as for the photon and the electromagnetic field, as explained in Chap. 6. The field operators are then constructed in terms of the annihilation and creation operators associated with each quantum oscillator and satisfy canonical relations, expressed in terms of commutators (for bosons) or anti-commutators (for fermions). Quantum states for the system are multi-particle states constructed as tensor products of the elementary oscillator states. They are completely characterized by occupation numbers, interpreted as the number of particles in each single-particle state, and generate the Fock–space of quantum states of the field. The Hamiltonian operator then describes the time evolution of a generic state of the system.

This route to the canonical quantization of fields, however, only works for free fields. In the presence of interactions, a basis of eigenstates of the Hamiltonian operator is in general not known. One has to give up the purpose of finding an exact solution to the canonical quantization problem and try to achieve a *perturbative* description of the interaction whenever this is feasible. It is useful in this respect, generalizing our discussion in Sect. 11.8, to write Lagrangian L of the interacting

---

[1] For fermionic fields it is more appropriate to talk about *anti-oscillators*, being them quantized using anti-commutators in order to reproduce the right statistics.

© Springer International Publishing Switzerland 2016
R. D'Auria and M. Trigiante, *From Special Relativity to Feynman Diagrams*,
UNITEXT for Physics, DOI 10.1007/978-3-319-22014-7_12

theory as the sum of a term L_0 describing the free fields and an interaction term $L_I \equiv L - L_0$:

$$L = L_0 + L_I. \tag{12.1}$$

A similar decomposition can be done for the Hamiltonian H of the system:

$$H = H_0 + H_I. \tag{12.2}$$

We shall assume the interaction term L_I not to involve time derivatives of the fields, so that we have: $H_I = -L_I$. If the interaction term H_I is "small", namely it is proportional to some small coupling constant λ, we can study its effect on the known solution to the free problem perturbatively. In other words we express the solution to the complete theory, i.e. field operators $\hat{\phi}$ and states[2] $|\psi\rangle$, in terms of the free field operators $\hat{\phi}_0$ and states $|\psi\rangle_0$ plus perturbative effects due to the interaction, which can be expanded in powers of λ and which vanish in the limit $\lambda \to 0$:

$$\hat{\phi}(x) = \hat{\phi}_0(x) + \sum_{n=1}^{\infty} \hat{\phi}_n(x)\,\lambda^n,$$

$$|\psi\rangle = |\psi\rangle_0 + \sum_{n=1}^{\infty} |\psi\rangle_n\,\lambda^n. \tag{12.3}$$

The example which we shall be mostly concerned with, is the interaction between an electron (or, in general, a charged fermion) and the electromagnetic field, which was dealt with in Sect. 11.8 of Chap. 11. In this case H_I has the following form (see Eq. (11.285) of Chap. 11):

$$H_I = -\int d^3\mathbf{x}\, e\, A_\mu\, \overline{\psi}\, \gamma^\mu\, \psi. \tag{12.4}$$

where $A_\mu(x)$ and $\psi(x)$ are the photon and the electron fields respectively, $e = -|e| < 0$ the electron charge. The dimensionless coupling constant associated with the electromagnetic interaction is the *fine structure constant* $\lambda = \alpha = \frac{e^2}{4\pi\hbar c} \sim \frac{1}{137}$, which is small and thus allows a perturbative analysis.

In this chapter we wish to give a concise account of the relativistic-covariant perturbation theory developed, for quantum electrodynamics, by Feynman, Dyson, Schwinger and Tomonaga, which generalizes the familiar perturbative analysis in non-relativistic quantum theory to a framework in which relativistic covariance is manifest at all perturbative orders. This approach provides a powerful and simple diagrammatic technique for computing amplitudes of scattering or decay processes, as well as perturbative corrections to generic physical observables: Each order λ^n

---

[2]In this book we often denote states by $|\psi\rangle$. The reader should, however, bear in mind that the Greek letter ψ in this symbol *has no relation* to fermion fields, generically denoted by $\psi(x)$.

term in a perturbative expansion is described in terms of diagrams made of basic building blocks called *propagators* and *vertices*.

The starting point of this analysis is to express the complete Lagrangian and Hamiltonian in terms of the free fields $\phi_0(x)$, namely in terms of fields evolving according to H_0 and quantized as operators acting on the Fock space of free field states. Such Hamiltonian should be itself regarded as the first term of a perturbative expansion and by no means describes the total energy of the interacting system. Similarly, the constants appearing in this Hamiltonian (coupling constants and masses) should be regarded just as the first term of a perturbative expansion in λ which yields the measured values of the corresponding physical quantities.

As we shall see, however, in the computation of the amplitudes of interaction processes, while everything is consistent to the lowest-order approximation, to second order divergent quantities occur, which seem to spoil the perturbative program. It turns out, however, that if we express these divergent amplitudes in terms of the *physical* (i.e. measurable) parameters, and not in terms of the *bare* ones (coupling constants, masses, charges of the free theory), then the amplitudes become finite, while the bare parameters, which are devoid of physical meaning in the interacting theory, become infinite. The technique to arrive at such finite results is usually referred to as *renormalization program* and we shall give of it just simple examples to second order in the S-matrix perturbative expansion. A complete proof that this procedure actually works to any order in perturbation theory can be found in the references given at the end of the chapter. It should also be said that the roles of the renormalization program and of the so-called *renormalization-group* flow related to it, are actually of foremost importance in modern physics. Indeed, as it became apparent from the later developments of quantum field theory, the renormalization technique is more than just a technical procedure for making an interacting theory predictive and thus experimentally testable, and has a profound bearing on the understanding of several physical phenomena.

Before entering into the mathematical details of this analysis let us discuss the relativistic-invariant description of scattering and decay processes.

12.2 Kinematics of Interaction Processes

If there were no interaction term H_I in the Hamiltonian, a system originally prepared in an eigenstate of the free Hamiltonian $H = H_0$, $|\psi, t = 0\rangle = |E\rangle_0$, and describing free particles with definite momenta and total energy E, will stay (in the Schroedinger picture) in the same state ever after $|\psi, t\rangle = e^{-\frac{i}{\hbar}\hat{H}_0 t}|E\rangle = e^{-\frac{i}{\hbar}Et}|E\rangle$. In the presence of an interaction $H_I \neq 0$, the eigenstates $|E\rangle_0$ of H_0 are no longer eigenstates of the complete Hamiltonian and a system initially prepared in $|E\rangle_0$ will in general evolve in time towards a different state. If we consider processes in which the interaction among the particles takes place in a small volume and during a short time-lapse, we can describe the states of the interacting particles long before ($t \to -\infty$) and long after ($t \rightarrow +\infty$) the interaction as free-particle states and express them in terms of eigenstates of H_0. We shall call these two asymptotic states as the states of incoming and outgoing particles, to be denoted by $|\psi_{in}\rangle$ and $|\psi_{out}\rangle$ respectively. They belong to the Fock space of free-particle states. We should think of the incoming and outgoing (in the far past and future respectively) particles as being so far apart from one another as not to feel their reciprocal action.[3] This picture would not be consistent with describing the corresponding states as those of particles with definite momenta, since momentum eigenstates are completely delocalized in space and time. We should instead think of $|\psi_{in}\rangle$ and $|\psi_{out}\rangle$ as combinations of momentum eigenstates describing *wave packets* with definite width ℓ, approaching each other before the interaction and departing from one another after it. Each wave packet will have a momentum which is indeterminate within a cubic element $\Delta^3\mathbf{p} \sim \hbar^3/\ell^3$ about a central value $\mathbf{p}$. We assume ℓ to be large enough for the probability of each process not to vary appreciably within $\Delta^3\mathbf{p}$ but to depend only on the mean values $\mathbf{p}_i$ of the momenta of each wave packet.

Let us now make some general remarks about the number of independent kinematic variables describing an isolated system of interacting particles in relation to its symmetry properties. Consider a process involving a total number N of particles (which include both the incoming and the outgoing ones), each particle being described by a 4-momentum p_i^μ, $i = 1, \ldots, N$, a polarization and a rest-mass m_i. Long before and after the interaction, for each free-particle we can write the *mass-shell* condition $p_i^2 = p_i^\mu p_{i\mu} = m_i^2 c^2$ (also called *on-shell* condition). Thus each free particle state is described by the three components of its linear momentum $\mathbf{p}_i$ and its polarization (i.e. as we have learned in the previous chapters, a free-particle is defined by an irreducible representation of the Poincaré group). If the particles are scalars, the total state of the system is therefore defined by $3N$ variables. Poincaré symmetry however reduces the number of independent variables. Invariance under space-time translations (a given process should look the same if observed in different places of our universe at different times) implies the conservation of the total 4-momentum, which amounts to 4 conditions on the $3N$ variables, cutting the number of independent ones down to $3N - 4$. The physics of the process is also invariant under rotations and boosts of the frame of reference (i.e. Lorentz invariance), though the description of the system in terms of the $3N - 4$ variables is not. If we are to achieve a Lorentz-invariant description of the process, we need to find a maximal number of independent combinations of the $3N - 4$ which are not affected by Lorentz transformations of the frame of reference (Lorentz-invariant quantities). We have already taken into account N of them, namely the rest-masses m_i^2, with the mass-shell condition. The remaining Lorentz-invariant quantities are obtained by requiring generic functions of the $3N - 4$ variables to be invariant under each of the six independent infinitesimal Lorentz transformations. This implies six further conditions which reduce the number of variables to a total of $3N - 10$ Lorentz-invariant quantities. The above

---

[3]This would not be true if, during the interaction process, bound states of particles are formed. The final system at $t \rightarrow +\infty$ would not consist in this case of free-particles only. We shall not consider interactions which allow the creation of bound states.

counting therefore accounts for the 10 conserved Noether charges associated with each Lorentz symmetry generator, see Sect. 8.8, which reduce the number of independent momentum components to $3N - 10$. For particles with spin, this number should be further multiplied by the number of spin states.

12.2.1 Decay Processes

Each elementary decay process consists of a single particle decaying into two or more particles, like, for instance, a neutron which decays into a proton, an electron and an anti-neutrino:

$$n \rightarrow p^+ + e^- + \bar{\nu}_e. \tag{12.5}$$

Consider a system of identical unstable particles prepared in a same initial state $|\psi_{in}\rangle$ at $t \rightarrow -\infty$. With the passing by of time a number of the initial particles will decay. If $N(t) \gg 1$ is the number of particles in a small volume dV^4 at a time t, so that $\rho(t) = \frac{N(t)}{dV}$ is the corresponding particle density, and if $dN(in) \ll N(t)$ denotes the number of these particles decaying between t and $t + dt$, we can write the probability of a decay per unit time as follows:

$$\frac{dP(in)}{dt} = \frac{dN(in)}{N(t)\, dt} = \frac{dN(in)}{\rho(t)\, dV\, dt}. \tag{12.6}$$

Experimentally one finds that this quantity is a constant, depending only on the initial state $|\psi_{in}\rangle$ and related to the probability of a single decay event to occur. Such constant is expressed in terms of a *decay width* $\Gamma(in)$, which has the dimension of an energy, divided by $\hbar$:

$$\frac{dN(in)}{\rho(t)\, dV\, dt} = \frac{\Gamma(in)}{\hbar}. \tag{12.7}$$

When computed in the rest-frame of the particle, the inverse of the above quantity gives the *mean life-time* $\tau \equiv \frac{\hbar}{\Gamma(in)}$, which is a characteristic feature of the particle itself. The mean life-time of an isolated neutron, for instance, is about 15 min, while that of a muon μ^- is of the order of 10^{-6} s (see Chap. 1).

Let us discuss now the relativistic covariance of Eq. (12.7). Suppose the quantities in (12.7) are referred to an inertial RF S and let us consider the same decay process as described from a different inertial RF S' (primed quantities being referred to the latter). The space-time volume element $dV\, dt$ is Lorentz-invariant, and so is the number of events contained therein: $dV\, dt = dV'\, dt'$, $dN = dN'$. Equation (12.7)

---

[4] Here we denote by dV a volume which is infinitesimal but still macroscopic in size, so as to contain a considerable number of particles.

implies that the product $\rho\,\Gamma(in)$ is Lorentz-invariant: $\rho\,\Gamma(in) = \rho'\,\Gamma(in')$, where in' refers to the initial state $|\psi'_{in}\rangle$ of the decaying particles as seen from S'.

Consider now a process in which a particle of rest mass M decays into a final system of n particles of rest masses $m_1, m_2, \ldots, m_n$. As outlined above, we consider a statistical sample consisting of identical decaying particles prepared in a same state with definite momentum and energy E_{in} (in the rest-mass frame we have $E_{in} = Mc^2$). We wish to study the probability for the decays to yield outgoing particles in a certain quantum state $|\psi_{out}\rangle$. Let us characterize $|\psi_{out}\rangle$ by a complete system of observables, which include the momenta of the outgoing particles $\mathbf{q}_i$, $i = 1, \ldots, n$, and other (discrete) quantities like the spin, which we collectively denote by α, so that $|\psi_{out}\rangle = |\alpha, \mathbf{q}_1, \ldots, \mathbf{q}_n\rangle$. If each outgoing particle is thought of as contained in a finite box of volume[5] V_i, $i = 1, \ldots, n$, in which it is quantized, the momenta are discrete as well and, denoting by $dN(in;\ \alpha, \mathbf{q}_1, \ldots, \mathbf{q}_n)$ the number of decay events, within $dV\,dt$, yielding particles in the asymptotic state $|\psi_{out}\rangle$, the probability per unit time of observing the n outgoing particles in the final state $|\alpha, \mathbf{q}_1, \ldots, \mathbf{q}_n\rangle$ reads:

$$\frac{d}{dt}P(in;\ \alpha, \mathbf{q}_1, \ldots, \mathbf{q}_n) = \frac{dN(in;\ \alpha, \mathbf{q}_1, \ldots, \mathbf{q}_n)}{\rho\,dV\,dt} = \frac{1}{\hbar}\,\Gamma(in;\ \alpha, \mathbf{q}_1, \ldots, \mathbf{q}_n).$$

$$(12.8)$$

Since, as already observed, $\frac{dN(in;\ \alpha, \mathbf{q}_1, \ldots, \mathbf{q}_n)}{dV\,dt}$ is Lorentz-invariant, also the product, which we shall denote by $\hat{\Gamma}$, of the partial-width $\Gamma(in;\ \alpha, \mathbf{q}_1, \ldots, \mathbf{q}_n)$ times ρ must have the same property. This implies that, in changing the reference frame from S to S' we have:

$$\hat{\Gamma}(in;\ \mathbf{q}_1, \ldots, \mathbf{q}_n) = \rho\,\Gamma(in;\ \mathbf{q}_1, \ldots, \mathbf{q}_n)$$
$$= \rho'\,\Gamma'(in',\ \mathbf{q}'_1, \ldots, \mathbf{q}'_n) = \hat{\Gamma}(in',\ \mathbf{q}'_1, \ldots, \mathbf{q}'_n), \qquad (12.9)$$

The total probability per unit time of the decay event is obtained by summing the partial probability (12.8) over all the final states

$$\frac{d}{dt}P(in) = \frac{\Gamma(in)}{\hbar} = \sum_{\mathbf{q}_1\ldots\mathbf{q}_n}\sum_{\alpha}\frac{d}{dt}P(in;\ \alpha, \mathbf{q}_1, \ldots, \mathbf{q}_n)$$

$$= \sum_{\mathbf{q}_1\ldots\mathbf{q}_n}\sum_{\alpha}\frac{1}{\hbar\rho}\,\hat{\Gamma}(in;\ \alpha, \mathbf{q}_1, \ldots, \mathbf{q}_n). \qquad (12.10)$$

As usual, in the limit of large volume V we may replace the sum over momenta by an integral: $\sum_{\mathbf{q}} = \int \frac{d^3q\,V}{(2\pi\hbar)^3}$ and write

---

[5]The normalization volumes V_i, here and in the following, should be thought of as having a microscopic size, given by the width of the wave packet describing the particle. It should not be confused with the macroscopic volume element dV in which the decay events occur.

$$\frac{d}{dt}P(in) = \frac{\Gamma(in)}{\hbar}$$

$$= \int \sum_\alpha \frac{1}{\hbar \rho} \hat{\Gamma}(in;\, \alpha,\, \mathbf{q}_1, \dots, \mathbf{q}_n) \frac{d^3\mathbf{q}_1\, V_1}{(2\pi\hbar)^3} \cdots \frac{d^3\mathbf{q}_n\, V_n}{(2\pi\hbar)^3}. \qquad (12.11)$$

Notice that each factor $\frac{d^3\mathbf{q}_i\, V_i}{(2\pi\hbar)^3}$ is Lorentz-invariant. The integrand in (12.11) now represents the probability per unit time of observing the final particles with momenta contained within elementary cubic volumes $d^3\mathbf{q}_i$ about some average values $\mathbf{q}_i$, $i = 1, \dots, n$. Let us stress that the momenta $\mathbf{q}_i$ are referred to the final state of the system at $t \to \infty$ in which the particles are infinitely far apart and in this state V_i represents the average volume occupied by the wave-packet associated with the ith particle. We can then define for each produced particle a density (number of ith-type particles per unit volume) $\rho_i = \frac{1}{V_i}$. According to our discussion in Sect. 9.5 of Chap. 9, we can, for each particle, refer the definition of the volume V_i to a RF S_{0i} in which the product of twice the volume times the energy has a certain value c_{0i} ($2\, V_{0i}\, E_{0i} = c_{0i}$) in the appropriate units and write:

$$\rho_i = \frac{1}{V_i} = \frac{2\, E_i}{2\, V_0\, E_{0i}} = \frac{2\, E_i}{c_{0i}}. \qquad (12.12)$$

The same can be done for the density ρ of the initial particle, by setting $\rho = 2\, E_{in}/c_0$, E_{in} being its energy. The normalization factors c_0, c_{0i} have dimension $(Newton) \times (length)^4$ and are relativistically invariant since they are defined in a specific frame of reference. As we shall see, these constants will finally drop out of the expressions for any physical quantity. Equation (12.11) will then have the following form:

$$\frac{dP(in)}{dt} = \frac{\Gamma(in)}{\hbar} = \frac{c_0}{2\, E_{in}} \int \frac{1}{\hbar} \hat{\Gamma}(in;\, \mathbf{q}_1, \dots, \mathbf{q}_n)\, d\Omega_{\mathbf{q}_1} \dots d\Omega_{\mathbf{q}_n}, \qquad (12.13)$$

where we have introduced the following Lorentz-invariant measures: $d\Omega_{\mathbf{q}_i} \equiv \frac{d^3\mathbf{q}_i\, c_{0i}}{(2\pi\hbar)^3\, 2\, E_i}$. The final momenta and energies are related by the mass-shell condition: $E_i^2 - c^2\, |\mathbf{q}_i|^2 = m_i^2\, c^4$. The above equation provides a Lorentz-covariant description of a decay process. The integration in (12.13) is performed over all possible final momenta of the outgoing particles, which are constrained by the energy-momentum conservation condition, being our system isolated:

$$P_{in}^\mu = P_{out}^\mu, \qquad (12.14)$$

where $P_{in}^\mu \equiv (\frac{1}{c}\, E_{in},\, \mathbf{p}_{in})$, is the energy-momentum vector of the decaying particle and $P_{out}^\mu \equiv \sum_{i=1}^n q_i^\mu$ is the total final energy-momentum of the system. We can take condition (12.14) into account in the integration by factoring out of $\hat{\Gamma}$ a $\delta^4(P_{in} - P_{out})$, that is redefining:

$$\hat{\Gamma} \rightarrow \hat{\Gamma}\,(2\pi\hbar)^4\,\delta^4(P_{in}-P_{out}). \tag{12.15}$$

Equation (12.13) will then read:

$$\frac{dP(in)}{dt} = \frac{\Gamma(in)}{\hbar} = \frac{c_0}{2\,E_{in}} \int \frac{1}{\hbar}\,\hat{\Gamma}(in;\,\mathbf{q}_1,\ldots,\mathbf{q}_n)\,d\Phi^{(n)}, \tag{12.16}$$

where $d\Phi^{(n)}$ is the n-particle relativistically invariant measure in phase space and is defined as:

$$\begin{aligned}
d\Phi^{(n)} &\equiv (2\pi\hbar)^4\,\delta^4(P_{in}-P_{out})\,\frac{d^3\mathbf{q}_1}{(2\pi\hbar)^3}\,V_1\cdots\frac{d^3\mathbf{q}_n}{(2\pi\hbar)^3}\,V_n \\
&= (2\pi\hbar)^4\,\delta^4(P_{in}-P_{out})\,\frac{d^3\mathbf{q}_1\,c_{01}}{(2\pi\hbar)^3\,2\,E_1}\cdots\frac{d^3\mathbf{q}_n\,c_{0n}}{(2\pi\hbar)^3\,2\,E_n}.
\end{aligned} \tag{12.17}$$

It is manifestly Lorentz-invariant since $\delta^4(P_{in}-P_{out})$ is. In Eq. (12.16), the kinematic analysis of the process, i.e. all the implications of the conservation laws, is encoded in the integration over $d\Phi^{(n)}$, and is separated from the dynamics of the process, which depends on the nature of the interaction involved, which is described by $\hat{\Gamma}$. This latter quantity, being Lorentz-invariant, should depend on the $3N - 10 = 3n - 7$ Lorentz-invariant variables associated with the system. For a particle decaying into two particles, $n = 2$ and $3n - 7$ is negative, meaning that all the kinematical variables are fixed by the symmetry of the system and the mass-shell condition, so that $\hat{\Gamma}$ will only depend on the rest-masses (i.e. it is a constant).

A same particle may decay into different systems of particles, defining different *decay channels*. For instance the neutral pion π^0, which is a neutral particle about 270 times as heavy as the electron, decays, most of the times, into two photons, $\pi^0 \rightarrow 2\gamma$. However about 1 % of them decay into an electron, a positron and a photon, $\pi^0 \rightarrow e^+ + e^- + \gamma$. There are other decay channels which are much rarer. We can define for each channel a decay width $\Gamma(in;\ channel)$, given by Eq. (12.16), where the sums and integrals on the right-hand-side are over all possible states of the decay products in the given channel. The total decay width $\Gamma(in)$, yielding the probability of decay per unit time, will then be given by the sum of the widths associated with each channel: $\Gamma(in) = \sum_{channel}\Gamma(in;\ channel)$. The relative probability associated with each channel can be characterized by a *branching ratio* $BR(in;\ channel) \equiv \frac{\Gamma(in;\ channel)}{\Gamma(in)}$, which tells us how likely is the corresponding decay to occur. For the neutral pion we have:

$$BR(\pi^0 \rightarrow 2\gamma) = \frac{\Gamma(\pi^0 \rightarrow 2\gamma)}{\Gamma(in)} \sim 99\,\%,$$

$$BR(\pi^0 \rightarrow e^+ + e^- + \gamma) = \frac{\Gamma(\pi^0 \rightarrow e^+ + e^- + \gamma)}{\Gamma(in)} \sim 1\,\%, \tag{12.18}$$

all other channels having BR less than 10^{-3}. The mean life-time of a neutral pion is about 10^{-16} s.

12.2.2 Scattering Processes

Consider a process in which particles are projected at a fixed target. If the incident particle comes close enough to the target particle at rest, the two will feel the interaction and be scattered or produce new particles. This happens if the *impact parameter*, i.e. the distance between the initial line of motion and the line parallel to it through the target particle, is "small enough". Depending on the nature of the interaction and on the energy of the incident particle, we can define an effective area about the target particle, on the plane perpendicular to the initial line of motion, so that if the incident particle crosses this area interaction takes place, otherwise the states of motion of the two particles remain practically unperturbed. This area is called *cross-section σ* of the interaction.

Let ρ_1 and ρ_2 be the densities of the incident and target particles respectively in the *laboratory frame S_0* in which the latter are at rest, and let **v** be the relative velocity of the two colliding particles in S_0 (that is the velocity of the incident particle). The number of incident particles colliding with a single target particle during a short time lapse dt is given by the number of particles which pass through the corresponding cross-sectional area σ: $\rho_1 v \sigma dt$, where $v = |\mathbf{v}|$ and $\rho_1 v$ is the *flux* of the incident particles. Multiplying this number by the number $\rho_2 dV$ of target particles in a small volume dV, we find the number dN of collisions taking place in dV during dt. The number of events per unit time and volume then reads:

$$\frac{dN(in)}{dV \, dt} = \rho_1 \rho_2 v \, \sigma(in). \tag{12.19}$$

The cross section σ is then defined as *the number of collision events for each scatterer, per unit flux of the incident beam and unit time*. We can also define the probability $dP(in)$ of a single event between t and $t + dt$ as the number of events per target particle, that is: $dP(in) = dN(in)/N_{target} = dN(in)/(\rho_2 dV)$. Equation (12.19) can also be written in the following way:

$$\frac{dP(in)}{dt} = (\rho_1 v) \sigma, \tag{12.20}$$

providing an alternative definition of cross section as the *probability of the scattering event per unit flux of the incident beam and unit time*.

Since the left-hand side of Eq. (12.19) is Lorentz-invariant, we wish to write the right-hand side in terms of Lorentz-invariant quantities as well. Suppose each collision produces n particles with rest-masses $m_1, \ldots, m_n$. We can consider, just as we did for the decays, the number of events $dN(in; \alpha, \mathbf{q}_1, \ldots, \mathbf{q}_n)$ which produce particles in a final state $|\psi_{out}\rangle$ characterized by momenta contained within an elementary momentum space volume $d^3 \mathbf{q}_i$ about an average value $\mathbf{q}_i$, $i = 1, \ldots, n$, and certain values of the remaining (discrete) quantum numbers α: $|\psi_{out}\rangle = |\alpha, \mathbf{q}_1, \ldots, \mathbf{q}_n\rangle$. To this end we write the cross section σ associated with this final state in terms of a *density* function Σ_{lab} times the invariant measure on the phase space $d\Phi^{(n)}$, which accounts for all the kinematic constraints:

$$\frac{dN(in; \; \alpha, \mathbf{q}_1, \ldots, \mathbf{q}_n)}{dV \, dt} = \rho_1 \, \rho_2 \, v \, \Sigma_{lab}(in; \; \alpha, \mathbf{q}_1, \ldots, \mathbf{q}_n) \, d\Phi^{(n)}. \qquad (12.21)$$

The above formula is still not Lorentz-invariant since we are in the reference frame S_0 in which the target particle is at rest. Let us now move to a generic RF S in which the incident and target particles have velocities $\mathbf{v}_1$, $\mathbf{v}_2$ and four-momenta p_1^μ, p_2^μ respectively. Let M_1, M_2 denote the rest-masses of the two interacting particles: $M_1^2 c^2 = p_1 \cdot p_1$, $M_2^2 c^2 = p_2 \cdot p_2$. Recalling the discussion in Sect. 9.5 and Eq. (9.141) therein, the densities ρ_1, ρ_2 in S can be expressed in terms of their corresponding values $\rho_1^{(0)}$, $\rho_2^{(0)}$ in the rest-frames of the two particles, through the γ-factors: $\rho_i = \rho_i^{(0)} \gamma_i$, $\gamma_i = (1 - \frac{v_i^2}{c^2})^{-\frac{1}{2}}$, $v_i = |\mathbf{v}_i|$, $i = 1, 2$. Let us show that the quantity $\rho_1 \, \rho_2 \, v$ in S_0 can be expressed with respect to S in the following Lorentz-invariant fashion

$$\rho_1^{(0)} \, \rho_2^{(0)} \sqrt{\frac{(p_1 \cdot p_2)^2}{M_1^2 M_2^2 c^2} - c^2}, \qquad (12.22)$$

as we can show using the properties $\frac{p_i^0}{M_i c} = \gamma_i$, $\frac{\mathbf{p}_i}{M_i} = \gamma_i \mathbf{v}_i$:

$$\rho_1^{(0)} \, \rho_2^{(0)} \sqrt{\frac{(p_1 \cdot p_2)^2}{M_1^2 M_2^2 c^2} - c^2} = \rho_1^{(0)} \, \rho_2^{(0)} \gamma_1 \gamma_2 c \sqrt{\left(1 - \frac{\mathbf{v}_1 \cdot \mathbf{v}_2}{c^2}\right)^2 - \frac{1}{\gamma_1^2 \gamma_2^2}}$$

$$= \rho_1 \, \rho_2 \sqrt{|\mathbf{v}_1 - \mathbf{v}_2|^2 - \frac{1}{c^2} \left(v_1^2 \, v_2^2 - (\mathbf{v}_1 \cdot \mathbf{v}_2)^2\right)}$$

$$= \rho_1 \, \rho_2 \sqrt{|\mathbf{v}_1 - \mathbf{v}_2|^2 - \frac{1}{c^2} |\mathbf{v}_1 \times \mathbf{v}_2|^2}$$

$$= \rho_1 \, \rho_2 f(\mathbf{v}_1, \mathbf{v}_2), \qquad (12.23)$$

where we have used the property $|\mathbf{v}_1 \times \mathbf{v}_2|^2 = v_1^2 v_2^2 - (\mathbf{v}_1 \cdot \mathbf{v}_2)^2$ and we have defined $f(\mathbf{v}_1, \mathbf{v}_2) \equiv \sqrt{|\mathbf{v}_1 - \mathbf{v}_2|^2 - \frac{1}{c^2} |\mathbf{v}_1 \times \mathbf{v}_2|^2}$. If the collision is head-on, the two initial velocities are collinear (i.e. $\mathbf{v}_1 \times \mathbf{v}_2 = 0$) and $f(\mathbf{v}_1, \mathbf{v}_2)$ is the modulus of the relative velocity: $f(\mathbf{v}_1, \mathbf{v}_2) = |\mathbf{v}_1 - \mathbf{v}_2|$. In the laboratory frame $\mathbf{v}_2 = 0$, $\mathbf{v}_1 = \mathbf{v}$ and the above expression yields $\rho_1 \, \rho_2 \, v$. Formula (12.23) is more general and also applies to the case in which one of the two particles (say particle 1) is massless, so that $v_1 = c$. In this case the corresponding rest-frame does not exist but Eq. (12.23) yields $\rho_1 \rho_2 (1 - v_2 \cos(\theta))$, where θ is the angle between $\mathbf{v}_1$ and $\mathbf{v}_2$. If both particles are massless, the frame S_0 does not exist and Eq. (12.23) gives $\rho_1 \rho_2 (1 - \cos(\theta))$.

Equation (12.21) can now be written in a fully Lorentz-invariant way:

$$\frac{dN}{dV\,dt}(in;\ \alpha,\ \mathbf{q}_1, \ldots, \mathbf{q}_n) = \rho_1 \rho_2 f(\mathbf{v}_1,\ \mathbf{v}_2)\, \Sigma(in;\ \alpha,\ \mathbf{q}_1, \ldots, \mathbf{q}_n)\, d\Phi^{(n)},$$

$$= \frac{(2E_1)\,(2E_2)}{c_{01}\,c_{02}} f(\mathbf{v}_1,\ \mathbf{v}_2)\, \Sigma(in;\ \alpha,\ \mathbf{q}_1, \ldots, \mathbf{q}_n)\, d\Phi^{(n)},$$

$$\tag{12.24}$$

where we have written $\rho_i = 1/V_i = 2E_i/c_{0i}$ and $\Sigma(in;\ \alpha,\ \mathbf{q}_1, \ldots, \mathbf{q}_n)$ is a Lorentz-invariant function which equals Σ_{lab} in the laboratory frame S_0. The quantity $\Sigma(in;\ \alpha,\ \mathbf{q}_1, \ldots, \mathbf{q}_n)\, d\Phi^{(n)}$ is also called *differential cross section $d\sigma$* and characterizes the probability that the scattering process, starting from a given initial state, yields final particles with momenta contained within infinitesimal neighborhoods $d^3\mathbf{q}_i$ of $\mathbf{q}_i$:

$$d\sigma(in;\ \alpha,\ \mathbf{q}_1, \ldots, \mathbf{q}_n) = \Sigma(in;\ \alpha,\ \mathbf{q}_1, \ldots, \mathbf{q}_n)\, d\Phi^{(n)}$$

$$= \frac{1}{\rho_1 \rho_2 f(\mathbf{v}_1,\ \mathbf{v}_2)} \frac{dN}{dV\,dt}(in;\ \alpha,\ \mathbf{q}_1, \ldots, \mathbf{q}_n)$$

$$= \frac{1}{\rho_1 f(\mathbf{v}_1,\ \mathbf{v}_2)} \frac{dP}{dt}(in;\ \alpha,\ \mathbf{q}_1, \ldots, \mathbf{q}_n). \tag{12.25}$$

In the above formula summation (or integration) over all quantities referred to the final state which $\Sigma(in;\ \alpha,\ \mathbf{q}_1, \ldots, \mathbf{q}_n)$ does not depend on is understood. It is useful, when explicitly calculating of $d\sigma$, to eliminate the delta-function δ^4 in the expression of $d\Phi^{(n)}$, by first integrating over one of the final linear momenta, and then over the total energy of the system. We shall illustrate this below when computing $d\Phi^{(2)}$.

The total cross section of the process is the sum of the differential cross sections over all possible states of the final n particles:

$$\sigma(in) = \sum_\alpha \int d\sigma(in;\ \alpha,\ \mathbf{q}_1, \ldots, \mathbf{q}_n) = \sum_\alpha \int \Sigma(in;\ \alpha,\ \mathbf{q}_1, \ldots, \mathbf{q}_n)\, d\Phi^{(n)}.$$

The Lorentz-invariant quantity $\Sigma(in;\ \alpha,\ \mathbf{q}_1, \ldots, \mathbf{q}_n)$, should only depend, aside from the rest-masses of the particles, on the $3N - 10 = 3n - 4$ remaining Lorentz-invariant variables associated with the system, and on the polarizations of the particles in the initial and final states. If the scattering produces two outgoing particles, $n = 2$, of momenta q_1^μ, q_2^μ, and rest-masses μ_1, μ_2, starting from two colliding particles with momenta p_1^μ, p_2^μ and rest masses m_1, m_2, the system is described by only two independent Lorentz-invariant variables (aside from the rest-masses which we consider as constants). It is useful to express them in terms of the Mandelstam variables s, t, u:

$$s \equiv (p_1 + p_2)^2 = (q_1 + q_2)^2, \quad t \equiv (p_1 - q_1)^2 = (p_2 - q_2)^2,$$
$$u \equiv (p_1 - q_2)^2 = (p_2 - q_1)^2, \tag{12.26}$$

which are manifestly Lorentz-invariant, though not independent. The relation between s, t, u is readily found by expressing them as follows:

$$s = (m_1^2 + m_2^2)c^2 + 2p_1 \cdot p_2, \quad t = (m_1^2 + \mu_1^2)c^2 - 2p_1 \cdot q_1,$$
$$u \equiv (m_1^2 + \mu_2^2)c^2 - 2p_1 \cdot q_2. \tag{12.27}$$

We then find:

$$\begin{aligned} s + t + u &= (3\,m_1^2 + m_2^2 + \mu_1^2 + \mu_2^2)c^2 + 2p_1 \cdot (p_2 - q_1 - q_2) \\ &= (m_1^2 + m_2^2 + \mu_1^2 + \mu_2^2)c^2. \end{aligned} \tag{12.28}$$

Therefore we can choose as independent Lorentz-invariant variables describing the system any two of s, t, u, the third Mandelstam variable being fixed in terms of them by (12.28).

In the center of mass frame $p_1 = (E_1/c, \mathbf{p})$, $p_2 = (E_2/c, -\mathbf{p})$, $q_1 = (E_1'/c, \mathbf{q})$, $q_2 = (E_2'/c, -\mathbf{q})$, $p_1 + p_2 = ((E_1 + E_2)/c, \mathbf{0})$ and thus $s = (E_1 + E_2)^2/c^2$, so that $c\sqrt{s}$ is the total energy. Consider now an elastic collision ($m_1 = \mu_1$, $m_2 = \mu_2$) in the center of mass frame. In this case $E_1 + E_2 = E_1' + E_2'$ implies $|\mathbf{p}| = |\mathbf{q}|$, from which it follows that $E_1 = E_1'$ and $E_2 = E_2'$. We then find, after some algebra:

$$t = -2\,|\mathbf{p}|^2\,(1 - \cos\theta) = -4\,|\mathbf{p}|^2\,\sin^2\frac{\theta}{2}, \tag{12.29}$$

$$u = -2\,|\mathbf{p}|^2\,(1 + \cos(\theta)) + \frac{(E_1 - E_2)^2}{c^2} = -4\,|\mathbf{p}|^2\,\cos^2\left(\frac{\theta}{2}\right)$$

$$+\,\frac{(E_1 - E_2)^2}{c^2}, \tag{12.30}$$

where θ is the angle between $\mathbf{p}$ and $\mathbf{q}$. The variable $-t$ represents the norm of the momentum $\Delta\mathbf{p} = \mathbf{p} - \mathbf{q}$ transferred during the process.

12.3 Dynamics of Interaction Processes

In the description we have given in last section of decay and scattering processes, we have encoded the dynamics of the event, namely the details of the interaction, in the Lorentz-invariant functions $\hat{\Gamma}$ and Σ, separating it from the kinematics, which is captured by the phase-space element $d\Phi^{(n)}$. In this Section we are going to express these quantities in terms of the interaction Hamiltonian H_I.

12.3.1 Interaction Representation

As anticipated in the introduction, in perturbation theory the Hamiltonian of the interacting system is computed on free fields, namely on fields evolving according to H_0.

In the *Schrödinger picture*, see Sect. 9.3.2, operators, including the Hamiltonian, are constant while states $|\psi(t)\rangle_S$ evolve in time according to the Schroedinger equation:

$$i\hbar \frac{\partial}{\partial t}|\psi(t)\rangle_S = \hat{H}^{(S)}|\psi(t)\rangle_S = (\hat{H}_0 + \hat{H}_I^{(S)})|\psi(t)\rangle_S. \qquad (12.31)$$

Both $\hat{H}_0$ and $\hat{H}_I^{(S)}$ can be expressed in terms of Hamiltonian-density operators

$$\hat{H}_0 = \int d^3\mathbf{x}\, \widehat{\mathcal{H}}_0, \quad \hat{H}_I^{(S)} = \int d^3\mathbf{x}\, \widehat{\mathcal{H}}_I^{(S)}, \qquad (12.32)$$

$\widehat{\mathcal{H}}_0$ and $\widehat{\mathcal{H}}_I^{(S)}$ being functions of the field-operators and their derivatives computed at some fixed time $t = t_0 \ll 0$ and thus not evolving (occasionally, in what follows, we shall explicitly take $t_0 = -\infty$). The reference instant t_0 is chosen long before the interaction process occurs, so that the particles are consistently regarded as free and described in terms of their free-field operators $\hat{\phi}_0$[6]:

$$\widehat{\mathcal{H}}_0 = \widehat{\mathcal{H}}_0(\hat{\phi}_0(t_0, \mathbf{x}), \partial_\mu\hat{\phi}_0(t_0, \mathbf{x})), \quad \widehat{\mathcal{H}}_I^{(S)} = \widehat{\mathcal{H}}_I^{(S)}(\hat{\phi}_0(t_0, \mathbf{x}), \partial_\mu\hat{\phi}_0(t_0, \mathbf{x})). \quad (12.33)$$

Clearly, the Schroedinger picture does not provide a relativistically-covariant description of the interaction since the free-field operators are all computed at $t = t_0$.

The time-evolution of states was described, in Sect. 9.3.2 of Chap. 9, in terms of a time-evolution operator $U(t, t_0)$, defined by property (9.73):

$$|\psi; t\rangle_S = U(t, t_0)|\psi; t_0\rangle_S. \qquad (12.34)$$

Let us recall the main properties of $U(t, t_0)$ discussed in Sect. 9.3.2. The inverse of $U(t, t_0)$ is the operator which maps the state at t back to t_0: $U(t, t_0)^{-1} = U(t_0, t)$. Substituting (12.34) into (12.31) we find that $U(t, t_0)$ is solution to the following equation

$$i\hbar \frac{d}{dt}U(t, t_0) = \hat{H}^{(S)}U(t, t_0), \qquad (12.35)$$

---

[6]Strictly speaking, identifying the fields of the particles long before the interaction process as free is not correct: As we shall see when discussing quantum electrodynamics, although not interacting with each other, charged particles interact with their own electromagnetic field. The effect of this self-interaction will be discussed and taken into account in Sects. 12.7 and 12.8.

with the initial condition $U(t_0, t_0) = 1$. From hermiticity of $\hat{H}^{(S)}$ it follows that $U(t, t_0)$ is unitary. Indeed let us first show that $U(t, t_0)^\dagger U(t, t_0)$ is constant:

$$
\begin{aligned}
\frac{d}{dt} (U(t, t_0)^\dagger U(t, t_0)) &= \left(\frac{d}{dt} U(t, t_0)^\dagger \right) U(t, t_0) + U(t, t_0)^\dagger \left(\frac{d}{dt} U(t, t_0) \right) \\
&= \frac{i}{\hbar} \left(U(t, t_0)^\dagger \hat{H}^{(S)\dagger} \right) U(t, t_0) + U(t, t_0)^\dagger \left(-\frac{i}{\hbar} \hat{H}^{(S)} U(t, t_0) \right) \\
&= \frac{i}{\hbar} U(t, t_0)^\dagger \hat{H}^{(S)} U(t, t_0) - \frac{i}{\hbar} U(t, t_0)^\dagger \hat{H}^{(S)} U(t, t_0) = 0.
\end{aligned}
$$

$$(12.36)$$

Being constant this operator should be equal to its value at $t = t_0$, namely $U(t, t_0)^\dagger U(t, t_0) = 1$, and thus $U(t, t_0)$ is unitary.

In the *Heisenberg picture* states $|\psi\rangle_H$ are constant while operators evolve in time. The relation between states in the Heisenberg and in the Schroedinger pictures is defined by the time-evolution operator: $|\psi\rangle_H = U(t, t_0)^\dagger |\psi(t)\rangle_S$. Similarly the operators $\hat{\mathcal{O}}(t)_H$ and $\hat{\mathcal{O}}$ in the two representations are related to one another in such a way that their mean values on the states is the same: $\hat{\mathcal{O}}(t)_H = U(t, t_0)^\dagger \hat{\mathcal{O}} U(t, t_0)$. As pointed out in the introduction, the fundamental problem which motivates the perturbative approach is that an exact solution to either Eq. (12.31) or (12.35) is not known. In order to develop a Lorentz-covariant perturbation theory, it is convenient to work in the *interaction representation* which is somewhat in between the Schroedinger and the Heisenberg picture. In this representation operators, which depend on the free-field operators, evolve according to $\hat{H}_0$, while states evolve according to $\hat{H}_I^{(S)}$. Let us introduce the time-evolution operator $U_0(t, t_0)$ associated with the free-field theory and thus satisfying the equation:

$$
i \hbar \frac{d}{dt} U_0(t, t_0) = \hat{H}_0 U_0(t, t_0). \tag{12.37}
$$

According to our discussion of Sect. 9.3.2, Chap. 9, being $\hat{H}_0$ a constant operator, the above equation is easily integrated: $U_0(t, t_0) = U_0(t - t_0) = e^{-\frac{i}{\hbar} \hat{H}_0 (t-t_0)}$. The states $|\psi(t)\rangle_I$ and operators $\hat{\mathcal{O}}(t)$ in the interaction picture are related to those in the Schroedinger representation as follows[7]:

$$
\begin{aligned}
|\psi(t)\rangle_I &\equiv U_0(t, t_0)^\dagger |\psi(t)\rangle_S = e^{\frac{i}{\hbar} \hat{H}_0 (t-t_0)} |\psi(t)\rangle_S, \\
\hat{\mathcal{O}}(t) &= e^{\frac{i}{\hbar} \hat{H}_0 (t-t_0)} \hat{\mathcal{O}} e^{-\frac{i}{\hbar} \hat{H}_0 (t-t_0)}.
\end{aligned} \tag{12.38}
$$

It is straightforward to verify that the state $|\psi(t)\rangle_I$ satisfies Eq. (9.76):

---

[7] Note that, for the sake of notational simplicity, we have simply denoted by $\hat{H}_0$ the free Hamiltonian in the Schroedinger representation as well as in the interaction picture.

$$i\hbar \frac{d}{dt}|\psi(t)\rangle_I = \hat{H}_I(t)\,|\psi(t)\rangle_I, \tag{12.39}$$

where $\hat{H}_I(t) \equiv e^{\frac{i}{\hbar}\hat{H}_0(t-t_0)}\hat{H}_I^{(S)}e^{-\frac{i}{\hbar}\hat{H}_0(t-t_0)}$ is the interaction Hamiltonian in the interaction representation. It non-trivially depends on time since $\hat{H}_I$ and $\hat{H}_0$ do not commute. In particular, if we compute in this picture the density $\widehat{\mathcal{H}}_I(t)$ associated with the interaction Hamiltonian, we see that it is expressed in terms of the free-field operators, and their derivatives, computed in a generic space-time point $x^\mu = (ct, \mathbf{x})$ and thus has a Lorentz-covariant expression:

$$\begin{aligned}\widehat{\mathcal{H}}_I(t) &\equiv e^{\frac{i}{\hbar}\hat{H}_0(t-t_0)}\widehat{\mathcal{H}}_I(\hat{\phi}_0(t_0,\mathbf{x}), \partial_\mu\hat{\phi}_0(t_0,\mathbf{x}))\,e^{-\frac{i}{\hbar}\hat{H}_0(t-t_0)} \\ &= \widehat{\mathcal{H}}_I(\hat{\phi}_0(t,\mathbf{x}), \partial_\mu\hat{\phi}_0(t,\mathbf{x})),\end{aligned}$$

where we have used the property of free-field operators of evolving according to $\hat{H}_0$:

$$e^{\frac{i}{\hbar}\hat{H}_0(t-t_0)}\hat{\phi}_0(t_0,\mathbf{x})\,e^{-\frac{i}{\hbar}\hat{H}_0(t-t_0)} = \hat{\phi}_0(t,\mathbf{x}). \tag{12.40}$$

12.3.2 The Scattering Matrix

In perturbation theory the solution to the evolution Eq. (12.39) is sought for in the form of a series expansion in the small coupling-parameter λ which $H_I(t)$ is proportional to. This expansion is to be determined by successive approximations.

Let us now define a time-evolution operator $U_I(t, t_0)$ for states in the interaction representation:

$$|\psi(t)\rangle_I = U_I(t, t_0)\,|\psi(t_0)\rangle_I, \tag{12.41}$$

satisfying the initial condition $U_I(t_0, t_0) = \mathbf{1}$. Substituting in (12.39), we find for $U_I(t, t_0)$ the following equation:

$$i\hbar\frac{d}{dt}U_I(t, t_0) = \hat{H}_I(t)\,U_I(t, t_0), \tag{12.42}$$

which is analogous in the interaction picture to Eq. (12.35) in the Schroedinger representation. Since $\hat{H}_I(t)$ is hermitian $(\hat{H}_I(t) = \hat{H}_I(t)^\dagger)$, we can apply the same argument used for U_0, see Eq. (12.36), to prove that U_I is unitary: $U_I(t, t_0)\,U_I(t, t_0)^\dagger = U_I(t, t_0)^\dagger\,U_I(t, t_0) = \mathbf{1}$. From this it follows that $U_I(t, t_0)^{-1} = U_I(t_0, t) = U_I(t, t_0)^\dagger$.

Let us now seek for a solution to the above equation in the form of a series expansion in λ:

$$U_I(t, t_0) = \sum_{k=0}^{\infty} U_k(t, t_0), \tag{12.43}$$

where each term $U_k(t, t_0)$ is proportional to λ^k. For $\lambda = 0$ there is no interaction and thus the interaction picture coincides with the Heisenberg one in which the states do not evolve in time, implying that the first term in the above expansion is the identity matrix $U_{k=0}(t, t_0) = \mathbf{1}$. The initial condition on $U_I(t, t_0)$ then implies on the other terms: $U_{k>0}(t_0, t_0) = \mathbf{0}$. Substituting the expansion (12.43) in (12.42), recalling that $\hat{H}_I(t)$ is proportional to λ, and equating the coefficients of the same power in the coupling-constant, we find the following iterative relation:

$$\frac{d}{dt} U_k(t, t_0) = -\frac{i}{\hbar} \hat{H}_I(t) U_{k-1}(t, t_0)$$

$$\Rightarrow U_k(t, t_0) = -\frac{i}{\hbar} \int_{t_0}^{t} dt_1 \, \hat{H}_I(t_1) U_{k-1}(t_1, t_0).$$

The above equation is formally solved for each k as follows:

$$U_k(t, t_0) = \left(-\frac{i}{\hbar}\right)^k \int_{t_0}^{t} dt_1 \int_{t_0}^{t_1} dt_2 \ldots \int_{t_0}^{t_{k-1}} dt_k \, \hat{H}_I(t_1) \hat{H}_I(t_2) \ldots \hat{H}_I(t_k). \quad (12.44)$$

It is convenient to write the above composite integral in a form in which all integrals are computed between t_0 and t. To this end, let us consider, for the sake of simplicity, the second order term $U_{k=2}(t, t_0)$:

$$U_2(t, t_0) = -\frac{1}{\hbar^2} \int_{t_0}^{t} dt_1 \int_{t_0}^{t_1} dt_2 \, \hat{H}_I(t_1) \hat{H}_I(t_2). \quad (12.45)$$

In the above expression the integration variables are in the following order $t_1 > t_2$. Let us define the *chronological operator* T as follows:

$$T[\hat{H}_I(t_1) \hat{H}_I(t_2)] \equiv \begin{cases} \hat{H}_I(t_1) \hat{H}_I(t_2) & \text{if } t_1 > t_2 \\ \hat{H}_I(t_2) \hat{H}_I(t_1) & \text{if } t_2 > t_1. \end{cases} \quad (12.46)$$

From the above definition we see that, if we compute the double integral of $T[\hat{H}_I(t_1) \hat{H}_I(t_2)]$ over the square in the (t_1, t_2)-plane defined by $t_0 \le t_1 \le t$, $t_0 \le t_2 \le t$, we find

$$\int_{t_0}^{t} dt_1 \int_{t_0}^{t} dt_2 \, T[\hat{H}_I(t_1) \hat{H}_I(t_2)] = \int_{t_0}^{t} dt_1 \int_{t_0}^{t_1} dt_2 \, \hat{H}_I(t_1) \hat{H}_I(t_2)$$

$$+ \int_{t_0}^{t} dt_2 \int_{t_0}^{t_2} dt_1 \, \hat{H}_I(t_2) \hat{H}_I(t_1).$$

Being t_1, t_2 integration variables, the two terms on the right-hand side are equal and we can then write:

$$U_2(t,\, t_0) = -\frac{1}{\hbar^2} \int_{t_0}^{t} dt_1 \int_{t_0}^{t_1} dt_2\, \hat{H}_I(t_1)\, \hat{H}_I(t_2)$$

$$= -\frac{1}{2\,\hbar^2} \int_{t_0}^{t} dt_1 \int_{t_0}^{t} dt_2\, T[\hat{H}_I(t_1)\, \hat{H}_I(t_2)]. \qquad (12.47)$$

Similarly we define the chronological operator on a generic k-fold product of $\hat{H}_I(t)$ operators at different times, as the operator which rearranges the factors so that the instants at which the operators are computed, decrease in reading the product from left to right (*time-ordered product*):

$$T[\hat{H}_I(t_{i_1})\, \hat{H}_I(t_{i_2}) \dots \hat{H}_I(t_{i_k})] \equiv \hat{H}_I(t_1)\, \hat{H}_I(t_2) \dots \hat{H}_I(t_k), \qquad (12.48)$$

where $t_1 > t_2 > \cdots > t_k$. Generalizing our discussion for the $k = 2$ case, we can convince ourselves that, since the left hand side of (12.48) is symmetric in the interchange of the factors, it contributes $k!$ equal terms when integrated over $t_1, t_2, \dots, t_k$, each varying from t_0 to t, so that the correct expression for $U_k(t, t_0)$ is obtained by dividing this integral by $k!$. Definition (12.48) then allows us to write $U_k(t,\, t_0)$ in (12.44) as follows:

$$U_k(t,\, t_0) = \frac{1}{k!} \left(-\frac{i}{\hbar}\right)^k \int_{t_0}^{t} dt_1 \int_{t_0}^{t} dt_2 \dots \int_{t_0}^{t} dt_k\, T[\hat{H}_I(t_1)\, \hat{H}_I(t_2) \dots \hat{H}_I(t_k)].$$

$$(12.49)$$

Note that, if the values of the operator $\hat{H}_I(t)$ at different times commuted (that is if $[\hat{H}_I(t),\, \hat{H}_I(t')] = 0$), there would be no issue of time-ordering and thus no need of using the T-operator. In this case the right-hand side of Eq. (12.49) would have a simple form in terms of the kth power of a single integral

$$U_k(t,\, t_0) = \frac{1}{k!} \left(-\frac{i}{\hbar}\right)^k \left(\int_{t_0}^{t} dt'\, \hat{H}_I(t')\right)^k, \qquad (12.50)$$

and the series (12.43) is easily summed to an exponential:

$$U_I(t,\, t_0) = \exp\left(-\frac{i}{\hbar} \int_{t_0}^{t} dt'\, \hat{H}_I(t')\right). \qquad (12.51)$$

In general, however, $[\hat{H}_I(t), \hat{H}_I(t')] \neq 0$ and the correct solution to Eq. (12.42) is not given by the above exponential but rather by the following formal expansion

$$U_I(t, t_0) = T\left[\exp\left(-\frac{i}{\hbar}\int_{t_0}^{t} dt'\,\hat{H}_I(t')\right)\right]$$

$$= \sum_{k=0}^{\infty}\frac{1}{k!}\left(-\frac{i}{\hbar}\right)^k \int_{t_0}^{t} dt_1 \int_{t_0}^{t} dt_2 \ldots \int_{t_0}^{t} dt_k\, T[\hat{H}_I(t_1)\,\hat{H}_I(t_2)\ldots\hat{H}_I(t_k)],$$

$$(12.52)$$

where the symbol $T[\exp(\ldots)]$ represents the prescription that the integrand in each multiple integral originating from the expansion of the exponential should be time-ordered. This means that, when acting by means of $U_I(t, t_0)$ on a state $|\psi(t_0)\rangle_I$, the values of the operator $\hat{H}_I(t)$ at earlier times should be applied to it *before* those computed at later times.

Let $|\psi(t)\rangle_I$ describe the state of the system at the time t. Long before the interaction, the system consists of free-particles described by the state $|\psi_{in}\rangle$ (we shall often omit the subscript "I" of the interaction representation), so that:

$$|\psi_{in}\rangle = \lim_{t\to-\infty}|\psi(t)\rangle_I = |\psi(-\infty)\rangle_I. \tag{12.53}$$

At a time t, the state $|\psi(t)\rangle_I$ can be formally expressed in terms of $|\psi_{in}\rangle$ using the time-evolution operator U_I:

$$|\psi(t)\rangle_I = U_I(t, -\infty)\,|\psi_{in}\rangle. \tag{12.54}$$

Long after interaction the system is described by the free-particle state $|\psi(+\infty)\rangle_I$, related to the initial state as follows:

$$|\psi(+\infty)\rangle_I = U_I(+\infty, -\infty)\,|\psi_{in}\rangle = \mathbf{S}\,|\psi_{in}\rangle, \tag{12.55}$$

where we have defined the *scattering matrix* $\mathbf{S}$ (*S*-matrix) as

$$\mathbf{S} \equiv U_I(+\infty, -\infty) = T\left[\exp\left(-\frac{i}{\hbar}\int_{-\infty}^{+\infty} dt'\,\hat{H}_I(t')\right)\right]$$

$$= \sum_{n=0}^{+\infty}\left(\frac{-i}{\hbar}\right)^n \frac{1}{n!}\int_{-\infty}^{+\infty} dt_1 \ldots \int_{-\infty}^{+\infty} dt_n\, T\left[\hat{H}_I(t_1)\ldots\hat{H}_I(t_n)\right]. \tag{12.56}$$

If we now use Eq. (12.32) in the interaction representation, we can express $\mathbf{S}$ in a Lorentz-invariant fashion, in terms of the interaction Hamiltonian density $\widehat{\mathcal{H}}_I(t, \mathbf{x}) = \widehat{\mathcal{H}}_I(x)$:

$$\mathbf{S} \equiv T\left[\exp\left(-\frac{i}{c\,\hbar}\int\limits_{-\infty}^{+\infty} d^4x\,\widehat{\mathcal{H}}_I(x)\right)\right]$$

$$= \sum_{n=0}^{+\infty}\left(\frac{-i}{c\,\hbar}\right)^n\frac{1}{n!}\int\limits_{-\infty}^{+\infty} d^4x_1\dots\int\limits_{-\infty}^{+\infty} d^4x_n\,T\left[\widehat{\mathcal{H}}_I(x_1)\dots\widehat{\mathcal{H}}_I(x_n)\right]. \qquad (12.57)$$

Just as U_I, $\mathbf{S}$ is a unitary operator acting on the Fock space of free-particle states:

$$\mathbf{S}\mathbf{S}^\dagger = \mathbf{1}, \qquad (12.58)$$

and it encodes the information about the interaction.

   In general one is interested in the probability of finding the system, after the interaction, in a free-particle state $|\psi_{out}\rangle$. If the particles are initially prepared in a state $|\psi_{in}\rangle$, this probability $P(in;\ out)$ reads

$$P(in;\ out) = \frac{|\langle\psi_{out}|\psi(+\infty)\rangle|^2}{\langle\psi_{out}|\psi_{out}\rangle\langle\psi(+\infty)|\psi(+\infty)\rangle} = \frac{|\langle\psi_{out}|\mathbf{S}|\psi_{in}\rangle|^2}{\langle\psi_{out}|\psi_{out}\rangle\langle\psi_{in}|\psi_{in}\rangle}, \qquad (12.59)$$

where we have used Eq. (12.55). The *transition amplitude* $A(in;\ out)$ (also called, for scattering processes, *scattering amplitude*) is then given by the matrix element of $\mathbf{S}$ between the initial and final states. Let us define an operator $\mathbf{T}$ so that $\mathbf{S} = \mathbf{1} + i\,\mathbf{T}$. The identity operator only contributes to the transition amplitude when the initial and final states coincide, namely when there is no interaction. Excluding this case we can write

$$A(in;\ out) \equiv \langle\psi_{out}|\mathbf{S}|\psi_{in}\rangle = i\,\langle\psi_{out}|\mathbf{T}|\psi_{in}\rangle. \qquad (12.60)$$

Since the system is isolated, the total four-momentum is conserved. If we think of the initial and final states as consisting of plane waves of total four-momenta P_{in}^μ, P_{out}^μ, respectively, we can factor out of the matrix element (12.60) a delta function implementing this constraint

$$\langle\psi_{out}|\mathbf{T}|\psi_{in}\rangle = (2\,\pi\,\hbar)^4\,\delta^4(P_{out} - P_{in})\,\langle\psi_{out}|\mathscr{T}|\psi_{in}\rangle, \qquad (12.61)$$

As previously emphasized, if the interacting particles were described by plane waves (i.e. eigenstates of the momentum operator), it would not even make sense to talk about "initial" and "final" states. We should always think of the process as of an interaction between wave-packets moving with linear momenta which are narrowly distributed about some average value, so that we can suppose the matrix element of $\mathbf{T}$ not to vary appreciably within the momentum intervals associated with each particle.

   To make contact with our discussion in Sects. 12.2.1 and 12.2.2 let us express the probability of an interaction process per unit time in terms of $\mathbf{S}$-matrix elements.

We start considering a scattering process of two particles of masses m_1, m_2, which yield a number of outgoing particles. The initial state describes two wave-packets of momenta narrowly distributed about two average values $\bar{\mathbf{p}}_1$, $\bar{\mathbf{p}}_2$. Therefore using Fock space representation it has the form

$$|\psi_{in}\rangle = |\psi_1\rangle |\psi_2\rangle,$$
$$|\psi_i\rangle = \int d\Omega_{\mathbf{p}} \sum_r f_i(\mathbf{p}, r) |\mathbf{p}, r\rangle, \quad i = 1, 2, \tag{12.62}$$

where $f_i(\mathbf{p}, r)$ is the weight of each $|\mathbf{p}, r\rangle$ contributing to the wave-packet $|\psi_{in}\rangle$, and where, as usual, $d\Omega_{\mathbf{p}} \equiv \frac{d^3\mathbf{p}_i}{(2\pi\hbar)^3} V_i$. The one-particle states $|\psi_i\rangle$ correspond to the following positive-energy solutions to the Klein-Gordon equation, which, for type-(a) bosons (complex scalar field and electromagnetic field) and type-(c) fermions, respectively, read:

$$\phi_i(x) = \langle 0|\hat{\phi}(x)|\psi_i\rangle = \int \frac{d^3\mathbf{p}}{(2\pi\hbar)^3} \hbar \sqrt{\frac{V_i}{2E_{\mathbf{p}}}} f_i(\mathbf{p}) e^{-\frac{i}{\hbar} p \cdot x}$$
$$= \hbar \int \frac{d\Omega_{\mathbf{p}}}{\sqrt{2E_{\mathbf{p}} V_i}} f_i(\mathbf{p}) e^{-\frac{i}{\hbar} p \cdot x},$$

$$A_\mu(x) = \langle 0|\hat{A}_\mu(x)|\psi_i\rangle = c\hbar \int \frac{d\Omega_{\mathbf{p}}}{\sqrt{2E_{\mathbf{p}} V_i}} \sum_{r=1}^2 f_i(\mathbf{p}, r) \varepsilon_\mu(\mathbf{p}, r) e^{-\frac{i}{\hbar} p \cdot x}, \tag{12.63}$$

$$\psi_i^\alpha(x) = \langle 0|\hat{\psi}^\alpha(x)|\psi_i\rangle = \int d\Omega_{\mathbf{p}} \sqrt{\frac{mc^2}{E_{\mathbf{p}} V_i}} \sum_{r=1}^2 f_i(\mathbf{p}, r) u(\mathbf{p}, r)^\alpha e^{-\frac{i}{\hbar} p \cdot x}, \tag{12.64}$$

We wish now to relate the Fourier coefficients f_i of the wave packets to the corresponding particle density. Consider, for instance, the case in which the incoming particles are described by either a complex scalar field, or a Dirac field.[8] For each of them we can define a conserved current $j_{i\mu} = (\rho_i, \frac{1}{c}\mathbf{j}_i)$, which, for a scalar and a fermion, respectively, reads:

$$j_{i\mu} = i\frac{c}{\hbar} (\phi_i^*(x)\partial_\mu\phi_i(x) - \partial_\mu\phi_i^*(x)\phi_i(x)),$$
$$j_{i\mu} = \overline{\psi}_i \gamma_\mu \psi_i, \tag{12.65}$$

no summation over i. The time-components $\rho_i = j_i^0$ of the current are positive definite, being constructed out of positive-energy solutions, and thus can be consistently regarded as one-particle densities. We are now considering a single interaction event and therefore, being ρ_i related to a single-particle state, it will be normalized as

---

[8]The final relations we are going to derive apply to the photon field as well.

follows[9]: $\int d^3\mathbf{x}\,\rho_i = 1$. This choice, using Eq. (12.65), implies the following normalization for the one-particle states, as the reader can easily verify[10]:

$$1 = \int d^3\mathbf{x}\,\rho_i = \int d\Omega_{\mathbf{p}} \sum_r |f_i(\mathbf{p}, r)|^2 = \langle \psi_i | \psi_i \rangle. \tag{12.66}$$

The above normalization is used also for the outgoing particles, so that $\langle \psi_{in} | \psi_{in} \rangle = \langle \psi_{out} | \psi_{out} \rangle = 1$.

In the following, for the sake of simplicity, we shall limit ourselves to two incoming spin 0 particles, described by complex fields $\phi_i(x)$, although the final relation between the transition probability, the differential cross section and the S-matrix elements, straightforwardly extends to spin 1/2 and 1 particles. Being $\phi_i(x)$ positive energy solutions, $\rho_i > 0$ can be consistently regarded as single-particle densities. We now observe that, substituting (12.63) inside (12.65) we find:

$$j_{i\mu}(x) = i\,\frac{c}{\hbar} \int \frac{d\Omega_{\mathbf{p}}}{\sqrt{2E_{\mathbf{p}}V_i}}\,\frac{d\Omega_{\mathbf{p}'}}{\sqrt{2E_{\mathbf{p}'}V_i}}\,\hbar^2$$

$$\times \left(f_i(\mathbf{p})^* f_i(\mathbf{p}')\,(-i)\,\frac{p'_{i\mu}}{\hbar}\,e^{-\frac{i}{\hbar}(p'-p)\cdot x} - i f_i(\mathbf{p}) f_i(\mathbf{p}')^*\,\frac{p'_{i\mu}}{\hbar}\,e^{\frac{i}{\hbar}(p'-p)\cdot x} \right)$$

$$\approx 2\,\frac{c}{\hbar^2}\,\bar{p}_{i\mu}\,|\phi(\mathbf{x})|^2, \tag{12.67}$$

where we have used the property that $f_i(\mathbf{p})$ are narrowly peaked about average values $\bar{\mathbf{p}}_i$ and thus we have approximated, in the integral, $p'_{i\mu},\,p_{i\mu}$ with $\bar{p}_{i\mu}$.[11] Using the same approximation, and restricting to the time component $\rho_i = j_i^0$ of the currents, we can set $p'^0 \approx p^0 \approx E_{\bar{\mathbf{p}}}/c$, so that the energy factors arising from the time-derivatives cancel against the factor $1/\sqrt{E_{\mathbf{p}}E_{\mathbf{p}'}}$, and we can write the densities ρ_i in the following form:

$$\rho_i(x) \approx \frac{1}{V_i} \int d\Omega_{\mathbf{p}}\,d\Omega_{\mathbf{p}'}\,f_i(\mathbf{p})^*\,f_i(\mathbf{p}')\,e^{-\frac{i}{\hbar}(p'-p)\cdot x}, \tag{12.68}$$

Let $|\psi_{out}\rangle$ describe a system of free outgoing wave packets whose momenta are also narrowly distributed about some average values, the average final total momentum being $\bar{P}_{out\,\mu}$. Let us now write the matrix element of $\mathbf{T}$ in plane waves components

---

[9]In Sects. 12.2.1 and 12.2.2, in contrast to the present section, we were considering collections of particles decaying or interacting (i.e. colliding beams) and thus the densities ρ were referred to multi-particle systems.

[10]Recall our choice of normalization for the single-particle momentum eigenstates: $\langle \mathbf{p}, r | \mathbf{p}', r' \rangle = \frac{(2\pi\hbar)^3}{V}\,\delta^3(\mathbf{p} - \mathbf{p}')\,\delta_{rr'}$.

[11]For fermionic fields, in the same approximation, we would find $j_{i\mu}(x) \approx \frac{\bar{p}_{i\mu}}{mc}\,\overline{\psi}_i(x)\,\psi_i(x)$.

$$\langle\psi_{out}|\mathbf{T}|\psi_{in}\rangle = \int d\Omega_{\mathbf{p}_1}\,d\Omega_{\mathbf{p}_2}\,f_1(\mathbf{p}_1)f_2(\mathbf{p}_2)\,\langle\psi_{out}|\mathbf{T}|\mathbf{p}_1\rangle|\mathbf{p}_2\rangle$$

$$= \int d\Omega_{\mathbf{p}_1}\,d\Omega_{\mathbf{p}_2}\,f_1(\mathbf{p}_1)f_2(\mathbf{p}_2)\,(2\pi\hbar)^4\delta^4(\bar{P}_{out\,\mu} - p_{1\,\mu} - p_{2\,\mu})$$

$$\times\langle\psi_{out}|\mathscr{T}|\mathbf{p}_1\rangle|\mathbf{p}_2\rangle. \tag{12.69}$$

To evaluate the transition probability, we need to compute the squared modulus of the above amplitude:

$$|\langle\psi_{out}|\mathbf{T}|\psi_{in}\rangle|^2 = \int d\Omega_{\mathbf{p}_1}\,d\Omega_{\mathbf{p}_2}\,d\Omega_{\mathbf{p}_1'}\,d\Omega_{\mathbf{p}_2'}\,f_1(\mathbf{p}_1)f_1^*(\mathbf{p}_1')f_2(\mathbf{p}_2)f_2^*(\mathbf{p}_2')$$

$$\times(2\pi\hbar)^8\,\delta^4(\bar{P}_{out\,\mu} - p_{1\,\mu} - p_{2\,\mu})\,\delta^4(\bar{P}_{out\,\mu} - p_{1\,\mu}' - p_{2\,\mu}')$$

$$\times\langle\psi_{out}|\mathscr{T}|\mathbf{p}_1\rangle|\mathbf{p}_2\rangle\langle\psi_{out}|\mathscr{T}|\mathbf{p}_1'\rangle|\mathbf{p}_2'\rangle^*. \tag{12.70}$$

Let us now approximate the matrix elements $\langle\psi_{out}|\mathscr{T}|\mathbf{p}_1\rangle|\mathbf{p}_2\rangle$, $\langle\psi_{out}|\mathscr{T}|\mathbf{p}_1'\rangle|\mathbf{p}_2'\rangle$ with the corresponding value computed on $\bar{\mathbf{p}}_i$, $\langle\psi_{out}|\mathscr{T}|\bar{\mathbf{p}}_1\rangle|\bar{\mathbf{p}}_2\rangle$. Writing $\delta^4(\bar{P}_{out\,\mu} - p_{1\,\mu} - p_{2\,\mu})\,\delta^4(\bar{P}_{out\,\mu} - p_{1\,\mu}' - p_{2\,\mu}')$ as $\delta^4(\bar{P}_{out\,\mu} - p_{1\,\mu} - p_{2\,\mu})\,\delta^4(p_{1\,\mu} + p_{2\,\mu} - p_{1\,\mu}' - p_{2\,\mu}')$ and expressing the second delta function as follows

$$(2\pi\hbar)^4\,\delta^4(p_{1\,\mu} + p_{2\,\mu} - p_{1\,\mu}' - p_{2\,\mu}') = \int d^4x\,e^{-\frac{i}{\hbar}(p_1+p_2-p_1'-p_2')\cdot x}, \tag{12.71}$$

the expression (12.70) can be recast in the form

$$|\langle\psi_{out}|\mathbf{T}|\psi_{in}\rangle|^2 = \left(\int d^4x \int d\Omega_{\mathbf{p}_1}\,d\Omega_{\mathbf{p}_2}\,d\Omega_{\mathbf{p}_1'}\,d\Omega_{\mathbf{p}_2'}\,f_1(\mathbf{p}_1)f_1^*(\mathbf{p}_1')\right.$$

$$\left.\times f_2(\mathbf{p}_2)f_2^*(\mathbf{p}_2')\,e^{-\frac{i}{\hbar}(p_1+p_2-p_1'-p_2')\cdot x}\right)$$

$$\times(2\pi\hbar)^4\,\delta^4(\bar{P}_{out\,\mu} - \bar{p}_{1\,\mu} - \bar{p}_{2\,\mu})\,|\langle\psi_{out}|\mathscr{T}|\bar{\mathbf{p}}_1\rangle|\bar{\mathbf{p}}_2\rangle|^2$$

$$= \left(\int d^4x\,\rho_1(x)\,V_1\,\rho_2(x)\,V_2\right)$$

$$\times(2\pi\hbar)^4\,\delta^4(\bar{P}_{out\,\mu} - \bar{p}_{1\,\mu} - \bar{p}_{2\,\mu})\,|\langle\psi_{out}|\mathscr{T}|\bar{\mathbf{p}}_1\rangle|\bar{\mathbf{p}}_2\rangle|^2, \tag{12.72}$$

where we have used the approximated expressions (12.68). Finally, differentiating both sides of Eq. (12.72) with respect to time and using the definition (12.59) of transition probability we find:

$$\frac{dP}{dt}(in;\,out) = c\left(\int d^3\mathbf{x}\,\rho_1(x)\,V_1\,\rho_2(x)\,V_2\right)$$

$$\times(2\pi\hbar)^4\,\delta^4(\bar{P}_{out\,\mu} - \bar{p}_{1\,\mu} - \bar{p}_{2\,\mu})\,|\langle\psi_{out}|\mathscr{T}|\bar{\mathbf{p}}_1\rangle|\bar{\mathbf{p}}_2\rangle|^2. \tag{12.73}$$

If we consider ρ_1 uniform in the region where particle 2 is localized (i.e. the region in which ρ_2 is non vanishing) we can write in the above expression $\int d^3\mathbf{x}\rho_2 = 1$.

If the final state $|\psi_{out}\rangle = |\alpha, \mathbf{q}_1, \ldots, \mathbf{q}_n\rangle$ describes n outgoing particles with definite momenta $\mathbf{q}_1, \ldots, \mathbf{q}_n$ and characterized by discrete quantum numbers which collectively denoted by α, we would rewrite Eq. (12.73) in the following form:

$$\frac{dP}{dt}(in; \alpha, \mathbf{q}_1, \ldots, \mathbf{q}_n) = c \, \rho_1(x) \, V_1 \, V_2 \, |\langle \alpha, \mathbf{q}_1, \ldots, \mathbf{q}_n | \mathscr{T} | \bar{\mathbf{p}}_1 \rangle | \bar{\mathbf{p}}_2 \rangle|^2 \, d\Phi^{(n)}.$$

$$(12.74)$$

where $d\Phi^{(n)}$, defined in (12.17), is, as usual, the phase space element which depends on the uncertainties $d^3\mathbf{q}_\ell$ in the definition of each final momenta and accounts for the kinematic constraints. Using Eq. (12.25) we can write the differential cross section:

$$d\sigma = \frac{1}{\rho_1 f(\mathbf{v}_1, \mathbf{v}_2)} \frac{dP}{dt}(in; \alpha, \mathbf{q}_1, \ldots, \mathbf{q}_n)$$

$$= c \, \frac{V_1 \, V_2}{f(\mathbf{v}_1, \mathbf{v}_2)} \, |\langle \alpha, \mathbf{q}_1, \ldots, \mathbf{q}_n | \mathscr{T} | \bar{\mathbf{p}}_1 \rangle | \bar{\mathbf{p}}_2 \rangle|^2 \, d\Phi^{(n)}, \qquad (12.75)$$

where, for head-on collisions, $f(\mathbf{v}_1, \mathbf{v}_2) = |\mathbf{v}_1 - \mathbf{v}_2|$.

Along the same lines we can derive the expression of a decay probability per unit time in terms of matrix elements of $\mathbf{T}$. In this case the initial state $|\psi_{in}\rangle = |\psi_1\rangle$ would describe a single particle with rest mass M and average four momentum $\bar{p}^\mu = (\frac{1}{c} E_{\bar{\mathbf{p}}}, \bar{\mathbf{p}})$. The reader can easily verify that

$$\frac{dP}{dt}(in; \alpha, \mathbf{q}_1, \ldots, \mathbf{q}_n) = c \left(\int d^3\mathbf{x} \, \rho(x) \, V \right) |\langle \alpha, \mathbf{q}_1, \ldots, \mathbf{q}_n | \mathscr{T} | \bar{\mathbf{p}} \rangle|^2 \, d\Phi^{(n)}$$

$$= c \, V \, |\langle \alpha, \mathbf{q}_1, \ldots, \mathbf{q}_n | \mathscr{T} | \bar{\mathbf{p}} \rangle|^2 \, d\Phi^{(n)}$$

$$= \frac{V}{\hbar} \, \hat{\Gamma}(in; \mathbf{q}_1, \ldots, \mathbf{q}_n) \, d\Phi^{(n)}, \qquad (12.76)$$

where $d\Phi^{(n)}$ is the phase-space element describing the n outgoing particles, ρ is the density associated with the decaying particle, V the corresponding volume and we have used the definition of the differential width $\hat{\Gamma}$ given in (12.16). The above equation allows to express $\hat{\Gamma}$ in terms of $\mathbf{T}$-matrix elements.

Note that in Eqs. (12.75) and (12.76), for scattering and decay processes respectively, each incoming and outgoing particle contributes a volume factor: In Eq. (12.75) V_i, $i = 1, 2$, for each colliding particle, in Eq. (12.76) the normalization volume V associated with the decaying particle, and V_ℓ, $\ell = 1, \ldots, n$, for each particle produced, from the definition of $d\Phi^{(n)}$. It is useful, as we did in the previous sections, to express each volume factor in terms of the energy of the corresponding particle by a suitable choice of the normalization: $V_i = c_{0i}/(2 E_{\mathbf{p}_i})$ (incoming particles), $V = c_0/(2 E_{\mathbf{p}})$ (decaying particle), and $V_\ell = c_{0\ell}/(2 E_{\mathbf{q}_\ell})$ (outgoing particles). As pointed out in Sect. 12.2.1, the dimensionful normalization coefficients c_{0k}, k running over all the incoming and outgoing particles, should finally drop out of the expression for the cross section or the decay width, since these physical quantities ought not depend on the specific normalization used. They indeed cancel against analogous coefficients originating from $|\langle\alpha,\mathbf{q}_1,\ldots,\mathbf{q}_n|\mathscr{T}|in\rangle|^2$, ($|in\rangle$ being $|\bar{\mathbf{p}}_1\rangle|\bar{\mathbf{p}}_2\rangle$ for scattering processes or $|\bar{\mathbf{p}}\rangle$ for decays). Let us motivate this by anticipating part of the forthcoming discussion on the structure of the scattering amplitude. Consider the contribution to the amplitude $\langle\alpha,\mathbf{q}_1,\ldots,\mathbf{q}_n|\mathscr{T}|in\rangle$, of a generic term in the perturbative expansion (12.57). It will be expressed as the matrix element between the initial and final states of products of a number of Hamiltonian densities $\widehat{\mathcal{H}}_I(x)$ computed in different points:

$$\langle\alpha,\mathbf{q}_1,\ldots,\mathbf{q}_n|\widehat{\mathcal{H}}_I(x_1)\ldots\widehat{\mathcal{H}}_I(x_m)|in\rangle. \tag{12.77}$$

As we shall see in the Sect. 12.3.6, the above matrix element can be written as a sum of terms, each containing a number of propagators, depending on m, which do not contribute volume factors, and factors of the form $\langle\mathbf{p}|\hat{\Phi}|0\rangle$, for each incoming and outgoing particle, Φ being the particle field and $\mathbf{p}$ its momentum.[12] These terms will contribute a factor $\prod_k \frac{1}{\sqrt{V_k}} \propto \prod_k \frac{1}{\sqrt{c_{0k}}}$, k running over the total number N of incoming and outgoing particles. Therefore the squared modulus of the matrix element precisely contains inverse normalization factors which cancel against those originating from the normalization volumes in the formula (12.75), so that the final expression for $d\sigma$ does not depend on them. This implies that we can forget about the normalization factors and safely replace in all formulas the normalization volumes V by $\frac{1}{2E}$. This is done by replacing the matrix element $\langle\mathscr{T}\rangle$ by a rescaled one $\langle\mathscr{T}'\rangle$ defined as follows:

$$\langle\mathscr{T}'\rangle = \prod_{k=1}^{N} \sqrt{V_k\,2E_k}\,\langle\mathscr{T}\rangle, \tag{12.78}$$

where the multiplicative factor on the right hand side does the job of replacing the normalization volumes V_k (one for each incoming and outgoing particle) in the expression of $\langle\mathscr{T}\rangle$ by $1/(2E_k)$. With this position Eqs. (12.75) and (12.76) will read:

$$d\sigma = c\,\frac{1}{4\,E_{\mathbf{p}_1}\,E_{\mathbf{p}_2}f(\mathbf{v}_1,\mathbf{v}_2)}\,|\langle\alpha,\mathbf{q}_1,\ldots,\mathbf{q}_n|\mathscr{T}'|\bar{\mathbf{p}}_1\rangle|\bar{\mathbf{p}}_2\rangle|^2\,d\Phi^{(n)}, \tag{12.79}$$

$$\frac{dP}{dt}(in;\,\alpha,\mathbf{q}_1,\ldots,\mathbf{q}_n) = \frac{1}{2\,\hbar\,E_{\mathbf{p}}}\,\hat{\Gamma}(in;\,\mathbf{q}_1,\ldots,\mathbf{q}_n)\,d\Phi^{(n)}$$

$$= \frac{c}{2\,E_{\mathbf{p}}}\,|\langle\alpha,\mathbf{q}_1,\ldots,\mathbf{q}_n|\mathscr{T}'|\bar{\mathbf{p}}\rangle|^2\,d\Phi^{(n)}, \tag{12.80}$$

---

[12]This will be shown in detail when proving Wick's theorem. Each matrix element $\langle\mathbf{p}|\hat{\Phi}|0\rangle$ contains a factor $\sqrt{V}$ coming from the expansion of $\hat{\Phi}$ in terms of creation and annihilation operators ($a^\dagger$, a, respectively), and a factor $\frac{1}{V}$ coming from the vacuum expectation value (v.e.v.) $\langle 0|aa^\dagger|0\rangle$,
(Footnote 12 continued)
which equals $\langle 0|[a,a^\dagger]|0\rangle = [a,a^\dagger]$ for bosons and $\{a,a^\dagger\}$ for fermions. The matrix element $\langle\mathbf{p}|\hat{\Phi}|0\rangle$ will thus contribute a factor $\frac{1}{\sqrt{V}}$.

where now $d\Phi^{(n)} = (2\pi\hbar)^4 \, \delta^4(\bar{P}_{out\,\mu} - \bar{p}_{1\,\mu} - \bar{p}_{2\,\mu}) \prod_{\ell=1}^{n} d\,\Omega_{\mathbf{q}_\ell}$ and $\Omega_{\mathbf{q}_\ell} = \frac{d^3 q_\ell}{(2\pi\,\hbar)^3 \, 2 E_{\mathbf{q}_\ell}}$.

The formula for the cross section, in particular, can be made manifestly invariant by recalling, from Eq. (12.23), that

$$f(\mathbf{v}_1, \mathbf{v}_2) = \frac{1}{\gamma_1 \, \gamma_2 \, M_1 \, M_2 \, c} \sqrt{(\bar{p}_1 \cdot \bar{p}_2)^2 - M_1^2 \, M_2^2 \, c^4}.$$

This allows to recast Eq. (12.79) in the following Lorentz-invariant form

$$d\sigma = \frac{1}{4 \, c^2 \sqrt{(\bar{p}_1 \cdot \bar{p}_2)^2 - M_1^2 \, M_2^2 \, c^4}} \, |\langle \alpha, \mathbf{q}_1, \dots, \mathbf{q}_n | \mathscr{T}' | \bar{\mathbf{p}}_1 \rangle | \bar{\mathbf{p}}_2 \rangle |^2 \, d\Phi^{(n)}.$$

$$(12.81)$$

If we are in the laboratory frame in which particle 2 is a target particle at rest, the above formula reduces to

$$d\sigma = \frac{1}{4 \, c^3 \, M_2 \, |\bar{\mathbf{p}}_1|} \, |\langle \alpha, \mathbf{q}_1, \dots, \mathbf{q}_n | \mathscr{T}' | \bar{\mathbf{p}}_1 \rangle | \bar{\mathbf{p}}_2 \rangle |^2 \, d\Phi^{(n)}. \qquad (12.82)$$

Upon replacing $V \to 1/(2E)$, the dimension of $|\langle \mathscr{T}' \rangle|^2$ for a scattering process involving two incoming and n outgoing particles is

$$\left[|\langle \mathscr{T}' \rangle|^2 \right] = [length]^{3n-2} \times [energy]^{n+2}. \qquad (12.83)$$

The reader is invited to re-derive Eqs. (12.81) and (12.82) in the more general case in which the incoming particles are generic bosonic fields (like a spin 1 photon) or fermionic fields (like an electron or a positron).

We have reduced the problem of studying, at a perturbative level, an interaction process to that of computing **S**-matrix (or, equivalently, **T**-matrix) elements, which encode the dynamics of the process itself. In what follows we shall introduce, in the framework of quantum electrodynamics, a graphical method, originally developed by Feynman, for computing these matrix elements.

Let us briefly elaborate on the issue of symmetry and its implications for interaction processes. Consider a transformation implemented on the space of states by a unitary operator U. Let U be a symmetry of the free theory which remains a symmetry also in the presence of interaction. In light of our discussion in Sect. 11.8, it will in particular commute with the interaction Hamiltonian: $[U, \hat{H}_I] = 0$. We also require the vacuum $|0\rangle$ of the interacting theory, which we assume to be unique, to be invariant under U: $U|0\rangle = |0\rangle$. It follows that U commutes with **S**:

$$[U, \mathbf{S}] = 0 \quad \Leftrightarrow \quad U^\dagger \mathbf{S} U = \mathbf{S}. \qquad (12.84)$$

As a consequence of this $\langle out|\mathbf{S}|in\rangle = \langle out|U^\dagger \mathbf{S} U|in\rangle = \langle out'|\mathbf{S}|in'\rangle$, where $|in'\rangle \equiv U|in\rangle$ and $|out'\rangle \equiv U|out\rangle$. This implies that the transition probabilities in (12.59) between the original and the transformed states be the same:

$$P(in; out) = P(in'; out').$$ (12.85)

As an example we can consider the Poincaré symmetry, which, as often stressed, encodes the fundamental assumption of homogeneity and isotropy of space-time, and which has been our guiding principle for constructing a relativistic theory. Invariance under a generic Poincaré transformation implies:

$$U^{\dagger}(\Lambda, x_0)SU(\Lambda, x_0) = S.$$ (12.86)

If the symmetry, on the other hand, involves time-reversal T, U is *antiunitary* and thus:

$$U^{\dagger}SU = S^{\dagger},$$ (12.87)

since U commutes, in the expansion of S, with all the $\hat{H}_I$ factors, but switches i into $-i$. Consequently we can write

$$\langle out|S|in \rangle = \langle out|S\,(in) \rangle = \langle U\,S\,(in)|U\,(out) \rangle = \langle S^{\dagger}\,U\,(in)|U\,(out) \rangle$$
$$= \langle in'|S|out' \rangle,$$ (12.88)

that is

$$P(in; out) = P(out'; in').$$ (12.89)

In other words one of the effects of a transformation which involves time reversal, is to invert the roles of the initial and final states, as it was to be expected.

12.3.3 Two-Particle Phase-Space Element

Before moving to dynamics and the computation of amplitudes, let us calculate, in the new conventions, the phase space element $d\Phi^{(2)}$ associated with two final particles of rest-masses μ_1, μ_2:

$$d\Phi^{(2)} = c\,\delta(E_{tot} - E_1' - E_2')\,\delta^3(\mathbf{p}_{tot} - \mathbf{q}_1 - \mathbf{q}_2)\frac{d^3\mathbf{q}_1 d^3\mathbf{q}_2}{(2\pi\hbar)^2\,4E_1'E_2'},$$ (12.90)

where $E_1' = E_{\mathbf{q}_1}$, $E_2' = E_{\mathbf{q}_2}$. We can integrate over $\mathbf{q}_2$ and then move to the center of mass frame where $\mathbf{q}_1 = \mathbf{q} = -\mathbf{q}_2$, obtaining

$$d\Phi^{(2)} = c\,\delta(E_{tot} - E_1' - E_2')\frac{|\mathbf{q}|^2\,d|\mathbf{q}|\,d\Omega}{(2\pi\hbar)^2\,4E_1'E_2'},$$ (12.91)

where we have written $d^3\mathbf{q} = |\mathbf{q}|^2 \, d|\mathbf{q}| \, d\Omega$, $d\Omega$ being the solid angle element[13]: $d\Omega = 2\pi \sin(\theta) d\theta$. Note now that E_1' and E_2' are not independent since $|\mathbf{q}|^2 = \frac{1}{c^2}(E_1')^2 - \mu_1^2 c^2 = \frac{1}{c^2}(E_2')^2 - \mu_2^2 c^2$. We wish to integrate (12.91) in the total energy $E_1' + E_2'$, in order to get rid of the remaining delta-function. To this end we use $|\mathbf{q}| d|\mathbf{q}| = E_1' dE_1'/c^2 = E_2' dE_2'/c^2$ and write:

$$|\mathbf{q}| d|\mathbf{q}| = \frac{E_1' |\mathbf{q}| d|\mathbf{q}| + E_2' |\mathbf{q}| d|\mathbf{q}|}{E_1' + E_2'} = \frac{E_1' E_2'}{c^2} \frac{d(E_1' + E_2')}{E_1' + E_2'}. \tag{12.92}$$

Integration of (12.91) over the total energy then yields:

$$d\Phi^{(2)} = \int \delta(E_{tot} - E_1' - E_2') \frac{|\mathbf{q}| \, d\Omega}{16(\pi\hbar)^2 c} \frac{d(E_1' + E_2')}{E_1' + E_2'} = \frac{|\mathbf{q}| \, d\Omega}{16(\pi\hbar c)^2 \sqrt{s}}, \tag{12.93}$$

We can express now $|\mathbf{q}|^2$ in terms of s by solving[14] $\sqrt{s} = \sqrt{\mu_1^2 c^2 + |\mathbf{q}|^2} + \sqrt{\mu_2^2 c^2 + |\mathbf{q}|^2}$:

$$|\mathbf{q}|^2 = \frac{(\mu_1^2 - \mu_2^2)^2 c^4 - 2s(\mu_1^2 + \mu_2^2)c^2 + s^2}{4s}. \tag{12.94}$$

Equation (12.93) then becomes:

$$d\Phi^{(2)} = \frac{\sqrt{(\mu_1^2 - \mu_2^2)^2 c^4 - 2s(\mu_1^2 + \mu_2^2)c^2 + s^2}}{32(\pi\hbar c)^2 s} d\Omega. \tag{12.95}$$

If, for example, $\mu_1 = \mu_2 = \mu$, we have $E_1' = E_2' = \sqrt{s} c/2$, $|\mathbf{q}|^2 = \frac{s}{4} - \mu^2 c^2$ and the above formula yields:

$$d\Phi^{(2)} = \frac{1}{32(\pi\hbar c)^2} \sqrt{1 - \frac{4\mu^2 c^2}{s}} \, d\Omega. \tag{12.96}$$

It is useful to totally express the phase-space element in terms of invariant quantities, so that the resulting formula can be easily specialized to the frame of reference in which a given process is most conveniently studied. To this end let us consider the

---

[13] We have supposed the dynamics of the process not to depend on the azimuthal angle φ, which is reasonable for an isolated system of interacting particles: In the case of a head-on collision both θ and φ are referred to the common direction of the two incident particles in the CM frame.

[14] For an elastic collision between two particles of rest masses m_1, m_1, we have $\mu_1 = m_1, \mu_2 = m_2$ and $|\mathbf{p}| = |\mathbf{q}|$. Using Eq. (12.94) one can easily show that the factor $\sqrt{(p_1 \cdot p_2)^2 - m_1^2 m_2^2 c^4}$ in the formula (12.81) for the cross-section can be alternatively be written as $\sqrt{s}|\mathbf{p}| = E|\mathbf{p}|/c, E = E_1 + E_2$ being the total energy of the system.

expression (12.29) for t. For given total energy of the system, in the CM frame, $|\mathbf{p}| = |\mathbf{q}|$ is fixed, so that θ is the only variable t depends on. Differentiating both sides we find:

$$d(-t) = -2\,|\mathbf{q}|^2 d\cos(\theta) = \frac{|\mathbf{q}|^2}{\pi}\,d\Omega. \tag{12.97}$$

Substituting the above expression in (12.95), and using (12.94) once again, we find

$$d\Phi^{(2)} = \frac{1}{8(\pi\hbar c)^2}\,\frac{d(-t)}{\sqrt{(\mu_1^2 - \mu_2^2)^2\,c^4 - 2s\,(\mu_1^2 + \mu_2^2)\,c^2 + s^2}}. \tag{12.98}$$

12.3.4 The Optical Theorem

Let us now discuss an important consequence of the unitarity property (12.58) of the S-matrix, known as the *optical theorem*. Consider a scattering between two particles of rest masses m_1, m_2, and let us use the short-hand notation of denoting by S_{fi} and $\mathscr{T}_{fi}$ the matrix elements $\langle \Psi_{out}|\mathbf{S}|\psi_{in}\rangle$, $\langle \Psi_{out}|\mathscr{T}|\psi_{in}\rangle$ (the subscripts f and i stand for final and initial state respectively). Let us now compute the matrix element of both sides of Eq. (12.58) between $|\psi_{in}\rangle$ and $|\psi_{out}\rangle$ and write $\mathbf{SS}^\dagger = \sum_n \mathbf{S}|n\rangle\langle n|\mathbf{S}^\dagger$, where $\{|n\rangle\}$ is a complete set of states in the Fock space. We can then rewrite Eq. (12.58) in components as follows:

$$\sum_n S_{fn}S_{in}^* = \delta_{fi}, \tag{12.99}$$

where we have used the property $\langle \Psi_{out}|\mathbf{1}|\psi_{in}\rangle = \delta_{fi}$ and denoted by S_{fn} and S_{in} the elements $\langle \Psi_{out}|\mathbf{S}|n\rangle$ and $\langle \Psi_{in}|\mathbf{S}|n\rangle$, respectively. Recall now the definition of the matrix element $\mathscr{T}_{ab}$ between two states a and b: $S_{ab} = \delta_{ab} + i\,(2\pi\hbar)^4\delta^4(P_a - P_b)\,\mathscr{T}_{ab}$, P_a, P_b being the total 4-momenta in the two states. Equation (12.99) can then be recast in the following form:

$$i(2\pi\hbar)^4\delta^4(P_{out} - P_{in})\left(\mathscr{T}_{fi} - \mathscr{T}_{if}^*\right)$$
$$+ \sum_n (2\pi\hbar)^8\delta^4(P_{out} - P_n)\delta^4(P_{in} - P_n)\mathscr{T}_{fn}\mathscr{T}_{in}^* = 0, \tag{12.100}$$

where we have used the property of $\{|n\rangle\}$ of being a complete set of states, $\sum_n \delta_{fn}\delta_{in} = \delta_{fi}$. Writing $\delta^4(P_{out} - P_n)\delta^4(P_{in} - P_n) = \delta^4(P_{out} - P_{in})\delta^4(P_{out} - P_n)$, we can deduce from Eq. (12.100) the following equation:

$$\mathscr{T}_{fi} - \mathscr{T}_{if}^* = \sum_n i(2\pi\hbar)^4\delta^4(P_{out} - P_n)\,\mathscr{T}_{fn}\mathscr{T}_{in}^*, \tag{12.101}$$

where all the kinematical quantities are subject to the constraint $P_{in} = P_{out}$. Note that, since $\mathcal{T}_{ab}$ is proportional to the (small) coupling constant λ, by (12.101), its lowest order component in λ is hermitian: $\mathcal{T}_{fi} = \mathcal{T}_{if}^*$.

Suppose now that the kinematical constraints only allow elastic processes. This means that $\delta^4(P_{in} - P_n)\mathcal{T}_{ni}$ is different from zero only for states $|n\rangle$ describing two (free) particles of rest masses m_1, m_2. In the CM frame the final state is totally defined by the scattering angle θ between $\mathbf{q}$ and $\mathbf{p}$. The value $\theta = 0$, in particular, corresponds to the *forward scattering* in which the initial and final states coincide $|\psi_{in}\rangle = |\psi_{out}\rangle$ (i.e. $i = f$ and $\mathbf{p} = \mathbf{q}$). In this case Eq. (12.101) reads:

$$2\,\mathrm{Im}(\mathcal{T}_{ii}) = \sum_n (2\pi\hbar)^4 \delta^4(P_{out} - P_n)\,|\mathcal{T}_{in}|^2. \tag{12.102}$$

Observe now that we can replace the sum over the intermediate states n, by the integral over the momenta $\mathbf{q}_1$, $\mathbf{q}_2$ of the corresponding two particles and the sum $\sum_{pol.(n)}$ over their polarizations:

$$\sum_n \rightarrow \sum_{pol.(n)} \int \frac{d^3\mathbf{q}_i}{(2\pi\hbar)^3}\, V_i. \tag{12.103}$$

Equation (12.102) will then read:

$$2\,\mathrm{Im}(\mathcal{T}_{ii}) = \sum_{pol.(n)} \int |\mathcal{T}_{in}|^2\, d\Phi^{(2)}. \tag{12.104}$$

Note that the left hand side is proportional to the total cross-section of the elastic scattering

$$\sigma_t(in) \equiv \int d\sigma(in;\, n) = \frac{c\,V_1\,V_2}{f(\mathbf{v}_1,\, \mathbf{v}_2)} \sum_{pol.(n)} \int |\mathcal{T}_{in}|^2\, d\Phi^{(2)}, \tag{12.105}$$

where we have used Eq. (12.75). Equation (12.104) can now be written in the following form:

$$2\,\mathrm{Im}(\mathcal{T}_{ii}) = \frac{f(\mathbf{v}_1,\, \mathbf{v}_2)}{c\,V_1\,V_2}\, \sigma_t. \tag{12.106}$$

We have not performed the replacement $V \rightarrow 1/(2E)$ yet. This is done by writing the left hand side in terms of $\mathcal{T}_{ii}'$, defined in Eq. (12.78):

$$\mathcal{T}_{ii} = \frac{1}{V_1\, V_2\, 4\, E_1\, E_2}\, \mathcal{T}_{ii}'. \tag{12.107}$$

The normalization volumes cancel and we end up with:

$$\text{Im}(\mathcal{T}_{ii}') = 2\,c^2\,\sqrt{(p_1 \cdot p_2)^2 - m_1^2 m_2^2\,c^4}\,\sigma_t = 2\,c\,E\,|\mathbf{p}|\,\sigma_t, \qquad (12.108)$$

where E is the total energy in the CM frame and we have used the comment in footnote 14. Equation (12.108) directly descends from the unitarity property of $\mathbf{S}$ and relates the imaginary part of the forward scattering amplitude to the total cross section of the process. It describes the content of the *optical theorem*.

12.3.5 Natural Units

Our analysis would simplify considerably if we could get rid of all the factors $\hbar$, c occurring in our formulas. This can be done by an appropriate redefinition of the units of measure: Being $\hbar$, c dimensionful universal constants, we can consistently relate the units of length (m), time (s) and mass (kg) to one another so that in the new system of units (referred to as *natural units*) $\hbar = c = 1$.[15] This can be done, for instance, by choosing length (or mass) as the only fundamental quantity and by defining in terms of it the units of mass (or length) and time. Let us denote by $c, \hbar$ the measures (numbers) of c and $\hbar$, respectively, in the standard system of units:

$$c = \text{c}\ (\text{m/s}), \quad \hbar = \hbar\ \left(\text{kg}\,(\text{m})^2/\text{s}\right),$$
$$\text{c} = 2.997925 \times 10^8, \quad \hbar = 1.054589 \times 10^{-34}.$$

When expressing meter, second, kg in natural units, we have:

$$\text{c}\,(\text{m/s}) = 1, \quad \hbar\ \left(\text{kg}\,(\text{m})^2/\text{s}\right) = 1.$$

The above equations can be solved in any one of the three units. Choosing for instance the meter as independent unit:

$$1\,\text{s} = \text{c}\,(\text{m}) \approx 3 \times 10^8\,(\text{m}), \quad 1\,\text{kg} = \frac{\text{c}}{\hbar}\,(\text{m}^{-1}) \approx 2.84 \times 10^{42}\,(\text{m}^{-1}),$$
$$1\,\text{J} = \frac{1}{\text{c}\hbar}\,(\text{m}^{-1}) \approx 3.16 \times 10^{25}\,(\text{m}^{-1}),\ 1\,\text{eV} \approx 5.07 \times 10^6\,(\text{m}^{-1}).$$

We see that in this system of units mass dimensions is inverse to length dimension while time has the same dimension as length:

---

[15] We are familiar with a similar choice for heat and energy: being the *mechanical equivalent of heat* a universal constant (4.186 J/cal) we can think of its value as just related to the different operational definitions used for heat and work, so that measuring the two equivalent quantities in the same standard units, namely setting 1cal = 4.186 J, this constant equals one.

By the same token, in the (rationalized) Heaviside-Lorentz system of units, see footnote 1 of Chap. 5, the unit of measurement for the electric charge is defined in terms of the units of length, mass and time by requiring that the vacuum permittivity ε_0 be one.

$$[time] = [length], \quad [mass] = [energy] = \frac{1}{[length]}.$$

In particular the mass m of a particle has the same value as its rest energy mc^2 and as the inverse Compton wavelength of the particle $mc/\hbar$. The electric charge, which in the Heaviside-Lorentz system had dimension of $(mass)^{\frac{1}{2}} (length)^{\frac{1}{2}} \times (time)^{-1}$, in natural units is *dimensionless*. The measures of the elementary electric charge and of the electron mass are

$$e \approx 0.303, \quad \alpha = \frac{e^2}{4\pi} \approx \frac{1}{137},$$
$$m_e \approx 9.109 \times 10^{-31} \text{ kg} = m_e c^2 \approx 0.511 \text{ MeV} \approx 2.59 \times 10^{12} \text{ (m}^{-1}).$$

The dimensions of bosonic and fermionic fields are:

$$[\phi] = \frac{1}{[length]}, \quad [\psi] = \frac{1}{[length]^{\frac{3}{2}}}, \tag{12.109}$$

while action is dimensionless. The Fourier expansion of a field operator in these units will have the following simpler form:

Complex scalar field :
$$\hat{\phi}(x) = \hat{\phi}_{(+)}(x) + \hat{\phi}_{(-)}(x),$$
$$\hat{\phi}_{(+)}(x) = \int \frac{d^3\mathbf{p}}{(2\pi)^3 \, 2E_\mathbf{p}} \, a_\mathbf{p} \, e^{-ip\cdot x},$$
$$\hat{\phi}_{(-)}(x) = \int \frac{d^3\mathbf{p}}{(2\pi)^3 \, 2E_\mathbf{p}} \, b_\mathbf{p}^\dagger \, e^{ip\cdot x},$$

Real vector field :
$$\hat{A}_\mu(x) = \hat{A}_{(+)\,\mu}(x) + \hat{A}_{(-)\,\mu}(x),$$
$$\hat{A}_{(+)\,\mu}(x) = \int \frac{d^3\mathbf{p}}{(2\pi)^3 \, 2E_\mathbf{p}} \sum_r a(\mathbf{p}, r)\, \varepsilon_\mu(\mathbf{p}, r) \, e^{-ip\cdot x},$$
$$\hat{A}_{(-)\,\mu}(x) = \int \frac{d^3\mathbf{p}}{(2\pi)^3 \, 2E_\mathbf{p}} \sum_r a(\mathbf{p}, r)^\dagger \, \varepsilon_\mu(\mathbf{p}, r)^* \, e^{ip\cdot x},$$

Spin–1/2 field:
$$\hat{\psi}(x) = \hat{\psi}_{(+)}(x) + \hat{\psi}_{(-)}(x),$$
$$\hat{\psi}_{(+)}(x) = \int \frac{d^3\mathbf{p}}{(2\pi)^3 \, 2E_\mathbf{p}} \sqrt{2m} \sum_{r=1}^{2} c(\mathbf{p}, r)\, u(\mathbf{p}, r) \, e^{-ip\cdot x},$$
$$\hat{\psi}_{(-)}(x) = \int \frac{d^3\mathbf{p}}{(2\pi)^3 \, 2E_\mathbf{p}} \sqrt{2m} \sum_{r=1}^{2} d(\mathbf{p}, r)^\dagger \, v(\mathbf{p}, r) \, e^{ip\cdot x},$$

where, as explained in the previous Section, we have replaced V by $1/(2\,E)$. In the sequel, we shall often label the helicity states of the photon by the index $i = 1, 2$, and denote the corresponding (complex) polarization vectors by $\varepsilon_\mu(\mathbf{p}, i)$, not to confuse it with the analogous index r of the fermion field.

Let us also recall the expressions for the Feynman propagators, which we shall use in the following sections:

<u>Complex scalar field :</u>

$$D_F(x - y) = \langle 0|T[\hat{\phi}(x)\hat{\phi}^\dagger(y)]|0\rangle = \int \frac{d^4p}{(2\pi)^4}\, D_F(p)\, e^{-ip\cdot(x-y)},$$

$$D_F(p) = \frac{i}{p^2 - m^2 + i\epsilon},$$

<u>Real, massless vector field :</u>

$$D_{F\mu\nu}(x - y) = \langle 0|T[\hat{A}_\mu(x)\hat{A}_\nu(y)]|0\rangle$$

$$= \int \frac{d^4p}{(2\pi)^4}\, D_{F\mu\nu}(p)\, e^{-ip\cdot(x-y)},$$

$$D_{F\mu\nu}(p) = -\frac{i}{p^2 + i\epsilon}\left(\eta_{\mu\nu} - (1 - \alpha)\frac{p_\mu p_\nu}{p^2}\right),$$

where α should not be mistaken for the fine structure constant. It is the constant associated with the choice of gauge fixing for the photon field.

<u>Spin–1/2 field:</u>

$$S_F(x - y)^\alpha{}_\beta = \langle 0|T[\hat{\psi}^\alpha(x)\overline{\hat{\psi}}_\beta(y)]|0\rangle$$

$$= \int \frac{d^4p}{(2\pi)^4}\, S_F(p)^\alpha{}_\beta\, e^{-ip\cdot(x-y)},$$

$$S_F(p) = \frac{i}{\not{p} - m + i\epsilon}. \tag{12.110}$$

From now on we shall use the natural units. The factors $\hbar$, c in any formula can be eventually restored by straightforward dimensional arguments.

12.3.6 The Wick's Theorem

In what follows we shall be interested in computing **S**-matrix elements between the initial and final states. This requires evaluating matrix elements of time-ordered products of interaction Hamiltonians, of the form (12.77). To this end it will be useful to express a time ordered product of free-field operators in terms of normal ordered products of the same operators. This is the content of Wick's theorem.

Let us introduce the notion of *contraction* between field-operators. If $\varphi_1(x), \varphi_2(x)$ denote generic (bosonic or fermionic) field operators (we shall omit, from now on, the hat over the symbols of field operators), the contraction $\overline{\varphi_1(x_1)\, \varphi_2(x_2)}$ is defined as follows:

$$\overline{\varphi_1(x_1)\, \varphi_2(x_2)} \equiv \langle 0|T\varphi_1(x_1)\, \varphi_2(x_2)|0\rangle. \tag{12.111}$$

Clearly to the v.e.v. on the right hand side only products of annihilation and creation operators of the *same kind* (i.e. associated with the same field) contribute, and thus the contraction is non vanishing only if $\varphi_2 = \varphi_1^\dagger$. A non vanishing contraction thus coincides with the Feynman propagator:

$$\overline{\phi(x_1)\phi^\dagger(x_2)} = D_F(x_1 - x_2),$$
$$\overline{\psi^\alpha(x_1)\bar{\psi}_\beta(x_2)} = S_F(x_1 - x_2)^\alpha{}_\beta,$$
$$\overline{A_\mu(x_1)A_\nu(x_2)} = D_{F\mu\nu}(x_1 - x_2). \tag{12.112}$$

If we denote by $\varphi_{(+)}$ and $\varphi_{(-)}$ the positive and negative energy parts of φ, proportional to the annihilation, creation operators respectively, we can write the contraction, or Feynman propagator, as a commutator (anti-commutator for fermionic fields) of field operators.[16] Suppose first $x_1^0 > x_2^0$

$$\overline{\varphi(x_1)\, \varphi^\dagger(x_2)} = \langle 0|\varphi(x_1)\, \varphi^\dagger(x_2)|0\rangle = \langle 0|\varphi_{(+)}(x_1)\, \varphi_{(-)}^\dagger(x_2)|0\rangle \tag{12.113}$$
$$= \langle 0|\left[\varphi_{(+)}(x_1),\, \varphi_{(-)}^\dagger(x_2)\right]_\pm |0\rangle = \left[\varphi_{(+)}(x_1),\, \varphi_{(-)}^\dagger(x_2)\right]_\pm,$$

where we have used the property that $\varphi_{(+)}|0\rangle = \langle 0|\varphi_{(-)} = 0$ and the fact that the commutator (or anti-commutator for fermions), of two field operators is a complex number. Similarly for $x_2^0 > x_1^0$ we find

$$\overline{\varphi(x_1)\, \varphi^\dagger(x_2)} = \pm\langle 0|\varphi^\dagger(x_2)\, \varphi(x_1)|0\rangle = \pm\left[\varphi_{(+)}^\dagger(x_2),\, \varphi_{(-)}(x_1)\right]_\pm, \tag{12.114}$$

the lower sign, here and in the following, refers to the case of two (components of) fermionic fields. From the definition of time ordering it follows that the contraction of two fermionic operators is odd with respect to the inversion of their order:

$$\overline{\psi(x_1)\psi(x_2)} = -\overline{\psi(x_2)\psi(x_1)}.$$

Let us now define the contraction between two operators which are not adjacent within a product:

---

[16]For the sake of simplicity we shall also denote the commutator and anti-commutator by $[\cdot,\, \cdot]_+, [\cdot,\, \cdot]_-$, respectively. That is: $[\cdot,\, \cdot]_+ = [\cdot,\, \cdot], [\cdot,\, \cdot]_- = \{\cdot,\, \cdot\}$.

$$\varphi_1(x_1)\varphi_2(x_2)\cdots\varphi_{k-1}(x_{k-1})\varphi_1^\dagger(x_k)\varphi_{k+1}(x_{k+1})\cdots$$

$$= \pm\left(\varphi_1(x_1)\,\varphi_1^\dagger(x_k)\right)\ \varphi_2(x_2)\cdots\varphi_{k-1}(x_{k-1})\,\varphi_{k+1}(x_{k+1})\cdots \quad (12.115)$$

where the minus sign only occurs if φ_1 is fermionic and, in bringing $\varphi_1^\dagger$ in front of the product, it has crossed an odd number of fermionic fields, as in the following case:

$$\psi_1(x_1)\psi_2(x_2)\overline{\psi}_1(x_3) = -\psi_1(x_1)\overline{\psi}_1(x_3)\ \psi_2(x_2). \quad (12.116)$$

Wick's Theorem A time-ordered product of field-operators can be written as a sum of normal ordered products as follows

$$T[\varphi_1(x_1)\ldots\varphi_n(x_n)] = :\varphi_1(x_1)\ldots\varphi_n(x_n): +$$

$$+ \sum_{\substack{\text{single}\\\text{contraction}}} :\varphi_1(x_1)\cdots\cdots\varphi_n(x_n): +$$

$$+ \sum_{\substack{\text{two}\\\text{contractions}}} :\varphi_1(x_1)\cdots\cdots\cdots\cdots\varphi_n(x_n): +$$

$$+\cdots \quad (12.117)$$

where the final ellipses represent terms with a higher number of contractions.

Before proving it, as an example, let us apply the theorem to a four-fermion product (one should think of each field below as a generic component of a Dirac spinor, so that $\overline{\psi}(x_1)$ should be intended as $\overline{\psi}_\alpha(x_1)$, $\psi(x_2)$ as $\psi^\beta(x_2)$, $\overline{\psi}(x_3)$ as $\overline{\psi}_\gamma(x_3)$ and $\psi(x_4)$ as $\psi^\sigma(x_4)$, with no contraction in general among the indices, which are suppressed for the sake of notational simplicity):

$$T[\overline{\psi}(x_1)\psi(x_2)\overline{\psi}(x_3)\psi(x_4)]$$

$$=:\overline{\psi}(x_1)\psi(x_2)\overline{\psi}(x_3)\psi(x_4): + :\overline{\psi}(x_1)\psi(x_2)\overline{\psi}(x_3)\psi(x_4)$$

$$: + :\overline{\psi}(x_1)\psi(x_2)\overline{\psi}(x_3)\psi(x_4): +\overline{\psi}(x_1)\psi(x_2)\overline{\psi}(x_3)\psi(x_4)$$

$$: + :\overline{\psi}(x_1)\psi(x_2)\overline{\psi}(x_3)\psi(x_4): + :\overline{\psi}(x_1)\psi(x_2)\overline{\psi}(x_3)\psi(x_4)$$

$$: + :\overline{\psi}(x_1)\psi(x_2)\overline{\psi}(x_3)\psi(x_4):$$

$$=:\overline{\psi}(x_1)\psi(x_2)\overline{\psi}(x_3)\psi(x_4): -S_F(x_2 - x_1): \overline{\psi}(x_3)\psi(x_4):$$

$$: -S_F(x_4 - x_1): \psi(x_2)\overline{\psi}(x_3): + S_F(x_2 - x_3): \overline{\psi}(x_1)\psi(x_4):$$

$$: -S_F(x_4 - x_3): \overline{\psi}(x_1)\psi(x_2): + S_F(x_2 - x_1)\,S_F(x_4 - x_3)$$

$$-S_F(x_4 - x_1)\,S_F(x_2 - x_3). \quad (12.118)$$

In the above derivation we have used the properties $\overline{\psi\psi} = \overline{\psi\psi} = 0$. We shall prove Wick's theorem by induction. Let us first prove it for $n = 2$ and use, for the sake of simplicity, the short-hand notation $\varphi_i \equiv \varphi_i(x_i)$:

$$T[\varphi_1\,\varphi_2] = :\varphi_1\,\varphi_2: +\overline{\varphi_1\,\varphi_2}. \tag{12.119}$$

From Eqs. (12.113) and (12.114), if φ_1 and φ_2 commute, their contraction is zero and the creation and annihilation operators in the time ordered product can be rearranged to obtain a normal ordered expression, so that Eq. (12.119) is trivially satisfied. If the two field operators do not commute, namely if $\varphi_2 = \varphi_1^\dagger$, we start considering the case $x_1^0 > x_2^0$ and expand the left hand side into products of the positive and negative energy components of the two fields:

$$
\begin{aligned}
T[\varphi_1\,\varphi_2] = \varphi_1\,\varphi_2 &= \varphi_{1\,(+)}\,\varphi_{2\,(+)} + \varphi_{1\,(+)}\,\varphi_{2\,(-)} + \varphi_{1\,(-)}\,\varphi_{2\,(+)} + \varphi_{1\,(-)}\,\varphi_{2\,(-)} \\
&= \varphi_{1\,(+)}\,\varphi_{2\,(+)} \pm \varphi_{2\,(-)}\,\varphi_{1\,(+)} + \varphi_{1\,(-)}\,\varphi_{2\,(+)} + \varphi_{1\,(-)}\,\varphi_{2\,(-)} \\
&\quad + \left[\varphi_{1\,(+)},\,\varphi_{2\,(-)}\right]_\pm =: \varphi_1\,\varphi_2: +\overline{\varphi_1\,\varphi_2},
\end{aligned}
\tag{12.120}
$$

where, in order to obtain a normal ordered expression, we had to swap the positions of $\varphi_{1\,(+)}$ and $\varphi_{2\,(-)}$ in second term of the second line, and this has produced a commutator/anti-commutator (the lower sign, as usual, refers to the case of two fermionic fields). Then we have used Eq. (12.113). If, on the other hand, $x_1^0 < x_2^0$ we have

$$
\begin{aligned}
T[\varphi_1\,\varphi_2] = \pm\varphi_2\,\varphi_1 &= \pm\left(\varphi_{2\,(+)}\,\varphi_{1\,(+)} + \varphi_{2\,(+)}\,\varphi_{1\,(-)} + \varphi_{2\,(-)}\,\varphi_{1\,(+)} + \varphi_{2\,(-)}\,\varphi_{1\,(-)}\right) \\
&= \pm\left(\varphi_{2\,(+)}\,\varphi_{1\,(+)} \pm \varphi_{1\,(-)}\,\varphi_{2\,(+)} + \varphi_{2\,(-)}\,\varphi_{1\,(+)} + \varphi_{2\,(-)}\,\varphi_{1\,(-)}\right) \\
&\quad \pm\left[\varphi_{2\,(+)},\,\varphi_{1\,(-)}\right]_\pm =: \varphi_1\,\varphi_2: +\overline{\varphi_1\,\varphi_2},
\end{aligned}
\tag{12.121}
$$

Suppose now the theorem holds for the product of n fields, let us prove it for $n + 1$. We start from a time-ordered product of the form $T[\varphi\varphi_1\ldots\varphi_n]$, where $\varphi \equiv \varphi(x)$. With no loss of generality, we can assume $x^0 > x_1^0, \ldots, x_n^0$, so that

$$
\begin{aligned}
T[\varphi\varphi_1\ldots\varphi_n] = \varphi\,T[\varphi_1\ldots\varphi_n] = \varphi\,\Big[&:\varphi_1\ldots\varphi_n: + \\
&+ \sum_{\substack{\text{single}\\\text{contraction}}} :\varphi_1\cdots\overline{\cdots\cdots}\varphi_n: + \\
&+ \sum_{\substack{\text{two}\\\text{contractions}}} :\varphi_1\cdots\overline{\cdots\cdots}\,\overline{\cdots\cdots}\varphi_n: + \cdots \Big],
\end{aligned}
\tag{12.122}
$$

where we have applied Wick's theorem to $T[\varphi_1\ldots\varphi_n]$. It is useful now to write $\varphi = \varphi_{(+)} + \varphi_{(-)}$ and to insert each component $\varphi_{(\pm)}$ inside the normal ordered products within square bracket. Since $\varphi_{(-)}$ contains a creation operator and multiplies the normal ordered terms to the left, it can be moved inside the normal order symbol, since the resulting product is already normal ordered: $\varphi_{(-)} : \varphi_1 \ldots \varphi_n :=: \varphi_{(-)}\varphi_1 \ldots \varphi_n :.$ This is not the case for $\varphi_{(+)}$, which contains an annihilation operator and thus should be moved to the right of all the creation operators in a normal ordered product in order for the resulting expression to be in normal order as well. Every time φ is moved past a field to the right, a commutator (or anti-commutator) is produced. Considering, for simplicity, only bosonic fields, we can write

$$
\begin{aligned}
\varphi_{(+)} : \varphi_{i_1} \ldots \varphi_{i_k} : &=: \varphi_{i_1} \ldots \varphi_{i_k} : \varphi_{(+)} + : [\varphi_{(+)}, \varphi_{i_1}]\varphi_{i_2} \ldots \varphi_{i_k} \\
&\quad : + : \varphi_{i_1}[\varphi_{(+)}, \varphi_{i_2}] \ldots \varphi_{i_k} : + \ldots : \varphi_{i_1}\varphi_{i_2} \ldots [\varphi_{(+)}, \varphi_{i_k}] \\
&=: \varphi_{(+)}\varphi_{i_1}\varphi_{i_2} \ldots \varphi_{i_k} : + : \overline{\varphi\varphi_{i_1}}\varphi_{i_2} \ldots \varphi_{i_k} \\
&\quad : + : \overline{\varphi\varphi_{i_1}\varphi_{i_2}} \ldots \varphi_{i_k} : + \cdots + : \overline{\varphi\varphi_{i_1}\varphi_{i_2} \ldots \varphi_{i_k}} :, \quad (12.123)
\end{aligned}
$$

where we have used the properties $: \varphi_{i_1} \ldots \varphi_{i_k} : \varphi_{(+)} =: \varphi_{(+)}\varphi_{i_1}\varphi_{i_2} \ldots \varphi_{i_k} :$ and $: \overline{\varphi\varphi_i} = [\varphi_{(+)}, \varphi_i]$ (recall that $x^0 > x_i^0$). The reader is invited to explicitly verify Eq. (12.123) in the simple case of three bosonic fields $\varphi, \varphi_1, \varphi_2$, by writing the expression of the normal ordered products in terms of the positive and negative-energy components of the fields, and to generalize the above derivation to the case of fermionic fields. We can now apply Eq. (12.123) to the product of $\varphi_{(+)}$ with each normal ordered term within square brackets in Eq. (12.122). The contractions in Eq. (12.123) yield all the missing contractions involving φ, which are needed to write Wick's formula (12.117) for the case of the $n + 1$ fields $\varphi, \varphi_1, \ldots, \varphi_n$. From the first term, for instance, indeed get : $\varphi_{(+)}\varphi_1 \ldots \varphi_n$: which sums up with: $\varphi_{(-)}\varphi_1 \ldots \varphi_n$: to give : $\varphi\varphi_1 \ldots \varphi_n$:, plus all the terms with single contractions involving φ. This completes the proof of the theorem.

We shall be interested in applying Wick's theorem to write a generic time ordered product

$$
T[\widehat{\mathcal{H}}_I(x_1) \ldots \widehat{\mathcal{H}}_I(x_n)], \quad (12.124)
$$

of the interaction Hamiltonian, in terms of normal ordered quantities. Each operator $\widehat{\mathcal{H}}_I(x)$ consists of a normal ordered product of field operators computed in the same point x. The whole product (12.124), however, is not normal ordered. In other words we have a time-ordered product of normal-ordered groups $\widehat{\mathcal{H}}_I(x)$ of field operators. As a corollary of Wick's theorem, we can prove that, when applying Eq. (12.117), the contractions between fields within a same normal ordered group (i.e. a same factor $\widehat{\mathcal{H}}_I(x)$) do not contribute. Let us prove this in the simple case of a normal ordered group : $\varphi(x)\varphi^\dagger(x)$: consisting of two fields and apply Wick's theorem to a product of the form $T[: \varphi(x)\varphi^\dagger(x) : \varphi_1 \ldots \varphi_n]$. Since the two fields $\varphi(x), \varphi^\dagger(x)$ are computed at the same time, we can write their product as a time-ordered one, and apply Eq. (12.119):

$$
\varphi(x)\varphi^\dagger(x) = T[\varphi(x)\varphi^\dagger(x)] =: \varphi(x)\varphi^\dagger(x) : + \overline{\varphi(x)\varphi}^\dagger(x). \quad (12.125)
$$

We can now write

$$T[: \varphi(x)\varphi^\dagger(x) : \varphi_1 \ldots \varphi_n] = T[\varphi(x)\overbrace{\varphi^\dagger(x)\varphi_1 \ldots \varphi_n}] - \overbrace{\varphi(x)\varphi^\dagger(x)}\, T[\varphi_1 \ldots \varphi_n].$$

Applying Wick's theorem to the time ordered terms on the right hand side, we see that the second term precisely cancels against the one containing the contraction $\overbrace{\varphi(x)\varphi^\dagger(x)}$ from the first one, which therefore does not appear in the final expression.

Let us now consider the problem of evaluating the matrix elements of **S** (or equivalently of **T**) between an initial state $|in\rangle = |\mathbf{p}_1, r_1; \ldots; \mathbf{p}_k, r_k\rangle$ describing k incoming particles with momenta $\mathbf{p}_1, \ldots, \mathbf{p}_k$ and spin components $r_1, \ldots, r_k$ (the $k = 1$ case corresponds to a decay process), and a final state $|\mathbf{q}_1, s_1; \ldots; \mathbf{q}_n, s_n\rangle$ describing n outgoing particles with momenta $\mathbf{q}_1, \ldots, \mathbf{q}_n$ and spin components $s_1, \ldots, s_n$. Denoting generically by a_i, $a_i^\dagger$ the annihilation and creation operators associated with the ith particle, respectively, we can write

$$|\mathbf{p}_1, r_1; \ldots; \mathbf{p}_k, r_k\rangle = \prod_{i=1}^{k} a_i(\mathbf{p}_i, r_i)^\dagger |0\rangle,$$

$$|\mathbf{q}_1, s_1; \ldots; \mathbf{q}_k, s_k\rangle = \prod_{\ell=1}^{n} a_\ell(\mathbf{q}_\ell, s_\ell)^\dagger |0\rangle, \tag{12.126}$$

so that

$$\langle \mathbf{q}_1, s_1; \ldots; \mathbf{q}_k, s_k | \mathbf{T} | \mathbf{p}_1, r_1; \ldots; \mathbf{p}_k, r_k\rangle = \langle 0 | \prod_{\ell=1}^{n} a_\ell(\mathbf{q}_\ell, s_\ell)\, \mathbf{T} \prod_{i=1}^{k} a_i(\mathbf{p}_i, r_i)^\dagger |0\rangle. \tag{12.127}$$

If we apply Wick's theorem to each term in the perturbative expansion (12.57) of the **S** matrix, we see that the only terms which contribute to the above matrix element are those containing for each incoming particle an annihilation operator on the right to match the corresponding creation operator acting on the vacuum, and, for each outgoing particle, a creation operator on the left, to match the corresponding annihilation operator to the left of **T**. In this way, the creation and annihilation operators, for each particle, would combine into non-vanishing matrix elements of the form $\langle 0 | a_j\, a_j^\dagger | 0 \rangle \neq 0$. Therefore the terms in the expression of **T** contributing to the amplitude of the process have the general (normal-ordered) form

$$\prod_{\ell=1}^{n} a_\ell(\mathbf{q}_\ell, s_\ell)^\dagger \prod_{i=1}^{k} a_i(\mathbf{p}_i, r_i) \times \text{(contractions)}, \tag{12.128}$$

the number of contractions depending on the order in λ of the term in the perturbative expansion (12.57). In the next section we shall review, within the theory of quantum electrodynamics, the Feynman rules for computing the contributions, of different order, to the amplitude of a given process.

12.4 Quantum Electrodynamics and Feynman Rules

Quantum electrodynamics (QED) is the quantum field theory describing the interaction of electrons and/or positrons, the quanta of the Dirac field, with photons, the quanta of the electromagnetic field. The Lagrangian density is obtained from Eq. (10.228) of Chap. 10 by adding the term describing the free Maxwell field $A_\mu(x)$, and reads:

$$
\begin{aligned}
\mathcal{L} &= \overline{\psi}\,(i\,\slashed{D} - m)\psi - \frac{1}{4}\, F_{\mu\nu}\, F^{\mu\nu} = \overline{\psi}\,(i\,\slashed{\partial} - m)\psi - \frac{1}{4}\, F_{\mu\nu}\, F^{\mu\nu} + A_\mu J^\mu \\
&= \mathcal{L}_0 + \mathcal{L}_I,
\end{aligned}
\tag{12.129}
$$

where, as usual, $\slashed{D} \equiv \gamma^\mu D_\mu$, $D_\mu \equiv \partial_\mu - ie A_\mu$ being the covariant derivative, and $J^\mu(x) \equiv e\,\overline{\psi}(x)\gamma^\mu\,\psi(x)$ is the conserved electric current.[17] The interaction Hamiltonian is $\mathcal{H}_I(x) = -\mathcal{L}_I = -A_\mu J^\mu$ and the corresponding operator is obtained by replacing the fields in its expression by the corresponding *free* operators, and by normal ordering the resulting products[18]:

$$
\widehat{\mathcal{H}}_I(x) \equiv -e\, : \overline{\psi}(x)\gamma^\mu\,\psi(x)\, A_\mu(x) : .
\tag{12.130}
$$

If we want to describe electromagnetic interaction processes involving not just an electron but also other charged fermion particles, like muons for example, we would need to include in the definition of $\mathcal{L}$ the corresponding kinetic term and electric current. For instance, to include a fermion with charge q and field $\psi_q(x)$, the electric current is to be defined as: $J_\mu(x) = e\,\overline{\psi}(x)\gamma^\mu\,\psi(x) + q\,\overline{\psi}_q(x)\gamma^\mu\,\psi_q(x)$.

Let us start considering for the time being interaction processes involving electrons, positrons and the electromagnetic fields. We need to compute the S-matrix elements between the initial and final states. The scattering matrix $\mathbf{S}$ is defined perturbatively in the coupling constant e, according to Eq. (12.57), the nth-order term $\mathbf{S}_{(n)}$ having the form:

$$
\mathbf{S}_{(n)} = \frac{(-i)^n}{n!} \int_{-\infty}^{+\infty} d^4x_1 \ldots \int_{-\infty}^{+\infty} d^4x_n\, T\left[\widehat{\mathcal{H}}_I(x_1) \ldots \widehat{\mathcal{H}}_I(x_n)\right].
\tag{12.131}
$$

As pointed out in the previous section, each of these terms is a time-ordered product of normal-ordered quantities $\widehat{\mathcal{H}}_I(x)$. In order to compute the contribution to a given process of $\mathbf{S}_{(n)}$, we would need to apply Wick's theorem in order to express each time-ordered product in terms of normal-ordered terms. It is then useful to represent each of these terms by a diagram, which will considerably simplify the task of computing the corresponding contribution to the amplitude. We associate with each factor $\widehat{\mathcal{H}}_I(x_i)$,

---

[17]In order to restore the $\hbar$ and c factors in the covariant derivative, we simply need to replace $e \to \frac{e}{\hbar c}$, as the reader can easily verify.

[18]For the sake of simplicity, we shall suppress the hats on the symbol of the field operators.

$i = 1, \ldots, n$, in the integral (12.131) a point in the diagram, localized in x_i^μ, called *vertex*. The plane of the diagram thus represents space-time, with just one spatial direction. Which of the two is the time direction will then depend on the specific process we are going to consider and will not be specified for the time being. The operator $\widehat{\mathcal{H}}_I(x_i)$ consists of three field-operators computed in x_i: $\psi(x_i)$ which destroys an electron in x_i or creates a positron in the same point and $\bar{\psi}(x_i)$ which creates an electron or destroys a positron in x_i. The former will be represented by a solid line ending in the vertex and directed towards it, the latter by a solid line originating in the vertex and directed outwards. Finally the operator $A_\mu(x_i)$ creates or destroys a photon in x_i. It will be represented by an undirected dashed line ending in x_i. In this way, we have associated with each factor in the integral (12.131) a vertex with three lines. The Wick expansion of the integrand in (12.131) will contain normal-ordered terms in which two or more operators belonging to different $\widehat{\mathcal{H}}_I$ factors are contracted (recall that contractions between operators in the same $\widehat{\mathcal{H}}_I$-factor do not enter the Wick expansion). These terms are graphically represented by drawing the three line-vertices for each factor and connecting the lines corresponding to the contracted operators. Contracted operators are then represented graphically by lines connecting two vertices: $\overline{\psi^\alpha(x_i)\bar{\psi}_\beta(x_j)}$ by a line connecting x_i and x_j and oriented from x_j to x_i; $\overline{A_\mu(x_i)A_\nu(x_j)}$ by a dashed, undirected line joining x_i to x_j. The reason why the latter line is not oriented is due to the symmetry of the photon propagator $D_{F\mu\nu}(x_i - x_j)$ with respect to an exchange of x_i and x_j: $D_{F\mu\nu}(x_i - x_j) = D_{F\mu\nu}(x_j - x_i)$. Each of these internal lines describes a *virtual* particle propagating between the two vertices. By virtual particle we mean a particle whose momentum does not satisfy the *on-shell* condition: $p^2 - m^2 = 0$ for the electron, $k^2 = 0$ for the photon, k^μ being the photon 4-momentum.

The normal ordered terms, besides the contractions, will also contain un-contracted field operators, which we shall refer to as *free*. These operators will be represented by lines extending from the vertex in which they are computed to infinity: The line representing $\psi(x)$ will originate at infinity and end in x. In the matrix element between initial and final state, it will contribute only if the initial state contains a free electron or if the final state contains a free positron. In these two cases $\psi(x)$ will destroy the incoming electron or create the outgoing positron in x, respectively. In the first case, if the electron state is $|\mathbf{p}_{e^-}, r\rangle = c^\dagger(\mathbf{p}_{e^-}, r)|0\rangle$, $\psi(x)$ would give the following non-vanishing contribution to the amplitude[19]:

---

[19]Recall that, in the light of our comments below Eq. (12.76), we have replaced everywhere the normalization volume of each particle with $1/(2E)$ so that, for instance, $[c(\mathbf{p}, r), c^\dagger(\mathbf{q}, s)]_- = (2\pi)^3 \, 2E \, \delta_{rs} \, \delta^3(\mathbf{p} - \mathbf{q})$. Note that, had we kept the normalization volumes, the calculation below would yield $\langle 0|\psi(x)|\mathbf{p}, r\rangle = \sqrt{\frac{m}{E_\mathbf{p} V}} \, u(\mathbf{p}, r) \, e^{-ip\cdot x}$, contributing a factor $1/\sqrt{V}$ to the amplitude, as anticipated in our discussion below Eq. (12.76).

$$\langle 0|\psi(x)|\mathbf{p}_{e^-}, r\rangle = \int d\Omega_{\mathbf{q}} \sqrt{2m} \sum_{s=1}^{2} u(\mathbf{q}, s) \langle 0|c(\mathbf{q}, s) c^\dagger(\mathbf{p}_{e^-}, r)|0\rangle e^{-iq \cdot x}$$

$$= \int d\Omega_{\mathbf{q}} \sqrt{2m} \sum_{s=1}^{2} u(\mathbf{q}, s) \langle 0|[c(\mathbf{q}, s), c^\dagger(\mathbf{p}_{e^-}, r)]_- |0\rangle_{e^-} e^{-iq \cdot x}$$

$$= \sqrt{2m}\, u(\mathbf{p}_{e^-}, r)\, e^{-ip_{e^-} \cdot x}. \tag{12.132}$$

Similarly, in the second case, if the final positron state is $|\mathbf{p}_{e^+}, r\rangle = d^\dagger(\mathbf{p}_{e^+}, r)|0\rangle$, the field $\psi(x)$ will contribute the following non-vanishing quantity:

$$\langle \mathbf{p}_{e^+}, r|\psi(x)|0\rangle = \sqrt{2m}\, v(\mathbf{p}_{e^+}, r)\, e^{-ip_{e^+} \cdot x}. \tag{12.133}$$

By the same token we can show that a free $\overline{\psi}(x)$ operator contributes only to those processes with an incoming positron, which will be destroyed in x, or with an outgoing electron, which will be created in x by the same operator. It will be represented by a line originating in the vertex x and extending towards infinity. In these two cases $\overline{\psi}(x)$ will therefore, contribute the following matrix elements to the amplitude:

$$\langle 0|\overline{\psi}(x)|\mathbf{p}_{e^+}, r\rangle = \sqrt{2m}\, \bar{v}(\mathbf{p}_{e^+}, r)\, e^{ip_{e^+} \cdot x},$$

$$\langle \mathbf{p}_{e^-}, r|\overline{\psi}(x)|0\rangle = \sqrt{2m}\, \bar{u}(\mathbf{p}_{e^-}, r)\, e^{ip_{e^-} \cdot x}. \tag{12.134}$$

We can also have a process with an electron in both the initial and final states. In this case the normal-ordered product $: \overline{\psi}_\alpha(y)\, \psi^\beta(x) :$ will contribute the following quantity to the amplitude:

$$\langle \mathbf{q}_{e^-}, r| : \overline{\psi}_\alpha(y)\, \psi^\beta(x) : |\mathbf{p}_{e^-}, r\rangle = \langle 0|c(\mathbf{q}_{e^-}, r)\, \overline{\psi}_{(-)\alpha}(y)\, \psi^\beta_{(+)}(x) c^\dagger(\mathbf{p}_{e^-}, r)|0\rangle$$

$$= \langle 0| \left[c(\mathbf{q}_{e^-}, r), \overline{\psi}_{(-)\alpha}(y) \right]_- \left[\psi^\beta_{(+)}(x), c^\dagger(\mathbf{p}_{e^-}, r) \right]_- |0\rangle$$

$$= \langle \mathbf{q}_{e^-}, r|\overline{\psi}_\alpha(y)|0\rangle \langle 0|\psi^\beta(x)|\mathbf{p}_{e^-}, r\rangle, \tag{12.135}$$

where we have used the property that $\langle 0|c\, c^\dagger c\, c^\dagger|0\rangle = \langle 0|[c, c^\dagger]_- [c\, c^\dagger]_- |0\rangle = \langle 0|c\, c^\dagger |0\rangle \langle 0|c\, c^\dagger |0\rangle$.

Finally a free photon field operator $A_\mu(x)$ can either destroy an incoming photon or create an outgoing one in x, giving the following contributions to the amplitude in these two cases respectively:

$$\langle 0|A_\mu(x)|\mathbf{k}, i\rangle = \varepsilon_\mu(\mathbf{k}, i)\, e^{-ik \cdot x},$$

$$\langle \mathbf{k}, i|A_\mu(x)|0\rangle = \varepsilon_\mu(\mathbf{k}, i)^*\, e^{ik \cdot x}, \tag{12.136}$$

where $i = 1, 2$ labels the transverse polarizations of the photon. Graphically a free $A_\mu(x)$ operator is represented by an infinite, undirected, dashed line extending from infinity to the vertex x. A same diagram can therefore describe a variety of processes:

A solid line extending from infinity to a vertex x (oriented towards it), can either describe an incoming electron destroyed in x or an outgoing positron created in x, and similarly a solid line originating in x and ending at infinity can either describe an incoming positron destroyed in x or an outgoing electron generated in the same point. The direction on a fermion line is thus not related to the direction of motion, but rather to the flow of the electron charge. Similarly a dashed line stretching from x to infinity can either describe a photon destroyed in x (i.e. absorbed) or a photon created at the same point. This ambiguity is due to the fact that a same normal ordered term in the Wick expansion of $\mathbf{S}_{(n)}$ will in general contribute to different processes. The lines of a diagram which originate or end at infinity are called *external legs*. When we consider specific processes, we shall identify the external legs with incoming or outgoing particles, thus fixing the time direction in the graph.

Consider the lowest order term $\mathbf{S}_{(1)}$:

$$\mathbf{S}_{(1)} = ie \int d^4x \; : \overline{\psi}(x)\gamma^\mu \psi(x) A_\mu(x) : . \tag{12.137}$$

The integrand is already normal-ordered and all field operators are free. It is represented by the vertex in Fig. 12.1 with three external legs. It may describe a process in which an electron decays in x into an electron and a photon (photon emission), or the analogous decay of a positron, or an electron and a positron annihilating and giving rise to a photon etc. All these processes, although having a non vanishing amplitude, cannot occur because of kinematical reasons. Consider, for instance, an electron with momentum p which emits a photon with momentum k, ending up in a free state with momentum p'. The momentum conservation implies:

$$p = p' + k. \tag{12.138}$$

Computing the norm of both sides and using the on-shell conditions $p'^2 = p^2 = m^2$, $k^2 = 0$, we find $m^2 = m^2 + 2k \cdot p'$, namely $k \cdot p' = 0$. In the rest frame of the initial electron this condition reads $k^0 p'^0 - \mathbf{k} \cdot \mathbf{p}' = k^0 (p'^0 + |\mathbf{p}'|)$, where we have used the properties $\mathbf{k} = -\mathbf{p}'$ and $k^0 = |\mathbf{k}|$. Since $p'^0 + |\mathbf{p}'| > 0$, momentum conservation implies $k^0 = 0$, namely that there is no final photon and that the initial electron stays still. Thus the probability of the process is suppressed by the phase space element $d\Phi$, which is non vanishing only for the trivial process with $k^0 = 0$.

Fig. 12.1 A vertex in QED

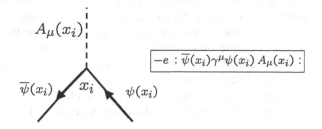

Let us now consider the second order term $\mathbf{S}_{(2)}$:

$$\mathbf{S}_{(2)} = -\frac{e^2}{2!} \int d^4x\, d^4y\, T[:\overline{\psi}(x)\gamma^\mu\psi(x)A_\mu(x)::\overline{\psi}(y)\gamma^\nu\psi(y)A_\nu(y):].$$

$$(12.139)$$

Wick expanding the integrand we find the following terms:

(1) $:\overline{\psi}(x)\gamma^\mu\psi(x)A_\mu(x)\,\overline{\psi}(y)\gamma^\nu\psi(y)A_\nu(y):$

(2) $:\overline{\psi}(x)\gamma^\mu\psi(x)\overline{A_\mu(x)\,\overline{\psi}(y)}\gamma^\nu\psi(y)A_\nu(y):$

$\qquad =:\overline{\psi}(x)\gamma^\mu\psi(x)\,\overline{\psi}(y)\gamma^\nu\psi(y):\,D_{F\mu\nu}(x-y)$

(3) $:\overline{\psi}(x)\gamma^\mu\overline{\psi(x)\,A_\mu(x)\,\overline{\psi}(y)}\gamma^\nu\psi(y)A_\nu(y):$

$\qquad =:\overline{\psi}(x)\gamma^\mu\,S_F(x-y)\,\gamma^\nu\psi(y)\,A_\mu(x)\,A_\nu(y):$

(4) $:\overline{\psi}(x)\gamma^\mu\psi(x)A_\mu(x)\,\overline{\psi}(y)\gamma^\nu\psi(y)A_\nu(y):$

$\qquad =:\overline{\psi}(y)\gamma^\mu\,S_F(y-x)\,\gamma^\nu\psi(x)\,A_\mu(y)\,A_\nu(x):$

(5) $:\overline{\psi}(x)\gamma^\mu\psi(x)A_\mu(x)\,\overline{\psi}(y)\gamma^\nu\psi(y)A_\nu(y):$

$\qquad = -S_F(y-x)^\alpha{}_\beta(\gamma^\mu)^\beta{}_\sigma\,S_F(x-y)^\sigma{}_\delta\,(\gamma^\nu)^\delta{}_\alpha\,:A_\mu(x)\,A_\nu(y):$

$\qquad = -\text{Tr}\left(S_F(y-x)\,\gamma^\mu\,S_F(x-y)\,\gamma^\nu\right)\,:A_\mu(x)\,A_\nu(y):$

(6) $:\overline{\psi}(x)\gamma^\mu\psi(x)A_\mu(x)\,\overline{\psi}(y)\gamma^\nu\psi(y)A_\nu(y):$

$\qquad =:\overline{\psi}(x)\gamma^\mu\,S_F(x-y)\gamma^\nu\psi(y):\,D_{F\mu\nu}(x-y)$

(7) $:\overline{\psi}(x)\gamma^\mu\psi(x)A_\mu(x)\,\overline{\psi}(y)\gamma^\nu\psi(y)A_\nu(y):$

$\qquad =:\overline{\psi}(y)\gamma^\mu\,S_F(y-x)\,\gamma^\nu\psi(x):\,D_{F\mu\nu}(y-x):$

(8) $:\overline{\psi}(x)\gamma^\mu\psi(x)A_\mu(x)\,\overline{\psi}(y)\gamma^\nu\psi(y)A_\nu(y):$

$\qquad = -\text{Tr}\left(S_F(y-x)\,\gamma^\mu\,S_F(x-y)\,\gamma^\nu\right)\,D_{F\mu\nu}(y-x):$ (12.140)

Note that the terms (3) and (4) only differ for the exchange $x \leftrightarrow y$, and thus, upon integrating over the positions of the two vertices, give an equal contribution. The same holds for (7) and (6). All the diagrams corresponding to these terms are illustrated in Fig. 12.2. We see that the diagram representing the first term is disconnected and describes two separate, single-vertex, processes occurring in x and y, each of these are forbidden by the conservation of 4-momentum, as explained above. The second term may describe an electron-electron scattering, in which one of the two incoming electrons emits a virtual photon in x, which is absorbed by the second incoming electron in y. Similarly it may also describe an electron-positron or a positron-positron

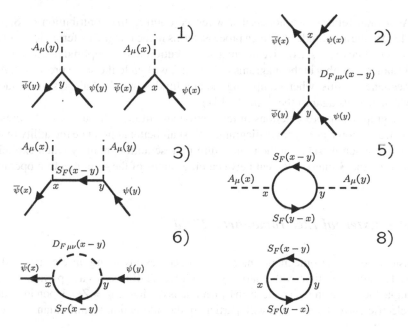

Fig. 12.2 Diagrams of $S_{(2)}$

scattering.*Thus in QED the electric interaction between two charged particles is described in terms of an exchange of virtual photons between them.* Finally the same diagram can describe the annihilation of an incoming electron and a positron in a point x, which produces a virtual photon decaying in y into a new electron-positron pair. The diagram (3) has several interpretations as well. It may describe an electron emitting two photons in y and x respectively, or absorbing a photon in y and emitting one in x, or vice-versa, or finally absorbing two photons in x and y respectively. It may also describe analogous emission/absorption processes by a positron moving from x to y (opposite to the flow of the electron charge). Note that, between two consecutive emissions/absorptions, the electron or positron is described by its propagator, namely it is virtual. Diagram (5) may describe a photon which produces a couple of virtual electron and positron in y, which annihilate in x to produce a final photon. Diagram (6) may describe an electron emitting a virtual photon in y and re-absorbing it in x. In other words the electron interacts with itself. Such process represents then a *self-interaction* of an electron or a positron. Finally (8) is a vacuum diagram: The initial and final states are both empty and at some point y, for instance, a virtual electron-positron couple and a photon are created from the vacuum and then destroyed in some other point x. Diagrams (1)–(3) do not contain *loops* and are thus called *tree diagrams*. The remaining diagrams, on the contrary, contain *loops*. As we shall see in Sect. 12.7, the corresponding amplitudes ar plagued by infinities. Note that in all the above diagrams there is no discontinuity in the orientation of the fermion lines. This is a general property which is related to the conservation of the electric charge and thus to the continuity of its flow in a given process.

As a final remark, we notice that when computing the contribution of $S_{(n)}$ in (12.131) to the amplitude of a given process, terms in the integral differing in the order of the n vertices $x_1, \ldots, x_n$ give an equal contribution to the amplitude. Exchanging, for instance, x with y in the diagrams in Fig. 12.2 will yield the same processes. As a consequence of this, when computing the amplitudes, we shall always find a factor $n!$ which cancels against the $\frac{1}{n!}$ in (12.131).

The graphical representations of interaction amplitudes, discussed in the present section, are known as Feynman diagrams. We shall better appreciate the utility of this technique when working in the momentum representation, namely when explicitly computing the S-matrix element between eigenstates of the 4-momentum operator.

12.4.1 External Electromagnetic Field

Consider now electromagnetic interaction processes involving an electron and an other fermionic particle with charge q, which we shall refer to as "particle q" (for example the scattering of an electron by a nucleus of charge $q = Ze$). Upon including the electric current associated with particle q, the interaction Hamiltonian will read

$$\widehat{\mathcal{H}}_I(x) \equiv - : [e\, \overline{\psi}(x)\gamma^\mu\, \psi(x) + q\, \overline{\psi}_q(x)\gamma^\mu\, \psi_q(x)]\, A_\mu(x) : . \qquad (12.141)$$

Suppose particle q is massive enough for its state not to change during the process:

$$|\psi_{in}\rangle = |in_e\rangle|in_q\rangle, \quad |\psi_{out}\rangle = |out_e\rangle|out_q\rangle = |out_e\rangle|in_q\rangle,$$

in other words $|out_q\rangle = |in_q\rangle$. For these processes we need not have a second quantized description of particle q in terms of a field operator, since we do not need to destroy an initial state and to create a new final one. Indeed, the only terms in the Wick expansion of S at all orders, which contribute to the amplitude are those having, as free operators, only ψ which destroys the electron in the initial state, $\overline{\psi}$ which creates an electron in the final state, ψ_q which destroys the incoming particle q and $\overline{\psi}_q$ which creates an outgoing one in the *same* state. The terms should not contain any external A_μ, since neither $|\psi_{in}\rangle$ nor $|\psi_{out}\rangle$ contains photons. This implies that all the photon fields are to be computed between vacuum states, namely they only contribute in contractions (i.e. propagators). The lowest order term contributing to the scattering amplitude is $S_{(2)}$:

$$S_{(2)} = \frac{(i)^2}{2!}\, e\, q \int d^4x\, d^4y : \left[\overline{\psi}(x)\gamma^\mu\, \psi(x)\overline{\psi}_q(y)\gamma^\nu\, \psi_q(y)\, D_{F\mu\nu}(x-y) \right.$$

$$\left. + (x \leftrightarrow y) \right] := (i)^2\, e\, q \int d^4x\, d^4y : \overline{\psi}(x)\gamma^\mu\, \psi(x)\overline{\psi}_q(y)\gamma^\nu\, \psi_q(y) : D_{F\mu\nu}(x-y).$$

$$(12.142)$$

where we have used the fact that, by virtue of the parity of $D_F^{\mu\nu}(x-y)$, the two terms in square bracket give an equal contribution to the integral. Computing the matrix element of $\mathbf{S}_{(2)}$ between the initial and final states, we find:

$$\langle\psi_{out}|\mathbf{S}_{(2)}|\psi_{in}\rangle = -e \int d^4x d^4y \langle out_e| : \overline{\psi}(x)\gamma^\mu \psi(x) : |in_e\rangle \, D_{F\mu\nu}(x-y) \, J_q^\nu(y),$$

(12.143)

where

$$J_q^\nu(y) \equiv \langle in_q| : \overline{\psi}_q(y)\gamma^\nu \psi_q(y) : |in_q\rangle,$$

(12.144)

is the classical current associated with particle q in the state $|in_q\rangle$. Recall now, from the definition of the Green's function D_F, that:

$$A_\mu^{ext}(x) \equiv i \int d^4y D_{F\mu\nu}(x-y) J_q^\nu(y),$$

(12.145)

is the classical electromagnetic field generated by the current J_q^ν. In our problem it represents the electromagnetic field generated by a particle whose state is unperturbed by the interaction process and will be referred to as an *external field*. The amplitude (12.143) can the be recast in the following first order form:

$$\langle\psi_{out}|\mathbf{S}_{(2)}|\psi_{in}\rangle = \langle out_e| \left(ie \int d^4x : \overline{\psi}(x)\gamma^\mu \psi(x) : A_\mu^{ext}(x) \right) |in_e\rangle$$

$$= \langle out_e| \left(-i \int d^4x \widehat{\mathcal{H}}_I^{ext}(x) \right) |in_e\rangle,$$

(12.146)

where

$$\widehat{\mathcal{H}}_I^{ext}(x) = -e : \overline{\psi}(x)\gamma^\mu \psi(x) : A_\mu^{ext}(x).$$

(12.147)

We have shown that, in all interaction processes in which particle q is just a "spectator", its effect on the electron can be accounted for by means of the external field A_μ^{ext} it generates. This is done by adding to the QED Hamiltonian describing just the electron and the electromagnetic field, the corresponding interaction term, generalizing thus the definition of the interaction Hamiltonian

$$\widehat{\mathcal{H}}_I'(x) \equiv \widehat{\mathcal{H}}_I(x) + \widehat{\mathcal{H}}_I^{ext}(x) = -e : \overline{\psi}(x)\gamma^\mu \psi(x) \, (A_\mu(x) + A_\mu^{ext}(x)) : .$$

This amounts in turn to redefining the electromagnetic potential in the QED Lagrangian (12.129) as the sum $A_\mu(x) + A_\mu^{ext}(x)$ of the field operator $A_\mu(x)$ and the external field $A_\mu^{ext}(x)$. Let us stress that $A_\mu^{ext}(x)$ is a classical field and not an

Fig. 12.3 Interaction with an external field

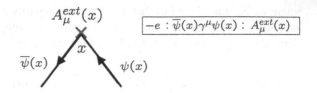

operator, namely it is a number and thus acts as the identity on the Fock space of free photons. Therefore the interaction term $\widehat{\mathcal{H}}_I^{ext}(x)$ contains just two field operators, $\psi, \overline{\psi}$. Graphically it will be represented by a 2-line vertex, with the external field being represented by a cross, as in Fig. 12.3.

12.5 Amplitudes in the Momentum Representation

12.5.1 Möller Scattering

Let us start considering a specific process describing the scattering between two electrons (labeled by 1, 2 respectively):

$$e^- + e^- \longrightarrow e^- + e^-. \tag{12.148}$$

The initial state describes the incoming electrons with momenta $\mathbf{p}_1$, $\mathbf{p}_2$ and polarizations r_1, r_2, respectively. The final momenta and polarizations of the two electrons are $\mathbf{q}_1$, $\mathbf{q}_2$ and s_1, s_2 respectively:

$$|\psi_{in}\rangle = |\mathbf{p}_1, r_1\rangle|\mathbf{p}_2, r_2\rangle,$$
$$|\psi_{out}\rangle = |\mathbf{q}_1, s_1\rangle|\mathbf{q}_2, s_2\rangle. \tag{12.149}$$

We shall compute the amplitude of the process to lowest order, namely the matrix element of $\mathbf{S}_{(2)}$ between the initial and final states. The only term contributing to the amplitude is the one described by the diagram (2) in Fig. 12.2, so that:

$$\langle\psi_{out}|\mathbf{S}_{(2)}|\psi_{in}\rangle = \frac{(ie)^2}{2!} \int d^4x d^4y$$
$$\times \Bigg[\langle\mathbf{q}_1, s_1|\langle\mathbf{q}_2, s_2| : \overline{\psi}(x)\gamma^\mu\psi(x)\,\overline{\psi}(y)\gamma^\nu\psi(y) :$$
$$|\mathbf{p}_1, r_1\rangle|\mathbf{p}_2, r_2\rangle\,D_{F\mu\nu}(x-y) \Bigg]. \tag{12.150}$$

We can convince ourselves that the only term in the normal product which contributes to the matrix element is the one of the form $c^\dagger c^\dagger cc$, since we need to destroy the two incoming electrons and to create the two outgoing ones. Let us explicitly compute the corresponding matrix element, bearing in mind that the two $c^\dagger$ operators come from the $\overline{\psi}$ fields, while the two c operator originate from the ψ fields. We write the initial and final states in terms of creation operators acting on the vacuum:

$$|\mathbf{p}_1, r_1\rangle|\mathbf{p}_2, r_2\rangle = c(\mathbf{p}_1, r_1)^\dagger c(\mathbf{p}_2, r_2)^\dagger|0\rangle,$$
$$|\mathbf{q}_1, s_1\rangle|\mathbf{q}_2, s_2\rangle = c(\mathbf{q}_1, s_1)^\dagger c(\mathbf{q}_2, s_2)^\dagger|0\rangle. \qquad (12.151)$$

There is a peculiarity about this kind of processes which involve identical particles in the initial and final states: There is an overall sign ambiguity in the amplitude due to the choice of the order in which the creation operators are written in (12.151). We can write:

$$\langle\psi_{out}| : \overline{\psi}(x)\gamma^\mu\psi(x)\,\overline{\psi}(y)\gamma^\nu\psi(y) : |\psi_{in}\rangle$$
$$= \int d\Omega_q d\Omega_p d\Omega_{q'} d\Omega_{p'}\, 4m^2 \sum_{s,r,s',r'}$$
$$\times \Bigg[-\bar{u}(\mathbf{q}, s)\gamma^\mu u(\mathbf{p}, r)\,\bar{u}(\mathbf{q}', s')\gamma^\nu u(\mathbf{p}', r')$$
$$\times \langle 0|c(\mathbf{q}_2, s_2)c(\mathbf{q}_1, s_1)c(\mathbf{q}, s)^\dagger c(\mathbf{q}', s')^\dagger c(\mathbf{p}, r)c(\mathbf{p}', r')c(\mathbf{p}_1, r_1)^\dagger c(\mathbf{p}_2, r_2)^\dagger|0\rangle$$
$$\times e^{-i[(p-q)\cdot x+(p'-q')\cdot y]} \Bigg], \qquad (12.152)$$

the minus sign on the second line is due to the definition of normal order for fermions. To compute the v.e.v. of the eight creation/annihilation operators, we compute a single state of the form $c\,c\,c^\dagger\,c^\dagger|0\rangle$, which is clearly proportional to the vacuum. To this end we move each c operator to the right until it annihilates the vacuum, at each step an anti-commutator being produced

$$c(\mathbf{p}, r)c(\mathbf{p}', r')c(\mathbf{p}_1, r_1)^\dagger c(\mathbf{p}_2, r_2)^\dagger|0\rangle$$
$$= c(\mathbf{p}, r)\{c(\mathbf{p}', r'),\, c(\mathbf{p}_1, r_1)^\dagger\}c(\mathbf{p}_2, r_2)^\dagger|0\rangle - c(\mathbf{p}, r)\,c(\mathbf{p}_1, r_1)^\dagger c(\mathbf{p}', r')\,c(\mathbf{p}_2, r_2)^\dagger|0\rangle$$
$$= \{c(\mathbf{p}', r'),\, c(\mathbf{p}_1, r_1)^\dagger\}\{c(\mathbf{p}, r),\, c(\mathbf{p}_2, r_2)^\dagger\}|0\rangle - c(\mathbf{p}, r)\,c(\mathbf{p}_1, r_1)^\dagger\{c(\mathbf{p}', r')\,c(\mathbf{p}_2, r_2)^\dagger\}|0\rangle$$
$$= \{c(\mathbf{p}', r'),\, c(\mathbf{p}_1, r_1)^\dagger\}\{c(\mathbf{p}, r),\, c(\mathbf{p}_2, r_2)^\dagger\}|0\rangle - \{c(\mathbf{p}, r)\,c(\mathbf{p}_1, r_1)^\dagger\}\{c(\mathbf{p}', r')\,c(\mathbf{p}_2, r_2)^\dagger\}|0\rangle. \qquad (12.153)$$

By the same token we prove that

$$\langle 0|c(\mathbf{q}_2, s_2)c(\mathbf{q}_1, s_1)c(\mathbf{q}, s)^\dagger c(\mathbf{q}', s')^\dagger = \langle 0|\{c(\mathbf{q}_1, s_1),\, c(\mathbf{q}, s)^\dagger\}\{c(\mathbf{q}_2, s_2),\, c(\mathbf{q}', s')^\dagger\}$$
$$- \langle 0|\{c(\mathbf{q}_2, s_2),\, c(\mathbf{q}, s)^\dagger\}\{c(\mathbf{q}_1, s_1),\, c(\mathbf{q}', s')^\dagger\}. \qquad (12.154)$$

The scalar product between the states in (12.153) and (12.154) gives rise to four terms, each of the form of a product of four anti-commutators:

$\langle 0|c(\mathbf{q}_2, s_2)c(\mathbf{q}_1, s_1)c(\mathbf{q}, s)^\dagger c(\mathbf{q}', s')^\dagger c(\mathbf{p}, r)c(\mathbf{p}', r')c(\mathbf{p}_1, r_1)^\dagger c(\mathbf{p}_2, r_2)^\dagger|0\rangle =$

$= \{c(\mathbf{q}_1, s_1), c(\mathbf{q}, s)^\dagger\}\{c(\mathbf{q}_2, s_2), c(\mathbf{q}', s')^\dagger\}\{c(\mathbf{p}', r'), c(\mathbf{p}_1, r_1)^\dagger\}\{c(\mathbf{p}, r), c(\mathbf{p}_2, r_2)^\dagger\}$

$- \{c(\mathbf{q}_1, s_1), c(\mathbf{q}, s)^\dagger\}\{c(\mathbf{q}_2, s_2), c(\mathbf{q}', s')^\dagger\}\{c(\mathbf{p}', r'), c(\mathbf{p}_2, r_2)^\dagger\}\{c(\mathbf{p}, r), c(\mathbf{p}_1, r_1)^\dagger\}$

$- \{c(\mathbf{q}_2, s_2), c(\mathbf{q}, s)^\dagger\}\{c(\mathbf{q}_1, s_1), c(\mathbf{q}', s')^\dagger\}\{c(\mathbf{p}', r'), c(\mathbf{p}_1, r_1)^\dagger\}\{c(\mathbf{p}, r), c(\mathbf{p}_2, r_2)^\dagger\}$

$+ \{c(\mathbf{q}_2, s_2), c(\mathbf{q}, s)^\dagger\}\{c(\mathbf{q}_1, s_1), c(\mathbf{q}', s')^\dagger\}\{c(\mathbf{p}', r'), c(\mathbf{p}_2, r_2)^\dagger\}\{c(\mathbf{p}, r), c(\mathbf{p}_1, r_1)^\dagger\}.$

Each anti-commutator provides a delta function on the momenta times a delta function on the polarizations. Substituting the above expansion in the integral (12.152), for each term the integration over the momenta and the summation over the polarizations disappear:

$\langle \psi_{out}| : \overline{\psi}(x)\gamma^\mu \psi(x)\,\overline{\psi}(y)\gamma^\nu \psi(y) : |\psi_{in}\rangle$

$= 4m^2 \Big[-\bar{u}(\mathbf{q}_1, s_1)\gamma^\mu u(\mathbf{p}_2, r_2)\,\bar{u}(\mathbf{q}_2, s_2)\gamma^\nu u(\mathbf{p}_1, r_1)\,e^{-i[(p_2-q_1)\cdot x+(p_1-q_2)\cdot y]}$

$+ \bar{u}(\mathbf{q}_2, s_2)\gamma^\mu u(\mathbf{p}_2, r_2)\,\bar{u}(\mathbf{q}_1, s_1)\gamma^\nu u(\mathbf{p}_1, r_1)\,e^{-i[(p_2-q_2)\cdot x+(p_1-q_1)\cdot y]}$

$+ \bar{u}(\mathbf{q}_1, s_1)\gamma^\mu u(\mathbf{p}_1, r_1)\,\bar{u}(\mathbf{q}_2, s_2)\gamma^\nu u(\mathbf{p}_2, r_2)\,e^{-i[(p_1-q_1)\cdot x+(p_2-q_2)\cdot y]}$

$- \bar{u}(\mathbf{q}_2, s_2)\gamma^\mu u(\mathbf{p}_1, r_1)\,\bar{u}(\mathbf{q}_1, s_1)\gamma^\nu u(\mathbf{p}_2, r_2)\,e^{-i[(p_1-q_2)\cdot x+(p_2-q_1)\cdot y]} \Big].$

$$(12.155)$$

Note that the first and the fourth term, as well as the second and the third ones within square brackets, are obtained from one another by exchanging x and y. They will then give equal contributions to the integral (12.150), producing a factor 2 which cancels against the $1/2!$. The first term in square brackets describes the electron $(\mathbf{p}_1, r_1)$ which is destroyed in y where the electron $(\mathbf{q}_2, s_2)$ is created. This transition is due to the emission of a virtual photon in y, which is absorbed by the electron $(\mathbf{p}_2, r_2)$, causing its transition to the state $(\mathbf{q}_1, s_1)$. In the second term the roles of the two final electrons is interchanged. The Feynman diagram representation of these two contributions to the amplitude are represented in Fig. 12.4. Both these diagrams have the same geometry represented in Fig. 12.2, (2). Since however we are now considering a specific process, it is useful to identify the external legs so as to identify in the plane of the graph the time direction: The plane of the picture represents space-time and time flows from right to left. The reader should however bear in mind that Feynman diagrams are not a graphical representation of the actual time evolution of the interacting system. They are just a graphical tool for constructing the contributions to the amplitude of a given process to all orders.

Substituting the above result in Eq. (12.150) we find

$\langle \psi_{out}|\mathbf{S}_{(2)}|\psi_{in}\rangle = (ie)^2 \int d^4x\, d^4y\, 4m^2$

$\times \Big[\big(-\bar{u}(\mathbf{q}_1, s_1)\gamma^\mu u(\mathbf{p}_2, r_2)\,\bar{u}(\mathbf{q}_2, s_2)\gamma^\nu u(\mathbf{p}_1, r_1)\,e^{-i[(p_2-q_1)\cdot x+(p_1-q_2)\cdot y]}$

$+ \bar{u}(\mathbf{q}_2, s_2)\gamma^\mu u(\mathbf{p}_2, r_2)\,\bar{u}(\mathbf{q}_1, s_1)\gamma^\nu u(\mathbf{p}_1, r_1)\,e^{-i[(p_2-q_2)\cdot x+(p_1-q_1)\cdot y]} \big)$

$\times \int \frac{d^4p}{(2\pi)^4}\,\tilde{D}_{F\mu\nu}(p)\,e^{-ip\cdot(x-y)} \Big],$

$$(12.156)$$

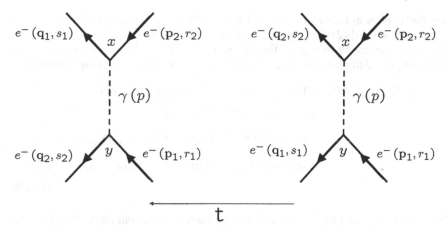

Fig. 12.4 Two 2nd-order contributions to the $e^- - e^-$ scattering amplitude

where we have written the photon propagator in momentum space. Let us notice that an incoming electron in a state $(\mathbf{p}, r)$ contributes a factor $u(\mathbf{p}, r)\, e^{-ip\cdot x_v}$ to the integrand in the expression of the amplitude, x_v^μ being the location of the vertex in which the electron state is destroyed. An outgoing electron created at x_v^μ is a state $(\mathbf{p}, r)$ contributes a factor $\bar{u}(\mathbf{p}, r)\, e^{ip\cdot x_v}$. In general with every particle annihilated or created, at a point x_v^μ is associated a characteristic factor $e^{-ip\cdot x_v}$, or $e^{ip\cdot x_v}$, respectively, p^μ being the corresponding 4-momentum. Therefore, in Eq. (12.156) the exponentials $e^{-ip\cdot x}$, $e^{ip\cdot y}$ signal the creation of a virtual photon in y with momentum p (being the photon virtual $p^2 \neq 0$), and its destruction in x. Let us now perform the integrations over the space-time positions x and y of the two vertices. The integrand depends on these variables only through the exponential factors. Let us consider the first term in Eq. (12.156). The integrals for the two vertices yield the following delta functions:

$$\int d^4x\, e^{-i\,(p_2 - q_1 + p)\cdot x} = (2\pi)^4\, \delta^4(p_2 - q_1 + p),$$

$$\int d^4y\, e^{-i\,(p_1 - q_2 - p)\cdot y} = (2\pi)^4\, \delta^4(p_1 - q_2 - p). \qquad (12.157)$$

We find, upon integration, a delta function which implements the conservation of 4-momentum at each vertex: At x the sum of the momenta p_2 and p of the incoming electron and photon equals the momentum q_1 of the outgoing electron; Similarly at y we have $p_1 = q_2 + p$. If we now perform the integration over the momentum of the virtual photon, we end up with a single delta-function implementing the conservation of 4-momentum for the whole process, as a consequence of the invariance of the system under global space-time translations:

$$\int \frac{d^4p}{(2\pi)^4} (2\pi)^8\, \delta^4(p_2 - q_1 + p)\, \delta^4(p_1 - q_2 - p) = (2\pi)^4\, \delta^4(p_1 + p_2 - q_1 - q_2).$$

Similarly, as far as the second term in Eq. (12.156) is concerned, the integration over x and over y yield $(2\pi)^4 \delta^4(p_2 - q_2 + p)$ and $(2\pi)^4 \delta^4(p_1 - q_1 - p)$, respectively, and upon integrating over p we find the same factor $(2\pi)^4 \delta^4(p_1 + p_2 - q_1 - q_2)$. Factoring this delta function out, we finally end up with the following amplitude

$$
\begin{aligned}
\langle \psi_{out} | S_{(2)} | \psi_{in} \rangle &= i \langle \psi_{out} | T_{(2)} | \psi_{in} \rangle = i \, (2\pi)^4 \delta^4(p_1 + p_2 - q_1 - q_2) \\
&\times \langle \psi_{out} | \mathscr{T}'_{(2)} | \psi_{in} \rangle = (2\pi)^4 \delta^4(p_1 + p_2 - q_1 - q_2) \, (ie)^2 \, 4m^2 \\
&\times \Big(-\bar{u}(\mathbf{q}_1, s_1) \gamma^\mu u(\mathbf{p}_2, r_2) \, \bar{u}(\mathbf{q}_2, s_2) \gamma^\nu u(\mathbf{p}_1, r_1) \, \tilde{D}_{F\mu\nu}(p_2 - q_1) \\
&\quad + \bar{u}(\mathbf{q}_2, s_2) \gamma^\mu u(\mathbf{p}_2, r_2) \, \bar{u}(\mathbf{q}_1, s_1) \gamma^\nu u(\mathbf{p}_1, r_1) \, \tilde{D}_{F\mu\nu}(p_1 - q_1) \Big) .
\end{aligned}
$$

$$(12.158)$$

We have thus found the second order contribution to the amplitude $\langle \psi_{out} | \mathscr{T}' | \psi_{in} \rangle$ entering the formula (12.74) for the probability per unit time of the process. Let us now insert in $\langle \psi_{out} | \mathscr{T}'_{(2)} | \psi_{in} \rangle$ the explicit expression of the photon propagator in momentum space. Consider the first term within brackets in Eq. (12.158):

$$
\begin{aligned}
&\bar{u}(\mathbf{q}_1, s_1) \gamma^\mu u(\mathbf{p}_2, r_2) \, \bar{u}(\mathbf{q}_2, s_2) \gamma^\nu u(\mathbf{p}_1, r_1) \, \tilde{D}_{F\mu\nu}(p) \Big|_{p=p_2 - q_1} \\
&= \left[\bar{u}(\mathbf{q}_1, s_1) \gamma^\mu u(\mathbf{p}_2, r_2) \frac{(-i)}{p^2 + i\epsilon} \left(\eta_{\mu\nu} - (1-\alpha) \frac{p_\mu p_\nu}{p^2} \right) \right. \\
&\qquad \times \bar{u}(\mathbf{q}_2, s_2) \gamma^\nu u(\mathbf{p}_1, r_1) \Big] \Big|_{p=p_2 - q_1} .
\end{aligned}
$$

$$(12.159)$$

Let us show that the $p_\mu p_\nu$ term in the propagator does not contribute to the above quantity by using the Dirac equation for the incoming and outgoing electrons (let us recall, for completeness, the same equations for the positron states as well):

$$
(\slashed{q} - m) u(\mathbf{q}, r) = 0, \quad \bar{u}(\mathbf{q}, r) (\slashed{q} - m) = 0, \tag{12.160}
$$

$$
(\slashed{q} + m) v(\mathbf{q}, r) = 0, \quad \bar{v}(\mathbf{q}, r) (\slashed{q} + m) = 0. \tag{12.161}
$$

Using $p = p_2 - q_1$, it is easy to show that $\bar{u}(\mathbf{q}_1, s_1) \slashed{p} u(\mathbf{p}_2, r_2) = 0$:

$$
\begin{aligned}
\bar{u}(\mathbf{q}_1, s_1) \slashed{p} u(\mathbf{p}_2, r_2) &= \bar{u}(\mathbf{q}_1, s_1) \slashed{p}_2 u(\mathbf{p}_2, r_2) - \bar{u}(\mathbf{q}_1, s_1) \slashed{q}_1 u(\mathbf{p}_2, r_2) \\
&= (m - m) \, \bar{u}(\mathbf{q}_1, s_1) u(\mathbf{p}_2, r_2) = 0.
\end{aligned}
$$

$$(12.162)$$

By the same token we prove that $\bar{u}(\mathbf{q}_2, s_2) \slashed{p} u(\mathbf{p}_1, r_1) = 0$. We can then conclude that:

$$
\begin{aligned}
i \langle \psi_{out} | \mathscr{T}'_{(2)} | \psi_{in} \rangle &= (ie)^2 \, 4m^2 \\
&\times \Big(-\bar{u}(\mathbf{q}_1, s_1) \gamma^\mu u(\mathbf{p}_2, r_2) \frac{-i}{(p_2 - q_1)^2} \bar{u}(\mathbf{q}_2, s_2) \gamma_\mu u(\mathbf{p}_1, r_1) \\
&\quad + \bar{u}(\mathbf{q}_2, s_2) \gamma^\mu u(\mathbf{p}_2, r_2) \frac{-i}{(p_1 - q_1)^2} \bar{u}(\mathbf{q}_1, s_1) \gamma_\mu u(\mathbf{p}_1, r_1) \Big) .
\end{aligned}
$$

$$(12.163)$$

As previously pointed out, there is an ambiguity in the overall sign, while the relative sign between the two terms in brackets is fixed and physically relevant.

12.5.2 A Comment on the Role of Virtual Photons

The treatment of the Möller scattering has shown that the interaction between electrons to second order in the fine structure constant can be viewed as due to the exchange of a *virtual photon* between the two electrons.[20] We recall that by virtual photon we mean a photon whose momentum $k = p_2 - p_1$ does not satisfy the mass-shell condition, $k^2 \neq 0$, and which can thus be interpreted as a massive particle with $m^2 = k^2$. While for a real photon only the two transverse polarizations are physical, when a *virtual* photon is exchanged, all four polarization vectors $\varepsilon_\mu^{(\lambda)}(\mathbf{k})$, $\lambda = 0, 1, 2, 3$ contribute to the amplitude. It is then interesting to see what is the role of the time-like and longitudinal photons $\varepsilon_\mu^{(0)}(\mathbf{k})$, $\varepsilon_\mu^{(3)}(\mathbf{k})$ in the interpretation of the process.

Let us refer for concreteness to the second diagram of the Möller scattering whose lowest order amplitude is given by the second term of Eq. (12.163). We observe that the sum over the indices μ of the gamma-matrices is due to the $\eta_{\mu\nu}$ factor of the photon propagator $\tilde{D}_{F\mu\nu} = -i\eta_{\mu\nu}(k^2 + i\epsilon)^{-1}$, which in turn comes from the completeness relation (11.229). Therefore the amplitude corresponding to the second term of Eq. (12.163) could have been alternatively written as

$$\bar{u}(\mathbf{q}_2, s_2)\gamma^\mu u(\mathbf{p}_2, r_2) \left[\sum_{\lambda=0}^{3} \varepsilon_{(\lambda)\mu}(\mathbf{k})\varepsilon_\nu^{(\lambda)}(\mathbf{k}) \right] \frac{-i}{(p_1 - q_1)^2} \bar{u}(\mathbf{q}_1, s_1)\gamma^\nu u(\mathbf{p}_1, r_1).$$

$$(12.164)$$

For a virtual photon we must take as polarization vectors a set which for $k^2 \to 0$ reduces to the set used for a real photon in (11.231) and (11.232). As seen in Sect. 11.7 of Chap. 11, such set is obtained by simply replacing the longitudinal vector $\varepsilon_\mu^{(3)}(\mathbf{k})$ of Eq. (11.232) with

$$\varepsilon_\mu^{(3)}(\mathbf{k}) = \frac{k_\mu - \eta_\mu(k \cdot \eta)}{\sqrt{(k \cdot \eta)^2 - k^2}}.$$

$$(12.165)$$

Let us now decompose the sum appearing in the completeness relation (11.229) into the sum over $\lambda = \sigma = 0, 3$, corresponding to the exchange of timelike and longitudinal photons and the sum over the transverse polarizations $\lambda = \sigma = 1, 2$. In particular, using Eq. (12.165) for $\varepsilon_\mu^{(3)}(\mathbf{k})$ and the value $\varepsilon_\mu^{(0)}(\mathbf{k}) = \eta_\mu$ of Eq. (11.234), we have

---

[20]The same interpretation is of course also true for the interaction between electron and positron in Bhabha scattering, see Sect. 12.5.3. For the sake of definiteness and simplicity we shall refer the considerations of this subsection to the Möller scattering.

$$\varepsilon_\mu^{(0)}(\mathbf{k})\varepsilon_\nu^{(0)}(\mathbf{k}) - \varepsilon_\mu^{(3)}(\mathbf{k})\varepsilon_\nu^{(3)}(\mathbf{k}) = \eta_\mu\eta_\nu - \frac{[k_\mu - \eta_\mu(k\cdot\eta)][k_\nu - \eta_\nu(k\cdot\eta)]}{(k\cdot\eta)^2 - k^2}.$$

Since we are interested in the contribution to the amplitude of the $\lambda = 0, 3$ polarizations, we substitute the right hand side of this expression into Eq. (12.164) with the sum restricted to the values $\lambda = 0$, $\lambda = 3$. Since, as already seen in the previous section, the terms proportional to k^μ do not contribute by virtue of gauge invariance[21] we obtain

$$-\frac{i}{k^2}\bar{u}(\mathbf{q}_2, s_2)\gamma^\mu u(\mathbf{p}_2, r_2)\bar{u}(\mathbf{q}_1, s_1)\gamma^\nu u(\mathbf{p}_1, r_1) \times \eta_\mu\eta_\nu\left(1 - \frac{(k\cdot\eta)^2}{(k\cdot\eta)^2 - k^2}\right)$$

$$= -i\bar{u}(\mathbf{q}_2, s_2)\gamma^0(\mathbf{p}_2, r_2)\frac{1}{k^2 - (k^0)^2}\bar{u}(\mathbf{q}_1, s_1)\gamma^0 u(\mathbf{p}_1, r_1)$$

$$= iu^\dagger(\mathbf{q}_2, s_2)u(\mathbf{p}_2, r_2)\frac{1}{|\mathbf{k}|^2}u^\dagger(\mathbf{q}_1, s_1)u(\mathbf{p}_1, r_1), \qquad (12.167)$$

where in the second step we have used the explicit value $\eta = (1, 0, 0, 0)$ valid in the Lorentz frame where $\varepsilon_\mu^{(1,2)}(\mathbf{k})$ are transverse (see Sect. 11.7 of Chap. 11). We now observe that $u^\dagger(\mathbf{q}_2, s_2)u(\mathbf{p}_2, r_2)$ and $u^\dagger(\mathbf{q}_1, s_1)u(\mathbf{p}_1, r_1)$ are the Fourier transform in the momentum space of the charge densities, while $\frac{1}{|\mathbf{k}|^2}$ is the Fourier transform of $1/(4\pi r)$. It follows that Eq. (12.167) represents an "instantaneous" Coulomb interaction between the two electrons. Adding the sum over the two transverse polarizations $\lambda = 1, 2$ the result is that the interaction between the two electrons is given by transverse "waves" plus an instantaneous Coulomb interaction.

12.5.3 Bhabha and Electron-Muon Scattering

Let us now consider the scattering between an electron e^- and a positron e^+ (first studied by H. Bahbha in 1936):

$$e^- + e^+ \longrightarrow e^- + e^+. \qquad (12.168)$$

Let the momenta and polarizations of the electron and positron be $(\mathbf{p}_-, r_-)$, $(\mathbf{p}_+, r_+)$ in the initial state, and $(\mathbf{q}_-, s_-)$, $(\mathbf{q}_+, s_+)$ after the interaction, respectively:

---

[21]In this special case this can be also seen directly. Indeed

$$k_\mu\bar{u}(\mathbf{q}_2, s_2)\gamma^\mu u(\mathbf{p}_2, r_2) = \bar{u}(\mathbf{q}_2, s_2)(p_2 - q_2)_\mu\gamma^\mu u(\mathbf{p}_2, r_2)$$
$$= -m\bar{u}(\mathbf{q}_2, s_2)u(\mathbf{p}_2, r_2) + m\bar{u}(\mathbf{q}_2, s_2)u(\mathbf{p}_2, r_2) = 0, \qquad (12.166)$$

and similarly for the other factor of Eq. (12.167).

$$|\psi_{in}\rangle = |\mathbf{p}_+, r_+\rangle|\mathbf{p}_-, r_-\rangle = d(\mathbf{p}_+, r_+)^\dagger c(\mathbf{p}_-, r_-)^\dagger|0\rangle,$$

$$|\psi_{out}\rangle = |\mathbf{q}_+, s_+\rangle|\mathbf{q}_-, s_-\rangle = d(\mathbf{q}_+, s_+)^\dagger c(\mathbf{q}_-, s_-)^\dagger|0\rangle,$$

$$\langle\psi_{out}| = \langle 0|c(\mathbf{q}_-, s_-)d(\mathbf{q}_+, s_+), \tag{12.169}$$

where we have used the property that, if A, B are two operators $(AB)^\dagger = B^\dagger A^\dagger$. Note that we have represented the initial state as resulting from the action on the vacuum of the electron creation operator followed by that of the positron, and we have used the same (conventional) ordering of creation operators for the final state. This will fix the overall sign of the amplitude, in contrast to the previous case in which identical particles where present in the initial and final states and the overall sign was ambiguous.

As for the electron-electron scattering, the second order contribution to the amplitude will come from term (2) in the Wick expansion (12.140) of $\mathbf{S}_{(2)}$ (represented by diagram (2) of Fig. 12.2):

$$\langle\psi_{out}|\mathbf{S}_{(2)}|\psi_{in}\rangle = \frac{(ie)^2}{2!}\int d^4x d^4y$$

$$\times\Big[\langle\mathbf{q}_+, s_+|\langle\mathbf{q}_-, s_-| : \overline{\psi}(x)\gamma^\mu\psi(x)\,\overline{\psi}(y)\gamma^\nu\psi(y) : |\mathbf{p}_+, r_+\rangle|\mathbf{p}_-, r_-\rangle$$

$$\times D_{F\mu\nu}(x - y)\Big]. \tag{12.170}$$

Only terms of the form $d^\dagger c^\dagger cd$ in the normal ordered product will contribute to the matrix element, with a term proportional to:

$$\langle 0|c(\mathbf{q}_-, s_-)d(\mathbf{q}_+, s_+)\, d^\dagger c^\dagger cd\, d(\mathbf{p}_+, r_+)^\dagger c(\mathbf{p}_-, r_-)^\dagger|0\rangle$$

$$= \{d(\mathbf{q}_+, s_+), d^\dagger\}\{d, d(\mathbf{p}_+, r_+)^\dagger\}\{c(\mathbf{q}_-, s_-), c^\dagger\}\{c, c(\mathbf{p}_-, r_-)^\dagger\}.$$

Each anti-commutator in the above expression provides a delta function on the momenta times a delta on the polarizations. These deltas single out, in the expansion of the field operators, the term with the same momentum and polarization as the corresponding external state: For instance $\{d(\mathbf{q}_+, s_+), d^\dagger\}$ will single out in the expansion of the ψ operator containing $d^\dagger$, the term proportional to $v(\mathbf{q}_+, s_+)$; $\{d, d(\mathbf{p}_+, r_+)^\dagger\}$ the term proportional to $\bar{v}(\mathbf{p}_+, r_+)$ in the expansion of the $\overline{\psi}$ field containing d, and so on. Since $d^\dagger$ may come either from $\psi(x)$ or from $\psi(y)$ and d from $\overline{\psi}(x)$ or from $\overline{\psi}(y)$, there are, in total, four such terms. Consider the contributions in which $d, d^\dagger$ originate from field operators computed in the same vertex. There are two of them, related by an exchange of the two vertices $x \leftrightarrow y$, which then give equal contributions to the integral (12.170). Each of them describes a positron and electron exchanging a virtual photon, as illustrated in Fig. 12.5a. Note that, in the corresponding Feynman diagram, the direction of motion for the positron is opposite to the orientation on the corresponding external leg, as is represented by an arrow parallel to it. The reason is that the arrow on an external fermionic leg indicates the flow of negative charge (electron charge), which is clearly opposite to the flow of the positron charge. One of them contributes to the integrand in (12.170) a term of the form:

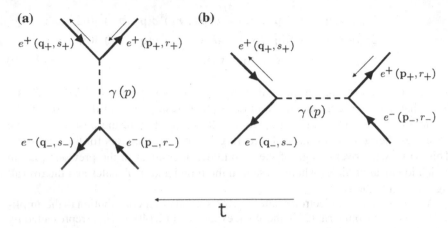

Fig. 12.5 Two 2nd-order contributions to the $e^- - e^+$ scattering amplitude: The *diffusion*, (a), and the annihilation, (b) diagrams

$$-\bar{v}(\mathbf{p}_+,\, r_+)\gamma^\mu v(\mathbf{q}_+,\, s_+)\, \bar{u}(\mathbf{q}_-,\, s_-)\gamma^\nu u(\mathbf{p}_-,\, r_-)\, e^{-i\,(p_+-q_+)\cdot x}\, e^{-i\,(p_--q_-)\cdot y},$$

to be contracted with the photon propagator, the other a similar term with $x \leftrightarrow y$. The minus sign in the above expression originates from the definition of normal ordering for fermions: $:d\,d^\dagger\,c^\dagger\,c := -d^\dagger\,c^\dagger\,c\,d.$

Consider now the two terms in which $d^\dagger$ and d originate from field operators computed in different vertices. They are also related by an exchange of the two vertices and thus give equal contributions to the integral (12.170). Each of them describes a process in which the incoming electron and positron lines converge on a same vertex, where the two particles are both destroyed (by c and d, respectively). They annihilate, producing a virtual photon which propagates up to the second vertex where it originates the couple of outgoing electron and positron (created by $c^\dagger$ and $d^\dagger$, respectively), see Fig. 12.5b. This is thus an annihilation process rather than a diffusion one. Its contribution to the integrand in (12.170) is a term of the form:

$$\bar{u}(\mathbf{q}_-,\, s_-)\gamma^\mu v(\mathbf{q}_+,\, s_+)\, \bar{v}(\mathbf{p}_+,\, r_+)\gamma^\nu u(\mathbf{p}_-,\, r_-)\, e^{-i\,(p_-+p_+)\cdot y}\, e^{i\,(q_++q_-)\cdot x},$$

to be contracted with the photon propagator. The integration over x and y yields the conservation of 4-momentum at each vertex. We have thus found two distinct contributions to this integral, one describing a diffusion and an other an annihilation process. In the former case the momentum of the photon is $p = p_- - q_- = q_+ - p_+$, while in the latter $p = p_- + p_+ = q_+ + q_-$ (the sign of p is irrelevant since the integral is invariant upon changing $p \to -p$ and $x \leftrightarrow y$).

Upon integration over x and y and the photon momentum p we end up with a single delta function implementing the conservation of the total momentum $p_- + p_+ = q_+ + q_-$. By factoring this delta function out, just as we did in the case of the electron-electron scattering we derive the expression for the matrix element of $\mathscr{T}'_{(2)}$:

$$i \langle \psi_{out} | \mathscr{T}'_{(2)} | \psi_{in} \rangle = (ie)^2 \, 4m^2$$

$$\times \left(-\bar{v}(\mathbf{p}_+, \, r_+) \gamma^\mu v(\mathbf{q}_+, \, s_+) \frac{-i}{(p_- - q_-)^2} \bar{u}(\mathbf{q}_-, \, s_-) \gamma_\mu u(\mathbf{p}_-, \, r_-) \right.$$

$$\left. + \bar{u}(\mathbf{q}_-, \, s_-) \gamma^\mu v(\mathbf{q}_+, \, s_+) \frac{-i}{(p_- + p_+)^2} \bar{v}(\mathbf{p}_+, \, r_+) \gamma_\mu u(\mathbf{p}_-, \, r_-) \right),$$

$$(12.171)$$

where we have used the properties $\bar{v}(\mathbf{p}_+, \, r_+)(\not{p}_+ - \not{q}_+)v(\mathbf{q}_+, \, s_+) = 0$ and $\bar{v}(\mathbf{p}_+, \, r_+)$ $(\not{p}_+ + \not{p}_-)u(\mathbf{p}_-, \, r_-) = 0$ which descend from Eqs. (12.161) and (12.160).

Consider now an electron-muon scattering:

$$e^- + \mu^- \longrightarrow e^- + \mu^-. \qquad (12.172)$$

The interaction Hamiltonian is obtained by writing the electric current as the sum of the electron and the muon ones, as in Eq. (12.141), in which "particle q" (which however now is no longer a "spectator") is the muon ($q = e = -|e| < 0$):

$$\widehat{\mathcal{H}}_I(x) \equiv -e : [\overline{\psi}(x)\gamma^\mu \, \psi(x) + \overline{\psi}_{(\mu)}(x)\gamma^\mu \, \psi_{(\mu)}(x)] A_\mu(x) : . \qquad (12.173)$$

We shall indicate the quantities associated with the muon by a subscript (μ), not to be confused with a 4-vector index. Let the initial and final electron states be $(\mathbf{p}_1, r_1)$, $(\mathbf{q}_1, s_1)$, while the initial and final muon states be $(\mathbf{p}_2, r_2)$, $(\mathbf{q}_2, s_2)$, respectively:

$$|\psi_{in}\rangle = |\mathbf{p}_1, \, r_1\rangle |\mathbf{p}_2, \, r_2\rangle = c(\mathbf{p}_1, \, r_1)^\dagger \, c_{(\mu)}(\mathbf{p}_2, \, r_2)^\dagger |0\rangle,$$

$$|\psi_{out}\rangle = |\mathbf{q}_1, \, s_1\rangle |\mathbf{q}_2, \, s_2\rangle = c(\mathbf{q}_1, \, s_1)^\dagger \, c_{(\mu)}(\mathbf{q}_2, \, s_2)^\dagger |0\rangle. \qquad (12.174)$$

The second order contribution to the amplitude, see Eq. (12.142), reads:

$$\langle \psi_{out} | \mathbf{S}_{(2)} | \psi_{in} \rangle = (ie)^2 \int d^4x d^4y$$

$$\times \langle \mathbf{q}_1, \, s_1 | \langle \mathbf{q}_2, \, s_2 | : \overline{\psi}(x)\gamma^\mu \, \psi(x)\overline{\psi}_{(\mu)}(y)\gamma^\nu \, \psi_{(\mu)}(y) : |\mathbf{p}_1, \, r_1\rangle |\mathbf{p}_2, \, r_2\rangle$$

$$\times D_{F\mu\nu}(x - y). \qquad (12.175)$$

In the expansion of the normal product in creation and annihilation operators, there is just one term contributing to the matrix element, of the form $c^\dagger c^\dagger_{(\mu)} c_{(\mu)} c$: The incoming muon can only be destroyed by $\psi_{(\mu)}(y)$ and the outgoing one only be created by $\overline{\psi}_{(\mu)}(y)$. We have therefore just one term contributing to the amplitude, represented in Fig. 12.6. This situation should be contrasted with the electron-electron case, in which the initial and final states consisted of identical particles and the independent diagrams contributing to the amplitude, modulo $x \leftrightarrow y$, were two, one obtained from the other by interchanging the external legs corresponding to the outgoing electrons.

Fig. 12.6 Two 2nd-order contribution to the $e^- - \mu^-$ scattering amplitude

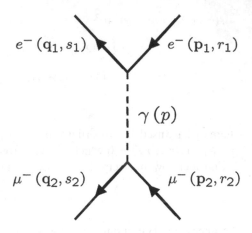

Using the property:

$$\langle 0|c_{(\mu)}(\mathbf{q}_2, s_2)c(\mathbf{q}_1, s_1)\, c^\dagger \, c^\dagger_{(\mu)} \, c_{(\mu)} \, c \, c(\mathbf{p}_1, r_1)^\dagger \, c_{(\mu)}(\mathbf{p}_2, r_2)^\dagger|0\rangle$$
$$= \{c_{(\mu)}(\mathbf{q}_2, s_2),\, c^\dagger_{(\mu)}\}\, \{c_{(\mu)},\, c_{(\mu)}(\mathbf{p}_2, r_2)^\dagger\}\, \{c(\mathbf{q}_1, s_1),\, c^\dagger\}\, \{c,\, c(\mathbf{p}_1, r_1)^\dagger\},$$

and integrating over the momenta and the positions of the vertices, we arrive at the following expression for the amplitude:

$$i\,\langle\psi_{out}|\mathscr{T}'_{(2)}|\psi_{in}\rangle = (ie)^2\,4m\,m_\mu$$
$$\times \left(\bar{u}_{(\mu)}(\mathbf{q}_2, s_2)\gamma^\nu u_{(\mu)}(\mathbf{p}_2, r_2)\frac{-i}{(p_1 - q_1)^2}\bar{u}(\mathbf{q}_1, s_1)\gamma_\nu u(\mathbf{p}_1, r_1)\right).$$

$$\text{(12.176)}$$

We leave as an exercise to the reader to show that the second order amplitude for the annihilation process, see Fig. 12.7,

$$e^- + e^+ \longrightarrow \mu^- + \mu^+, \tag{12.177}$$

Fig. 12.7 Two 2nd-order contribution to the $e^- + e^+ \to \mu^- + \mu^+$ amplitude

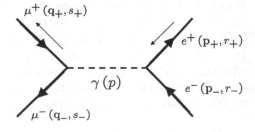

reads:

$$i \langle \psi_{out} | \mathcal{T}'_{(2)} | \psi_{in} \rangle = (ie)^2 \, 4m \, m_\mu$$

$$\times \left(\bar{u}_{(\mu)}(\mathbf{q}_-, s_-) \gamma^\nu v_{(\mu)}(\mathbf{q}_+, s_+) \frac{-i}{(p_+ + p_-)^2} \bar{v}(\mathbf{p}_+, r_+) \gamma_\nu u(\mathbf{p}_-, r_-) \right).$$

$$(12.178)$$

12.5.4 Compton Scattering and Feynman Rules

Let us now consider the interaction between electromagnetic radiation (photons) and matter, in particular a process in which a photon is scattered by an electron (Compton scattering):

$$\gamma + e^- \longrightarrow \gamma + e^-, \qquad (12.179)$$

in which the initial state consists of a photon γ in the state $(\mathbf{p}_1, i)$ ($i = 1, 2$ being its polarization) and an electron in the state $(\mathbf{p}_2, r)$, while the final state describes a photon and electron in the states $(\mathbf{q}_1, i')$, $(\mathbf{q}_2, s)$, respectively:

$$|\psi_{in}\rangle = |\mathbf{p}_1, i\rangle |\mathbf{p}_2, r\rangle = a(\mathbf{p}_1, i)^\dagger c(\mathbf{p}_2, r)^\dagger |0\rangle,$$

$$|\psi_{out}\rangle = |\mathbf{q}_1, i'\rangle |\mathbf{q}_2, s\rangle = a(\mathbf{q}_1, i')^\dagger c(\mathbf{q}_2, s)^\dagger |0\rangle. \qquad (12.180)$$

We shall compute the second-order amplitude of this process. The only terms in the Wick expansion (12.140) of $\mathbf{S}_{(2)}$ contributing to the amplitude are those containing two electromagnetic free fields and two electron free fields, namely terms (3) and (4), which, however, give an equal contribution upon integration over x and y. We can thus focus of (3) and write:

$$\langle \psi_{out} | \mathbf{S}_{(2)} | \psi_{in} \rangle = (ie)^2 \int d^4x d^4y$$

$$\times \langle \mathbf{q}_1, i' | \langle \mathbf{q}_2, s | : \overline{\psi}(x) \gamma^\mu S_F(x - y) \gamma^\nu \psi(y) A_\mu(x) A_\nu(y) : |\mathbf{p}_1, i\rangle |\mathbf{p}_2, r\rangle.$$

$$(12.181)$$

Expanding the free field in creation an annihilation operators, we can convince ourselves that the only terms contributing to the matrix element have the form: $c^\dagger a^\dagger c a$, which destroys the initial photon and electron (operators a, c, respectively) and creates the outgoing ones (operators $a^\dagger$, $c^\dagger$). Their non-vanishing contributions have the general form:

$$\langle 0 | c(\mathbf{q}_2, s) a(\mathbf{q}_1, i') c^\dagger a^\dagger c a a(\mathbf{p}_1, i)^\dagger c(\mathbf{p}_2, r)^\dagger |0\rangle$$

$$= [a(\mathbf{q}_1, i'), a^\dagger] [a, a(\mathbf{p}_1, i)^\dagger] \{c(\mathbf{q}_2, s), c^\dagger\} \{c, c(\mathbf{p}_2, r)^\dagger\}. \quad (12.182)$$

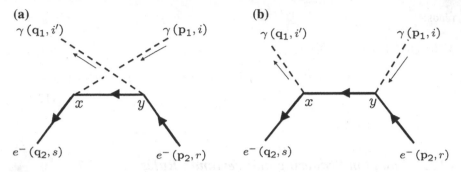

Fig. 12.8 Two 2nd-order contributions to the Compton scattering amplitude

Let us note, however, that there are two terms of the form $c^\dagger a^\dagger ca$: One in which a comes from $A_\mu(x)$ (and thus $a^\dagger$ from $A_\mu(y)$), the other in which a comes from $A_\mu(y)$ (and thus $a^\dagger$ from $A_\mu(x)$). The former describes a process in which the incoming electron emits the outgoing photon ($\mathbf{q}_1$, i') in y and absorbs the incoming one in x, see Fig. 12.8a, while in the latter the incoming photon ($\mathbf{p}_1$, i) is absorbed in y and the outgoing one emitted in x, see Fig. 12.8b. Using, for each term, Eq. (12.182), and eliminating, by integration, the delta functions arising from the commutators and anticommutators, we end up with:

$$\langle \psi_{out} | S_{(2)} | \psi_{in} \rangle = (ie)^2 \int d^4x d^4y \int \frac{d^4p}{(2\pi)^4} 2m$$

$$\times \left[\bar{u}(\mathbf{q}_2, s)\gamma^\mu \frac{i}{\not{p} - m} \gamma^\nu u(\mathbf{p}_2, r)\varepsilon_\mu(\mathbf{p}_1, i)\, \varepsilon_\nu(\mathbf{q}_1, i')^*\, e^{-i(p_2 - q_1 - p)\cdot y} \right.$$

$$\times\, e^{i(q_2 - p_1 - p)\cdot x} + \bar{u}(\mathbf{q}_2, s)\gamma^\mu \frac{i}{\not{p} - m} \gamma^\nu u(\mathbf{p}_2, r)\varepsilon_\nu(\mathbf{p}_1, i)\, \varepsilon_\mu(\mathbf{q}_1, i')^*$$

$$\left. \times e^{-i(p_1 + p_2 - p)\cdot y}\, e^{i(q_2 + q_1 - p)\cdot x} \right], \tag{12.183}$$

We note that under the exchange

$$\mathbf{p}_1 \leftrightarrows -\mathbf{q}_1; \quad \epsilon_\mu(\mathbf{p}_1, i) \leftrightarrows \epsilon_\mu(\mathbf{q}_1, i')^*,$$

the total matrix element remains invariant. This invariance is known as *crossing symmetry*, the graph (a) being referred to as the crossed term of graph (b). The integrations over x and y implement the conservation of momentum at each vertex, while the integration over the momentum p of the virtual electron yields the global delta function $(2\pi)^4 \delta^4(p_1 + p_2 - q_1 - q_2)$. The matrix element of $\mathscr{T}'_{(2)}$ reads:

$$i \langle \psi_{out} | \mathscr{T}'_{(2)} | \psi_{in} \rangle = (ie)^2 2m$$

$$\times \bar{u}(\mathbf{q}_2, s) \left[\gamma^\mu \frac{i}{\not{p}_2 - \not{q}_1 - m} \gamma^\nu + \gamma^\nu \frac{i}{\not{p}_1 + \not{p}_2 - m} \gamma^\mu \right] u(\mathbf{p}_2, r)$$

$$\times \varepsilon_\mu(\mathbf{p}_1, i)\, \varepsilon_\nu(\mathbf{q}_1, i')^*. \tag{12.184}$$

Let us now verify that the above result does not depend on our gauge choice for the electromagnetic potential, namely that it is not affected by a gauge transformation $A_\mu \to A_\mu + \partial_\mu \Lambda$. In momentum space a gauge transformation amounts to adding an unphysical component to ε_μ:

$$\varepsilon_\mu(\mathbf{p}, i) \longrightarrow \varepsilon_\mu(\mathbf{p}, i) + \chi(\mathbf{p}) p_\mu. \tag{12.185}$$

To show that such component gives no contribution to Eq. (12.184), let us replace any of the photon polarization vectors (e.g. $\varepsilon_\mu(\mathbf{p}_1, i)$) by the corresponding 4-momentum $p_{1\,\mu}$ and prove that the resulting expression is zero. It suffices to prove that the following quantity vanishes:

$$\bar{u}(\mathbf{q}_2, s)\left[\gamma^\mu \frac{i}{\not{p}_2 - \not{q}_1 - m}\gamma^\nu + \gamma^\nu \frac{i}{\not{p}_1 + \not{p}_2 - m}\gamma^\mu\right] u(\mathbf{p}_2, r)\, p_{1\,\mu}$$

$$= \bar{u}(\mathbf{q}_2, s)\left[\not{p}_1 \frac{i}{\not{p}_2 - \not{q}_1 - m}\gamma^\nu + \gamma^\nu \frac{i}{\not{p}_1 + \not{p}_2 - m}\not{p}_1\right] u(\mathbf{p}_2, r). \tag{12.186}$$

To this end let us write, in the first and second terms within square brackets:

$$\not{p}_1 = -(-\not{p}_1 + \not{q}_2 - m) + (\not{q}_2 - m) = -(\not{p}_2 - \not{q}_1 - m) + (\not{q}_2 - m)$$

$$\not{p}_1 = (\not{p}_1 + \not{p}_2 - m) - (\not{p}_2 - m), \tag{12.187}$$

respectively. As far as the first term is concerned, we can use $\bar{u}(\mathbf{q}_2, s)(\not{q}_2 - m) = 0$, while $(\not{p}_2 - \not{q}_1 - m)$ cancels against the electron propagator, yielding $-\bar{u}(\mathbf{q}_2, s)\gamma^\nu u(\mathbf{p}_2, r)$. Similarly, for the second term, we use the equation of motion $(\not{p}_2 - m)u(\mathbf{p}_2, r) = 0$, while $(\not{p}_1 + \not{p}_2 - m)$ cancels against the propagator, yielding $\bar{u}(\mathbf{q}_2, s)\gamma^\nu u(\mathbf{p}_2, r)$. Summing the two contributions we find for (12.186):

$$\bar{u}(\mathbf{q}_2, s)(-\gamma^\nu + \gamma^\nu)u(\mathbf{p}_2, r) = 0, \tag{12.188}$$

thus proving that the amplitude (12.184) is gauge invariant. This result extends to any amplitude with external photon fields and is required by consistency of the quantum theory.

Feynman Rules. From the above discussion we can formulate few simple graphical rules which allow us to write each perturbative contribution to the amplitude of a process. The order-n amplitude $i\langle \mathscr{T}'_{(n)}\rangle$ of a process is computed as follows:

- Write n three-leg vertices, of the form in Fig. 12.1, identify some of these legs with the incoming and outgoing fields (external legs) and connect all the remaining legs to one another (photon to photon, electron to electron) in all possible ways. In this way we write all possible n-vertex Feynman diagrams with the given external legs. Each diagram yields a contribution to the amplitude, which should be finally summed up over all the diagrams.
- In each diagram we associate an incoming electron or positron with the field $u(\mathbf{p}, r)$ or $\bar{v}(\mathbf{p}, r)$, respectively, while outgoing electrons and positrons contribute

fields $\bar{u}(\mathbf{p}, r)$ and $v(\mathbf{p}, r)$ respectively. Finally an incoming or outgoing photon contributes a polarization vector $\varepsilon_\mu(\mathbf{p}, i)$ or $\varepsilon_\mu(\mathbf{p}, i)^*$, respectively.

- Each vertex is associated with a factor $ie\,\gamma^\mu = ie\,((\gamma^\mu)^\alpha{}_\beta)$, where α can contract either an incoming positron $(\bar{v}_\alpha)$ or an outgoing electron $(\bar{u}_\alpha)$ field, while index β can contract either an outgoing positron v^β or an incoming electron u^β.
- 4-momentum is conserved at each vertex.
- Each internal fermion line contributes a propagator $\tilde{S}_F(p) = \frac{i}{\not{p}-m+i\varepsilon}$, while an internal photon line contributes a propagator $\tilde{D}_{F\mu\nu}(p)$ (the final expression for the amplitude does not depend on the gauge choice α), where p^μ is the momentum carried by the corresponding virtual particle.
- Diagrams differing by an exchange of two external legs corresponding to identical fermions contribute terms with a relative minus sign.

12.5.5 Gauge Invariance of Amplitudes

Different gauge choices for the photon field should not affect the physical predictions of the theory. This is indeed the case since, as we are going to show shortly, the S-matrix element defining the amplitude of a process is gauge invariant.

Consider a generic diagram, or set of diagrams, describing an interaction process. As we have already noted, fermion lines always form polygonal curves which consist of internal lines, contributing fermion propagators to the amplitude, between external ones describing incoming and outgoing fermions. The orientation on the fermionic line segments along a polygonal path, which represents the flow of the fermion charge, is continuous because of charge conservation. Their end points are vertices at which an internal or external photon line ends. The former contributes a photon propagator $D_{F\mu\nu}(k)$ to the amplitude, the latter a photon polarization vector $\epsilon_\mu(k, r)$. In both cases the index μ contracts the γ^μ matrix at the vertex. The gauge choice for a virtual photon, is encoded in the term $k_\mu k_\nu$ within $D_{F\mu\nu}(k)$, which, if the amplitude is to be gauge invariant, should not contribute to the S-matrix element. This was indeed shown to be the case for the Möller scattering. A gauge transformation on a real photon, on the other hand, induces a transformation of the amplitude which is obtained by replacing the polarization vector $\varepsilon_\mu(\mathbf{k}, r)$ with k_μ, according to (12.185). When discussing the Compton scattering amplitude, we have proven that such transformation is indeed ineffective. We have also shown that the final amplitude of the process is the sum of all diagrams in which a given (internal or external) photon line is attached to a different vertex of a fermion line, see Fig. 12.8 for the Compton scattering. In Fig. 12.9 this is illustrated for a generic fermion line with $n + 2$ vertices: The three diagrams represented in the picture are the contributions to the amplitude in which the photon with momentum k (k-photon) is attached to the $(i + 3)$th, $(i + 2)$th, and $(i + 1)$th vertices respectively; These are clearly part of the sum over the $n + 2$ diagrams in which the k-photon line ends in all possible vertices. The momenta p_i, $i = 0, \ldots, n + 1$, are fixed in terms of the momentum p_0 of the

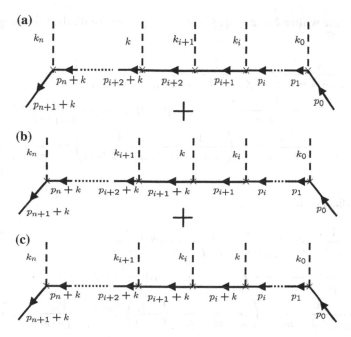

Fig. 12.9 Contributions to the amplitude of a process in which a photon with momentum k^μ ends in different vertices along a fermion line

incoming fermion and of the photon momenta k_i by the momentum conservation at each vertex:

$$p_i = p_0 + \sum_{\ell=0}^{i-1} k_\ell. \tag{12.189}$$

To prove gauge invariance in full generality, it suffices then to show that, replacing the k-photon polarization with the corresponding momentum k_μ (that is replacing it with a *gauge photon* $\varepsilon_\mu \propto k_\mu$) in each of the $n+2$ diagrams the sum of the resulting amplitudes is zero. Consider, for instance, the contribution from diagram a) of Fig. 12.9. In this case the matrix γ^μ in the $(i+3)$th vertex is contracted with k_μ, yielding:

$$\cdots \gamma^{\mu_{i+2}} \frac{1}{\slashed{p}_{i+2} + \slashed{k} - m} \slashed{k} \frac{1}{\slashed{p}_{i+2} - m} \gamma^{\mu_{i+1}} \frac{1}{\slashed{p}_{i+1} - m} \cdots$$

$$= \cdots \gamma^{\mu_{i+2}} \frac{1}{\slashed{p}_{i+2} + \slashed{k} - m} (\slashed{k} + \slashed{p}_{i+2} - m - (\slashed{p}_{i+2} - m)) \frac{1}{\slashed{p}_{i+2} - m} \gamma^{\mu_{i+1}} \cdots$$

$$= \cdots \gamma^{\mu_{i+2}} \left(\frac{1}{\slashed{p}_{i+2} - m} - \frac{1}{\slashed{p}_{i+2} + \slashed{k} - m} \right) \gamma^{\mu_{i+1}} \frac{1}{\slashed{p}_{i+1} - m} \cdots, \tag{12.190}$$

where we have written $\not{k} = \not{k} + \not{p}_{i+2} - m - (\not{p}_{i+2} - m)$. Similarly, from diagram (b) we have:

$$\cdots \frac{1}{\not{p}_{i+2} + \not{k} - m} \gamma^{\mu_{i+1}} \frac{1}{\not{p}_{i+1} + \not{k} - m} \not{k} \frac{1}{\not{p}_{i+1} - m} \gamma^{\mu_i} \frac{1}{\not{p}_i - m} \cdots$$

$$= \cdots \gamma^{\mu_{i+1}} \frac{1}{\not{p}_{i+1} + \not{k} - m} (\not{k} + \not{p}_{i+1} - m - (\not{p}_{i+1} - m)) \frac{1}{\not{p}_{i+1} - m} \gamma^{\mu_i} \cdots$$

$$= \cdots \frac{1}{\not{p}_{i+2} + \not{k} - m} \gamma^{\mu_{i+1}} \left(\frac{1}{\not{p}_{i+1} - m} - \frac{1}{\not{p}_{i+1} + \not{k} - m} \right) \gamma^{\mu_i} \frac{1}{\not{p}_i - m} \cdots ,$$
(12.191)

and from diagram (c):

$$\cdots \frac{1}{\not{p}_{i+1} + \not{k} - m} \gamma^{\mu_i} \frac{1}{\not{p}_i + \not{k} - m} \not{k} \frac{1}{\not{p}_i - m} \gamma^{\mu_{i-1}} \frac{1}{\not{p}_{i-1} - m} \cdots$$

$$= \cdots \gamma^{\mu_i} \frac{1}{\not{p}_i + \not{k} - m} (\not{k} + \not{p}_i - m - (\not{p}_i - m)) \frac{1}{\not{p}_i - m} \gamma^{\mu_{i-1}} \cdots$$

$$= \cdots \frac{1}{\not{p}_{i+1} + \not{k} - m} \gamma^{\mu_i} \left(\frac{1}{\not{p}_i - m} - \frac{1}{\not{p}_i + \not{k} - m} \right) \gamma^{\mu_{i-1}} \frac{1}{\not{p}_{i-1} - m} \cdots . \quad (12.192)$$

Note that the second term in (12.190) cancels against the first one in (12.191) and that the second term in (12.191) is canceled by the first one in (12.192). We can then convince ourselves that, summing all the $n + 2$ diagrams up, the contributions from the intermediate diagrams cancel out and we are thus left with the two contributions from the diagrams in which the k-photon ends in the first and last vertices. These read:

$$\cdots \gamma^{\mu_0} \frac{1}{\not{p}_0 + \not{k} - m} \not{k} u(p_0, r)$$

$$= \cdots \gamma^{\mu_0} \frac{1}{\not{p}_0 + \not{k} - m} (\not{p}_0 + \not{k} - m - (\not{p}_0 - m)) u(p_0, r), \quad (12.193)$$

$$\bar{u}(p_{n+1} + k, s) \not{k} \frac{1}{\not{p}_{n+1} - m} \gamma^{\mu_n} \cdots$$

$$= \bar{u}(p_{n+1} + k, s) (-(\not{p}_{n+1} - m) + \not{k} + \not{p}_{n+1} - m) \frac{1}{\not{p}_{n+1} - m} \gamma^{\mu_n} \cdots \quad (12.194)$$

The first term in Eq. (12.193) cancels against the second term from the next contribution, in which the k-photon ends in the second vertex, while the second term is zero by virtue of the Dirac equation: $(\not{p}_0 - m) u(p_0, r) = 0$. Similarly the first term in Eq. (12.194) cancels against the first term from the previous diagram, in which the k-photon ends in the one but last vertex, while the second term is zero by virtue of the Dirac equation.

This argument equally applies to all the fermion lines in the diagrams of a process, showing that a gauge transformation on a generic photon field does not affect the total amplitude.

12.5.6 Interaction with an External Field

Let us end this subsection by considering the interaction of an electron with an external electromagnetic field A_μ^{ext}, analyzed in Sect. 12.4.1. This process is described within QED by replacing, in the interaction term of the Hamiltonian, $A_\mu \to A_\mu + A_\mu^{ext}$. Let us write the first order term in the amplitude in the non-relativistic limit. Let the incoming and outgoing electron states be $(\mathbf{p}, r)$ and $(\mathbf{q}, s)$, respectively. Let us restore, for this analysis only, all the $\hbar$, c factors, as well as the normalization volume V_e of the electron. The lowest-order S-matrix element reads:

$$\langle \psi_{out} | \mathbf{S}_{(1)} | \psi_{in} \rangle = \langle \psi_{out} | \frac{ie}{\hbar c} \int d^4x : \bar{\psi}(x) \gamma^\mu \psi(x) : A_\mu^{ext}(x) | \psi_{in} \rangle. \qquad (12.195)$$

Let us write the external field in Fourier components:

$$A_\mu^{ext}(x) = \int \frac{d^4k}{(2\pi\hbar)^4} \tilde{A}_\mu^{ext}(k)\, e^{-\frac{i}{\hbar} k \cdot x}, \qquad (12.196)$$

where k^μ is a 4-momentum. In the Lorentz gauge we then have $\partial^\mu A_\mu^{ext} \Leftrightarrow k^\mu \tilde{A}_\mu^{ext} = 0$. The matrix element in (12.195) is readily computed by writing the electron field operators and the initial and final states in terms of creation and annihilation operators. As usual the integration over x yields the conservation of momentum at the vertex: $q = p + k$. Integrating out all the delta functions we find:

$$\langle \psi_{out} | \mathbf{S}_{(1)} | \psi_{in} \rangle = i\, \frac{e}{\hbar c} \left(\frac{mc^2}{\sqrt{E_\mathbf{p} E_\mathbf{q}}\, V_e} \right) \bar{u}(\mathbf{p} + \mathbf{k}, s) \gamma^\mu u(\mathbf{p}, r) \tilde{A}_\mu^{ext}(k). \qquad (12.197)$$

In the non-relativistic limit we retain only terms of order less than two in the ratio $\frac{v}{c}$. We can then approximate the energy of the electron with its rest energy $E_\mathbf{q} \approx E_\mathbf{p} \approx mc^2$. In this limit $k^0 = E_\mathbf{q} - E_\mathbf{p} \approx 0$, namely $k^\mu \approx (0, \mathbf{k})$. Defining $\mathcal{A} \equiv \tilde{A}_\mu^{ext}(k)\,\gamma^\mu$ and writing $u(\mathbf{p}, r) \approx \frac{\not{p} + mc}{2mc} u(\mathbf{0}, r)$, where $u(\mathbf{0}, r) \equiv (\varphi_r, \mathbf{0})$, we find

$$\langle \psi_{out} | \mathbf{S}_{(1)} | \psi_{in} \rangle \approx i\, \frac{e}{\hbar c\, V_e}\, \frac{\bar{u}(\mathbf{0}, s)(\not{p} + \not{k} + mc)\mathcal{A}(\not{p} + mc)u(\mathbf{0}, r)}{4m^2 c^2}. \qquad (12.198)$$

It is useful to write the field strength $F_{\mu\nu}$ of the external electromagnetic field in momentum space as well:

$$F_{\mu\nu} = \int \frac{d^4k}{(2\pi\hbar)^4}\, \tilde{F}_{\mu\nu}(k)\, e^{-\frac{i}{\hbar}k\cdot x}\,; \quad \tilde{F}_{\mu\nu}(k) = -\frac{i}{\hbar}\,(k_\mu \tilde{A}_\nu^{ext} - k_\nu \tilde{A}_\mu^{ext}),$$

(12.199)

and to introduce the Fourier transforms $\tilde{\mathbf{E}}(k)$, $\tilde{\mathbf{B}}(k)$ of the electric and magnetic fields, respectively:

$$\tilde{E}_i \equiv \tilde{F}_{i0}, \quad \tilde{B}_i \equiv \frac{1}{2}\, \epsilon_{ijk}\, \tilde{F}^{jk}.$$

(12.200)

We wish now to rewrite the matrix $(\not{p} + \not{k} + mc)\not{A}(\not{p} + mc)$ using the properties:

$$\not{A}\not{p} = 2\,(\tilde{A}^{ext} \cdot p) - \not{p}\not{A}, \quad \not{k}\not{A} = (k \cdot \tilde{A}^{ext}) + \frac{1}{2}\,[\not{k},\,\not{A}]$$

$$= (k \cdot \tilde{A}^{ext}) + k_\mu \tilde{A}_\nu^{ext}\, \gamma^{\mu\nu},$$

(12.201)

which directly descend from the gamma matrix algebra (recall that $\gamma^{\mu\nu} \equiv \frac{1}{2}\,[\gamma^\mu, \gamma^\nu]$). We then find

$$\begin{aligned}
(\not{p} + \not{k} + mc)\not{A}(\not{p} + mc) &= \not{p}\not{A}\not{p} + \not{k}\not{A}(\not{p} + mc) + mc\,\{\not{A},\,\not{p}\} + m^2 c^2\,\not{A} \\
&= [2\,(\tilde{A}^{ext} \cdot p) + (k \cdot \tilde{A}^{ext}) + k_\mu \tilde{A}_\nu^{ext}\, \gamma^{\mu\nu}]\,(\not{p} + mc) \\
&= [2\,(\tilde{A}^{ext} \cdot p) + k_\mu \tilde{A}_\nu^{ext}\, \gamma^{\mu\nu}]\,(\not{p} + mc),
\end{aligned}$$

(12.202)

where we have used properties (12.201), the on-shell condition $p^2 = m^2 c^2$ and the Lorentz gauge condition $k \cdot \tilde{A}^{ext} = 0$. We can now rewrite the amplitude (12.197):

$$\langle \psi_{out}|\mathbf{S}_{(1)}|\psi_{in}\rangle \approx i\, \frac{e}{4\,\hbar c\, V_e\, m^2\, c^2}\, \times$$

$$\times\, \bar{u}(\mathbf{0}, s)[2\,(\tilde{A}^{ext} \cdot p) + \frac{i\hbar}{2}\tilde{F}_{\mu\nu}\gamma^{\mu\nu}]\,(\not{p} + mc)u(\mathbf{0}, r) = i\, \frac{e}{4\,\hbar c\, V_e\, m^2\, c^2}$$

$$\times\, \bar{u}(\mathbf{0}, s)[2\,(\tilde{A}^{ext} \cdot p) - i\,\hbar\tilde{E}_i\,\gamma^{0i} + \frac{i\hbar}{2}\,\epsilon_{ijk}\,\gamma^{ij}\tilde{B}^k]\,(\not{p} + mc)u(\mathbf{0}, r).$$

(12.203)

Let us now write the matrix $\not{p} + mc$:

$$\not{p} + mc = mc(1 + \gamma^0) - \mathbf{p}\cdot\boldsymbol{\gamma} = \begin{pmatrix} 2mc\,\mathbf{1}_2 & -\mathbf{p}\cdot\boldsymbol{\sigma} \\ \mathbf{p}\cdot\boldsymbol{\sigma} & \mathbf{0}_2 \end{pmatrix}.$$

(12.204)

in the non-relativistic limit, the off-diagonal blocks are subleading, so we shall only consider the diagonal ones. The matrix element between $\bar{u}(0, s)$ and $u(0, r)$ belongs to the upper-diagonal block, and, being γ^{0i} off-diagonal, the term in (12.203) containing the electric field is subleading in the non-relativistic limit. To lowest order in $\frac{v}{c}$ we find:

$$
\begin{aligned}
\langle \psi_{out} | \mathbf{S}_{(1)} | \psi_{in} \rangle &\approx \frac{i}{\hbar c V_e}\, \varphi_s^\dagger \left[\frac{e}{mc} (\tilde{A}^{ext} \cdot p) + \frac{ie\hbar}{4\,mc}\, \epsilon_{ijk}\, \gamma^{ij} \tilde{B}^k \right] \varphi_r \\
&= -\frac{i}{\hbar c V_e}\, \varphi_s^\dagger \left[e\,\tilde{V}(k) - \frac{e}{mc}\, p^i \tilde{A}_i^{ext}(k) - \frac{e}{mc}\, \mathbf{s} \cdot \tilde{\mathbf{B}}(k) \right] \varphi_r \\
&= -\frac{i}{\hbar} \int d^4x \langle \psi_{out} | \widehat{\mathcal{H}}_I^{ext}(x) | \psi_{in} \rangle ,
\end{aligned}
\tag{12.205}
$$

where $\tilde{V}(k) = -\tilde{A}_0^{ext}(k)$ is the Fourier transform of the electrostatic potential, $\mathbf{s} = \hbar\,\boldsymbol{\sigma}/2$ is the spin vector. To derive Eq. (12.205) we have used the property

$$
\frac{i\hbar}{2}\, \epsilon_{ijk}\, \gamma^{ij} \tilde{B}^k = \hbar \begin{pmatrix} \tilde{B}_k \sigma^k & \mathbf{0}_2 \\ \mathbf{0}_2 & \tilde{B}_k \sigma^k \end{pmatrix} .
\tag{12.206}
$$

The quantity between square brackets in Eq. (12.205) can be compared with the analogous quantity appearing on the right hand side of Eq. (10.223) of Chap. 10. Since $\mathbf{s} = \frac{\hbar}{2}\boldsymbol{\sigma}$, we see that, excluding the kinetic term $\frac{1}{2m}\mathbf{p}^2$, the expressions of the two interaction Hamiltonians coincide at first order in e. The last term in square brackets is the coupling term $-\boldsymbol{\mu}_s \cdot \tilde{\mathbf{B}}$ of the electron spin to the external magnetic field, where the magnetic moment associated with the spin is usually written in the form

$$
\boldsymbol{\mu}_s = \frac{ge}{2mc}\, \mathbf{s}.
\tag{12.207}
$$

g being the so called electron *g-factor*. Comparing this definition with the corresponding term in Eq. (12.205), we see that for the electron $g = 2$, to lowest order in the perturbative expansion (classical value). We also note that, setting $\mathbf{s} = \frac{\hbar}{2}\boldsymbol{\sigma}$, this value coincides with the result given in Eqs. (10.224) and (10.226), namely the gyromagnetic ratio $\frac{e}{mc}$ is twice as large as the one related to the orbital angular momentum.

The exact amplitude is obtained by summing to (12.205) all higher order corrections $\langle \mathbf{S}^{(n)} \rangle$. In particular the term in $\langle \mathbf{S}^{(2)} \rangle$, described by the diagram in Fig. 12.10, will be computed in Sect. 12.8.6 and will provide an important test of the theory against experiments.

Fig. 12.10 First perturbative
contribution to the electron
anomalous magnetic moment

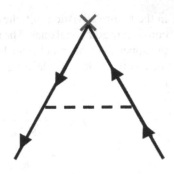

12.6 Cross Sections

We analyze here two instances of interaction processes: the Bahbha and the Compton
scattering.

12.6.1 The Bahbha Scattering

Let us first compute the cross section for the electron-positron scattering (12.168), to
lowest order in α. The Lorentz-invariant variables describing the process are, aside
from the electron (positron)-mass m, the cross scalar products among the 4-momenta
p_+^μ, p_-^μ, q_+^μ, q_-^μ, all of which can be expressed in terms of the three Mandelstam
variables s, t, u defined in (12.26):

$$p_+ \cdot p_- = q_+ \cdot q_- = \frac{s - 2m^2}{2}, \quad p_+ \cdot q_+ = p_- \cdot q_- = \frac{2m^2 - t}{2},$$

$$p_+ \cdot q_- = q_+ \cdot p_- = \frac{2m^2 - u}{2}. \tag{12.208}$$

Equation (12.28) in this case implies $s + t + u = 4\,m^2$. The explicit form of t, u in
the CM frame is given by Eqs. (12.29) and (12.30):

$$t = -4\,|\mathbf{p}|^2 \sin^2\left(\frac{\theta}{2}\right), \quad u = -4\,|\mathbf{p}|^2 \cos^2\left(\frac{\theta}{2}\right), \tag{12.209}$$

being $E_1 = E_2 = E_1' = E_2' = E = \sqrt{m^2 + |\mathbf{p}|^2}$ and $s = 4\,E^2$.
Let us now use Eq. (12.81) to write, in the CM frame:

$$d\sigma = \frac{1}{2\sqrt{s(s - 4m^2)}} |\langle \mathscr{T}_{(2)}' \rangle|^2 \, d\Phi^{(2)} = \frac{1}{64\,\pi^2\,s} |\langle \mathscr{T}_{(2)}' \rangle|^2 \, d\Omega, \tag{12.210}$$

where we have used the property $\sqrt{(p_+ \cdot p_-)^2 - m^4} = \frac{1}{2}\sqrt{s(s - 4m^2)}$ and the general form (12.96) of $d\Phi^{(2)}$. We shall consider the simpler case in which the incoming particles are not polarized and the spin states of the final particles are not measured. The probability per unit time is then computed by averaging the one referred to distinct polarizations of the electron-positron system, over the initial spin state and summing over the final ones. This amounts, in Eq. (12.210), to define:

$$|\langle \mathscr{T}'_{(2)}\rangle|^2 = \frac{1}{4} \sum_{s_\pm, r_\pm} |\langle \mathbf{q}_+, s_+|\langle \mathbf{q}_-, s_-|\mathscr{T}'_{(2)}|\mathbf{p}_+, r_+\rangle|\mathbf{p}_-, r_-\rangle|^2, \quad (12.211)$$

where the factor $1/4$ is related to the average over the four distinct polarization states of the initial electron-positron system. Let us now use our previous result (12.171) for the scattering amplitude and write:

$$|\langle \mathscr{T}'_{(2)}\rangle|^2 = 4\,e^4\,m^4 \sum_{s_\pm, r_\pm} \left[AA^* + BB^* + AB^* + BA^*\right], \quad (12.212)$$

where the terms

$$A = -\frac{1}{t}\bar{v}(\mathbf{p}_+, r_+)\gamma^\mu v(\mathbf{q}_+, s_+)\,\bar{u}(\mathbf{q}_-, s_-)\gamma_\mu u(\mathbf{p}_-, r_-),$$

$$B = \frac{1}{s}\bar{u}(\mathbf{q}_-, s_-)\gamma^\mu v(\mathbf{q}_+, s_+)\,\bar{v}(\mathbf{p}_+, r_+)\gamma_\mu u(\mathbf{p}_-, r_-), \quad (12.213)$$

are referred to the diagrams in Fig. 12.5a (diffusion) and b (annihilation), respectively. Now, using the gamma-matrix properties $(\gamma^\mu)^\dagger \gamma^0 = \gamma^0 \gamma^\mu$ and $(\gamma^0)^\dagger = \gamma^0$, one can easily show that

$$(\bar{u}_1 \gamma^\mu u_2)^* = u_2^\dagger (\gamma^\mu)^\dagger (\gamma^0)^\dagger u_1 = u_2^\dagger \gamma^0 \gamma^\mu u_1 = \bar{u}_2 \gamma^\mu u_1, \quad (12.214)$$

and similarly $(\bar{v}_1 \gamma^\mu u_2)^* = \bar{u}_2 \gamma^\mu v_1$, $(\bar{v}_1 \gamma^\mu v_2)^* = \bar{v}_2 \gamma^\mu v_1$, so that we can write

$$A^* = -\frac{1}{t}\bar{v}(\mathbf{q}_+, s_+)\gamma^\mu v(\mathbf{p}_+, r_+)\,\bar{u}(\mathbf{p}_-, r_-)\gamma_\mu u(\mathbf{q}_-, s_-),$$

$$B^* = \frac{1}{s}\bar{v}(\mathbf{q}_+, s_+)\gamma^\mu u(\mathbf{q}_-, s_-)\,\bar{u}(\mathbf{p}_-, r_-)\gamma_\mu v(\mathbf{p}_+, r_+). \quad (12.215)$$

Next we need to recall the formulas (10.182) and (10.183) of Chap. 10, for the projectors on the positive and negative-energy solutions of the Dirac equation.[22]

---

[22]In the case of polarized fermions, there would be no summation over the spin states and we should use, for each particle, the expressions in Sect. 10.6.3 for the projector on the corresponding polarization:

$$u(\mathbf{p}, r)\,\bar{u}(\mathbf{p}, r) = \frac{(\not{p} + m)}{4m}\,(1 + \varepsilon_r \gamma^5 \not{n}), \quad v(\mathbf{p}, s)\,\bar{v}(\mathbf{p}, s) = \frac{(\not{p} - m)}{4m}\,(1 - \varepsilon_s \gamma^5 \not{n}).$$

$$\sum_r u(\mathbf{p}, r)^\alpha \bar{u}(\mathbf{p}, r)_\beta = \frac{(\not{p} + m)^\alpha{}_\beta}{2m},$$

$$\sum_r v(\mathbf{p}, r)^\alpha \bar{v}(\mathbf{p}, r)_\beta = \frac{(\not{p} - m)^\alpha{}_\beta}{2m}, \qquad (12.216)$$

to rewrite the AA^* term in (12.212) as follows:

$$\sum_{r_\pm, s_\pm} AA^* = \frac{1}{16\,m^4\,t^2} \mathrm{Tr}\left(\gamma^\mu(\not{q}_+ - m)\gamma^\nu(\not{p}_+ - m)\right) \mathrm{Tr}\left(\gamma_\mu(\not{p}_- + m)\gamma_\nu(\not{q}_- + m)\right).$$

To prove the above formula it is useful to write, in the product AA^*, the spinor indices explicitly. By the same token we find:

$$\sum_{r_\pm, s_\pm} BB^* = \frac{1}{16\,m^4\,s^2} \mathrm{Tr}\left(\gamma^\mu(\not{q}_+ - m)\gamma^\nu(\not{q}_- + m)\right) \mathrm{Tr}\left(\gamma_\mu(\not{p}_- + m)\gamma_\nu(\not{p}_+ - m)\right),$$

$$\sum_{r_\pm, s_\pm} AB^* = -\frac{1}{16\,m^4\,st} \mathrm{Tr}\left(\gamma^\mu(\not{q}_+ - m)\gamma^\nu(\not{q}_- + m)\gamma_\mu(\not{p}_- + m)\gamma_\nu(\not{p}_+ - m)\right),$$

$$\sum_{r_\pm, s_\pm} BA^* = -\frac{1}{16\,m^4\,st} \mathrm{Tr}\left(\gamma^\mu(\not{q}_+ - m)\gamma^\nu(\not{p}_+ - m)\gamma_\mu(\not{p}_- + m)\gamma_\nu(\not{q}_- + m)\right).$$

To compute the above traces we need to recall from Appendix G the following gamma-matrix identities:

$$\mathrm{Tr}(\gamma^\mu\gamma^\nu) = 4\,\eta^{\mu\nu},$$
$$\mathrm{Tr}(\gamma^\mu\gamma^\nu\gamma^\rho\gamma^\sigma) = 4(\eta^{\mu\nu}\eta^{\rho\sigma} + \eta^{\mu\sigma}\eta^{\rho\nu} - \eta^{\mu\rho}\eta^{\nu\sigma}),$$
$$\mathrm{Tr}(\gamma^{\mu_1} \cdots \gamma^{\mu_{2k+1}}) = 0,$$
$$\gamma^\mu \not{A} \gamma_\mu = -2\not{A},$$
$$\gamma^\mu \not{A}\not{B} \gamma_\mu = 4\,(A \cdot B),$$
$$\gamma^\mu \not{A}\not{B}\not{C} \gamma_\mu = -2\not{C}\not{B}\not{A}. \qquad (12.217)$$

The first of the above identities is proven by writing $\gamma^\mu\gamma^\nu = 2\eta^{\mu\nu} - \gamma^\mu\gamma^\nu$, computing the trace of both sides and using the cyclic property of the trace $\mathrm{Tr}(\gamma^\mu\gamma^\nu) = \mathrm{Tr}(\gamma^\nu\gamma^\mu)$. Similarly the second is proven by shifting γ^σ to the left, past the other three gamma-matrices, and then using again the cyclic property of the trace $\mathrm{Tr}(\gamma^\mu\gamma^\nu\gamma^\rho\gamma^\sigma) = \mathrm{Tr}(\gamma^\sigma\gamma^\mu\gamma^\nu\gamma^\rho)$. Finally the property that the trace of an odd number of gamma-matrices is zero is easily proven using the properties $(\gamma^5)^2 = \mathbf{1}$, $\gamma^\mu\gamma^5 = -\gamma^5\gamma^\mu$ of the γ^5-matrix:

$$\mathrm{Tr}(\gamma^{\mu_1} \cdots \gamma^{\mu_{2k+1}}) = \mathrm{Tr}(\gamma^{\mu_1} \cdots \gamma^{\mu_{2k+1}}(\gamma^5)^2) = \mathrm{Tr}(\gamma^5\gamma^{\mu_1} \cdots \gamma^{\mu_{2k+1}}\gamma^5)$$
$$= -\mathrm{Tr}(\gamma^{\mu_1} \cdots \gamma^{\mu_{2k+1}}). \qquad (12.218)$$

We shall also need the general identities (G.25), (G.26) of Appendix G.

Now the reader can easily derive the following formulas:

$$\text{Tr}\left(\gamma^\mu(\not{p}\pm m)\gamma^\nu(\not{q}\pm m)\right) = 4\left[p^\mu q^\nu + p^\nu q^\mu - \left(p\cdot q - m^2\right)\eta^{\mu\nu}\right],$$

$$\text{Tr}\left(\gamma^\mu(\not{p}\pm m)\gamma^\nu(\not{q}\mp m)\right) = 4\left[p^\mu q^\nu + p^\nu q^\mu - \left(p\cdot q + m^2\right)\eta^{\mu\nu}\right],$$

which are needed, together with (G.25), to derive, after some algebra, the following expressions:

$$\sum_{r_\pm,s_\pm} AA^* = \frac{1}{m^4\,t^2}\left(2\,(p_+\cdot p_-)(q_+\cdot q_-) + 2\,(q_+\cdot p_-)(p_+\cdot q_-)\right.$$

$$\left. + t\,(q_+\cdot p_+) + t\,(q_-\cdot p_-) + t^2\right) = \frac{1}{m^4\,t^2}\left(\frac{s^2+u^2}{2} + 4\,m^2\,(t-m^2)\right),$$

$$\sum_{r_\pm,s_\pm} BB^* = \frac{1}{m^4\,s^2}\left(\frac{t^2+u^2}{2} + 4\,m^2\,(s-m^2)\right),$$

$$\sum_{r_\pm,s_\pm} AB^* = \sum_{r_\pm,s_\pm} BA^* = \frac{2}{m^4\,st}(p_+\cdot q_-)\left(p_+\cdot q_- + 2m^2\right)$$

$$= \frac{2}{m^4\,st}\left(m^2 - \frac{u}{2}\right)\left(3m^2 - \frac{u}{2}\right). \tag{12.219}$$

Inserting the above result in Eq. (12.210) we find a general formula for the cross section:

$$\frac{d\sigma}{d\Omega} = \frac{\alpha^2}{s}\left[\frac{1}{t^2}\left(\frac{s^2+u^2}{2} + 4\,m^2\,(t-m^2)\right) + \frac{1}{s^2}\left(\frac{t^2+u^2}{2} + 4\,m^2\,(s-m^2)\right)\right.$$

$$\left. + \frac{4}{st}\left(m^2 - \frac{u}{2}\right)\left(3m^2 - \frac{u}{2}\right)\right], \tag{12.220}$$

where, as usual, $\alpha = e^2/(4\pi)$. The first and second terms in square brackets originate from the squared norm of the diffusion and annihilation terms in the amplitude, respectively, while the third is a cross product.

Consider now the non-relativistic limit in which $E \sim m$ ($s \sim 4m^2$) and we neglect terms of the order $|\mathbf{p}|^2/m^2$ (like t/m^2 and u/m^2). In this limit we see that the second and third terms on the right hand side of (12.220) are subleading, that is the annihilation amplitude does not contribute to the cross section, which then reduces to[23]:

$$\frac{d\sigma}{d\Omega} = \left(\frac{\alpha}{2\,\mu v^2}\right)^2 \frac{1}{\sin^4\left(\frac{\theta}{2}\right)}, \tag{12.221}$$

---

[23] Note that, in order to restore the $\hbar$, c factors, we would need to multiply Eq. (12.220) by $\hbar^2$, while Eq. (12.221) needs no $\hbar$, c factors to be restored.

where $v = |\mathbf{p}|/\mu = 2\,|\mathbf{p}|/m$ is the relative velocity, being $\mu = m/2$ the reduced mass of the system. Since only the diffusion diagram contributes in the non-relativistic limit, we would have obtained the same result for the scattering of an electron off any charge-e particle. Equation (12.221) reproduces the classical result obtained by E. Rutherford when studying the scattering of alpha-particles off heavy nuclei.

12.6.2 The Compton Scattering

Let us now analyze the Compton scattering (12.179). We shall denote the quantities associated with the photon and the electron by 1 and 2, respectively. The initial and final 4-momenta of the photon are written as follows:

$$p_1 = (\omega,\, \mathbf{p}_1),\quad q_1 = (\omega',\, \mathbf{q}_1), \tag{12.222}$$

where $|\mathbf{p}_1| = \omega$, $|\mathbf{q}_1| = \omega'$, ω, ω' being the angular frequencies of the incoming and outgoing electromagnetic plane-waves. We shall analyze the scattering in the laboratory frame in which the electron is initially at rest: $\mathbf{p}_2 = 0$. The angular variables are referred to the direction of the incident photon. In particular we shall denote by θ the photon diffusion angle, namely of the angle between $\mathbf{q}_1$ and $\mathbf{p}_1$.

Recalling that, for physical photons, $p_1^2 = q_1^2 = 0$, we have:

$$s - m^2 = 2\,p_1 \cdot p_2 = 2\,q_1 \cdot q_2 = 2m\omega,\quad u - m^2 = -2\,p_1 \cdot q_2$$
$$= -2\,p_2 \cdot q_1 = -2m\omega'. \tag{12.223}$$

Let us now use the conservation of the total 4-momentum and write $q_2 = p_1 + p_2 - q_1$. Computing the squared norm of both sides we find

$$0 = p_1 \cdot p_2 - p_1 \cdot q_1 - p_2 \cdot q_1 = m\,(\omega - \omega') - \omega\omega'\,(1 - \cos(\theta)),$$

from which we find

$$\frac{1}{\omega'} - \frac{1}{\omega} = \frac{1}{m}\,(1 - \cos(\theta)). \tag{12.224}$$

Let us now express the phase-space element in terms of photon quantities. To this end let us write the t variable as follows

$$t = 2\,m^2 - 2\,p_2 \cdot q_2 = -2\,p_1 \cdot q_1 = -2\,\omega\omega'\,(1 - \cos(\theta)) = -2\,m\,(\omega - \omega'),$$

where we have used Eq. (12.224). For a given ω, t will depend on the variable ω', related to θ by (12.224). From the above equation we find $dt = 2m\,d\omega'$. Using Eq. (12.224) we can write $d\omega' = \omega'^2\,d\cos(\theta)/m$ and thus:

$$d(-t) = -2m \, d\omega' = -2\,\omega'^2 \, d\cos(\theta) = \frac{\omega'^2}{\pi} \, \Omega. \tag{12.225}$$

We can now write the phase space element substituting the above expression for $d(-t)$ in Eq. (12.98) and identifying $\mu_2 = m$, $\mu_1 = 0$:

$$d\Phi^{(2)} = \frac{1}{8\pi} \frac{d(-t)}{s - m^2} = \frac{\omega'^2}{16\pi^2 \, m\,\omega} d\Omega. \tag{12.226}$$

The differential cross-section, to lowest order, reads:

$$d\sigma = \frac{1}{4\,p_1 \cdot p_2} |\langle \mathscr{T}'_{(2)}\rangle|^2 \, d\Phi^{(2)} = \frac{1}{16\pi} \frac{d(-t)}{(s-m^2)^2} |\langle \mathscr{T}'_{(2)}\rangle|^2$$

$$= \frac{1}{(8\pi \, m)^2} \left(\frac{\omega'}{\omega}\right)^2 |\langle \mathscr{T}'_{(2)}\rangle|^2 \, d\Omega. \tag{12.227}$$

Let us now evaluate $|\langle \mathscr{T}'_{(2)}\rangle|^2$. As usual we consider unpolarized initial particles and we do not measure the spin states of the outgoing electron and photon. This implies that the probability per unit time should be summed over the final polarizations and averaged over the initial ones, which amounts, in (12.227), to write

$$|\langle \mathscr{T}'_{(2)}\rangle|^2 = \frac{1}{4} \sum_{i,i',r,s} |\langle \mathbf{q}_1, i'|\langle \mathbf{q}_2, s| \mathscr{T}'_{(2)} |\mathbf{p}_1, i\rangle |\mathbf{p}_2, r\rangle|^2$$

$$= e^4 \, m^2 \sum_{i,i',r,s} \Big[\varepsilon_\mu(\mathbf{p}_1, i) \, \varepsilon_\nu(\mathbf{q}_1, i')^* \varepsilon_\rho(\mathbf{p}_1, i)^* \, \varepsilon_\sigma(\mathbf{q}_1, i')$$

$$\times \bar{u}(\mathbf{q}_2, s) \left(\gamma^\mu \frac{1}{\not{p}_2 - \not{q}_1 - m} \gamma^\nu + \gamma^\nu \frac{1}{\not{p}_1 + \not{p}_2 - m} \gamma^\mu\right) u(\mathbf{p}_2, r)$$

$$\times \bar{u}(\mathbf{p}_2, r) \left(\gamma^\sigma \frac{1}{\not{p}_2 - \not{q}_1 - m} \gamma^\rho + \gamma^\rho \frac{1}{\not{p}_1 + \not{p}_2 - m} \gamma^\sigma\right) u(\mathbf{q}_2, s) \Big]. \tag{12.228}$$

Consider the quantity $R_{\mu\nu} \equiv \sum_{i=1}^{2} \varepsilon_\mu(\mathbf{p}, i)\varepsilon_\nu(\mathbf{p}, i)^*$ in the Coulomb gauge where $\varepsilon_\mu = (0, \boldsymbol{\varepsilon})$. This tensor has only spatial components, $R_{00} = R_{0i} = R_{i0} = 0$, which, being $\boldsymbol{\varepsilon}$ transverse to the direction of propagation $\mathbf{n} \equiv \mathbf{p}/|\mathbf{p}|$, read:

$$R_{ij} = \delta_{ij} - n_i n_j. \tag{12.229}$$

The reader may easily verify that $R_{\mu\nu}$ can be written in the form:

$$R_{\mu\nu} = -\eta_{\mu\nu} + \chi_\mu p_\nu + \chi_\nu p_\mu, \tag{12.230}$$

where $\chi_0 = \frac{1}{2\omega}$, $\chi_i = -n_i/(2\,\omega)$. The last two terms on the right hand side of (12.230) can be reabsorbed by a gauge transformation of ε_μ and do not contribute to (12.228) since, as shown in Sect. 12.5, the amplitude of the process is gauge invariant and thus the contraction of the photon momentum with any of the gamma-matrices γ^μ, γ^ν, γ^ρ, γ^σ is zero (see Eqs. (12.186) and (12.188)). We can thus replace in Eq. (12.228) the sum over the photon polarizations, $\sum_{i=1}^{2} \varepsilon_\mu(\mathbf{p}, i)\varepsilon_\nu(\mathbf{p}, i)^*$, by $-\eta_{\mu\nu}$. Equation (12.228) can now be recast in the following compact form:

$$|\langle \mathscr{T}'_{(2)} \rangle|^2 = \frac{e^4}{4} \operatorname{Tr} \left[(\slashed{p}_2 + m) \left(\gamma^\mu \frac{\slashed{p}_2 - \slashed{q}_1 + m}{u - m^2} \gamma^\nu + \gamma^\nu \frac{\slashed{p}_1 + \slashed{p}_2 + m}{u - s^2} \gamma^\mu \right) \right.$$
$$\left. \times (\slashed{p}_2 + m) \left(\gamma_\nu \frac{\slashed{p}_2 - \slashed{q}_1 + m}{u - m^2} \gamma_\mu + \gamma_\mu \frac{\slashed{p}_1 + \slashed{p}_2 + m}{u - s^2} \gamma_\nu \right) \right],$$

$$(12.231)$$

where we have used the property $(\slashed{p} + m)(\slashed{p} - m) = p^2 - m^2$. Using the identities (G.25) and (G.25), we find the following useful formulas

$$\operatorname{Tr}\left[(\slashed{q}_2 + m)\gamma^\mu (\slashed{p}_2 - \slashed{q}_1 + m)\gamma^\nu (\slashed{p}_2 + m)\gamma_\nu (\slashed{p}_2 - \slashed{q}_1 + m)\gamma_\mu \right]$$
$$= 8 \left[4m^4 - (s - m^2)(u - m^2) + 2m^2(u - m^2) \right],$$
$$\operatorname{Tr}\left[(\slashed{q}_2 + m)\gamma^\mu (\slashed{p}_2 + \slashed{p}_1 + m)\gamma^\nu (\slashed{p}_2 + m)\gamma_\nu (\slashed{p}_2 + \slashed{p}_1 + m)\gamma_\mu \right]$$
$$= 8 \left[4m^4 - (s - m^2)(u - m^2) + 2m^2(s - m^2) \right],$$
$$\operatorname{Tr}\left[(\slashed{q}_2 + m)\gamma^\mu (\slashed{p}_2 - \slashed{q}_1 + m)\gamma^\nu (\slashed{p}_2 + m)\gamma_\mu (\slashed{p}_2 + \slashed{p}_1 + m)\gamma_\nu \right]$$
$$= \operatorname{Tr}\left[(\slashed{q}_2 + m)\gamma^\mu (\slashed{p}_2 + \slashed{p}_1 + m)\gamma^\nu (\slashed{p}_2 + m)\gamma_\mu (\slashed{p}_2 - \slashed{q}_1 + m)\gamma_\nu \right]$$
$$= 8 m^2 [4m^2 + (s - m^2) + (u - m^2)].$$

$$(12.232)$$

Expanding the right hand side of (12.231) and using the above identities we find:

$$|\langle \mathscr{T}'_{(2)} \rangle|^2 = 8 e^4 \left[m^4 \left(\frac{1}{u - m^2} + \frac{1}{s - m^2} \right)^2 + m^2 \left(\frac{1}{u - m^2} + \frac{1}{s - m^2} \right) \right.$$
$$\left. - \frac{1}{4} \left(\frac{s - m^2}{u - m^2} + \frac{u - m^2}{s - m^2} \right) \right]$$
$$= 8 e^4 \left[\frac{m^2}{4} \left(\frac{1}{\omega'} - \frac{1}{\omega} \right)^2 + \frac{m}{2} \left(\frac{1}{\omega} - \frac{1}{\omega'} \right) + \frac{1}{4} \left(\frac{\omega}{\omega'} + \frac{\omega'}{\omega} \right) \right],$$

$$(12.233)$$

where we have used Eq. (12.223). Next we use Eq. (12.224) to rewrite the above equation in the following form:

$$|\langle \mathscr{T}'_{(2)} \rangle|^2 = 2 e^4 \left[\frac{\omega}{\omega'} + \frac{\omega'}{\omega} - \sin^2(\theta) \right].$$

$$(12.234)$$

Substituting the above result in (12.227) we obtain the following formula for the differential cross-section:

$$\frac{d\sigma}{d\Omega} = \frac{r_e^2}{2} \left(\frac{\omega'}{\omega}\right)^2 \left[\frac{\omega}{\omega'} + \frac{\omega'}{\omega} - \sin^2(\theta)\right], \tag{12.235}$$

where $r_e \equiv \frac{\alpha}{m} = \frac{e^2}{4\pi m}$ $(= \frac{e^2}{4\pi mc^2})$ is the *classical radius of the electron*. The above formula was originally found by O. Klein and Y. Nishina in 1929, and by I.E. Tamm in 1930.

In the limit of low-energy photons, in which $\omega \ll m$, neglecting terms of order $\frac{\omega}{m}$, we can approximate, by virtue of Eq. (12.224), ω' with ω and write:

$$\frac{d\sigma}{d\Omega} \approx \frac{r_e^2}{2} \left(1 + \cos^2(\theta)\right). \tag{12.236}$$

The approximation becomes exact if we compute the total cross-section at threshold, that is when $\omega, \omega' \to 0$. Since

$$\lim_{\omega,\omega'\to 0} \frac{\omega}{\omega'} = 1$$

by integration in $d\Omega$ we easily find

$$\sigma_{(thr.)} = \frac{8}{3}\pi r_e^2 = \frac{8}{3}\pi \left(\frac{e^2}{4\pi m}\right)^2. \tag{12.237}$$

12.7 Divergent Diagrams

So far we have considered amplitudes and cross-sections corresponding to *tree-diagrams*, that is diagrams where no closed lines (*loops*) appear. They correspond to the terms from (1)–(4) of the second order S-matrix $\mathbf{S}^{(2)}$ given in Eq. (12.140) and to the corresponding diagrams in Fig. 12.2. The amplitudes from (5)–(8) instead involve loops, as it is apparent from the same figure. In particular diagrams (5) and (6) in Fig. 12.2 (which represents the equal contributions of the terms (6) and (7) in Eq. (12.140)) refer to transitions between initial and final states consisting of a single particle with the same quantum numbers. They are referred to as *self-energy* transitions. Considering first the *electron self-energy diagram* corresponding to the terms (6) and (7). The S-matrix element contributing to the process is read from Eq. (12.140)[24]:

---

[24] The factor $\frac{1}{2}$ is canceled by the sum of the two identical terms (6) an (7) of Eq. (12.140).

$$(ie)^2 \int d^4x d^4y \; : \bar\psi(x)\gamma^\mu S_F(x-y)\gamma^\nu \psi(y) : D_{F\mu\nu}(x-y)$$

$$= -i \int d^4x d^4y \; : \bar\psi(x)\Sigma(x-y)\psi(y) :, \tag{12.238}$$

where we have defined[25]:

$$\Sigma(x-y) \equiv -ie^2 \, \gamma^\mu S_F(x-y)\gamma^\nu D_{F\mu\nu}(x-y)$$

$$= -ie^2 \int \frac{d^4q}{(2\pi)^4} \frac{d^4k}{(2\pi)^4} \, \gamma^\mu \frac{1}{\slashed{q}-m} \gamma^\nu \frac{1}{k^2} e^{-i(q+k)\cdot(x-y)}$$

$$= \int \frac{d^4p}{(2\pi)^4} \, \Sigma(p)\, e^{-ip\cdot(x-y)}. \tag{12.239}$$

and $p = q + k$. The Fourier transform $\Sigma(p)$ of $\Sigma(x-y)$ reads

$$\Sigma(p) \equiv -ie^2 \int \frac{d^4k}{(2\pi)^4} \, \gamma^\mu \frac{1}{\slashed{p}-\slashed{k}-m} \gamma^\nu \frac{1}{k^2}. \tag{12.240}$$

Computing the $S_{(2)}$ term (12.238) between an incoming and outgoing electron states with momenta P_i and P_f, the space-time integrals yield a delta-function $\delta^4(P_f - P_i)$ implementing the total momentum conservation ($P_f = P_i$), times a second delta-function which sets $P_i = q + k = p$. We find:

$$i\langle \mathscr{T}' \rangle = -2mi\bar u(p,r)\Sigma(p)u(p,r). \tag{12.241}$$

Notice that, in contrast to the tree-amplitudes, here the delta functions are not enough to eliminate all the momentum integrals and we are left with the integral (12.240) over the photon momentum k. This is a common feature of diagrams containing loops. The function $\Sigma(p)$ describes the presence of the kind of loop in Fig. 12.11a, in the momentum space representation of an amplitude and only depends on the inflowing momentum p. We see that the integral representing $\Sigma(p)$ is *linearly divergent* as $k \to \infty$ since there are four powers of k in the numerator and three in the denominator. Divergencies arising for high values of the four-momentum circulating in a loop are also called *ultraviolet divergences*.

As for the *photon self-energy*, corresponding to diagram (5) of Fig. 12.2, the term in $S_{(2)}$ contributing to it can be written as follows:

$$-i \int d^4x d^4y \; : A_\mu(x)A_\nu(y) : \varPi^{\mu\nu}(x-y), \tag{12.242}$$

---

[25]Here and in the following we shall omit, for the sake of simplicity, the integration prescription defined by the infinitesimal term $i\epsilon$ in the Feynman propagators.

Fig. 12.11 **a** Electron self-energy diagram; **b** Photon self-energy diagram; **c** Second order vacuum–vacuum transition

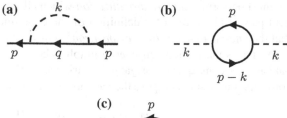

where we have defined

$$\Pi^{\mu\nu}(x-y) \equiv ie^2 \, \mathrm{Tr}\left(\gamma^\mu S_F(x-y)\gamma^\nu S_F(y-x)\right) = \int \frac{d^4k}{(2\pi)^4} \, \Pi^{\mu\nu}(k) \, e^{-ik\cdot(x-y)},$$

where we have denoted by k the difference between the momenta of the internal fermions, see Fig. 12.11b. The Fourier transform $\Pi_{\mu\nu}(k)$ of $\Pi_{\mu\nu}(x-y)$ reads:

$$\Pi^{\mu\nu}(k) = -ie^2 \int \frac{d^4p}{(2\pi)^4} \, \mathrm{Tr}\left(\gamma^\mu \frac{1}{\not{p}-m}\gamma^\nu \frac{1}{\not{p}-\not{k}-m}\right). \qquad (12.243)$$

Computing $S_{(2)}$ between two single-photon states, and factoring out the delta function which implements the total momentum conservation, we find:

$$i \langle \mathbf{k}, i | \mathscr{T}' | \mathbf{k}, i \rangle = -i \, \varepsilon_\mu(k, i)^* \, \Pi^{\mu\nu}(k) \, \varepsilon_\nu(k, i). \qquad (12.244)$$

Just like $\Sigma(p)$, $\Pi^{\mu\nu}(k)$ represents the presence, in the momentum representation of an amplitude, of a loop (in this case a fermion loop), and only depends on the inflowing photon momentum k. It is referred to as the *vacuum polarization tensor*. We see that $\Pi^{\mu\nu}(k)$ exhibits a *quadratic ultraviolet divergence*. As we shall discuss in the following, the presence of infinities is a general feature of the perturbative expansion when closed lines, that is *loops*, appear in a diagram, since the integration over the virtual particles circulating in the loop makes in general the integral divergent. We note that the existence of transitions between initial and final one-particle states implies that the one-particle states are *not stable*, since

$$U(+\infty, -\infty)|\mathbf{p}, s\rangle \neq |\mathbf{p}, s\rangle, \qquad (12.245)$$
$$U(+\infty, -\infty)|\mathbf{k}, i\rangle \neq |\mathbf{k}, i\rangle. \qquad (12.246)$$

Furthermore in our case the matrix elements are divergent. As we shall see in Sect. 12.8.2 to dispose of the linear divergence in the electron self-energy graph we must perform a *mass renormalization* which allows us to absorb the linear divergent part of (12.240) in the definition of the experimental mass, thereby ensuring that the *one-particle electron (or positron) states* (12.245) are stable.

As far as the photon self-energy graph is concerned, we can show that it must vanish as a consequence of gauge invariance of the S-matrix. To show this it is convenient to first decompose the vacuum polarization tensor in Eq. (12.243) as

$$\Pi_{\mu\nu}(k) = \Pi_{\mu\nu}(0) + \Pi_{\mu\nu}^{(1)}(k), \tag{12.247}$$

so that

$$\Pi_{\mu\nu}^{(1)}(k) = \Pi_{\mu\nu}(k) - \Pi_{\mu\nu}(0) \to 0$$

as $k \to 0$. Let us prove that the quadratic divergence of $\Pi_{\mu\nu}(k)$ is entirely contained in $\Pi_{\mu\nu}(0)$. To this end consider the general operator expansion[26]

$$\frac{1}{A+B} = \frac{1}{A} - \frac{1}{A}B\frac{1}{A} + \frac{1}{A}B\frac{1}{A}B\frac{1}{A} + \dots$$

and apply it to the propagator $(\not{p} - \not{k} - m)^{-1}$ with $A = \not{p} - m$ and $B = -\not{k}$ We obtain

$$\frac{1}{(\not{p} - \not{k} - m)} = \frac{1}{(\not{p} - m)} + \frac{1}{(\not{p} - m)}\not{k}\frac{1}{(\not{p} - m)} + \dots \tag{12.248}$$

Inserting this into Eq. (12.243) we find

$$\Pi_{\mu\nu}(0) = -ie^2 \int \frac{d^4p}{(2\pi)^4} \operatorname{Tr}\left(\gamma^\mu \frac{1}{\not{p} - m} \gamma^\nu \frac{1}{\not{p} - m}\right), \tag{12.249}$$

while $\Pi_{\mu\nu}^{(1)}(k)$ is given by the sum of the other terms in the expansion each of which features extra powers of p in the denominator implying that the divergence of $\Pi_{\mu\nu}^{(1)}(k)$ is at worst linear.

Since $\Pi_{\mu\nu}(k)$ is a rank-two Lorentz tensor depending only of k, by Lorentz covariance we may set

$$\Pi_{\mu\nu}(0) = A\eta_{\mu\nu},$$
$$\Pi_{\mu\nu}^{(1)}(k) = C(k^2)k^2\eta_{\mu\nu} + D(k^2)k_\mu k_\nu. \tag{12.250}$$

---

[26]The expansion is easily derived from the identity

$$\frac{1}{A+B} = \frac{1}{A}(A+B-B)\frac{1}{A+B} = \frac{1}{A} - \frac{1}{A}B\frac{1}{A+B}.$$

where A is a constant. On the other hand, gauge invariance of the $S_{(2)}$ term in (12.242) implies

$$k^\mu \Pi_{\mu\nu}(k) = 0. \qquad (12.251)$$

This can be shown by gauge-transforming the polarization vectors in (12.244) according to (12.185). This induces the following extra pieces in the amplitude[27]:

$$\varepsilon^\mu(x)\Pi_{\mu\nu}(k)k^\nu\chi + \varepsilon^\nu(x)\Pi_{\mu\nu}(k)k^\mu\chi + \chi^2 k^\mu k^\nu \Pi_{\mu\nu}.$$

Requiring them to vanish, by gauge invariance, Eq. (12.251) follows. Using the decomposition (12.250), we have

$$k^\mu \Pi_{\mu\nu}(k) = k_\nu \left(A + C(k^2)k^2 + D(k^2)k^2 \right) = 0.$$

For $k^2 = 0$, with $k_\mu \neq 0$, we find

$$A = 0.$$

Hence for arbitrary k

$$C(k^2) = -D(k^2).$$

Since the term containing the quadratic divergence A vanishes the expression of the vacuum polarization tensor reduces to

$$\Pi_{\mu\nu}(k) = \Pi_{\mu\nu}^{(1)}(k) = C(k^2)(\eta_{\mu\nu}k^2 - k_\mu k_\nu). \qquad (12.252)$$

Finally, inserting this result in the matrix element $\langle \mathbf{k}, i | \mathscr{T}' | \mathbf{k}, i \rangle$ in (12.244), we find

$$i \langle \mathbf{k}, i | \mathscr{T}' | \mathbf{k}, i \rangle = -i\,\varepsilon_\mu(k, i)\, C(k^2)(\eta^{\mu\nu}k^2 - k^\mu k^\nu)\,\varepsilon_\mu(k, i) = 0, \qquad (12.253)$$

as a result of the mass-shell and transversality conditions: $k^2 = 0$, $k^\mu\varepsilon_\mu = 0$.

We conclude that the vanishing of the photon self-energy for the second order S-matrix elements ensures that *one-particle photon states (12.246) are stable.*

Let us note that the result $A = 0$, was obtained by requiring gauge invariance of the S-matrix. Actually if one computes the trace $\Pi^\mu{}_\mu(0)$ directly from the integral expression (12.243) one finds

$$A = \frac{1}{4}\Pi^\mu{}_\mu(0) = -\frac{i}{4}e^2 \int \frac{d^4p}{(2\pi)^4} \mathrm{Tr}\left(\gamma^\mu \frac{(\not{p} + m)}{p^2 - m^2}\gamma_\mu \frac{(\not{p} + m)}{p^2 - m^2} \right)$$

$$= -ie^2 \int \frac{d^4p}{(2\pi)^4} \frac{(-2p^2 + 4m^4)}{(p^2 - m^2)^2}, \qquad (12.254)$$

---

[27]Note that this relation is obtained by gauge-transforming the polarization vectors in (12.244) according to (12.185), and requiring the variation of the amplitude to vanish.

where we have used $\mathrm{Tr}(\gamma^\mu) = 0$. The result is that A does not vanish, but diverges quadratically so that gauge invariance appears to be violated. It is however essential that *no gauge invariance property be lost in a consistent theory*. Actually it can be shown that there are different ways of manipulating the divergent integral, one of them reestablishing the $A = 0$ result. From this point of view we can say that the gauge invariance of the theory must be the guiding principle in defining divergent ill-defined integrals and thus consistency of the quantum theory implies $A = 0$.

Finally we consider the last term of the second order amplitude (12.140), namely the term (8) which is associated with the matrix element $\langle 0|S^{(2)}|0\rangle$. Diagrammatically it is a graph with no external line, see Fig. 12.11c, consisting of just propagators, and it is referred to as a *vacuum–vacuum transition*. The amplitude is readily calculated to be

$$\langle 0|S_{(2)}|0\rangle = ie^2 (2\pi)^4 \, \delta^{(4)}(0) \int \frac{d^4p}{(2\pi)^4} \frac{d^4q}{(2\pi)^4} \, \mathrm{Tr}\left[\gamma^\mu \frac{1}{p\!\!\!/ - m}\gamma_\mu \frac{1}{q\!\!\!/ - m}\right] \frac{1}{k^2},$$

where $k = q - p$, and corresponds, in momentum space, to the graph in Fig. 12.11c. We have two sources of infinities in the above expression,: One given by the $\delta^{(4)}(0)$ factor due to the absence of external lines in the diagram, implying that the matrix element is proportional to the four-dimensional volume in space-time; The other infinity shows up in the double integral which is ultraviolet divergent in p as well as in q. Actually we may simply ignore this diagram along with all vacuum–vacuum transition amplitudes, of any order in the perturbative expansion. For example at fourth order we may have the vacuum diagrams in Fig. 12.12a. To show that the sum of all these diagram is physically irrelevant, we recall that the S-matrix elements describe the evolution of a state vector from $t = -\infty$ to $t = +\infty$ in the inter-action picture (it is a mapping between asymptotic free-particle states). Under this transformation the vacuum state must remain invariant. Let us denote by

$$C = \langle 0|S|0\rangle,$$

the sum of *all the vacuum–vacuum transitions to all orders in perturbation theory*. Conservation of the four-momentum p^μ implies that the S-matrix can only map the

Fig. 12.12 a Fourth order vacuum–vacuum transition; **b** Disconnected fourth order graph for the Compton scattering

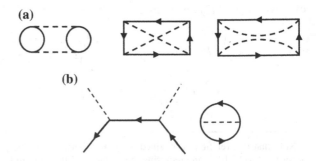

vacuum state, which has $p^\mu = 0$, into itself. Therefore

$$\mathbf{S}|0\rangle = C|0\rangle = \langle 0|\mathbf{S}|0\rangle|0\rangle.$$

However unitarity of the S-matrix implies

$$\langle 0|\mathbf{S}\mathbf{S}^\dagger|0\rangle = \langle 0|CC^*|0\rangle = \langle 0|0\rangle = 1 \;\Rightarrow\; |C|^2 = 1.$$

The conclusion is that C is just a phase factor and can be disregarded.

Note that each Feynman diagram can be accompanied by a set of vacuum graphs. For example at fourth order we may have the disconnected graph in Fig. 12.12b and so on at any order in perturbation theory. Since in any disconnected diagram the S-matrix element is the product of the matrix elements of the disconnected parts, we conclude that the constant C appears as an overall multiplicative phase factor in the S-matrix. If S' is the S-matrix with all the disconnected diagrams omitted CS' is the full S-matrix differing from S by a trivial phase factor. It follows that all the disconnected Feynman diagrams can be omitted in studying the perturbative expansion.

12.8 A Pedagogical Introduction to Renormalization

In Sect. 12.7 we have shown that the last three diagrams of Fig. 12.10 are expressed in terms of divergent integrals. Aside from the divergences associated with vacuum–vacuum transitions (which, as we have seen, can be disregarded because their effect is of multiplying any S-matrix element by a same phase factor), the divergence associated with the photon self-energy transitions (vacuum polarization) was shown to vanish on the grounds of gauge invariance. On the other hand, the divergence associated with the electron self-energy graph was found to be somewhat "serious" in that there seems to be no simple and consistent way to eliminate it.[28]

Actually the treatment of the aforementioned divergences was given for matrix elements $\langle out|\mathbf{S}^{(2)}|in\rangle$ of $\mathbf{S}^{(2)}$ between *initial and final single-particle states obeying the equations of motion of the free theory* (on-shell particles), namely[29]

$$(\not{p} - m)u(p, s) = 0 \tag{12.255}$$

$$k^2 = 0; \quad \varepsilon \cdot k = 0. \tag{12.256}$$

---

[28]Note that the problem of the electron self-energy already exists in the classical theory of the electron. Indeed, either one assumes the electron to be a point particle without structure, in which case the total energy of the electron together its associated field is infinite; or one assumes a finite electron radius, in which case it should explode as a consequence of the internal charge distribution.

[29]Here and in the following we shall refer, for simplicity, only to electron wave functions $u(p, s)$, to electron lines and so on. However all our analysis equally applies to the electron antiparticle, the positron, as well as to, any other charged lepton, like muons and tau mesons.

Photon and electron self-energies are just an example of diagrams containing *loops*. As already mentioned, the presence of loops in a Feynman diagram entails an integration over the momentum k of the virtual particle circulating in the loop, and this, in general, implies an ultraviolet divergence of the integral when the $k \to \infty$. Thus, when we consider higher order terms in the perturbative expansion many more ultraviolet divergences (actually infinitely many) show up in the computation of the S-matrix. This tells us that the Feynman rules for the computation of the amplitudes are in some sense incomplete since they do not tell us what to do with divergent integrals when computing amplitudes beyond the lowest tree-level. It turns out, however, that if we express amplitudes in terms of the *physical measurable* parameters of the theory, namely the mass and the coupling constant, the amplitudes become finite.

Let us make this statement more precise. It must be observed that when we consider higher order terms in perturbation theory, the parameters m and e appearing in the Lagrangian do not represent the experimental values of mass and coupling constant, as it was anticipated in the introduction. For example the electron experimental mass is defined as the expectation value of the Hamiltonian (the energy operator) when the one-particle electron state has zero three-momentum. This has to be computed, to the order of precision required, using

$$m_{exp} = \left. \frac{\langle \mathbf{p}, s | \hat{H} | \mathbf{p}, s \rangle}{\langle \mathbf{p}, s | \mathbf{p}, s \rangle} \right|_{\mathbf{p}=0},$$

$|\mathbf{p}, s\rangle$ and $\hat{H}$ being the states and Hamiltonian operator of the complete interacting theory, the former being perturbatively expressed in terms of free states in (12.3). Notice that in no situation an electron state can be identified with a free state in the Fock space, of the kind we have been using so far in our analysis: the higher order terms in the expansion (12.3) are always present. The reason for this is that an electron is never isolated since it always interacts at least with its own electromagnetic field, and its self interaction contributes to the perturbative expansion (12.3).

Similarly the coupling constant, the physical charge of the electron, should be defined as the quantity which appears in an experimental result. For example the charge may be defined as the parameter that appears in the Compton scattering cross-section at threshold. Therefore to any order in the computation, the result must be given by the formula (12.237) with e replaced by e_{exp}.

When these definitions are implemented, the parameters e, m entering the original Lagrangian can be expressed in terms of the physical ones by relations of the form

$$e = e(e_{exp}),$$
$$m = m(m_{exp}, e_{exp}).$$

In the following it will be convenient to rename e_0 and m_0 the parameters entering the Lagrangian,, called the *bare* parameters, while the physical measurable values of the coupling constant and mass will be denoted by e and m, respectively, so that the relations (12.257) take the form

$$e_0 = e_0(e),$$
$$m_0 = m_0(m, e).$$

As we shall see in the sequel, these relations actually contain divergent quantities. This means that, since the experimental mass and charge are obviously finite, the parameters e_0, and m_0 entering the Lagrangian, in terms of which the Feynman rules were constructed, must be themselves infinite in order to obtain finite results for the physical parameters. Therefore e_0 and m_0 are *not observable*. By eliminating e_0, m_0 in terms of m, e, and by suitable redefinition of the fields, all higher order amplitudes turn out to be finite.

The technique used to handle the divergences appearing in perturbation theory is called *renormalization*. By means of it the divergences can be isolated and reinterpreted as unobservable redefinitions of the mass, coupling constants and field operators of the theory.

Let us observe that the renormalization program requires manipulations of infinite quantities, given in terms of divergent integrals showing up in the perturbative expansion. This raises many questions of mathematical consistency, not all of them having a clear answer. However, even if its mathematical formulation may seem somewhat unsatisfactory, from a pragmatic point of view the renormalization program is fully justifiable, since by means of it we are able to extract finite results which are found to be in remarkable agreement with experiments. We shall give examples of that in the last section.

Renormalization is therefore a necessary route in order to extract physical verifiable predictions from the quantum theory of fields.

The complete renormalization program is rather complicated and its full exposition is beyond the scope of this book. The key point however is that, as it will be shown in the next section, in quantum electrodynamics all the divergences appear-

Fig. 12.13 One-loop divergent diagrams in QED

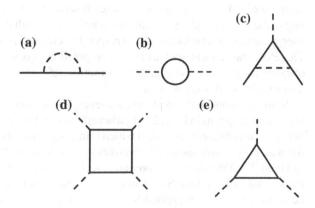

Fig. 12.14 a Second order
electron self-energy parts in
fourth order diagrams;
b Second order photon
self-energy parts in fourth
order diagrams

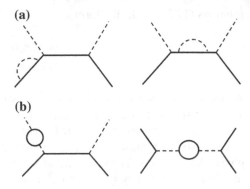

ing in higher order corrections are associated with a *limited number* of diagrams, represented in Fig. 12.13a–c, namely the self-energy diagrams *considered as parts of larger higher-order graphs,* together with the *vertex-part* diagram to be defined below. This means that, once we are able to consistently eliminate the divergences associated with this limited number of diagrams, all the divergences of the theory, at least in principle, can be eliminated. In this case we say that the theory, QED in our case, is *renormalizable.*

By *self-energy part* or *self-energy insertion* of a larger diagram we mean a portion of the graph which, if cut off from the rest, is a self energy diagram of the kinds illustrated in Fig. 12.11a, b. It is important at this stage to distinguish between *self-energy parts* and *self-energies,* computed for the second order amplitudes in Sect. 12.7. Indeed, in the latter case the amplitude was taken between external *on-shell states,* that is between states obeying the equations of motion of a free particle, while the former describe just parts of the amplitude associated with the larger diagram. They can be viewed themselves as self-energy amplitudes, whose external lines however, may not describe on-shell particles, but rather be internal lines of the larger graph, represented by propagators. For example in the Compton scattering we may have, at fourth order, the diagrams in Fig. 12.14a. We see that in these diagrams there is a second-order 1-loop electron self-energy inserted in an external and internal electron line of the larger graph. Note that, in the latter case, both the lines attached to the self-energy part are internal and thus correspond to electron propagators in the amplitude. Therefore the self-energy part is not computed between *external one-particle states* obeying the free-particle equations of motion $(\not{p} - m)u(p, s) = 0$, i.e. the inflowing momentum is off-shell $p^2 \neq m^2$.

Similarly, considering a photon self-energy insertion, we may have at fourth order the 1-loop diagrams in Fig. 12.14b, where the second-order self-energy part is inserted between two photon lines, one of which is internal and thus describes a virtual photon, for which the mass-shell and transversality conditions $k^2 = 0$, $\varepsilon \cdot k = 0$ are not satisfied. It follows that our computations of the self-energies given in Sect. 12.7 for the second-order 1-loop S-matrix elements between external *on-shell states* should be reconsidered when applied to self-energy parts. In particular the proof that the (divergent) 1-loop *self-energy* of the photon is zero does not apply when we have a photon *self-energy part* since the conditions $k^2 = \varepsilon \cdot k^\mu = 0$ used in the proof do not hold. The same is true *a fortiori* for the 1-loop electron self-energy considered in Sect. 12.7, where we have seen that it is actually divergent even if it refers to an amplitude between external states. We shall show in the following that both the photon and electron self-energy insertions can be made finite by *mass renormalization*, *coupling constant renormalization* and *field renormalization*.

Furthermore, when we consider diagrams that are parts of larger ones, besides the self-energy graphs, a further divergent contribution comes into play, namely the *vertex part*, whose 1-loop diagram, (second order in the coupling constant), is given in Fig. 12.13c. This diagram exists only when it is *part* of a larger diagram since when the external electron and photon lines are on the mass-shell momentum conservation cannot be satisfied. This justifies why it was not considered when discussing S-matrix elements between external states.[30]

12.8.1 Power Counting and Renormalizability

In this section we show that QED is a renormalizable theory, by which we mean, as mentioned earlier, that only a limited number of amplitudes is divergent. In particular, to one-loop, the divergent amplitudes are those associated with the self-energies and vertex insertions discussed in the previous section, see Fig. 12.13a–c. This justifies the assertion made in the previous section that the consideration of the self-energies and vertex parts are actually sufficient to show that in QED all the divergences can be disposed of.

We have so far restricted our attention to diagrams containing just one loop. A generic diagram may however contain various loops. Let us define the *superficial degree of divergence D_G* of a diagram as a number which signals, if non-negative, the presence in the amplitude of divergent integrals. We observe that in each Feynman graph there is an integration in $\frac{d^4 p}{\not{p} - m}$ for each *internal* fermion (electron) line and an integration $\frac{d^4 k}{k^2}$ for each *internal* boson (photon) line, which contributes three and two units, respectively, to the degree of divergence of the amplitude, Furthermore, at each vertex we have a δ^4-function expressing conservation of the momenta flowing in and out of the vertex. This eliminates four momentum integrations at each vertex (i.e. an integral in $d^4 p$). However one of the momentum δ^4-functions just implements the conservation of the total momentum and thus is ineffective in eliminating momentum integrals. Taking this into account, we may define the *superficial degree of divergence* to be given by

$$D_G \equiv 3F_i + 2B_i - 4(V - 1). \tag{12.257}$$

---

[30]There are in principle further divergences associated with the so-called photon-photon system, Fig. 12.13d and the three-photon vertex, Fig. 12.13e. Such divergences are however armless (see below.).

where F_i and B_i are the number of internal fermion and boson lines, respectively, and V is the number of vertices. Here we are using the words "fermion" and "boson" instead of electron and photon because our considerations in general apply to any theory containing bosons and fermions. What actually characterizes a theory is the *interaction vertex*. For QED the interaction Hamiltonian density is $\mathcal{H}_I = -e\bar{\psi}(x)\gamma^\mu\psi(x)A_\mu(x)$, so that at each vertex there are two fermion and one boson lines. Let us denote F_i, F_e and B_i, B_e the number of internal and external lines. Since an external line is connected to one vertex and an internal line connects two vertices, it is easy to see that[31]

$$2F_i + F_e = 2V,$$
$$2B_i + B_e = V. \tag{12.258}$$

If we solve the above equations in F_i and B_i and substitute the result into Eq. (12.257) we find

$$D_G = 4 - \frac{3}{2}F_e - B_e. \tag{12.259}$$

We conclude that the degree of divergence does not depend on the number of vertices and internal lines, but only on the number of external lines. In particular we see that $D_G > 0$ only for a limited number of diagrams. In general when this happens we say that the theory is *renormalizable*. Therefore QED is renormalizable.

On the other hand the renormalizability property is related to the physical dimension of the coupling constant. Let us first recall from Sect. 12.3.5 that, using natural units, the fermions have dimension $\frac{3}{2}$ in mass units and the bosons dimension 1. Furthermore the action of a theory is dimensionless, so that the Lagrangian density has dimension (in mass) $[M^4]$. On the other hand, if from each vertex f fermionic and b bosonic lines originate, respectively, we must have that the dimension of the coupling constant g in front of the interaction Lagrangian density is

$$[\lambda] = [M^{4-\nu}] = [M^{d_\lambda}],$$

where $\nu = \frac{3}{2}f + b$ and $d_\lambda = 4 - \nu$ is the mass-dimension of λ. In particular for quantum electrodynamics we find $\nu = \frac{3}{2}f + b = 4$, so that the coupling constant $\frac{e}{\sqrt{4\pi}}$, $(\frac{e}{\sqrt{4\pi\hbar c}}$ in the usual units) is indeed *dimensionless*. For a general theory we can generalize Eqs. (12.258) as follows

$$2F_i + F_e = f V$$
$$2B_i + B_e = b V. \tag{12.260}$$

---

[31]To derive these relations, one can cut each internal fermion line of a diagram into two parts. The total number of lines so obtained should be twice the number of vertices. In this counting however, each internal line contributes two units (i.e. a total of $2F_i$ units) and each external ones a single unit (i.e. a total of F_e units). A similar argument applies to the boson lines.

Substituting the values of F_i and B_i in Eq. (12.257) we find

$$
D_G = bV - B_e + \frac{3}{2}(fV - F_e) - 4V + 4
$$

$$
= V\left(b + \frac{3}{2}f - 4\right) + 4 - \frac{3}{2}F_e - B_e = -d_\lambda V - \frac{3}{2}F_e - B_e + 4,
$$

$$(12.261)$$

where we have used the definition $d_\lambda = 4 - \frac{3}{2}f - b$ of the mass-dimension of λ. We then have the following cases:

If $b + \frac{3}{2}f < 4$, that is if $d_\lambda > 0$, as the perturbative order V increases, D_G decreases and amplitudes are finite. We say in this case that the theory is *super-renormalizable*.

If $b + \frac{3}{2}f = 4$, so that $d_\lambda = 0$, the coupling constant is dimensionless. In this case D_G is independent of V and the theory is *renormalizable*. The divergences occurring in the infinitely many loop diagrams, as we shall illustrate in the sequel for the case of quantum electrodynamics, can be disposed of by adding a finite number of *counterterms* to the Lagrangian, which amounts to a redefinition of the parameters of the theory (*renormalization*).

If $b + \frac{3}{2}f > 4$, $d_\lambda < 0$, the theory is *non-renormalizable* since, by increasing the order of the diagram, that is the number of the vertices, the degree of divergence also increases. This time, however, in order to dispose of the divergences occurring to each order in the coupling constant, counterterms should be added to the original Lagrangian, whose functional dependence on the fields and their (higher-order) derivatives, would in general depend on the corresponding power of the coupling constant. Therefore, in contrast to the renormalizable theories, in the non-renormalizable ones the divergences cannot be cured through the redefinition of a finite number of parameters but infinitely many counterterms need be added to the Lagrangian. An example of a non-renormalizable theory is the quantized Einsten gravity, in which the interactions are all expressed in terms of a fundamental coupling constant, which is proportional to the square root of Newton's constant and thus has dimension of a length in natural units.

After this general discussion, let us come back to the case of the QED. To one-loop order besides the three divergent diagrams (a–c) of Fig. 12.13, corresponding to the electron and photon self-energy parts and vertex part discussed before, which are of second-order in the coupling constant, there is also at one loop a divergent fourth-order diagram, represented in Fig. 12.13d, which is referred to as the photon-photon system and an order-three three-photon vertex, see Fig. 12.13e. Applying Eq. (12.257) to diagrams (a)–(c) we immediately conclude by power counting that the electron self-energy part is linearly divergent ($D_G = 1$), the photon self-energy part is quadratically divergent ($D_G = 2$) and the vertex part is logarithmically divergent ($D_G = 0$).

As far as the photon-photon system is concerned, it is logarithmically divergent, while the three-photon vertex is linearly divergent $D_G = 1$. However, an explicit evaluation of the former diagram, shows that the coefficient of its divergent part is exactly zero. Therefore we shall disregard this diagram in the following. As far as the three-photon vertex is concerned, it is zero being odd under charge conjugation and thus would violate the charge conjugation symmetry (recall from Chap. 11 that the photon is odd under charge conjugation: $\eta_C = -1$). Actually, by the same argument, one can show that all diagrams with an odd number of external photons is zero (Furry's theorem).

Renormalizability of QED means that all the divergences appearing in the perturbative expansions can be eliminated. As previously anticipated, a complete account of the full renormalization program to all orders is rather heavy and complicated and would be outside the scope of this pedagogical introduction. In the following we shall therefore limit ourselves to apply the renormalization program to the self-energy and vertex parts given in Fig. 12.13 which correspond to 1-loop insertions and are therefore of *second order in the coupling constant*. We believe that even in this restricted framework the main ideas used in the full renormalization program, to all orders, can be understood.

Thus far we have been dealing with divergences as if they were well defined quantities. Actually, in order to make sense of divergent integrals, and their manipulations, it is important, as a first step, to make such divergent integrals finite by some *regularization* procedure. The general procedure is the following. One first separates the divergent integral into two parts,[32] where the first part is still divergent, but the divergence is entirely contained in a set of divergent *constants*, that is in a set of integrals which do not depend on the external momenta, the second part instead is completely finite and, in general, will depend on the external momenta. To show how this separation can be made we quote the following simple example.[33] Let us consider the following integral

$$\sigma(p) = \int\limits_0^\infty \frac{dk}{k+p},$$

which is logarithmically divergent. If we differentiate with respect to p we obtain

$$\sigma'(p) = -\int\limits_0^\infty \frac{dk}{(k+p)^2} = -\frac{1}{p}.$$

Therefore
$$\sigma(p) = -\log p + c.$$

We have thus separated the divergent part of $\sigma(p)$, given by the constant c, from its finite part. Analogously, from the linearly divergent integral

$$\sigma(p) = \int\limits_0^\infty \frac{k\,dk}{k+p},$$

---

[32] Actually, beyond 1-loop, there are divergences that require a more careful treatment than just separation into a divergent and a finite part (overlapping divergences). We can neglect them, since we are going to discuss only 1-loop self-energy and vertex insertions which cannot give rise to this kind of divergences.

[33] See Weinberg's book [13].

by the same procedure, we obtain

$$\sigma(p) = a + bp + p \log p,$$

where a and b are divergent constants. In general for a divergence of order D of the integral we obtain a polynomial in the external momenta of degree $D - 1$ whose constant coefficients are divergent plus a *finite part*. Actually the given decomposition is equivalent to the first terms of a Taylor expansion of the integral in the external momenta. However we must pay attention to the fact that the separation between a divergent part and a finite part is *not uniquely defined*. Indeed we may always change the value of the finite part by adding a constant to it and subtracting the same constant from the divergent part. In order to have a uniquely defined expansion we must therefore add some requirement dictated by physical considerations.

The previous examples are given in terms of one-dimensional integrals. Coming back to the divergent four-dimensional integrals arising from loop integration, in order to manipulate the "constant" divergent integrals they need to be *regulated*, that is made finite, by a some convenient *regularization scheme*.

There are several regularization schemes which do the work and would be worth discussing, since they allow to compute the explicit form of the divergence. However, being their treatment rather technical, it would be outside the limited discussion of the renormalization that we plan to present. We shall therefore avoid entering the detail of the regularization procedures. We can just give a simple example of how regularization can be achieved for the linearly divergent integral (12.240). In this case can use the so called Pauli–Villars scheme of regularization by modifying the photon propagator in the integral as follows

$$\eta_{\mu\nu} \frac{-i}{k^2} \;\rightarrow\; \eta_{\mu\nu} \left(\frac{-i}{k^2} - \frac{-i}{k^2 - \Lambda^2} \right) = -i\eta_{\mu\nu} \frac{-\Lambda^2}{k^2 (k^2 - \Lambda^2)}.$$

The integral (12.240) becomes

$$\Sigma^{(reg)}(p) = -ie^2 \int \frac{d^4k}{(2\pi)^4} \, \gamma^\mu \frac{1}{(\not{p} - \not{k}) - m} \gamma_\mu \left[\frac{-\Lambda^2}{k^2 (k^2 - \Lambda^2)} \right]. \qquad (12.262)$$

We see that $\Sigma^{(reg)}(p)$ is finite as long as Λ^2 is kept finite. If we separate the integral into a divergent (in the $\Lambda^2 \rightarrow \infty$ limit) and a finite part, as we shall do in Sect. 12.8.2, the finite part remains the same when $\Lambda^2 \rightarrow \infty$, while the divergent regulated part becomes infinite only when $\Lambda^2 \rightarrow \infty$.

This example shows how regularization allows us to manipulate quantities which become divergent only when the regularization is removed. In the following, with abuse of notation, we shall call these regulated quantities "divergent", but it must be kept in mind that they are in fact regulated. Only once regularization has been performed the renormalization program allows us to separate the divergent part of an integral from its finite part and to prove that the entire divergence can be eliminated by appropriate redefinitions of the mass and coupling constant of the theory. This procedure is referred to as *mass and coupling constant renormalization*, and will be discussed in the next sections.

Since our process of elimination of the infinities will require a redefinition of the parameters and of the fields entering the Lagrangian, we will rewrite the initial Lagrangian, in terms of which the Feynman rules are defined, as follows:

$$\mathcal{L}_0 = \overline{\psi}_0 \, (i\partial\!\!\!/ - m_0)\psi_0 - \frac{1}{4} F_{0\mu\nu} \, F_0^{\mu\nu} + e_0 A_{0\mu} \, \overline{\psi}_0 \gamma^\mu \psi_0. \qquad (12.263)$$

Therefore, in this notation, the fields $\psi(x)$, $A_\mu(x)$ and the coupling constant e appearing in all the formulas written so far, should be intended as $\psi_0(x)$, $A_{0\mu}(x)$ and e_0, respectively.

12.8.2 The Electron Self-Energy Part

Consider an internal electron line of a Feynman graph. When adding higher order contributions to the amplitude of the same process, we will have to consider a diagram which differs from the initial one only in the insertion of a self-energy part in the electron line. In summing the contributions from the two diagrams, all the rest factorizes while the propagator associated with the internal line is replaced by the following sum, see Fig. 12.15:

$$S_F(x - y) + \int d^4x_1 d^4x_2 \, S_F(x - x_1)[-i\Sigma(x_1 - x_2)]S_F(x_2 - y),$$

where $\Sigma(x_1 - x_2)$ was defined in (12.239). Since the effect of the higher order contribution is accounted for by replacing the propagator $S_F(x - y)$ in the original amplitude with the above sum, the second term on the right hand side can be seen as a second order correction to the free propagator. In momentum space this correction reads,

$$S_F(p) + S_F(p) \left[-i\Sigma(p)\right] S_F(p), \qquad (12.264)$$

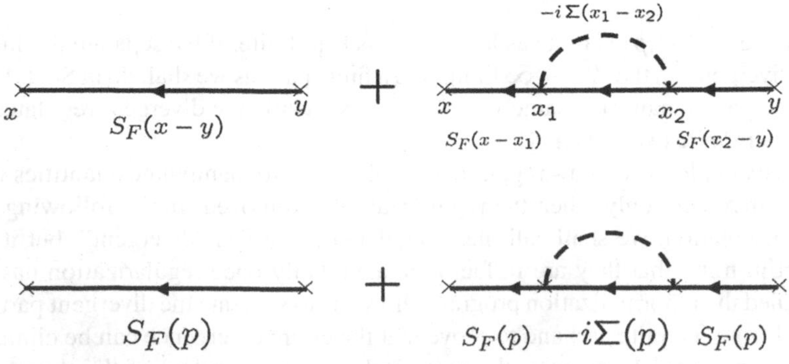

Fig. 12.15 Corrected electron propagator by insertion of a second-order self-energy part

where

$$S_F(p) = i \, (p\!\!\!/ - m_0)^{-1} \equiv i \frac{p\!\!\!/ + m_0}{p^2 - m_0^2},$$

and $-i\Sigma(p)$ is given in Eq. (12.240). Diagrammatically the 1-loop corrected propagator is given in Fig. 12.15.[34]

The correction (12.264) to the free propagator can be improved by considering the so called *chain approximation*. In this procedure one considers higher order corrections to the free propagator arising as a (infinite) sum of all the graphs obtained as *chains of 1-loop insertions* as in Fig. 12.16. The improved correction S_F' to the propagator is then

$$
\begin{aligned}
S_F'(p) &= S_F(p) + S_F(p)[-i\Sigma(p)]S_F(p) \\
&\quad + S_F(p)[-i\Sigma(p)]S_F(p)[-i\Sigma(p)]S_F(p) + \cdots \\
&= S_F(p) \left(\frac{1}{1 + i\Sigma(p)\, S_F(p)} \right) \\
&= \frac{i}{p\!\!\!/ - m_0} \frac{1}{1 - \Sigma(p)(p\!\!\!/ - m_0)^{-1}} = \frac{i}{p\!\!\!/ - m_0 - \Sigma(p)}.
\end{aligned}
\tag{12.265}
$$

Note that each correction term in the above sequence is two orders (in the coupling constant) higher than the preceding one, since each self-energy insertion in the chain is of second-order.[35]

To proceed we apply the considerations of the last section to separate the divergent part of $\Sigma(p)$ from its finite part. As we have noted earlier, the expansion of a divergent

---

[34]Note that if (12.264) were taken between external (on-shell) electron lines, we would recover the electron self-energy computed in Sect. 12.7, namely Eq. (12.241).

[35]Actually we could make the chain approximation (12.264) exact if we would consider each electron self energy insertion not restricted to one-loop order. This can be done by introducing the concept of one particle irreducible (1PI) diagram. A diagram is 1PI if it cannot be disconnected by cutting one internal line. Thus we may consider a *self-energy diagram* which has contributions from 1PI diagram only, like the three fourth order diagrams of Fig. 12.17a, while the graph (b), being reducible, would not contribute. The reason for selecting only 1PI diagrams is that the reducible diagrams can always be decomposed in 1PI diagrams *without further integration*, and therefore if we can take care of the divergences of the 1PI diagrams, we automatically take care of the general diagram. Let us denote the correction to the free propagator due to the sum of all possible 1PI self-energy diagrams by $-i\Sigma^*(p)$, see Fig. 12.18a. The correction (12.264) becomes

$$S_F(p) + S_F(p) \left(-i\Sigma^*(p) \right) S_F(p). \tag{12.266}$$

If we now perform the chain expansion as in (12.265) but with $-i\Sigma(p\!\!\!/)$ replaced by $-i\Sigma^*(p)$, we obtain the exact propagator in the form

$$S_F'(p) = \frac{i}{p\!\!\!/ - m_0 - \Sigma^*(p)}, \tag{12.267}$$

see Fig. 12.18b. In the following however we will limit ourself to consider the approximation (12.264) where only the 1-loop integral $\Sigma(p\!\!\!/)$, lowest order approximation of $\Sigma^*(p)$, appears.

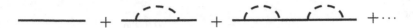

Fig. 12.16 Correction to the electron propagator in the chain approximation

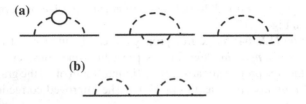

(a)

(b)

Fig. 12.17 Examples of fourth-order corrections to the electron propagator: **a** one-particle irreducible; **b** one-particle reducible

(a)

$$-i\Sigma^*(p) \equiv -i\Sigma(p) + \cdots$$

(b)

$$S_F \qquad S_F[-i\Sigma^*]S_F \qquad S_F[-i\Sigma^*]S_F[-i\Sigma^*]S_F$$

Fig. 12.18 **a** Definition of $-i\Sigma^*(p)$ as the sum of the corrections to the electron propagators due to all 1PI diagrams; **b** Exact propagator

integral into a polynomial in the external momenta with divergent coefficients plus a finite remainder is equivalent to a Taylor series expansion, truncated to the first divergent terms plus a finite remainder. Let us apply this technique to the divergent integral

$$\Sigma(p) = -ie_0^2 \int \frac{d^4k}{(2\pi)^4} \gamma^\mu \frac{1}{\not{p} - \not{k} - m_0} \gamma_\mu \frac{1}{k^2}. \tag{12.268}$$

By differentiation with respect to the *external* momentum p, we increase the power of k in the denominator by one unit making the result only logarithmically divergent. Through a second differentiation we obtain a finite, that is convergent, integral. In our case we have then a Taylor expansion truncated to first order in p plus a finite remainder. Taking into account that by Lorentz invariance $\Sigma(p)$ can only be function of $\not{p}$ and p^2, we expand $\Sigma(p)$ in powers of $(\not{p} - m)$ where m is *arbitrary*:

$$\Sigma(p) = \delta m + B(\not{p} - m) + \Sigma^{(c)}(p). \tag{12.269}$$

Here δm and B are divergent constants given by

$$\delta m = \Sigma(p)\big|_{p^2=m^2}, \quad B = \frac{1}{4} \gamma^\mu \frac{\partial \Sigma}{\partial p^\mu}\bigg|_{p^2=m^2}.$$

while $\Sigma^{(c)}(p)$ is convergent and satisfies

$$\Sigma^{(c)}(p) = \gamma^\mu \frac{\partial \Sigma^{(c)}}{\partial p^\mu} = 0 \quad \text{for } p^2 = m^2. \tag{12.270}$$

We see that the entire divergence of $\Sigma(p)$ is contained in the infinite constants δm and B.

We can now insert the result (12.269) into the expression on the right hand side of (12.265), obtaining

$$\frac{i}{p\!\!\!/ - m_0 - \Sigma(p)} = \frac{i}{p\!\!\!/ - m_0 - \delta m - B(p\!\!\!/ - m) - \Sigma^{(c)}(p)}. \tag{12.271}$$

We see that the pole of the improved propagator $S'_F(p)$, which defines the mass of the particle, is no longer at $p^2 = m_0^2$. If we choose the arbitrary parameter m to satisfy:

$$m_0 + \Sigma(m) = m, \tag{12.272}$$

where $\Sigma(m) \equiv \Sigma(p)\big|_{p^2=m^2} = \delta m$, Eq. (12.271) yields

$$S'_F(p) = \frac{i}{(p\!\!\!/ - m)(1 - B) - \Sigma^{(c)}(p)}. \tag{12.273}$$

Recalling that $\Sigma^{(c)}$ vanishes for $p^2 = m^2$, m becomes the mass of the particle, which is shifted from its original value m_0, the shift being proportional to the divergent quantity $\delta m \equiv \Sigma(m)$. Since δm is divergent we conclude that the *bare mass* m_0 present in the original Lagrangian must be divergent as well, in order for the physical mass m to be finite.[36] The mass renormalization given by the mass shift (12.272) provides the removal of the divergent term $\delta m = \Sigma(m)$ from the corrected propagator, but it still depends on the infinite constant B.[37] As it is apparent from the (12.273), this infinite

---

[36] Naively one could think that the separation of the physical mass into the bare mass m_0 and the mass-shift $\delta m = \Sigma(m)$ would correspond to the separation of the electron mass into a "mechanical" and a "electromagnetic" mass. However such separation is devoid of physical meaning since it cannot be observed. We also note that the process of mass renormalization is not a peculiarity of field theory. For example when an electron moves inside a solid it has a renormalized mass m^*, also called effective mass, which is different from the mass measured in the absence of the solid, i.e. the bare mass m_0. However, differently from our case, the effective and bare mass can be measured separately, while in field theoretical case m_0 cannot be measured.

[37] We observe that this term would give a vanishing contribution if we had an external on-shell state instead of the propagator in (12.264) since the term $B(p\!\!\!/ - m)$ in Eq. (12.269) is zero on the free electron wave function.

constant changes the residue at the pole from its original value i to

$$i(1 - B)^{-1}.$$

To dispose of the divergent constant B we observe that neglecting higher order terms in e_0^2 we may write

$$\Sigma^{(c)}(p) \simeq \Sigma^{(c)}(p)(1 - B),$$

that is

$$\Sigma^{(c)}(p) \simeq \Sigma^{(c)}(p)(1 - B) = \Sigma^{(c)}(p)Z_2^{-1},$$

where

$$Z_2 \equiv (1 - B)^{-1}. \tag{12.274}$$

Equation (12.273) can be recast in the following form

$$S'_F(p) = i\frac{Z_2}{(\not{p} - m) - \Sigma^{(c)}(p)}.$$

We see that the expression multiplying Z_2 is completely finite. On the other hand, the multiplicative constant Z_2 can be reabsorbed in a redefinition of the electron field, namely by defining a renormalized physical field $\psi(x)$ in terms of a *bare* unphysical one $\psi_0(x)$ as follows:

$$\psi_0 = Z_2^{\frac{1}{2}} \psi. \tag{12.275}$$

Recalling indeed the definition (12.110) of the Feynman propagator and its Fourier transform, we have

$$\begin{aligned} S'_F &= \int d^4\xi \, e^{ip\cdot\xi} \langle 0| \, T(\psi(y + \xi)\bar{\psi}(y))|0\rangle \\ &= Z_2^{-1} \int d^4\xi \, e^{ip\cdot\xi} \, \langle 0|T(\psi_0(y + \xi)\bar{\psi}_0(y))|0\rangle \\ &= i\frac{1}{(\not{p} - m) - \Sigma^{(c)}}. \end{aligned} \tag{12.276}$$

so that, when written in terms of the renormalized mass m and the renormalized field ψ, the corrected propagator is completely finite. The renormalization of the bare field into the physical field by the divergent constant Z_2 given in Eq. (12.275) is usually referred to as the *wave function renormalization*.[38]

At the Lagrangian level, we can give an interpretation of the renormalization procedure as the addition of *counterterms* to the original Lagrangian $\mathcal{L}_0$. Indeed, taking into account Eqs. (12.272) and (12.275) we have

---

[38]Recall that a one-particle state and its wave function $\psi(x)$ is related to the quantum fields $\hat{\psi}$ by $\langle 0|\hat{\psi}(x)|a\rangle$, see for example Eq. (12.64) for a boson particle.

$$\mathcal{L}_0^{Dirac} = \overline{\psi}_0 \, (i\partial\!\!\!/ - m_0)\psi_0 = Z_2 \, \overline{\psi} \, (i\partial\!\!\!/ - m)\psi + Z_2\overline{\psi}\psi\delta m. \qquad (12.277)$$

Therefore the Dirac Lagrangian written in terms of the physical mass and fields is

$$\begin{aligned}
\mathcal{L}^{Dirac} &= \overline{\psi} \, (i\partial\!\!\!/ - m)\psi \\
&= \mathcal{L}_0^{Dirac} - (Z_2 - 1) \, \overline{\psi} \, (i\partial\!\!\!/ - m)\psi - Z_2\overline{\psi}\psi\delta m. \qquad (12.278)
\end{aligned}$$

One can verify that applying the Feynman rules to these countertcrms the mass m_0 acquires the correction (12.272) while the logarithmically divergent part B is subtracted from $\Sigma(p)$.

12.8.3 The Photon Self-Energy

Let us now consider the photon self-energy graph. We perform the same steps as in the case of the electron self-energy graph. A photon self-energy insertion in an internal photon line defines a second order correction to the photon propagator. In the coordinate representation the second-order corrected photon propagator reads:

$$\begin{aligned}
D'_{F\mu\nu}(x-y) &= D_{F\mu\nu}(x-y) \\
&\quad + \int d^4x_1 d^4x_2 D_{F\mu\rho}(x-x_1)[-i\,\Pi^{\rho\sigma}(x_1-x_2)]\,D_{F\mu\rho}(x_2-y),
\end{aligned}$$

and is represented diagrammatically in Fig. 12.19. In the momenutm representation the corrected propagator reads:

$$D'_{F\mu\nu}(k) = D_{F\mu\nu}(k) + D_{F\mu\rho}(k)[-i\Pi_{\rho\sigma}(k)]D'_{F\sigma\nu}(k). \qquad (12.279)$$

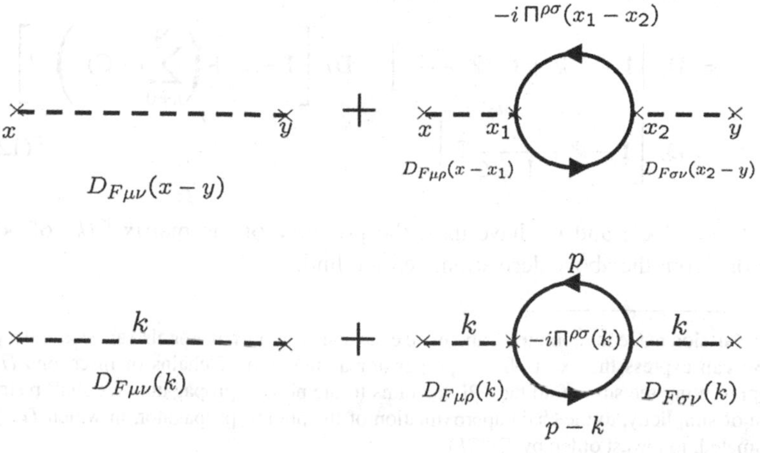

Fig. 12.19   Second order correction to the photon propagator

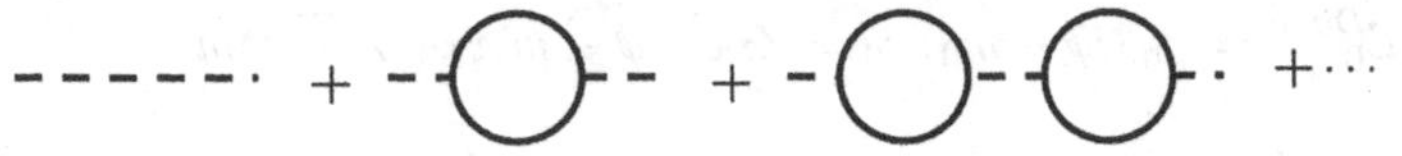

Fig. 12.20 Correction to the propagator in the chain approximation

As in the electron case, the self-energy diagram is just part of a larger graph and the inflowing momentum k is off mass-shell. For this reason $\Pi_{\rho\sigma}(k)$ is non-vanishing and is expressed by the divergent integral in Eq. (12.243), which is of second order in the charge e_0.

Just as for the electron case, we limit ourselves to the chain approximation and consider all the higher order corrections to the propagator originating from chains of self-energy insertions, see Fig. 12.20.[39] Performing the sum over the chain of 1-loop diagrams is, however, somewhat more complicated than in the case of the electron self-energy, because of the tensor indices carried by $\Pi_{\mu\nu}$. We may proceed as follows. Denote by $\mathbf{D}_F(k)$ and $\mathbf{\Pi}(k)$ the 4×4 matrices $D_{F\mu\nu}(k)$ and $\Pi^{\rho\sigma}(k)$. Define now the projector $\mathbb{P}(k) = (\mathbb{P}(k)^\mu{}_\nu)$:

$$\mathbb{P}(k)^\mu{}_\nu \equiv \delta^\mu_\nu - \frac{k^\mu k_\nu}{k^2}. \tag{12.280}$$

The reader can easily verify that $\mathbb{P}(k)^n = \mathbb{P}(k)$. From the general form (12.252) of the vacuum polarization tensor found in Sect. 12.7 and the expression of $D_{F\mu\nu}(k)$ in (12.110), it follows that:

$$- i\, \Pi^{\rho\sigma}(k)\, D_{F\sigma\nu}(k) = -C(k^2)\, \mathbb{P}(k)^\rho{}_\nu. \tag{12.281}$$

The corrected propagator in the chain approximation reads:

$$\begin{aligned}
\mathbf{D}'_F &= \mathbf{D}_F + \mathbf{D}_F[-i\mathbf{\Pi}]\mathbf{D}_F + \mathbf{D}_F[-i\mathbf{\Pi}]\mathbf{D}_F[-i\mathbf{\Pi}]\mathbf{D}_F + \cdots \\
&= \mathbf{D}_F\left[1 - i\,\mathbf{\Pi}\,\mathbf{D}_F + (-i\,\mathbf{\Pi}\,\mathbf{D}_F)^2 + \cdots \right] \\
&= \mathbf{D}_F\left[1 - C\,\mathbb{P} + C^2\,\mathbb{P} + \cdots \right] = \mathbf{D}_F\left[1 - \mathbb{P} + \left(\sum_{n=0}^{\infty}(-C)^n \right) \mathbb{P} \right] \\
&= \mathbf{D}_F\left[1 - \mathbb{P} + \frac{1}{1+C}\,\mathbb{P} \right],
\end{aligned} \tag{12.282}$$

where $C = C(k^2)$ and we have used the property of the matrix $\mathbb{P}(k)$ of being a projector. From the above derivation we then find:

---

[39]The discussion made in footnote 35 about the exact electron propagator also applies to the photon case. We can express the exact photon propagator as the sum of chains of insertions $\Pi^{*\mu\nu}(k)$ each representing the sum of all the 1PI diagrams to the photon propagator. We shall restrict, for the sake of simplicity, to the chain approximation of the photon propagator, in which $\Pi^{*\mu\nu}(k)$ is approximated, to lowest order, by $\Pi^{\mu\nu}(k)$.

$$D'_{F\mu\nu}(k) = -i\frac{\eta_{\mu\nu}}{k^2}\left(\frac{1}{1+C} + \frac{C}{1+C}\frac{k^\mu k_\nu}{k^2}\right). \tag{12.283}$$

In light of the discussion made in Sect. 12.5.5 about the gauge invariance of S-matrix elements, we can disregard the $k_\mu k_\nu$ terms in $\mathbf{D}'_F$ since they give a vanishing contribution to the S-matrix. Therefore the non trivial part of $D'_{F\mu\nu}(k)$ reduces to

$$D'_{F\mu\nu}(k) = -\frac{i\eta_{\mu\nu}}{k^2\left(1+C(k^2)\right)}. \tag{12.284}$$

The important point is the fact that, *assuming that $C(k^2)$ has no pole at $k^2 = 0$*, the pole of the photon propagator is not shifted with respect to the tree diagram level, namely it is located at $k^2 = 0$. Therefore *no mass renormalization is needed for the photon self-energy part.* Recalling our discussion in Sect. 12.7, this absence of renormalization is due both to gauge invariance which implies the vanishing of the quadratically divergent part $\Pi_{\mu\nu}(0) = A\eta_{\mu\nu}$, and to the assumption of regularity of $C(k^2)$ at $k^2 = 0$.

We now have to eliminate the further divergent term $C(k^2)$ from $D'_{F\mu\nu}(k)$ the residue being now given by $(1+C(0))^{-1}$. This can be done exactly as in the case of the electron self-energy part. We first observe that the quantity

$$\Pi^{(c)}(k^2) = k^2 C(k^2) - k^2 C(0) = k^2 C^{(R)}(k^2), \tag{12.285}$$

where $C^{(R)}(k^2) \equiv C(k^2) - C(0)$, must be finite since $C(k^2)$ is logarithmically divergent. In fact $\Pi_{\mu\nu}(k)$ is given by a quadratically divergent integral and $C(0)$ is the coefficient of k^2 in its expansion around $k^2 = 0$

$$k^2 C(k^2) = k^2 C(0) + \Pi^{(c)}(k^2). \tag{12.286}$$

Therefore the divergence is entirely contained in $C(0)$. We fix the ambiguity in (12.285) alluded to in Sect. 12.8.2, assuming $\Pi^{(c)}(0) = 0$, which can always be done by shifting a constant from $C(0)$ to $\Pi^{(c)}(k^2)$. Therefore the corrected propagator takes the form

$$D'_{F\mu\nu}(k) = -\frac{i\eta_{\mu\nu}}{k^2\left(1+C(0)\right) + \Pi^{(c)}(k^2)}. \tag{12.287}$$

We see that the residue changes by the factor $Z_3 \equiv (1+C(0))^{-1}$. Moreover, neglecting higher order terms, we may also write

$$\Pi^{(c)}(k^2) \simeq \Pi^{(c)}(k^2)(1+C(0)) = \Pi^{(c)}(k^2)Z_3^{-1},$$

so that Eq. (12.287) becomes

$$D'_{F\mu\nu}(k) = -\frac{i\eta_{\mu\nu}Z_3}{k^2 + \Pi^{(c)}(k^2)}. \tag{12.288}$$

where the expression multiplying Z_3 is completely finite. As for the electron self-energy, the factor Z_3 can be now by reabsorbed by the *photon wave function renormalization*, namely by setting

$$A_{0\,\mu} = Z_3^{\frac{1}{2}} A_\mu \qquad (12.289)$$

Indeed recalling Eqs. (12.110) we have

$$
\begin{aligned}
D'_{\mu\nu F} &= \int d^4\xi \, e^{ip\cdot\xi} \langle 0| \, T(A_\mu(y+\xi)A_\nu(y))|0\rangle \\
&= Z_3^{-1} \int d^4\xi \, e^{ip\cdot\xi} \langle 0|T(A_{0\mu}(y+\xi)A_{0\nu}(y))|0\rangle \\
&= -i\eta_{\mu\nu} \frac{1}{k^2 + \Pi^{(c)}(k^2)} = -i\,\eta_{\mu\nu} \frac{1}{k^2(1+C^{(R)}(k^2))}. \qquad (12.290)
\end{aligned}
$$

Similarly to what we did for the electron self-energy, the photon wave function renormalization (12.289) can be interpreted at the Lagrangian level, as the addition of a counterterm. Indeed we can write for the electromagnetic free Lagrangian density

$$\mathcal{L}_{F_0^2} = -\frac{1}{4} F_0^{\mu\nu} F_{0\,\mu\nu} = -\frac{1}{4} Z_3 F^{\mu\nu} F_{\mu\nu}. \qquad (12.291)$$

Therefore the electromagnetic Lagrangian density in terms of the physical renormalized fields is

$$\mathcal{L}_{F^2} = -\frac{1}{4} F^{\mu\nu} F_{\mu\nu} = \mathcal{L}_{F_0^2} + \frac{1}{4}(Z_3 - 1)F^{\mu\nu} F_{\mu\nu}. \qquad (12.292)$$

The change in the photon propagator given by the self-energy insertion is referred to as *vacuum polarization*. The *vacuum polarization* is a physical measurable effect. Indeed, let us consider for example the Möller scattering. We have seen in Sect. 12.5.2 that, in a specific Lorentz frame, we can separate the interaction due to the exchange of transverse photon from the one due to the exchange of longitudinal and timelike photons, the latter resulting in a instantaneous Coulomb potential energy, whose Fourier transform is[40]

$$e_0 \, V(|\mathbf{k}|) = \frac{e_0^2}{|\mathbf{k}|^2}.$$

When the self-energy insertion is taken into account, we have to replace the lowest order photon propagator $D_{F\mu\nu}$ with the new propagator $D'_{F\mu\nu}$ given in Eq. (12.288). This implies that the *vacuum polarization* changes the Coulomb law as follows

$$\frac{e^2}{|\mathbf{k}|^2} \rightarrow \frac{e^2}{|\mathbf{k}|^2 \, [1 + C^{(R)}(-|\mathbf{k}|^2)]}.$$

---

[40]With respect to Eq. (12.163) we have replaced the coupling constant e with e_0 since the amplitude was computed to lowest order in the coupling constant.

where we have defined $e^2 = Z_3 e_0^2$ and we have set $k^2 = -|\mathbf{k}|^2$ since we are in the non-relativistic limit in which $k^0 = 0$. The factor $(1 + C^{(R)}(-|\mathbf{k}|^2))$ behaves much like a dielectric constant $\epsilon(\mathbf{k})$ since, as we show below, it reduces the effective charge (in absolute value) 'seen' at a given $|\mathbf{k}|$, as $|\mathbf{k}|$ decreases (i.e. as the distance from the charge increases), in the same way as it happens for charges in a dielectric material. Pictorially we may say that the vacuum polarization creates electron-positron virtual pairs circulating in the loop with a resulting partial screening of the electric charge, as it happens for a charge in a polar dielectric material. The actual value of $\Pi^{(c)}(-|\mathbf{k}|^2))$ can be computed explicitly by appropriate regularization of $\Pi_{\mu\nu}(k^2)$. One finds that for $\mathbf{k}^2 \ll m^2$ (the threshold for the pair production $e^+ e^-$)

$$\frac{e^2}{|\mathbf{k}|^2 [1 + C^{(R)}(-|\mathbf{k}|^2)]} \simeq \frac{e^2}{|\mathbf{k}|^2} \left(1 + \frac{\alpha}{15\pi} \frac{|\mathbf{k}|^2}{m^2} \right). \tag{12.293}$$

By Fourier transforming to configuration space we have

$$eV(\mathbf{x}) = \int \frac{d^3\mathbf{k}}{(2\pi)^3} e^{i\mathbf{k}\cdot\mathbf{x}} \frac{-e^2}{|\mathbf{k}|^2[1 + C^{(R)}(-|\mathbf{k}|^2)]} \simeq -\frac{e^2}{4\pi r} - \frac{\alpha}{15\pi} \frac{e^2}{m^2} \delta^{(3)}(\mathbf{x}),$$

where $r = |\mathbf{x}|$. This change indicates that the electromagnetic force becomes stronger at small distances.[41] This effect can be measured in hydrogen-like atoms, where the wave function is non-zero at the origin for s-waves. In fact this produces a shift of the $2s_{\frac{1}{2}}$ level given by

$$\Delta E = \int d^3\mathbf{x} \, |\psi(\mathbf{x})|^2 \left(-\frac{\alpha}{15\pi} \frac{e^2}{m^2} \delta^{(3)}(\mathbf{x}) \right) = -\frac{4\alpha^2}{15m^2} |\psi(0)|^2,$$

and using $|\psi(0)|^2 = \frac{\alpha^3 m^3}{8\pi}$ for the $2s$ state, we get

$$\Delta E = -1.123 \times 10^{-7} \text{ eV}.$$

This change has in particular the effect of removing the degeneration between the $2s_{\frac{1}{2}}$ and $2p_{\frac{1}{2}}$ levels. As will be discussed in the last section, the Lamb shift also removes the degeneration with a much larger correction. The agreement between theory and experiments, however, is good enough to verify the shift due to the vacuum polarization.

---

[41] The seeming singularity due to the presence of the delta function is actually due to our approximation $|\mathbf{k}|^2 \ll m^2$. In general the correction will be smooth and strongly peaked around $\mathbf{x} = 0$.

12.8.4 The Vertex Part

We now discuss the third divergent diagram, namely the *vertex part* whose 1-loop second order graph is shown in Fig. 12.21, together with the tree-level vertex. The element of $\mathbf{S}_{(3)}$ contributing to the amplitude reads:

$$
(ie_0)^3 \int d^4x d^4y d^4z : \bar{\psi}_0(x)\gamma^\mu S_F(x-y)\gamma^\nu S_F(y-z)\gamma^\rho \psi_0(z) :
$$

$$
\times D_{F\mu\rho}(x-z) A_{0\nu}(y) = ie_0 \int d^4x d^4y d^4z : \bar{\psi}_0(x)\, \Lambda^\nu(x,z|y)\, \psi_0(z) : A_{0\nu}(y),
$$

$$(12.294)$$

where we have defined the vertex part connecting the three external legs as:

$$
\Lambda^\nu(x,z|y) \equiv (ie_0)^2\, \gamma^\mu S_F(x-y)\gamma^\nu S_F(y-z)\gamma^\rho D_{F\mu\rho}(x-z).
$$

Using the explicit form (12.110) of the propagators in momentum representation we can write:

$$
\Lambda^\nu(x,z|y) = (ie_0)^2 \int \frac{d^4q'}{(2\pi)^4} \frac{d^4q}{(2\pi)^4} \frac{d^4k'}{(2\pi)^4}
$$

$$
\times \gamma^\mu \frac{i}{q\!\!\!/' - m}\gamma^\nu \frac{i}{q\!\!\!/ - m}\gamma_\mu \frac{-i}{(k')^2} e^{i(q'-q)\cdot(y-x)} e^{i(q+k')\cdot(z-x)}.
$$

$$(12.295)$$

Changing integration variables from q, q', k' to $k = q' - q$, $p = q + k'$ and q the above expression simplifies to:

$$
\Lambda^\nu(x,z|y) = \int \frac{d^4p}{(2\pi)^4} \frac{d^4k}{(2\pi)^4} \Lambda^\mu(p+k,p) e^{ik\cdot(y-x)} e^{ip\cdot(z-x)}, \quad (12.296)
$$

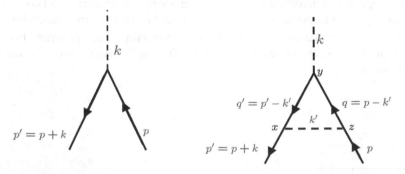

Fig. 12.21 Third order vertex loop

where

$$\Lambda^\mu(p+k,p) \equiv -ie_0^2 \int \frac{d^4q}{(2\pi)^4} \gamma^\mu \frac{1}{\not{k}+\not{q}-m} \gamma^\nu \frac{1}{\not{q}-m} \gamma_\mu \frac{1}{(p-q)^2}. \qquad (12.297)$$

When the operator (12.294) is computed between on-shell states, conservation of the total momentum sets the corresponding amplitude to zero, as explained in Sect. 12.4. Consequently the third order term (12.294) can only contribute to the amplitude of a larger process, in which at least one of the three external legs is an internal line of the corresponding Feynman diagram. This means that at least one of the fields $\bar\psi_0(x)$, $\psi_0(z)$, $A_{0\mu}(y)$ in (12.294) is contracted with some other one within a higher order S-matrix term, to make a propagator. Another process to which such term may contribute is the interaction of an electron with an external field, in which case $A_{0\mu}(y)$ is to be replaced by $A_{0\mu}^{ext}(y)$. In this case the one loop vertex diagram contributes to the amplitude a term of the form

$$2ime_0 \, V^\mu(p',p) A_{0\mu}^{ext}(k) = 2ime_0 \bar u(p') \Lambda^\mu(p',p) u(p) A_{0\mu}^{ext}(k), \qquad (12.298)$$

where $V^\mu(p',p) \equiv \bar u(p') \Lambda^\mu(p',p) u(p)$, while p and $p' = p + k$ are the momenta of the incoming and outgoing electrons, respectively. We see that the above term has the same form as the tree vertex contribution (12.197) except for the presence of $\Lambda^\mu(p',p)$ instead of γ^μ. Similarly, if the photon of momentum k is a virtual photon within a larger graph, the current $V^\mu(p',p)$ will have to be contracted with the corresponding photon propagator $D_{F\mu\nu}(k)$. According to our discussion in Sect. 12.5.5, gauge invariance with respect to the incoming photon of momentum $k = p' - p$ requires the current V^μ to be conserved (i.e. divergenceless), namely:

$$k_\mu V^\mu(p',p) = (p' - p)_\mu V^\mu(p',p) = 0. \qquad (12.299)$$

When summing all the contributions to a given amplitude coming from S-matrix terms of orders differing by two units, we will have to sum contributions from two diagrams differing just in the substitution of a tree vertex by a one loop vertex. Adding up the two terms amounts to effectively replacing in the lowest order one:

$$\gamma^\mu \rightarrow \Gamma^\mu(p',p),$$
$$\Gamma^\mu(p',p) \equiv \gamma^\mu + \Lambda^\mu(p',p). \qquad (12.300)$$

The quantity $\Lambda^\mu(p',p)$ then represents a second order correction to a vertex, whose integral expression in (12.297) has a logarithmic divergence for large values of the integration variable q, representing the momentum of a virtual electron. The matrix $\Gamma^\mu(p',p)$ is referred to as the *second order corrected vertex*. There are other corrections to the vertex, obtained by inserting self-energy parts in the legs of the three diagram. These are in principle accounted for by using the exact propagators for the electrons and the photon.

Expanding $\Lambda^\mu(p', p)$ in p, p'

$$\Lambda_\mu(p', p) = L\gamma_\mu + \Lambda_\mu^f(p', p), \tag{12.301}$$

where $L\gamma^\mu = \Lambda_\mu(0, 0)$ and one can isolate the divergent part L, which is a constant, from the finite remainder $\Lambda_\mu^f(p', p)$. The second order corrected vertex $\Gamma^\mu(p', p)$ consequently splits as follows:

$$\Gamma_\mu(p', p) = (1 + L)\gamma_\mu + \Lambda_\mu^f(p', p). \tag{12.302}$$

The ambiguity in the definition of L is fixed as follows. Let us first show that, on the general grounds of Lorentz covariance, if $p' = p$, the current $V^\mu(p, p) = \bar{u}(p)\Lambda^\mu(p, p)u(p)$ is proportional, through a constant, to $\bar{u}(p)\gamma^\mu u(p)$. By Lorentz covariance we can indeed convince ourselves that $\Lambda^\mu(p, p)$, which is a spinorial matrix depending on p, can only be combination of the matrices $p^\mu \mathbf{1}$ and γ^μ. Using then the property[42]

$$\bar{u}(p)\gamma^\mu u(p) = \frac{p^\mu}{m}\,\bar{u}(p)u(p), \tag{12.304}$$

we conclude that

$$V^\mu(p, p) = \bar{u}(p)\Lambda^\mu(p, p)u(p) = f_0\,\bar{u}(p)\gamma^\mu u(p), \tag{12.305}$$

f_0 being a constant. Actually, using Lorentz covariance and the gauge invariance condition (12.299), one can show that the current $V^\mu(p', p)$ can only have the following general form

$$V^\mu(p', p) = \bar{u}(p')\left(F_1(k^2)\,\gamma^\mu + F_2(k^2)\,\gamma^{\mu\nu}\,k_\nu\right)u(p), \tag{12.306}$$

where $k = p' - p$.[43] It follows that $V^\mu(p, p) = F_1(0)\bar{u}(p)\gamma^\mu u(p)$, so that $F_1(0) = f_0$. We now fix the ambiguity in L by requiring $L = f_0$, which implies

---

[42]To show this use the general on-shell identity (which only holds on-shell):

$$\bar{u}(p')\gamma^\mu u(p) = \frac{1}{2m}\left[\bar{u}(p')\gamma^\mu\,\not{p}u(p) + \bar{u}(p')\not{p}'\gamma^\mu u(p)\right]$$
$$= \bar{u}(p')\left[\frac{p'^\mu + p^\mu}{2m} - \gamma^{\mu\nu}\frac{(p'_\nu - p_\nu)}{2m}\right]u(p), \tag{12.303}$$

where we have written $\gamma^\mu\gamma^\nu = \eta^{\mu\nu} + \gamma^{\mu\nu}$, and $\gamma^{\mu\nu}$ being defined as $[\gamma^\mu, \gamma^\nu]/2$.

[43]See Weinberg's book [13] for a general derivation of this formula. There the most general form of $\Gamma^\mu(p', p)$ is written in terms of γ-matrices, p and p'. The number of independent terms reduces considerably upon using the Dirac equation $\not{p}u(p) = mu(p)$ $(\bar{u}(p')\not{p}' = m\,\bar{u}(p'))$ and the identity (12.303). By further implementing the gauge invariance condition (12.299) the final expression boils down to the one in Eq. (12.306).

$$\bar{u}(p, s')\Lambda^f_\mu(p, p)u(p, s) = 0. \tag{12.307}$$

Let us observe that the vertex $\Gamma_\mu(p', p)$ contains in general the coupling constant e_0 and a factor $Z_2 Z_3^{\frac{1}{2}}$ originating from the wave function renormalization of the electron and photon fields

$$\mathcal{L}^I_0 = e_0\, \bar{\psi}_0 \gamma^\mu \psi_0 A_{0\,\mu} = e_0 Z_2 Z_3^{\frac{1}{2}}\, \bar{\psi} \gamma^\mu \psi A_\mu. \tag{12.308}$$

We conclude that the logarithmic divergence in the vertex part correction can be absorbed in a *charge (or coupling constant) renormalization* as follows

$$e_0 Z_2 Z_3^{\frac{1}{2}} \to e_0 Z_2 Z_3^{\frac{1}{2}} (1 + L) = e, \tag{12.309}$$

where e defines the physical renormalized coupling constant. Setting $1 + L = Z_1^{-1}$ we rewrite (12.309) as follows

$$e = e_0 Z_2 Z_3^{\frac{1}{2}} Z_1^{-1}. \tag{12.310}$$

We now show an important identity between the vertex function $\Gamma_\mu(p', p)$ and the propagator $S'_F(p)$. The identity, referred to as *Ward identity*, is

$$\Gamma_\mu(p, p) = i\frac{\partial S'^{-1}_F(p)}{\partial p^\mu} \tag{12.311}$$

and, as we shall presently show, it is a consequence of the gauge invariance of the theory. The identity is trivially satisfied by the tree level vertex γ_μ and the free propagator $i\,(p\!\!\!/ - m)^{-1}$. To next order, using Eqs. (12.265) and (12.301), we can rewrite the Ward identity as follows

$$\Lambda_\mu(p, p) = -\frac{\partial \Sigma(p)}{\partial p^\mu} \tag{12.312}$$

The proof (to second order) can be done by exploiting the fact that, in the presence of a *constant* external electromagnetic field $A^{ext}_{0\mu}$, the electron self-energy part is modified as follows

$$-i\Sigma(p) \to -i\Sigma(p) + ie_0 \tilde{A}^{ext}_{0\mu} \Lambda_\mu(p, p) + \ldots, \tag{12.313}$$

where the right hand side represents a power series in the constant $\tilde{A}^{ext}_\mu$ and the second term represents a single interaction with $\tilde{A}^{ext}_\mu$. Note that, since the external field is constant, it transfers zero momentum, so that its Fourier transform is non-zero only for $k = 0$: $\tilde{A}^{ext}_{0\mu} = \tilde{A}^{ext}_{0\mu}(k = 0)\, \delta^4(k)$. Diagrammatically we can represent (12.313) as in Fig. 12.22. On the other hand, gauge invariance requires that the interaction with

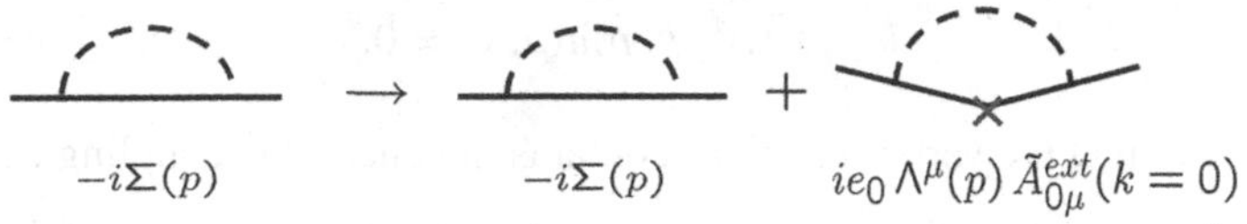

$$-i\Sigma(p) \qquad\qquad -i\Sigma(p) \qquad\qquad ie_0\,\Lambda^\mu(p)\,\tilde{A}^{ext}_{0\mu}(k=0)$$

Fig. 12.22 Insertion of an external field in an electron line at zero momentum transfer $k = 0$

the external field can be obtained by performing the *minimal coupling substitution*:

$$p_\mu \to p_\mu + e_0 A^{ext}_\mu. \tag{12.314}$$

Therefore we also have

$$\Sigma(p) \to \Sigma(p) + e_0 \tilde{A}^{ext}_\mu \left.\frac{\partial \Sigma(p)}{\partial p^\mu}\right|_{\tilde{A}^{ext}_\mu=0} + \dots \tag{12.315}$$

Comparison of (12.313) and (12.315) gives the Ward identity (12.312).

An important consequence of the Ward identity is that the wave function and vertex renormalization constants are equal

$$Z_1 = Z_2 \ \Leftrightarrow \ L = -B, \tag{12.316}$$

where B was defined in (12.269). To show this we compute the right hand side of the Ward identity (12.312) using Eq. (12.269) while on the left hand side we substitute Eq. (12.301). We obtain

$$L\gamma_\mu + \Lambda^f_\mu(p,p) = -\gamma_\mu B - \frac{\partial}{\partial p^\mu}\Sigma^{(c)}(p).$$

We now sandwich this relation between external on-shell states and find

$$\bar{u}(p,s')\gamma_\mu u(p,s)L = -B\bar{u}(p,s')\gamma_\mu u(p,s). \tag{12.317}$$

where we have used Eq. (12.307) and the fact that $\frac{\partial}{\partial p^\mu}\Sigma^{(c)}(p)$ vanishes for $p\!\!\!/ = m$. Recalling the definition (12.274) and that $Z_1 \equiv (1+L)^{-1}$, we immediately obtain Eq. (12.316).

The equality (12.316) implies that the coupling constant renormalization (12.309) reduces to

$$e = Z_3^{\frac{1}{2}} e_0 \to e_0 A_{0\mu} = e A_\mu. \tag{12.318}$$

The cancelation between the electron and photon wave function renormalization constants has been shown to work *at one loop* level (second order in the coupling constant). Actually the implementation of the full renormalization program reveals that the cancelation between the constants Z_1 and Z_2 is valid at all orders of the perturbation theory. It follows that these renormalizations are in fact spurious. This result is of fundamental importance. Indeed, generalizing to the electromagnetic interaction of other charged particles, it implies that *the electromagnetic coupling is universal.*

The interpretation of the coupling constant renormalization follows the usual lines. Starting from Eqs. (12.308) and (12.309) we have

$$\mathcal{L}_0^I = e_0\,\bar{\psi}_0\gamma^\mu\psi_0\,A_{0\,\mu} = Z_1 e\,\bar{\psi}\gamma^\mu\psi\,A_\mu \tag{12.319}$$

Therefore the physical renormalized interaction Lagrangian density can be written as

$$\mathcal{L}^I = e\,\bar{\psi}\gamma^\mu\psi\,A_\mu = \mathcal{L}_0^I + e(1 - Z_1)\,\bar{\psi}\gamma^\mu\psi\,A_\mu. \tag{12.320}$$

12.8.5 One-Loop Renormalized Lagrangian

We can now summarize the results of the previous sections writing down the relation between the bare Lagrangian density (12.263) we started from and the physical renormalized Lagrangian density $\mathcal{L}$. Adding Eqs. (12.278), (12.292) and (12.320) we find

$$
\begin{aligned}
\mathcal{L}_0 &= \bar{\psi}_0\,(i\partial\!\!\!/ - m_0)\psi_0 - \frac{1}{4}\,F_{0\,\mu\nu}\,F_0^{\mu\nu} + e_0\,A_{0\,\mu}\,\bar{\psi}_0\gamma^\mu\psi_0 \\
&= \mathcal{L} + \Delta\mathcal{L}
\end{aligned}
\tag{12.321}
$$

where

$$\mathcal{L} = \bar{\psi}\,(i\,\partial\!\!\!/ - m)\psi - \frac{1}{4}F^{\mu\nu}F_{\mu\nu} + e\,\bar{\psi}\gamma^\mu\psi\,A_\mu, \tag{12.322}$$

and

$$
\Delta\mathcal{L} = (Z_2 - 1)\,\bar{\psi}\,(i\,\partial\!\!\!/ - m)\psi + Z_2\bar{\psi}\psi\delta m + \\
\frac{1}{4}(Z_3 - 1)F^{\mu\nu}F_{\mu\nu} - e(1 - Z_1)\psi\gamma^\mu\bar{\psi}A_\mu. \tag{12.323}
$$

The relation between the bare fields and parameters and the physical ones is given by

$$\psi_0 = Z_2^{\frac{1}{2}}\psi; \quad A_{0\mu} = Z_3^{\frac{1}{2}}A_\mu$$

$$m_0 = m - \delta m; \quad e_0 = Z_1 Z_2^{-1} Z_3^{-\frac{1}{2}} e = Z_3^{-\frac{1}{2}} e \tag{12.324}$$

Note that the added terms in $\Delta\mathcal{L}$ have exactly the same structure as the terms present in the original Lagrangian $\mathcal{L}_0$.

The conclusion is that in order to have finite two-point Green's functions, that is propagators, and vertex functions we must start from a Lagrangian whose fields and parameters are not the physical fields and parameters, but are the unphysical, formally infinite bare quantities defined by Eq. (12.324). This has been shown at one-loop level or, equivalently, at second order for the self-energy and vertex insertions. In the general theory of renormalization one proves that the results obtained at one-loop level are sufficient to render finite the diagrams to any order in the perturbative expansion.

12.8.6 The Electron Anomalous Magnetic Moment

We have seen that the removal of the divergences from the second- order self-energy and vertex parts of a larger diagram is achieved by separating the divergent from the finite parts of the amplitude, the former being reabsorbed in the mass, coupling constant and wave-function (field) renormalization. The finite parts, on the other hand, give a well defined contribution to the amplitude and the result of its computation can be compared with experiment.[44]

In this subsection we want to give an important example of this finite contribution in a specific case, namely the (second-order) correction to the scattering of an electron by an external field A_μ^{ext}. This will allow us to compute the *anomalous magnetic moment of the electron* and compare the result with experiment.

Let us start with the first-order computation of the scattering amplitude of an electron in the external field A_μ^{ext}. It was computed in Sect. 12.5, Eq. (12.197), with the result

$$\langle\psi_{out}|\mathbf{S}_{(1)}|\psi_{in}\rangle = i\,\frac{e}{\hbar c}\left(\frac{mc^2}{\sqrt{E_\mathbf{p}E_{\mathbf{p}'}}\,V_e}\right)\bar{u}(\mathbf{p}',s)\gamma^\mu u(\mathbf{p},r)\,\tilde{A}_\mu^{ext}(k), \qquad (12.325)$$

where $k = p' - p$. Let us now consider the second order correction to the vertex part whose diagram is given in Fig. 12.23. We know that the vertex correction is given by the right hand side of Eq. (12.301), where the entire (logarithmic) divergence is contained in the constant L and can be reabsorbed in the coupling constant renormalization via $Z_1 = (1 + L)^{-1}$. Hence Λ_μ^f represents an observable effect. We are thus confronted with the explicit computation of Λ_μ^f. The computation of this integral is not trivial and we shall only quote the result. If the electron is supposed on the mass-shell, $p'^2 = p^2 = m^2$, and if the momentum transfer k^μ is small one obtains

---

[44]Corrections given by the finite parts of loop diagrams are often referred to as *radiative corrections*.

Fig. 12.23 Radiative
correction to the A_μ^{ext} field

$$\bar{u}(p')\Lambda_\mu^f(k)u(p) = \frac{\alpha}{2\pi}\,\bar{u}(p')\left[-\frac{1}{2mc}\gamma_{\mu\nu}k^\nu + \frac{2k^2}{3m^2c^2}\,\gamma_\mu\left(\ln\frac{m}{\lambda_{min}} - \frac{3}{8}\right)\right]u(p).$$

$$(12.326)$$

Comparing the above formula with the general expression in Eq. (12.306) we can identify the invariant functions $F_1(k^2)$, $F_2(k^2)$ in the latter with the following quantities:

$$F_1(k^2) = L + \frac{\alpha}{3\pi}\frac{k^2}{m^2c^2}\left(\ln\frac{m}{\lambda_{min}} - \frac{3}{8}\right),$$

$$F_2(k^2) = -\frac{\alpha}{4\pi mc}, \qquad\qquad (12.327)$$

where we have used the identification of L with $F_1(k^2 = 0) = f_0$. We see that only F_1 is divergent, the divergence being in L and is reabsorbed in the charge renormalization, while F_2 is finite and gives the correction to the electron magnetic moment, as we shall show.

The constant λ_{min} in (12.326) is a fictitious photon mass that has been introduced in order to avoid the divergence of the integral for small k, known as the "*infrared catastrophe*". In fact to obtain the previous result the photon propagator has been modified as follows

$$-i\frac{1}{k^2} \rightarrow -i\frac{1}{k^2 + \lambda_{min}^2}. \qquad\qquad (12.328)$$

This modification obviously entails that the amplitude (12.326) diverges when we let the photon mass go to zero giving rise to the so-called *infrared catastrophe*. Let us shortly comment on this point, since this kind of *infrared* divergence occurs quite often when computing Feynman diagrams. Actually this *infrared divergence* has nothing to do with the ultraviolet one present in $\Lambda_\mu(p', p)$ which was included in the definition of L. Its origin lies in the fact that considering an electromagnetic interaction process, we are asking a wrong question, namely: What is the amplitude of electron scattering with the emission of *no photon*? Now, in any scattering experiment, the electrons can radiate photons whose energy and momentum is sufficiently small to be undetected by the experimental apparatus. If the apparatus has an energy resolution E_m, then photons with energy $E < E_m$ will remain undetected. When the amplitude for the *soft photon* emission is combined with the infrared divergent amplitude, the divergence disappears.

Coming back to the our second order amplitude, we see that the first order amplitude (12.325) is changed as follows

$$\langle \psi_{out} | \mathbf{S}_{(1)} + \mathbf{S}_{(3)} | \psi_{in} \rangle$$

$$= i \frac{e}{\hbar c} \left(\frac{mc^2}{\sqrt{E_{\mathbf{p}} E_{\mathbf{p}'}} V_e} \right) \bar{u}(\mathbf{p}', s) \left(\gamma^\mu + \Lambda^{f\mu}(p', p) \right) u(\mathbf{p}, r) \tilde{A}_\mu^{ext}(k), \qquad (12.329)$$

where $\Lambda^{f\mu}(p', p)$, the finite remainder of the second-order vertex part, is the *radiative correction* to the first-order electron scattering. This is not the only correction to the first-order scattering. There is a further correction arising from the vacuum polarization graph of Fig. 12.23. One can show that the external field will get replaced by

$$A_\mu^{(ext)}(k) \rightarrow A_\mu^{(ext)}(k) \left(1 - \frac{\alpha}{15\pi} \frac{k^2}{m^2 c^2} \right),$$

which amounts just to adding $-\frac{1}{5}$ to $-\frac{3}{8}$ in the last term of Eq. (12.326).

We now show that the first term of Eq. (12.326), depending on the function F_2, computed at zero momentum transfer ($k^2 = 0$), represents the effect of an *anomalous electron magnetic moment* to the amplitude. To this end let us rewrite the current $\bar{u}(p')\gamma^\mu u(p)$ in the three-level part (12.197) of (12.329) using Eq. (12.303). As shown in Sect. 12.5.6 by evaluating the non-relativistic limit of the tree amplitude, the term contributing to the magnetic coupling is the one proportional to $\gamma^{\mu\nu} k_\mu \tilde{A}_\nu^{ext}$ which has the following form:

$$i \frac{1}{\hbar c V_e} \frac{e}{2mc} \bar{u}(p') \gamma^{\mu\nu} k_\mu \tilde{A}_\nu^{ext}(k) u(p),$$

where we have used the non-relativistic approximation $E_{\mathbf{p}} \sim E_{\mathbf{p}'} \sim mc^2$. The factor $e/(mc) = ge/(2mc)$ represents the gyromagnetic ratio that we have computed earlier. If we add the second order correction represented by the first term in Eq. (12.326) we end up with

$$i \frac{1}{\hbar c V_e} \frac{e}{2mc} (1 - 2mc F_2) \bar{u}(p') \gamma^{\mu\nu} k_\mu \tilde{A}_\nu^{ext}(k) u(p).$$

We see that the gyromagnetic ratio has acquired a correction of the form:

$$\frac{e}{mc} \rightarrow \frac{e}{mc}\,(1 - 2mc\,F_2) = \frac{2e}{2mc}\left(1 + \frac{\alpha}{2\pi}\right) = \frac{ge}{2mc}, \qquad (12.330)$$

corresponding to a corrected g-factor:

$$g = 2\left(1 + \frac{\alpha}{2\pi}\right).$$

This result was first obtained by Schwinger in 1948. The quantum deviation

$$\Delta\mu = \frac{(g-2)e}{2\,mc}\,\mathbf{s},$$

of the electron magnetic moment from its classical value, due to perturbative corrections, is usually referred to as the *electron anomalous magnetic moment*. Nowadays the very high precision measurements [10] of $g - 2$ provide the most stringent tests of QED (the agreement between theory and experiment is to within ten parts in a billion).[45]

There is another experimental result which is successfully predicted by quantum electrodynamics, and is worth mentioning without entering into heavy technical details. It is the splitting of the $2s_{1/2}$ and $2p_{1/2}$ levels in hydrogen atom, which was first measured by Lamb and Retherford in 1947 and is known as the *Lamb shift*. Indeed one can interpret the correction Λ_μ^f appearing in Eq. (12.326) as a modification of the effective field $\tilde{A}_\mu^{ext}$ seen by the electron, in our case $\tilde{A}_\mu^{ext}$ reducing to the Coulomb potential. This modification produces a splitting of the $2s_{1/2}$ and $2p_{1/2}$ levels, but it still depends on the λ_{min} cutoff present in Eq. (12.326). However if one takes into account the contribution from emission and absorption of virtual photons of momenta less than λ_{min}, then the dependence from λ_{min} cancels out. The final result gives for the splitting a value 1052.01 Mc/s. By improved theoretical calculations the value is raised to 1057.916. This agrees with the experimental value of with an accuracy of 10^{-5}.

12.8.7 References

For further reading see Refs. [3], [8, vol. 4], [9, 13].

---

[45]Since in order to test QED predictions for higher order corrections to a given quantity (like the $g - factor$), a high-precision determination of the coupling constant α is needed, one uses the QED formulas to experimentally determine α. QED is then tested by comparing the values of α determined from different experiments.

Appendix A
The Eötvös' Experiment

Let us consider two bodies, with inertial masses m_I e m_I', and suppose we attach them to the ends of a torsion pendulum as in Fig. A.1. We denote by ℓ and ℓ' the distances of the masses from the center of suspension. Let z and x be directed vertically and southwards, respectively; on the left part of Fig. A.1 these directions have been drawn at a particular point P of the terrestrial surface, the y direction being the normal passing through P corresponding to the west–east direction. Note that the centrifugal force m_I **a** due to the rotation of the earth forms an angle θ with the vertical direction equal to the latitude of P, while the gravitational force m_G **g** is directed towards the center of the earth.

On the right part of the Fig. A.1 we have drawn the torsion pendulum, and the centrifugal forces m_I **a** and m_I' **a** have been decomposed along the x and z axes.

The centrifugal forces acting in the x direction give rise to a momentum along the vertical direction z given by:

$$\tau_z = m_I \, a_x \, \ell - m_I' \, a_x \, \ell'. \tag{A.1}$$

On the other hand, equilibrium in the east–west direction requires the vanishing of τ_x, so that we may write:

$$(m_G \, g - m_I \, a_z)\, \ell = \left(m_G' \, g - m_I' \, a_z\right) \ell'. \tag{A.2}$$

If we now substitute the value of ℓ' given by (A.2) into Eq. (A.1) we find:

$$\tau_z = m_I \, a_x \, \ell \, g \, \frac{\left(\frac{m_G'}{m_I'} - \frac{m_G}{m_I}\right)}{g \frac{m_G'}{m_I'} - a_z}. \tag{A.3}$$

This component τ_z, if non vanishing, should be balanced by the momentum exerted by the torsion of the rod to which the pendulum is suspended. Experimentally no torsion momentum is observed, and therefore we must have: $\tau_z = 0$, that is:

© Springer International Publishing Switzerland 2016
R. D'Auria and M. Trigiante, *From Special Relativity to Feynman Diagrams*,
UNITEXT for Physics, DOI 10.1007/978-3-319-22014-7

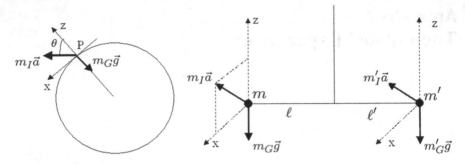

Fig. A.1 The Eotvös' experiment

$$\frac{m'_G}{m'_I} = \frac{m_G}{m_I}.$$ (A.4)

It follows that the ratio between inertial and gravitational masses does not depend on the particular body we are considering. Choosing the same unit for their measure we conclude that the two masses are indeed equal.

Appendix B
The Newtonian Limit of the Geodesic Equation

In this section we show that in the non-relativistic limit $v \ll c$, by further assuming the gravitational field to be *weak* and *stationary*, the geodesic equation (3.56) reduces to the Newton equation of a particle in a gravitational field. We recall from Chap. 3 that the metric field $g_{\mu\nu}(x)$ is the generalization of the Newtonian potential, and the statement that the gravitational field be *weak* and *stationary* is expressed by conditions (3.61) and (3.62), computing all quantities to first order in v/c and h.

We first rewrite Eq. (3.56) by splitting the coordinate index μ into $\mu = 0$ and $\mu = i(i = 1, 2, 3)$:

$$\frac{d^2(ct)}{d\tau^2} + \Gamma^0_{00}\left(\frac{d(ct)}{d\tau}\right)^2 + 2\,\Gamma^0_{0i}\frac{d(ct)}{d\tau}\frac{dx^i}{d\tau} + \Gamma^0_{ij}\frac{dx^i}{d\tau}\frac{dx^j}{d\tau} = 0, \tag{B.1}$$

$$\frac{d^2x^i}{d\tau^2} + \Gamma^i_{00}\left(\frac{d(ct)}{d\tau}\right)^2 + \Gamma^i_{jk}\frac{dx^k}{d\tau}\frac{dx^j}{d\tau} + 2\,\Gamma^i_{0j}\frac{dct}{d\tau}\frac{dx^j}{d\tau} = 0. \tag{B.2}$$

Since

$$\left(\frac{dx^i}{d\tau}\right) \Big/ \left(\frac{dx^0}{d\tau}\right) = \frac{v^i}{c}, \tag{B.3}$$

one recognizes that the condition $v/c \ll 1$ makes the last two terms of both equations negligible, so that Eqs. (B.1) and (B.2) become:

$$\frac{1}{c}\frac{d^2t}{d\tau^2} + \Gamma^0_{00}\left(\frac{dt}{d\tau}\right)^2 = 0 \tag{B.4}$$

$$\frac{1}{c^2}\frac{d^2x^i}{d\tau^2} + \Gamma^i_{00}\left(\frac{dt}{d\tau}\right)^2 = 0. \tag{B.5}$$

Taking into account that the time derivative of $g_{\mu\nu}$ is zero for a stationary field, from (3.59) we find:

© Springer International Publishing Switzerland 2016
R. D'Auria and M. Trigiante, *From Special Relativity to Feynman Diagrams*,
UNITEXT for Physics, DOI 10.1007/978-3-319-22014-7

$$\Gamma^0_{00} = \frac{1}{2} g^{0\rho} (-\partial_\rho g_{00} + 2 \partial_0 g_{\rho 0}) = -\frac{1}{2} (\eta^{0\rho} - h^{0\rho}) \partial_\rho g_{00} + O(h^2)$$

$$= -\eta^{00} \partial_0 h_{00} + O(h^2) = O(h^2) \simeq 0, \tag{B.6}$$

$$\Gamma^i_{00} = -\frac{1}{2} g^{ij} \partial_j g_{00} = -\frac{1}{2} (\eta^{ij} - h^{ij}) \partial_j h_{00} + O(h^2)$$

$$= \frac{1}{2} \partial_j h_{00} + O(h^2), \tag{B.7}$$

where we have taken into account Eqs. (3.61), (3.62), the fact that $\eta^{ij} = -\delta^{ij}$, and the inverse of relation (3.61), namely:

$$g^{\mu\nu} = \eta^{\mu\nu} - h^{\mu\nu} + O(h^2). \tag{B.8}$$

Equation (B.4) implies, taking into account (B.6)

$$\frac{dt}{d\tau} = const. \tag{B.9}$$

so that $\frac{d^2 x^i}{d\tau^2} = \left(\frac{dt}{d\tau}\right)^2 \frac{d^2 x^i}{dt^2}$. By virtue of (B.9) and (B.7), Eq. (B.5) becomes:

$$\frac{d^2 x^i}{dt^2} = -\frac{c^2}{2} \partial_i h_{00}, \tag{B.10}$$

where the minus sign on the right hand side originates from the metric. This is exactly Newton's equation of a particle in a gravitational field if we identify the Newtonian potential $\phi(\mathbf{x})$ with h_{00} as follows:

$$\frac{\phi}{c^2} = \frac{1}{2} h_{00}. \tag{B.11}$$

Indeed, with such identification, Eq. (3.64) can be rewritten as:

$$\frac{d^2 x^i}{dt^2} = -\partial_i \phi. \tag{B.12}$$

Furthermore, from the previous equations, we also see that in the limit of non-relativistic, weak and static field we can write:

$$g_{00} = 1 + h_{00} = 1 + 2 \frac{\phi}{c^2}. \tag{B.13}$$

Appendix C
The Twin Paradox

The so called twin paradox is the seemingly contradictory situation arising from a naive application of the *time dilation* phenomenon discussed in Chap. 1 to the following conceptual experiment.

Let A e B be two twins which are initially both at rest on earth. Suppose the twin B makes a journey on a high speed spaceship with constant velocity $\mathbf{v}$ and then comes back to earth meeting again the twin A. Let S be the frame of reference on earth and S' the one attached to the spaceship. If $\varDelta t$ is the time duration, relative to the earth's system S, of the total journey of B, if we were to naively apply the special relativity formulas given in Chap. 1, and since the two events (departure of B from A and final meeting of the two twins) occur in the same place relative to S', the corresponding time $\varDelta t'$ elapsed in the spaceship frame S' is related to $\varDelta t$ by the time dilation relation $\varDelta t = \varDelta t' \, \gamma(v)$. It follows that the twin B must be *younger* than the twin A when they meet again. This result appears to be paradoxical, since from the principle of relativity it follows that it is the same thing to consider B traveling with velocity $\mathbf{v}$ with respect to A or A traveling with velocity $-\mathbf{v}$ with respect to B. Since time dilation depends on v^2, considering B at rest and A traveling, it should be also possible to argue that A be younger than B. This puzzling result can be easily seen not to be correct if we recall that the special relativity effects can be applied only to frames of reference in *relative uniform motion*. If the two twins are to meet again to find out who is the younger, the spaceship system S' must invert its motion in order to come back to earth and therefore there is a part of its motion which is *accelerated* with respect to S. The situation is therefore not *symmetrical* since the S frame always remains *inertial*, while the frame S' is *non-inertial* during the inversion of its motion. There is thus no logical contradiction in saying that B is younger than A.

Even if the analysis of the twin paradox can be made entirely within the framework of special relativity we shall give its solution by applying the principle of equivalence discussed in Chap. 3 and showing that in both reference systems S and S' the twin B is younger than the twin A.

We shall perform the computation to the first order in $\frac{v^2}{c^2}$ and we shall denote by t_1, t_3, t_2 the time durations of the forth and back journeys and the inversion of

© Springer International Publishing Switzerland 2016
R. D'Auria and M. Trigiante, *From Special Relativity to Feynman Diagrams*,
UNITEXT for Physics, DOI 10.1007/978-3-319-22014-7

motion, respectively. In the frame of reference S' the corresponding times lapses will be denoted by t_1', t_2', t_3'.

- Let us first compute the total time duration of the journey from the point of view of the twin A, that is relative to the frame of reference S.
  The B twin in the frame S', measures a total duration of the journey $t' = t_1' + t_2' + t_3'$, while A measures $t = t_1 + t_2 + t_3$ where:

$$t_1' = t_1 \sqrt{1 - v^2/c^2} \simeq t_1 \left(1 - \frac{1}{2} v^2/c^2\right), \tag{C.1}$$

$$t_3' = t_3 \sqrt{1 - v^2/c^2} \simeq t_1 \left(1 - \frac{1}{2} v^2/c^2\right), \tag{C.2}$$

$$t_2' \simeq 0 \quad t_2 \simeq 0, \tag{C.3}$$

where we have set $t_2' = t_2 \simeq 0$ since the time of turnaround of S', from the point of view of the *inertial* frame S, can be neglected compared with t_1 and t_3. Note that the times t_i' are *proper times* since B is at rest in S'.
Setting $t_1 = t_3$ the total duration of the journey of B from the point of view of A is:

$$t' = 2 t_1 \left(1 - \frac{1}{2} \frac{v^2}{c^2}\right). \tag{C.4}$$

Thus, if we take $v = 9 \times 10^7$ m/s and $t_1 = 20\, years$, and if the two twins were, say, 22 years old when B departed, as they meet again after the trip, their age difference will be $t_1 \frac{v^2}{c^2} \approx 2\, years$: A will be 62 and B 60. It is instructive to derive Eq. (C.4) from geometric considerations, see Fig. C.1.

Fig. C.1 World-lines of the two twins in a space-time diagram

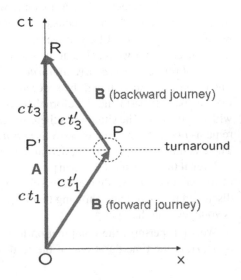

Let us plot on a space-time diagram, relative to S, the trajectories (world-lines) of the two twins. Let the points O and R in the diagram be the events in which they depart and meet again, respectively. The twin A is at rest in S and thus its world-line is vertical, directed along the time direction. Suppose, for the sake of simplicity, that the twin B moves forth and back along the x-axis, so as to describe, in the diagram, two segments: One, OP, with positive slope $\frac{\Delta x}{c\Delta t} = \frac{v}{c} > 0$ during the forward journey, and an other, PR, with slope $-\frac{v}{c} < 0$, during the backward journey. The *lengths* of the two world-lines, divided by c, measure the proper-time intervals relative to A and B (i.e. the times measured by A and B, respectively) between the two events O and R. Since A is at rest in S, its proper time interval is $t = \frac{|OR|}{c} = t_1 + t_2 + t_3 \simeq 2t_1$. As for B, its proper time interval is

$$t' = t'_1 + t'_2 + t'_3 \simeq 2t'_1 = \frac{2}{c}|OP|. \tag{C.5}$$

From the diagram one would naively conclude that $t' > t$ since the length of the trajectory of B appears to be greater than that of A. Recall, however, that we are in Minkowski space and that lengths are measured with the Lorentzian signature for the metric. As a consequence, in contrast to the Pythagorean theorem which holds in Euclidean geometry, the squared length of the hypotenuse of the right triangle OPP' is given by the *difference* of the squared lengths of the catheti, instead of the sum (in other words the hypotenuse is shorter than each of the catheti):

$$|OP| = \sqrt{|OP'|^2 - |PP'|^2} = \sqrt{c^2 t_1^2 - \Delta x^2} = ct_1\sqrt{1 - \frac{v^2}{c^2}}, \tag{C.6}$$

where we have used $\Delta x = v\,t_1$. Substituting the above result in (C.5), and expanding the square root to the first order in v^2/c^2, we find (C.4).

- Let us now compute the duration of the journey from the point of view of B himself (frame of reference S').

In this case t_1 e t_3 are *proper times*, being the twin A at rest with respect to the earth's frame of reference S, which is now moving relative to B, and we have

$$t'_1 = \frac{t_1}{\sqrt{1 - v^2/c^2}} \simeq t_1\left(1 + \frac{1}{2}v^2/c^2\right), \tag{C.7}$$

$$t'_3 = \frac{t_3}{\sqrt{1 - v^2/c^2}} \simeq t_3\left(1 + \frac{1}{2}v^2/c^2\right).$$

Let us now compute t'_2 which now, as opposite to the previous analysis, cannot be neglected: We are indeed now in a non-inertial frame of reference and, as we shall see below, it will turn out to be proportional to t_1.

Indeed, during the turnaround of the spaceship, there is an acceleration field $g = \frac{2v}{t_2}$ with respect to earth (directed towards the earth itself). According to the equivalence principle, we can interpret this acceleration as due to an equivalent gravitational potential with strength $\phi = g h$ where $h = v t_1'$; using Eq. (3.75) one obtains:

$$t_2' = t_2 \left(1 - \frac{g h}{c^2}\right) = t_2 \left(1 - \frac{2vh}{t_2 c^2}\right) = t_2 - \frac{2v^2}{c^2}t_1' = t_2 - 2\frac{v^2}{c^2}t_1 + O\left(\frac{v^4}{c^4}\right),$$

where we made use of (C.7) implying $t_1' = t_1 + O(v^2/c^2)$. The final result is therefore:

$$t' = t_1' + t_2' + t_3' \simeq 2 t_1 \left(1 + \frac{1}{2}\frac{v^2}{c^2}\right) + t_2 - 2\frac{v^2}{c^2}t_1$$

$$\simeq 2t_1 \left(1 - \frac{1}{2}\frac{v^2}{c^2}\right), \tag{C.8}$$

where we have used $t_2 \ll t_1$, see (C.3).

We see that Eq. (C.8) coincides with (C.4). We conclude that from both the points of view of A and B the time elapsed for the twin B is shorter than the time elapsed for the twin A. In other words, after the journey the twin B is younger than the twin A.

Appendix D
Jacobi Identity for Poisson Brackets

We show that given three dynamical variables $f(p, q)$, $g(p, q)$, $h(p, q)$ their Poisson brackets obey the *Jacobi identity*, namely:

$$\{f_1, \{f_2, f_3\}\} + \{f_2, \{f_3, f_1\}\} + \{f_3, \{f_1, f_2\}\} = 0, \tag{D.1}$$

where we have renamed $f(p, q)$, $g(p, q)$, $h(p, q)$ of the text (see Eq. (7.39) of Sect. 8.3) with the more convenient notation $f_1(p, q), f_2(p, q), f_3(p, q)$.

Let us compute $\{f_1, \{f_2, f_3\}\}$:

$$
\begin{aligned}
\{f_1, \{f_2, f_3\}\} &= \frac{\partial f_1}{\partial q_i} \frac{\partial}{\partial p_i} \left[\frac{\partial f_2}{\partial q_j} \frac{\partial f_3}{\partial p_j} - \frac{\partial f_2}{\partial p_j} \frac{\partial f_3}{\partial q_j} \right] - \frac{\partial f_1}{\partial p_i} \frac{\partial}{\partial q_i} \left[\frac{\partial f_2}{\partial q_j} \frac{\partial f_3}{\partial p_j} - \frac{\partial f_2}{\partial p_j} \frac{\partial f_3}{\partial q_j} \right] \\
&= \frac{\partial f_1}{\partial q_i} \left[\frac{\partial^2 f_2}{\partial p_i \partial q_j} \frac{\partial f_3}{\partial p_j} + \frac{\partial^2 f_3}{\partial p_i \partial p_j} \frac{\partial f_2}{\partial q_j} - \frac{\partial^2 f_2}{\partial p_i \partial p_j} \frac{\partial f_3}{\partial q_j} - \frac{\partial^2 f_3}{\partial p_i \partial q_j} \frac{\partial f_2}{\partial p_j} \right] \\
&\quad - \frac{\partial f_1}{\partial p_i} \left[\frac{\partial^2 f_2}{\partial q_i \partial q_j} \frac{\partial f_3}{\partial p_j} + \frac{\partial^2 f_3}{\partial q_i \partial p_j} \frac{\partial f_2}{\partial q_j} - \frac{\partial^2 f_2}{\partial q_i \partial p_j} \frac{\partial f_3}{\partial q_j} - \frac{\partial^2 f_3}{\partial q_i \partial q_j} \frac{\partial f_2}{\partial p_j} \right],
\end{aligned}
$$

where sum over the repeated indices i, j is understood.

Considering the terms which are bilinear in the first derivatives with respect to the two $q_i's$ we have:

$$\frac{\partial f_1}{\partial q_i} \left[\frac{\partial f_2}{\partial q_j} \frac{\partial^2 f_3}{\partial p_i \partial p_j} - \frac{\partial f_3}{\partial q_j} \frac{\partial^2 f_2}{\partial p_i \partial p_j} \right]. \tag{D.2}$$

Adding to this expression the analogous terms coming from the second and third term of the identity (D.1) which are simply obtained by cyclic permutations of 1, 2, 3, we see that the total contribution sum up to zero:

© Springer International Publishing Switzerland 2016
R. D'Auria and M. Trigiante, *From Special Relativity to Feynman Diagrams*,
UNITEXT for Physics, DOI 10.1007/978-3-319-22014-7

$$\frac{\partial f_1}{\partial q_i}\left[\frac{\partial f_2}{\partial q_j}\frac{\partial^2 f_3}{\partial p_i \partial p_j} - \frac{\partial f_3}{\partial q_j}\frac{\partial^2 f_2}{\partial p_i \partial p_j}\right]$$

$$+ \frac{\partial f_2}{\partial q_i}\left[\frac{\partial f_3}{\partial q_j}\frac{\partial^2 f_1}{\partial p_i \partial p_j} - \frac{\partial f_1}{\partial q_j}\frac{\partial^2 f_3}{\partial p_i \partial p_j}\right]$$

$$+ \frac{\partial f_3}{\partial q_i}\left[\frac{\partial f_1}{\partial q_j}\frac{\partial^2 f_2}{\partial p_i \partial p_j} - \frac{\partial f_2}{\partial q_j}\frac{\partial^2 f_1}{\partial p_i \partial p_j}\right] = 0.$$

The same of course would happen if we considered all the other terms bilinear in the first derivatives with respect to two $p_i's$ and to one q_i and one p_i. Therefore the total sum is identically zero.

Appendix E
Induced Representations and Little Groups

E.1 Representation of the Poincaré Group

The single particle states $|p, r\rangle$ are constructed as a basis of a (infinite dimensional) space $V^{(c)}$ supporting a unitary, irreducible representation of the Poincaré group. This construction is effected through the method of *induced representations*: We start defining the single particle states $|\bar{p}, r\rangle$ in a *fixed reference frame* S_0, where the four momentum is a *standard one* $p^\mu = \bar{p}^\mu$. These states differ by the internal degree of freedom, labeled by r, related to the spin of the particle and which is acted on by the little group $G^{(0)} \subset SO(1, 3)$ of the momentum $\bar{p} \equiv (\bar{p}^\mu)$ (spin group), consisting of the Lorentz transformations $\Lambda^{(0)}$ which leave $\bar{p}$ inert:

$$\Lambda^{(0)} \in G^{(0)} \Leftrightarrow \Lambda^{(0)\mu}{}_\nu \bar{p}^\nu = \bar{p}^\mu. \tag{E.1}$$

A transformation $\Lambda^{(0)}$ of $G^{(0)}$ is implemented on the states $|\bar{p}, r\rangle$ by a unitary operator $U(\Lambda^{(0)})$ which then maps $|\bar{p}, r\rangle$ into an eigenstate of the four-momentum corresponding to the same eigenvalue $\bar{p}$. The vector $U(\Lambda^{(0)})|\bar{p}, r\rangle$ has then to be a linear combination of the basis elements $|\bar{p}, r\rangle$ through a matrix $\mathcal{R} \equiv (\mathcal{R}^r{}_s)$:

$$U(\Lambda^{(0)})|\bar{p}, r\rangle = \mathcal{R}(\Lambda^{(0)})^s{}_r |\bar{p}, s\rangle. \tag{E.2}$$

Such matrix $\mathcal{R}(\Lambda^{(0)})$ defines a (unitary) representation $\mathcal{R}$ of $G^{(0)}$ which characterizes the spin of the particle. For a massive particle $m^2 \neq 0$, $G^{(0)} = SU(2)$, see Sect. E.2, and $\mathcal{R}$ has dimension $2s + 1$ (that is $r = 1, \ldots, 2s + 1$), s being the spin of the particle (in units $\hbar$); for a massless particle, $m^2 = 0$, $G^{(0)}$ is effectively $SO(2)$, generated by the helicity operator, see Sect. E.2, and $r = 1, 2$ labels the helicity state. Proper Lorentz transformations do not alter the eigenvalue of the helicity, as proven in Sect. 9.4.2.

A state $|p, r\rangle$, corresponding to a generic four momentum $p \equiv (p^\mu)$ is defined by acting on $|\bar{p}, r\rangle$ with the Lorentz boost Λ_p which relates S_0 to the RF S in which the momentum of the particle is p: $p = \Lambda \bar{p}$. If $U(\Lambda)$ is the unitary transformation

© Springer International Publishing Switzerland 2016
R. D'Auria and M. Trigiante, *From Special Relativity to Feynman Diagrams*,
UNITEXT for Physics, DOI 10.1007/978-3-319-22014-7

implementing a Lorentz transformation Λ on the states, $|p, r\rangle$ is then defined as:

$$|p, r\rangle = |\Lambda_p \bar{p}, r\rangle \equiv U(\Lambda_p)|\bar{p}, r\rangle. \tag{E.3}$$

The above relation defines $|p, r\rangle$ and $U(\Lambda_p)$ at the same time. Equations (E.2) and (E.3) allow to define the action of a generic Lorentz transformation Λ on the states $|p, r\rangle$ through a corresponding unitary operator $U(\Lambda)$. Suppose Λ transforms p into p': $p' = \Lambda p$. We can then write:

$$U(\Lambda)|p, r\rangle = U(\Lambda) U(\Lambda_p)|\bar{p}, r\rangle = U(\Lambda_{p'}) \left(U(\Lambda_{p'})^{-1} U(\Lambda) U(\Lambda_p) \right) |\bar{p}, r\rangle$$

$$= U(\Lambda_{p'}) U \left(\Lambda_{p'}^{-1} \Lambda \Lambda_p \right) |\bar{p}, r\rangle, \tag{E.4}$$

where $\Lambda_{p'}$ is the Lorentz boost connecting $\bar{p}$ to p'. Note now that the transformation $\Lambda^{(0)} \equiv \Lambda_{p'}^{-1} \Lambda \Lambda_p$ first maps $\bar{p}$ into p, then p into p' and finally p' back into $\bar{p}$. It therefore belongs to the little group $G^{(0)}$ of $\bar{p}$ and thus its action on $|\bar{p}, r\rangle$ is defined in (E.2). We then find:

$$U(\Lambda)|p, r\rangle = \mathcal{R}^{r'}{}_r U(\Lambda_{p'})|\bar{p}, r'\rangle = \mathcal{R}^{r'}{}_r |\Lambda p, r'\rangle, \tag{E.5}$$

where now the rotation matrix $\mathcal{R}$, associated with $\Lambda^{(0)}$, depends on both Λ and p: $\mathcal{R} = \mathcal{R}(\Lambda, p)$. If Λ is a simple boost, the corresponding rotation $\mathcal{R}(\Lambda, p)$ is called *Wigner rotation*.

The action of a Poincaré transformation (Λ, x_0) on $|p, r\rangle$ then reads:

$$e^{-\frac{i}{\hbar} x_0^\mu \hat{P}_\mu} U(\Lambda)|p, r\rangle = \mathcal{R}^{r'}{}_r e^{-\frac{i}{\hbar} x_0^\mu \hat{P}_\mu} |\Lambda p, r'\rangle = \mathcal{R}^{r'}{}_r e^{-\frac{i}{\hbar} x_0 \cdot (\Lambda p)} |\Lambda p, r'\rangle.$$

As mentioned in Chap. 9, the procedure illustrated here for constructing the unitary, infinite dimensional representation of the Poincaré group on single particle states starting from the (finite-dimensional) representation of the spin group is called *method of induced representations*.

Having defined the single particle states $|p, r\rangle$ and the action of Poincaré transformations on them, let us prove general properties that were used, or simply mentioned, in Sect. 9.4.1.

- The little group $G_p^{(0)}$ of a generic momentum p, defined in Eq. 9.109, is related to $G^{(0)}$ through conjugation by Λ_p: $G_p^{(0)} = \Lambda_p G^{(0)} \Lambda_p^{-1}$. To see this we first observe that with each element $\Lambda^{(0)}$ of $G^{(0)}$, defined by the property $\Lambda^{(0)} \bar{p} = \bar{p}$, we can associate a unique transformation $\Lambda_p^{(0)}$ in the little group $G_p^{(0)}$ of p, whose effect consists in a first boost to the RF S_0 in which the four-momentum is $\bar{p}$, followed by the transformation $\Lambda^{(0)}$ which leaves $\bar{p}$ inert, and then a second boost back to the initial frame in which the momentum is p: $\Lambda_p^{(0)} = \Lambda_p \Lambda^{(0)} \Lambda_p^{-1}$. We easily verify that $\Lambda_p^{(0)}$ so defined leaves p invariant:

$$\Lambda_p^{(0)} p = \left(\Lambda_p \, \Lambda^{(0)} \, \Lambda_p^{-1}\right) p = \Lambda_p \, \Lambda^{(0)} \, \bar{p} = \Lambda_p \bar{p} = p, \qquad \text{(E.6)}$$

which implies that $\Lambda_p^{(0)} \in G_p^{(0)}$. Similarly, given an element $\Lambda_p^{(0)} \in G_p^{(0)}$ we can construct the unique element $\Lambda^{(0)} = \Lambda_p^{-1} \, \Lambda_p^{(0)} \, \Lambda_p$ in $G^{(0)}$. This proves that little groups corresponding to four-momenta with the same mass squared, are conjugated to one another, and thus share the same structure, though being represented by different matrices. The one between $G_p^{(0)}$ and $G^{(0)}$ is the same kind of relation, that we have called isomorphism in footnote 12 of Chap. 9, which exists between the little group $O(1, 3)$ of the origin (Lorentz group), and that of a generic space-time point x, $O(1, 3)_x$, and implies that the two groups realize the same symmetry.

- In order for the representation U of the Poincaré group on the single-particle states $|p, r\rangle$ to be irreducible, the representation $\mathcal{R}$ of the spin group $G^{(0)}$ has to be irreducible as well. Indeed, if $\mathcal{R}$ were reducible, there would be a proper subset of states in $\mathcal{S}_0$, denoted by $|\bar{p}, s\rangle_0$ which is stable with respect to the action of $G^{(0)}$. The states $|p, s\rangle_0 = U(\Lambda_p) |\bar{p}, s\rangle_0$ span a proper subspace $V_0^{(c)}$ of the full Hilbert space $V^{(c)}$ which is stable with respect to the Lorentz group. This is easily shown by applying a generic Lorentz transformation $U(\Lambda)$ to $|p, s\rangle_0$, as in Eq. (E.5): The corresponding $G^{(0)}$ transformation $\Lambda^{(0)} \equiv \Lambda_{p'}^{-1} \Lambda \Lambda_p$ will act on $|\bar{p}, s\rangle_0$ mapping it into a combination of states in the same $G^{(0)}$-invariant subspace. The action of $\Lambda_{p'}$ on such combination will therefore still be in $V_0^{(c)}$. Thus the full representation of the Lorentz group would be reducible.

Consequently $\mathcal{R}$ is the $(2s+1)$-dimensional representation of the spin group SU(2) for massive particles, while it is the one-dimensional representation defined by a given value of the helicity for massless particles.

E.2 Little Groups

The *little group* of a four-momentum vector $p = (p^\mu)$ was defined in Sect. 9.4 as the set of all the Lorentz transformations $\Lambda_p^{(0)}$ leaving p invariant, namely satisfying Eq. (9.109). Such set is indeed a group, as the reader can easily verify. Let us construct the little group $G^{(0)}$ of the standard four-momentum $\bar{p}$. Writing Eq. (9.109) for an infinitesimal transformation (4.171) we find[1]:

$$\Lambda^{(0)} \bar{p} \approx \left(1 + \frac{i}{2\,\hbar} \delta\theta_{\rho\sigma} J^{\rho\sigma}\right) \bar{p} = \bar{p} \; \Rightarrow \; \delta\theta_{\rho\sigma} J^{\rho\sigma} \bar{p} = 0, \qquad \text{(E.7)}$$

from which we deduce that $G^{(0)}$ is generated by those combinations $\delta\omega_{\rho\sigma} J^{\rho\sigma}$ of the Lorentz generators $J^{\rho\sigma}$ which annihilate $\bar{p}$. Let us consider the different cases:

---

[1]Recall that $J^{\rho\sigma} = -i\hbar \, L^{\rho\sigma}$, $L^{\rho\sigma}$ being defined in (4.170).

$\mathbf{m^2 > 0}$: In this case we can choose the standard RF $\mathcal{S}_0$ as the rest frame of the particle in which $\bar{p} = (mc, 0, 0, 0)$. Equation (E.7) the implies the following conditions on the infinitesimal generators:

$$\begin{pmatrix} 0 & \delta\theta_{0,1} & \delta\theta_{0,2} & \delta\theta_{0,3} \\ \delta\theta_{0,1} & 0 & -\delta\theta_{1,2} & -\delta\theta_{1,3} \\ \delta\theta_{0,2} & \delta\theta_{1,2} & 0 & -\delta\theta_{2,3} \\ \delta\theta_{0,3} & \delta\theta_{1,3} & \delta\theta_{2,3} & 0 \end{pmatrix} \begin{pmatrix} 1 \\ 0 \\ 0 \\ 0 \end{pmatrix} = \begin{pmatrix} 0 \\ 0 \\ 0 \\ 0 \end{pmatrix} \Rightarrow \delta\theta_{0\mu} = 0, \quad \text{(E.8)}$$

that is the infinitesimal generators of $G^{(0)}$ read:

$$\frac{i}{2\hbar} \delta\theta_{ij} J^{ij} = \frac{i}{\hbar} \delta\theta_i J^i, \quad \text{(E.9)}$$

having defined $\delta\theta_i = -\epsilon_{ijk} \delta\theta^{jk}/2$. We conclude that $G^{(0)}$ is the rotation group SO(3). When we consider the action of these generators on states, $\hat{J}_i$ also contains the spin-component $\hat{S}_i$, which can act on bi-dimensional representations (as it is the case for spin 1/2 particles). Since SO(3) has no such representation, it is appropriate to say that $\hat{J}_i$ generate the spin group SU(2).

$\mathbf{m^2 = 0}$: The standard four-momentum vector can be chosen to be $\bar{p} = (E, E, 0, 0)/c$. Equation (E.7) the implies:

$$\begin{pmatrix} 0 & \delta\theta_{0,1} & \delta\theta_{0,2} & \delta\theta_{0,3} \\ \delta\theta_{0,1} & 0 & -\delta\theta_{1,2} & -\delta\theta_{1,3} \\ \delta\theta_{0,2} & \delta\theta_{1,2} & 0 & -\delta\theta_{2,3} \\ \delta\theta_{0,3} & \delta\theta_{1,3} & \delta\theta_{2,3} & 0 \end{pmatrix} \begin{pmatrix} 1 \\ 1 \\ 0 \\ 0 \end{pmatrix} = \begin{pmatrix} 0 \\ 0 \\ 0 \\ 0 \end{pmatrix} \Rightarrow \begin{cases} \delta\theta_{01} = 0 \\ \delta\theta_{0a} = -\delta\theta_{1a}, \end{cases} \quad \text{(E.10)}$$

where $a = 2, 3$. The generators of $G^{(0)}$ consist in $J^{23} = -J^1$ which generates rotations about the direction X of motion, and the following two matrices:

$$N^a \equiv J^{0a} - J^{1a}. \quad \text{(E.11)}$$

From the commutation relations among the $J^{\rho\sigma}$·s we deduce:

$$[J^{23}, N^2] = -i\hbar N^3; \quad [J^{23}, N^3] = i\hbar N^2; \quad [N^2, N^3] = 0. \quad \text{(E.12)}$$

A group generated by three generators J^{23}, N^a with the above commutation relations is denoted by ISO(2) and contains an SO(2) subgroup generated by J^{23} and a two-parameter subgroup of translations generated by N^a. It is the group of congruences on the Euclidean plane E_2.

Defining $N^{\pm} = N^2 \pm i N^3$ and the helicity matrix $\Gamma \equiv J^1 = -J^{23}$ we find:

$$[\Gamma, N^{\pm}] = \pm\hbar N^{\pm}; \quad [N^+, N^-] = 0. \quad \text{(E.13)}$$

Consider now the action of the operators $\hat{\Gamma}$, $\hat{N}^\pm$ on the states $|\bar{p}, r\rangle$. In going from the 4×4 matrix representation of these operators, to their representation on states, the commutation structure (E.13) is preserved. Moreover $\hat{N}^+ = (\hat{N}^-)^\dagger$, while $\hat{\Gamma}$ is hermitian and can thus be diagonalized. Suppose it has an eigenvalue $\hbar s$ on $|\bar{p}, s\rangle$ (the state vectors being normalized to one). Note that the operators $\hat{N}^+$, $\hat{N}^-$ behave as creation and annihilation operators in the sense that, using (E.13) one can easily verify the following:

$$\hat{N}^+ |\bar{p}, s\rangle = \alpha_0 |\bar{p}, s + 1\rangle; \quad \hat{N}^- |\bar{p}, s + 1\rangle = \alpha_0^* |\bar{p}, s\rangle, \tag{E.14}$$

α_0 being some complex number. If we continue applying those operators we can construct infinitely many states $|\bar{p}, s + k\rangle$:

$$\hat{N}^+ |\bar{p}, s + k\rangle = \alpha_k |\bar{p}, s + k + 1\rangle; \quad \hat{N}^- |\bar{p}, s + k + 1\rangle = \alpha_k^* |\bar{p}, s + k\rangle, \tag{E.15}$$

Note that $\hat{N} \equiv \hat{N}^+ \hat{N}^- = \hat{N}^- \hat{N}^+$ is positive definite and:

$$\hat{N} |\bar{p}, s + k\rangle = \hat{N}^- \hat{N}^+ |\bar{p}, s + k\rangle = |\alpha_k|^2 |\bar{p}, s + k\rangle = \hat{N}^+ \hat{N}^- |\bar{p}, s + k\rangle$$
$$= |\alpha_{k-1}|^2 |\bar{p}, s + k\rangle; \quad k = \ldots, -2, -1, 0, 1, 2, \ldots, \tag{E.16}$$

from which we deduce that $|\alpha_{k-1}|^2 = |\alpha_k|^2 = |\alpha|^2$. If we require the system to have finitely many spin states, corresponding to its internal degrees of freedom, some state should be annihilated by $\hat{N}^1$, which implies $\alpha_k = 0$ for some k, and thus $\alpha = 0$. We conclude that $\hat{N}^\pm$ and $\hat{N}$ must be zero on any state (consequently also $\hat{N}^a$ are zero): The only generator of the little group which has non trivial action on the states is the helicity operator $\hat{\Gamma}$ generating the SO(2) subgroup of ISO(2). The condition that the single particle state transform in an irreducible representation of SO(2) further implies that there can be just two helicity states:

$$\hat{\Gamma} |\bar{p}, \pm s\rangle = \pm \hbar s |\bar{p}, \pm s\rangle, \tag{E.17}$$

s being the spin of the particle.

$\mathbf{m^2 < 0}$: Let us just mention this case which corresponds to an *unphysical* particle called *tachyon* which moves faster than light: $\frac{v^2}{c^2} = \frac{|\mathbf{p}|^2 c^2}{E^2} > 1$. The standard four-momentum vector can be chosen to be $\bar{p} = (0, p^1, 0, 0)$. Clearly Eq. (E.7) is solved by a 4×4 matrix $\mathbf{A}$ obtained from $\delta\theta_{\rho\sigma} J^{\rho\sigma}$ by deleting the second row and the second column. It generates Lorentz transformations in the three-dimensional subspace of M_4 spanned by the coordinates (ct, y, z) and orthogonal to the X-axis. This space is a three-dimensional Minkowski space M_3 with a metric $\eta = \mathrm{diag}(+1, -1, -1)$ and the corresponding symmetry subgroup of the Lorentz group is therefore $G^{(0)} = $ SO(1, 2).

Appendix F
SU(2) and SO(3)

The group SU(2) is the group of all 2×2 unitary matrices with unit determinant (also called *special unitary matrices*). Let $\mathbf{S} = (S^r{}_s)$ be a generic element of the group. By definition $\mathbf{S}^\dagger \mathbf{S} = \mathbf{1}_2$ and $\det(\mathbf{S}) = 1$. From our general discussion of unitary matrices, it follows that, we can write $\mathbf{S}$, in a neighborhood of the identity, as the exponential of i times a hermitian matrix $\mathbf{A}$ as follows:

$$\mathbf{S} = e^{i\mathbf{A}} \quad \Rightarrow \quad \mathbf{A}^\dagger = \mathbf{A}. \tag{F.1}$$

From the matrix property $\det(\mathbf{S}) = \exp(i \operatorname{Tr}(\mathbf{A}))$, it follows that, being $\mathbf{S}$ special, $\mathbf{A}$ should be traceless. The most general 2×2 hermitian traceless matrix has the form:

$$\mathbf{A} = \begin{pmatrix} a & b-ic \\ b+ic & -a \end{pmatrix} = b\,\sigma^1 + c\,\sigma^2 + a\,\sigma^3, \tag{F.2}$$

where σ^i are the Pauli matrices, defined as:

$$\sigma^1 = \begin{pmatrix} 0 & 1 \\ 1 & 0 \end{pmatrix}; \quad \sigma^2 = \begin{pmatrix} 0 & -i \\ i & 0 \end{pmatrix}; \quad \sigma^3 = \begin{pmatrix} 1 & 0 \\ 0 & -1 \end{pmatrix}. \tag{F.3}$$

The Pauli matrices therefore form a basis for 2×2 hermitian traceless matrices, and thus a basis of the algebra of infinitesimal generators of SU(2). The reader can verify that these three matrices satisfy the following relations:

$$\sigma^i \sigma^j = \delta^{ij}\,\mathbf{1}_2 + i\,\epsilon^{ijk}\,\sigma^\kappa. \tag{F.4}$$

In particular, we can choose as basis elements the matrices $s_i \equiv \hbar\sigma^i/2$ which satisfy the following commutation relations:

$$[s_i,\, s_j] = i\hbar\,\epsilon_{ijk}\,s_k, \tag{F.5}$$

© Springer International Publishing Switzerland 2016
R. D'Auria and M. Trigiante, *From Special Relativity to Feynman Diagrams*,
UNITEXT for Physics, DOI 10.1007/978-3-319-22014-7

as it can be easily verified using (F.4). Note that the three matrices s_i satisfy the same commutation relations as the components $\hat{M}_i$ of the orbital angular momentum, which generate the group SO(3) of rotation in the three-dimensional Euclidean space. These two groups share therefore the same structure in a neighborhood of the identity element (they are *locally isomorphic*). For this reason the spin is sometimes improperly referred to as an internal angular momentum. The two groups are however globally different and this reflects in the fact that SU(2) has representations (the even-dimensional ones) which SO(3) does not have.

Let us illustrate the relationship between SU(2) and SO(3) in some more detail. We define a mapping between elements of the two groups as follows. Consider an element (2×2 complex matrix) $\mathbf{S} = (S^r{}_s), r, s = 1, 2$, of SU(2) and its adjoint action on the Pauli matrices: $\mathbf{S}^{-1} \sigma_i \mathbf{S} = \mathbf{S}^\dagger \sigma_i \mathbf{S}, i, j = 1, 2, 3$. Since the Pauli matrices form a basis for hermitian traceless matrices, resulting matrix is still hermitian traceless:

$$(\mathbf{S}^\dagger \sigma_i \mathbf{S})^\dagger = \mathbf{S}^\dagger \sigma_i^\dagger \mathbf{S} = \mathbf{S}^\dagger \sigma_i \mathbf{S}, \quad \mathrm{Tr}(\mathbf{S}^\dagger \sigma_i \mathbf{S}) = \mathrm{Tr}(\mathbf{S}\mathbf{S}^\dagger \sigma_i) = \mathrm{Tr}(\sigma_i) = 0.$$

Therefore $\mathbf{S}^\dagger \sigma_i \mathbf{S}$ can be expanded in the basis (σ_i). Let us denote by $R[\mathbf{S}]_i{}^j$ the components along σ_i of $\mathbf{S}^\dagger \sigma_i \mathbf{S}$:

$$\mathbf{S}^\dagger \sigma_i \mathbf{S} = R[\mathbf{S}]_i{}^j \sigma_j. \tag{F.6}$$

Since $\mathbf{R}[\mathbf{S}] \equiv (R[\mathbf{S}]_i{}^j)$ is a 3×3 matrix, we have thus defined a correspondence which maps a 2×2 matrix $\mathbf{S}$ of SU(2) into a 3×3 matrix $\mathbf{R}[\mathbf{S}]$. We want to show first that this correspondence is a homomorphism, namely that $R[\mathbf{S}_1 \mathbf{S}_2]_i{}^j = R[\mathbf{S}_1]_i{}^k R[\mathbf{S}_2]_k{}^j$:

$$\begin{aligned}(\mathbf{S}_1 \mathbf{S}_2)^\dagger \sigma_i (\mathbf{S}_1 \mathbf{S}_2) &= \mathbf{S}_2^\dagger (\mathbf{S}_1^\dagger \sigma_i \mathbf{S}_1) \mathbf{S}_2 = R[\mathbf{S}_1]_i{}^k (\mathbf{S}_2^\dagger \sigma_k \mathbf{S}_2) \\ &= R[\mathbf{S}_1]_i{}^k R[\mathbf{S}_2]_k{}^j \sigma_j = (\mathbf{R}[\mathbf{S}_1] \mathbf{R}[\mathbf{S}_2])_i{}^j \sigma_j.\end{aligned} \tag{F.7}$$

Let us prove now that the matrix $\mathbf{R}[\mathbf{S}]$ is real by computing the hermitian-conjugate of both sides of Eq. (F.6) and using the property that the left hand side is hermitian:

$$\mathbf{S}^\dagger \sigma_i \mathbf{S} = (\mathbf{S}^\dagger \sigma_i \mathbf{S})^\dagger = (R[\mathbf{S}]_i{}^j)^* \sigma_j. \tag{F.8}$$

Since the components associated with any vector (in this space vectors are hermitian matrices!) are unique, comparing (F.8)–(F.6) we find: $(R[\mathbf{S}]_i{}^j)^* = R[\mathbf{S}]_i{}^j$. Using the first of properties (10.66), we can write

$$R[\mathbf{S}]_i{}^j = \frac{1}{2} \mathrm{Tr}[(\mathbf{S}^\dagger \sigma_i \mathbf{S}) \sigma_j], \tag{F.9}$$

Finally let us show that the matrix $\mathbf{R}[\mathbf{S}]$ is orthogonal. To this end we use the general property of homomorphisms that: $\mathbf{R}[\mathbf{S}^{-1}] = \mathbf{R}[\mathbf{S}]^{-1}$ and write

$$R[\mathbf{S}]^{-1}{}_i{}^j = R[\mathbf{S}^\dagger]_i{}^j = \frac{1}{2} \mathrm{Tr}[(\mathbf{S} \sigma_i \mathbf{S}^\dagger) \sigma_j] = \frac{1}{2} \mathrm{Tr}[(\mathbf{S}^\dagger \sigma_j \mathbf{S}) \sigma_i] = R[\mathbf{S}]_j{}^i,$$

where we have used the cyclic property of the trace. We conclude that $\mathbf{R}[\mathbf{S}]^{-1} = \mathbf{R}[\mathbf{S}]^T$, which means that $\mathbf{R}[\mathbf{S}] \in O(3)$. Let us show that $\mathbf{R}[\mathbf{S}] \in SO(3)$, namely that $\det(\mathbf{R}[\mathbf{S}]) = 1$. To show this let us use the property that $\sigma_1 \sigma_2 \sigma_3 = i \, \mathbf{1}_2$. Then, from unitarity of $\mathbf{S}$ it follows that:

$$\mathbf{1}_2 = \mathbf{S}^\dagger \, \mathbf{S} = -i \, \mathbf{S}^\dagger \, \sigma_1 \sigma_2 \sigma_3 \, \mathbf{S} = -i \, (\mathbf{S}^\dagger \, \sigma_1 \, \mathbf{S}) \, (\mathbf{S}^\dagger \, \sigma_2 \, \mathbf{S}) \, (\mathbf{S}^\dagger \, \sigma_3 \, \mathbf{S})$$
$$= -i \, (R[\mathbf{S}]_1{}^i \, \sigma_i) \, (R[\mathbf{S}]_2{}^j \, \sigma_j) \, (R[\mathbf{S}]_3{}^k \, \sigma_k) = -i \, R[\mathbf{S}]_1{}^i \, R[\mathbf{S}]_2{}^j \, R[\mathbf{S}]_3{}^k \, (\sigma_i \, \sigma_j \, \sigma_k).$$
$$\text{(F.10)}$$

Now use the following property of the Pauli matrices

$$\sigma_i \, \sigma_j \, \sigma_k = i \, \epsilon_{ijk} \, \mathbf{1}_2 + \delta_{ij} \, \sigma_k - \delta_{ik} \, \sigma_j + \delta_{jk} \, \sigma_i, \qquad \text{(F.11)}$$

which follows from (F.4), to rewrite $\sigma_i \, \sigma_j \, \sigma_k$. Note that the terms with the δ matrix do not contribute because of the orthogonality property of $\mathbf{R}$: $R[\mathbf{S}]_k{}^i \, R[\mathbf{S}]_\ell{}^j \, \delta_{ij} = \sum_{i=1}^3 R[\mathbf{S}]_k{}^i \, R[\mathbf{S}]_\ell{}^i = \delta_{k\ell}$, which is zero if $k \neq \ell$. The only term in $\sigma_i \, \sigma_j \, \sigma_k$ which contributes to the summation is $i \, \epsilon_{ijk} \, \mathbf{1}_2$, and therefore we can rewrite Eq. (F.10) as follows:

$$R[\mathbf{S}]_1{}^i \, R[\mathbf{S}]_2{}^j \, R[\mathbf{S}]_3{}^k \, \epsilon_{ijk} \, \mathbf{1}_2 = \mathbf{1}_2. \qquad \text{(F.12)}$$

We recognize in the sum $R[\mathbf{S}]_1{}^i \, R[\mathbf{S}]_2{}^j \, R[\mathbf{S}]_3{}^k \, \epsilon_{ijk}$ the expression of the determinant of a matrix in terms of its entries and therefore we conclude that:

$$\det(\mathbf{R}[\mathbf{S}]) = 1, \qquad \text{(F.13)}$$

namely that $\mathbf{R}[\mathbf{S}] \in SO(3)$. We have thus defined a homomorphism between $SU(2)$ and $SO(3)$:

$$\mathbf{S} \in SU(2) \xrightarrow{\ \mathbf{R}\ } \mathbf{R}[\mathbf{S}] \in SO(3). \qquad \text{(F.14)}$$

This homomorphism is two-to-one. Indeed, the matrix $\mathbf{S}$ which corresponds to a given orthogonal one $\mathbf{R}[\mathbf{S}]$ is defined modulo a sign: $\mathbf{R}[\mathbf{S}] = \mathbf{R}[-\mathbf{S}]$. In a neighborhood of the identity of $SU(2)$, the correspondence is therefore one-to-one and thus the two groups are called *locally isomorphic*.

The fact that s_i and the 3×3 $SO(3)$-generators $\mathbf{M}_i$, defined in (4.131), have the same commutation relations allows to write the correspondence $\mathbf{R}$ as a mapping between the element $\mathbf{S} = e^{\frac{i}{\hbar} \theta^i s_i}$ of $SU(2)$ and $e^{\frac{i}{\hbar} \theta^i \mathbf{M}_i}$ of $SO(3)$ defined above in the following way:

$$\mathbf{R}[e^{\frac{i}{\hbar} \theta^i s_i}] = e^{\frac{i}{\hbar} \theta^i \mathbf{M}_i} \in SO(3), \qquad \text{(F.15)}$$

as it can be easily verified for infinitesimal transformations ($\theta^i \ll 1$).

Appendix G
Gamma Matrix Identities

We collect in this Appendix the most useful formulae used for the manipulation of gamma-matrices. All the following relations are actually a consequence of the defining anticommutation rules (10.61), namely

$$\gamma^\mu\gamma^\nu + \gamma^\nu\gamma^\mu = 2\eta^{\mu\nu} \qquad \mu, \nu = 0, 1, 2, 3, \tag{G.1}$$

where we recall that $\eta^{\mu\nu} \equiv \mathrm{diag}(+1, -1, -1, -1)$. Let us first observe that, since the matrix representation is four-dimensional, from (G.1) it follows

$$\eta_{\mu\nu}\gamma^\mu\gamma^\nu \equiv \gamma^\mu\gamma_\mu = 4. \tag{G.2}$$

Let us suppose that we have an expression of the type

$$\gamma^\mu \left(\gamma^\nu\gamma^\rho\gamma^\sigma \cdots \right) \gamma_\mu.$$

Using several times the anticommutation rules (G.1), the two γ_μ can be put side by side, and we find the following formulae:

$$\gamma^\mu\gamma^\nu\gamma_\mu = -2\gamma^\nu, \tag{G.3}$$
$$\gamma^\mu\gamma^\nu\gamma^\rho\gamma_\mu = 4\eta^{\nu\rho},$$
$$\gamma^\mu\gamma^\nu\gamma^\rho\gamma^\sigma\gamma_\mu = -2\gamma^\sigma\gamma^\rho\gamma^\nu.$$

The first is readily proven by writing $\gamma^\rho\gamma^\mu = -\gamma^\mu\gamma^\rho + 2\eta^{\mu\rho}$.

As for the second we write

$$\gamma_\mu\gamma^\rho\gamma^\sigma\gamma^\mu = \gamma_\mu\gamma^\rho \left(-\gamma^\mu\gamma^\sigma + 2\eta^{\mu\sigma}\right) = 2\gamma^\rho\gamma^\sigma + 2\gamma^\sigma\gamma^\rho = 4\eta^{\rho\sigma}.$$

© Springer International Publishing Switzerland 2016
R. D'Auria and M. Trigiante, *From Special Relativity to Feynman Diagrams*,
UNITEXT for Physics, DOI 10.1007/978-3-319-22014-7

In an analogous way, using the previous result, we have for the third identity

$$\gamma_\mu \gamma^\rho \gamma^\sigma \gamma^\tau \gamma^\mu = \gamma_\mu \gamma^\rho \gamma^\sigma \left(-\gamma^\mu \gamma^\tau + 2\eta^{\mu\tau} \right) = -4\eta^{\rho\sigma} \gamma^\tau + 2\gamma^\tau \gamma^\rho \gamma^\sigma$$
$$= -4\eta^{\rho\sigma} \gamma^\tau + 2\gamma^\tau \left(-\gamma^\sigma \gamma^\rho + 2\eta^{\rho\sigma} \right) = -2\gamma^\tau \gamma^\sigma \gamma^\rho.$$

In most applications the indices of the gamma matrices are contracted with four-vectors. Introducing the notation

$$\slashed{a} = \gamma \cdot a = \gamma^\mu a_\mu, \tag{G.4}$$

and using (G.1) we have, for example

$$\slashed{a}\slashed{b} + \slashed{b}\slashed{a} = 2a \cdot b; \quad \slashed{a}\slashed{a} = a \cdot a. \tag{G.5}$$

The formulae (G.3) take the following form

$$\gamma^\mu \slashed{a} \gamma_\mu = -2\slashed{a} \tag{G.6}$$
$$\gamma^\mu \slashed{a}\slashed{b} \gamma_\mu = 4a \cdot b \tag{G.7}$$
$$\gamma^\mu \slashed{a}\slashed{b}\slashed{c} \gamma_\mu = -2\slashed{c}\slashed{b}\slashed{a}. \tag{G.8}$$

Consider now the matrix γ^5 defined in (10.194) and define the following matrices:

$$\gamma^{\mu\nu} = \gamma^{[\mu} \gamma^{\nu]} = \frac{1}{2} \left[\gamma^\mu, \gamma^\nu \right] \equiv -i\,\sigma^{\mu\nu} ; \quad \gamma^{\mu\nu\rho} = \gamma^{[\mu} \gamma^\nu \gamma^{\rho]}, \tag{G.9}$$

where $\sigma^{\mu\nu}$ was defined in (10.98). We may easily prove the following duality relations

$$\gamma^5 \gamma_\mu = -\frac{i}{3!} \epsilon_{\mu\nu\rho\sigma} \gamma^{\nu\rho\sigma},$$
$$\gamma^5 \gamma_{\mu\nu} = -\frac{i}{2} \epsilon_{\mu\nu\rho\sigma} \gamma^{\rho\sigma},$$
$$\gamma^5 \gamma_{\mu\nu\rho} = i\,\epsilon_{\mu\nu\rho\sigma} \gamma^\sigma. \tag{G.10}$$

As for the first one, multiplying both sides of (10.194) by γ_3 to the right and using the property $\gamma^3 \gamma_3 = \mathbf{1}$, we find:

$$\gamma^5 \gamma_3 = i\gamma^0 \gamma^1 \gamma^2 = i\,\epsilon_{0123}\, \gamma^0 \gamma^1 \gamma^2 = -i\,\epsilon_{3012}\, \gamma^0 \gamma^1 \gamma^2, \tag{G.11}$$

where we have used $\epsilon_{0123} = 1$. By Lorentz covariance, the above relation implies the first of Eqs. (G.10). As far as the second equation is concerned, we further multiply Eq. (G.11) to the right by γ_2 and find:

$$\gamma^5 \gamma_3 \gamma_2 = i\gamma^0 \gamma^1 = -i\,\epsilon_{3201}\, \gamma^0 \gamma^1. \tag{G.12}$$

Covariantizing the above equation, the second of Eqs. (G.10) follows. By a similar argument, the third of those equations can also be proven.

Let us now consider the following set of 16 matrices

$$\Gamma_A = \left\{ 1, \gamma^5, \gamma^\mu, \gamma^5 \gamma^\mu, \gamma^{\mu\nu} \right\}; \quad A = 1, \ldots, 16, \tag{G.13}$$

where, when A labels the matrices $\gamma^{\mu\nu}$, we consider only the six independent couples (μ, ν) with $\mu < \nu$, so as to avoid repetitions: $A = (\mu, \nu) = \{(0, 1), (0, 2), (0, 3), (1, 2), (1, 3), (2, 3)\}$.

Note that for each value of A:

$$\gamma^A \gamma_A = \varepsilon_A \, 1, \quad \text{(no summation over } A)$$

where $\varepsilon_A = 1$ for $\Gamma_A = \{1, \gamma^\mu, \gamma^5\}$ and $\varepsilon_A = -1$ for $\Gamma_A = \{\gamma^{\mu\nu}, \gamma^5 \gamma^\mu\}$. The following properties hold:

$$\Gamma^A \neq 1 \Rightarrow \text{Tr}(\Gamma^A) = 0 \; ; \quad \frac{1}{4} \text{Tr}(\Gamma^A \Gamma_B) = \varepsilon_A \, \delta_B^A, \tag{G.14}$$

where $\varepsilon_A = \pm 1$. For instance $\text{Tr}(\gamma^{\mu\nu} \gamma_{\rho\sigma}) = -8 \, \delta_{\rho\sigma}^{\mu\nu} == -4 \, (\delta_\rho^\mu \delta_\sigma^\nu - \delta_\sigma^\mu \delta_\rho^\nu)$. In this case $A = (\mu\nu)$, $B = (\rho\sigma)$ and $\varepsilon_A = -1$, being $\delta_B^A = 2 \, \delta_{\rho\sigma}^{\mu\nu}$.

The proof that all the Γ^A, s, except 1, are traceless is based on the observation that the trace of the product of two anticommuting matrices is zero. Indeed from the invariance of the trace of a matrix product under a cyclic permutation of the matrices, we have

$$AB = -BA \rightarrow \text{Tr}(AB) = 0.$$

Now $\gamma^5 \gamma^\mu$ and $\gamma^{\mu\nu}$ are already in this form. As for γ^5 it suffices to write

$$\gamma^5 = i \gamma^0 \gamma^1 \gamma^2 \gamma^3 = -i \gamma^1 \gamma^2 \gamma^3 \gamma^0.$$

Taking the trace of the above γ^μ-matrix products, it immediately follows that this trace must vanish. On the other hand we can observe from (10.194) that the explicit for of γ^5 in the Pauli basis is traceless. Clearly this product is basis-independent, since a change of basis amounts to a conjugation of γ^5 or any other matrix by a non singular one U, and such conjugation does not affect the value of the trace: $\text{Tr}(U^{-1} \gamma^5 U) = \text{Tr}(U U^{-1} \gamma^5) = \text{Tr}(\gamma^5) = 0$. A similar argument applies to the γ^μ matrices, which are traceless in the Pauli basis, and thus are traceless in any other basis.

We can regard the Γ^A as vectors in a vector space and define among them a symmetric scalar product $(\cdot, \cdot)$ as follows: $(\Gamma^A, \Gamma^B) \equiv \text{Tr}(\Gamma^A \Gamma^B)$. Being the Γ^A mutually orthogonal with respect to this scalar product, they are linearly independent. Indeed if we consider a generic combination of them

$$\sum_1^{16} c_A \Gamma^A = 0,$$

upon multiplication of both sides by Γ_B and taking the trace, we obtain:

$$\sum_1^{16} c_A \operatorname{Tr}(\Gamma^A \Gamma_B) = 4\varepsilon_B c_B = 0 \;\to\; c_B = 0 \;\; \forall B.$$

A generic 4×4 matrix is defined by its 16 elements, which implies that the vector space consisting of all the 4×4 matrices is 16-dimensional. The 16 linearly independent matrices $\{\Gamma^A\}$ form therefore a basis for this space, i.e. a complete set of matrices in terms of which any other matrix M can be expressed as a unique linear combination:

$$M = \sum_{A=1}^{16} C_A \,\Gamma^A; \quad C_A = \frac{\varepsilon_A}{4}\operatorname{Tr}(M\Gamma^A). \tag{G.15}$$

In Chap. 12 we shall often need to compute traces of products of gamma matrices. Let us derive some useful properties of these traces. We start defining the following quantities:

$$T^{\mu_1\mu_2\cdots\mu_n} \equiv \frac{1}{4}\operatorname{Tr}(\gamma^{\mu_1}\gamma^{\mu_2}\cdots\gamma^{\mu_n}). \tag{G.16}$$

These are Lorentz-invariant tensors. To prove this let us observe that the γ^μ matrices, if written in components $(\gamma^\mu)^\alpha{}_\beta$, can be viewed as a mixed Lorentz-tensor, with two contravariant indices μ, α in the fundamental and spinorial representations, respectively, and one covariant spinorial index β. As a Lorentz tensor, it is invariant sice, if we simultaneously apply to all its indices a Lorentz transformation Λ, it remains unchanged:

$$(\gamma^\mu)^\alpha{}_\beta \xrightarrow{\Lambda} \Lambda^\mu{}_\nu \, S(\Lambda)^\alpha{}_{\alpha'} S(\Lambda)^{-1\,\beta'}{}_\beta \,(\gamma^\nu)^{\alpha'}{}_{\beta'} \;\Leftrightarrow$$

$$\Leftrightarrow\; \gamma^\mu \xrightarrow{\Lambda} \Lambda^\mu{}_\nu \, S(\Lambda)\gamma^\nu \, S(\Lambda)^{-1} = \gamma^\mu, \tag{G.17}$$

where we have used (10.88) with $\Lambda \to -\Lambda$. If we apply the above property to each gamma-matrix in $T^{\mu_1\mu_2\cdots\mu_n}$ we find:

$$T^{\mu_1\mu_2\cdots\mu_n} \equiv \frac{1}{4}\operatorname{Tr}(\gamma^{\mu_1}\gamma^{\mu_2}\cdots\gamma^{\mu_n})$$

$$= \Lambda^{\mu_1}{}_{\nu_1}\cdots\Lambda^{\mu_n}{}_{\nu_n}\frac{1}{4}\operatorname{Tr}(S\gamma^{\nu_1}S^{-1}S\gamma^{\nu_2}S^{-1}\cdots S\gamma^{\nu_n}S^{-1})$$

$$= \Lambda^{\mu_1}{}_{\nu_1}\cdots\Lambda^{\mu_n}{}_{\nu_n}\,T^{\nu_1\nu_2\cdots\nu_n}, \tag{G.18}$$

where we have used the invariance of the trace under conjugation of the gamma-matrices by $S = S(\Lambda)$. This proves the Lorentz-invariance of the tensors $T^{\mu_1\mu_2\cdots\mu_n}$. These tensors should therefore be expressed in terms of the only invariant tensor of the full Lorentz group $O(1, 3)$, namely $\eta_{\mu\nu}$. Since every tensor made up in terms of the metric tensor is necessarily of even order, the trace of the product of an odd number of gamma matrices is zero.

From the anticommutation relations (G.1), using the cyclic property of the trace we easily find

$$T^{\mu\nu} \equiv \frac{1}{4}\,\mathrm{Tr}(\gamma^\mu\gamma^\nu) = \eta^{\mu\nu}. \tag{G.19}$$

Let us now consider the trace of the product of four gamma matrices. We show that

$$T^{\mu_1\mu_2\mu_3\mu_4} = \eta^{\mu_1\mu_2}\,\eta^{\mu_3\mu_4} - \eta^{\mu_1\mu_3}\,\eta^{\mu_2\mu_4} + \eta^{\mu_1\mu_4}\,\eta^{\mu_2\mu_3}. \tag{G.20}$$

By successive steps we bring γ^{μ_1} to the right end of the product. In the first step, using (G.1) we find

$$T^{\mu_1\mu_2\mu_3\mu_4} = 2\eta^{\mu_1\mu_2}\,T^{\mu_3\mu_4} - T^{\mu_2\mu_1\mu_3\mu_4} = 2\eta^{\mu_1\mu_2}\,\eta^{\mu_3\mu_4} - T^{\mu_2\mu_1\mu_3\mu_4}. \tag{G.21}$$

As a second step we (anti)commute γ^{μ_1} with γ^{μ_3} on the right hand side of (G.21), and so on until after the last anticommutation we find the tensor $T^{\mu_2\mu_3\mu_4\mu_1}$, which equals to $T^{\mu_1\mu_2\mu_3\mu_4}$ by the cyclic identity of the trace. Putting together the results of the successive commutations we recover Eq. (G.20).

The same iterative procedure can be applied to any number of gamma matrices. As a further example consider the trace of six gamma matrices. We can write

$$\begin{aligned}
T^{\mu_1\mu_2\mu_3\mu_4\mu_5\mu_6} &= \eta^{\mu_1\mu_2}\,T^{\mu_3\mu_4\mu_5\mu_6} - \eta^{\mu_1\mu_3}\,T^{\mu_2\mu_4\mu_5\mu_6} \\
&\quad + \eta^{\mu_1\mu_4}\,T^{\mu_2\mu_3\mu_5\mu_6} - \eta^{\mu_1\mu_5}\,T^{\mu_2\mu_3\mu_4\mu_6} + \eta^{\mu_1\mu_6}\,T^{\mu_1\mu_2\mu_3\mu_4},
\end{aligned}$$

and the four-index tensors can be reduced using (G.21). In general, using the property

$$T^{\mu_1\mu_2\cdots\mu_n} = \eta^{\mu_1\mu_2}\,T^{\mu_3\cdots\mu_n} - \eta^{\mu_1\mu_3}\,T^{\mu_2\cdots\mu_n} + \cdots + \eta^{\mu_1\mu_n}\,T^{\mu_2\cdots\mu_{n-1}}, \tag{G.22}$$

a generic rank tensor can be reduced to combinations of products of η-matrices.

In actual computations the Lorentz indices of the gamma matrices are contracted with four vectors a_μ, so that we typically have to evaluate expressions like

$$(a_1 a_2 \ldots a_n) \equiv \frac{1}{4}\,\mathrm{Tr}(\slashed{a}_1\slashed{a}_2 \ldots \slashed{a}_n). \tag{G.23}$$

In that case formula (G.22) implies:

$$(a_1 a_2 \ldots a_n) = (a_1 \cdot a_2)\,(a_3 \ldots a_n) - (a_1 \cdot a_3)\,(a_2 \ldots a_n) + \cdots + (a_1 \cdot a_n)\,(a_2 \ldots a_{n-1}). \tag{G.24}$$

Using the properties discussed above we can also prove the following identities which will be useful when computing cross sections in Chap. 12:

$$\text{Tr}\left[(\slashed{A}+a)\gamma^\mu(\slashed{B}+b)\gamma^\nu(\slashed{C}+c)\gamma_\mu(\slashed{D}+d)\gamma_\nu\right]$$
$$= -32(A\cdot C)(B\cdot D) + 16\,ab(C\cdot D) + 16\,ac(B\cdot D) + 16\,ad(B\cdot C)$$
$$+16\,bc(A\cdot D) + 16\,bd(A\cdot C) + 16\,cd(A\cdot B) - 32\,abcd, \tag{G.25}$$

$$\text{Tr}\left[(\slashed{A}+a)\gamma^\mu(\slashed{B}+b)\gamma^\nu(\slashed{C}+c)\gamma_\nu(\slashed{D}+d)\gamma_\mu\right]$$
$$= 16\left[(A\cdot D)(B\cdot C) + (A\cdot B)(C\cdot D) - (A\cdot C)(B\cdot D)\right]$$
$$- 32\left[ab(C\cdot D) + ad(B\cdot C) + bc(A\cdot D) + cd(A\cdot B)\right]$$
$$+ 64\,ac(B\cdot D) + 16\,bd(A\cdot C) + 64\,abcd. \tag{G.26}$$

Appendix H
Simultaneity and Rigid Bodies

In this Appendix we shall discuss the relativity of the concept of *simultaneity*, giving some simple examples and illustrating its main consequences. We shall also prove the meaninglessness, in relativity, of the classical concept of *rigid body*.

Let us first discuss simultaneity.

According to the invariance of the speed of light and its independence of the source speed, two events taking place at two points P_1, P_2 are defined to be *simultaneous* when two light rays departing from them arrive at the *same instant* in the midpoint between P_1 and P_2. As we have seen in Chap. 1, the simultaneity of two events occurring at different points, observed in a frame S is not in general observed in a different inertial frame S', that is (see Eq. (1.70)):

$$\Delta t = 0 \rightarrow \Delta t' = -\gamma \frac{\beta}{c} \Delta x \neq 0. \tag{H.1}$$

This result can be made intuitive in the following way.

Let S be the observer at rest on a railway, that is relative to the earth (considered as an inertial frame) and S' be an observer on a train of length $l' = A'B'$ moving at a velocity V in the x direction.[2] Let the two events be the lighting of two light bulbs in A' and B', of coordinates x'_A and x'_B in S', at the two ends of the train (B' at the front, A' at the rear). In the instants the two events occur, A' and B' coincide with two points A, B along the railway, of coordinates x_A and x_B in S, respectively. We suppose the two events to be *simultaneous* in S. That means that the two light signals reach simultaneously the midpoint C of the length $l = AB$, as measured along the railway, of coordinate $x_C = (x_B - x_A)/2$. In S the two events therefore occur at the same time: $t_A = t_B$.

In the S' frame connected to the train the signal coming from the end A' reaches the midpoint C' of the train, of coordinate $x_{C'} = (x'_B - x'_A)/2$, *later* than that sent from B', since during the light propagation the train has shifted its position in the direction of motion. It follows that in the S' frame the event "lighting of the bulb" in B' precedes

---

[2]As always we suppose that the two frames are in the standard configuration.

© Springer International Publishing Switzerland 2016
R. D'Auria and M. Trigiante, *From Special Relativity to Feynman Diagrams*,
UNITEXT for Physics, DOI 10.1007/978-3-319-22014-7

the corresponding event in A', i.e. the two events are therefore no longer simultaneous and $t'_A < t'_B$. This is consistent with Eq. (H.1) since $t'_B - t'_A = -\gamma \frac{\beta}{c} (x_B - x_A)$ is negative being $x_B > x_A$.

Note that this result, which seems paradoxical from the point of view of classical mechanics, is due to the fact that we are used to consider the propagation of light as instantaneous. However, if we were to use the Galileo transformations of classical mechanics, we could find a similar result by assuming the speed of light to be finite. Using Galileian transformation laws we would write:

$$c'_A = c - V \; ; \qquad c'_B = c + V, \tag{H.2}$$

where c'_A and c'_B are the speeds of the light signals from the two endpoints of the train in S', propagating in opposite directions, while c is the corresponding common speed in the frame S connected to the earth. It is then obvious that, since $c'_B > c'_A$, in the frame S' the signals from A' reaches the midpoint of the train later than that from B'. From the classical viewpoint, however, this would have no bearing on the simultaneity of the two events since the observer in S', taking into account the different speeds of the two signals, would find A' and B' to be simultaneous just as they are for the observer in S. This is not the case in special relativity, where $c'_B = c'_A = c$ and different transformation laws should be used.

We also note that the fact that the coincidence of the times of two events is relative to the reference frame, is quite analogous to the fact that the occurrence of two events at the same place is relative to the frame, a well accepted fact in everyday life. For example, if a bulb is lit and later turned off on a train, the two events happen in the same point in S', but in different places in S.

Finally we observe that if we consider the same process from the point of view of an observed located in C', at rest in S', the point B' is seen to coincide with B before A' is seen to coincide with A. That means that the observer perceives the length of the train $l' = A'B'$ larger then the length $l = AB$ on the railway, i.e. he observes a *contraction* of the length l. This shows, as already observed in Chap. 1, a close connection between simultaneity and length-contraction.

Let us now discuss the meaninglessness of the concept of *rigid body* in special relativity.

As we mentioned at the end of Sect. 1.4, the concept of *rigid body*, being related to the invariance of distances in classical mechanics, looses its meaning in a relativistic theory. This conceptual limitation can be made apparent by considering accelerated bodies, and we shall take as an example the circular motion of the hand of a clock (the second-hand, for example).

Indeed suppose we have a clock in an inertial frame S. An observer in this frame sees the clock hands, which are *straight segments*, moving with uniform circular motion (which is indeed accelerated) around the center of the clock face. As we shall show below, the same hand is seen *curved* by a second inertial observer O' in a reference frame S'.

Indeed let us take a point P on the rotating hand whose coordinates in the S-frame are x, y. We have

$$
\begin{aligned}
x(t) &= r \cos \omega t, \\
y(t) &= r \sin \omega t,
\end{aligned}
\tag{H.3}
$$

r being the of P from the center along the hand and ω its angular velocity. Using Lorentz transformations the same point in the frame S' (standard configuration between S' and S being understood) has the following time-dependence of its coordinates x', y'

$$
\begin{aligned}
x'(t') &= \gamma \, (r \, \cos \omega t - V t), \\
y'(t') &= \gamma \, (r \, \sin \omega t - V t).
\end{aligned}
\tag{H.4}
$$

where t' is related to t by:

$$
t' = \gamma \left(t - \frac{V}{c^2} r \cos \omega t \right) \equiv f(r, t).
\tag{H.5}
$$

To express t in terms of t' we have to solve the transcendental equation $f(r, t) = t'$ in order to write

$$
t - t(r, t').
$$

Though we cannot give $t(r, t')$ explicitly, we can Taylor-expand it in r and write

$$
t = \frac{t'}{\gamma} + \frac{V}{c^2} r \, \cos[\omega \, t(r, t')] = \frac{t'}{\gamma} + \frac{V}{c^2} r \, \cos \left(\frac{\omega t'}{\gamma} \right) + O(r^2).
\tag{H.6}
$$

Substituting this expansion in (H.4) and expanding $x'(r, t')$ and $y'(r, t')$ in r, the presence of quadratic terms in this expansion implies that the clock hand is a *curved line* in S'.

The meaninglessness of the concept of rigid body in relativity is also apparent in the famous *pole and hole paradox* first described by Rindler [15].

References

N.B. The chapters indicated between square brackets after each reference, refer to the chapters of the present book.

1. P.G. Bergmann, *Introduction to the Theory of Relativity* (Dover publications, New York, 1976) [Chaps. 1, 2, 3]
2. J.F. Cornwell, *Group Theory in Physics. An Introduction* (Academic Press, San Diego, 1997) [Chaps. 4, 7]
3. A. Di Giacomo, *Lezioni di Fisica Teorica* (Edizioni ETS, Pisa, 1992) [Chaps. 10, 11, 12]
4. P.A.M. Dirac, *The Principles of Quantum Mechanics* (Clarendon Press-Oxford, Oxford, 1958) [Chap. 9]
5. H. Georgi, *Lie Algebras in Particle Physics* (Westview Press, Boulder, 1999) [Chaps. 4, 7]
6. H. Goldstein, C. Poole, J. Safko, *Classical Mechanics* (Addison Wesley, Boston, 2001) [Chap. 8]
7. C. Itzinksons, J.B. Zuber, *Quantum Field Theory* (McGraw-Hill Education, New York, 1980) [Chaps. 10, 11, 12]
8. L.D. Landau, E.M. Lifshitz, *Course of Theoretical Physics, Mechanics*, vol. 1, *The Classical Theory of Fields*, vol. 2, *Quantum Mechanics*, vol. 3, *Quantum Electrodynamics*, vol. 4, (Butterworth Heinemann, Elsevier Science, Amsterdam, 1976) [Chaps. 8, 9, 10, 11, 12]
9. D. Lurié, *Particles and Fields* (Interscience Publishers, Wiley, New York, 1968) [Chaps. 10, 11, 12]
10. B. Odom, D. Hanneke, B. DUrso, G. Gabrielse, Phys. Rev. Lett. **97**, 030801 (2006), [Chap. 12]
11. W. Rindler, *Relativity* (Oxford University Press, Oxford, 2001) [Chap. 1, 2, 3]
12. S. Weinberg, *Gravitation and Cosmology-Principles and Applications of the General Theory of Relativity* (Wiley, New York, 1972) [Chap. 1, 2, 3]
13. S.L. Weinberg, *The Quantum Theory of Fields*, vol.1 (Cambridge University Press, Cambridge, 1995) [Chap. 10, 11, 12]
14. B.G. Wybourne, *Classical Groups for Physicists* (Wiley, New York, 1973) [Chap. 4, 7]
15. W. Rindler, Length contraction paradox. Am. J. Phys. **29** (1961) 365–6, [Appendix H]

© Springer International Publishing Switzerland 2016   591
R. D'Auria and M. Trigiante, *From Special Relativity to Feynman Diagrams*,
UNITEXT for Physics, DOI 10.1007/978-3-319-22014-7

Index

© Springer International Publishing Switzerland 2016
R. D'Auria and M. Trigiante, *From Special Relativity to Feynman Diagrams*,
UNITEXT for Physics, DOI 10.1007/978-3-319-22014-7

Index

601

Printed in the United States
By Bookmasters